DATE DUE			

Isozymes

IV

Genetics and Evolution

Organizing Committee and Editorial Board

CLEMENT L. MARKERT
Chairman and Editor
Department of Biology
Yale University
New Haven, Connecticut

BERNARD L. HORECKER
Roche Institute of Molecular Biology
Nutley, New Jersey

ELLIOT S. VESELL
Department of Pharmacology
Pennsylvania State University
College of Medicine
Hershey, Pennsylvania

JOHN G. SCANDALIOS
Department of Biology
University of South Carolina
Columbia, South Carolina

GREGORY S. WHITT
Provisional Department of Genetics and Development
University of Illinois
Urbana, Illinois

The Third International Conference on Isozymes
Held at Yale University, New Haven, Connecticut
April 18-20, 1974

Isozymes

IV

Genetics and Evolution

EDITED BY

Clement L. Markert
Department of Biology
Yale University

ACADEMIC PRESS New York San Francisco London 1975

A Subsidiary of Harcourt Brace Jovanovich, Publishers

ACADEMIC PRESS RAPID MANUSCRIPT REPRODUCTION

ACADEMIC PRESS, INC.
111 Fifth Avenue, New York, New York 10003

United Kingdom Edition published by
ACADEMIC PRESS, INC. (LONDON) LTD.
24/28 Oval Road, London NW1

Library of Congress Cataloging in Publication Data
Main entry under title:

Isozymes.

CONTENTS: 1. Molecular structure.–2. Physiological structure.–3. Developmental biology. [etc.]
1. Isoenzymes–Congresses. I. Markert, Clement Lawrence, (date) II. Title. [DNLM: 1. Isoenzymes –Congresses. W3 IN182A 1974i / QU135 1587 1974i]
QP601.I75 574.1'925 74-31288
ISBN 0-12-472704-2 (v.4)

PRINTED IN THE UNITED STATES OF AMERICA

Contents

CONTENTS

CONTENTS

CONTENTS

CONTENTS

Contributors

Irene Abraham Department of Biology, Yale University, New Haven, Connecticut 06520

Julian Adams Division of Biological Sciences, The University of Michigan, Ann Arbor, Michigan 48104

R. W. Allard Department of Genetics, University of California, Davis, California 95616

Sally Allen Division of Biological Sciences, The University of Michigan, Ann Arbor, Michigan 48104

Fred W. Allendorf Northwest Fisheries Center, National Marine Fisheries Service, National Oceanic and Atmospheric Administration, Seattle, Washington 98112

Norman Arnheim Department of Biochemistry, State University of New York, Stony Brook, New York 11790

George S. Bailey Department of Biochemistry, University of Otago, Dunedin, New Zealand

James N. Baptist University of Texas System Cancer Center, M.D. Anderson Hospital and Tumor Institute, Houston, Texas 77025

Gunhild Beckman Department of Medical Genetics, University of Umeå, 901 85 Umeå, Sweden

Lars Beckman Department of Medical Genetics, University of Umeå, 901 85 Umeå, Sweden

B. N. Billeck Canada Department of the Environment, Fisheries and Marine Service, Freshwater Institute, Winnipeg, Manitoba, Canada R3T 2N6

Alan H. Brush Biological Sciences Group, University of Connecticut, Storrs, Connecticut 06268

John M. Burns Museum of Comparative Zoology, Harvard University, Cambridge, Massachusetts 02138

Mary Ann Butler University of Texas Health Science Center, Graduate School of Biomedical Sciences, Texas Medical Center, Houston, Texas 77025

H. L. Carson Department of Genetics, University of Hawaii, Honolulu, Hawaii 96822

Arthur Chovnick Biological Sciences Group, The University of Connecticut, Storrs, Connecticut 06268

J. W. Clayton Canada Department of the Environment, Fisheries and Marine Service, Freshwater Institute, Winnipeg, Manitoba, Canada R3T 2N6

M. T. Clegg Division of Biological and Medical Sciences, Brown University, Providence, Rhode Island 02912

M. M. Cooper Department of Biology, Western Michigan University, Kalamazoo, Michigan 49001

Linda L. Darga Department of Anatomy, Wayne State University Medical School, Detroit, Michigan 48202

Gladys E. Deibler National Institutes of Health, Bethesda, Maryland 20014

Howard Dene Department of Biology, Wayne State University, Detroit, Michigan 48202

Winifred W. Doane Department of Biology, Yale University, New Haven, Connecticut 06520

G. Drummond Department of Pharmacology, New York University, New York, New York 10016

Wolfgang Engel Institut für Humangenetik und Anthropologie der Universität, D-7800 Freiburg, West Germany

Robert E. Ferrell Department of Human Genetics, University of Michigan Medical School, Ann Arbor, Michigan 48104

Jerome F. Fredrick The Research Laboratories, Dodge Chemical Company, Bronx, New York 10469

Roger E. Ganschow The Children's Hospital Research Foundation, Elland and Bethesda Avenues, Cincinnati, Ohio 45229

Charles T. Garten, Jr. Savannah River Ecology Laboratory, Drawer E, Aiken, South Carolina 29801

William Gelbart Biological Sciences Group, University of Connecticut, Storrs, Connecticut 06268

Ian Gibson School of Biological Sciences, University of East Anglia, Norwich, Norfolk, England

Morris Goodman Department of Anatomy, Wayne State University School of Medicine, Detroit, Michigan 48201

Sheldon I. Guttman Department of Zoology, Miami University, Oxford, Ohio 45056

J. L. Hamerton Division of Genetics, Department of Pediatrics, University of Manitoba, Winnipeg, Manitoba, Canada

Harry Harris Galton Laboratory, University College, London, England

Samuel H. Hori Department of Zoology, Faculty of Science, Hokkaido University, Sapporo, Japan

Ronald Humphrey University of Texas System Cancer Center, M.D. Anderson Hospital and Tumor Institute, Houston, Texas 77025

P. Ihssen Ontario Ministry of Natural Resources, Research Branch, Maple, Ontario, Canada

M. Jacobs Laboratory of Plant Genetics, Vrije Universiteit Brussel, B-1640 Sint-Genesius-Rode, Belgium

Alan Jamieson Ministry of Agriculture, Fisheries, and Food, Fisheries Laboratory, Lowestoft, Suffolk, England

W. E. Johnson Department of Biology, Western Michigan University, Kalamazoo, Michigan 49001

A. L. Kahler Department of Genetics, University of California, University of California, Davis, California 95616

Tsutomu Kamada Department of Zoology, Faculty of Science, Hokkaido University, Sapporo, Japan

K. Y. Kaneshiro Department of Genetics and Entomology, University of Hawaii, Honolulu, Hawaii 96822

R. C. Karn Department of Medical Genetics, Indiana University Medical Center, Indianapolis, Indiana 46202

Donald W. Kaufman Savannah River Ecology Laboratory, Drawer E, Aiken, South Carolina 29801

Richard K. Koehn Department of Ecology and Evolution, State University of New York, Stony Brook, New York 11790

Ann L. Koen C. S. Mott Center for Human Growth and Development, Wayne State University School of Medicine, Detroit, Michigan 48201

R. L. Kirk Department of Human Biology, The John Curtin School of Medical Research, Canberra ACT Australia 2601

M. M. Kolar Department of Biology, Yale University, New Haven, Connecticut 06520

L. M. Lehrner Department of Zoology, Indiana University, Bloomington, Indiana 47401

H. B. LéJohn Department of Microbiology, University of Manitoba, Winnipeg, Manitoba, R3T 2N2, Canada

Alan L. Lin Department of Environmental Physiology, Virginia Institute of Marine Science, Gloucester Point, Virginia 23062

Seth Lubega Howard University, Washington, D.C. 20001

Phyllis J. McAlpine Division of Genetics, Department of Pediatrics, University of Manitoba, Winnipeg, Manitoba, Canada

Margaret McCarron Biological Sciences Group, University of Connecticut, Storrs, Connecticut 06268

G. M. Malacinski Department of Zoology, Indiana University, Bloomington, Indiana 47401

C. L. Markert Department of Biology, Yale University, New Haven, Connecticut 06520

Russell E. Martenson National Institutes of Health, Bethesda, Maryland

Takeo Maruyama Center for Demographic and Population Genetics, University of Texas, Houston, Texas 77025

Thomas S. Matney University of Texas Health Science Center, Graduate School of Biomedical Sciences, Texas Medical Center, Houston, Texas 77025

Genji Matsuda Department of Biochemistry, Nagasaki University School of Medicine, Nagasaki, Japan

CONTRIBUTORS

Martha R. Matteo Biology Department, University of Massachusetts, Boston, Massachusetts 02125

Bernard P. May Northwest Fisheries Center, National Marine Fisheries Service, National Oceanic and Atmospheric Administration, Seattle, Washington 98112

A. Donald Merritt Department of Medical Genetics, Indiana University Medical Center, Indianapolis, Indiana 46202

J. N. Miceli Department of Pharmacology, New York University, New York, New York 10016

Roger Milkman Department of Zoology, University of Iowa, Iowa City, Iowa 52242

Yoshikatsu Mochizuki Department of Zoology, Faculty of Science, Hokkaido University, Sapporo, Japan

T. Mohandas Division of Genetics, Department of Pediatrics, University of Manitoba, Winnipeg, Manitoba, Canada

G. William Moore Department of Anatomy, Wayne State University School of Medicine, Detroit, Michigan 48201

E. J. Murgola University of Texas System Cancer Center, M.D. Anderson Hospital and Tumor Institute, Houston, Texas 77025

Ken Nozawa Department of Variation Research, Primate Research Institute, Kyoto University, Inuyama, Aichi-ken, 484, Japan

Yoshiko Okura Department of Variation Research, Primate Research Institute, Kyoto University, Inuyama, Aichi-ken, 484, Japan

William R. A. Osborne Department of Human Genetics, University of Michigan Medical School, Ann Arbor, Michigan 48104

Janardan Pandey Biological Sciences Group, The University of Connecticut, Storrs, Connecticut 06268

Sarah B. Pipkin Howard University, Washington, D.C. 20001

Jane H. Potter Howard University, Washington, D.C. 20001

Jeffrey R. Powell Department of Biology, Yale University, New Haven, Connecticut 06520

Dennis A. Powers Department of Biology, Johns Hopkins University, Baltimore, Maryland 21218

Dianne Powers Department of Biology, Johns Hopkins University, Baltimore, Maryland 21218

William Prychodko Department of Biology, Wayne State University, Detroit, Michigan 48202

Paul R. Ramsey Department of Biology, Presbyterian College, Clinton, South Carolina 29325

M. E. Richardson Department of Zoology, University of Texas, Austin, Texas 78712

R. H. Richardson Department of Zoology, University of Texas, Austin, Texas 78712

Barnett B. Rosenblum Department of Medical Genetics, Indiana University Medical Center, Indianapolis, Indiana 46202

Yoshikazu Sado Department of Zoology, Faculty of Science, Hokkaido University, Sapporo, Japan

Stanley N. Salthe Department of Biology, Brooklyn College – CUNY, Brooklyn, New York 11210

John G. Scandalios Genetics Laboratory, Department of Biology, University of South Carolina, Columbia, South Carolina 29208

Jörg Schmidtke Institut für Humangenetik und Anthropologie, der Universität, D-7800 Freiburg, West Germany

F. Schwind Laboratory of Plant Genetics, Vrije Universiteit Brussel, B-1640 Sint-Genesius-Rode, Belgium

Robert K. Selander Department of Biology, University of Rochester, Rochester, New York 14627

J. B. Shaklee Department of Zoology, University of Illinois, Urbana, Illinois 61801

Takayoshi Shotake Department of Variation Research, Primate Research Institute, Kyoto University, Inuyama, Aichi-ken 484, Japan

Michael J. Siciliano University of Texas System Cancer Center, M.D. Anderson Hospital and Tumor Institute, Houston, Texas 77025

Charles F. Sing Department of Human Genetics, University of Michigan Medical School, Ann Arbor, Michigan 48104

Michael H. Smith Savannah River Ecology Laboratory, Drawer E, Aiken, South Carolina 29801

P. E. Smouse Department of Human Genetics, University of Michigan, Ann Arbor, Michigan 48104

Eugenia Springer University of Maryland, College Park, Maryland

John J. Stegeman Department of Biology, Woods Hole Oceanographic Institution, Woods Hole, Massachusetts 02543

W. W. M. Steiner Department of Genetics, University of Hawaii, Honolulu, Hawaii 96822

Clyde Stormont Department of Reproduction, School of Veterinary Medicine, University of California, Davis, California 95616

Yoshiko Suzuki Department of Reproduction, School of Veterinary Medicine, University of California, Davis, California 95616

R. Szabadi Department of Pharmacology, New York University, New York, New York 10016

Robert J. Tanis Department of Human Genetics, University of Michigan Medical School, Ann Arbor, Michigan 48104

Richard E. Tashian Department of Human Genetics, University of Michigan Medical School, Ann Arbor, Michigan 48104

Alan R. Templeton Society of Fellows, University of Michigan Medical School, Ann Arbor, Michigan 48104

Diane Thomas University of Texas Health Science Center, Graduate School of Biomedical Sciences, Texas Medical Center, Houston, Texas 77025

Lim Soo Thye Department of Biochemistry, University of Otago, Dunedin, New Zealand

Philip L. Townes Departments of Anatomy and Pediatrics, University of Rochester School of Medicine and Dentistry, Rochester, New York 14642

D. N. Tretiak Canada Department of the Environment, Fisheries and Marine Service, Freshwater Institute, Winnipeg, Manitoba, Canada R3T 2N6

Fred M. Utter Northwest Fisheries Center, National Marine Fisheries Service, National Oceanic and Atmospheric Administration, Seattle, Washington 98112

R. C. Vrijenhoek Department of Zoology, Rutgers University, New Brunswick, New Jersey 08903

J. R. Wall Department of Biology, George Mason University, Fairfax, Virginia 22030

Sarah W. Wall Department of Biology, George Mason University, Fairfax, Virginia 22030

Jewell C. Ward Department of Medical Genetics, Indiana University Medical Center, Indianapolis, Indiana 46202

Marion C. Watt University of Texas System Cancer Center, M.D. Anderson Hospital and Tumor Institute, Houston, Texas 77025

W. W. Weber, Department of Pharmacology, University of Michigan, Ann Arbor, Michigan 48104

Mark L. Weiss Department of Anatomy, Wayne State University Medical School, Detroit, Michigan 48202

Lowell R. Weitkamp Departments of Anatomy and Pediatrics, University of Rochester School of Medicine and Dentistry, Rochester, New York 14642

G. S. Whitt Department of Zoology, University of Illinois, Urbana, Illinois 61801

Nöel P. Wilkins Department of Zoology, University College, Galway, Ireland

Christopher Wills Department of Biology, University of California, San Diego, La Jolla, California 92037

Ulrich Wolf Institut für Humangenetik und Anthropologie der Universität, D-7800 Freiburg, West Germany

David A. Wright Biology Department, M.D. Anderson Hospital and Tumor Institute, Houston, Texas 77025

Toshio Yamauchi Department of Biology, M.D. Anderson Hospital and Tumor Institute, Houston, Texas 77025

Tsuneyuki Yamazaki National Institute of Genetics, Mishima 411, Japan

Satoshi Yonezawa Department of Zoology, Faculty of Science, Hokkaido University, Sapporo, Japan

Akira Yoshida Department of Biochemical Genetics, City of Hope National Medical Center, Duarte, California 91010

Paul L. Zubkoff Department of Environmental Physiology, Virginia Institute of Marine Science, Gloucester Point, Virginia 23062

Preface

Isozymes are now recognized, investigated, and used throughout many areas of biological investigation. They have taken their place as an essential feature of the biochemical organization of living things. Like many developments in the biomedical sciences, the field of isozymes began with occasional, perplexing observations that generated questions which led to more investigation and, finally, with the application of new techniques, to a clear recognition and appreciation of a new dimension of enzymology.

The area of isozyme research is only about 15 years old but has been characterized by an exponential growth. Since the recognition in 1959 that isozymic systems are a fundamental and significant aspect of biological organization, many thousands of papers have been published on isozymes. Several hundred enzymes have already been resolved into multiple molecular forms, and many more will doubtless be added to the list. In any event, it is now the responsibility of enzymologists to examine every enzyme system for possible isozyme multiplicity.

Two previous international conferences have been held on the subject of isozymes, both under the sponsorship of the New York Academy of Sciences—the first in February 1961, and the second in December 1966. And now, after a somewhat longer interval, the Third International Conference was convened in April 1974, at Yale University. For many years, a small group of investigators has met annually to discuss recent advances in research on isozymes. They have published a bulletin each year and have generally helped to shape the field; in effect, they have been a standing committee for this area of research. From this group emerged the decision to hold a third international conference, and an organizing committee of five was appointed to carry out the mandate for convening a third conference. This Third International Conference was by far the largest of the three so far held with 224 speakers representing 21 countries, and organized into nine simultaneous sessions for three days on April 18, 19, and 20, 1974. Virtually every speaker submitted a manuscript for publication, and these total almost 4,000 pages. The manuscripts have been collected into four volumes entitled, *I. Molecular Structure; II. Physiological Function; III. Developmental Biology; and IV. Genetics and Evolution.* The oral reports at the Conference and the submitted manuscripts cover a vast area of biological research. Not every manuscript fits precisely into one or another of the four volumes, but the most appropriate assignment has been made wherever possible. The quality of the volumes and the success of the Conference must be credited to the participants and to the organizing committee. The scientific community owes much to them.

Acknowledgments

I would like to acknowledge the help of my students and my laboratory staff in organizing the Conference and in preparing the volumes for publication. I am grateful to my wife, Margaret Markert, for volunteering her time and talent in helping to organize the Conference and in copy editing the manuscripts.

Financial help for the Conference was provided by the National Science Foundation, the National Institutes of Health, International Union of Biochemistry, Yale University, and a number of private contributors:

Private Contributors

American Instrument Company
Silver Spring, Maryland 20910

Canalco
Rockville, Maryland 20852

CIBA-GEIGY Corporation
Ardsley, New York

Gelman Instrument Company
Ann Arbor, Michigan 48106

Gilford Instrument Laboratories, Inc.
Oberlin, Ohio

Hamilton Company
Reno, Nevada 89502

Kontes Glass Company
Vineland, New Hampshire 08360

The Lilly Research Laboratories
Indianapolis, Indiana

Merck Sharp & Dohme
West Point, Pennsylvania

Miles Laboratories, Inc.
Elkhart, Indiana

New England Nuclear
Worcester, Massachusetts 01608

Schering Corporation
Bloomfield, New Jersey

Smith Kline & French Laboratories
Philadelphia, Pennsylvania

Warner-Lambert Company
Morris Plains, New Jersey

GENES, ISOZYMES, AND EVOLUTION

JOHN G. SCANDALIOS
Genetics Laboratory, Department of Biology
University of South Carolina, Columbia, S.C. 29208

INTRODUCTION

Mutation and gene recombination are the major processes that generate the wealth of genetic variability that underlies the evolutionary process. Since Garrod formulated the concept of inborn errors of metabolism, it has been appreciated that genetic variation can be expressed via altered enzyme activity. Under the influence of Beadle's "one gene-one enzyme hypothesis" many geneticists have studied and further clarified the relationships between the genome and biochemical phenotype.

Although enzyme multiplicity was observed and described many years ago (Theorell, 1940) the biological significance of this phenomenon was not appreciated until fairly recently. With the advent of new and more refined biochemical tools in recent years, there has been an increasing awareness that enzyme multiplicity is the rule rather than the exception, as previously thought. Along with this knowledge came an accumulation of experimental data from various sources clearly pointing to the fact that multiple molecular forms of enzymes were not all physical artifacts as earlier thought by the physical chemists, but that in many instances these multiple forms were products of specific genes; furthermore, this phenomenon presented the investigator with a powerful tool to probe a large variety of biological problems. Markert and Møller (1959) coined the term _isozyme_ as an operational definition to encompass the multiple enzyme forms in biological systems. The term has since been widely adopted in the genetic and biochemical literature because of its convenience and because it sharply points up the generality and biological significance of multiple forms of enzymes. Since the in-depth definition and classification of isozymes are dealt with in some detail by a number of authors in these Proceedings, I will not dwell on that point, but I do refer the reader to the recent report on nomenclature of isozymes by the Subcommittee on Multiple Molecular Forms of Enzymes of the IUPAC-IUB (1971).

To date, a large number of isozyme systems have been investigated in a wide variety of organisms across the evolutionary scale. The elucidation of a particular isozyme system frequently involves both genetic and biochemical methodology. From such studies it has become apparent that isozymes may be generated in cells in a variety of ways. Generally speaking, isozymes can be products of different genetic sites, or they may result from secondary alterations in the structure of single polypeptide species.

One genetic mechanism by which isozymes can be generated is by gene duplication with subsequent mutations at daughter and parental loci. Two genes are capable of generating a variety of different isozymic forms if a number of functional multimers can be formed. In cases where homomultimers can be synthesized from subunits encoded in a single gene, the presence of allelic variants of this gene may give rise to hybrid heteromultimeric isozymes (composed of subunits from each of two allelic genes). Hybrid isozymes can also be generated by non-allelic gene interactions (e.g., LDH in animal systems and the catalase heteromultimers generated by subunits from the non-allelic genes Ct_1 and Ct_2 during maize development).

Multiple forms of enzymes may also arise by non-genetic chemical or physical means. This category of isozymes however, is discussed elsewhere (Markert and Whitt, 1968; Scandalios, 1974) and by several authors in these Proceedings.

In this brief overview, I shall attempt to stress with examples when possible, the usefulness of isozymes whose genetics is well understood as effective markers in studies of various aspects of genetics and evolution.

The use of isozymes as genetic markers has increased dramatically over the last decade. Isozymes offer a number of important advantages over more conventional morphological markers. Isozyme variants frequently occur spontaneously and seldom produce obviously deleterious effects. Variant alleles are generally codominant making it possible to easily and positively identify heterozygotes as well as homozygotes, and to monitor conveniently the time of expression of the parental alleles in the heterozygote. Since isozymes represent specific gene products, variants are more likely to represent single gene lesions than are complex morphological markers. Aside from the obvious utility of isozymes to the protein chemist concerned with problems of enzyme structure and function, they have been used by geneticists to estimate the degree of genetic polymorphism present in populations, genetic differences within and between species, differential gene expression during development, gene dosage effects on enzyme structure and function, intra- and intergenic complementation and heterosis, and the evolution of genes and organisms.

The area of population genetics has in many ways been "revitalized" by the discovery of large amounts of electrophoretically detectable genetic variation within natural populations. In *Drosophila pseudoobscura*, it has been estimated that 30% of the structural gene loci within any population are polymorphic, and an average individual is likely to be heterozygous at about 12% of its loci (Lewontin and Hubby, 1966); in man, of 71 loci examined 28% show polymorphism and it has been estimated that 6.7% heterozygosity exists per individual (Harris and Hopkinson, 1972); and in plants Allard and his associates (1971) have estimated that 31%-54% of the loci examined in *Avena sp*. are polymorphic.

The vast amounts of genetic isozyme polymorphism in diverse groups of organisms has provoked several hypotheses in a search for an explanation. The so called "neutral" theory of genic polymorphism has been expounded best by Kimura (1968). This hypothesis suggests that most of the observed variation is physiologically irrelevant and that the alleles are adaptively equivalent and their variation random. The alternate or "selection" hypothesis asserts that the polymorphism observed within populations is a consequence of balancing forms of natural selection, especially heterosis. These hypotheses and recent developments regarding them have eloquently been discussed in recent reviews by Lewontin (1973) and Johnson (1973). Recent experimental evidence (Johnson, 1974; Wills, 1974; Wills and Nichols, 1973; Scandalios et al., 1972; Felder and Scandalios, 1971) supports the notion that polymorphism among enzyme loci is not selectively neutral, but is related to physiological and metabolic function, and that the allelic isozymes may have subtle but physiologically significant differences. Data from our own investigations with maize catalase and alcohol dehydrogenase clearly show that hybrid isozymes generated by allelic or non-allelic interactions have "improved" physico-chemical properties when compared to the parental molecules; this may be metabolically advantageous to the individual carrying them (Scandalios et al. 1972; Felder and Scandalios, 1973). Resolution of the issues between these two theories is a major and fruitful preoccupation of population geneticists today and has led to much experimental and theoretical work. That the evolutionary role of enzyme polymorphisms may be to provide metabolic flexibility in a changeable environment is becoming more apparent; however, more experimental verification is in order.

Mutations at duplicated loci coding for a specific enzyme may lead to altered enzymatic functions and bring about the evolution of new enzymes as exemplified by the phosphagen kinases (Watts and Watts, 1968).

Isozymes are effectively being employed in studies to ascertain phylogenetic relationships. Hart (1970) has successfully used the isozymes of alcohol dehydrogenase of hexaploid wheat to demonstrate that in nullisomic-tetrasomic lines the genes coding for ADH subunits are located on chromosomes 4A, 4B, and 4D--the homoeologous chromosomes from each of the three diploid progenitors.

In studies of regulation of gene function, isozymes have been used by a number of investigators as markers and have proven to be ideal gene products for the analysis of gene expression during development and differentiation of eukaryotic cells.

The ability to lose or gain specific biochemical characteristics is a common feature of cellular development and differentiation. Qualitative and quantitative changes in protein (enzyme) activity are two principal alterations encountered in the life cycle of most organisms. The appearance of new or increased enzyme activities in a developing organism may result from the *de novo* synthesis of the enzyme molecule or from the activation of pre-existing enzyme precursors. Although the exact mechanisms by which higher organisms regulate the differential expression of macromolecules are not yet understood, it is certain that such changes require regulation at the gene level or in the pathway between the gene and its final product, the functional enzyme. Isozymes are thus an expression of the differentiation of cells, and a detailed analysis of their changing patterns and properties during development may lead to some understanding of basic genetic and metabolic mechanisms underlying cellular differentiation.

There are now a number of cases in both plants and animals where isozymes have been successfully employed as markers to study factors controlling differential gene expression. The most recent reviews on the subject are those by Vesell and Fritz (1971), Whitt, Miller, and Shaklee (1973), and Scandalios (1974). Additionally, a number of papers in these Proceedings deal with the topic of developmental genetics and differential gene expression in a variety of organisms.

The fact that subunits of multimeric isozyme variants can frequently interact to generate heteromultimeric species of enzymes in heterozygotes renders them excellent markers in somatic cell genetic analysis; in such studies, isozymes can be used to ascertain co-expression of two variant genes within a single cell (Ruddle, 1969). Until recently, before isozymes were effectively employed in genetic analysis, human genetic analysis was largely restricted to family and population studies - a rather limited approach which generated only meager data regarding autosomal linkage relationships (Stern, 1973). With the advent of such techniques as somatic cell hybridization and the utilization of isozymes as gene markers much progress has been made toward mapping the human genome.

The above are but a few of the areas and problems where isozymes have and are being used effectively in efforts to answer basic genetic questions. Upon examining the four volumes emanating from this conference the wide applicability of isozymes to biological problems will become apparent. Additionally, on comparing this conference to the previous two in the series, it becomes obvious that there has been a welcome departure from heavy reliance upon electrophoresis as the sole, though still the simplest, technique for resolution of isozymes. In recent years there has been a deeper examination of isozymes by both the new and the older conventional tools of the protein chemist leading to the discovery of subtle but crucial differences between isozymes. In fact, when we consider that most estimates of genetic enzyme polymorphism are based solely on electrophoretic differences, then we must conclude that in spite of the large amount of variability reported, we are grossly underestimating the amount of genetic variation. For example, mutations which insert or delete neutral amino acids from specific polypeptide chains are not likely to be detected by differences in electrophoretic mobility. Yet such mutations may lead to altered physicochemical properties of the molecules and affect cellular metabolism so as to alter the adaptive ability of the organism that carries one or another of the variants, or both as in heterozygotes.

Based on the results presented at this conference, it is obvious that isozymes will be used even more extensively in the future and will help to resolve many old questions and will also surely generate new and exciting questions in genetics and evolution.

ACKNOWLEDGEMENT

Work from the author's laboratory was supported by the U.S. Atomic Energy Commission under contract A T (38-1)-770.

REFERENCES

Allard, R.W. and A.L. Kahler 1971. Allozyme polymorphisms in plant populations. *Stadler Symp.* (Univ. Mo.) 3:9-24.

Felder, M.R. and J.G. Scandalios 1971. Effects of homozygosity and heterozygosity on certain properties of genetically defined electrophoretic variants of alcohol dehydrogenase isozymes in maize. *Molec. Gen. Genetics* 111:317-326.

Felder, M.R., J.G. Scandalios, and E. Liu 1973. Purification and partial characterization of two genetically defined alcohol dehydrogenase isozymes in maize. *Biochem. Biophys. Acta.* 318:149-159.

Harris, H. and D.A. Hopkinson 1972. Average heterozygosity per locus in man: an estimate based on the incidence of enzyme polymorphisms. *Ann. Hum. Genet.* 36:9-20.

Hart, G.E. 1970. Evidence for triplicate genes for alcohol dehydrogenase in hexaploid wheat. *Proc. Nat. Acad. Sci. U.S.A.* 66:1136-1141.

IUPAC-IUB Commission on Biological Nomenclature 1971. The nomenclature of multiple molecular forms of enzymes recommendations. *Arch. Biochem. Biophys.* 147:1-3.

Johnson, G.B. 1973. Enzyme polymorphism and biosystematics: The hypothesis of selective neutrality. *Ann. Rev. Ecol. System* 4:93-116.

Johnson, G.B. 1974. Enzyme polymorphism and metabolism. *Science* 184:28-37.

Kimura, M. 1968. Genetic variability maintained in a finite population due to mutational production of neutral and nearly neutral isoalleles. *Genet. Res.* 11:247-269.

Lewontin, R.C. and J.L. Hubby 1966. A molecular approach to the study of genetic heterozygosity in natural populations. II. Amount of variation and degree of heterozygosity in natural populations of *Drosophila pseudoobscura*. *Genetics* 54:595-609.

Lewontin, R.C. 1973. Population genetics. *Ann. Rev. Genet.* 7:1-17.

Markert, C.L. and F. Møller 1959. Multiple forms of enzymes: Tissue, ontogenetic, and species specific patterns. *Proc. Nat. Acad. Sci. U.S.A.* 45:753-763.

Markert, C. L. and G.S. Whitt 1968. Molecular varieties of isozymes. *Experientia* 24:977-991.

Ruddle, F.H. 1969. Enzyme and genetic linkage studies employing human and mouse somatic cells. In: *Problems in Biology: RNA in Development*. E. Hanly, ed. Univ. Utah Press, Salt Lake City.

Scandalios, J.G., E. Liu, and M. Campeau 1972. The effects of intragenic and intergenic complementation on catalase structure and function in maize: A molecular approach to heterosis. *Arch. Biochem. Biophys.* 153:695-705.

Scandalios, J.G. 1974. Isozymes in development and differentiation. *Ann. Rev. Plant Physiol.* 25:225-258.

Stern, C. 1973. *Principles of Human Genetics.* W.H. Freeman and Co., San Francisco.

Vesell, E.S. and P.J. Fritz 1971. Factors affecting the activity, tissue disstribution, synthesis and degradation of isozymes. In: *Enzyme Synthesis and Degradation in Mammalian Systems.* M. Rechcigl, ed. Univ. Park Press, Maryland.

Watts, R.L. and D.C. Watts 1968. Gene evolution and the evolution of enzymes. *Nature* 217:1125-1130.

Whitt, G.S., E.T. Miller, and J.B. Shaklee 1973. Developmental and biochemical genetics of lactate dehydrogenase isozymes in fishes. In: *Genetics and Mutagenesis of Fish.* J.H. Schroder, ed. Springer-Verlag, Berlin.

Wills, C. and L. Nichols 1973. How genetic background masks single-gene heterosis in *Drosophila. Proc. Nat. Acad. Sci. U.S.A.* 69:323-325.

Wills, C.J. 1975. Selective pressures on isozymes in *Drosophila. IV. Isozymes: Genetics and Evolution.* C. L. Markert, ed., Academic Press, New York.

ISOZYMES AND NON-DARWINIAN EVOLUTION: A RE-EVALUATION

JEFFREY R. POWELL
Department of Biology
Yale University
New Haven, Connecticut 06520

ABSTRACT. The controversy over the importance of stochastic processes in molecular evolution is reviewed. Experimental data relevant to this controversy are obtained from three different sources: isozyme techniques (especially electrophoresis), amino acid sequences, and nucleic acid hybridization studies. The interrelationship of data obtained by these techniques is discussed. The conclusion is reached that, in the main, genetic variation revealed by isozyme techniques is adaptively significant. However, electrophoresis can detect only about 0.1% of the possible nucleotide substantions in a eukaryote genome. The other two sources of information reveal that certain selective constraints are operating on the other 99.9% of the potential variation; however, no conclusion can be drawn whether drift or selection is the predominant force in evolutionary change of this material. Thus the view that natural populations are highly polymorphic for adaptively significant genetic variation is not necessarily incompatible with the view that most evolutionary changes on the nucleotide level are a result of random fixation of neutral mutations.

INTRODUCTION

Traditionally there have been two approaches to biological problems. One is to reduce biologically complex phenomena to relatively simple physio-chemical explanations. This approach has been called reductionist or Cartesian. Other biologists feel that not all biological phenomena can be reduced to physio-chemical explanations, especially when dealing with populations of individuals. They endeavor to discover and study population phenomena which are unique to biological systems and not reducible to chemistry. This latter approach has been called compositionist or Darwinian. (See Nagel, 1961, Simpson, 1964, and Dobzhansky, 1966 and 1968 for further discussion.) That these two approaches are complements rather than alternatives has become increasingly apparent in the last ten years or so. With the growing interest of biochemists in molecular evolution and the increasing use of biochemical techniques by population biologists, there is expanding common ground and understanding between these two groups of biologists.

Three advances are primarily responsible for bridging the gap between the biochemical approach and the population approach. First is the application of isozyme techniques to the study of amounts and patterns of genetic variation in natural and experimental populations. The second is the increase of data available for amino acid sequences of a variety of proteins in many species (Dayhoff, 1969). The third advance is the use of DNA-DNA and DNA-RNA hybridization techniques to determine the types of DNA present in genomes (e.g. repetitive vs. unique) and to compare DNA homologies among species. The purpose of this paper is to very briefly summarize the data relevant to evolutionary theory obtained from these three approaches. The importance of the interrelationship of these approaches will be emphasized especially as it relates to the problem of non-Darwinian evolution.

CLASSICAL VS. BALANCE VIEW AND NON-DARWINIAN EVOLUTION

A major problem of population and evolutionary genetics has been to determine how much genetic variation exists in natural populations. As is often the case, two opposite views have evolved in trying to answer this question. These two views have been called the classical theory of the genetic structure of populations and the balance theory (Dobzhansky, 1970). The classical view maintains that there is a best fit genotype (the "wild-type") which is homozygous at most of its loci for alleles that yield the highest fitness. Selection will act to fix these optimal alleles in the population. Polymorphism and heterozygosity at a locus are rare. Diagrammatically this view has been represented as follows (Dobzhansky, 1970; Wallace, 1968):

$$\frac{A_1\ B_1\ C_1\ D_1\ E_2\ F_1\ G_1\ H_1\ I_1}{A_1\ B_1\ C_1\ D_1\ E_1\ F_1\ G_1\ H_1\ I_1}$$

A fundamental reason for maintaining this view is related to the problem of genetic load: too much genetic variability leads to unbearable genetic loads (Crow and Kimura, 1963; Kimura and Ohta, 1971b; Wallace, 1970).

The alternative view can be diagrammed as follows:

$$\frac{A_1\ B_3\ C_2\ D_1\ E_4\ F_6\ G_1\ H_1\ I_3}{A_2\ B_1\ C_1\ D_1\ E_1\ F_2\ G_3\ H_1\ I_2}$$

In words, polymorphism and heterozygosity are thought to be the rule at most loci, and complete homozygosity or monomorphism is the exception. The notion of an ideal genotype for all environments at all times does not exist. This great amount of heterozygosity is thought to be maintained by various forms of balancing selection, and thus the name balance theory. This selection may be heterotic, frequency-dependent, or some other as yet undiscovered selective mechanism which maintains polymorphism. (See Lewontin 1974 for further discussion and history of these two views.)

With the introduction of isozyme techniques to the study of natural populations (Lewontin and Hubby, 1966; Johnson et al, 1966; Harris, 1966), the balance view appears to have been vindicated. An average of approximately 40% of the protein loci studied in sexually outcrossing species are polymorphic for electrophoretically detectable alleles. (See Johnson 1973 and Selander and Johnson 1973 for recent reviews.) As mentioned below, electrophoresis detects only a fraction of all possible amino acid differences. Assuming these loci are an unbiased sample of all loci, it can be concluded that most loci are polymorphic and heterozygosity is high.

The situation is not quite so simple, however. What remains to be proven is whether these electrophoretically separable proteins confer different fitnesses on the individuals in which they are functioning. In other words, are these differences adaptively significant? Is this the main basis of adaptive evolution? Or is it possible that as far as natural selection is concerned $A_1 \equiv A_2$, $B_3 \equiv B_1$, $C_2 \equiv C_1$, etc.? We would then have to turn back to a variant of the classical view. What the isozyme studies showed was that the loci were polymorphic but not necessarily functionally polymorphic. At about the same time as electrophoresis was first being widely applied to population studies, a theory was being propounded which claimed that most amino acid substitutions have occurred in evolution by genetic drift (King and Jukes, 1969; Kimura, 1968). These authors maintained that many mutations which result in amino acid changes in proteins don't affect the fitnesses of the carriers of the mutations. These mutations would then lead to selectively equivalent alleles and have been called "neutral mutations." If alleles are selectively equivalent, random processes become the most important factor in determining their frequencies: fixation and loss of alleles are stochastic events. The protein polymorphisms detected by electrophoresis are thought to represent a phase in this random evolutionary process (Kimura and Ohta, 1971a). Because in this view the major force of evolutionary change is not natural selection, this theory has been labeled (or mis-labeled) non-Darwinian evolution(King and Jukes, 1969). Regardless of its appropriateness, the term has become widely accepted and will be used here.

THE POPULATION GENETICIST'S APPROACH

Population geneticists have been preoccupied with determining whether drift of selection controls the allelic isozyme (allozyme) frequencies in populations. If the isozyme polymorphisms can be shown to be balanced polymorphisms maintained by selection then the balance view of populations is more correct. If the polymorphisms prove to be effectively neutral, controlled by drift, then the classical picture remains tenable. The classical view has thus evolved (drifted?) to a neo-classical view (Lewontin, 1973) dependent upon the validity of the non-Darwinian theory. Briefly, I will review the types of evidence revealed by studies on the selective significance and mechanisms of maintenance of isozyme polymorphisms. This is not meant to be a complete review of all such work.

Heterosis or overdominance is the best known mechanism able to maintain stable balanced polymorphism. One way of detecting heterosis at a locus is to observe a significant excess of heterozygotes over Hardy-Weinberg expectations after selection has occurred. This is often difficult for several reasons (Powell, 1974a). Nevertheless heterozygote excesses have been observed at a number of electrophoretically polymorphic loci (Richmond and Powell, 1970; Powell, 1974a; Marshall and Allard, 1970) and it is unlikely that these excesses were caused by linkage effects (Powell, 1974b). An ingenious variation on this theme is the work of Wills and Nichols (1971) in which heterozygote excess is only detected when the environment is conditioned with the substrate of the enzyme being observed. This is particularly interesting in light of Parsons' (1971) theory of extreme environment heterosis. Yamazaki (1972) has, however, pointed out a possible flaw in the Wills and Nichols experiment. (See, also, Wills, this volume.)

Schwartz and Laughner (1969) have demonstrated a molecular mechanism for heterosis for electrophoretic variants of the alcohol dehydrogenase locus of maize. One allele codes for a subunit of *Adh* which is stable but relatively inactive. Another allele codes for a subunit which is very active but unstable. The heteropolymers which exist in heterozygotes are both active and stable.

Frequency-dependent selection can also maintain a stable polymorphism. Kojima (1971) and his colleagues have been able to show that in some cases the fitnesses of carriers of electrophoretic alleles change as the frequencies of the alleles in the population change. The fitness of an allele is inversely proportional to its frequency.

What may be considered a special case of frequency-dependent selection is multiple-niche selection. In the absence of heterosis, a stable polymorphism can be maintained if more than one niche is available to a population and alternative genotypes have different fitnesses in the different environments (Levene, 1953). A prediction from this theory is that more genetic variability should be maintained in populations in variable environments than in populations in relatively uniform and constant environments. In an experiment in which the variability of the environment was experimentally controlled, Powell (1971) has shown that the amount of electrophoretically detectable genetic variation maintained in populations of *Drosophila willistoni* is directly proportional to the degree of environmental diversity experienced by the populations. McDonald and Ayala (1974) have confirmed these results using *D. pseudoobscura*. Bryant (1974) has shown a similar relationship between electrophoretic variation in natural populations and temporal environmental variation.

Another way of investigating selection on polymorphisms is to examine geographic patterns of allele frequencies. According to the non-Darwinian theory, allele frequencies should show no systematic correlations among isolated populations. Contrary to this prediction protein polymorphisms in Drosophila do show very strong correlations among widely separated populations (Prakash et al, 1969) and even among populations isolated by hundreds of miles of water (Ayala et al, 1971). Furthermore, according to the non-Darwinian theory, when geographic variation is found it should not be systematically correlated with environmental factors. Allele frequency clines such as found by Koehn (1969) are the expected outcome of selection, not drift. This is especially true when an environmental gradient (such as temperature) can be correlated with differential activity of the electrophoretic alleles, as is the case in Koehn's work. The study by Rockwood-Sluss et al (1973) correlates variation in allele frequencies at several loci with variation in the concentration of certain chemicals in the host plant of *Drosophila pachea*. The works of Allard and his colleagues demonstrate a very strong relationship between multi-locus electrophoretic genotypes and the degree of dryness of the environment of *Avena barbata* (Clegg and Allard,

1972; Hamrick and Allard, 1972).

Another approach to show that electrophoretic variants are not selectively equivalent is to demonstrate that there exist biochemical differences between allelic isozymes which may have important physiological consequences. This has been done rather thoroughly for several human variants (Harris, 1971) as well as for some electrophoretic alleles in *Drosophila* and Lepidoptera (G. B. Johnson, 1974 and personal communication). Similarly, Johnson (1974) has been able to show a strong correlation between the degree of polymorphism at a locus and the physiological function of the enzyme encoded by the locus. Loci coding for enzymes which act on substrates originating external to the organism have the highest degree of variation followed by regulatory enzymes follwed by non-regulatory enzymes acting on substrates produced internally. Such a regular pattern is highly unlikely if this variation is adaptively insignificant.

All of this evidence, taken together, has led many workers to conclude that under most circumstances selection is the major force maintaining electrophoretically detectable genetic polymorphisms in populations. Thus the balance view of population genetics is confirmed. Does this conclusion then rule out the non-Darwinian view? Not necessarily. The question which must be asked is whether electrophoresis detects all possible mutations. If not, is it selecting for types of mutations likely to be most strongly selected? The answer to the first question is obviously no. If one takes a random sequence of nucleotides coding for a protein, one can calculate the consequences of various base changes and estimate the relative frequency of each type of change. The following types of changes can be distinguished (figures based on an extrapolation of calculations by Holmquist et al, 1972):

I. Base changes, no change in amino acid sequences coded for: 32%

II. Changes in amino acids, no change in net electric charge of protein: 46%

III. Changes in amino acids which change charges of proteins: 22%

So far we have only been discussing type III mutants which make up only about 1/5 of the possible mutations. The stimulating recent work of Bernstein et al (1973) is very interesting in this context. By using differential heat denaturation techniques, they have been able to show that type II mutants are prevalent in natural populations. These preliminary data indicate that the selection story may not be nearly so clear for this type of mutant. (Unfortunately, the genetic tests to show that these heat-sensitive alleles are truly Mendelian variants were not reported in this paper.) More studies employing these newer techniques must be done before much can be concluded from them. There are other approaches which may allow us to determine something about the selective significance of type I and II mutations.

AMINO ACID SEQUENCE STUDIES

In dealing with amino acid sequence data we are no longer dealing with existing polymorphisms, but rather with the end result of amino acid substitutions which have occurred since species diverged from common ancestors. According to the non-Darwinian theory, the existing polymorphisms are merely a stage in the stochastic replacement process which produces the observed divergent amino acid sequences. Approximately 2/3 of all amino acid substitutions are the result of type II mutations. Therefore when we gain insights into the selective significance or restrictions of amino acid substitutions in general, we are dealing primarily with type II mutants. There are four conclusions from amino acid sequence comparisons which bear on the problem of non-Darwinian evolution.

First, some amino acid sites in a protein molecule vary considerably from species to species, while other sites are very conservative. This observation has led to the "covarion" theory (Fitch and Markowitz, 1970; Fitch 1972) which states that for any given molecule a certain limited number of codons may vary and still produce a protein which is functioning acceptably. These sites are called concomitantly variable codons or covarions. By extrapolating the percent sites varied among different numbers of species to a single species, the percent of sites which belong to the covarion class in a single species can be determined. This turns out to be about 10% for cytochrome-c (Fitch and Markowitz, 1970) and 25% for the β-δ hemoglobin group (Fitch 1972). Milkman (1974) has reasoned that the covarion class must be a good deal smaller per individual allele than these estimates. Thus only a minority of sites are free to vary in any species and produce an observable polymorphism. An important caveat is that sites which are covarions in one species are not necessarily covarions in another. If one of the covarions changes then the selective constraints on other sites almost always change: some sites which are free to vary are no longer, others which were under strong selective constraints are now freed from these constraints (Fitch, 1972). This will be important later.

The second point is that different proteins evolve at different rates. For example fibrinopeptide A has evolved some 15-20 times faster than cytochrome-c. Histones are extremely slow in changing, only two amino acid differences exist between peas and cattle (Smith et al, 1970). Doubtless, histones are exceptional molecules. If one considers only covarions, on the other hand, then the apparent disparity among molecules in rates of evolutionary change disappears. That is, if instead of considering the rate of substitution for all sites we restrict our attention to rates of substitution per covarion, then the rates for many molecules (excluding histones) appear remarkably uniform (Fitch, 1972). This is very inportant as the non-Darwinian theory predicts that the rate of substitution of neutral changes should equal the mutation rate to neutral mutations (Kimura and Ohta, 1971b). If such mutations occur at about the same rate per covarion in all molecules then the rate of change should be approximately the same in all molecules. This is observed at least for some molecules.

A third prediction from non-Darwinian theory is that rates of neutral substitutions in proteins should be constant throughout evolutionary time. This is "probably the strongest evidence for the theory" (Kimura and Ohta, 1971a). Such constancy is found for a variety of proteins if one calculates rate of change over relatively long periods of time (30 million years or more). However substitutions may be occurring by selection quite rapidly and irregularly. Only when such substitutions are averaged over long periods of time do they appear to occur at a uniform rate. "The entire argument is based upon a confusion between an average and a constant" (Stebbins and Lewontin, 1972). Furthermore a somewhat odd finding falls out of such studies: the rate of change is relatively constant over sidereal time rather then generation time (Jukes and Holmquist, 1972). Thus for a protein to be evolving at the same rate over sidereal time in bacteria and man, the per generation mutation rate would have to be 5 orders of magnitude higher in man than in bacteria. This seems biologically unreasonable.

The fourth piece of evidence concerns the chemical nature of the substituted amino acids. If the covarions of a protein are under little or no selective constraints, then any amino acid should be able to substitute for another. That is, just by considering the nature of the genetic code, one can calculate probabilities of each amino acid substitution irrespective of the chemical nature of the amino acids involved. Have amino acid substitutions occurred independently of their chemical natures? Clarke (1970a) has shown a strong correlation between the chemical nature of the amino acids being replaced and the chemical nature of the amino acids doing the replacing. Thus even within the group of sites known as covarions, there are selective constraints on what substitutions are permissible. Does this mean then that these substitutions have occurred by selection? Not necessarily. One might predict that mutations to codons coding for amino acids of similar chemical properties are the most likely to code for selectively equivalent alleles. Furthermore, Pelc (1965) has pointed out that a single base change in a codon usually produces a codon coding for an amino acid of very similar chemical properties to the original amino acid. Two or more base changes are required to code for an amino acid of radically different chemical properties. This may, in part, account for the observation of Clarke.

These four findings lead to a tentative conclusion. If amino acid substitutions are occurring by random drift of neutral mutations then they are subject to two restrictions: (1) only a minority of sites are involved (covarions) and (2) only chemically similar amino acids can be substituted at these sites. This is not to imply that substitutions which conform to these restrictions are necessarily caused by random drift, but only that substitutions of this type are more likely to be neutral. Thus we have concluded something about type II mutations. Is it possible to say anything about type I mutations, i.e. synonymous mutations?

As pointed out independently by Richmond (1970) and Clarke (1970b) even synonymous mutations may cause a fitness change in an allele. Different t-RNA's recognizing different codons for the same amino acid may have different efficiencies or be in different concentrations in certain cells. Thus changing the t-RNA's needed to build a protein may cause a fitness change. Is there any experimental evidence for this? The only work of which I am aware bearing on this question is that of McCarthy and Farquhar (1972) on histones. As mentioned above, the amino acid sequences of histones have remained remarkably constant over immensely long evolutionary times. Thus there was probably strong selection to maintain the particular amino acid sequence. However, if selection cannot detect synonymous mutations in the third base of a codon ("wobble" position) then the DNA coding for histones should have drifted apart at this position. McCarthy and Farquhar (1972) have compared histone mRNA from a variety of species by very sensitive techniques of DNA-RNA hybridization. The degree of divergence in nucleotide sequence of the genes coding for histones is much less than predicted if third position changes are neutral. Also, within a genome there are multiple copies of histone genes; if synonymous mutations are neutral, then there should be some amount of drifting apart of the histone genes within a genome. As far as McCarthy and Farquhar could detect, there is no intragenomic divergence of histone mRNA's. (Their technique could detect 1% or less nucleotide differences.) Thus at least for histone-coding genes, there appear to be selective constraints even on type I mutations.

CONSIDERATIONS OF TOTAL DNA

So far I have been dealing with only a small part of the total DNA of an organism, that part coding for amino acid sequences of proteins. Let us consider briefly what might be the relative selective constraints on the total DNA of an organism.

The first consideration concerns the functions of different parts of the genome. I will distinguish six classes of DNA each having a different function (or no function at all):

A. "Junk DNA," having no function, evolutionary relic or mistake.

B. Non-transcribed, but serving a function such as structure.

C. Transcribed, not translated; e.g. DNA coding for nucleus restricted RNA.

D. Transcribed, translated, producing a functional protein; e.g. enzymes.

E. DNA involved in gene regulation; e.g. recognition sites for polymerases or controlling molecules.

F. DNA is transcribed and RNA is final functioning product; e.g. rRNA, tRNA.

So far we have been only been dealing with class D DNA. Can we make a prediction as to the relative selective constraints on each class and thus predict the probability of truly neutral mutations occurring in each class? I would like to suggest that the degree of selective constraint is in the order listed, i.e. "junk DNA" has the least constraints and DNA coding for functioning RNA molecules has the most constraints. I will briefly review some evidence to support this contention.

Almost by definition, it is easy to conclude that nonfunctional DNA (class A) should have essentially no selective constraints. The proportion of the DNA in this class or even the existence of such DNA may be questioned, but arguments have been proferred for its existence (Ohno, 1972).

There is some evidence for ranking class C as having fewer constraints than the classes below it. Shearer and McCarthy (1970) have compared the divergence of nucleus-restricted RNA and cytoplasmic RNA of the mouse with the rat, hamster, guinea pig, and rabbit. Nucleus-restricted RNA was more divergent in these species than cytoplasmic RNA. This implies that DNA coding for RNA which never leaves the nucleus, and is thus not translated, is freer to change than DNA coding for RNA which is translated in the cytoplasm.

Because rRNA cistrons (rDNA) are relatively easy to isolate, much work has been done on their evolution (Birnstiel et al, 1972; McCarthy and Farquhar, 1972). This has been done primarily by the technique of nucleic acid hybridization. The following general picture has emerged: rDNA evolves more slowly than the bulk of DNA. Within the rDNA cistron the area coding for the functioning subunits (23S and 16S in prokaryotes and 28S and 18S in eukaryotes) evolves more slowly than the DNA coding for the segment of the rRNA precursor molecule which is cleaved off and not functional. This later DNA would be class C DNA i.e. transcribed DNA but the RNA acts only as a spacer between the functional RNA's of the cistron. Within the area of the rRNA cistron encoding for the functional subunits, evolution has been remarkably slow. Among species the greatest differences occur in the length of spacer DNA. The relative conservatism of rRNA cistrons holds true for mitochondrial rDNA as well (Dawid, 1972). As with histone genes, multiple copies of rDNA cistrons are known to occur within a genome. No differences have been detected among cistrons within a genome. An internal repair mechanism has been postulated to account for this intragenomic indentity of multiple gene copies (Callan, 1967). The reason rDNA is so conservative is probably due to the fact that selection can act on every base in the region coding for functioning rRNA. There are no "wobble" positions as in DNA coding for proteins.

This sketchy evidence does not prove the ranking listed above, but it is all consistent with the ranking. Nothing has been said about gene control sequences (class E) as nothing is really known about them. The ranking here is however plausible: mutations which result in a somewhat less than optimally functioning protein are more permissible than mutations in regulatory sites which result in no molecule being produced or at the wrong time or in wrong concentration.

How much of the DNA falls into each class? This is impossible to answer with precision. In the context of the isozyme studies what is most relevant to the non-Darwinian controversy is how much of the DNA is coding for proteins. Various estimates have been given for the number of structural genes in eukaryotes. I will use a number which is probably correct at least to an order of magnitude. Let us assume that the average mammalian genome contains 3×10^4 structural genes each coding for an average of 200 amino acids per gene. There are on the average about 6.5 x 10^9 DNA nucleotides per diploid nucleus in mammals. This means that about 0.25% of the nucleotides are directly involved in coding for structural genes or 0.5% if one considers that the complementary strand of the DNA directly involved is also under the same selective constraints. The exact figures may not be correct, but the conclusion that the great majority of the DNA is not structural genes is inescapable. Another way of arriving at the same conclusion is to realize that 93% of known enzyme activities are present in both prokaryotes and mammals (Britten and Davidson, 1971) and yet there has been approximately a 650 fold increase in the amount of DNA from bacteria to mammals (Sparrow et al, 1972).

CONCLUSIONS

What has this excursion into molecular biology told us about the non-Darwinian controversy? First, workers using isozyme techniques are studying only a small fraction of the total genome. If all the proteins of an organism were subjected to electrophoretic analysis only about 20% of the mutations in about 0.5% of the DNA - or 0.1% of the total possible nucleotide substitutions - would be detected. I have tried to review briefly the evidence that this 0.1% of the possible variation is usually subject to natural selection. To determine the true extent of polymoprhism or genetic divergence among taxa, population geneticists would want to know the DNA sequences of individuals in a population. Also one would want to know the function of each area of the DNA molecule to be able to interpret the importance of base differences. If a technique could be devised to sequence DNA at the remarkable rate of 1 base/sec it would take 4 months to sequence a bacterium and about 100 years for a mammal (Hoyer and Roberts, 1967). Obviously this is not a practical way to proceed. Realizing the incompleteness of the electrophoretic approach other approaches must be sought. The heat-stability tests of Bernstein et al (1973) or the subunit interaction studies of MacIntyre (1972) are attempts in this direction and should reveal something about type II mutations. Nucleic acid hybridization is another fruitful approach.

One of the implications of the considerations here concerns the nature of the argument between selectionists and neutralists. Kimura (1968) argued that the substitutional load would be too great if all DNA base changes occur by selection. Some experimental population geneticists argue that certain isozyme polymorphisms are selectively maintained. Are these two notions incompatible? I think not. Much of the polemics between the two groups has occurred on two different levels - is it surprising then that neither side has convinced the other of the validity of their stance? They are often arguing about two different things. Furthermore, no one claims that all evolutionary changes occur by selection or all by drift. Any observer of the organic world realizes that adaptive evolution has occurred. As pointed out by Crow (1972) the argument is similar to the nature-nurture argument. Once the necessary statistical procedures were worked out and experiments done, it was recognized that the argument is really a non-argument. It seems to me that the balance view of population genetics can be true and still leave room for considerable non-Darwinian evolution. Little is known of the adaptive nature of the 99.9% of the potential variation and evolutionary changes below the resolving power of electrophoretic isozyme techniques. As outlined above there are some constraints on this variation, but how much?

If, as has been argued, the variation detected by isozyme techniques is adaptively significant,why worry about the "evolutionary noise" occurring below the resolving power of our technique? As evolutionists aren't we interested in the stuff which evolves adaptations? First it isn't clear that isozyme techniques detect all the adaptively important variation. Almost certainly it doesn't. Consider for example the importance of regulation of gene activity in adaptive evolution. Even if we could detect and study all the adaptively significant variation, there are still reasons to be concerned with neutral changes. Neutral changes may affect adaptive evolution. For example let us consider Fitch's covarion theory and assume that substitutions at the variable sites are indeed neutral. After one site has gone through an amino acid substitution by drift, the selective constraints on several other sites will be changed. This could radically change the adaptive potential of the molecule if adaptive changes become necessary. Or let us assume that a species becomes divided into two isolated populations; each population started with a certain amino acid, A, at a crucial site in a molecule. One population by chance has a neutral substitution of A→B while the other population remains A. Now assume a drastic environmental shift occurs which requires amino acid C at the position. The codon at this crucial site is such that one base change changes the codon from A→B and another single base changes B→C; A→C requires two base changes. The probability that the population which still has A fixed will have the necessary double mutation is very low. It may well go extinct. On the other hand the population which by chance has already fixed B, will stand a much better chance of getting the single mutation necessary to get to C and will probably survive. Chance determined which population could made an adaptive shift.

Thus, evolution is not controlled by either stochastic processes or deterministic processes. It is controlled by a combination of both. Some evolutionary geneticists have recognized this fact for many decades (Wright, 1931; Dobzhansky, 1937). With the recent advances in evolutionary theory brought about by isozyme and other biochemical techniques, the importance of considering the interaction of evolutionary forces has been greatly emphasized.

ACKNOWLEDGMENTS

I thank the following people for constructive criticism of an early draft of this paper: Th. Dobzhansky, R. Milkman, M. Watson, B. Wilcox, H. Wistrand, R. Richmond and G. Johnson.

REFERENCES

Ayala, F. J., J. R. Powell, and Th. Dobzhansky 1971. Polymorphisms in continental and island populations of *Drosophila willistoni*. *Proc. Natl. Acad. Sci. USA* 68: 2480-2483.

Bernstein, S. C., L. H. Throckmorton, and J. L. Hubby 1973. Still more variability in natural populations. *Proc. Natl. Acad. Sci. USA* 70: 3928-3931.

Birnstiel, M. L., M. Chipchase, and J. Speirs 1972. The ribosomal RNA cistrons. *Prog. Nuc. Acid Res. and Mol. Biol.* 11: 351-389.

Britten, R. J. and E. H. Davidson 1971. Repetitive and non-repetitive DNA sequences and a speculation on the origins of evolutionary novelty. *Quart. Rev. Biol.* 46: 111-138.

Britten, R. J. and D. E. Kohne 1968. Repeated sequences in DNA. *Science* 161: 529-540.

Bryant, E. H. 1974. On the adaptive significance of enzyme polymorphisms in relation to environmental variability. *Am. Nat.* 108: 1-19.

Callan, H. J. 1967. The organization of genetic units in chromosomes. *J. Cell Science* 2: 1-11.

Clarke, B. 1970a. Selective constraints on amino-acid substitutions during the evolution of proteins. *Nature* 228: 159-160.

Clarke, B. 1970b. Darwinian evolution and proteins. *Science* 168: 1009-1011.

Clegg, M. T. and R. W. Allard 1972. Patterns of genetic differentiation in the slender wild oat species *Avena barbata*. *Proc. Natl. Acad. Sci. USA* 69: 1820-1824.

Crow, J. F. 1972. Darwinian and non-Darwinian evolution. *Proc. Sixth Berkeley Symp. Math. Stats. and Prob.* 5: 1-22.

Crow, J. F. and M. Kimura 1963. The theory of genetic loads. *Proc. XI Int. Cong. Genetics* 3: 495-506.

Dawid, I. B. 1972. Evolution of mitochondrial DNA sequences in *Xenopus*. *Devel. Biol.* 29: 139-151.

Dayhoff, M. 1969. *Atlas of protein sequence*. Nat. Biomed. Res. Found.; Silver Springs, Md.

Dobzhansky, Th. 1937. *Genetics and the origin of species*. Columbia University Press, New York.

Dobzhansky, Th. 1966. Are naturalists old-fashioned? *Am. Nat.* 100: 541-550.

Dobzhansky. Th. 1968. On Cartesian and Darwinian aspects of biology. *The Graduate Journal* 8: 99-117.

Dobzhansky, Th. 1970. *Genetics of the evolutionary process*. Columbia University Press, New York.

Fitch, W. M. 1972. Does the fixation of neutral mutations form a significant part of observed evolution in proteins? In *Evolution of Genetic Systems*, Brookhaven Symp. Biol. No. 23. H. H. Smith, ed. pp. 186-216.

Fitch, W. M. and E. Markowitz 1970. An improved method for determining codon variability in a gene and its application to the rate of fixation of mutations in evolution. *Biochem. Genet.* 4: 579-593.

Hamrick, J. L. and R. W. Allard 1972. Microgeographic variation in allozyme frequencies in *Avena barbata*. *Proc. Natl. Acad. Sci. USA* 69: 2100-2104.

Harris, H. 1966. Enzyme polymorphism in man. *Proc. Roy. Soc.* London, B 164: 298-310.

Harris, H. 1971. Polymorphism and protein evolution: the neutral mutation - random drift hypothesis. *J. Med. Genet.* 8: 444-452.

Holmquist, R., C. Cantor, and T. Jukes 1972. Improved procedures for comparing homologous sequences in molecules of proteins and nucleic acids. *J. Mol Biol.* 64: 145-161.

Hoyer, B. H. and R. B. Roberts 1967. Studies of DNA homology by the DNA-agar technique. In *Molecular Genetics*, Vol. II. J. H. Taylor, ed. Academic Press, New York.

Johnson, F. M., C. Kanapi, R. H. Richardson, M. R. Wheeler, and W. S. Stone 1966. An analysis of polymorphisms among isozyme loci in dark and light *Drosophila ananassae* strains from American and Western Samoa. *Proc. Natl. Acad. Sci. USA* 56: 119-125.

Johnson, G. B. 1973. Enzyme polymorphisms and biosystematics: the hypothesis of selective neutrality. *Ann. Rev. Ecol. Syst.* 4: 93-116.

Johnson, G. B. 1974. Enzyme polymorphism and metabolism. *Science* 184: 28-37.

Jukes, T. H. and R. Holmquist 1972. Estimation of evolutionary

changes in certain homologous polypeptide chains. *J. Mol. Biol.* 64: 163-179.
Kimura, M. 1968. Evolutionary rate at the molecular level. *Nature* 217: 624-626.
Kimura, M. and T. Ohta 1971a. Protein polymorphism as a phase of molecular evolution. *Nature* 229: 467-469.
Kimura, M. and T. Ohta 1971b. *Theoretical aspects of population genetics.* Princeton University Press, Princeton, N. J.
King, J. L. and T. H. Jukes 1969. Non-Darwinian evolution. *Science* 164: 788-798.
Koehn, R. K. 1969. Esterase heterogeneity: dynamics of a polymorphism. *Science* 163: 943-944.
Kojima, K. I. 1971. Is there a constant fitness for a given genotype? No! *Evolution* 25: 281-285.
Levene, H. 1953. Genetic equilibrium when more than one ecological niche is available. *Am. Nat.* 87: 331-333.
Lewontin, R. C. 1973. Population genetics. *Ann. Rev. Genet.* 7: 1-17.
Lewontin, R. C. 1974. *The Genetic Basis of Evolutionary Change.* Columbia University Press, New York.
Lewontin, R. C. and J. L. Hubby 1966. A molecular approach to the study of genic heterozygosity in natural populations. II. Amount of variation and degree of heterozygosity in natural populations of *Drosophila pseudoobscura.* *Genetics* 54: 595-609.
MacIntyre, R. J. 1972. Multiple alleles and gene divergence in natural populations. In *Evolution of Genetic Systems*, Brookhaven Symp. Biol. No. 23. H. H. Smith, ed. pp. 144-185.
Marshall, D. R. and R. W. Allard 1970. Maintenance of isozyme polymorphisms in natural populations of *Avena barbata.* *Genetics* 66: 393-399.
McCarthy, B. J. and M. N. Farquhar 1972. The rate of change of DNA in evolution. In *Evolution of Genetic Systems*, Brookhaven Symp. Biol. No. 23. H. H. Smith, ed. pp. 1-43.
McDonald, J. F. and F. J. Ayala 1974. Genetic response to environmental heterogeneity. *Nature*, in press.
Milkman, R. 1974. How many covarions per species? *Biochem. Genet.* 11: 181-182.
Nagel, E. 1961. *The structure of science.* Harcourt-Brace, New York.
Ohno, S. 1972. So much "junk" DNA in our genome. In *Evolution of Genetic Systems*, Brookhaven Symp. Biol. No. 23. pp. 366-370.
Parsons, P. A. 1971. Extreme-environment heterosis and genetic loads. *Heredity* 26: 579-583.

Pelc, S. R. 1965. Correlation between coding-triplets and amino acids. *Nature* 207: 597-599.

Powell, J. R. 1971. Genetic polymorphisms in varied environments. *Science* 174: 1035-1036.

Powell, J. R. 1974a. Heterosis at an enzyme locus of Drosophila: evidence from experimental populations. *Heredity* 32: 105-108.

Powell, J. R. 1974b. Interaction of genetic loci: the effect of linkage disequilibrium on Hardy-Weinberg expectations. *Heredity* 32: 151-158.

Prakash, S., R. C. Lewontin, and J. L. Hubby 1969. A molecular approach to the study of genic heterozygosity in natural populations. IV. Patterns of genic variation in central, marginal and isolated populations of *Drosophila pseudoobscura*. *Genetics* 61: 841-858.

Richmond, R. C. 1970. Non-Darwinian evolution: a critique. *Nature* 225: 1025-1028.

Richmond, R. C. and J. R. Powell 1970. Evidence of heterosis associated with an enzyme locus in a natural population of *Drosophila*. *Proc. Natl. Acad. Sci. USA* 67: 1264-1267.

Rockwood-Sluss, E. S., J. S. Johnston, and W. B. Heed 1973. Allozyme genotype-environment relationships. I. Variation in natural populations of *Drosophila pachea*. *Genetics* 73: 135-146.

Schwartz, D. and W. J. Laughner 1969. A molecular basis for heterosis. *Science* 166: 626-627.

Selander, R. K. and W. E. Johnson 1973. Genetic variation among vertebrate species. *Ann. Rev. Ecol. Syst.* 4: 75-91.

Shearer, R. W. and B. J. McCarthy 1970. Related base sequences in DNA of simple and complex organisms. IV. Evolutionary divergence of base sequence in mouse L-cell cytoplasmic and nucleus-restricted RNA. *Biochem. Genet.* 4: 395-408.

Simpson, G. G. 1964. *This view of life*. Harcourt Brace, New York.

Smith, E. L., R. J. DeLange, and J. Bonner 1970. Chemistry and biology of histones. *Physiol. Rev.* 50: 159-170.

Sparrow, A. H., H. J. Price, and A. G. Underbrink 1972. A Survey of DNA content per cell and per chromosome of prokaryotic and eukaryotic organisms: some evolutionary considerations. In *Evolution of Genetic Systems*, Brookhaven Symp. Biol. No. 23. H. H. Smith, ed. pp. 451-494.

Stebbins, G. L. and R. C. Lewontin 1972. Comparative evolution at the levels of molecules, organisms, and populations. *Proc. Sixth Berkeley Symp. Math. Stats.* 5: 23-42.

Wallace, B. 1968. *Topics in population genetics*. Norton, New York.

Wallace, B. 1970. *Genetic Load: its biological and conceptual*

aspects. Prentice-Hall, Englewood Cliffs, N. J.
Wills, C. and L. Nichols 1971. Single-gene heterosis in *Drosophila* revealed by inbreeding. *Nature* 233: 123-125.
Wright, S. 1931. Evolution in Mendelian populations. *Genetics* 6: 111-178.
Yamazaki, T. 1972. Detection of single gene effect by inbreeding. *Nature* 240: 53-54.

GENETIC POPULATION STRUCTURE AND BREEDING SYSTEMS

ROBERT K. SELANDER AND DONALD W. KAUFMAN[1]
Department of Biology, University of Rochester
Rochester, New York 14627

ABSTRACT. Electrophoretically demonstrable allozymes (allelic isozymes) have been employed as markers to study microgeographic and macrogeographic genetic population structure in hermaphroditic snails having different breeding systems. In *Helix aspersa*, an obligate outcrosser, intracolony genetic variability is high (~73% of regional diversity is within colonies) and there is strong local intercolony heterogeneity, particularly where migration is limited. Demes occupy an area of ~500 square meters and have an effective breeding size of ~15 individuals. The pattern of variation among introduced North American populations is generally consistent with a stochastic model of differentiation. *Rumina decollata*, a habitual self-fertilizer, is a complex of strongly inbred, monogenic strains, one of which has colonized North America. In Montpellier, France, where the species is native, a study of two strongly differentiated strains differing in micro-habitat distribution demonstrated that, in mixed colonies, an average of ~9% of individuals are of mixed genome. The frequency of outcrossing was crudely estimated at 15%. That the strains remain monogenic despite the potential for introgression suggests strong coadaptation of their respective genomes. Possible ecological correlates of breeding systems are discussed in relation to the differences shown by *Helix* and *Rumina*. This and other studies of population structure emphasize the convergence between animals and plants in strategies of ecogenetic adaptation involving regulation of the breeding system.

INTRODUCTION

Relationships between the breeding system and the amount and organization of genetic variability in natural populations have been extensively studied in plants (Grant, 1958; Baker, 1959; Stebbins, 1960; Allard et al., 1968; Jain, 1969; and papers in Baker and Stebbins, 1965). But this aspect of population genetics has been relatively neglected by zoologists, in part because there is little variety in the breeding

---

[1]Present address: Savannah River Ecology Laboratory, Drawer E, Aiken, South Carolina 29801.

systems of the higher and genetically better known groups of animals. Among animals with separate sexes, outcrossing is, of course, obligatory, and only occasionally has parthenogenesis developed as an alternate reproductive mode (White, 1970). But in hermaphroditic animals, including many large groups of invertebrates (Ghiselin, 1969) and even some lower vertebrates (Harrington and Kallman, 1968), self-fertilization (autogamy) is possible. We might therefore expect some degree of convergence with plants in strategies of adaptation involving organization of genetic variation through control of the breeding system.

In recent years, we have been employing electrophoretic techniques to analyze allelic isozymes (allozymes) in order to study the genetic population structure of snails having different breeding systems (Selander and Kaufman, 1973a, 1974a,b,c; Selander and Hudson, 1974; Kaufman and Selander, 1974). In this paper, we have summarized our findings relative to two hermaphroditic species, one, *Helix aspersa*, an obligate outcrosser, and another, *Rumina decollata*, a facultative but habitual self-fertilizer. Both species are terrestrial pulmonates, both are native to Europe, and both are superb colonizers, having established themselves widely in North America and other parts of the world. A comparison of level and distribution of genetic variability in these species demonstrates the profound influence of the breeding system on population structure and raises important questions regarding the adaptive strategies underlying variation in the breeding system.

Because of the generally low vagility of land snails and their tendency to occur in discrete colonies that may be founded by small numbers of individuals, stochastic processes should play an important role in determining gene frequencies in populations. Our studies of snail populations secondarily have provided an opportunity to assess the effects of genetic drift and the founder effect, processes neglected or even dismissed by many evolutionists and ecological geneticists.

RESULTS

POPULATION STRUCTURE IN HELIX ASPERSA

The brown snail is native to Great Britain and Europe. It was introduced to California from France in 1859, and by 1930 had spread over much of the state, where it inhabits urban and rural gardens and orchards. Only very locally has it "escaped" from cultivated habitats into native vegetation types. Largely through the agency of nursery stock, subintroductions of *Helix* to Texas and other western states have occurred.

Although *Helix* is a functional hermaphrodite, it is generally believed to be autosterile (Fretter and Graham, 1964), and our genetic data are wholly consistent with this notion. Populations are characterized by a large amount of variability at structural gene loci; an electrophoretic analysis of enzymes encoded by 17 loci in several populations yielded an average estimate of heterozygosity of .20, which is equivalent to that recorded in the more highly variable species of *Drosophila* (Selander and Kaufman, 1973b). In our extensive analyses of population structure, we have assayed allozymic variation in four types of enzymes, each represented by two isozymes encoded by separate loci: malate dehydrogenase (Mdh-1, Mdh-2), leucine aminopeptidase (Lap-1, Lap-2), glutamic-oxaloacetic transaminase (Got-1, Got-2), and phosphoglucomutase (Pgm-1, Pgm-2). At each of the eight loci, alleles are codominant; there are no "null" alleles.

Microgeographic structure. Populations of *Helix* are highly subdivided, and local intercolony heterogeneity in allele frequencies can be very strong, particularly where migration is restricted. To analyze population structure on a microgeographic scale, we collected all individuals on two adjacent city blocks in Bryan, Texas (Figure 1). Our efforts yielded 2,218 snails, distributed in 43 colonies (Selander and Kaufman, 1974a).

Within colonies, there is no significant net deviation in heterozygote proportions at any locus. For the four loci that are highly polymorphic in Bryan (Mdh-1, Lap-1, Lap-2, and Pgm-1), the net value of $F = (H_{exp} - H_{obs})/H_{exp}$ is -.00829 (P>.4) for the 43 colonies. Hence the data provide no evidence of inbreeding within colonies.

There is considerable genetic heterogeneity among colonies within blocks, and major interblock differences in allele frequencies are apparent at all polymorphic loci (Fig. 2). To measure intercolony variation in allele frequency, we have employed Wright's F_{ST} statistic, the standardized genetic variance;

$$F_{ST} = \frac{s^2}{\bar{p}(1-\bar{p})},$$

where s^2 is the weighted variance of allele p (corrected, where appropriate, for sampling variance). Values of F_{ST} are rather uniform for all loci on either block; for Block A, mean F_{ST} = .027, and for Block B, F_{ST} = .041; the mean for the two blocks is .0337. Relatively large shifts in allele frequencies occurring on either block across an alley running between North Texas and

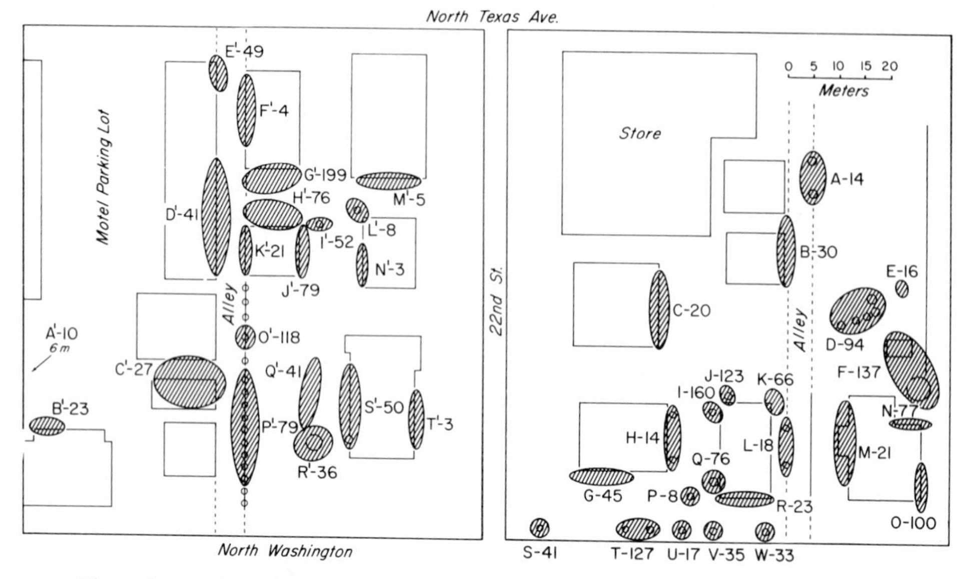

Fig. 1. Distribution and size of colonies of *Helix aspersa* on adjacent city blocks in Bryan, Texas. Block A (left): 20 colonies, designated A' to T'. Block B (right): 23 colonies, designated A to W. Numbers following letters indicate numbers of snails collected. Width of 22nd Street is not to scale.

North Washington avenues (Fig. 2) presumably reflect reduced movement of snails between colonies.

By visual inspection of maps of allele frequencies similar to those shown in Figure 2, we have identified seven homogeneous (and apparently panmictic) units or demes on each block (colonies of less than 10 individuals were omitted); for example, colonies G', H', I', J', and K' form one such unit. The area occupied by the larger demes (those consisting of several colonies) is roughly 500 square meters (equivalent to a circle with a diameter of 25 meters). That colonies less distant than 25 meters are likely to be homogenous in allele frequencies is also demonstrated in Figure 3, in which values of Nei's (1972) coefficient of genetic identity (I, based on five polymorphic loci) for all pairs of colonies on Block B are plotted against geographic distance between colony pairs. Up to about 25 meters, mean genetic identity between colonies is high and independent of distance.

For the two blocks, the harmonic mean number of adults per panmictic unit is 15. This number, which is an estimate of the effective breeding size of populations, is sufficiently

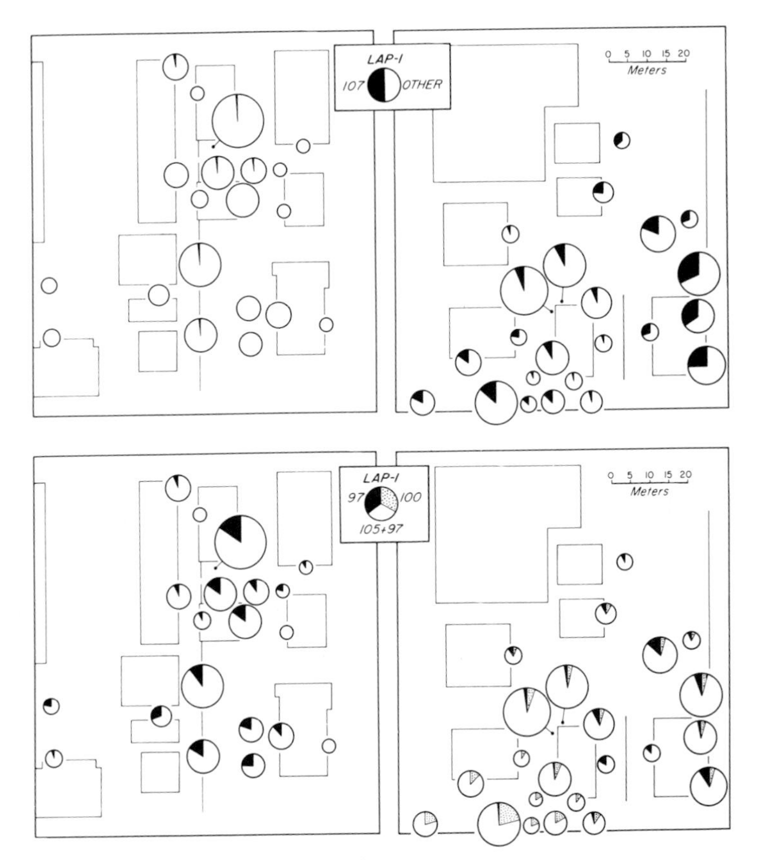

Fig. 2. Allele frequencies at Lap-1 locus in colonies of *Helix aspersa* on adjacent city blocks in Bryan, Texas. Area of circle proportional to number of individuals in colony (see Figure 1).

small that the effects of genetic drift should be evident . And indeed, there is considerable heterogeneity among colonies, even within single blocks, the pattern of which does not correspond to any visible features of the habitat.

According to stochastic models, intercolony variance should be larger for small colonies than for large colonies. As shown in Table I, there is a significant direct relationship between colony size and the magnitude of I, based on five loci. A comparable relationship in the snail *Cepaea nemoralis* in France was cited by Lamotte (1951) as evidence that intercolony differentiation is caused, in part at least, by genetic drift. However, Cain and Sheppard (1954) rejected this interpretation, pointing out that, because the

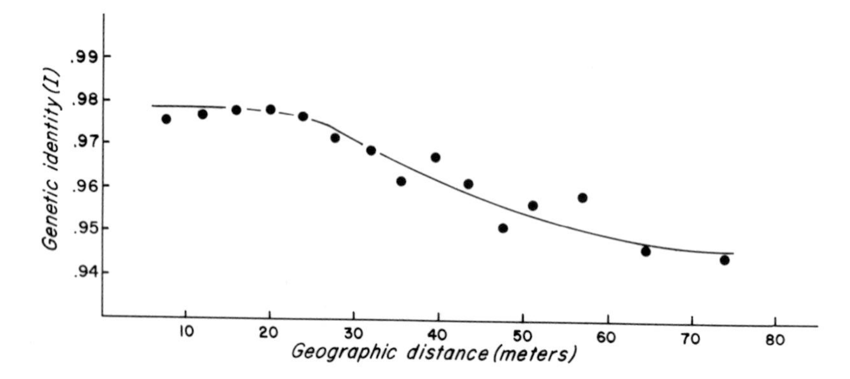

Fig. 3. Mean genetic identity of pairs of colonies of *Helix aspersa* on Block B as a function of mean geographic distance between pairs.

TABLE I

MEAN GENETIC IDENTITY BETWEEN PAIRS OF COLONIES OF *HELIX*

Colony Size	Block A[1]	Block B[1]
Small		
Number of colonies	8	10
Median size	7	5
Mean identity ($\bar{I}$)	.9392	.9589
Medium		
Number of colonies	7	9
Median size	41	16
Mean identity ($\bar{I}$)	.9756	.9751
Large		
Number of colonies	5	4
Median size	79	52
Mean identity ($\bar{I}$)	.9908	.9745

[1] Significance of variation in $\underline{I}$: Block A, $F_{(2, 56)} = 16.43$ ($P \ll .001$). Block B, $F_{(2,84)} = 7.75$ ($P < .001$).

smaller colonies sampled by Lamotte occupied smaller areas and therefore presumably experienced lesser degrees of spatial variation in habitat, the average degree of habitat difference between small colonies was greater. Hence, the relationship could be attributed to selection. In the case of *Helix* at Bryan, however, variation in diversity of habitat cannot be a significant factor, for there is no relationship between the area occupied by a colony and its size (Fig. 1).

Models of random differentiation predict that estimates of F_{ST} derived from covariances of allele pairs will match those calculated directly from allele frequencies, and that correlations of frequencies of allele pairs calculated from mean frequencies will be similar to observed correlations (Nei, 1965). These conditions are satisfied for allele pairs at the two multi-allele loci (Lap-1 and Lap-2) (Table II), and thus the analysis is consistent with the thesis that there is a strong random component in the differentiation of the colonies.

TABLE II

STANDARDIZED VARIANCE (ESTIMATED FROM COVARIANCE) AND CORRELATION OF ALLELE FREQUENCIES AMONG *HELIX* COLONIES

Area and allele pair	F_{ST}	Correlation (r) Observed	Correlation (r) Expected	P
Block A (20 colonies)				
Lap-1: 105-97	.023	-.996	-.982	.03
Lap-2: 100-84	.023	-.813	-.733	.56
Block B (23 colonies)				
Lap-1: 107-105	.091	-.883	-.711	.11
Lap-2: 100-84	.021	-.542	-.652	.44

Macrogeographic structure. Much of our effort has involved a comparison of macrogeographic variation in *Helix aspersa* in western North America, where the species has been living for only about 60 generations, and in France and Italy, where it is native (Selander and Kaufman, 1974b,c). Three levels of variation have been considered: (1) within populations on single blocks; (2) between populations on different blocks within cities (intracity); and (3) between populations in different cities (intercity). Our approach has been to apportion the total genetic diversity into various components and to compare F_{ST} estimates within and between cities. The analysis is based on over 5000 individuals in 164 local samples, each representing a group of snails collected in a single garden, churchyard, citrus grove, or other similar setting over an area no greater than 1000 square yards.

Before presenting the results of our analysis, we should note that, in the qualitative sense of allelic representation, North American populations are not genetically depauperate compared with those of Europe. Either the inoculum established in California (near San Jose) in 1859 was large enough to have included most of the alleles occurring in European populations or the diversity of the original population subsequently was augmented by other introductions.

To apportion genic diversity, we have followed Lewontin (1972) in employing the Shannon-Weaver information index. The results of our analysis for three regions are shown in Table III. Three points are noteworthy: First, a surprisingly large part (between 69% and 78%) of the total diversity for regions is within populations on single blocks. Second, for North American regions the between cities component of diversity is only slightly greater than the between populations-within cities component; this means that the division of the population into cities has only a slightly greater effect on diversity than does the division of the population into blocks within cities. Third, the between populations-within cities component of diversity is significantly smaller in France than in North America (8% versus 13%).

TABLE III

APPORTIONMENT OF GENIC DIVERSITY IN *HELIX*[1]

	Number of		Proportion of diversity		
Region	Populations	Cities	Within populations	Between populations within cities	Between cities
California	109	34	.7187	.1342	.1473
Other states	12	4	.6925	.1300	.1775
S. France	19	4	.7780	.0805	.1415

[1] Mean values for 6 loci.

A similar picture emerges from a comparison of intracity and intercity standardized variances in California and southern France (Table IV). In California, mean intracity F_{ST} (.116) is only slightly smaller than the mean intercity $\overline{F_{ST}}$ (.162). However, in France, intracity F_{ST} is considerably smaller, .030. Why are populations within cities less genetically diverse in France than in North America? One possibility is that colony sites are environmentally more uniform in France than in California. However, our field

TABLE IV

STANDARDIZED VARIANCES OF ALLELE FREQUENCIES IN *HELIX*

Locus	Mean F_{ST} California[1] Intracity	California[1] Intercity	France[2] Intracity	France[2] Intercity
Mdh-1	.131	.163	.008	.165
Lap-1	.115	.146	.019	.159
Lap-2	.101	.140	.022	.116
Got-1	.112	.145	.054	.082
Got-2			.058	.044
Pgm-1	.094	.211	.037	.325
Pgm-2	.144	.167	.015	.018
Mean	.1161	.1620	.0305	.1298

[1] 109 samples for intracity variance, 34 for intercity.

[2] 19 samples for intracity variance, 9 for intercity.

observations suggest a more likely explanation involving variation in the amount of migration between colonies. In North America, *Helix* has a more strongly insular population structure, being confined almost entirely to well-watered gardens (or, locally, to citrus groves) in urban areas; it does not occupy hedgerows, parkways, or riparian habitats that could serve as avenues of migration between blocks. Hence, gene flow between the rather widely scattered populations is limited, being largely dependent upon transfer, intentional or otherwise, by man. In France, however, *Helix* is less restricted in habitat distribution, being found wherever there is plant cover, including abandoned fields, streamsides, hedgerows, and parkways. Hence, we believe there is a lesser degree of isolation between the large garden populations that we sampled.

An unusually strong insular pattern of distribution of *Helix* was found in the vicinity of Rome, Italy, where the species has an extremely spotty distribution, occurring in small ($N < 50$), highly disjunct colonies separated by distances averaging several miles. For a set of six colonies collected near Rome, mean F_{ST} (over 7 loci) = .301. Most colonies were fixed at one or more loci, and F_{ST} estimates for individual loci are even more heterogeneous than those for southern France, varying from .002 for Pgm-1 to .546 for Pgm-2 . While it may be argued that this high level of genetic heterogeneity reflects unusually large environmental (and, hence, selective) variation among colony sites, we are inclined to attribute the increased genetic variance to genetic drift in the almost complete absence of migration . In marginal environments such as that near Rome, it is likely that a high turnover rate among the small colonies, owing to an increased probability of extinction per unit time, enhances drift by reducing the effective population number (Selander and Kaufman, 1974b).

An analysis and discussion of macrogeographic genetic variation in terms of selective and stochastic models is beyond the scope of this report (see Selander and Kaufman, 1974b,c). It will suffice here to note that patterns of variation at most loci studied in the North American populations are generally consistent with expectations of a stochastic model. However, the Mdh-1 locus shows conspicuous clinal variation in allele frequencies, which is not readily compatible with the stochastic model. And in Europe F_{ST} estimates are strongly heterogeneous over loci and there are major regional differences in allele frequencies at most loci. Whether these non-random patterns of variation are the result of selection on the enzyme loci themselves or on the segments of chromosomes that they mark is problematical.

POPULATION STRUCTURE IN RUMINA DECOLLATA

Rumina decollata is native to the Mediterranean region of Europe and North Africa. It was introduced to South Carolina (and perhaps elsewhere in eastern North America) before 1822, from a stock originating in Europe; by 1915 it had spread westward to Texas and had colonized Mexico, Bermuda and Cuba; and by 1966 it was established in Arizona and southern California. Unlike *Helix*, *Rumina* has managed to invade riparian and other native habitats in North America, although it occurs most commonly in gardens and fields in urban areas.

Our interest in the ecogenetics of *Rumina* was initially stimulated by the discovery that North American populations are a monogenic strain; all individuals from 33 localities were homozygous and allelically identical at 25 structural gene loci assayed electrophoretically (Selander and Kaufman, 1973a). Subsequently, we were able to show that *Rumina* in its native range is a complex of monogenic strains generated by a breeding system of facultative self-fertilization (Selander et al., 1974). Nine local populations sampled in France and Tunisia in 1972 represented seven distinctive strains.

Recent field work has demonstrated that colonies composed of two or more strains are common and that the genetic system of *Rumina* is broadly similar to that of strongly selfing plants in which most or all genetic variance is between strongly inbred lines. Here we will limit our discussion to the ecogenetic relationships of two strains in the Montpellier area of southern France (Selander and Hudson, 1974). Other aspects of our work on *Rumina* in Europe and Africa are being reported elsewhere.

Macrogeographic variation. In the Montpellier area, as in much of southern France, two monogenic strains of *Rumina* are widely distributed. Colonies may consist of a single strain or a mixture of the two, with or without a small proportion of individuals of mixed genome. One strain (designated RFAF by Selander and Kaufman, 1973a) has a gray body (with a black dorsal stripe) and a pale yellow foot, while the other (corresponding to the UMF strain identified earlier) has a black body and an olive-gray foot. These color characters are continuously variable and have a multifactorial genetic basis. At 26 enzyme loci assayed, the dark and light strains share the same allele at 13 and are fixed for different alleles at the other 13. Thus they are as genetically divergent as the average pair of sibling species of *Drosophila* (Ayala et al., 1970; Zouros, 1973; Wagner and Selander, 1974) and rodents (Selander and Johnson, 1973). Further evidence of genetic differentiation between the strains is provided by laboratory studies now in progress: the dark strain lays a larger number of eggs per clutch, is generally less active, and grows at a slower rate under laboratory conditions.

The dark strain is the majority type in the Montpellier region. Among 21 local populations sampled, the dark strain was the exclusive type in 12, and at only one site was the light strain the exclusive type. At the remaining eight sites, both strains occurred in the same population. An analysis of these collections demonstrated that there is no heterozygosity in the Montpellier region apart from that proximally generated by outcrossing between dark and light strains.

For detailed studies of population structure, we selected five loci (Lap-1, Est-6, Est-10, Est-11, and 6-Pgd) at which dark and light genomes are distinctive. Alleles fixed in the strains are as follows:

	Lap-1	Est-6	Est-10	Est-11	6-Pgd
Dark	100	100	95	90	83
Light	95	95	100	100	95

Employing the letters "F" (= fast-migrating) and "S" (slow-migrating) in reference to electrophoretic mobility of allozymes, we can represent the dark and light genomes, respectively, as FFSSS and SSFFF (see Table 6).

Microgeographic variation. The genetic and ecologic relationships of the strains of *Rumina* were studied intensively in a park and adjacent residential gardens on the Boulevard des Arceaux in Montpellier (Fig. 4). A systematic search

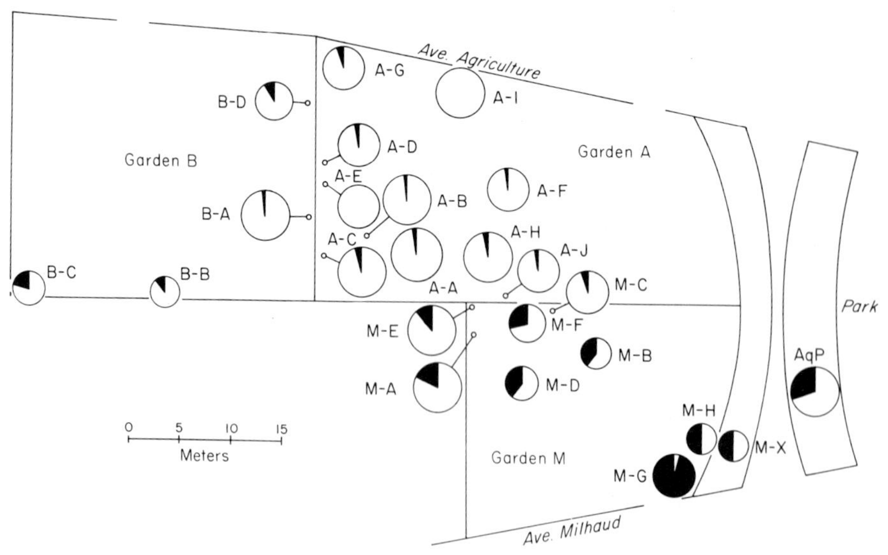

Fig. 4. Distribution, size, and strain composition of 24 colonies of *Rumina decollata* in study area on Boulevard des Arceaux, Montpellier. Area of circle proportional to colony size; black and white areas proportional to frequencies of dark and light individuals, respectively.

over the area yielded 2280 individuals, representing 24 colonies (Table V).

TABLE V

COMPOSITION OF *RUMINA* COLONIES IN THE BOULEVARD DES ARCEAUX STUDY AREA

Colony	Number of individuals		Percentage of genomes		
	Collected	Electrophoresed	Dark	Light	Mixed
AqP	113	52	33	63	4
M-X	10	10	50	50	0
M-A	110	62	18	74	8
M-B	5	5	40	60	0
M-C	74	19	5	58	37
M-D	19	19	39	50	11
M-E	142	71	11	75	14
M-F	41	38	29	55	16
M-G	58	30	97	3	0
M-H	4	4	50	50	0
A-A	510	51	2	98	0
A-B	109	42	2	93	5
A-C	175	23	4	87	9
A-D	77	40	2	95	3
A-E	83	30	0	100	0
A-F	77	42	2	88	10
A-G	91	46	4	83	13
A-H	110	37	3	92	5
A-I	137	55	0	84	16
A-J	56	54	2	94	4
B-A	208	94	1	85	14
B-B	9	9	11	67	22
B-C	25	25	20	72	8
B-D	37	32	9	81	9
Total	2280	890			
Mean					
Weighted			12.6	78.3	9.0
Unweighted			18.1	73.2	8.6

In two colonies (A-E and A-I), no individuals of "pure" dark genome were collected, but in one of these (A-I), 16% of individuals were of mixed genome, indicating the former presence of one or more dark individuals. The light genome was the predominant type in most colonies, but in one (M-G), 97% of individuals were dark. Over all, the unweighted average proportion of light genomes per colony was 78%; 18% were the dark genome; and 9% were of mixed genome. Proportions of individuals having mixed genomes varied from 0% (in six colonies) to 37% (M-C).

Unweighted mean values of F for the 23 colonies in which both strains were represented are homogeneous over loci (range, .752 to .788), and the mean is .775. The genetic composition of a representative colony is shown in Table VI.

TABLE VI

COMPOSITION OF *RUMINA* COLONY M-E IN BOULEVARD DES ARCEAUX STUDY AREA

Number of individuals	Genotype at indicated locus				
	Lap-1	Est-6	Est-10	Est-11	6-Pgd
8	F	F	S	S	S
53	S	S	F	F	F
1	S	S	F	FS	FS
1	F	FS	S	F	FS
1	FS	S	FS	S	F
1	F	F	FS	FS	FS
1	F	S	FS	FS	FS
1	FS	S	FS	FS	FS
4	FS	FS	FS	FS	FS

If we assume that the probability of outcrossing is equal within and between strains and that the system is in equilibrium, estimates of the proportion of outcrossing may be derived from the F values (Nei and Syakudo, 1958; Vasek, 1968). For the 23 colonies in which both strains were present, the mean value over loci is .150. However, both assumptions are perhaps unwarranted. As shown in Figure 5, crossing between the strains occurs more frequently in colonies where the dark strain is relatively rare than in those in which it is common. This striking frequency-dependent effect suggests that, when given the opportunity, individuals tend preferentially to outcross with members of their own strain. Additionally, it is unlikely that the populations on the Boulevard des Arceaux are in equilibrium with respect to the generation of heterozygosity through outcrossing and its loss through selfing. This is suggested by the large proportion of heterozygotes in individuals of mixed genome (Table VI). In experimental studies of breeding now in progress, we anticipate finding variation in the outcrossing/selfing ratio dependent upon environment, colony composition, and genotype, as shown previously for plants (e.g., Horovitz and Harding, 1972; Hamrick and Allard, 1972).

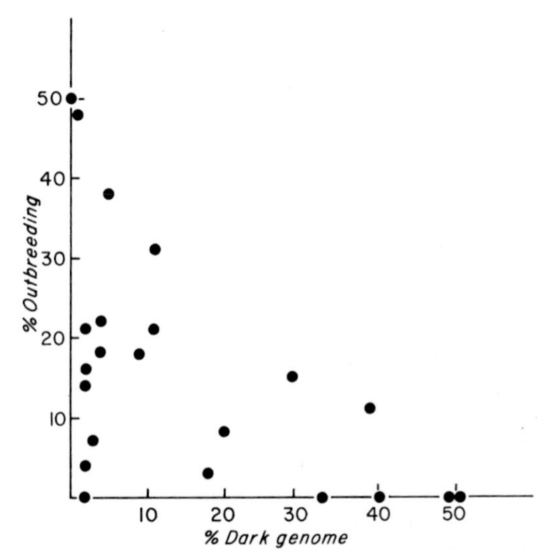

Fig. 5. Relationship between proportion of individuals of dark genome and outcrossing frequency (estimated from F) in colonies of *Rumina decollata* in Boulevard des Arceaux study area, Montpellier.

In the Montpellier area and elsewhere, there is a conspicuous inter-strain difference in habitat association. Colonies having a high proportion of dark genomes tend to occur in relatively protected, mesic habitats (under logs, boards, and rocks), whereas colonies of the light strain tend to occupy more open, xeric situations. In areas such as the Boulevard des Arceaux gardens, where both types of habitat are occupied by the "appropriate" strain, local intercolony heterogeneity in allele frequencies is very large. For the 24 colonies, we have calculated an average F_{ST} of .294 over the five loci studied. This compares with a value of .03 obtained for colonies of *Helix* on individual blocks in Bryan, where the area is approximately four times that of the study area in Montpellier.

DISCUSSION

Several generalizations are beginning to emerge from population genetic studies of snails and other animals in which the breeding system is variable. For example, a comparison of *Helix* and *Rumina* demonstrates that no one adaptive ecogenetic strategy is employed by all colonizing species. In colonizing North America, *Helix aspersa* has maintained essentially all the genic variability characteristic of its native European populations, with the potential for adaptive response to geographically variable environmental factors. At one locus studied (Mdh-1), clinal patterns of variation suggesting response to environmental selective factors are already apparent after only 60 generations of exposure to the new environment. In *Helix*, as in *Cepaea* (Murray, 1964), multiple mating and sperm storage may minimize the effects of population bottlenecks and fluctuations in reducing variability. Additionally, *Helix* is relatively vagile as snails go, so that intracolony variation often may be augmented by migration. It well may be that the relatively high mobility is one factor ultimately responsible for the breeding system of *Helix*, which seems designed to maximize outcrossing and genic recombination. In any event, genic variability is high in most local colonies, and the same alleles tend to be distributed throughout the range of the species.

An extremely different picture is presented by *Rumina decollata*, North American populations of which apparently are descended from one or more individuals of a single monogenic strain. In the apparent absence of genetic variability, adaptation of the North American populations to local microniches and regional patterns of variation in environmental factors presumably depends entirely on phenotypic flexibility and plasticity. Parenthetically, we may note that the success of *Rumina* in North America suggests that a high level of genetic variability is not essential to the survival of populations, at least for the short term. But in animals, as in plants (Baker, 1953; Stebbins, 1960), continued self-fertilization probably is an evolutionary dead-end.

A comprehensive discussion of theories relating to the adaptive significance of close inbreeding cannot be attempted here for want of space (see reviews by Allard et al., 1968; Carson, 1967). Most theories fall into one of two groups, those that consider selfing as little more than a provision for colonization or adjustment to low population densities (Baker, 1955; Ghiselin, 1969), and those which view it as part of a total ecogenetic strategy of adaptation to particular patterns of environmental heterogeneity. One of the earlier attempts to interpret selfing per se as an adaptive response to specific habitat conditions was made by Stebbins (1958), who postulated that there is a selective advantage to colonizing populations that can rapidly build up large numbers of genetically uniform, well-adapted individuals in temporary habitats. This and related themes were developed by Allard et al. (1968), who argued that the structure of inbreeding populations of plants results from an integration of the breeding system into the totality of genetic factors affecting population structure. Thus the breeding system is regarded as one of a series of interacting components functioning to regulate genetic variability. Recent theoretical contributions to our understanding of the adaptive significance of breeding systems also have been made by Lewontin (1965), Levins (1968), and Roughgarden (1972).

If *Rumina* is typical of self-fertilizing animals, our results confirm the thesis of Allard et al. (1968) that even strongly inbred species generally retain much genetic variability, distributed largely among inbred lines but capable of being released through occasional outcrossing. As noted elsewhere (Selander and Hudson, 1974), the breeding system and population structure of *Rumina* are in many respects remarkably similar to those of strongly selfing grasses, particularly the slender wild oat species *Avena barbata*. Allard and his associates (Allard et al., 1972; Hamrick and Allard, 1972; Clegg and Allard, 1972) have shown that strains of *Avena barbata* marked by black and gray lemma morphs are homozygous for coadapted blocks of alleles differentially adapted to xeric and mesic habitats. Selection for this arrangement reduces the potential distribution of genotypes so that those conferring high fitness occur with great frequency . Although we as yet have no direct experimental evidence of genetic coadaptation for the dark and light strains of *Rumina*, circumstantial evidence is provided by the fact of their continued monomorphism at structural gene loci, notwithstanding the potential for introgression resulting from outbreeding.

The ultimate adaptive "reason(s)" for the striking difference in the genetic systems of *Rumina* and *Helix* is not entirely apparent. However, we suspect that any factor sufficient to profoundly influence the genetic system of species must be a major and, in all probability, a conspicuous feature in the life of the organism. One feature is obvious: compared with *Helix* (and indeed with most other land snails), *Rumina* is strikingly sedentary. Once founded, a colony is likely to persist in the same spot, with stable microenvironmental conditions, for many generations, the progeny of the founders encountering the same set of conditions to which their parents were exposed. We suspect that the predictability of the environment has permitted *Rumina* to abandon recombination and to develop specialized genomes especially suited to specific sets of environmental conditions. Population fitness is maximized by a close matching of genotypes to microenvironments (Selander and Hudson, 1974).

In closing, we should like to comment on the relative influences of stochastic and deterministic factors on the population structure of snails. Cain and Sheppard (1950) and a host of other workers associated with the British school of ecological genetics (see Ford, 1971) have convincingly demonstrated that visual and physiological selection is involved in the maintenance of color and banding polymorphism in *Cepaea* (review in Jones, 1973). But stochastic processes have been neglected or even dismissed by these workers, although drift continues to be invoked as an important factor in snail population structure on the basis of statistical evidence (Lamotte, 1959; Goodhart, 1963; Komai and Emura, 1954; and Hickson, 1972). Further understanding is unlikely to come from a polarization of views regarding drift. The essence of the problem was identified by Lamotte: evidence that certain selective factors are influencing the genetic composition of populations does not mean that the observed differences between local colonies and populations are explicable as consequences of these factors. Conversely, by invoking stochastic processes to explain patterns of variation in *Helix*, we do not argue that selection is unimportant. Because, as noted by Dobzhansky (1970), the theoretically desirable aim of evolutionary investigations is to quantify respective contributions of different factors to gene frequency change, as well as their interactions, we believe that theories of population genetics neglecting drift are unrealistic.

ACKNOWLEDGEMENTS

Our research has been supported by NSF grant GB-37690 and NIH grant GM-20731. D.W. Kaufman was a Postdoctoral Fellow on NIH training grant 5T01-00337.

REFERENCES

Allard, R.W., G.R. Babbel, M.T. Clegg, and A.L. Kahler 1972. Evidence for coadaptation in *Avena barbata*. *Proc. Nat. Acad. Sci.* 69:3043-3048.

Allard, R.W., S.K. Jain, and P.L. Workman 1968. The genetics of inbreeding populations. *Adv. Genet.* 14:55-131.

Ayala, F.J., C.A. Mourão, S. Pérez-Salas, R. Richmond, and T. Dobzhansky 1970. Enzyme variability in the *Drosophila willistoni* group. I. Genetic differentiation among sibling species. *Proc. Nat. Acad. Sci.* 67:225-232.

Baker, H.G. 1953. Race formation and reproductive method in flowering plants. *Symp. Soc. Exp. Biol.* 7:114-143.

Baker, H.G. 1955. Self-compatibility and establishment after "long-distance" dispersal. *Evolution* 9:347-348.

Baker, H.G. 1959. Reproductive methods as factors in speciation in flowering plants. *Cold Spring Harbor Symp. Quant. Biol.* 24:177-191.

Baker, H.G. and G.L. Stebbins (eds.) 1965. *The Genetics of Colonizing Species*. Academic Press, New York.

Cain, A.J. and P.M. Sheppard 1950. Selection in the polymorphic land snail *Cepaea nemoralis*. *Heredity* 4:275-294.

Cain, A.J. and P.M. Sheppard 1954. Natural selection in *Cepaea*. *Genetics* 39:89-116.

Carson, H.L. 1967. Inbreeding and gene fixation in natural populations. In: *Heritage from Mendel*, R.A. Brink and E.D. Styles (eds.), University of Wisconsin Press, Madison, Wisconsin. pp. 281-308.

Clegg, M.T. and R.W. Allard 1972. Patterns of genetic differentiation in the slender wild oat species *Avena barbata*. *Proc. Nat. Acad. Sci.* 69:1820-1824.

Dobzhansky, T. 1970. *Genetics of the Evolutionary Process*. Columbia Univ. Press, New York.

Ford, E.B. 1971. *Ecological Genetics*, Third ed., Chapman and Hall, London.

Fretter, V. and A. Graham 1964. Reproduction. In: *Physiology of Mollusca, Vol. 1*, K.M. Wilbur and C.M. Yonge (eds), Academic Press, New York. pp. 127-164.

Ghiselin, M.T. 1969. The evolution of hermaphroditism among animals. *Quart. Rev. Biol.* 44:189-208.

Goodhart, C.B. 1963. 'Area effects' and non-adaptive variation between populations of *Cepaea* (Mollusca). *Heredity* 18:459-465.

Grant, V. 1958. The regulation of recombination in plants. *Cold Spring Harbor Symp. Quant. Biol.* 23:337-363.

Hamrick, J.L. and R.W. Allard 1972. Microgeographical variation in allozyme frequencies in *Avena barbata*. *Proc. Nat. Acad. Sci.* 69:2100-2104.

Harrington, R.W., Jr. and K.D. Kallman 1968. The homozygosity of clones of the self-fertilizing hermaphroditic fish *Rivulus marmoratus* Poey (Cyprinodontidae, Atheriniformes). *Amer. Nat.* 102:337-343.

Hickson, T.G.L. 1972. A possible case of genetic drift in colonies of the land snail *Theba pisana*. *Heredity* 29: 177-190.

Horovitz, A. and J. Harding 1972. Genetics of *Lupinus*. V. Intraspecific variability for reproductive traits in *Lupinus nanus*. *Bot. Gaz.* 133:155-165.

Jain, S.K. 1969. Comparative ecogenetics of two *Avena* species occurring in central California. In: *Evolutionary Biology, Vol. 3*, T.H. Dobzhansky et al. (eds), Appleton-Century-Crofts, New York. pp. 73-118.

Jones, J.S. 1973. Ecological genetics and natural selection in molluscs. *Science* 182:546-552.

Kaufman, D.W. and R.K. Selander 1974. *Studies of genetic variation in Mesodon.* (in preparation).

Komai, T. and S. Emura 1954. A study of population genetics on the polymorphic land snail *Bradybaena similaris*. *Evolution* 9:400-418.

Lamotte, M. 1951. Recherches sur la structure génétique des populations naturelles de *Cepaea nemoralis (L)*. *Bull. Biol. Suppl.* 35:1-239.

Lamotte, M. 1959. Polymorphism of natural populations of *Cepaea nemoralis*. *Cold Spring Harbor Symp. Quant. Biol.* 24:65-84.

Levins, R. 1968. *Evolution in Changing Environments*. Princeton Univ. Press, Princeton, New Jersey.

Lewontin, R.C. 1965. Selection for colonizing ability. In: *The Genetics of Colonizing Species*, H.G. Baker and G.L. Stebbins (eds.), Academic Press, New York. pp. 77-91.

Lewontin, R.C. 1972. The apportionment of human diversity. In: *Evolutionary Biology, Vol. 6*, T. Dobzhansky et al. (eds), Appleton-Century-Crofts, New York. pp. 381-398.

Murray, J. 1964. Multiple mating and effective population size in *Cepaea nemoralis*. *Evolution* 18:283-291.

Nei, M. 1965. Variation and covariation of gene frequencies in subdivided populations. *Evolution* 19:256-258.

Nei, M. 1972. Genetic distance between populations. *Amer. Nat.* 106:283-292.

Nei, M. and K. Syakudo 1958. The estimation of outcrossing in natural populations. *Jap. J. Genet.* 33:46-51.

Roughgarden, J. 1972. Evolution of niche width. *Amer. Nat.* 106:683-718.

Selander, R.K. and R.O. Hudson 1974. Animal population structure under close inbreeding: the land snail *Rumina* in southern France. *Amer. Nat.* submitted.

Selander, R.K. and W.E. Johnson 1973. Genetic variation among vertebrate species. *Ann. Rev. Ecol. Syst.* 4:75-91.

Selander, R.K. and D.W. Kaufman 1973a. Self-fertilization and genetic population structure in a colonizing land snail. *Proc. Nat. Acad. Sci.* 70:1186-1190.

Selander, R.K. and D.W. Kaufman 1973b. Genic variability and strategies of adaptation in animals. *Proc. Nat. Acad. Sci.* 70:1875-1877.

Selander, R.K. and D.W. Kaufman 1974a. Genetic population structure in the brown snail *(Helix aspersa)*. I. Microgeographic variation. *Evolution* (in press).

Selander, R.K. and D.W. Kaufman 1974b. Genetic population structure in the brown snail (*Helix aspersa*). II. North American populations. In preparation.

Selander, R.K. and D.W. Kaufman 1974c. Genetic population structure in the brown snail (*Helix aspersa*). III. European populations. In preparation.

Selander, R.K., D.W. Kaufman, and R.S. Ralin 1974. Self-fertilization in the terrestrial snail *Rumina decollata*. *Veliger* 16:265-270.

Stebbins, G.L. 1958. Longevity, habitat, and release of genetic variability in the higher plants. *Cold Spring Harbor Symp. Quant. Biol.* 23:365-378.

Stebbins, G.L. 1960. The comparative evolution of genetic systems. In: *Evolution After Darwin, Vol. 1*, S. Tax (ed), Univ. Chicago Press, Chicago, Illinois. pp. 197-226.

Vasek, F.C. 1968. Outcrossing in natural populations: A comparison of outcrossing estimation methods. In: *Evolution and Environment*, E.T. Drake (ed), Yale Univ. Press, New Haven, Connecticut. pp. 369-385.

Wagner, R.P. and R.K. Selander 1974. Isozymes in insects and their significance. *Ann. Rev. Entomol.* 19:117-138.

White, M.J.D. 1970. Heterozygosity and genetic polymorphism in parthenogenetic animals. In: *Essays in Evolution and Genetics in Honor of Theodosius Dobzhansky*, M.K. Hecht and W.C. Steere (eds), Appleton-Century-Crofts, New York. pp. 237-262.

Zouros, E. 1973. Genic differentiation associated with the early stages of speciation in the *mulleri* group of *Drosophila*. *Evolution* 27:601-621.

ISOZYMES IN EVOLUTIONARY SYSTEMATICS

JOHN M. BURNS
Museum of Comparative Zoology
Harvard University
Cambridge, Massachusetts 02138

ABSTRACT. Evolutionary biologists exploiting electrophoretic methods of isozyme separation have shown that most natural populations of sexually reproducing organisms maintain considerable genetic polymorphism which becomes extensively altered in the course of speciation. To a systematist, this means that isozymic characters should be particularly powerful at low taxonomic levels.

Three different ways of using isozymic characters are considered, and concrete examples of their application to low-level systematic problems are treated in detail. In all cases, isozymic characters are used in combination with other, more conventional, characters rather than alone.

(i) A single, extraordinarily variable dimeric esterase polymorphism controlled by 3 to as many as 25 or 30 alleles works both in distinguishing and in clustering North American species belonging to the difficult butterfly genus *Colias*.

(ii) Two isozymes of phosphoglucomutase serve as natural markers for release experiments that conclusively demonstrate premating isolation between two slightly differentiated allopatric populations of Mediterranean mosquitoes of the genus *Aedes*.

(iii) Close analysis of geographic variation in one esterase and five dehydrogenases, as well as in the color of the dewlap (a structure used in territorial and courtship displays and in species discrimination), shows that one species of Haitian *Anolis* lizard actually comprises three ecologically equivalent, parapatric species.

Evolutionary biologists have exploited electrophoretic methods of isozyme separation to show, in natural populations of sexually reproducing organisms, (i) that numerous genetic loci are normally polymorphic, (ii) that some individual loci are exceedingly variable, and (iii) that most species--even sibling species--are genetically well-differentiated from one another. To a systematist, this means that electrophoretic characters should be particularly powerful at low taxonomic levels, i.e., in analyzing population differentiation, distinguishing species, and clustering them into species groups.

I will consider three ways of working with isozymic characters in solving low-level systematic problems: (i) using one highly variable enzyme both to distinguish and to cluster species within a difficult genus; (ii) using one enzyme as a marker in testing for the existence of premating isolation between related populations; and (iii) using a number of enzymes to distinguish sibling species (the most obvious approach). In all cases, isozymic characters are used in combination with other, more conventional, characters rather than alone.

One highly variable enzyme and a host of sulfur butterflies *(Colias)*

A variable enzyme polymorphism controlled by 3 to as many as 25 or 30 alleles is a sensitive tool for systematic and evolutionary studies of related natural populations.

This remarkable tool was foreshadowed when 127 adults of *Colias eurytheme* from Austin, Texas, were individually electrophoresed and their dimeric esterase patterns compared: no less than 13 different homopolymers were clearly distinguished, and at least 13 alleles were therefore thought to be involved (Johnson and Burns, 1966). Genetic analysis launched with five wild females each known from spermatophore dissection (Burns, 1968) to have mated but once showed, however, that the allelic series must be longer: of 20 different alleles that could appear in a total of five single-pair matings, 13 actually did; in other words, on close investigation, just 10 individuals (rather than 127) were found to carry 13 different alleles for the dimeric esterase (Burns and Johnson, 1967).

In a sample of 85 central Connecticut adults of *C. eurytheme* and its closest relative, *C. philodice*, the dimeric esterase was as polymorphic as in the central Texas *eurytheme* (Burns and Johnson, 1967). Natural populations of these pierid butterflies appeared to consist largely of individuals heterozygous for the dimeric esterase and different from one another. Several population comparisons later, it was conservatively estimated that, in each of these species, about 25 alleles for the dimeric esterase commonly coexist at one point in space and time (Burns, 1970).

In other butterflies, the richest isozyme variation known stems from 9- to 14-allele polymorphisms of a dimeric esterase in the lycaenid *Hemiargus isola* (Burns and Johnson, 1971). Originally, the dimeric esterase of *Colias* was arbitrarily called EST E (Johnson and Burns, 1966), whereas that of *Hemiargus* was designated ES-D (for ES[terase]-D[imeric]) (Burns and Johnson, 1971). The more informative ES-D symbol is now applied to *Colias*, as well.

To make better taxonomic sense of *Colias*--a refractory

genus in which species limits and groups are often indistinct--I have undertaken long-term comparative analysis of ES-D polymorphism in natural populations. Chiefly holarctic in distribution (with minor intrusions into South America, Africa, and India), *Colias* includes about 60 species, of which one-fourth occur in North America. Even these are still poorly understood, although much studied biologically in the last few decades.

North American species of *Colias* show three principal patterns of larval foodplant selection: (i) various genera of herbaceous legumes (Leguminosae) such as *Astragalus, Lathyrus, Lupinus, Thermopsis, Trifolium, Vicia, Medicago,* and *Melilotus;* (ii) *Salix* (Salicaceae); and (iii) *Vaccinium* (Ericaceae). Relations between the *Vaccinium*-eaters and the *Salix*-eaters are felt by most workers to be especially close and confusing.

On the basis of their ES-D variation, North American species of *Colias* fall into three discrete groups that coincide with clusters based on choice of foodplant: the *Vaccinium*-eaters, with relatively low levels of variation; the *Salix*-eaters, at an appreciably higher level; and the legume-eaters, at high to exceedingly high levels. With few exceptions, each species can be distinguished from all others by its pattern of ES-D variation alone, provided sizable samples are compared (Fig. 1). Interspecific differences can be conveyed in part by describing the number of different *Es-d* alleles normally maintained in a population, as well as the width and location of the zone that the isozymes controlled by those alleles occupy on an electrophoretic gel. That is done in Table I, where the mobilities of ES-D isozymes are expressed in average R_F values. (R_F is the distance from the origin that an isozyme has moved relative to the distance moved by a buffer front that produces a brownish line across the gel.)

In the *Vaccinium*-eaters *(C. interior, C. palaeno, C. pelidne,* and *C. behrii)*, variation is low, both because there are only 3 to 5 *Es-d* alleles in individual natural populations and because the ES-D isozymes have a very narrow mobility range. The *Salix*-eater *(C. scudderii)* usually maintains more alleles (7), and the isozymes span a perceptibly wider range than they do in any of the *Vaccinium*-eating species. Indeed, the ES-D mobility range of *scudderii* is almost equal, in width and in location, to that of all four *Vaccinium*-eaters together--but it is still narrow. Although, in view of this general similarity, the *Salix*-eater might appear to belong with the *Vaccinium*-eaters, it does not: in all populations of *scudderii* examined, two *Es-d* alleles are common (one ranging in frequency from 0.29 to 0.55 and the other, from 0.31 to 0.51); and both these major alleles control isozymes with mobilities

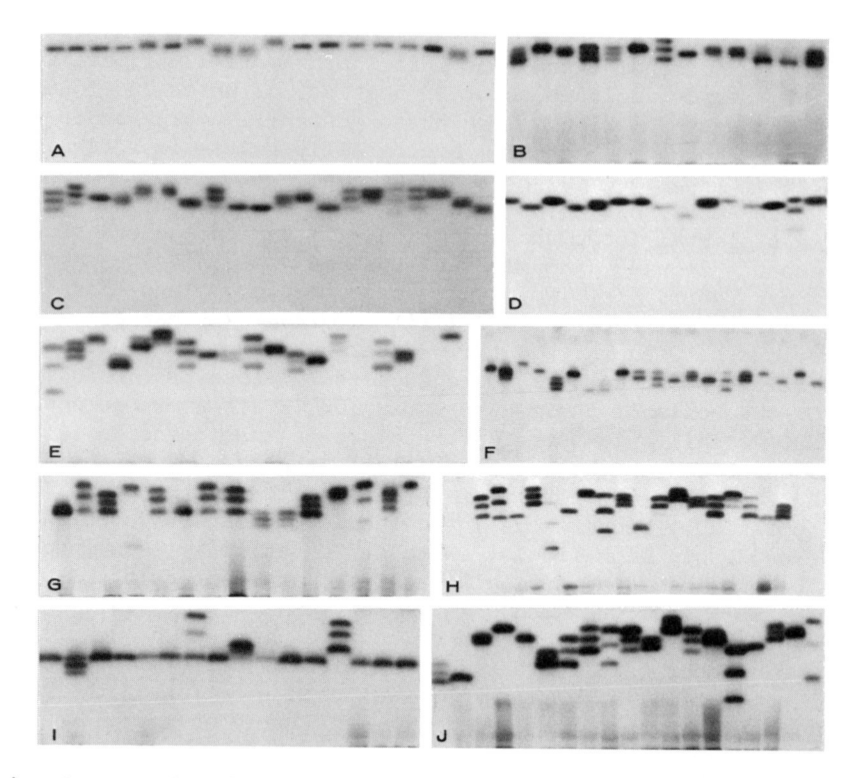

Fig. 1. Starch gels showing electrophoretically detectable isozymes of a dimeric esterase (ES-D) in samples from local populations of 10 North American species of *Colias* butterflies. A. *C. behrii*, 18♂ from Elizabeth Lake, Sierra Nevada, 2900 m, Tuolumne Co., California, July 31, 1970. B. *C. scudderii*, 11♂ 2♀ from willow bogs NE of Battle Lake, Sierra Madre, 2800 m, Carbon Co., Wyoming, August 21, 1968. C. *C. interior*, 20♂ from Marion, 1220 m, Flathead Co., Montana, July 28, 1972. D. *C. pelidne*, 9♂ 6♀ from 12.7 km E of Macks Inn, 2000 m, Fremont Co., Idaho, August 5, 1972. E. *C. alexandra*, 6♂ from Pine Flat, La Sal Mountains, 2500 m, Montrose Co., Colorado and San Juan Co., Utah, June 27, 1971; and 11♂ 2♀ from Hannagan Meadows, 2750 m, Greenlee Co., Arizona, June 26, 1971. F. *C. meadii*, 20♂ from Cumberland Pass, 3660 m, Gunnison Co., Colorado, July 28, 1967. G. *C. harfordii*, 8♂ 7♀ from Laguna Lakes, Laguna Mountains, 1650 m, San Diego Co., California, June 18, 1968. H. *C. eurytheme*, 18♂ from Davis, 15 m, Yolo Co., California, September 10, 1970. I. *C. occidentalis*, 16♂ from Canyon Creek Canyon, Ochoco Mountains, 1370 m, Crook Co.,

Fig. 1 legend continued
Oregon, July 10, 1970. J. *C. philodice*, 19♂ from Durham, 75 m, Middlesex Co., Connecticut, October 16, 1968.

---

distinct from the mobilities of any isozymes occurring in the *Vaccinium*-eaters.

In the seven legume-eaters studied (*C. meadii*, *C. chrysomelas*, *C. occidentalis*, *C. alexandra*, *C. harfordii*, *C. philodice*, and *C. eurytheme*), variation is high both because most species have extremely long series of *Es-d* alleles and because the ES-D isozymes of all these species have a wide to extremely wide mobility range. In the utmost cases (*eurytheme* and *philodice*), there are at least 25 (perhaps 30) *Es-d* alleles in single populations, and the ES-D zone spans half the total distance from origin to front. The ES-D isozymes of all other species of *Colias* investigated have mobilities that fall well within this wide range, with a pronounced clustering toward (but not at) the fast end. Isozymes of the less variable species (the *Vaccinium*-eaters and the *Salix*-eater) are all relatively fast moving; and the bulk of the isozymes of *eurytheme* and *philodice* are crowded in the upper half of the total mobility range. [*Hemiargus isola*, with a similarly rich ES-D polymorphism (i.e., with numerous alleles controlling isozymes having a wide mobility range), is also a legume-eater that attacks a broad array of species in various genera (Burns and Johnson, 1971).]

The *Vaccinium*-eaters provide strong indirect evidence that selection molds ES-D polymorphism. *Es-d* allelic variation is strikingly conservative among these species. *Colias palaeno*, which is holarctic in distribution, and *interior*, *pelidne*, and *behrii*, which are strictly nearctic, tend to replace one another geographically (Fig. 2), although the mutual spatial relations of *interior* and *pelidne* are not well understood.

Colias interior, with a nearly transcontinental distribution, has been sampled in both eastern and western North America (Fig. 2) and *Es-d* allelic frequencies compared over short distances, over long distances, across partial barriers, and across absolute gaps in distribution, as well as in time. In the eastern United States, samples come (i) from the southern edge (Stratton, Vermont) of the main range, (ii) from a point (Otsego Lake, Michigan) that lies within the main range but in an area now cut off by water barriers (Lake Michigan and Lake Huron), and (iii) from a population (Seneca Creek, West Virginia) occurring as a glacial relict at high elevations in the southern Appalachian Mountains, where it has clearly been isolated from other segments of the species population

TABLE I

Species	Number of populations examined	Number of different alleles in a single population	Approximate mobility range of ES-D isozymes (R_F units)	Approximate number of R_F units spanned
Vaccinium-eaters:				
behrii	2	3	.48 - .53	6
interior	8	4, 5	.50 - .57	8
palaeno	1	5	.45 - .55	11
pelidne	1	5	.42 - .52	11
Salix-eaters:				
scudderii	5	7 (3)*	.44 - .57	14
Legume-eaters:				
meadii	6	10 - 13 (4)*	.35 - .54	20
chrysomelas	1	4	.32 - .52	21
occidentalis	1	6	.32 - .56	25
alexandra	11	14 - 17 (3)*	.33 - .58	26
harfordii	3	14 - 17	.35 - .59	25
philodice & *eurytheme*	9	ca. 25	.15 - .63	49

*In parentheses, the reduced number of alleles in a peripherally isolated population.

Dimeric esterase (ES-D) variation in North American populations of *Colias*.

for a few thousand years. In all of these samples there are four *Es-d* alleles, of which three are common and one is relatively rare (Table II). The frequencies of the common alleles show no more than mild fluctuations in time and space; and, most remarkably, the rare allele consistently occurs at frequencies of 0.03 to 0.06--even in the widely disjunct montane isolate! Moreover, in the western United States, where five samples have been taken at much closer intervals in the northern Rocky Mountains of Idaho and Montana (Fig. 2), the same four *Es-d* alleles are maintained at frequencies similar to those in the east (even the rare allele persists at frequencies of 0.02 to 0.09) and a fifth allele (which is common in both *palaeno* to the north and *pelidne* to the south) is added, also at low frequencies of 0.02 to 0.10 (Table II).

Colias behrii has a highly restricted range--far removed from all other *Vaccinium*-eaters--at elevations above 3000 m in the Sierra Nevada of California. Clearly a glacial relict, it is the least polymorphic for ES-D of all species of *Colias*. It has just three *Es-d* alleles, one of which is unusually common. Their frequencies appear stable in space and time.

In the *Vaccinium*-eaters as a whole, ES-D polymorphism is exceedingly conservative (for *Colias* butterflies). All sampled populations of all four species share a single allele, $Es\text{-}d^{.517}$, which is always a common allele and usually the most common one. Three of the four *Vaccinium*-eaters share two alleles, $Es\text{-}d^{.535}$ and $Es\text{-}d^{.500}$, controlling isozymes with mobilities equally faster and slower than that of $Es\text{-}d^{.517}$. Three species have one appreciably frequent allele that is essentially or actually missing in other species: for *interior*, this is $Es\text{-}d^{.570}$, controlling the fastest moving of all *Vaccinium*-eater isozymes; for *pelidne*, $Es\text{-}d^{.420}$, controlling the slowest; and for *behrii*, $Es\text{-}d^{.483}$ (Table II).

Collectively, the main alleles of the *Vaccinium*-eaters form an arithmetic series ranging from $Es\text{-}d^{.483}$ to $Es\text{-}d^{.570}$. Mobility of the autodimers controlled by the six alleles in this series increases by a constant amount, as if some standard unit difference in net surface charge of the molecule were repeatedly involved.

In view of their geographic distribution (Fig. 2), larval foodplant *(Vaccinium)*, and ES-D variation (Table II), *C. palaeno*, *C. interior*, *C. pelidne*, and *C. behrii* undoubtedly are a compact phylogenetic group--some current taxonomic opinion to the contrary notwithstanding.

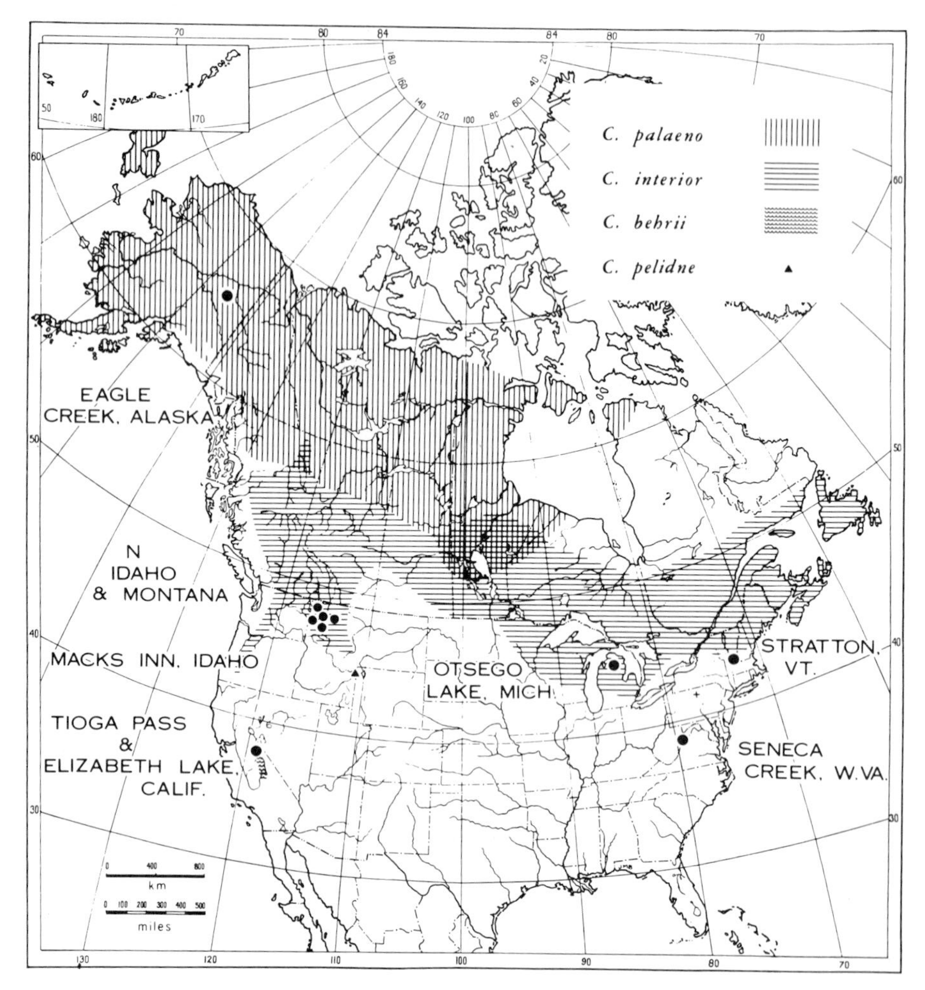

Fig. 2. Geographic distribution of *Vaccinium*-eating species of *Colias* in North America (modified from Hovanitz, 1950). The known ranges of three largely allopatric species (*C. palaeno*, *C. interior*, and *C. behrii*) are shown and those local populations sampled for electrophoresis are marked by dots and named. [A small cross in Pennsylvania marks the point at which, in low mountains, a single fresh male of *C. interior* was collected in 1968 (Shapiro, 1969); it undoubtedly represents a relict population.] For the fourth *Vaccinium*-eater, *C. pelidne*, no attempt is made to indicate total range, which is both uncertain and complexly related to that of *C. interior*. The population of *C. pelidne* sampled for electrophoresis is

Fig. 2 legend continued
marked by a triangle and named. See Table II for data on *Es-d* allelic frequencies in all of these populations.

Two forms of one enzyme and two mosquitoes *(Aedes)*

Coluzzi and Bullini (1971) elegantly used a single electrophoretically detectable enzyme difference to establish beyond doubt the existence of two sibling species within the *mariae* complex of the mosquito genus *Aedes*. Isozymes at the phosphoglucomutase locus provided natural markers for release experiments designed to test premating isolation in slightly differentiated allopatric populations.

The allopatric populations in question are *zammitii* of the eastern Mediterranean and *mariae* of the western Mediterranean. Their scant morphologic differences are only statistically demonstrable. The mosquitoes themselves mate freely with each other in cages. Although F_1 hybrid males are partially sterile, F_1 hybrid females are not; and hybrids of each sex are vigorous. Both mosquitoes breed identically in rock pools along Mediterranean coasts and have a generation time of about two weeks.

A massive release experiment just south of Rome mixed *zammitii* from eastern Italy with *mariae* from western Italy. These two populations had been shown to be monomorphic for different isozymes of PGM. Individuals of each population produced a distinctive electrophoretic pattern, and hybrids between them gave a combined pattern. On July 14, 1970, about 25,000 larvae and pupae of *zammitii* were collected at a point on the east coast and released on the west coast in an area where the *mariae* population had been reduced by a comparable amount to make roughly equal numbers of the two species. The release area was an "ecologic island" comprising less than one km of rocky coast enclosing about 18 breeding pools and flanked on both ends by sandy beaches. From this area, fourth instar larvae and pupae were sampled in modest numbers on July 15, 19, 22, and 26 and weekly thereafter until October 11. Adults emerging in the laboratory from these samples were electrophoresed and identified by their PGM pattern. Altogether, 2536 specimens of *Aedes* were so screened: 1414 were *zammitii*, 1112 were *mariae*, and a mere 10 were hybrids.

It was concluded that *A. mariae* and *A. zammitii* had chanced, in allopatry, to evolve nearly foolproof premating isolating mechanisms and should, therefore, be regarded as separate species.

TABLE II. ALLELIC FREQUENCIES FOR ES-D IN NORTH AMERICAN POPULATIONS OF 4 *Vaccinium*-EATING SPECIES OF *COLIAS*

Population	Year sampled	Sample size	Frequency of Es-d alleles							
			.420	.483	.500	.517	.535	.553	.570	other
C. palaeno										
ALASKA: Eagle Creek	1970	20			.27	.55	.08	.08		.02
C. interior										
VT.: Stratton	1968	61				.17	.30	.04	.49	
" "	1970	54				.21	.40	.06	.33	
W. VA.: Seneca Creek	1968	15				.40	.43	.03	.13	
MICH.: Otsego Lake	1968	31				.29	.55	.03	.13	
MONT.: Marion	1972	100			.08	.36	.27	.07	.22	
" nr. Libby	1972	64	.01		.08	.38	.27	.09	.16	
" nr. DeBorgia	1972	31			.10	.44	.27	.02	.18	
IDAHO: nr. Naples	1972	51			.08	.44	.27	.06	.15	
" nr. Athol	1972	43			.02	.40	.12	.03	.43	
C. pelidne										
IDAHO: nr. Macks Inn	1972	86	.07	.006	.43	.49				.006
C. behrii										
CALIF.: Tioga Pass	1968	60		.04		.83	.13			
" " "	1970	84		.02		.87	.11			
" Elizabeth Lake	1970	56		.04		.79	.17			

Six enzymes and three lizards (*Anolis*)

The dewlap of male anoles functions in territorial and courtship displays and in species discrimination. In use, this collapsible fold of loose skin projects conspicuously from the throat in the midsagittal plane like a fan. Color pattern of the dewlap usually differs less within species than it does between coexisting species.

However, two West Indian anoles--*Anolis distichus* and *A. brevirostris*--show striking geographic variation in dewlap color. Each of the 8 subspecies of *A. distichus* occurring on mainland Hispaniola differs strongly in dewlap color from all adjacent *distichus* subspecies (with one exception), and dewlap color seems to intergrade between nearly all of these divergent adjacent subspecies (Schwartz, 1968). In the driest parts of Hispaniola, the widespread *A. distichus* gives way to its closest relative, *A. brevirostris*. By analyzing geographic variation both in dewlap color and in six enzymes, Webster and Burns (1973) demonstrated that 12 Haitian populations of "*brevirostris*" spanning about 150 km actually represent three ecologically equivalent, parapatric species.

Early in this analysis, in 1969, when "*brevirostris*" had been sampled from just three populations, dewlaps appeared to vary clinally over all of the region studied: they were wholly bright orange in population 2, basally orange with a broad white margin in population 6, and wholly pale with a basal tinge of warm pigment in population 11 (Fig. 3). But each of these samples had a unique set of esterase isozymes, and the northern sample was fully distinct from those to the south at one of the loci controlling lactate dehydrogenase. Accordingly, in 1970, "*brevirostris*" was sampled much more finely in space; and detailed electrophoretic comparison was expanded to six geographically varying enzymes: α-glycerophosphate dehydrogenase, 6-phosphogluconate dehydrogenase, and two isocitrate dehydrogenases--as well as LDH and Est.

Within the apparent cline in dewlap color, there are really two sharp discontinuities--between populations 3 and 4, on the one hand, and 7 and 8, on the other--but a new (true) cline that is relatively short emerges (Fig. 3). In general, dewlaps in populations 1 to 3 are uniform and vivid; in populations 4 to 7, bicolored; and in populations 8 to 12, uniform and dull. Only the bicolored central populations show clinal variation: in population 4, dewlaps are all white or, more often, white with a small light orange basal spot; from population 4 to 7, the average extent (and intensity) of basal orange increases at the expense of white until, in population 7, dewlaps phenotypically approach those in populations 1 to 3. It is almost as though populations 4 to 7 were a massive in-

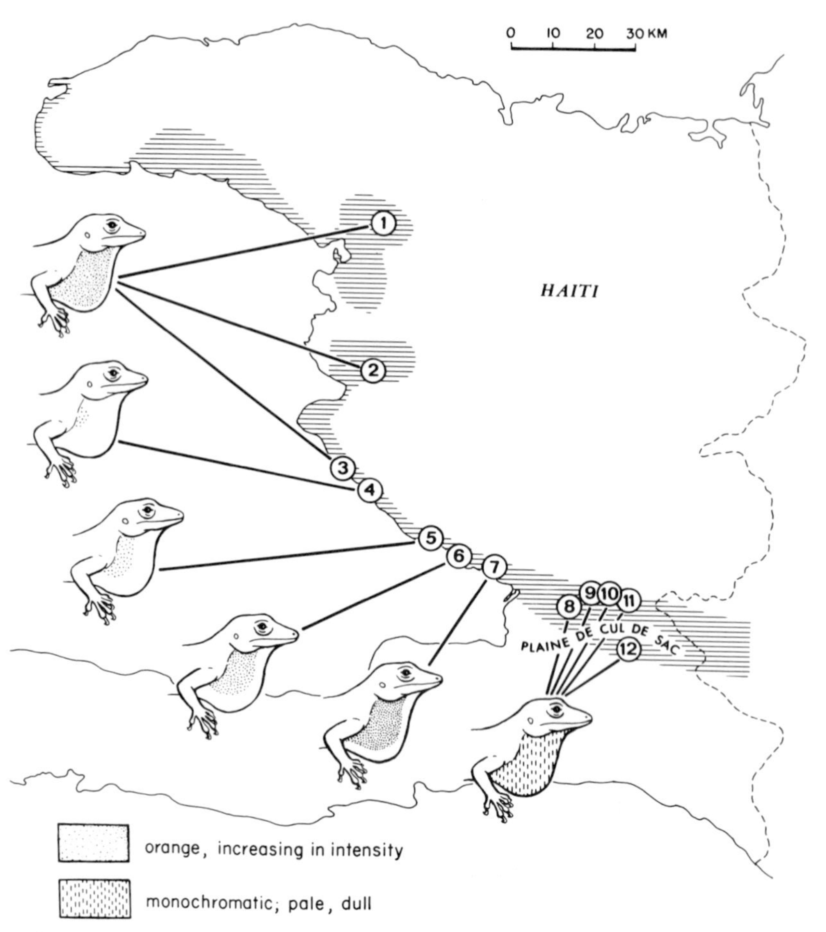

Fig. 3 Dewlap color patterns typical of "*Anolis brevirostris*" males in a dozen Haitian populations (numbered 1 to 12); hatching shows the linear geographic distribution of "*A. brevirostris*" in this region (after Webster and Burns, 1973).

version within an overall gradient in dewlap variation along linearly distributed "*brevirostris*" populations (Fig. 3).

On the basis of isozymic characters, these same lizard populations form three clear clusters. Electrophoretically detectable variants are remarkably uniform within populations 1 to 3 and 4 to 7 and 8 to 12 but are different between these groups. Table III shows this pattern of geographic variation for two of the enzymes studied. Taken together, data from all six enzymes reveal complete differentiation at four genetic loci between populations 3 and 4 and again between popula-

TABLE III

Population	Number lizards sampled	Frequency of αGPD alleles				Frequency of 6PGD alleles				
		a	b	c	d	a	b	c	d	e
1	51	.01		.99			.01	.99		
2	40	.09		.91				1.00		
3	82			.91	.09	.006		.994		
4	9			1.00					1.00	
5	36			1.00					1.00	
6	39			1.00					1.00	
7	34			1.00					1.00	
8	19		1.00						.82	.18
9	57		1.00						.84	.16
10	25		1.00						.68	.32
12	45		1.00						1.00	

Allelic frequencies for αGPD and 6PGD in 11 Haitian populations representing three sibling species within "*Anolis brevirostris*" (from Webster and Burns, 1973).

tions 7 and 8. Since this differentiation occurs over short distances in which significant barriers to movement are lacking, the differentiated populations must be reproductively isolated.

Variation in dewlaps perfectly supports the conclusion that "*brevirostris*" populations 1 to 12 comprise three separate species (A [populations 1 to 3], B [4 to 7], and C [8 to 12]). The two sharp discontinuities in dewlap color coincide with zones of contact between species A and B and species B and C. The steep cline in dewlap color within species B connects the divergent extremes that enhance visual differences where B meets its nearest relatives, A and C (Fig. 3).

ACKNOWLEDGMENTS

This work was supported in part by National Science Foundation grants GB-5935 and GB-37832 and by a grant from the William F. Milton Fund of Harvard University.

REFERENCES

Burns, J. M. 1968. Mating frequency in natural populations of skippers and butterflies as determined by spermatophore counts. *Proc. Natl. Acad. Sci. USA* 61: 852-859.

Burns, J. M. 1970. Highly polymorphic butterfly esterases. *Isozyme Bull.* 3: 49-50.

Burns, J. M. and F. M. Johnson 1967. Esterase polymorphism in natural populations of a sulfur butterfly, *Colias eurytheme*. *Science* 156: 93-96.

Burns, J. M. and F. M. Johnson 1971. Esterase polymorphism in the butterfly *Hemiargus isola:* stability in a variable environment. *Proc. Natl. Acad. Sci. USA* 68: 34-37.

Coluzzi, M. and L. Bullini 1971. Enzyme variants as markers in the study of pre-copulatory isolating mechanisms. *Nature* 231: 455-456.

Hovanitz, W. 1950. The biology of *Colias* butterflies. I. The distribution of the North American species. *Wasmann J. Biol.* 8: 49-75.

Johnson, F. M. and J. M. Burns 1966. Electrophoretic variation in esterases of *Colias eurytheme* (Pieridae). *J. Lepid. Soc.* 20: 207-211.

Schwartz, A. 1968. Geographic variation in *Anolis distichus* Cope (Lacertilia, Iguanidae) in the Bahama Islands and Hispaniola. *Bull. Mus. Comp. Zool.* 137: 255-309.

Shapiro, A. M. 1969. New distributional data on three northeastern United States butterflies. *J. Lepid. Soc.* 23: 265-269.

Webster, T. P. and J. M. Burns 1973. Dewlap color variation and electrophoretically detected sibling species in a Haitian lizard, *Anolis brevirostris*. *Evolution* 27: 368-377.

PREDICTING GENE FREQUENCIES IN NATURAL POPULATIONS: A TESTABLE HYPOTHESIS

DENNIS A. POWERS and DIANNE POWERS
Department of Biology
The Johns Hopkins University
Baltimore, Maryland 21218

ABSTRACT. Darwinian theory has been criticized because it is more often used to make retrodictions than predictions. The neutralist theory has become popular among molecular evolutionists because it leads to predictions and testable hypotheses. Our studies provide a testable hypothesis for predicting gene frequencies in natural populations from a selectionist viewpoint, utilizing enzymology and physical environmental parameters. We generate an equation, $q = \frac{E^n}{E_e^n + E^n}$, and test its predictive powers by employing the LDH-B locus in *Fundulus heteroclitus*. The results are consistent with selectionist predictions. The extent of selection or lack of it remains an open question that we will continue to test via a biochemical approach.

INTRODUCTION

Comparative biochemists and physiologists have provided considerable information concerning the molecular machinery for adapting to environmental stress (Prosser, 1973).

When confronted with stress, a species must either adapt or die. Because poikilotherms must adapt to thermal stress, efforts have been made to study their adaptive mechanisms (see review by Prosser, 1967). In general, poikilotherms modify their metabolic rate as their environmental temperature increases. Metabolic rate, in turn, is a function of a complex interconnected series of temperature-sensitive enzymes. The works of Hochachka et al. (1971), Somero (1969), Selander, Hunt, and Yang (1969), and Shaw (1965), and many others, have established that poikilothermic species are, generally, enzymatically adapted for their thermal environments.

Utilizing gel electrophoresis and histochemical staining, population geneticists have demonstrated a great deal of protein variation within populations of the same species (e.g., Prakash, et al., 1969). These data have stimulated discussion concerning the role of selection in maintaining protein polymorphisms and the role (if any) of environmental factors (e.g., temperature). In other words, are populations enzymatically adapted for their external environment?

Many investigators advocate a form of selection operating to balance the protein polymorphism (Jungck, 1971; Richmond, 1970; Powell, 1971; Prakash, 1969; Johnson, 1974; Ayala, 1970; Ayala, et al., 1971), but others (Kimura, 1968a, 1968b; Kimura and Crow, 1964; Arnheim and Taylor, 1969; King and Jukes; 1969; Kimura and Ohta, 1971; Jukes and King, 1971) have convincingly argued that observed polymorphisms are selectively neutral.

Since present estimates of mutation rates, genetic load, "effective" population size, and evolutionary rates are inexact, we have addressed our research efforts toward studying the structural and functional properties of allelic products in relation to environmental variables.

If a species is found over a large geographical range, the variant thermal conditions could exceed the limits of some individual's ability to adapt. Survival may be accomplished by physiological adaptation and/or population adjustments in the frequencies of temperature sensitive gene products. If the species survive through selective processes (i.e., of temperature sensitive loci), one would expect to find different gene frequencies of polymorphic, thermal sensitive, loci in populations living under different thermal regimes. Some studies have attributed genetic clines over geographic space to be related to changing thermal conditions (e.g., Schopf and Gooch, 1971; Johnson, et al., 1969; Koehn, et al., 1971; Koehn, 1969 and Merritt, 1972). A few have attempted to explain the clines by correlating an enzymic property of the phenotype with temperature. Two such studies were done on esterases in fish (Koehn, 1969 and Koehn, et al., 1971). Because of the differences in the molecular weights of the esterase "phenotypes" (see: Koehn, 1969), the non-specificity of esterases, the secondary esterase activity of several enzymes with different primary activity, the nature of the assay employed, the lack of genetic information and the thermal induction of different esterase phenotypes (Baldwin, et al., 1970), these works are difficult to interpret. In spite of the inherent difficulties, suggestions of these studies have provoked our interest. Merritt (1972) did not have genetic information nor a complete kinetic study but he did employ a specific enzyme (LDH; L-lactate NAD oxidoreductase, E.C. 1.1.1.27). In the light of Hochachaka and Somero's (1973) implication that Michaelis constants are ecospecific, Merritt observed that thermally induced changes in Michaelis constants were correlated with the rarity of one LDH phenotype in southern latitudes.

If temperature variation is a primary cause in selecting some enzyme phenotypes, one might be able to predict, *a priori*, gene frequencies in selected types of natural populations by using the functional parameters of the allozymes and thermal information from the locality in question. The present study was undertaken to explore the feasibility of this hypothesis.

Fundulus heteroclitus was chosen because they are available in large numbers and found along most of the east coast of the United States and Canada. An LDH-B sununit polymorphism was reported by Whitt (1969). We found (1973) LDH phenotypes to be genetic products that are inherited in a Mendelian manner. *Fundulus* are found in one of the steepest thermal gradients in the world having large seasonal variation in temperature at each locality (daily temperature measurements are availbale along most of the coast), the populations are relatively continuous from Canada to Florida, they are an inshore species so a linear model can be employed, they are easily raised and bred under laboratory conditions, and a wealth of physiological, embryological, and morphological work has been done on the species. Although several *Fundulus* enzyme systems are being studied in our laboratory, the kinetics of the liver LDH-B locus were the first to be examined in relation to temperature as a model system to test the predictability of allozymes in natural populations (see Discussion for the significance of this LDH to fish).

EXPERIMENTAL PROCEDURE

Fish were sampled during 1971, 1972 and 1973 by seining and baited killitraps. Samples to be used for population studies were frozen shortly after capture while samples for kinetic studies were kept alive. A series of studies indicated that tissues frozen between 0° and -100°C lost LDH activity exponentially with time. Freezing at -190°C was adequate to maintain 90% activity indefinitely. Fish used for kinetic studies were maintained alive in aquaria at 5°, 10°, 15°, 20°, 25°, and 30° C for two months before analysis. Other fish were analyzed directly from the field.

Mummichogs used for breeding studies were collected from the two following locations: (1) from Sandy Point on the Chesapeake Bay on October 15, 1972 and (2) from Woods Hole, Mass. during the summers of 1972 and 1973. Fish were maintained at a density of one fish per 2.5 gallons of 12.5% sea water at 20° ± 2°C with a photoperiod of 14 hours of light and 10 hours of darkness. The feeding schedule and other particulars will be described elsewhere.

Individual livers were weighed, then 0.25 ml of 0.1 M Tris HCl (pH 7.2) per 100 mg of tissue was added. The tissue was homogenized in buffer, on ice, in a glass tissue homogenizer. The homogenate was transferred to a centrifuge tube and centrifuged in a Sorvall RC-2B centrifuge at 10,000 x g for 30 minutes at 4°C. The supernatant was normally used directly for kinetic studies because partial purification did not modify the kinetic parameters and crude extracts allowed comparisons of relative LDH activity between individuals. The liver (or heart) LDH was devoid of LDH-A subunits (Powers, 1972, 1973). Once the kinetic analyses were completed protein concentration and phenotype were determined. Protein concentration was determined by the colorimetric method of Lowry et al. (1951) using bovine serum albumin as a standard reference while LDH phenotype was detected by starch gel electrophoresis and histochemical staining for LDH activity. Electrophoresis was done in 0.75 M Tris, 0.25 M citric acid buffer (pH 6.9) as described by Whitt (1969).

For kinetic studies, NAD and NADH were standardized in an automated Beckman spectrophotometer at 340 nm. The dual beam spectrophotometer was interfaced with a teletype so that data could be gathered for computer analysis. Temperature was recorded continuously inside the cuvettes and each sample was stirred with a micro-magnetic stirrer. The substrates (lactate and pyruvate) were standardized enzymatically by employing crystalline LDH and following the stoichiometry of NADH or NAD formation.

LDH activity from lactate to pyruvate was measured in 1.0 ml of 0.1 M Tris-HCl at various pH values with 1 mM NAD while activity from pyruvate to lactate was usually measured in 0.02 mM Imidazole buffer (pH 7.0) with 0.02% bovine serum albumin and 0.01 mM NADH. The latter concentration of NADH was held constant throughout the study. This is an important factor because the kinetics for the pyruvate to lactate conversion is complicated by the fact that the Michaelis constant (Km) for pyruvate increases with increasing NADH concentration. Furthermore, temperature corrections must be made for the buffers.

Initial velocities were expressed as moles of substrates/liter/hour/μg protein nitrogen or per gram of wet weight tissue. All determinations were made in triplicate and each phenotype was repeated 30-50 times under various acclimation regimes and during various times of the year. This procedure was followed to see if there were environmentally induced changes in the degree of expression for the LDH loci.

K_m was determined by Lineweaver-Burk plots using 16 different substrate concentrations for each of the following reaction temperatures: 0°, 5°, 10°, 15°, 20°, 25°, 30°, 35°, and 40° C. In addition, the B'B' phenotype was studied at 6°, 7°, 8°, 9°, 11°, 12°, 13°, and 14° C.

Heat denaturation studies on the various LDH phenotypes were done at 50° C for times ranging from 10 minutes to 1 hour with and without substrates and cofactors. The assay temperature was 25°C.

Blood and tissue substrate levels, pH and bicarbonate levels were measured by methods described elsewhere (Powers, 1974; Clark and Powers, 1973; Powers, 1973).

RESULTS

Allozymic velocities were found to differ with temperature. The homozygote (B'B') has a greater velocity per microgram of animal tissue than either of the other phenotypes (see Fig. 1). Since the kinetics were not significantly modified by initial purification we preferred to use the supernatant of fresh animal homogenates because this approach allowed comparisons between both individuals and phenotypes. Structural and functional studies on highly purified enzymes are currently underway.

The B'B' homozygote had an activation energy that changed drastically at about 12.4°C. Below 12.4°C the activation energy was about 50.7 kilocals per mole while above that temperature the value was 11.7 kilocals per mole. The sharp change in activation energy with temperature suggests a thermally induced structural rearrangement that directly affects the functional properties of the enzyme. The other homozygote (BB) and the heterozygote (BB') had values of 11.1 kilocals per mole over the whole temperature range (See fig. 1).

Using Michaelis-Menten kinetics, the equilibrium constant (K_m) was found to vary between the allozymes (see Fig. 2). The B'B' homozygote was the least changed over the whole temperature range. The heterozygote showed less change in K_m from -5°C to 20°C than the homozygotes. However, above 20°C the K_m increased rapidly presumably because of the breakdown in the enzyme-substrate complex (note legend of Fig. 2).

Heat stability studies (see Fig. 3) indicate the B'B' homozygote to be the most stable at 50°C, the BB homozygote to be the next most stable while the heterozygote (B'B) was the least stable. Since the heterozygote is a tetramer of non-identical subunits the large difference in heat stability suggests the heterozygote tetramer is dissociating into inactive subunits at 50°C. Although enzyme stability is dependent on subunit contacts, there is <u>no</u> evidence for subunit cooperativity. Stabilization of the tetramer with either substrate or cofactor greatly enhanced subunit stability. Substrate and cofactor also stabilized the tetramers of the other two phenotypes. While a weakening of subunit bonds in

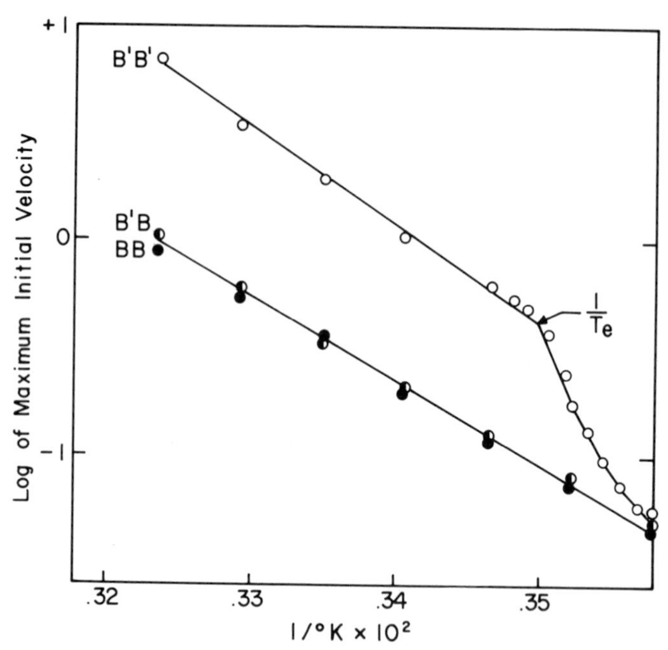

Fig. 1. The relationship between the maximum initial velocity and the inverse of the absolute temperature for the three allotypes. In this Arrhenius plot, the slope is equal to E/2.303R where E = activation energy, °K = absolute temperature, R = the universal gas constant.

---

heterozygotes may be detrimental at higher temperatures, it could conceivably be a positive factor at lower temperatures by affecting substrate affinity or other kinetic parameters.

Physiological studies. Studies on numerous physiological parameters as a function of thermal acclimation revealed some interesting results. Almost all parameters were affected by increased temperature (Table I). Muscle lactate levels were not greatly affected but blood lactate showed large variations between individuals and the differences were most pronounced at the warmer acclimation temperatures (i.e., >20° C) (See Table I) (also see Powers, 1974).

The results of the population studies can be seen in Fig. 6 and Table II. The significance of these results will be discussed in a later section.

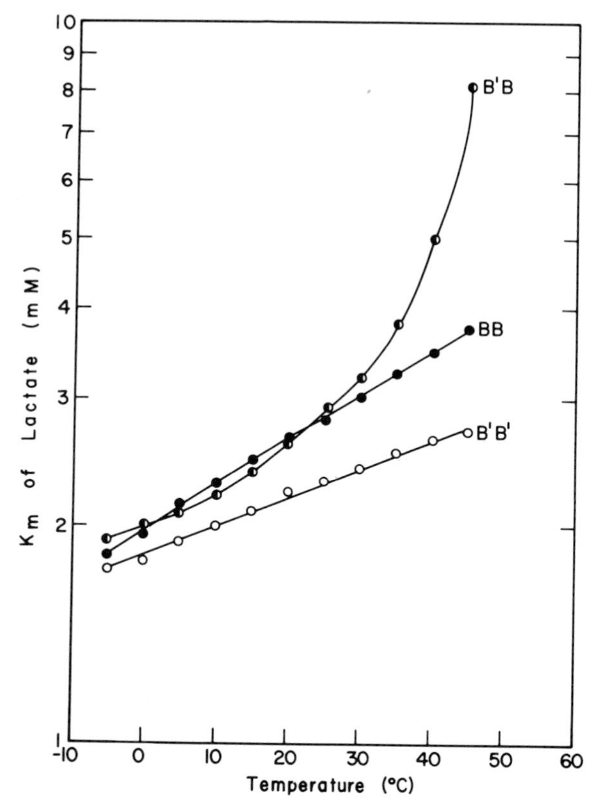

Fig. 2. Michaelis-Menten equilibrium constant (K_m) plotted against temperature (degrees centigrade). At lower temperatures the Km's are not significantly different.

DISCUSSION AND CONCLUSIONS

Physiological significance of LDH-B in fish. In 1972 we described a molecular strategy of adaptation for fast and slow water habitats that involved an evolutionary modification of the Bohr effect on fish hemoglobins (Powers, 1972). More recently we elucidated some of the subtle seasonally induced physiological adaptations of fish hemoglobins, their allosteric modifiers, total oxygen binding capacity and optimization of their internal molecular environment (Powers, 1974). During our current study we became aware of individual variation in the capacity to control blood pH both in constant and variable environments. The major contributing factor was the fish's ability to handle high levels of blood lactate resulting from bursts of muscular anaerobic metabolism induced by swimming stress or increased anaerobiosis provoked by either

TABLE I

PHYSIOLOGICAL ADAPTATION TO THERMAL STRESS

	Acclimation Temperature (^{o} C)			
	2°	5°	10°	15°
Environmental O_2 gms/100 ml H_2O	.0063	.0060	.0054	.0048
ATP/Hb (moles/mole)	1.82 ± .22	1.84 ± .13	1.87 ± .16	1.64 ± .12
Percent Hematocrit	28.4 ± 1.5	30.0 ± 1.5	32.6 ± 1.7	37.0 ± 3.0
Blood pH	-	7.39 ± .14	7.34 ± .05	7.25 ± .05
Blood Bicarbonate (mM/L)	-	13.1 ± 0.1	13.8 ± 0.2	14.0 ± 0.1
Blood Lactate mg/100 ml	-	8.2 ± 3.1	9.4 ± 2.7	15.0 ± 3.6
Muscle lactate μM/gram tissue	-	15.71 ± 0.49	15.41 ± 0.51	15.0 ± 0.62
Liver Glycogen (mg/g wet weight)	13.36 ± 11.24	50.49 ± 13.06	55.40 ± 11.93	67.41 ± 11.81
Muscle Glycogen (mg/g wet weight)	3.7 ± 2.01	4.20 ± 1.75	4.70 ± 1.01	4.41 ± .71
Blood Glucose (mg %)	200	100	75	70

TABLE I (cont.)

	Acclimation Temperature (^{o} C)		
	20°	25°	30°
Environmental O_2 gms/100 ml H_2O	.0043	.0038	.0037
ATP/Hb (moles/mole)	1.54 ± .12	1.38 ± .26	1.10 ± .17
Percent Hematocrit	44.0 ± 2.0	47 ± .16	50.1 ± 2.1
Blood pH	7.10 ± .14	6.90 ± .12	6.85 ± .04
Blood Bicarbonate (mM/L)	14.1 ± 0.3	14.5 ± 0.3	15.1 ± 0.2
Blood Lactate mg/100 ml	20.1 ± 8.1	23.0 ± 12.7	42.7 ± 10.9
Muscle lactate μM/gram tissue	14.14 ± 0.52	13.80 ± 0.70	13.29 ± 0.50
Liver Glycogen (mg/g wet weight)	87.14 ± 13.81	95.27 ± 25.2	115.22 ± 31.6
Muscle Glycogen (mg/g wet weight)	4.81 ± .91	4.80 ± 1.10	4.75 ± .67
Blood Glucose (mg %)	75	85	100

TABLE II

Phenotypic variation of the lactate dehydrogenase B locus in *Fundulus heteroclitus*, with expected proportions in parentheses, allelic frequencies and goodness-of-fit X^2 tests of observed and expected distributions.

Locality	LDH-B'B'	LDH-B'B	LDH-BB	N	q ± s.e.	X^2	E^+	$q^\#$
1. Halifax, N.S., Canada	0 (0.21)	9 (8.58)	87 (87.21)	96	.047 ± .015	.232	7.8	.045
2. Boothbay, Me.	0 (0)	0 (0)	58 (58)	58	0 ± .000	.000	7.8	.045
3. Gloucester, Mass.	11 (2.03)	21 (18.93)	43 (44.03)	75	.177 ± .033	.777	9.5	.147
4. Woods Hole, Mass. (1969)*	21 (17.78)	90 (96.44)	134 (130.78)	245	.269 ± .020	1.093		
Woods Hole, Mass. (1972)	6 (4.18)	24 (13.85)	48 (46.18)	78	.231 ± .048	.693	10.8	.287
Woods Hole, Mass. (1973)	11 (10.97)	62 (62.11)	88 (87.93)	161	.261 ± .031	.293		
5. Stony Brook, New York	2 (3.75)	26 (22.50)	32 (33.75)	60	.250 ± .040	1.452	11.4	.365
6. Jones Beach, New York	8 (9.18)	26 (23.63)	14 (15.19)	48	.438 ± .051	.485	11.9	.433
7. Keansburg, New Jersey	8 (7.44)	21 (22.12)	17 (16.44)	46	.402 ± .051	.118	12.0	.446

TABLE II (cont.)

Locality	LDH-B'B'	LDH-B'B	LDH-BB	N	q ± s.e.	X^2	E^+	$q^\#$
8. Atlantic Highlands, New Jersey	14 (16.12)	42 (37.76)	20 (22.12)	76	.461 ± .040	.957	12.1	.460
9. Point Pleasant New Jersey	29 (29.45)	36 (35.09)	10 (10.45)	75	.627 ± .039	.050	12.5	.513
10. Atlantic City, New Jersey	20 (20.67)	23 (21.65)	5 (5.67)	48	.656 ± .048	.185	13.3	.614
11. Cape May New Jersey	21 (21.11)	21 (20.78)	5 (5.11)	47	.670 ± .048	.005	13.8	.670
12. Woodland Beach, Del.	26 (24.1)	16 (19.8)	6 (4.1)	48	.708 ± .046	1.79	13.6	.648
13. Cedartown, Md.	173 (175.9)	118 (112.2)	15 (17.9)	306	.758 ± .017	.810	14.4	.728
14. Wachapreague, Virginia(1972)	108 (106.40)	56 (59.33)	10 (8.27)	174	.782 ± .018	.571	15.1	.786
Wachapreague, Virginia(1973)	60 (59.1)	33 (34.8)	6 (5.1)	99	.773 ± .030	.257		
15. Cape Henry, Virginia	11 (11.25)	8 (7.50)	1 (1.25)	20	.750 ± .068	.089	15.7	.825
16. Beaufort, North Carolina	106 (105.39)	23 (24.22)	2 (1.39)	131	.897 ± .019	.331	18.2	.968

TABLE II (cont.)

Locality	LDH-B'B'	LDH-B'B	LDH-BB	N	q ± s.e.	X^2	E^+	$q^\#$
17. Ladies Island, North Carolina	30 (30.01)	1 (0.98)	0 (0.01)	31	.984 ± .016	.008	20.8	.968
18. Savannah, Georgia	48 (48.00)	0 (0.00)	0 (0.00)	48	1.000 ± .000	.000	21.0	.970

* Data from Whitt (1969).

+ Taken from the latidude and Fig. 5 ($^\circ$ C).

Predicted gene frequency.

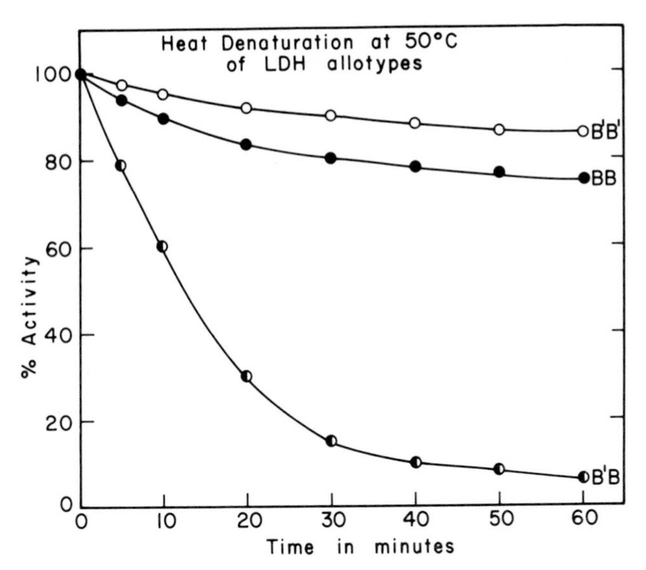

Fig. 3. Heat denaturation curves for the LDH-B allotypes at 50°C.

lowering oxygen and/or increasing water temperature. Individual variability in blood lactate suggested there might be a genetic component related to the ability to convert blood lactate to pyruvate (i.e., in the liver) for gluconeogenesis or the channeling of pyruvate into the citric acid cycle. The degree of lactate control was reflected in the ability to maintain blood pH within reasonable limits. When blood pH could not be adequately controlled, death often resulted presumably because fish could not bind oxygen at the gills. The Bohr effect, considered an advantage under normal circumstances, can prevent binding of oxygen at the gills when blood pH is inadequately controlled (details to be published elsewhere).

Temperature and/or oxygen induced changes in blood pH are the result of an emphasis on anaerobic metabolism causing increases in blood lactate. Since lactate levels increase with temperature of acclimation (see Table I), it is reasonable that the K_m of the liver LDH could increase with temperature (see Fig. 2) as long as the K_m was sufficiently low to maintain a reasonable affinity for the substrate so that blood pH could be adequately controlled. This argument is complicated by the fact that lowering blood pH, somewhat, with increasing temperature can be a beneficial phenomenon for optimizing the interaction between hemoglobin subunit cooperativity and its allosteric modifier (i.e., ATP) as we have pointed out earlier (Powers, 1972; Powers, 1974; Powers and Edmundson, 1972). But, as indicated in Table I, there is no significant change in erythrocyte ATP levels until the temperature of acclimation exceeds 20°C. Consequently, below 20°C the control of blood lactate, therefore, pH, could be critical to survival because lowering blood pH would reduce the ability to bind oxygen thereby making competition for available food or swimming away from a predator more difficult. We are currently exploring this possibility via selection experiments.

The Hypothesis. Our kinetic data suggests that the electrophoretically fast phenotype (B'B') would be at an advantage in warm water (i.e., above 12.4°C and at a disadvantage in colder water (i.e., below 12.4°C with possible equality below 4°C). Although we have preliminary evidence that suggests the heterozygote (BB') is biochemically superior to both BB and B'B' at low temperatures, we shall consider the heterozygote intermediate to the homozygotes between 0 and 20°C as suggested by figures 1 and 2. If the water temperature is fluctuating above and below 12.4°C (i.e., during the course of the year), then the net result is heterozygote superiority in a thermally changing environment (Levins and MacArthur, 1966). That is to say, if the heterozygote is intermediate to the homozygotes, selection for the B'B' homozygote in warmer waters and selection against that phenotype in cooler waters would result in a net heterozygote superiority over the combined temperature range. Therefore, a heterosis model will be used because of the changing thermal conditions of the environment. We shall generalize our predictive model on this basis, but it should be kept in mind that these are assumptions.

If q^2 and p^2 are the probabilities of two homozygous phenotypes, of a two allele polymorphism, that are affected by some environmental variable E such that:

$$\text{If:} \quad q^2 \propto W_{B'B'} \propto \frac{1}{S_2} \propto E^{c1} \text{ and } p^2 \propto W_{BB} \propto \frac{1}{S_1} \propto E^{c2}$$

(where: $W_{B'B'}$ = fitness of B'B', W_{BB} = fitness of BB, S_2 = selection coefficient of B'B' and S_1 = selection coefficient of BB, and $-\infty \leqslant c_2 \leqslant 0 \leqslant c_1 \leqslant +\infty$).

If the heterozygote is intermediate over a range ($E_i \rightarrow E_n$) that is in a fine grained changing environment (Levins and MacArthur, 1966), or B'B is more fit than both B'B' and BB for each E, then for the range of E the above equation could reflect a heterozygote superiority model. Therefore, when populations are in equilibrium with E, that is, when

$\frac{\Delta q}{\Delta t} = 0$, then: $\frac{q}{(1-q)} = \frac{S_1}{S_2}$ where: S_1 and S_2 are selection coefficients for p^2 and q^2 in a heterosis model and q is the <u>equilibrium gene frequency</u>.

Since: $\frac{1}{S_1} \propto E^{c_2}$ and $\frac{1}{S_2} \propto E^{c_1}$

Then: $S_1 = \frac{1}{k_2 E^{c_2}}$ and $S_2 = \frac{1}{k_1 E^{c_1}}$

Therefore: $\frac{S_1}{S_2} = \frac{k_1}{k_2} E^{c_1 - c_2} = KE^n$

(where: K and n are system specific constants and because $-\infty \leqslant c_2 \leqslant 0 \leqslant c_1 \leqslant +\infty$ then $0 < n < +\infty$).

Therefore: $\frac{S_1}{S_2} = \frac{q}{(1-q)} = KE^n$ (when: $\frac{\Delta q}{\Delta t} = 0$)

then: $q = \frac{KE^n}{1 + KE^n}$

One can evaluate K at the environment of equal adaptability or equality (E_e) (e.g., the temperature of allelic equivalence).

$$\left.\frac{q}{(1-q)}\right]_{E = E_e} = KE_e^n = 1$$

(where: $q = (1-q)$)

Therefore: $\frac{q}{(1-q)} = \frac{E^n}{E_e^n}$

$$qE_e^n = E^n - qE^n$$

$$\boxed{q = \frac{E^n}{E_e^n + E^n}} \text{ and, therefore: } p = \boxed{\frac{E_e^n}{E_e^n + E^n}}$$

The curve of the above equation can be seen in Fig. 6 for various values of n and an arbitrary $E_e = 0.5$ for illustration purposes.

Consequently one may characterize the above equation by:

$$\frac{dq}{dE} = \frac{nE^{n-1}}{E_e^n + E^n} - \frac{nE^{n-1}\,E^n}{(E_e^n + E^n)^2}$$

$$\frac{dq}{dE} = \frac{nE^{n-1}}{E_e^n + E^n} \quad \left(1 - \frac{E^n}{E_e^n + E^n}\right)$$

$$\boxed{\frac{dq}{dE} = \frac{n}{E}\, q(1-q) = \frac{n}{E}\, pq}$$

$$\frac{d^2q}{dE} = \frac{-nq(1-2)}{E^2} \quad \left\{1-n(1-2q)\right\}$$

Therefore the point of inflection (q_i) for Figure 6 would be:

$$q_i = \frac{n-1}{2n}$$

So: $\lim_{n \to \infty} q_i = 1/2$

Consequently, as n increases, the fitnesses in the different environmental ranges (i.e., above and below E_e) become greater until once n is very large the curve in Fig. 4 becomes a step function with essentially selection for only one homozygote in one environment and the other homozygote in the other environment. If this happened at many "critical" loci the populations could conceivably "initiate" speciation.

The first test. Since the environmental variable (E) measured in our kinetic study was temperature, we shall use temperature directly as the variable E in this initial attempt to test the predictive powers of the equation $q = \frac{E^n}{E_e^n + E^n}$. Our data suggest that the E_e for the mummichog LDH-B system is about 12.4°C (see text). One may solve for the exponent (n) by knowing the gene frequency from at least one natural population and an accurate temperature profile from that locality. Although accurate temperature data is available from Woods Hole, the gene frequency data (Whitt, 1969) was not gathered from the point of temperature measurement. Consequently, fish were typed from Wachapreague, Virginia during the spring of 1972 and fall of 1973. This site was chosen because there was a temperature probe where the fish were collected. It

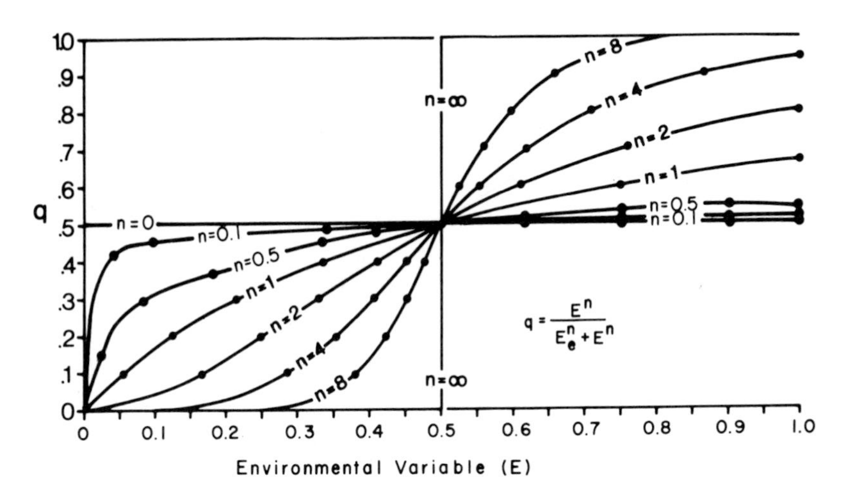

Fig. 4. Several representative curves for the equation: $q = \frac{E^n}{E_e^n + E^n}$. The value of 0.5 has been selected for E_e and several values of n are represented.

should be noted that there was no significant change in gene frequency during these years nor were there significant changes in the gene frequencies from Woods Hole, Mass. or Beaufort, N. C. (see Table II). Therefore, $\frac{\Delta q}{\Delta t} = 0$ which is necessary for equilibria models. Using one locality and $E_e = 12.4^oC$, the exponent for the LDH-B system was found to be: $n = 6.6$.

Since the yearly water temperature at a locality varies in a sinesoidal manner, the average water temperature was used as a single value that reflects relatively equal deviation in both directions. Undoubtedly, the extreme values are the most important to an animal but the average water temperature reflects the amount of time near those extremes. The water temperature varies considerably along the range of *Fundulus* and most localities show the same general degree of variation but the curves are shifted up or down depending on the locality (Anonymous, 1968). There is, of course, less variability at the upper and lower extremes and, because some measurements are in open ocean, some cited values are probably lower than those experienced by the fish. A temperature much above E_e (i.e., the temperature of phenotypic equivalence) indicates little, if any, time for selecting against the warm water phenotype (B'B') while a low temperature would reflect the opposite case. If $E = E_e$, it would indicate relatively equal

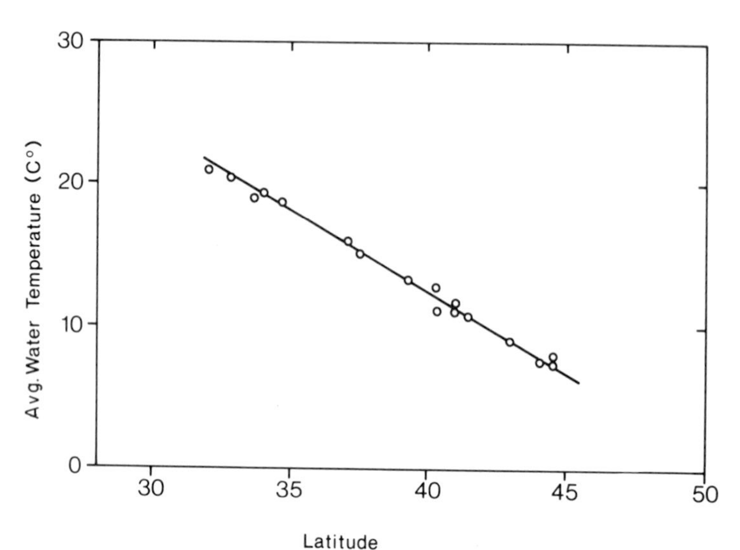

Fig. 5. Average water temperature of representative localities plotted against latitude of the localities in question.

time above and below the temperature of equivalence and should reflect relatively equal selection times, however the degree of selection may vary.

Unfortunately, temperatures are not always available from the sites where fish are collected. Therefore, the temperature was estimated from latitude of each locality and published oceanographic data (Anonymous, 1968; see Fig. 5). Since local thermal conditions vary, we must expect some error factor from this estimation. However, for a number of intuitively obvious biological reasons, we do not expect precision from our predictions. Rather, we expect a trend that will be predictable and therefore temperature estimates will suffice.

Using the equation $q = \frac{E^n}{E_e^n + E^n}$ and the temperature data, gene frequencies were estimated from 18 localities. The actual gene frequencies (see Table II) were found to correspond remarkably well with the predicted values.

In addition, when frequencies are plotted against average water temperature, (see Fig. 6) there is a sharp change in gene frequency around 12°C which is consistent with our enzyme kinetic data and our predictions. The actual selection scheme is undoubtedly much more complex than suggested here but it appears that temperature is a selective force for these LDH allotypes, and in this case anyway the data cannot be easily accounted for by selective neutrality.

While the concept of natural selection for advantageous

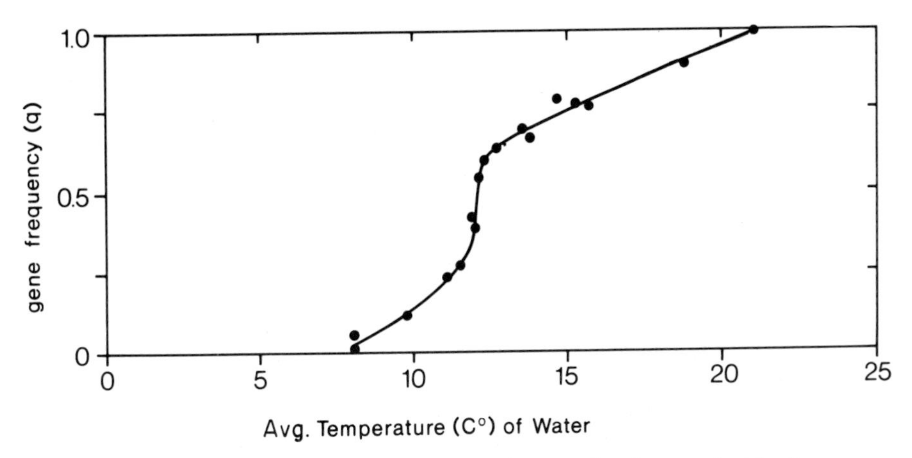

Fig. 6. Gene frequencies of natural populations plotted against the average water temperature of the locality from which they were taken. Temperature is estimated from Fig. 5.

genes is engrained in the thinking of modern biologists, we would be advised to remember that natural selection operates on the reproductive unit rather than on isolated molecules. A particular detrimental molecule could presumably be the determining component of a population's fitness but these are probably rare exceptions rather than the rule. Therefore, we must caution the reader not to extrapolate our data to absurd generalizations. We are not suggesting unigenetic selection. These fish are probably not controlled solely by their LDH phenotypes. It is the summation and synergistic interaction between fitnesses of numerous loci that determine survival.

So far, our predictive hypothesis appears useful. We are currently rigorously testing this hypothesis with 20 other enzyme systems, as well as examining the molecular basis for structural and functional variations in allotypes. Because of difficulties in determining the environment of equivalence (E_e) for protein systems and the limitations on the accurate determination of environmental parameters (E), the precision of predictive equations for natural situations are self limiting and, therefore, not universally applicable. However, the biochemical and physiological significance of selection at the phenotype level should transcend these difficulties. Therefore, future work will not attempt to predict absolute gene frequencies but will establish relationships between laboratory experiments and field studies.

ACKNOWLEDGEMENTS

This work was supported by grants: NSF (GB37584), NIH (GM00265-15), and EPRI (P407711). We thank Drs. R. Koehn, J. Mitton, and R. Lewontin for helping bridge the communication gap between biochemistry and population biology.

REFERENCES

Anonymous. 1968. Surface water temperatures at tide stations: Atlantic Coast., U. S. Dept. of Com. Coast and Geodetic Survey, Washington, D.C.

Arnheim, N. and C. E. Taylor 1969. Non-Darwinian evolution: consequences for neutral allelic variation. *Nature* 223: 900-901.

Ayala, F. J. 1970. Enzyme variability in the *Drosophila willistoni* group, I. Genetic differences among sibling species. *Proc. Nat. Acad. Sci.* USA 67: 225-232.

Ayala, F. J., J. R. Powell, and T. Dobzhansky 1971. Polymorphisms in continental island population. *Proc. Nat. Acad. Sci.* USA 68: 2480-2483.

Baldwin, J. and P. W. Hochachka 1970. Functional significance of isozymes in thermal acclimation: acetylcholinesterase from trout brain. *Biochem. J.* 116: 883-887.

Clark, V. and D. A. Powers 1973. Environmental adaptation of *F.heteroclitus* muscle and liver LDH's. *Biol. Bull.* 145: 428.

Hochachka, P. W. and J. K. Lewis 1971. Interacting effects of pH and temperature on the K_m values for fish LDH. *Comp. Biochem. Physiol.* 38: 925-933.

Johnson, G. B. 1974. Enzyme polymorphism and metabolism. *Science* 184: 28-37.

Johnson, F. M., H. E. Schaffer, F. E. Gillaspy, and E. S. Rockwood 1969. Isozyme genotype-environment relationships in natural populations of harvester ant, *P.barbatus*, from Texas. *Biochem. Genet.* 3: 429-450.

Jukes, T. H. and J. L. King 1971. Deleterious mutations and neutral substitutions. *Nature* 231: 114-115.

Jungck, J. R. 1971. Pre-Darwinian and Non-Darwinian evolution of proteins. *Molecular Biol.* 3: 307-318.

Kimura, M. 1968a. Genetic variability maintained in a finite population due to mutational production of neutral and nearly neutral isoalleles. *Genetic Res. Camb.* 11: 247-269.

Kimura, M. 1968b. Evolutionary rate at the molecular level. *Nature* 217: 624-626.

Kimura, M. and J. F. Crow 1964. The number of alleles that can be maintained in a finite population. *Genetics*

49: 725-738.

Kimura, M. and T. Ohta 1971. Protein polymorphism as a phase of molecular evolution. *Nature* 229: 467-469.

King, J. L. and T. H. Jukes 1969. Non-Darwinian evolution. *Science* 164: 788-798.

Koehn, R. K. 1969. Esterase heterogeneity: dynamics of a polymorphism. *Science* 163: 943-944.

Koehn, R. K., J. E. Perez, and R. B. Merritt 1971. Esterase enzyme function and genetical structure of populations of the freshwater fish, *Notropis stamineus*. *Am. Nat.* 105: 51-69.

Levins, R. and R. MacArthur 1966. The maintenance of genetic polymorphism in a spatially heterogeneous environment: variations on a theme by Howard Levene. *Am. Nat.* 100: 585-589.

Lowry, O. H., N. J. Rosebrough, A. L. Farr, and R. J. Randall 1951. Protein measurement with the folin phenol reagent. *J. Biol. Chem.* 193: 265-275.

Merritt, R. B. 1972. Geographic distribution and enzymatic properties of LDH allozymes in the fathead minnow, *Pimephalus promelas*. *Am. Nat.* 106: 173-184.

Powell, J. R. 1971. Genetic polymorphisms in varied environments. *Science* 174: 1035-1036.

Powers, D. A. 1972. Enzyme kinetics in predicting gene frequencies of natural populations. *Amer. Soc. of Ichth. and Herp.* 52: 77.

Powers, D. A. 1972. Hemoglobin adaptations for fast and slow water habitats in sympatric catostomid fishes. *Science* 177: 360-362.

Powers, D. A. 1973. Predicting gene frequencies in natural populations. II. The genetic and physiological basis of protein polymorphisms. *Biol. Bull.* 145: 540.

Powers, D. A. 1974. Structure-function and molecular ecology of fish hemoglobins. In: *Hemoglobins: Comparative Molecular Biology Models for the Study of Disease*. Ann. New York Acad. Sci., 241: 472-490.

Powers, D. A. and A. B. Edmundson 1972. Multiple hemoglobins of catostomid fish: I. Isolation and characterization of the isohemoglobins from *C.clarkii*. *J. Biol. Chem.* 247: 6694-6707.

Prakash, S. 1969. Genic variation in natural populations of *Drosophila persimilis*. *Proc. Nat. Acad. Sci.* USA. 62: 778-784.

Prakash, S., R. Lewontin, and J. Hubby 1969. A molecular approach to the study of genetic heterozygosity in natural populations. IV. Patterns of genic variation in central, marginal and isolated populations of *Drosophila pseudoobscura*. *Genetics* 61: 841-858.

Prosser, C. L. 1967. Molecular mechanisms of temperature adaptation. *Amer. Assoc. Adv. Sci.*, Washington, D. C.

Prosser, C. L. 1973. *Comparative Animal Physiology*. 3rd. Ed. W. B. Saunders Company, Philadelphia, Pa.

Richmond, R. C. 1970. Non-Darwinian evolution: A critique. *Nature* 225: 1025-1028.

Schopf, T. and J. J. Gooch 1971. Gene frequencies in a marine ectoproct: A cline in natural populations related to sea temperature. *Evolution* 25: 286-289.

Selander, R. K., W. G. Hunt, and S. Y. Yang 1969. Protein polymorphism and genetic heterozygosity in two European subspecies of the house mouse. *Evolution* 23: 379-390.

Shaw, C. R. 1965. Electrophoretic variations in enzymes. *Science* 149: 936-941.

Somero, G. N. 1969. Enzymic mechanisms of temperature compensation: immediate and evolutionary effect of temperature on enzymes of aquatic poikilotherms. *Am. Nat.* 103: 517-530.

Whitt, G. S. 1969. Homology of LDH genes: E gene function in the teleost nervous system. *Science* 166: 1156-1158

GENIC HETEROZYGOSITY AND POPULATION DYNAMICS IN SMALL MAMMALS

MICHAEL H. SMITH[1], CHARLES T. GARTEN, JR.[1], AND PAUL R. RAMSEY[2]

Savannah River Ecology Laboratory, Drawer E, Aiken, S.C. 29801[1] and Department of Biology Presbyterian College, Clinton, S.C. 29325[2]

ABSTRACT. Isozyme research and the development of electrophoretic techniques permit analysis of allozyme (allelic isozyme) variation and estimation of genic heterozygosity. Despite recent interest in the genetics of natural populations, relationships between vertebrate population dynamics and genic heterozygosity are relatively unexplored. Concurrent changes in small mammal behavior, genetics, and population numbers may be associated with temporal changes in a populations's level of heterozygosity. This hypothesis is supported by studies of heterosis and by correlations between reproductive effort, behavioral performance, morphological characters, and genic heterozygosity in oldfield mice (*Peromyscus polionotus*). Changes in allelic equitability, an index of genic heterozygosity, accompany fluctations in small mammal numbers. Heterozygosity appears to decline in increasing small mammal populations due to inbreeding and increases in declining populations due to the superior fitness of heterozygous individuals. Increased heterozygosity may confer a survival advantage in competitive environments. We predict that reproductive, behavioral, morphological, and physiological traits directly associated with individual fitness will exhibit correlations with genic heterozygosity in most vertebrates.

INTRODUCTION

Until recently, the importance of genetics as a factor influencing population processes received little attention. Studies that demonstrate genetic changes accompany fluctuations in small mammal numbers (Semeonoff and Robertson, 1968; Tamarin and Krebs, 1969; Berry and Murphy, 1970; Gaines and Krebs, 1971; Krebs, et al., 1973; Ramsey, 1973) have corroborated suggestions that fluctuations in numbers reflect changes in a populations's genetic composition (Birch, 1960). Two models of genetic regulation of populations have been proposed. Pimentel (1961) suggested an interspecific genetic feedback mechanism which attempts to explain changes in population density by the mutual evolution of predator-prey relationships. The assumptions of Pimentel's model limit its usefulness in explaining short-term changes in numbers (Lomnick, 1971). Chitty (1967) hypothesized an intraspecific model for vole populations whereby density-dependent selection for differing genotypes causes a decline in numbers. Data from vole populations appear to support Chitty's ideas (Krebs, 1970; Krebs et al., 1973), but the mechanisms linking demographic events to genetic changes in populations are unknown. Because of the many variables influencing population numbers, it would be naive to seek a single cause of fluctuations in densities (Christian and Davis, 1964; Aumann, 1965; Smith, 1971). Nevertheless, some genetic basis of attributes affecting population numbers, such as reproductive potential and survival, have been established (Chai, 1959; Roderick and Storer, 1961).

Genetic effects on population numbers could be expressed through individual loci or combinations of loci. Temporal changes in allele frequencies may reflect alterations in some pervasive attribute of the genome, like genic heterozygosity. Changes in heterozygosity could be one mechanism linking demography and genetics in small mammal populations. Some will regard this as an extreme view; however, we feel that a relationship between population dynamics and heterozygosity is implicated. The evidence comes from studies of heterosis and correlations between heterozygosity and the characteristics of individuals and populations.

Isozyme research and the development of starch-gel electrophoresis permit analysis of allozyme variation and estimation of genic heterozygosity (Lewontin and Hubby, 1966; Selander and Johnson, 1973). Genic heterozygosity exhibits geographic variation in oldfield mice (*Peromyscus polionotus*; Selander et al., 1971) and temporal variation of allelic frequencies occur in local populations (Ramsey, 1973). In addition, studies of life history (Smith, 1966), demography (Smith, 1971; Caldwell, 1964; Davenport, 1964; Ramsey, 1973) and behavioral performance (Garten, 1974) of oldfield mice point to the importance of genic heterozygosity in population regulation of this species. Our purpose is to elaborate on relationships between genic heterozygosity and population parameters in the oldfield mouse and to propose a general model for the effects of heterozygosity on natural small mammal populations.

To demonstrate that heterozygosity is important in influencing natural population fluctuations, it is necessary for us to show the following: (1) heterozygosity changes with time, (2) these changes are associated with changes in population numbers, and (3) there are mechanisms by which shifts in heterozygosity can directly influence population number. Both regional and local comparisons of population characteristics can be made from our data on oldfield mice.

REGIONAL COMPARISONS OF REPRODUCTION, DENSITY, AND GENIC HETEROZYGOSITY

Oldfield mice were collected from five areas along a north-south transect of the species distribution (Selander et al., 1971); these included areas in South Carolina, from Aiken to Columbia; middle Georgia, from just north of Statesboro to Reidsville; Florida-Georgia border, from Valdosta to Lake City; central Florida, from the Ocala National Forest and areas in the immediate vicinity; and finally south Florida, from Frostproof to just south of Lake Placid. Mice were collected every four to six weeks at each location by digging out their burrows (Smith, 1968). Collections were made at four locations for two years (1966-1968) and at the central Florida area for six years (1962-1968). Over 6,000 mice were dissected to obtain reproductive information. Sex ratios are approximately 1:1 in this species (Smith, 1967). The extensive sampling program and the large sample size allowed an accurate estimate of reproductive effort without the normal sampling biases. Detailed analysis of this reproductive data will be published elsewhere.

Data on genic heterozygosity were based on 745 *P. polionotus* collected at 30 locations which included the same areas as those used in the reproductive study. Genic analysis was by starch-gel electrophoresis, and heterozygosity estimates were based on 30 proteins encoded by 32 structural loci (Selander, et al., 1971). These included 7 hydrolases, 11 dehydrogenases, 1 oxidase, 3 isomerases and mutases, 2 transaminases, 1 peptidase, and 7 nonenzymatic proteins. Detailed discussion of this variation is given in Selander et al. (1971).

A linear positive relationship between the mean reproductive rate (number of offspring produced per adult female per month) and mean genic heterozygosity was observed across the five areas (Fig. 1). The relationship has a high degree of predictability, but correlation alone cannot be used to argue cause and effect. Inbreeding depression of reproductive rate, which is associated with decreased genic heterozygosity, is well documented (Barnett and Coleman, 1960; Falconer, 1960; Barnett, 1964) and supports the hypothesis that differential heterozygosity over geographical regions results in different reproductive rates.

Genic heterozygosity is positively correlated with the level of environmental variability (e.g., mean monthly temperature; Bryant, 1974). Oldfield mice in more variable environ-

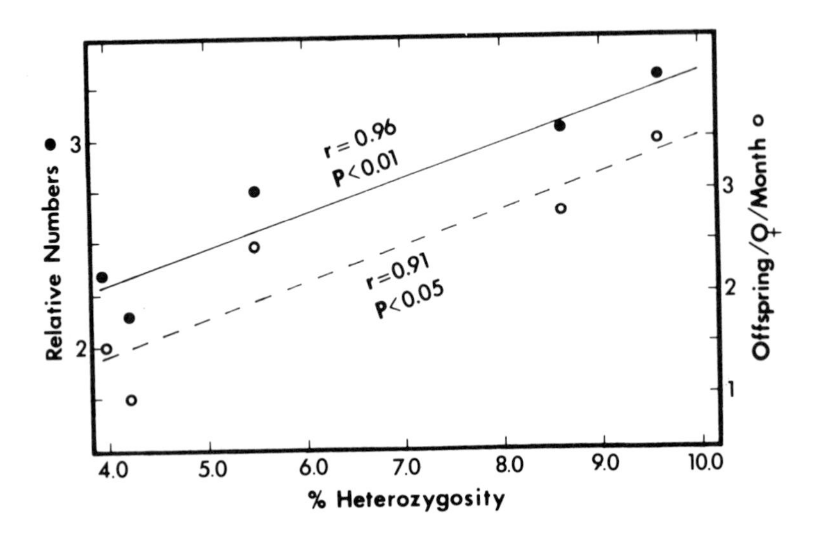

Fig. 1. Relative numbers of oldfield mice, expressed as numbers of mice captured per hour by digging burrows, and reproductive rates of adult female oldfield mice in relation to average per cent genic heterozygosity. The sample locations and their respective heterozygosity estimates are: middle Georgia (3.9), South Carolina (4.2), Florida-Georgia border (5.5), central Florida (8.6), and south Florida (9.6).

ments have the highest levels of genic heterozygosity and the highest reproductive rates. In addition, these populations also have the highest relative densities (Fig. 1). Selection for heterozygosity in response to environmental uncertainty could be the cause of the observed correlations. Environmental uncertainty could result in directional selection for higher reproductive performance independent of genic heterozygosity. However, our results, along with those of others, suggest a direct relationship between reproductive performance and genic heterozygosity. Environmental uncertainty probably increases reproductive performance by increasing heterozygosity.

Single gene effects can be important in determining reproductive performance or other characters related to individual fitness. However, most reproductive attributes, as well as behaviors and morphological characters affecting reproduction, are quantitative polygenic traits (Falconer, 1960; Fuller and Thompson, 1960). Overall heterozygosity may be more important than specific alleles or multiple alleles in affecting individual fitness (Lerner, 1954; Smathers, 1961). Testing this hypothesis requires techniques to assess the degree of genic heterozygosity in organisms. Electrophoretically detectable allelic variation yields a reliable index of genic variability or heterozygosity (Soulé et al., 1973; Selander and Kaufman, 1973) that can be used as a quantitative variable in studies of heterosis.

ADDITIONAL CORRELATES WITH GENIC HETEROZYGOSITY

Behavior is important to survival and reproductive fitness in small mammals. Aggressive mice have a survival advantage in some populations (Turner and Iverson, 1973), apparently because increased aggressiveness conveys an advantage in competition for food (Calhoun, 1963) and space (Mackintosh, 1973). Socially dominant mice also have a reproductive advantage over subordinate individuals in laboratory experiments (DeFries and McClearn, 1970). These observations indicate that aggressive and socially dominant mice have a high fitness value in competitive environments.

Correlations between behavior and genic heterozygosity have been observed in oldfield mice (Garten, 1974). Mice were collected by digging out burrows at Columbia, S.C., Statesboro, Ga., Crestview, Fla., Ocala, Fla., and Frostproof, Fla. Average per cent genic heterozygosity, based on 31 loci, was determined using starch-gel electrophoresis (Garten, 1974). Most of the sample locations, as well as the enzymes used in estimating heterozygosity, were the same as those used to determine relationships between heterozygosity, reproductive performance, and relative numbers.

In arena tests between pairs of adult male oldfield mice from the same location, accumulated attack time, or total time mice spend fighting, increased with increasing genic heterozygosity across the five samples (Fig. 2). When mice from different locations were matched in arena tests, the average number of times mice from a particular sample were socially dominant increased with increasing heterozygosity (Fig. 2). Socially dominant mice were recognized by their increased proneness to approach, attack, and aggressively groom their opponents. Subordinate mice exhibited avoidance, defensive, and submissive postures in response to actions by socially dominant mice. Aggressive oldfield mice also have an advantage in competition for food as suggested by a correlation between social dominance and food control ($r=0.66$; $P < 0.01$). Food control is the amount of time each mouse spends eating when a small food cube is dropped into the arena between the contestants.

Correlations were examined across geography because heterozygosity can be estimated more accurately for groups of mice than for individuals. The same trends in behavior were observed on an individual level. In 29 of 42 food-competition tests between mice from different locations, the mouse with more heterozygous loci ate longer than its less heterozygous opponent.

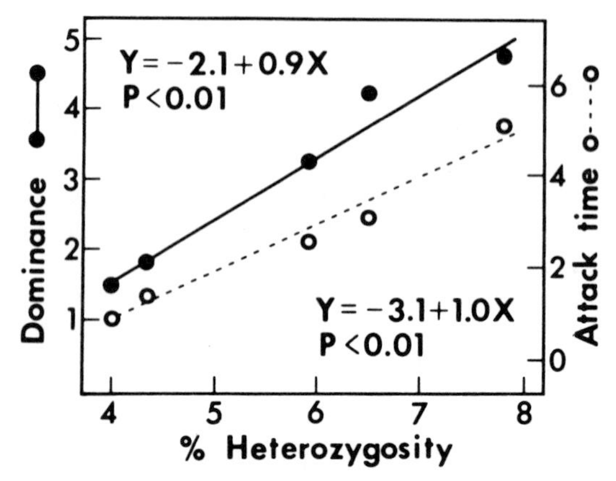

Fig. 2. Mean social dominance and mean accumulated attack time (sec) regressed against mean genic heterozygosity of mainland oldfield mice from five sample locations. The locations and their estimated average per cent genic heterozygosity are: Columbia, S.C. (4.0), Statesboro, Ga. (4.3), Frostproof, Fla. (5.9), Crestview, Fla. (6.5), and Ocala, Fla. (7.8).

The observed ratio of food control by more heterozygous vs less heterozygous mice (29:13) differs significantly from an expected 1:1 ratio ($\chi^2_{(1)} = 6.10$; $P < 0.05$).

As with reproductive performance and heterozygosity, cause and effect cannot be inferred from correlations between behavior and heterozygosity unless other supporting evidence exists. Relationships between small mammal behavior and genic heterozygosity are suggested by studies documenting heterosis for nonsocial behaviors like wheelrunning (Bruell, 1964a), trial and error learning (Abeelen, 1966; Smart, 1970) and exploration (Bruell, 1964b). Hybrid laboratory mice are also better competitors for limited food in contests with their inbred parents (Manosevitz, 1972). Inbreeding decreases aggressiveness and social dominance ability in chickens (Craig and Baruth, 1965). Hybrid sunfish also appear more aggressive than their parental species and this may confer an advantage in competition for food in natural populations (Whitt, et al., 1973). Bruell (1964a, 1964b) suggested that behavioral vigor would be associated with the proportion of heterozygous loci. Mechanisms underlying this phenomenon remain to be discovered.

Morphological characters like body size frequently influence behavior and reproduction. Litter size increases with increasing body size in oldfield mice (Caldwell and Gentry,

1965) and other small mammals (Beer and MacLeod, 1961; Keller and Krebs, 1970). Social dominance and competitive abilities were significantly correlated with body weight in male oldfield mice (r_s=0.39; $P<0.05$ and r_s=0.42; $P < 0.05$, respectively). Mean body weight increased with increasing numbers of heterozygous loci in male oldfield mice (r=0.98; $P < 0.001$; Garten, 1974). Body size is thus related to genic heterozygosity and probably influences levels of aggressive behavior and reproductive performance.

Aggressive behavior appears to be important in limiting small mammal numbers or in causing declines from peak densities (Krebs, 1970; Healey, 1967; Sadleir, 1965; Christian and Davis, 1964). Relationships between aggression and heterozygosity suggest that concurrent changes in behavior, gene frequencies, and population numbers may be linked by a behavior-genetic mechanism similar to that proposed by Chitty (1967). However, the mechanism is not linked to a limited number of genes, but rather is dependent upon heterotic effects across the entire genome.

IMPORTANCE OF HETEROZYGOSITY IN LOCAL POPULATIONS

Correlations between fitness and genic heterozygosity across geography do not necessarily imply that similar trends would be found within local populations. However, our data are suggestive of similar effects taking place at this level of organization. It seems appropriate to construct a model outlining the general sequence of events taking place during population fluctuations in the oldfield mouse. Evidence from other species will be examined to determine whether it is consistent with the general model.

Heterozygosity can be changed in a natural population in various ways. The number of polymorphic loci, number of alleles, and allele frequencies are important in changing heterozygosity levels. In a local population allele frequency is probably the most important variable influencing heterozygosity. Maximum heterozygosity is achieved when the alleles at a given locus occur in equal frequency. Temporal, spatial or biological subdivision could also have an effect on heterozygosity (Wahlund effect). For example, males and females or animals in different age cohorts could have different allele frequencies (Ramsey, 1973). Heterozygosity will increase with the reestablishment of panmixia in formerly subdivided populations.

Two measures of heterozygosity can be used as an index to genic variability. The average proportion of heterozygous loci per individual has been commonly used for this purpose (e.g. Selander, et al., 1971). Another measure, equitability, is defined as the observed number of heterozygotes divided by the maximum number of heterozygotes expected if the alleles have equal frequencies (Johnson and Feldman, 1973). The equitability index reflects changes in heterozygosity resulting from population subdivision and later panmixia and shifts in allele frequency. Equitability can vary between 0 and 1 and is a useful index of genic changes in a population.

Expected changes in small mammal populations during one annual population fluctuation can be summarized as in Table 1.

TABLE I

Relationships between numbers of animals, breeding structure, dispersal (E_t = emigration/unit time and I_t = immigration/unit time), selection for heterozygosity and equitability among the trappable population. There is a lag between changes in the breeding structure and equitability because of the time needed for maturation of young and for population turnover.

Relative Numbers	Breeding Structure	Dispersal $E_t \simeq I_t$	Selection for Heterozygosity	Allelic Equitability
Low-High	Inbreeding Outbreeding	Low-High	Low-High	High-Low
High-Low	Outbreeding Inbreeding	High-Low	High-Low	Low-High

Most of the information used to construct this model came from studies on the oldfield mouse. In this species, inbreeding occurs during the early parts of the increase phase, at which time selection for mice with higher levels of heterozygosity is close to zero or may even have a negative value. Indirect evidence for inbreeding comes from the rapid rise in the frequency of a light brown phenotype in the population at this time (Fig. 3). This phenotype results from homozygosity for an autosomal recessive allele (Dawson, et al., 1969), and its frequency may be used as an indicator of the degree of homozygosity resulting from inbreeding. When densities are low, mice do not disperse and the population becomes spatially static (Fig. 3; Smith, 1971; Briese and Smith, 1974). Another evidence for inbreeding during the increase phase is the high degree of assortative mating occurring in the population (Smith et al., 1972). At low densities and under favorable environmental conditions, it is probably advantageous for mice to remain in one place and to emphasize reproductive activities rather than movement or overt aggressive behaviors.

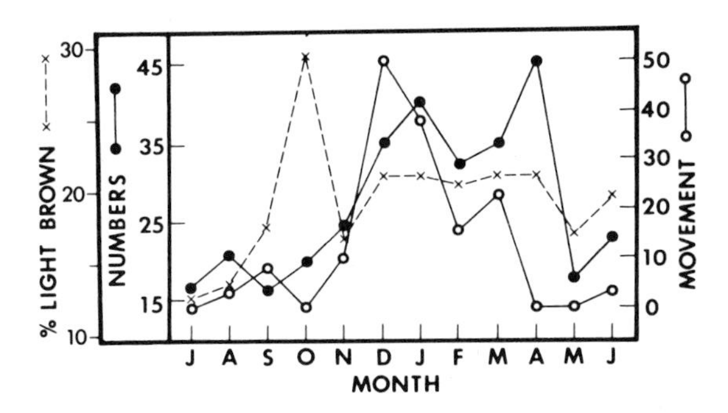

Fig. 3. Average oldfield mouse numbers (Smith, 1971), movement (emigration and immigration; Briese and Smith, 1974), and per cent light brown phenotype (bb) on a 3.6 ha oldfield for each month over a year. Movement is expressed as the number of mice captured leaving or entering the population by pitfall trapping along a 40 cm high drift fence surrounding the field.

At a later time during the increase phase, dispersal occurs and the population becomes relatively more outbred (Fig. 3). At this time equitability starts to increase and the relative proportion of homozygotes in the population declines. There will be a time delay before these changes in equitability are observed in the trappable adult population because of the length of the gestation (23-30 days) and nursing periods (21 days) for young mice (Carmon et al., 1963). Peak population numbers are first observed in the early winter when food is probably a limiting factor (Fig. 3; Smith, 1971). At this time of resource limitation, the more heterozygous-aggressive animals probably have an advantage over the more homozygous types. Scarring is seldom observed in this species so overt aggression is probably not common and competition must take a more subtle form.

There is a dramatic increase in equitability in both *Peromyscus polionotus* and *Microtus pennsylvanicus* as populations decline (Fig. 4). Four different loci provide the data for this trend and probably serve as markers of more extensive alterations in the genome. The shift from high equitability -low numbers to low equitability-high numbers must be quite rapid as most of the data points occupy terminal positions on the relationship (Fig. 4). Most dispersal occurs over a very

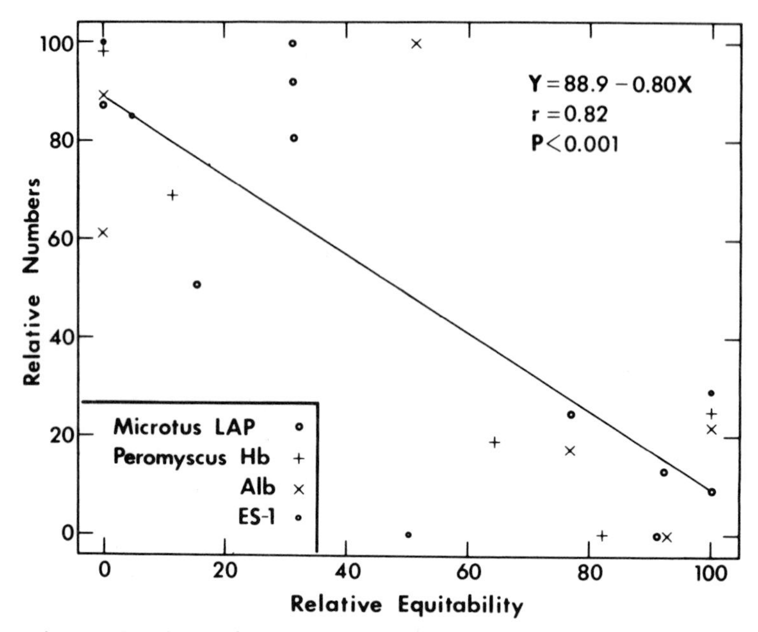

Fig. 4. Relationship between relative numbers of small mammals and relative allelic equitability. Standardization to a relative value was accomplished by subtracting the minimum value from the observed and dividing by the range and then multiplying by 100. *Microtus pennsylvanicus* data were calculated for a declining vole population from leucine amino peptidase allele frequencies (Krebs et al., 1973). *Peromyscus polionotus* data were calculated for declining (esterase-1) and increasing (hemoglobin and albumin) field populations (Ramsey, 1973).

short time period during the population fluctuation in the oldfield mouse and in most other species studied (Fig. 3; Briese and Smith, 1974). Pulsed dispersal could cause a relatively rapid shift in allelic equitability.

The decline phase in oldfield mouse populations is not caused by emigration (Garten and Smith, 1975), and the importance of aggressive behavior can only be speculated upon. There is a summer decline in reproductive rate that is probably a direct response of the population to high temperatures (Davenport, 1964; Smith, 1966). However, the populations start declining three to four months before the peak summer temperatures (Fig. 3). Predation has been suggested as the causative factor for population decline from peak winter numbers in the cotton rat (Schnell, 1968). For there to be a genetic-behavior feedback mechanism in the oldfield mouse, we must show selection pressures against the relatively homozygous animals during the decline period. The number of homozygous light brown mice did decline during the summer months of July and August (Fig. 3). With intensive predation the ratio of relatively heterozygous to homozygous mice should increase in local populations. Increased predation at the end of the spring may be due to the many species of snakes that come out of hibernation at this time and start feeding.

PREDICTIONS BASED ON THE TABULAR MODEL

The hypothetical model and its supporting evidence permit predictions about the role of genic heterozygosity in animal populations. Some of the predictions are in agreement with results already published. Other predictions are untested and require additional research before their general or specific applicability can be determined. We will discuss three general predictions concerning relationships between population ecology and genetics.

1) Reproductive performance and population numbers will be positively correlated with genic heterozygosity on a regional scale. This prediction follows from our results and those of others (Ayala, 1968; Beardmore, 1970; Berger, 1971; Redfield, 1973). Redfield (1973) reported a significant positive correlation between population density and the frequency of heterozygotes at an Ng locus among yearling blue grouse. Our calculations showed a positive regression of per cent breeding males in unfenced *Microtus pennsylvanicus* populations against allelic equitability as calculated from leucine amino peptidase allele frequencies (Krebs et al., 1973; $F_{1,7} = 10.8$; $P < 0.05$). In the same population, a regression of per cent lactating females against allelic equitability showed a positive slope. Therefore, reproductive activity appears to increase with increasing heterozygosity in vole populations.

Predictions from the model may not be confined to just small mammal populations. Laboratory experiments with *Drosophila* from natural and hybrid populations demonstrate those with greater variability have higher densities and reproductive rates (Ayala, 1968; Beardmore, 1970). Mean productivity increases with increasing genetic variance in *Drosophila melanogaster* (Beardmore, 1970). In addition, hybrid fruit fly populations increase to significantly higher numbers and exhibit significantly greater productivity than their parental populations (Ayala, 1968). Our calculations show population size increases with increasing allelic equitability in *Drosophila melanogaster* cultures (Es-6 locus; Beardmore, 1970). These cultures do not show the same trend as local mammal populations

(Fig. 4) because subdivision and inbreeding are probably unimportant in population cages for *Drosophila*.

Observed associations between local population numbers and genic heterozygosity may be similar to patterns across geography or they can be quite different depending on the breeding structure of the species. For example, temporally numbers may be negatively correlated with heterozygosity in local populations when population subdivision leads to inbreeding. This is exemplified by decreasing relative numbers with increasing relative equitability in small mammal populations (Fig. 4). Subdivision (Reimer and Petras, 1967; Selander, 1970), inbreeding (Rasmussen, 1964; Petras, 1967), and assortative mating (Smith et al., 1972) appear to be common phenomena in some small mammal populations.

Many population processes may be adapted to some particular mean level of heterozygosity ($\bar{H}$). Expected effects on population numbers with a change in heterozygosity (ΔH) might be a direct function of $\Delta H / \bar{H}$. Thus populations which show greater fluctuations in numbers might be expected to have lower levels of genic variability. For example, species such as *Sigmodon hispidus* and *Microtus* spp. should have lower average levels of genic heterozygosity than *P. leucopus*, *P. maniculatus* or *P. polionotus* based on information concerning the degree of population fluctuation (Odum, 1955; Terman, 1968; Smith, 1971). Although isozyme data are not available for most of these species, *S. hispidus* does show lower average heterozygosity than mainland populations of *P. polionotus* (Selander et al., 1971; Johnson et al., 1972).

2) Behavioral, morphological, and physiological traits associated with survival should show correlations with genic heterozygosity. Bruell (1964a) concluded that behaviors closely associated with fitness would exhibit heterotic inheritance. We expect that relationships between behavior, body size, and heterozygosity will exist whenever these characters directly or indirectly affect an individual's probability of survival. Survivorship itself will probably be significantly correlated with heterozygosity. For example, selection against homozygous genotypes in young hybrid fish populations leaves more heterozygous individuals (Fujino and Kang, 1968; Whitt et al., 1973; Avise and Smith, 1974). In declining and low density *Microtus ochrogaster* populations, male voles heterozygous at the transferrin locus survive better than homozygotes (Krebs, et al., 1973; Gaines and Krebs, 1971). Frelinger (1972) has shown increased survival for pigeons that are heterozygous for transferrin. Heterozygous animals are probably better equipped to survive due to their superior

behavioral, morphological, and physiological phenotype.

3) Density-dependent movement patterns are important in determining the level of genic heterozygosity in a population, and heterozygous individuals will exhibit a greater propensity to dispersal than less heterozygous animals. Differential predisposition to movement in small mammals (Howard, 1960) undoubtedly has a genetic basis. In *Microtus pennsylvanicus*, 89% of the females heterozygous at the transferrin locus are lost from increasing populations through dispersal (Krebs, et al., 1973). Although the exact relationship between dispersal and exploratory behavior is unknown, exploratory behavior increases with increasing genic heterozygosity in oldfield mice (Garten, 1974). Because of their superior fitness, heterozygous individuals are better equipped to survive the uncertainty of dispersal and have a better chance of successfully adapting to a new environment than less heterozygous animals. Ayala (1968) concluded that the success of a population in colonizing and adapting to a new environment is dependent on the amount of genetic variability in the initial population. Once heterozygous animals have settled in a new environment, they may find competition for food or space relaxed so that their chance of survival is increased and energetic or social limitations to reproduction are removed. Under such conditions, heterozygous individuals could maximize their fitness.

SUGGESTIONS FOR FUTURE WORK

Ideas presented in this paper are speculative and the general inferences we have made are admittedly tenuous in many cases. Our effort has not been to arouse confrontation, but to suggest a new direction for population biologists interested in the regulation of animal numbers. We feel that progress will need to be made in a number of areas before the generality of our predictions can be tested.

Techniques for observing additional isozyme systems must be developed from the body fluids of numerous species. Additional polymorphic protein systems are needed to chart changes in genic heterozygosity with changes in population numbers. A larger sample of "genetic markers" would enable investigators to more precisely examine relationships between fitness and heterozygosity at the individual level. In addition, estimating genic heterozygosity by examining allelic variation in blood plasma proteins makes nondestructive sampling of local populations possible. Multivariate analysis on the level of several local populations is needed to define the role of genetics in the regulation of vertebrate populations. Until a large amount of allelic variation can be nondestructively detected with electrophoretic techniques, geographic trends in morphology, reproduction, behavior, and genic heterozygosity should be examined in a number of vertebrate species. Behavioral observations should be made in conjunction with genetic and demographic measurements to expand our understanding of behavior-genetic relationships in vertebrates and the role of genic heterozygosity in animal populations.

ACKNOWLEDGEMENTS

Research was supported by Contracts AT(38-1)-310 and AT (38-1)-819 between the U.S. Atomic Energy Commission and the University of Georgia. We also appreciate the many helpful discussions with the staff and students of the Savannah River Ecology Laboratory.

REFERENCES

Abeelen, J.H.F. Van 1966. Effects of genotype on mouse behavior. *Anim. Behav.* 14:218-225.

Aumann, G.D. 1965. Microtine abundance and soil sodium levels. *J. Mammal.* 46:594-604.

Avise, J.C. and M.H. Smith 1974. Biochemical genetics of sunfish. I. Geographic variation and subspecific intergradation in the bluegill, *Lepomis macrochirus*. *Evolution* 28:42-56.

Ayala, F.J. 1968. Genotype, environment, and population numbers. *Science* 162:1453-1459.

Barnett, S.A. and E.M. Coleman 1960. "Heterosis" in F_1 mice in a cold environment. *Genet. Res., Camb.* 1:25-38.

Barnett, S.A. 1964. Heterozygosis and the survival of young mice in two temperatures. *Quart. J. Exp. Physiol.* 49: 290-296.

Beardmore, J. 1970. Ecological factors and the variability of gene-pools in *Drosophila*. In: *Essays in Evolution and Genetics in Honor of Theodosius Dobzhansky*. M.K. Hecht and W.C. Steere, eds. Appleton-Century-Crofts, New York, pp. 299-314.

Beer, J.R. and C.F. MacLeod 1961. Seasonal reproduction in the meadow vole. *J. Mammal.* 42:483-489.

Berger, E.M. 1971. A temporal survey of allelic variation in natural and laboratory populations of *Drosophila melanogaster*. *Genetics* 67:121-136.

Berry, R.J. and H.M. Murphy 1970. The biochemical genetics of an island population of the house mouse. *Proc. Roy. Soc. Lond. B.,* 176:87-103.

Birch, L.C. 1960. The genetic factor in population ecology. *Am. Nat.* 94:5-24.

Briese, L.A. and M.H. Smith 1974. Seasonal abundance and movement of nine species of small mammals. *J. Mammal.* 55: 615-629.

Bruell, J.H. 1964a. Heterotic inheritance of wheelrunning in mice. *J. Comp. Physiol. Psychol.* 58:159-163.

Bruell, J.H. 1964b. Inheritance of behavioral and physiological characters of mice and the problem of heterosis. *Am. Zool.* 4:125-138.

Bryant, E.H. 1974. On the adaptive significance of enzyme polymorphisms in relation to environmental variability. *Am. Nat.* 108:1-19.

Caldwell, L.D. 1964. An investigation of competition in natural populations of mice. *J. Mammal.* 45:12-30.

Caldwell, L.D. and J.B. Gentry 1965. Natality in *Peromyscus polionotus* populations. *Am. Midl. Nat.* 74:168-175.

Calhoun, J.B. 1963. *The Ecology and Sociology of the Norway Rat.* U. S. Government Printing Office, Washington.

Carmon, J.L., F.B. Golley, and R.G. Williams 1963. An analysis of the growth and variability in *Peromyscus polionotus*. *Growth* 27:247-254.

Chai, C.K. 1959. Lifespan in inbred and hybrid mice. *J. Hered.* 50:203-208.

Chitty, D. 1967. The natural selection of self-regulatory behaviour in animal populations. *Proc. Ecol. Soc. Aust.* 2:51-78.

Christian, J.J. and D.E. Davis 1964. Endocrines, behavior, and population. *Science* 146:1550-1560.

Craig, J.V., and R.A. Baruth 1965. Inbreeding and social dominance ability in chickens. *Anim. Behav.* 13:109-113.

Davenport, L.B., Jr. 1964. Structure of two *Peromyscus polionotus* populations in old-field ecosystems at the AEC Savannah River Plant. *J. Mammal.* 45:95-113.

Dawson, W.D., M.H. Smith, and J.L. Carmon 1969. A third independent occurrence of the brown mutant in *Peromyscus*. *J. Hered.* 60:286-288.

DeFries, J.C. and G.E. McClearn 1970. Social dominance and Darwinian fitness in the laboratory mouse. *Am. Nat.* 104:408-411.

Falconer, D.S. 1960. *Introduction to Quantitative Genetics.* Oliver and Boyd, London. 240 pp.

Frelinger, J.A. 1972. The maintenance of transferrin polymorphism in pigeons. *Proc. Nat. Acad. Sci. U.S.A.* 69: 326-329.

Fujino, K. and T. Kang 1968. Transferrin groups of tunas. *Genetics* 59:79-91.

Fuller, J.L. and W.R. Thompson 1960. *Behavior Genetics*. John Wiley and Sons, Inc., New York. 396 pp.

Gaines, M.S. and C.J. Krebs 1971. Genetic changes in fluctuating vole populations. *Evolution* 25:702-723.

Garten, Charles T., Jr. 1974. Relationships between behavior, genetic heterozygosity, and population dynamics in the oldfield mouse, *Peromyscus polionotus*. Unpublished *M.S. Thesis*, Univeristy Georgia, Athens. 88 pp.

Garten, C.T., Jr. and M.H. Smith 1974. Movement by oldfield mice and population regulation. *Acta Theriol.* 32: 513-514.

Healey, M.C. 1967. Aggression and self-regulation of population size in deer-mice. *Ecology* 48:377-392.

Howard, W.E. 1960. Innate and environmental dispersal of individual vertebrates. *Amer. Midl. Nat.* 63:152-161.

Johnson, G.B. and M.W. Feldman 1973. On the hypothesis that polymorphic enzyme alleles are selectively neutral. I. The evenness of allele frequency distribution. *Theoret. Popul. Biol.* 4:209-221.

Johnson, W.E., R.K. Selander, M.H. Smith, and Y.J. Kim 1972. Biochemical genetics of sibling species of the cotton rat (*Sigmodon*). *Studies in Genetics, 7* (Univ. Texas Publ. 7213):297-305.

Keller, B.L. and C.J. Krebs 1970. *Microtus* population biology. III. Reproductive changes in fluctuating populations of *M. ochrogaster* and *M. pennsylvanicus* in southern Indiana, 1965-1967. *Ecol. Monog.* 40-263-294.

Krebs, C.J. 1970. *Microtus* population biology: behavioral changes associated with the population cycle in *M. ochrogaster* and *M. pennsylvanicus*. *Ecology* 51:34-52.

Krebs, C.J., M.S. Gaines, B.L. Keller, J.H. Myers, and R.H. Tamarin 1973. Population cycles in small rodents. *Science* 179:35-41.

Lerner, I.M. 1954. *Genetic Homeostasis*. Oliver and Boyd, Edinburgh. 134 pp.

Lewontin, R.C. and J.L. Hubby 1966. A molecular approach to the study of genic heterozygosity in natural populations. II. Amount of variation and degree of heterozygosity in natural populations of *Drosophila pseudoobscura*. *Genetics* 54:595-609.

Lomnick, A. 1971. Animal population regulation by the genetic feedback mechanism: a critique of the theoretical model. *Am. Nat.* 105:413-421.

Mackintosh, J.H. 1973. Factors affecting the recognition of territory boundaries by mice (*Mus musculus*). *Anim. Behav.* 21:464-470.

Manosevitz, M. 1972. Behavioral heterosis: food competition in mice. *J. Comp. Physiol. Psychol.* 79:46-50.

Odum, E.P. 1955. An eleven year history of a *Sigmodon* population. *J. Mammal.* 36:368-378.

Petras, M.L. 1967. Studies of natural populations of *Mus*. I. Biochemical polymorphisms and their bearing on breeding structure. *Evolution* 21:259-274.

Pimentel, D. 1961. Animal population regulation by the genetic feedback mechanism. *Am. Nat.* 95:65-79.

Ramsey, P.R. 1973. Spatial and temporal variation in genetic structure of insular and mainland populations of *Peromyscus polionotus*. *Unpublished Ph.D. dissertation*, Univ. Ga. Athens. 103 pp.

Rasmussen, D.I. 1964. Blood group polymorphism and inbreeding in natural populations of the deer mouse *Peromyscus maniculatus*. *Evolution* 18:219-229.

Redfield, J.A. 1973. Demography and genetics in colonizing populations of blue grouse (*Dendragapus obscurus*). *Evolution* 27:576-592.

Reimer, J.D. and M.L. Petras 1967. Breeding structure of the house mouse, *Mus musculus*, in a population cage. *J. Mammal.* 48:88-99.

Roderick, T.S. and J.B. Storer 1961. Correlation between mean litter size and mean lifespan among 12 inbred strains of mice. *Science* 48:48-49.

Sadleir, R.M.F. 1965. The relationship between agonistic behaviour and population changes in the deermouse *Peromyscus maniculatus*. *J. Anim. Ecol.* 34:331-352.

Schnell, J.H. 1968. The limiting effects of natural predation on experimental cotton rat populations. *J. Wildl. Mgmt.* 32:698-711.

Selander, R.K. 1970. Behavior and genetic variation in natural populations. *Am. Zool.* 10: 53-66.

Selander, R.K., M.H. Smith, S.Y. Yang, W.E. Johnson, and J.B. Gentry 1971. Biochemical polymorphism and systematics in the genus *Peromyscus*. I. Variation in the old-field mouse (*Peromyscus polionotus*). *Studies in Genetics, 6* (Univ. Texas Publ. 7103):49-90.

Selander, R.K. and W.E. Johnson 1973. Genetic variation among vertebrate species. *Ann. Rev. Ecol. Syst.* 4:75-91.

Selander, R.K. and D.W. Kaufman 1973. Genic variability and strategies of adaptation in animals. *Proc. Nat. Acad. Sci. U.S.A.* 70:1875-1877.

Semenoff, R. and F.W. Robertson 1968. A biochemical and ecological study of plasma esterase polymorphism in natural populations of the field vole, *Microtus agrestis* L. *Biochem. Genet.* 1:205-222.

Smart, J.L. 1970. Trial-and-error behaviour of inbred and F_1 hybrid mice. *Anim. Behav.* 18:445-453.

Smathers, K.M. 1961. The contribution of heterozygosity at certain gene loci to fitness of laboratory populations of *Drosophila melanogaster*. *Am. Nat.* 95:27-37.

Smith, M.H. 1966. The evolutionary significance of certain behavioral, physiological, and morphological adaptations of the old-field mouse, *Peromyscus polionotus*. *Unpublished Ph.D. dissertation*, Univ. Fla., Gainesville. 187 pp.

Smith, M.H. 1967. Sex ratios in laboratory and field populations of the old-field mouse, *Peromyscus polionotus*. *Res. Pop. Ecol.* 9:108-112.

Smith, M.H. 1968. A comparison of different methods of capturing and estimating numbers of mice. *J. Mammal.* 49: 455-462.

Smith, M.H. 1971. Food as a limiting factor in the population ecology of *Peromyscus polionotus* (Wagner). *Ann. Zool. Fennici* 8:109-112.

Smith, M.H., J.L. Carmon, and J.B. Gentry 1972. Pelage color polymorphism in *Peromyscus polionotus*. *J. Mammal.* 53: 824-833.

Soule, M.E., S.Y. Yang, M.G.W. Weiler, and G.C. Gorman 1973. Island lizards: the genetic-phenetic variation correlation. *Nature* 242:191-193.

Tamarin, R.H. and C.J. Krebs 1969. *Microtus* population biology. II. Genetic changes at the transferrin locus in fluctuating populations of two vole species. *Evolution* 23: 183-211.

Terman, C.R. 1968. Population dynamics. In: *Biology of Peromyscus (Rodentia)*. J.A. King, ed., Special Publ. 2, Amer. Soc. Mammal, pp. 412-450.

Turner, B.N. and S.L. Iverson 1973. The annual cycle of aggression in male *Microtus pennsylvanicus*, and its relation to population parameters. *Ecology* 54:967-981.

Whitt, G.S., W.F. Childers, J. Tranquilli, and M. Champion 1973. Extensive heterozygosity at three enzyme loci in hybrid sunfish populations. *Biochem. Genet.* 8:55-72.

ISOZYME POLYMORPHISM MAINTENANCE MECHANISMS VIEWED FROM THE STANDPOINT OF POPULATION GENETICS

TSUNEYUKI YAMAZAKI and TAKEO MARUYAMA
National Institute of Genetics, Mishima 411, Japan and
Center for Demographic and Population Genetics
University of Texas at Houston, Houston, Texas 77025

ABSTRACT. All available data on isozyme polymorphism were collected from the published literature. These data were used to test several rival mechanisms which appear to be responsible for maintaining the genetic variation in natural populations. Five independent methods were used in the examination. They are: (1) The sum of heterozygotes due to alleles which have a specified gene frequency. In this method the geographical structure of a population is taken into account, but one must be able to distinguish between a mutant allele and the original allele. (2) Consistency between molecular evolution and the isozyme polymorphism. The former is amino acid substitution among homologous proteins of different species with a common origin. (3) A relationship between the heterozygosity and the number of alleles at each locus. (4) The distribution of heterozygosity with respect to gene frequency given by $4Nu\ Y^{-1}(1-Y)^{4Nu-1} \times 2Y(1-Y) \propto (1-Y)^{4Nu}$. In this method, one need not distinguish between mutant and original alleles, but populations are assumed to be panmictic. (5) Geographical uniformity of gene frequencies in a large structured population.

In all the five examinations, the theoretical expectations were obtained and compared with the appropriate quantities calculated from the data. Particular attention was paid to distinguishing between the neutral hypothesis and the selection hypotheses. The result of each of the tests supports the hypothesis that the random genetic drift of neutral mutants is the major source of isozyme polymorphisms.

Studies on various isozymes in man and other organisms have led to the discovery of a considerable number of electrophoretically demonstrable variant forms which are genetically determined. Extensive examinations of the genetic variation at a large number of loci and for a wide range of organisms have revealed that approximately 30 to 50 percent of all loci are polymorphic in natural populations and that about 10 percent of loci in each individual are heterozygous.

These values of polymorphic loci and of the heterozygosity are indeed very high. For instance, let us assume that there are 10,000 loci in our genome, then about 1000 of them have different alleles or are heterozygous. Naturally the finding of this large amount of genetic variability in natural populations led us to wonder about its cause and the mechanisms which maintain it. This is one of the most exciting and important problems in contemporary population genetics.

This question has been asked repeatedly since the publication of the influential papers by Lewontin and Hubby (1966) and by Harris (1966). Concurrently, a related but slightly different problem has arisen concerning molecular evolution, observed as amino acid substitutions in homologous proteins of different descendant species with common origin. Following the pioneering work of Zuckerkandl and Pauling (1965), Kimura (1968) was the first person who examined this question seriously from the viewpoint of population genetics, and he was led to the conclusion that the majority of the amino acid substitutions must be the result of the genetic random drift of selectively neutral mutants. This phenomenon is known as "non-Darwinian evolution" a term coined in another widely read paper by King and Jukes (1969).

The question we are asking here is not the cause of amino acid substitutions observed among homologous proteins, but the cause of the different forms of electrophoretically demonstrable variations existing in a single population or species. The two questions are different, but they are related, particularly in the view of Kimura and Ohta's (1971) hypothesis that the isozyme polymorphisms and the molecular evolution are not separate phenomena, but two aspects of a single process, namely, that the isozyme polymorphism is a phase of the transitional process of amino acid substitution in a population. They further argue that as in molecular evolution, the majority of isozyme polymorphisms are results of the random genetic drift of neutral mutants.

The question as to whether natural selection or the random genetic drift is the major force for the isozyme polymorphisms is by no means settled yet. Of course, both factors must be operating, but the question is, which of the two forces is playing the more important role, and if it is possible to measure, can we determine the relative importance of the two factors?

We would like to emphasize that our main concern is not to demonstrate an underlying mechanism for any particular locus or for a set of particular loci. Instead, we are concerned with the nature of all loci collectively. The data were collected from published literature, and they were examined in various ways in conjunction with theoretical expectations derived from population genetics theory. The literature from which the data were collected are given in Yamazaki and Maruyama (1975). Throughout this paper, "the data" refers to the collection of all the data, unless otherwise stated.

I. *HETEROZYGOTE DISTRIBUTION WITH RESPECT TO GENE FREQUENCY*

In this section we use one "invariant" relationship in an attempt to discriminate among rival hypotheses (Yamazaki and Maruyama, 1972 and 1974). A newly arisen mutant will eventually be fixed in the population or lost by extinction. During the process the gene will pass through various gene frequencies. It has been shown that, whatever the time that the mutant spends at a given frequency, the expected total number of mutant heterozygotes is a simple function independent of the population structure (Maruyama, 1972 and 1974). If the process has been going on for sufficient time to reach a steady state, the ergodic principle applies and we can regard observations of many mutants at one time as equivalent to observing a single mutant for a long time. If the mutant gene is neutral the total number of heterozygotes during the time the mutant is at frequency Y is proportional to 1-Y. If the mutant has a selective advantage or disadvantage, s in heterozygotes and 2s in homozygotes, then the frequency of heterozygotes is proportional to $2S(1-Y)S(1/2N)/s\ S(1)$ where $S(z)=1-\exp(-4Nsz)$ with N=the population size. The heterozygote distribution can be obtained also for a case where the heterozygote is favored.

A practical difficulty is that when we observe a heterozygote AA' we do not know whether A or A' is the mutant. However, if the frequency of A is Y, that of A' is 1-Y. We therefore reflect the gene frequency scale around the value of 0.5 so that heterozygotes correspond to the gene frequencies Y and 1-Y appear at the same abscissa when the data are graphed. The ordinate is the number of heterozygotes (Fig.1).

In cases 1 and 2 (neutral and favored mutant) the distribution of heterozygosity is constant with respect to the gene frequency (Y) in the whole population and is easily distinguished from case 3 (deleterious mutant) and case 4 (balanced selection) which have peaks at 0 and at 0.5, respectively. All the data which come from polymorphism surveys of which we are aware were used in the analysis.

The distribution of heterozygosity from the data are indicated by the dots in Figure 1. These utilize a total of 1530 alleles of 600 polymorphic loci from 16 different species. The data are clearly more consistent with the first two hypotheses than with either of the other two. (For more details, see Yamazaki and Maruyama, 1972 and 1974, and for the theory see Maruyama, 1972 and 1974).

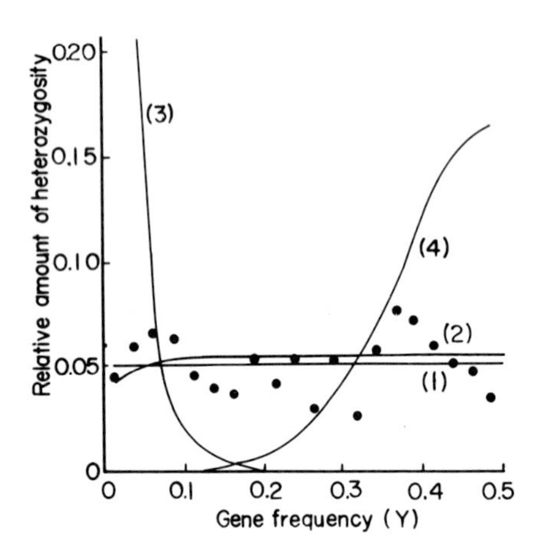

Fig. 1. Distribution patterns of heterozygosity. The curves indicate the theoretical expectations: (1) neutral; (2) advantageous (Ns=10); (3) deleterious (Ns= -10); (4) overdominance (Ns=10). The circles indicate the observed results. (The total area under each curve and the circles is unity). Y is the global gene frequency.

The curves (1) and (2), the neutral and the selectively advantageous cases, cannot be distinguished by this analysis. However, it is quite remarkable that the data are in good agreement with possibilities (1) and (2), while they are not consistent with possibilities of deleterious mutants or overdominance. Therefore, the result of this analysis suggests that almost all isozyme polymorphisms are the results of the random drift of neutral mutants or selectively advantageous mutants, but not overdominant or deleterious mutants. Mukai et al. (1973) have presented entirely different evidence leading to a similar conclusion.

The above analysis has been challenged. However, even if some of the apparent drawbacks of the analysis are taken into account, the same data still fit best to the theoretical expection based on the neutral assumption, but not to the heterotic case (see Ewens and Feldman, 1974 and Yamazaki and Maruyama, 1974).

II. *DISTINCTION BETWEEN* (1) *AND* (2)

Of course the preceding analysis alone does not settle the important question as to whether various kinds of balancing selection really contribute only a small fraction of genetic variation, and the question probably will remain unsolved for a long time to come. However, if this question of balancing polymorphism versus neutral or transient polymorphism can be considered tentatively settled, the next thing would be to distinguish between the two alternatives (1) and (2), i.e., the neutral versus advantageous mutant. This question seems to present no great problem, and the answer is, in fact, yes. In this section, as in the preceding, we again rely entirely on the quantities which are independent of the population structure. We need to define a few quantities.

Let θ and $1-\theta$ be the relative amounts of heterozygosity due to neutral polymorphisms and to selectively advantageous mutants. It was shown that a neutral mutant gene produces on the average 2N heterozygotes while it is segregating in the population of total size N, and an advantageous mutant produces 4N heterozygotes. (In heterozygosity, they are $2N/N=2$ and $4N/N=4$). These values of heterozygosity are invariant in relation to the geographical structure of the population (Maruyama, 1971 and 1972). Therefore, regardless of population structure, the relative rates of neutral and advantageous mutations are θ and $(1-\theta)/2$. We also know the ultimate fixation probability of each of these mutants: it is obviously $1/2N$ for a neutral mutant, and it is approximately equal to 2s for a mutant of advantage s (Haldane, 1927; Kimura, 1962). It is important to point out that fixation probability of an advantageous mutant is independent of the population structure (Maruyama, 1970 and 1972). Therefore the ratio of the number, E_n, of gene (amino acid) substitutions caused by random drift (neutrality), to the number, E_a, of those caused by selection is

$$\frac{E_n}{E_a} = \frac{\theta/2N}{\frac{2Ns(1-\theta)}{2N}} = \frac{\theta}{2Ns(1-\theta)} \qquad (1)$$

and rearranging

$$\theta = \frac{1}{1 + \frac{E_a}{2sN\ E_n}}$$

If $2sN\ E_n >> E_a$, $\theta \simeq 1$, which means that most of naturally occurring polymorphisms are due to neutral genes. For example, let us suppose that more than 5 percent of all evolutionary fixations are due to random fixation of neutral mutants and on the average $sN > 10$ for advantageous mutants; then neutral polymorphisms would outnumber transient polymorphisms (see Maruyama, 1972a).

Formula (1) has another important implicaton, namely if the fraction of neutral polymorphism, θ, is significantly less than unity, easily $2Ns(1 - \theta) >> 1$, and therefore $E_n/E_a << 1$. In other words, if polymorphisms due to transitional advantageous mutants represent a fair fraction of the total variation, most of the gene (amino acid) substitutions should be the result of selection. However this is quite implausible in view of the evidence for non-Darwinian evolution that a majority of evolutionary change observed as amino acid substitutions among homologous proteins of different species are the results of the random drift of neutral mutants (Kimura, 1968 and 1969; King and Jukes, 1969).

III. *HETEROZYGOSITY AND NUMBER OF ALLELES*

If a locus is polymorphic, we may ask how many alleles are segregating in a population? Are there any relationships between the two quantities? This question was first examined by Johnson (1972).

We need to introduce some symbols used in what follows. We denote by n_e the inverse of the homozygosity and by n_a the number of alleles at a locus under consideration. Johnson predicts that if all the alleles at loci in question are selectively neutral, the ratio, n_e/n_a, should decrease as n_a increases. Johnson has presented three different sets of data, all contradicting the above prediction, and has argued against the neutral hypothesis that most polymorphisms are the results of random genetic drift of neutral mutants. However his analysis was only qualitative and applied only to a small set of data.

We have extended the same examination to the same set of data used in Section I which amounts to far more than those used by Johnson. We have also provided the exact values expected theoretically of the ratio $\bar{n}_e/\bar{n}_a$ as a function of $\bar{n}_e$ and sample size (Fig. 2). The theoretical relationship between n_a and n_e is determined by first calculating $\bar{n}_a$ as a function of $\bar{n}_e$ and the sample size n (2n genes),

$$\bar{n}_a = (n_e - 1) \int_0^1 (1-x)^{\bar{n}_e -2} x^{-1} [1 - (1-x)^{2n} dx]$$

and then calculating $\bar{n}_e/\bar{n}_a$ (see Ewens, 1972). (A bar indicates the expectation of the quantity under consideration.) Unlike Johnson, we find that the observed data are in good accord with theory. Agreement between the theoretical and observed data is seen consistently in most of the individual sets of data, as well as in the average of all data combined. In Figure 2, the two theoretical relationships between $\bar{n}_a$ and $\bar{n}_e/\bar{n}_a$ come close to each other in spite of the large difference in their sample size. When gene frequency data on one locus are available from more than one observation (location), their average value is given in the figures. The graphs are, however, not altered significantly if each observation is presented independently.

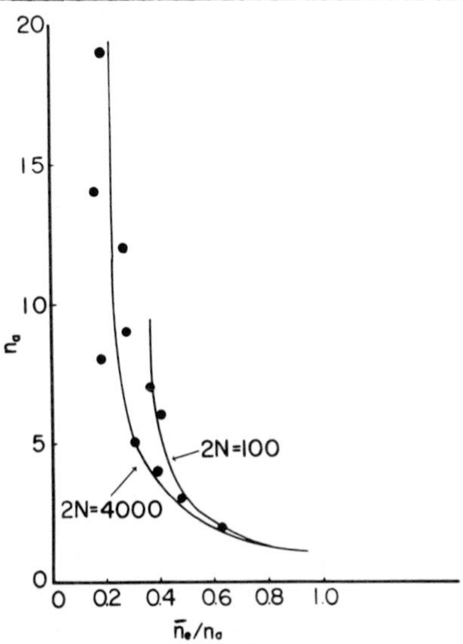

Fig. 2. The relationship of the homozygosity to the number of alleles, (all the data based on a sample of size greater than 100 genes are included). The curves indicate the theoretical expectations of their relationship. Each dot indicates the average coordinate of the analyzed data which belong to a given value of n_a.

Theoretically the curves in Figure 2 are very steep for $\bar{n}_a$ greater than three. Therefore, the direction (plus or minus) alone, particularly in that region, is not reliable enough to show whether or not the data agree with the theoretical expectation. It is more important to note that the observations come very close to the expected values irrespective of the direction itself. Although this was negative in most cases, we certainly came across some instances in which the direction of the observed slope seemed positive rather than negative. Nevertheless, the actual values in these cases are close to the theoretical ones and we believe they are not contrary to the neutral hypothesis. Our study provides both the exact expectation and the analysis based on more data. From Figure 2 it is clear that, overall, the data are consistent with theoretical considerations based on the neutral hypothesis (see Yamazaki and Maruyama, 1973a). It asserts also that most isozyme polymorphisms cannot be due to selection of various sorts, otherwise the relationship between n_e/n_a and n_a would be different from that given here. Kirby and Halliday (1973) have reported a similar result. This finding alone is of course insufficient to prove the hypothesis, but it invalidates Johnson's claim (1973 and 1974).

IV. *ANOTHER DISTRIBUTION OF HETEROZYGOSITY*

In Section I, we have presented an analysis of heterozygote distribution which is independent of the population structure. That analysis requires essentially that a mutant and original allele are distinguishable or that the mutation rate is low and therefore there are at most two high frequency alleles at a locus. There is another distribution of heterozygosities with respect to the gene frequencies. This distribution is, however, based on the assumption that the population is panmictic, and in turn, it does not require the assumptions imposed in the other analysis, i.e., the distinction between a mutant and original or the low mutation rate.

Assuming that every mutant is unique and selectively neutral, Kimura and Crow (1964) have shown that the number of alleles whose frequency is in a neighborhood of Y is proportional to $Y^{-1}(1-Y)^{4Nu-1}$, where N is the population size and u is the mutation rate. Therefore, if we multiply this formula by $Y(1-Y)$, we get the amount of heterozygosity, namely $(1-Y)^{4Nu}$ (2) which is proportional to the sum of the heterozygotes whose gene frequency is in a neighborhood of Y.

We have applied the same set of data used in Section I to formula (2), assuming that the populations are nearly panmictic and therefore the average mean frequency of each allele can be regarded as Y of the formula.

To determine the theoretical expectation, we need to know the value of 4Nu in formula (2). Kimura and Crow (1964) have shown that if f is the probability that two randomly chosen genes are identical by descent, $1/f = 1 + 4Nu$. The value of f can be calculated from data. It is known that the average value of f is rather invariant over a wide range of organisms, but it varies among different loci. The values for some loci are close to zero, while those for other loci are as high as 0.5, which is about the highest we found. Therefore, we present the pattern of the curve given by (2) for two extreme values of the parameter, 4Nu = 0.5 and 0 (Fig. 3). If the neutral hypothesis is correct, the data should appear in a reasonable range. The results are presented in Figure 3. As in Figure 1, since we are interested in the relative amount of heterozygosity, the curves and the dots in the figure are adjusted to make the areas under each of them unity. The data appear to be consistent with the theoretical expectation. Expectations based on other assumptions have different patterns. For example, that pattern based on the assumption that the majority of polymorphisms are maintained by some sort of balancing selection has a high peak in the neighborhood of Y = 1/2.

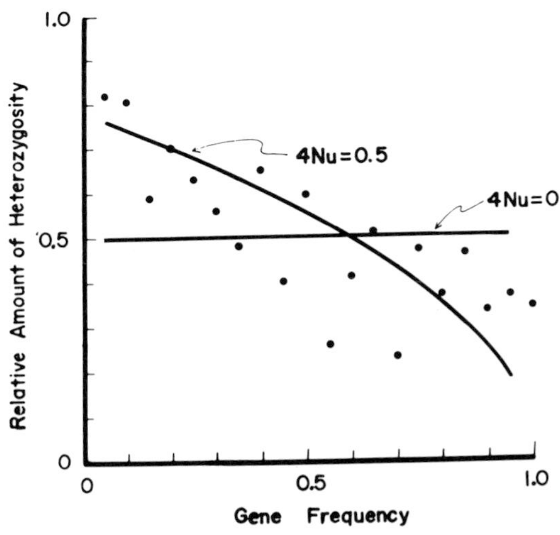

Fig. 3. Distribution of heterozygosity. The curves indicate the theoretical expectations based on the neutral hypothesis. The circles indicate the observed results. (The total area under each curve and the circles is unity).

In the analysis, we ignored the population structure, but the value of 4Nu was allowed to vary. The results of both analyses of this section and of Section I support the neutral hypothesis. We believe that an analysis of the distribution of heterozygotes has an advantage in this problem, because those alleles with a low frequency do not contribute appreciably to heterozygosity, while there may be a large number of rare mutants of deleterious effect, which can easily destroy the validity of an analysis based on the actual number of alleles. For example, suppose that there are ten alleles (A_1, A_2, A_3, ...) segregating at a locus, and that except A_1 and A_2 of both frequencies nearly 50 percent and of equal fitness, all the others are deleterious and of very low frequency. In a situation like this almost all the heterozygosity is due to A_1 and A_2 and therefore is neutral variation. The presence or absence of rare alleles A_3, A_4, ... has very little influence on the conclusion of an analysis based on the heterozygosity, while it is of great importance on the conclusion based on the actual number of alleles.

V. *POPULATION STRUCTURE AND GENE FREQUENCY*

There have been several attempts to use gene frequency data in a structured population to elucidate the underlying mechanisms that are responsible for maintaining isozyme polymorphisms. In advancing their neutral hypothesis, Kimura and Ohta (1971) have relied partly on a finding that in a two dimensional population of finite size the seemingly uniform distribution of gene frequencies over a large area is indeed possible with a surprisingly small amount of migration (see Kimura and Maruyama, 1971). It has been emphasized repeatedly that a truly infinite population and a large, but finite population are quite different in a fate of a neutral mutant (Maruyama, 1970 and 1971; Nei, 1975). When we consider a neutral mutant, it is almost certain that we need to take the size of population into account, and if we ignore the size, we may be led to a false conclusion due to the incorrect assumption.

For example, Bulmer (1973) has applied Malecot's (1967) formula for the probability of identity between genes of a given distance to some Drosophila data. While Malecot's formula used by Bulmer is valid only for an infinite population, he concluded that the values of the formula are far too small in comparison with those of the Drosophila data, and therefore that the data is incompatible with the neutral hypothesis. These conclusions are consequences of the incorrect assumption of the infinite population size. When a correct assumption and a corrected formula in which the assumption of finite population size is taken into account, then the values of the theoretical expectations are, in fact, in agreement with the data examined by Bulmer. (For details, see Maruyama and Kimura, 1974).

We would like to conclude the paper by pointing out that the collection of all available data of isozyme polymorphism is consistent with the neutral hypothesis under the five independent critical aspects tested, and that the data collectively are incompatible with selection schemes, particularly with the overdominance hypothesis.

ACKNOWLEDGEMENTS

We would like to thank Dr. K. Weiss for critical reading of the manuscript. This study was supported in part by a grant-in-aid from the Naito Foundation, Japan, and by U.S. Public Health Service Research Grants GM 19513 and GM 20293

REFERENCES

Bulmer, M.G. 1973. Geographical uniformity of protein polymorphisms. *Nature* 241:199-200.

Ewens, W.J. 1972. The sampling theory of selectively neutral alleles. *Theoret. Pop. Biol.* 3:87-112.

Ewens, W.J. and M.W. Feldman 1974. Analysis of neutrality in protein polymorphism. *Science* 183:446-448.

Haldane, J.B.S. 1927. A mathematical theory of natural and artificial selection. Part V. Selection and mutation. *Proc. Camb. Phil. Soc.* 23:838-844.

Harris, H. 1966. Enzyme polymorphism in man. *Proc. Roy. Soc.* B164:298-310.

Johnson, G.B. 1972. Enzyme polymorphisms: Evidence that they are not selectively neutral. *Nature, New Biol.* 237:170-171.

Johnson, G.B. 1974. Enzyme polymorphism and metabolism. *Science* 184:28-37.

Kimura, M. 1962. On the probability of fixation of mutant genes in a population. *Genetics* 47:713-719.

Kimura, M. 1968. Evolutionary rate at the molecular level. *Nature* 217:624-626.

Kimura, M. 1969. The rate of molecular evolution considered from the standpoint of population genetics. *Proc. Nat. Acad. Sci. U.S.A.* 63:1181-1188.

Kimura, M. and J.F. Crow 1964. The number of alleles that can be maintained in a finite population. *Genetics* 49:725-738.

Kimura, M. and T. Maruyama 1971. Patern of neutral polymorphism in a geographically structured population. *Genet. Res.* (Camb.) 18:125-131.

Kimura, M. and T. Ohta 1971. Protein polymorphism as a phase of molecular evolution. *Nature* 229:467-469.

King, J.L. and T.H. Jukes 1969. Non-Darwinian evolution. *Science* 164:788-798.

Kirby, G. and R. Halliday 1973. Another view of neutral alleles in natural populations. *Nature* 241:463-464.

Lewontin, R.C. and J.L. Hubby 1966. A molecular approach to the study of genic heterozygosity in natural populations. II. Amount of variation and degree of heterozygosity in natural populations of *Drosophila pseudoobscura*.

Genetics 54:595-609.

Malecot, G. 1967. Identical loci and relationship. *Proc. Fifth Berkeley Symp. Math. Stat. Prob.* 4:317-332.

Maruyama, T. 1970. On the fixation probability of mutant genes in a subdivided population. *Genet. Res.* (Camb.) 15:221-225.

Maruyama, T. 1970a. Effective number of alleles in a subdivided population. *Theoret. Pop. Biol.* 1:273-306.

Maruyama, T. 1971. An invariant property of a structured population. *Genet. Res.* (Camb.) 18:81-84.

Maruyama, T. 1971a. Analysis of population structure. II. Two-dimensional stepping stone models of finite length and other geographically structured populations. *Ann. Hum. Genet.* (Lond) 35:179-196.

Maruyama, T. 1972. Some invariant properties of a geographically structured finite population: Distribution of heterozygotes under irreversible mutation. *Genet. Res.* (Camb.) 20:141-149.

Maruyama, T. 1972a. A note on the hypothesis: Protein polymorphism as a phase of molecular evolution. *J. Molec. Evol.* 1:368-370.

Maruyama, T. 1974. A simple proof that certain quantities are independent of the geographical structure of population. *Theoret. Pop. Biol.* 5:148-153.

Maruyama, T. and M. Kimura 1974. Geographical uniformity of selectively neutral polymorphisms. *Nature* 249:30-32.

Mukai, T., R.A. Cardllino, T.K. Watanabe and J.F. Crow 1974. The genetic variance of viability and its components in a local population of *Drosophila melanogaster*. *Genetics* (in press).

Nei, M. 1975. *Molecular Population Genetics and Evolution.* North Holland and Elsevier.

Yamazaki, T. and T. Maruyama 1973. Evidence for the neutral hypothesis of protein polymorphism. *Science* 178:56-58.

Yamazaki, T. and T. Maruyama 1973a. Evidence that enzyme polymorphisms are selectively neutral. *Nature, New Biol.* 245:140-141.

Yamazaki, T. and T. Maruyama 1974. Analysis of neutrality in protein polymorphisms. *Science* 183:448.

Yamazaki, T. and T. Maruyama 1974a. Evidence that enzyme polymorphisms are selectively neutral, but blood group polymorphisms are not. *Science* 183:1091-1092.

Yamazaki, T. and T. Maruyama 1975. *Protein Polymorphism in Natural Populations.* (in preparation).

Zuckerkandl, E. and L. Pauling 1965. Evolutionary divergence and convergence in proteins. In Bryson and Vogel (eds.), *Evolving Genes and Proteins.* New York: Academic Press.

A SEARCH FOR THE GENETIC UNIT OF SELECTION

CHARLES F. SING[1] AND ALAN R. TEMPLETON[2]
Department of Human Genetics[1] and Society of Fellows[2]
University of Michigan Medical School, 1137 E. Catherine St.,
Ann Arbor, Michigan 48104

ABSTRACT. Natural populations of *Drosophila mercatorum* have all the properties common to other species of *Drosophila* and have an additional characteristic, parthenogenesis, which allows experimental control of factors which affect genetic variation. This species was used as the biological design to estimate the size of the genetic units whose fitnesses combine in a mathematically simple additive or multiplicative fashion to explain genotypic fitness. Deviations from random expectations of genotype frequencies in parthenogenetic progeny of heterozygous females could be attributed to selection because no factors other than meiosis affect the observations. The data collected gave evidence for true coadaptation involving nonadditive (nonmultiplicative) interactions between nonalleles. The unit of selection was found to be a function of the genetic state of the background. The greater the perturbation of the coadapted genotype by meiosis the larger the unit of genetic material which behaves as an additive (multiplicative) unit. Selective neutrality of allelic variation may be an artifact of our failure to measure the proper genetic unit.

INTRODUCTION

It is commonly accepted that selection operates on an individual's entire phenotype. Furthermore, it is the combination of effects of the physiologically meaningful genes which make up the genotype of the individual which contributes to the determination of the adaptive and fitness properties of his phenotype. Biochemical and developmental genetic studies indicate that the effect of each physiologically meaningful gene is combined with the effects of other such genes at various levels of organization to produce an integrated biological system, the organism. However, these findings are often ignored by population geneticists in their studies to determine the forces which maintain polymorphic variation at isozyme loci in natural populations of sexually reproducing diploid organisms. The possible answers from such investigations have been clearly stated by the alternate hypotheses (explanations) for genetic polymorphisms which have been put forth in the form of numerous theoretical models. In every case, a model is constructed so as to explain the intermediate allelic frequencies within a subpopulation and the dispersion of allelic frequencies among subpopulations in terms of the operation of the particular factors of interest. The neutralists have chosen to build models which explain allele frequency variation as a function of mutation rate, μ, and the demographic dimensions of the hypothesis space (Fig. 1A); migration (D_1), effective population size (D_2), and mating system (D_3). The variance of allele frequencies among subpopulations for a locus at equilibrium is defined as some function of variables which represent the four dimensions of the hypothesis space

$$\sigma_p^2 = f(\mu, D_1, D_2, D_3). \tag{1}$$

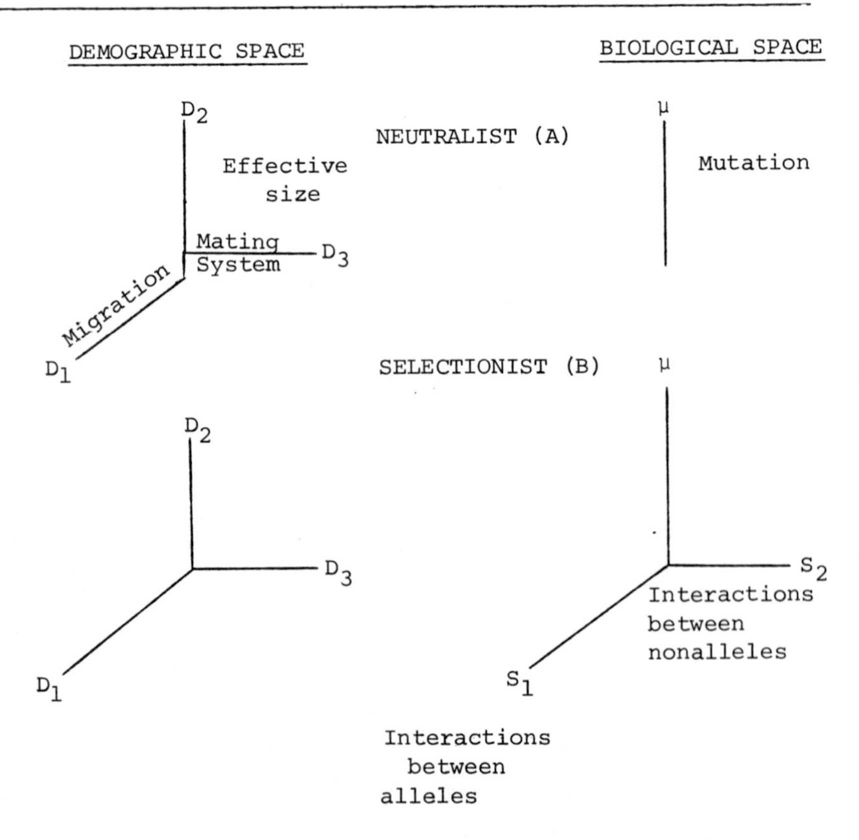

Fig. 1. The dimensions of the hypothesis space which have been considered to explain observed genetic variation.

Polymorphism within a subpopulation can be explained as a transitional state in the process of fixation of neutral or "near neutral" mutations and allele frequency variation among subpopulations is explained as a result of the process whose properties are defined by the appropriate choices of values of the variables μ, D_1, D_2, and D_3. Additional biological factors expand the hypothesis space (Fig. 1B) to include the differential biological properties of the isozyme variations. The selectionist school hypothesizes that in addition to mutation and the demographic factors, allelic variation is determined by the fitness interaction between alleles (S_1, Fig. 1B) or between the fitness values of alleles at the locus and those of nonalleles at other loci (S_2, Fig. 1B). The effect of the latter may also be modified by the linkage relationship, say R. The variance of allele frequency under this expanded hypothesis space becomes

$$\sigma_p^2 = f(\mu, D_1, D_2, D_3, S_1, S_2, R) \qquad (2)$$

Population genetic models have tended not to include all dimensions simultaneously but to be developed only in terms of those which give an adequate explanation for the observed allelic variation in nature. The plethora of such models is well documented by the Crow and Kimura (1970) text and the book by Ewens (1969). It is unfortunate for the science of population genetics that most workers conclude that a particular dimension of the hypothesis space is either sufficient or proved if the associated model is an adequate description of the observed natural variation.

The analysis of equations (1) and (2) using observations of allele frequencies from subdivisions of some natural population to identify the relevant dimensions is plagued by a number of major complications. First, it is not known if populations are in equilibrium and/or the values taken by the variables defining the hypothesis space change over time. Secondly, the effects of the many variables which determine allelic variation cannot yet be independently estimated using data sampled from natural populations so as to obtain the relative role of each of the dimensions of the hypothesis space. And, thirdly, the locus may not be the genetic unit upon which all the dimensions of the hypothesis space operate. Certain factors may affect a larger "chunk" of genetic material, hence, measures of this larger unit must be used to accurately predict or describe the evolutionary process as it is affected by that dimension. If certain dimensions of the hypothesis space affect a genetic unit larger than the locus, then models (such as (1) or (2)) which relate observations on single loci to those dimensions may lead to erroneous inferences about the factors which determine allelic variation. What then is the genetic unit which can be used to describe or predict the evolutionary process? Is it a gene locus, a "chunk"of chromosome, or an entire chromosome which is the proper unit whose behavior may be predicted by the dimensions of the hypothesis space suggested by population genetic models?

Each generation recombination and assortment reorganizes the array of genotypes which underlie the array of phenotypes exposed to selection. Because of this breakdown and reorganization of parental genomes it is obvious that the entire genotype may not be the genetic unit which measures the operation of selection. If the fitness of a phenotype is determined by a nonindependent relationship of contributions of individual loci, the genetic unit is probably determined by some balance between the cohesive forces of selection and linkage which retain genotype organization and the dispersive forces of meiosis and population demography.

Franklin and Lewontin (1970) showed by computer simulation that the effects of close physical linkage and selection (multiplicative model) can interact in such a way that it is necessary to define some genetic unit greater than a single locus to describe the process of micro-evolutionary change. Estimates of gametic disequilibrium in populations of sexually reproducing diploids (Sinnock and Sing, 1972; Zouros and Krimbas, 1973; Charlesworth and Charlesworth, 1973) are evidence of such organization. However, in all of these studies the factors -- linkage, epistasis, and joint sampling of non-alleles due to finite size or migration effects, were confounded because of restrictions on sampling which are inherent in studying natural populations. In this paper we review our work to estimate the size of the genetic unit upon which selection operates.

PARTHENOGENESIS AS A DESIGN FOR THE STUDY OF THE GENETIC UNIT OF SELECTION

There has been a general failure to distinguish the role of selection from the effects of the demographic factors in determining genotypic and allelic frequencies in natural populations. Statistical models and sampling designs have been inadequate and the values of the many variables which operate simultaneously that are measures of demographic and biological effects have not been estimable from allele frequency variation alone. Furthermore, auxiliary information on the selective factors in the environment has been lacking because of our ignorance of what those factors might be or how they might be measured. The reports of Powell and Maruyama in these proceedings represent the approach often taken by the selectionist and neutralist schools, respectively. Each selects a population genetic model which explains allelic variation in observed natural populations in terms of the dimensions of causation which represent his point of view and then shows that the available data fit that model. Recognizing that neither approach is designed to reject the other and that statistical models for analysis are not presently available to discriminate between alternate hypotheses given the kinds of data available from natural populations, we have chosen to select a biological design which follows the reductionist strategy in attacking complex systems. We have selected *Drosophila mercatorum* as an organism with all the properties common to other species of *Drosophila* but has one feature, parthenogenesis, which allows us to experimentally control the factors affecting genetic variation.

The ability of normally sexually reproducing *Drosophila mercatorum* to reproduce parthenogenetically provides a biological design which enables us to

1) control experimentally the demographic influences on genotype organization,
2) control the mode of reproduction so as to set up a variety of genotypic arrays whose frequencies depend only on the effects of meiosis and selection, and
3) utilize the multilocus genetic theory and associated statistical models developed for this mode of reproduction (Asher, 1970; Templeton, 1972).

We utilized this design to test for the operation of selection, and if such effects were present, to estimate the additive or multiplicative genetic unit which adequately describes the fitness differences among the parthenogenetically produced progeny.

Parthenogenetic reproduction in certain strains of *D. mercatorum* is almost exclusively by post meiotic gamete duplication followed by fusion of the cleavage nuclei to restore diploidy (Fig. 2). Gamete duplication yields only homozygous progeny. Two parthenogenetically reproducing lines, one established in 1965 (the O strain from flies collected on Oahu) and one established in 1961 (the S strain from flies collected in San Salvador) were used in the studies reported here. Templeton and Rothman (1973) estimated that the frequency of gamete duplication in these strains of *D. mercatorum* is 96%, with the remaining progeny a result of central or terminal fusion of ootids.

A survey of isozyme loci indicated that these two strains were fixed for different alleles at approximately 1/3 of their loci. Although each strain has reproduced parthenogenetically for nine years or more, the females have not lost the capacity

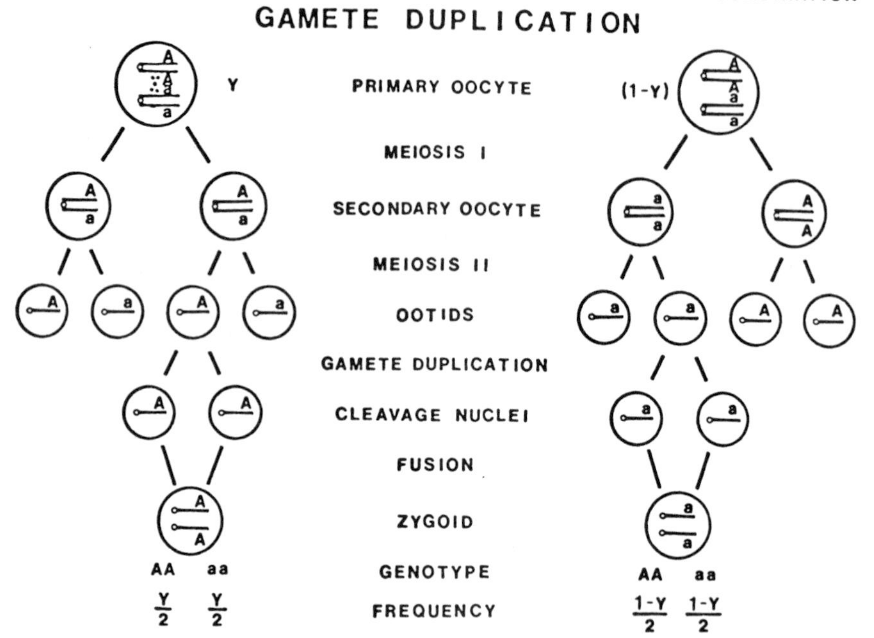

Fig. 2. A diagramatic representation of the meiotic events of gamete duplication.

for sexual reproduction. Using this feature, heterozygous females for specific marker loci were constructed by crossing a "bridge" S male (Carson et al., 1969 for backcrossing scheme used) with an O female. Figure 3 gives the mating scheme to produce a female which is heterozygous S-O for each of the four major chromosomes (2 acrocentric and 1 metacentric autosomes and an acrocentric "X" chromosome). The parthenogenetic progeny of this S-O female represent a spectrum of homozygous genotypes determined by assortment and recombination operating on 100% of the two parental genomes. To estimate the contribution of background reorganization on genotype distributions for marker loci, data from these progeny were compared to data from progeny produced parthenogenetically by females heterozygous for 60% and 40% of their genotype. (See Templeton, Sing and Thirtle, 1974, for details of the mating design).

The visible marker genotypes (Fig. 3) on each of the chromosomes assured the proper selection of treatment females for the three levels (100%, 60%, 40%) of perturbation of the O and S parental genomes. The data collected from these three groups of progeny were the six locus genotypes for the loci shown in Figure 4. Approximately 2000 progeny (from many

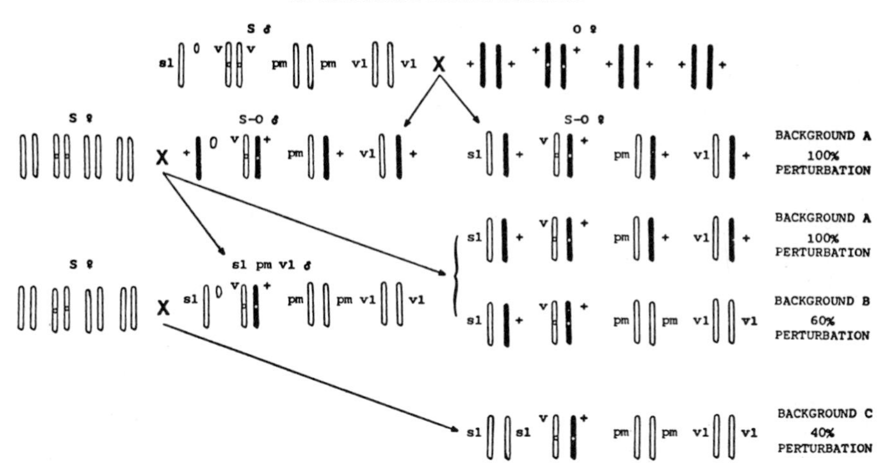

Fig. 3. The mating scheme used to generate the experimental treatments and a diagramatic representation of the chromosomal constitution of the S-O hybrids which produced the three treatment levels of perturbation. Each arm represents approximately 20% of the total genome.

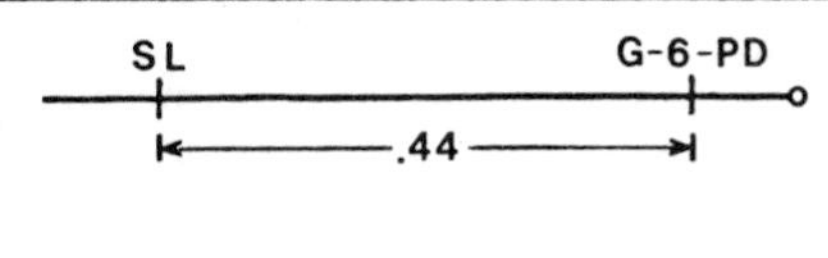

X Chromosome

EST B EST A XDH V

.26 .43 .27

Metacentric Autosome

Fig. 4. The genetic map of two of four major chromosomes of *D. mercatorum* used to study the genetic unit of selection.

genotypically identical females) were scored for these six loci for each perturbation treatment. Under the null hypothesis of selective neutrality of the genetic regions marked by allelic variations at these six loci, we expect an equal proportion of S and O segregants at each locus. Likewise, in the absence of selection, we expect the proportion of nonrecombinant and recombinant homozygous genotypes defined by the two, three, four, five or six loci to be a function of map distances only.

Furthermore, the null hypothesis predicts that the recovery of segregant genotypes for these markers will be independent of the level of perturbation of the parental genomes (100%, 60% or 40%) by meiosis.

Progeny were raised so as to minimize larval competition and were collected for isozyme analysis within 24 hours of eclosion. Deviations from assortment of genotypes as expected on the basis of the operation of meiosis and map distances between loci were taken as evidences for differential viability. We compared the frequencies of segregants for combinations of the six markers in the parthenogenic progeny with those observed in the sexually produced progeny from an identical heterozygous female, to insure that an excess of nonrecombinant genotypes was a reflection of true coadaptation of nonalleles to homozygosity in the O and S strains and not simply failure of the loci to recombine as expected (when in fact the null was true). In this way the heterozygous sexually produced progeny act as a control treatment in which the selective forces of homozygosity are minimized. Coadaptation is defined as the excess frequency of nonrecombinant parental O or S genomes in homozygous genotypes above and beyond that found in heterozygous genotypes. For our purposes we define the genetic unit of selection as that combination of regions marked by the loci being measured whose fitnesses combine in a mathematically simple additive (or multiplicative) fashion with other such units to produce an adequate statistical description of the genotypic frequencies observed in the parthenogenetic progenies.

SUMMARY OF RESULTS

A detailed description of the derivation of the experimental genotypes, the statistical models for analysis of data and the complete presentation of the data with the analyses are given by Templeton, Sing, and Thirtle (1974). Only a summary of the outcome of the study will be possible here.

Evidence for rejection of the null hypothesis of random assignment of different alleles to the O and S strains is of two kinds. First, the number of viable parthenogenetic offspring (daughters) per female decreases as the degree of perturbation, by recombination, of the O and S genomes increases. The average number of viable daughters per female was 1.63 for the 100% perturbation treatment, 5.36 for the 60% treatment, 10.25 for the 40% treatment and 14.54 for S females. Secondly, we tested the difference between the parthenogenetic and sexual progeny in the proportions of parental genotypes (O and S) with respect to all two through six locus combinations. The two and four locus results are representative and are given

TABLE I

PROPORTIONS OF O AND S PARENTAL GENOTYPES FOR ALL TWO LOCUS SYSTEMS IN THE SEXUAL CONTROLS, P_C, AND IN THE PARTHENOGENETIC TREATMENT GROUPS, P_E

Linkage Relationship of Markers	Markers in System	P_c	Perturbation Level 100% P_e	60% P_e	40% P_e
Same Chromosome Arm	EstA/EstB	.7383	.7609	.7756*	.7523
	v/XDH	.7292	.7695	.7713**	.7518
	sl/G-6-PD	.5809	.6139	.6357**	-
Same Chromosome, Different Arms	EstA/v	.5459	.6120*	.5346	.5425
	EstA/XDH	.5704	.6307	.5787	.5915
	EstB/v	.5373	.6468****	.5481	.5397
	EstB/XDH	.5473	.6450**	.5395	.5632
Unlinked	EstA/G-6-PD	.4971	.5606*	.5159	-
	EstB/G-6-PD	.5105	.5584	.5104	-
	v/G-6-PD	.5060	.5571	.4888	-
	XDH/G-6-PD	.4925	.5413	.4919	-
	EstA/sl	.5018	.5451	.5176	-
	EstB/sl	.5043	.5657*	.5066	-
	v/sl	.4877	.5942***	.4951	-
	SDH/sl	.5131	.5148	.5013	-

* P_c and P_e are significantly different at the 5% level.
** the 1% level.
*** the 0.1% level.
**** the 0.01% level.

TABLE II

PROPORTIONS OF O AND S PARENTAL GENOTYPES FOR ALL FOUR LOCUS SYSTEMS IN THE SEXUAL CONTROLS, P_C, AND IN THE PARTHENOGENETIC TREATMENT GROUPS, P_E

Linkage Relationship of Markers	Markers in System	P_c	Perturbation Level 100% P_e	60% P_e	40% P_e
A Pair of Loci on Each of Two Arms	(B A) (X v)	.3122	.4108***	.3557*	.3480*
	(B A) (sl G)	.2248	.3079**	.2641*	-
	(v X) (sl G)	.2085	.2904**	.2491*	-
1st Pair on Same Arm And Second Pair on Different Arms So That Three Arms Are Involved	(sl G) B X	.1575	.2638***	.1757	-
	(sl G) B v	.1465	.2840***	.1680	-
	(sl G) A v	.1554	.2613**	.1737	-
	(sl G) A X	.1627	.2350**	.1891	-
	(B A) v sl	.1997	.3265***	.2178	-
	(B A) X sl	.2082	.2906**	.2279	-
	(BA) X G	.1997	.3047***	.2246	-
	(BA) v G	.2375	.3079**	.2025	-
	(vX) B G	.1956	.3076***	.2028	-
	(vX) B sl	.1936	.3218***	.2094	-
	(vX) A sl	.2101	.3076***	.2155	-
	(vX) A G	.1985	.2863**	.2126	-

[1] ( ) implies enclosed markers on same arm.

* P_c and P_e are significantly different at the 5% level.

** the 1% level; *** the 0.1% level.

in Tables 1 and 2, respectively. In the 100% perturbation all two through six locus systems showed an excess of parental types and this excess grew larger as the number of loci (and proportion of marked genome) simultaneously considered increased. This can be seen by contrasting the results given in Tables 1 and 2. In progeny of the 60% and 40% treatment females, all significant excesses of nonrecombinant parental genotypes are confined to those combinations involving markers on the same arm. Therefore, when the spectrum of background genotypes was closer to the parental S genome, the genotypes marked by loci on different arms deviated less from expected proportions. With regard to sexual progeny, Table 1 demonstrates that for eight systems of unlinked pairs of loci all parental proportions are close to the *a priori* expected proportion of 0.5. None of the deviations are statistically significant from 0.5 at the 5% level of probability. Thus, the forces that operate on these same systems in the parthenogenetic progeny of the 100% perturbation treatment do not operate in the sexual controls. In addition to these analyses we tested for deviations of O and S single locus genotypes for each of the six loci from the *a priori* segregation ratio of 0.5. Despite the fact that there were large deviations in multilocus systems in the 100% treatment results, only one of the six loci, spotless, deviated significantly (5% level of probability) from 0.5 when considered as single locus systems. All frequency data from 2-6 locus analyses support the rejection of the null hypothesis that loci scattered throughout the parental O and S genomes are randomly fixed for functionally equivalent, "neutral" alleles. The magnitude of the deviation from this null hypothesis is greatest when the genotypic spectrum deviates farthest from the O or S genome.

THE GENETIC UNIT OF SELECTION

Analyses of the frequency data were undertaken to 1) estimate the additive (or multiplicative) fitness contributions of each region identified by a marker to multiple locus homozygote fitness and 2) estimate the non-additive (nonmultiplicative) deviation of observed fitnesses from expectations based on the estimated marginal locus fitnesses. Only two and three locus data were utilized because of sample size considerations. Goodness-of-fit to four fitness models was measured to determine the size of the genetic unit which behaves as an additive or multiplicative unit of genotypic fitness. The models are presented in Table 3 with a schematic outline of the decision rule for identifying whether the locus, the arm of a chromosome, or the entire marked region of the genome is the small-

TABLE III

MODELS FOR ESTIMATING THE GENETIC UNIT OF SELECTION
AND DECISION RULES FOR GOODNESS OF FIT TO DATA

I. Additivity among loci

II. Multiplicity among loci

III. Additivity between pair of loci on the same arm and a third locus

IV. Multiplicity between pair of loci on the same arm and a third locus

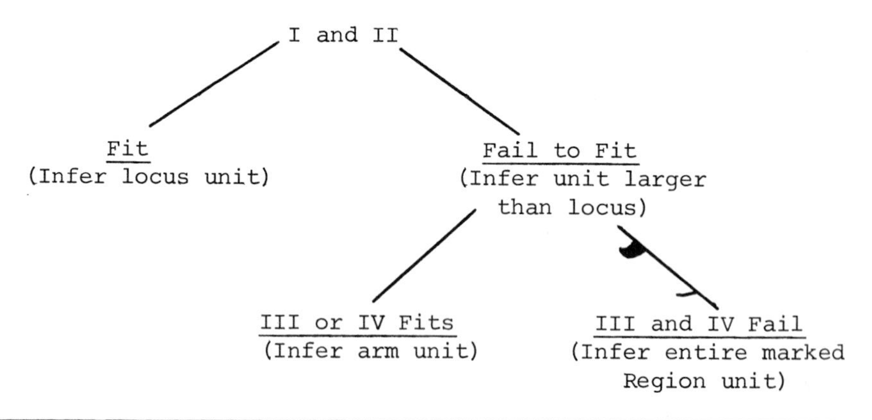

est statistical unit which behaves in a non-interactive way with other such units. Rejection of models I and II for two or three loci implies that the unit is larger than the region marked by a locus. Rejection of I and II for three loci but failure to reject II or IV implies that the entire marked region cannot be divided into additive (multiplicative) units and hence is defined as the genetic unit.

All analyses of fitness models using two-locus data from the 100% perturbation treatment gave a positive interaction term which was usually larger in magnitude than the additive effects. Of the 7 two locus systems showing a significant deviation from the null hypothesis, only the unlinked combination of spotless and Esterase A fit models I and II and hence the outcome of selection for only 1 of the 15 two locus combinations could be adequately described by marginal fitnesses For the 60% perturbation treatment all three of the two-locus systems on the same chromosome arm fail to fit models I or II, whereas the remaining combinations show that any selection operating can be explained by a non-interactive fitness model. Therefore, all interactions between regions marked by two loci are confined to markers on the same chromosome arm with none detected between loci on different arms. In the 40% perturbation experiment the same pattern of two-locus interactions was determined as for the 60% perturbation except that only one same arm pair (vermillion and Xdh) showed a statistically significant deviation from the null based on meiosis alone and for which the additive (multiplicative) model was rejected. The three-locus fitness analyses support the same conclusion. In the progeny of the 100% perturbation treatment the null hypothesis was rejected for 13 three locus combinations and of these only two fit model I, II, III, or IV. For the 60% perturbation experiment only one of the seventeen combinations showing significant selection had a fitness pattern such that the relationship between the pair and the third could not be explained by a chromosome arm additive (multiplicative) model.

INFERENCES

1) Selection has operated to determine the O and S genomes. The nonrandom combinations of nonalleles found in these two strains is supported by the depression in viability when the genotype deviates from either the O or S combination and the preferential survival of nonrecombinant vs recombinant progeny from the S-O hybrid females. These data represent evidence for true coadaptation involving nonadditive or nonmultiplicative interactions between nonalleles because no factors other than meiosis and selection are affecting the observations upon which inferences are made.

2) This study illustrates that selective neutrality of allelic variation at a locus may be an artifact of our failure to measure the proper genetic unit. The strongest selection was detected at the 100% level of perturbation, yet only one of the six markers showed significant marginal fitness differences among alleles. We have measured the degree to which an allele may be a functional part of a system of interacting nonalleles. It is likely that genotype organization contributes to determining the frequency behavior of an allele in natural populations of *D. mercatorum*. Studies are underway to evaluate the relative role of such organization when sex and population structure are added to the system studied here.

3) The unit of selection is a function of the genetic state of the background. Figure 5 illustrates the implied relationship between degrees of recombination among coadapted alleles

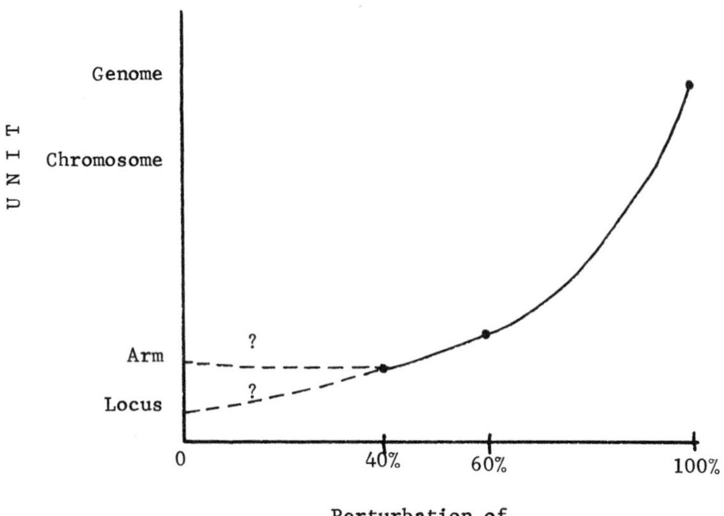

Fig. 5. The genetic unit of selection as a function of reorganization of a coadapted genotype.

and the genetic unit which behaves as an additive-multiplicative unit. Studies are in progress to determine whether the relationship intersects the "unit" axis at the level of the locus or if some larger "chunk" smaller than an arm retains the additive (multiplicative) property which has been so central to the conceptualization of population genetic models.

ACKNOWLEDGEMENT

This work was supported by NSF Grant GB 41278 and AEC Grant AT(11-1)-1552.

We wish to acknowledge the dedication and technical expertise contributed to this project by Barbara Thirtle.

REFERENCES

Asher, J.H., Jr. 1970. Parthenogenesis and genetic variability. II. One locus models for various diploid populations. *Genetics* 66:369-391.

Carson, H.L., I.Y. Wei, and J.A. Niederkorn, Jr. 1969. Isogenicity in parthenogenetic strains of *Drosophila mercatorum*. *Genetics* 63:619-628.

Charlesworth, Brian and Deborah Charlesworth 1973. A study of linkage disequilibrium in populations of *Drosophila melanogaster*. *Genetics* 73:351-359.

Crow, J.F. and M. Kimura 1970. *An Introduction to Population Genetics Theory*. Harper and Row, New York.

Ewens, W.J. 1969. *Population Genetics*. Methuen and Co. Ltd. London.

Franklin, I. and R.C. Lewontin 1970. Is the gene the unit of selection? *Genetics* 65:707-734.

Sinnock, Pomeroy and Charles F. Sing 1972. Analysis of multilocus genetic systems in Tecumseh, Michigan. II. Consideration of the correlation between nonalleles in gametes. *Amer. J. Hum. Genet.* 24:393-415.

Templeton, A.R. 1972. Statistical models of parthenogenesis. Ph.D. Thesis, University of Michigan, Ann Arbor.

Templeton, A.R. and E.D. Rothman 1973. The population genetics of parthenogenetic strains of *Drosophila mercatorum*. I. One locus model and statistics. *Theor. and Applied Genet.* 43:204-212.

Templeton, A.R., C.F. Sing, and B. Thirtle. The unit of selection in *Drosophila mercatorum*. I. The interaction of selection and meiosis in parthenogenetic strains. Submitted to *Genetics*.

Zouros, E. and C.B. Krimbas 1973. Evidence for linkage disequilibrium maintained by selection in two natural populations of *Drosophila subobscura*. *Genetics* 73:659-674.

MULTIPLE ALLELISM AND ISOZYME DIVERSITY IN HUMAN POPULATIONS

HARRY HARRIS
Galton Laboratory
University College
London

ABSTRACT. The causes of isozyme formation are conveniently classified into three main categories: 1) multiple gene loci; 2) multiple allelism; 3) "secondary" isozyme formation due to post-translational structural modifications. An essential difference between multiple allelism and the other categories of causes is that while the latter represent phenomena which are in general common to all members of a species, multiple allelism gives rise to isozyme differences between individual members of the species.

Two different questions concerning the isozyme diversity generated by multiple allelism are considered.

The first has to do with its population aspects. Much work has been done on the occurrence of so-called electrophoretic polymorphisms attributable to the occurrence of two or more "common" alleles at particular gene loci. Here, particular attention is directed to the incidence and distribution of "rare" alleles. These appear to be ubiquitous in human populations and to occur both at "polymorphic" and "non-polymorphic" loci. The results of a study extending over a wide range of loci are summarized. Estimates of the variation from locus to locus in the heterozygosities attributable to both "common" and "rare" alleles have been obtained.

The second question is concerned with the diversity of the isozyme patterns observed in heterozygotes. This in general depends on the subunit structure of the particular enzyme. An attempt is made to obtain a picture of the distribution of subunit structures among different enzymes from the isozyme patterns revealed by electrophoretic studies.

Multiple molecular forms of an enzyme, or isozymes, may be generated in different ways and it is convenient to classify these causes of isozyme formation into three main categories (Harris, 1969; Hopkinson and Harris, 1971).

1. The occurrence of multiple gene loci coding for structurally distinct polypeptide chains of the enzyme.

2. The occurrence of multiple allelism at a single locus, with different alleles determining structurally different versions of a particular polypeptide chain.

3. The occurrence of so-called 'secondary' isozyme formation due to post-translational modifications of the enzyme structure.

In practice two or more of these different types of cause are often operative in any particular case, so that the isozyme patterns observed are complex and usually require for their elucidation a combination of both biochemical and genetical methods of analysis.

In the present paper I propose to consider only the second of these categories, that is, isozyme formation due to multiple allelism. The essential difference between this category of causes and the others, is that while the latter represent phenomena which in general are common to all members of the species, multiple allelism by its very nature gives rise to differences between individual members of the species in the isozymes that they form. Thus multiple gene loci and post-translational changes as causes of isozymes, can be regarded as providing the basic genetic and biochemical framework on which individual diversity is generated by allelic differences.

Where multiple alleles occur at a locus coding for a polypeptide chain of a particular enzyme protein, the individuals homozygous for different alleles will as a rule show different isozymic forms of the enzyme, and heterozygous individuals will exhibit more complex isozyme patterns because the two different alleles they carry generally produce structurally different polypeptide products.

In theory a very large number of different alleles may be generated by separate mutations within the confines of a single gene. Thus from a typical gene containing a DNA sequence of say 900 bases, as many as 2700 different alleles may be generated by separate mutations involving single base alterations, since each base may be substituted by one of three others. About 70-75% of these different mutants may be expected to result in single aminoacid substitutions in the corresponding polypeptide chain, and this is a major source of isozyme diversity. But in addition a variety of other alterations in the polypeptide structure, such as deletions of parts of amino acid sequence, chain elongations and various types of rearrangement of the sequence, can be produced by other types of mutational event.

Because of gene mutations which have occurred in single individuals in earlier generations, many of these different alleles exist among living members of the species today, and result in individual diversity. The extent of such diversity in any given population depends, of course, on the incidence of the different alleles which oocur. And in recent years a great deal of information about this has been obtained.

ENZYME POLYMORPHISM AND AVERAGE HETEROZYGOSITY

Following the widespread application of enzyme electrophoresis to population genetic studies, it soom emerged that in both human and animal populations there exist at an unexpectedly high proportion of gene loci coding for the structure of enzymes and other proteins, two or more quite common alleles which result in isozyme differences detectable by electrophoresis.

A recent summary of the data in man is given in Table 1. It is based on electrophoretic enzyme surveys carried out in unrelated individuals of European origin (Harris and Hopkinson, 1972). In all it appeared that the enzyme products of some 71 different gene loci had been studied, and of these 20 showed electrophoretic polymorphism (i.e. 28%). For the purpose of this analysis a locus was regarded as polymorphic if there was evidence for two or more common alleles and the proportion of heterozygous individuals was 2% or more of the total population.

TABLE 1

ENZYME POLYMORPHISM IN EUROPEANS
(Harris and Hopkinson, 1972)

Number of loci screened	71
Number of loci showing electrophoretic polymorphism (i.e. >0.02 heterozygotes)	20
Percentage of polymorphic loci	28.2
Average heterozygosity per locus (detected electrophoretically)	0.067

A better idea of the extent of the isozymic diversity generated in this way is provided by estimating the average heterozygosity per locus. This is done by summing the actual values observed for the proportion of heterozygotes at each locus, and dividing by the total number of loci (i.e. 71). This gives an estimate for the average heterozygosity per locus of 0.067. So this particular set of data suggest that in man any single individual is likely to be heterozygous for common alleles giving rise to electrophoretic differences at about 7% of his loci coding for enzyme structure. Because of the many different allelic combinations at the various loci that can occur, it is likely that no two individuals, with the exception of monozygotic twins, are exactly alike in their isozymic constitutions.

Essentially similar results have of course also been obtained from electrophoretic surveys of enzymes and proteins in natural populations of a variety of other animal species. Data on nearly fifty different species have now been reported. In other vertebrates the average heterozygosities per locus do not differ very much from those found in man. But in invertebrate species they appear to be significantly higher (Selander and Kaufman, 1973).

It must be noted that these surveys are only concerned with alleles which determine electrophoretic differences. But only a proportion, probably no more than one-third, of all possible mutant alleles which produce structural changes in enzyme proteins, will be detectable in electrophorttic surveys. So the true heterozygosity is likely to be greater than these estimates suggest, and there are probably many isozyme differences which occur but which are not detected by the procedures presently used.

RARE ALLELES

In searching for polymorphism in any given population, the number of different unrelated individuals examined need usually amount to no more than a hundred or two, because the alleles involved are by definition relatively common. But to study rare alleles producing uncommon isozyme variants very much larger samples are required.

The point is illustrated by our own findings in the case of one of the loci which determines the enzyme phosphoglucomutase. Some years ago we showed that three common phenotypes could be readily identified and these were due to two common alleles with frequencies in European populations of about 0.76 and 0.24; the homozygous phenotypes occurring in about 56% and 6% of the population respectively and the heterozygous phenotype in about 36% of the population (Spencer et al., 1964). Subsequently in the course of examining this enzyme in a large number of different individuals a number of rare phenotypes were identified, which could be attributed to the occurrence of rare alleles in heterozygous combination with one or the other of the two common alleles, but only about one in a thousand people was found to be heterozygous for one or another of them (Hopkinson and Harris, 1966). The findings in 10,333 unrelated Europeans are illustrated in Figure 1. There were five different rare alleles, and the number of each among the 20,666 alleles

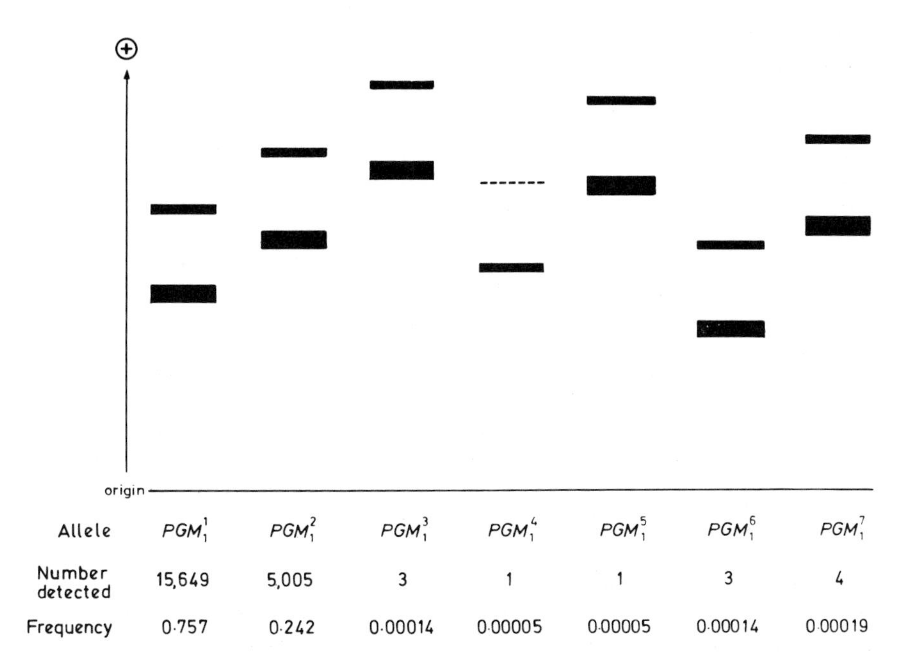

Fig. 1. Diagram of the electrophoretic isozyme pattern determined by seven different alleles at the phosphoglucomutase $\underline{PGM}_1$ locus. The incidence of the different alleles found in the course of screening 10,333 unrelated Europeans (i.e. 20,666 alleles) is indicated. Data of Harris et al (1974).

---

screened is indicated.

Similar rare electrophoretic variants had also been detected in the course of investigation of other enzymes, and recently (Harris, et al., 1973) we decided to put together the data which had accumulated over a period of about ten years in order to see whether we could obtain a general picture of the incidence of such rare alleles over a wide range of loci.

We found that we had reasonably reliable data on 43 different enzyme loci. The number of unrelated individuals examined varied for the different enzymes from just a few hundred up to more than 10,000, but the average per locus was about 2,500, and since the loci are probably all or nearly all autosomal this implies that on average we had screened the products of about 5,000 genes at each locus for electrophoretic variants. For the purpose of the analysis we defined a rare allele as one which had a gene frequency of less than 1 in 200 in the general population. In all, 56 different ones had been identified, and more than 80% of these appeared to have frequencies of less than

1 in 1,000, so the majority of these alleles are indeed extremely uncommon.

The tabulations of the detailed findings over the 43 loci (Harris et al., 1973) are rather extensive, so in order to illustrate the main features of the data I have simply summarized in Table 2 the results in just 12 of them. The first line for example gives the findings for the phosphoglucomutase locus ($\underline{PGM_1}$) which have already been mentioned (Fig. 1). The number of unrelated individuals tested was 10,333, i.e. 20,666 genes. Five different rare alleles were identified, and these occurred in 12 individuals who were heterozygous for one or another of them. So the combined incidence of rare heterozygotes was 1.16 per 1,000. At this locus there are also two common alleles, and the heterozygosity in the population due to these is 0.366, or 366 per 1,000.

In the surveys one or more rare alleles were detected at just over half the 43 loci studied. Obviously the chance of detecting such rare events will depend on the size of the sample. And indeed the number of individuals examined was on average much lower among the loci where no rare alleles were detected (1,300), than among the others (4,023). Also at several of the loci, although we failed to find rare variants in these surveys, such variants have in fact been found in other studies. So in general it seems not unlikely that such rare alleles actually occur at virtually all loci.

It is obvious that estimates of the frequencies of individual alleles must inevitably be very imprecise because most of them are extremely rare. It is more practical therefore in comparing different loci to consider the combined incidence of heterozygosity due to the rare alleles at each of the loci (penultimate column Table 2). These heterozygosities per locus turn out to vary considerably, extending over a 150-fold range even if we consider only those loci at which rare alleles were in fact discovered. To some extent this variation no doubt occurs for technical reasons, because we know that there is much variation in the discriminative power of the electrophoretic methods available for the different enzymes. Nevertheless we are inclined to think that the variation does to a significant degree reflect real differences between loci.

A case in point is locus PL which determines a specific alkaline phosphatase peculiar to the placenta and which shows a quite remarkable degree of allelic variation (Robson and Harris, 1967, Donald and Robson, 1973). In the course of examining placentae from 3,244 individuals, evidence for as many as 14 different alleles was obtained. Ten of these can be counted as rare alleles according to our definition and they account for as many as 23 heterozygotes per 1,000 individuals.

TABLE 2

INCIDENCE OF RARE ALLELES DETERMINING ELECTROPHORETIC ENZYME VARIANTS IN UNRELATED EUROPEANS Further details and also detailed data on 31 other loci are given in Harris, Hopkinson and Robson, (1973).

Enzyme	Locus	No. of unrelated individuals tested	No. of different rare alleles	No. of rare heterozygotes	Heterozygotes for rare alleles (per 1000)	Heterozygotes for common alleles (per 1000)
Phosphoglucomutase	$\underline{PGM_1}$	10,333	5	12	1.16	366
Phosphoglucomutase	$\underline{PGM_2}$	10,333	3	7	0.68	-
Glutamate-oxalate transaminase 'soluble'	$\underline{GOT_S}$	1,195	2	2	1.67	-
Glutamate-oxalate transaminase 'mitochondrial'	$\underline{GOT_M}$	1,195	0	0	-	33
Adenylate kinase	AK	6,760	1	1	0.15	77
Nucleoside phosphorylase	NP	1,542	2	2	1.30	-
Pyrophosphatase	PP	2,190	0	0	-	-
Alkaline phosphatase -placental	PL	3,244	10	76	23.43	502
Peptidase -B	PEP-B	7,041	3	15	2.13	-
Triosephosphate isomerase	TPI	1,750	2	2	1.17	-
Superoxide dismutase	$\underline{SOD_A}$	11,237	1	7	0.62	-
Inosine triphosphatase	ITP	641	0	0	-	-

The common alleles give rise to some 500 heterozygotes per 1,000 individuals. The number of different rare alleles found in this particular case and the degree of heterozygosity to which they give rise, makes us suspect that at this locus some unusual process may be responsible either for generating the allelic diversity or maintaining it at this high level.

Despite this variation in heterozygosity from locus to locus, it is nevertheless of some interest to estimate the average heterozygosity per locus due to rare alleles, over the full range of loci including those where no rare variants at all were discovered (Table 3). This average is 1.76 per 1,000 for all the loci. But it is reduced to 1.14 per 1,000 if we exclude PL. Thus the data as a whole suggest that on average for any single locus we may expect that between 1 and 2 individuals per 1,000 will be heterozygous for a rare allele determining an electrophoretic variant. If we suppose for example that there are say 30,000 loci coding for enzyme structure, each of us is likely to be carrying at least 30 such alleles, and it is extremely improbable that any two of us will, with the exception of monozygotic twins, have exactly the same combination.

It is also if interest to ask whether the so-called 'polymorphic' loci and 'non-polymorphic' loci differ in the incidence of rare alleles. The average heterozygosity per locus due to rare alleles is 2.61 per 1,000 for the 13''polymorphic' loci and 1.16 for the 'non-polymorphic' loci (Table 4). This is a considerable difference, but it is entirely accounted for by locus PL. If this locus is removed from the calculation the two values are almost the same.

The alleles which determine these rare electrophoretic enzyme variants must have originated by mutations occurring in a single individuals in earlier generations. The question therefore arises as to how many of the rare variants observed in these surveys are the consequence of fresh mutations. That is to say mutations which occurred in the germ line of one or other of the parents of individuals found to have the variant.

In the course of this work we have systematically carried out studies on the families of individuals showing rare variants whenever this was practicable, and an extensive body of pedigree data has been assembled. In fact no clear example of a fresh mutation has been found. But of course not all the family data is informative, because sometimes critical individuals could not be tested. However in 77 unrelated individuals who were heterozygous for one or another of these rare variants we have been able to test both the parents. And in each case, either the father or the mother of the individual with the variant showed the same variant. So none of these cases was due to a

TABLE 3

Average heterozygosity per locus for rare alleles determining electrophoretic enzyme variants. For data see Harris, Hopkinson, and Robson (1973). PL is the locus for placental alkaline phosphatase.

	Number of loci	Total No. of genes screened	Total No. of rare alleles	Heterozygotes per locus per 1000 individuals
All loci	43	231,508	204	1.76
All loci excluding PL	42	225,020	128	1.14

TABLE 4

Average heterozygosity per locus for rare alleles determining electrophoretic enzyme variants at polymorphic and non-polymorphic loci. For detailed data see Harris, Hopkinson and Robson (1973). PL is the locus for placental alkaline phosphatase.

	Number of loci	Total No. of genes screened	Total No. of rare alleles	Heterozygotes per locus per 1000 individuals
Non-polymorphic loci	30	135,750	79	1.16
Polymorphic loci	13	95,758	125	2.61
Polymorphic loci excluding PL	12	89,280	49	1.10

fresh mutation.

We can therefore conclude that the great majority of the rare variants observed in our population are not the products of fresh mutations, and the fraction attributable to fresh mutations must be very small - probably less than 1-2%.

DISTRIBUTION OF HETEROZYGOSITIES

From the data about both the common and the rare alleles at each of these 43 enzyme loci, it was possible to estimate for each locus the heterozygosity for alleles determining electrophoretic differences.

The distribution of these heterozygosities is shown in Figure 2. It has two very interesting features. The first is the very considerable variation in heterozygosity which occurs

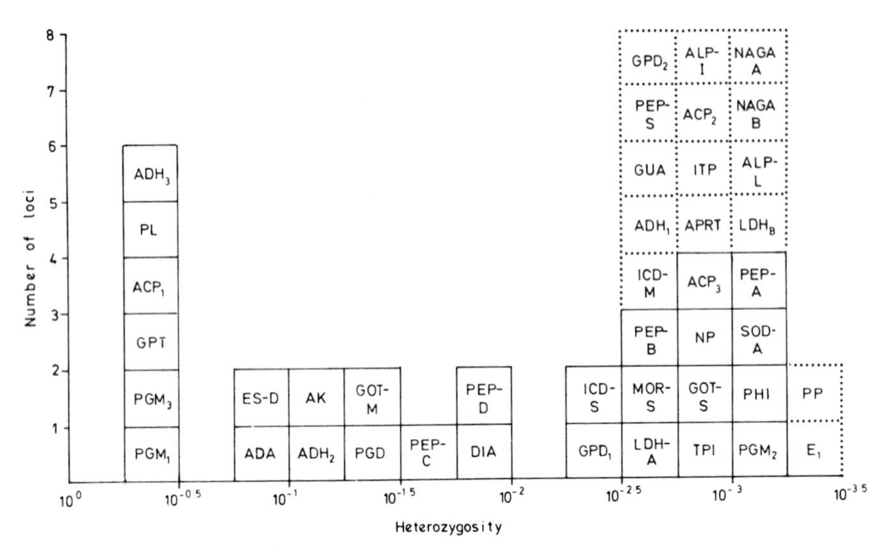

Fig. 2. Distribution of heterozygosities for alleles determining electrophoretic variants at 43 loci (Harris et al., 1973).

Each square represents a locus, the symbol for which is indicated. At some of the loci (indicated by dotted lines) no heterozygotes were detected in the surveys, but to include them in the distribution they have been placed in the position where they would have been if in fact one heterozygote had been observed. The true heterozygosities for these loci must therefore lie somewhat to the right of the position shown.

KEY: ADA, adenosine deaminase; ACP_1, ACP_2, ACP_3, acid phosphatase loci; ADH_1, ADH_2, ADH_3, alcohol dehydrogenase loci; AK, adenylate kinase; ALP-I, intestinal alkaline phosphatase; ALP-L liver alkaline phosphatase; APRT, adenine phosphoribosyl transferase; DIA, NADH diaphorase; E_1, serum cholinesterase; ES-D, esterase D; GPD_1, GPD_2, α-glycerophosphate dehydrogenase loci; GOT-S, GOT-M, glutamate-oxaloacetate transaminase loci; GPT, glutamate-pyruvate transaminase; GUA, guanase; ICD-S, ICD-M, isocitrate dehydrogenase loci; ITP, inosine triphosphatase; LDH_A, LDH_B, lactate dehydrogenase loci; MOR-S, NAD malate dehydrogenase (soluble); NAGA-A, NAGA-B, hexosaminidase loci; NP, purine nucleoside phosphorylase; PEP-A, PEP-B, PEP-C, PEP-D, PEP-S, peptidase loci; PGD, phosphogluconate dehydrogenase; PGM_1, PGM_2, PGM_3, phosphoglucomutase loci; PHI, phosphohexose isomerase; PL, placental alkaline phosphatase; PP, inorganic pyrophosphatase; SOD-A, superoxide dismutase; TPI, triosephosphate isomerase.

from locus to locus. This variation extends over a more than

500-fold range, from loci where some 50% of the population are heterozygotes, to loci where the incidence of heterozygotes is less than 1 in 1,000.

The other interesting feature is that there appears to be a clear tendency to bimodality, at least when the distribution is plotted on a log-scale. This, if confirmed, might suggest a real biological dichotomy between so-called 'polymorphic' loci with high heterozygosities, and so-called 'monomorphic' loci with low heterozygosities.

However it is not clear to what extent these apparent features of the distribution derive from the fact that it is only concerned with heterozygosity attributable to alleles whose enzyme products show electrophoretic differences. It is not known how far the occurrence of non-electrophoretically detectable variants would affect the overall picture.

SUBUNIT STRUCTURE

So far I have only considered the level of heterozygosity as a measure of isozyme diversity within a population. But the complexity of the isozyme patterns which occur in heterozygotes depend on the subunit structures of the enzyme proteins. This adds an extra dimension to the matter, and it is of some interest to see whether the extensive electrophoretic studies which have now been carried out on different enzymes for various reasons can, in this connection also, provide us with any general picture of what occurs.

If an enzyme protein is monomeric, one expects that the isozyme pattern in a heterozygote will simply represent a mixture of the characteristic isozymes seen in the two corresponding homozygotes. If however the enzyme protein is multimeric, and there are in homozygotes at least two identical polypeptide chains, then one may expect in heterozygotes to find 'hybrid' or 'heteromeric' isozymes in addition to the characteristic isozymes present in the two corresponding homozygotes. Thus in the simplest situation where the enzyme protein is dimeric and is made up of two identical polypeptide chains whose amino acid sequences are determined at a single gene locus, one expects to find three characteristic isozymes in a heterozygote, two of which are 'homodimers' as in the corresponding homozygotes, and the third is a 'heterodimer' containing polypeptide chains coded by each allele. If the enzyme protein is a trimer with three identical polypeptide subunits in homozygotes, then four characteristic isozymes are expected in heterozygotes, two of which are 'hybrids'. Similarly if the enzyme protein is a tetramer with four identical polypeptide subunits in the homozygote, then a pattern of five characteristic isozymes is expected in heterozygotes, three of them being 'hybrid' or 'heteromeric' forms.

Examples of each of these situations have indeed been observed, and very often it is possible to infer the probable subunit structure of the enzyme protein from the electrophoretic pattern observed in heterozygotes. There are however certain exceptions. For example in the case of multimeric enzymes determined by genes located on the X-chromosome, hybrid isozymes are not usually seen in tissue extracts from heterozygotes, because only one allele is on the active X-chromosome in any one cell and therefore as a rule the polypeptide products of the two alleles are not produced together in the same cell and cannot form a 'hybrid' isozyme. Other exceptions may arise if, for example, the dissociation of the multimeric enzyme happens to occur more rapidly than the rate of the electrophoretic separation, as is the case with haemoglobin; or if for stereochemical reasons the association of the two different polypeptides is restricted.

Simple electrophoretic studies in other situations may also provide information about the subunit structure of isozymes. For example where two or more different gene loci are concerned in determining polypeptides which can associate to form isozymes, the electrophoretic pattern in homozygous individuals can be informative. For instance the tetrameric structure of lactate dehydrogenase was originally suggested by the five banded isozyme patterns displayed in different tissues, due as it was subsequently shown to the formation of the distinctive A and B polypeptides determined by two separate gene loci.

Studies on somatic cell hybrids obtained by fusing tissue cultured cells from two different species represent yet another way in which the subunit structure of an enzyme may be inferred from the isozyme pattern. In such cases provided the enzyme has a different electrophoretic mobility in the two species and the enzyme is multimeric, hybrid isozyme formation is readily observed in the somatic cell hybrids. Numerous examples of this effect have now been found in studies on human-mouse and human-chinese hamster somatic cell hybrids (Ruddle, 1972). As a general rule the electrophoretic patterns are in essentials similar to those observed in heterozygotes in either of the species alone, and the findings complement one another very exactly. Exceptions however occur in the case of multimeric enzymes determined by genes located on the X-chromosome. Thus when somatic cell tissue extracts are examined for such enzymes in heterozygous females, no hybrid form is usually observed because only one or other of the two alleles present is on the so-called 'active' X-chromosome in any single cell. However somatic cell hybrids between two different species can contain active X-chromosomes from both species. So both types of polypeptide are produced and if the enzyme is multimeric hybrid isozymes are seen.

Further evidence supporting the subunit structure inferred by these methods can often also be obtained by in vitro hybridization experiments, again using electrophoresis to examine isozyme products. In such experiments the aim is to find conditions under which a multimeric enzyme can be dissociated into its subunits which can then be allowed to recombine to give enzymically active products. If a mixture of the enzymes from two different species, or from two different loci in the same species, or from two different alleles at a single locus in the same species, are then subjected to this procedure the appropriate hybrid isozymes can be formed. Such in vitro hybridization has now been accomplished in a number of cases and in each case the resulting isozyme pattern obtained has confirmed the subunit structure expected from isozyme studies in the various in vivo situations discussed above.

Because of the consistency of the electrophoretic findings for any one enzyme when two or more of these different approaches have been used, one is encouraged to see whether one can obtain from the available data a picture of the general distribution of subunit structures over a wide range of enzymes.

Table 5 lists a series of enzymes which have on the basis of electrophoretic studies of the isozymes been classified as monomers, dimers, trimers or teteramer. They are the products of 50 different gene loci. Nearly three quarters appear to be multimeric.

TABLE 5

A SERIES OF ENZYMES CLASSIFIED BY SUBUNIT STRUCTURE

The enzymes represent the products of fifty different loci. The classification is based on the electrophoretic patterns observed in heterozygotes, somatic cell hybrids, or in-vitro dissociation-recombination experiments.

MONOMERS

1) phosphoglucomutase ($\underline{PGM_1}$)
2) phosphoglucomutase ($\underline{PGM_2}$)
3) phosphoglucomutase ($\underline{PGM_3}$)
4) carbonic anhydrase ($\underline{CA\ I}$)
5) carbonic anhydrase ($\underline{CA\ II}$)
6) red cell acid phosphatase (ACP_1)
7) adenylate kinase
8) adenosine deaminase
9) NADH diaphorase

10) mannosephosphate isomerase
11) peptidase B
12) peptidase C
13) phosphoglycerate kinase

DIMERS

1) alcohol dehydrogenase ($\underline{ADH_1}$)
2) alcohol dehydrogenase ($\underline{ADH_2}$)
3) alcohol dehydrogenase ($\underline{ADH_3}$)
4) glycerol-3-phosphate dehydrogenase ($\underline{GPD_1}$)
5) glycerol-3-phosphate dehydrogenase ($\underline{GPD_2}$)
6) acid phosphatase ($\underline{ACP_2}$)
7) acid phosphatase (ACP_3)
8) glutamate oxalate transaminase (S)
9) glutamate oxalate transaminase (M)
10) malate dehydrogenase (S)
11) malate dehydrogenase (M)
12) adenine phosphoribosyl transferase
13) phosphohexose isomerase
14) phosphogluconate dehydrogenase
15) peptidase A
16) peptidase D
17) pyrophosphatase
18) glucose-6-phosphate dehydrogenase
19) superoxide dismutase (A)
20) enolase
21) esterase D
22) isocitrate dehydrogenase (S)
23) glutamate-pyruvate transaminase
24) 2,3 diphosphoglycerate mutase
25) inosine triphosphatase
26) placental alkaline phosphatase

TRIMERS

1) purine nucleoside phosphorylase

TETRAMERS

1) lactate dehydrogenase (A)
2) lactate dehydrogenase (B)
3) lactate dehydrogenase (C)
4) aldolase (A)
5) aldolase (B)
6) aldolase (C)
7) pyruvate kinase
8) superoxide dismutase (B)
9) serum cholinesterase
10) malic enzyme (S)

Since the enzymes in the table were selected only in that sufficiently precise electrophoretic techniques were available for the study of the isozymes formed by heterozygotes, somatic cell hybrids, or by multiple homozygous loci it is reasonable to regard them as representative of enzymes in general. If so, one can conclude that a dimeric structure is the most common form (52%). A tetrameric structure is not uncommon (20%) but a trimeric structure is rather unusual, since only one example was found in the 50 cases listed. In 26% of the cases no hybrid isozymes were detected and these are almost all probably monomers.

Proof of a particular subunit structure can of course best be obtained by comparing the molecular weights of the native enzyme and of its products after complete dissociation by, for example, denaturation in the presence of high concentrations of urea, guanidine or sodium dodecyl sulphate. This is often difficult because it involves the use of highly purified enzyme preparations whereas most of the methods discussed above can be carried out on crude tissue extracts or on only partially purified preparations.

Purine nucleoside phosphorylase provides an interesting example of a case where all these different methods have been applied. The original suggestion that this enzyme is a trimer came from electrophoretic studies on heterozygotes (Edwards, et al., 1971). This was supported by the findings in human-mouse somatic cell hybrids grown in tissue culture and by in vitro dissociation and recombination experiments using the human enzyme with the mouse enzyme and also the human enzyme with the bovine enzyme. In all cases a four banded isozyme pattern expected from a trimer was obtained. Further supporting evidence came from the observation that three moles of hypoxanthine appeared to bind one mole of the enzyme (Agarwal and Parks, 1969). The situation appeared however to be very unusual since at that time no certain example of an enzyme with a trimeric structure had been identified. Eventually however molecular size determinations by ultracentrifugation using the purified bovine enzyme was found to give a value of 84,000 which was in agreement with estimates obtained by gel filtration on non-purified human material; and molecular size determinations on the bovine enzyme dissociated into its subunits by treatment with sodium dodecyl sulphate gave a value of 28,000, just one third of the molecular size of the native enzyme (Edwards et al., 1973).

VARIATION IN ACTIVITY

Since the polypeptide products of different alleles at any given gene locus will as a rule differ in their structures, though often by only single amino acid substitutions, the isozymes in which they occur may often be expected to differ in their properties, and this will commonly be reflected in differences in their enzymic activities. Such differences in activity may arise either because the particular alteration in structure affects the active site and so alters the specific catalytic activity; or because the structural change alters the stability of the molecule so that it may be broken down in the cell at an altered rate; or because the structural change is associated with an alteration in the rate of synthesis. In heterozygotes where the two alleles because of such effects, make different contributions to the total activity, this is readily detected by the fact that the isozyme pattern observed electrophoretically is asymmetric. Such asymmetric isozyme patterns in heterozygotes are seen not infrequently and are usually easily distinguished from the so-called symmetric patterns which occur when the two alleles make equal contributions to the activity. Thus in the case of heterozygotes for a dimeric enzyme, the activities of the characteristic three isozymes observed will be in the ratio 1:2:1 if the two alleles contribute equally, but if one allele contributes say only half the activity of the other an asymmetric pattern will be observed and the ratio of activities may be roughly 4:4:1, although the actual ratio will in fact depend on exactly how the activity difference arises.

Where the altered polypeptide structure results in a severe reduction in enzyme activity this may be manifested in individuals homozygous for the particular allele, by a metabolic disorder consequent on the specific enzyme deficiency - a so-called 'inborn error of metabolism'. Sometimes indeed there may be a complete or virtually complete absence of the specific enzyme activity in such homozygotes. And it is of interest to note that in these cases the presence of the allele will not usually be detectable in the electrophoretic pattern seen in heterozygotes, although the total activity in such individuals will be somewhat reduced, generally to about half the normal level.

In general one can expect that the series of alleles which may occur at any given gene locus, will result in a wide variation in enzyme activities and this will be reflected in the isozyme patterns which are observed.

REFERENCES

Agarwal, R. P. and R. E. Parks 1969. Purine nucleoside phosphorylase from human erythrocytes. IV Crystallization and some properties. *J. Biol. Chem.* 244: 644.

Donald, L. J. and E. B. Robson 1973. Rare variants of placental alkaline phosphatase. *Ann. hum. Genet.*,Lond. 37: 303-

Edwards, Y. H., P. A. Edwards, D. A. Hopkinson 1973. A trimeric structure for mammalian purine nucleoside phosphorylase. *FEBS Letters* 32: 235-

Edwards, Y. H., D. A. Hopkinson and H. Harris 1971. Inherited variants of human nucleoside phosphorylase. *Ann. Hum. Genet.*, Lond. 34: 395-

Harris, H. 1969. Genes and isozymes. (Review Lecture). *Proc. Roy. Soc.* Lond. B. 174: 1-

Harris, H. and D. A. Hopkinson 1972. Average heterozygosity per locus in man: an estimate based on the incidence of enzyme polymorphism. *Ann. Hum. Genet.* Lond. 36: 9-

Harris, H., D. A. Hopkinson, and E. B. Robson 1973. The incidence of rare alleles determining electrophoretic variants: data on 43 enzyme loci in man. *Ann. Hum. Genet.*, Lond. 37: 237-

Hopkinson, D. A. and H. Harris 1966. Rare phosphoglucomutase phenotypes. *Ann. Hum. Genet.*, Lond. 30: 167-

Hopkinson, D. A. and H. Harris 1971. Recent work on isozymes in man. *Annual Review of Genetics*,5: 5-

Robson, E. B. and H. Harris 1967. Further studies on the genetics of placental alkaline phosphatase. *Ann. Hum. Genet.*, Lond. 30: 219-

Ruddle, F. H. 1972. Linkage analysis using somatic cell hybrids. *Advances in Human Genetics* 3: 173- Edit. H. Harris and K. Hirschhorn, Plenum Press, New York.

Selander, R. K. and D. W. Kaufman 1973. Genic variability and strategies of adaption in animals. *Proc. Nat. Acad. Sci., U. S. A.* 70: 1875-

Spencer, N., D. A. Hopkinson, and H. Harris 1964. Phosphoglucomutase polymorphism in man. *Nature*, Lond. 204: 742-

ISOZYME ANALYSIS OF SOMATIC CELL HYBRIDS: ASSIGNMENT OF THE PHOSPHOGLUCOMUTASE$_2$ (*PGM_2*) GENE LOCUS TO CHROMOSOME 4 IN MAN WITH DATA ON THE MOLECULAR STRUCTURE AND HUMAN CHROMOSOME ASSIGNMENTS OF SIX ADDITIONAL MARKERS

PHYLLIS J. McALPINE, T. MOHANDAS, and J. L. HAMERTON
Division of Genetics, Department of Pediatrics,
University of Manitoba and Health Sciences Children's Centre,
Winnipeg, Manitoba, CANADA

ABSTRACT. The analysis of isozyme patterns of lysates of human-rodent somatic cell hybrids provides a convenient method for detecting the presence of human gene products in gene assignment studies and for obtaining data on the molecular structure of these gene products. The expression of human fumarate hydratase (FH) and guanylate kinase (GuK) in Chinese hamster-human somatic cell hybrids was found to be dependent upon the presence of chromosome 1 from man. FH appears to be a tetramer composed for four identical subunits while GuK may exist in monomeric form. Evidence for the monomeric structure of phosphoglucomutase$_1$ (PGM_1) and peptidase C (Pep C) isozymes and for the dimeric structure of 6-phosphogluconate dehydrogenase (PGD) isozymes in man was obtained; *PGM_1*, *Pep C*, and *PGD* are also chromosome 1 markers in man. Data obtained have provided evidence that the isozymes related to the phosphoglucomutase$_2$ (*PGM_2*) and to the phosphoglucomutase$_3$ (*PGM_3*) gene loci in man are monomeric and that the *PGM_2* gene locus can be assigned to human chromosome 4. Regional mapping indicates that it can be excluded from the distal portion of the long arm of this chromosome. Further evidence of the dimeric structure of the cytoplasmic form of glutamic oxaloacetic transaminase (*GOT_1*) was obtained. The tentative assignments of the *PGM_3* gene locus to chromosome 6, the gene loci for the cytoplasmic form of GOT and a high Km form of fibroblast hexokinase to chromosome 10 and the gene locus for the dimeric form of superoxide dismutase to chromosome 21 in man have been confirmed.

INTRODUCTION

Mapping the human genome using interspecific somatic cell hybrids, in particular those formed from the fusion of human cells with hamster or mouse cells, has been well documented. Gene assignments are based on the establishment of positive correlations between the presence of precisely identified human chromosomes, or parts thereof, and the presence of detectable human gene products. Electrophoretic analysis of lysates of somatic cell hybrids has proved to be very useful for distinguishing gene products, notably enzymes of human and rodent origin on the basis of differences in their electrophoretic mobility.

In addition to identifying gene products which are human in origin, the analysis of isozyme patterns in somatic cell hybrids can provide information regarding the molecular structure of these enzymes. If the human and rodent gene products each exist as monomers, the isozyme patterns observed when the human gene is expressed resembles a superimposition of the human isozyme pattern on that of the rodent. However, in those instances in which the enzymes are polymeric, one or more isozymes with electrophoretic mobility intermediate between that of the human and rodent isozymes, in addition to bands corresponding to the human and rodent forms, may be observed when the structural gene(s) associated with the synthesis of the human monomers is(are) expressed. These intermediate bands are considered to be composed of both human and rodent subunits and are referred to as heteropolymers. The isozymes consisting only of identical human or rodent monomers are termed homopolymers. The number of human-rodent heteropolymers theoretically possible in such instances depends upon the number of structural gene loci which determine the synthesis of the monomers in each species as well as the number of subunits which combine to form an enzyme molecule. Because of gene dosage effects in somatic cell hybrids, human homopolymers frequently cannot be visualized in the gels, but the presence of one or more heteropolymers containing human and rodent subunits may be considered as evidence of the expression of a human gene. When the human gene is not expressed, an isozyme pattern identical with that of the rodent parent is expected.

MATERIALS AND METHODS

In our laboratory somatic cell hybrids have been obtained from fusion between a mutant hypoxanthine guanine phosphoribosyl transferase deficient Chinese hamster cell line (Gee et al., 1974) and normal human fibroblasts or lymphocytes as described previously (Douglas et al., 1973a, Hamerton et al., 1973). Following the fusion process, the cells were plated in thymidine-hypoxanthine-aminopterin-glycine (THAG) selective medium and allowed to proliferate until colonies were just visible. Only one colony per tissue culture vessel was isolated and analyzed, thereby attempting to ensure that each hybrid cell line was essentially independent in origin. All hybrid cell lines were analyzed by electrophoretic methods for approximately forty human gene products. The following enzymes were examined using the methods indicated: phosphoglucomutase (Spencer et al., 1964), peptidase C (Povey et al., 1972), guanylate kinase (Monn and Christiansen, 1972), and phosphopentomutase (Quick et al., 1972). Details of the methods of examination of 6-phosphogluconate dehydrogenase, fumarate hydratase, glutamic oxaloacetic transaminase, hexokinase, and superoxide dismutase will be reported elsewhere (McAlpine et al., 1974). Cytological analyses were carried out as reported previously (Douglas et al., 1973a) in order to identify the human chromosomes retained.

RESULTS AND DISCUSSION

CHROMOSOME 1 MARKERS

Data obtained from both somatic cell hybrid and from family studies have established that the 6-phosphogluconate dehydrogenase (*PGD*), phosphoglucomutase$_1$ (*PGM*$_1$) and peptidase C (*Pep C*) structural gene loci are located on chromosome 1 in man (Van Cong et al., 1971, Westerveld and Meera Khan, 1972, Ruddle et al., 1972, Hamerton et al., 1973 and Robson et al., 1973). *PGD* and *PGM*$_1$ have been localized to the distal third of the short arm of this chromosome (Douglas et al., 1973b, Burgerhout et al., 1973, 1974, and Jongsma et al., 1973) while *Pep C* is on the long arm (Burgerhout et al., 1973, 1974 and Jongsma et al., 1973). Evidence from this laboratory suggests that *Pep C* may be quite close to the distal tip of the long arm, in the (1)(q41→qter) region. More recently, the presence of human guanylate kinase (GuK) and fumarate hydratase (FH) have been reported to be correlated with the presence of human chromosome 1 (Meera Khan et al., 1974, Van Someren et al., 1974). Guanylate kinase, also referred to as guanosine monophosphate kinase (E.C.2.7.4.8), catalyzes the interconversion of guanosine 5'-monophosphate and guanosine 5'-diphosphate while fumarate hydratase (E.C.4.2.1.2) catalyses the conversion of fumarate to malate.

We have analyzed the PGD, PGM$_1$, Pep C, GuK, and FH isozyme patterns of nine independently derived somatic cell hybrid lines. Human PGM$_1$, Pep C, and GuK isozymes, and PGD and FH heteropolymers, presumably consisting of both human and hamster subunits, were observed in four of these cell lines. The remaining five cell lines had isozyme patterns identical with those of the hamster fibroblasts and, thus, did not express the human form of these five markers. Correlation of the expression of the human form of PGD, PGM$_1$, Pep C,

GuK, and FH with the human chromosome complements of these cell lines indicated that the human form of these markers was expressed only in those cell lines which had retained human chromosome 1. When chromosome 1 from man was not present, the human form of these markers was not expressed (Table 1).

As no evidence of the presence of human-hamster heteropolymers involving the gene products of either the human PGM_1 (Fig. 1) or the *Pep C* gene loci in somatic cell hybrids expressing these human gene loci was observed, the isozymes associated with the PGM_1 and *Pep C* gene loci apparently are single polypeptides. The monomeric structure of the human PGM_1 and Pep C isozymes is consistent with the isozyme patterns observed in tissues of individuals considered to be heterozygous at the PGM_1 (Spencer et al., 1964) or at the *Pep C* (Santachiara Benerecetti, 1970 and Povey et al., 1972) gene loci. PGM_1 and *Pep C* heterozygotes have isozyme patterns which resemble a simple mixture of the isozymes associated with each of the alleles present at the PGM_1 or at the *Pep C* gene loci respectively in these individuals.

The presence of a single PGD heteropolymer in those hybrid cell lines expressing the human *PGD* gene locus, indicating that human PGD is a dimer, is consistent with the isozyme patterns observed in cell lysates of individuals considered to be heterozygous at the *PGD* gene locus (Parr, 1966).

Family data regarding the inheritance of GuK and FH isozyme patterns have not been reported. However, the presence of the human form of these two markers in somatic cell hybrids only when human chromosome 1 was present suggests that the synthesis of these GuK and FH isozymes is genetically determined. The absence of human-hamster GuK heteropolymers in lysates of somatic cell hybrids which express human GuK indicates that this enzyme probably exists in monomeric form, and, hence, its synthesis would be expected to be related to a single gene locus.

FH from pig heart has been shown to be a tetramer composed of four apparently identical subunits (Kanarek et al., 1964). More recent studies based on isoelectrofocusing (Penner and Cohen, 1971) and column chromatography with substrate elution (Woodfin, 1975) have provided evidence for heterogeneity of pig heart FH, but whether this heterogeneity is genetically determined or results from post-transcriptional alterations of a single polypeptide is not certain. The FH isozyme patterns of human and the hamster fibroblasts obtained after electrophoresis using cellulose acetate (Cellogel, Chemetron, Italy) each consisted of a single isozyme, with the human band migrating more anodally than the hamster component. The FH isozyme patterns of hybrid cell lines retaining human

TABLE 1

CORRELATION OF THE PRESENCE OF HUMAN PGD, PGM_1, Pep C, FH, AND GuK WITH THE PRESENCE OF CHROMOSOME 1 FROM MAN

PGD	Expression of human markers PGM_1	Pep C	FH	GuK	Human chromosome 1	Number of independent cell lines
+	+	+	+	+	+	4
-	-	-	-	-	-	5

+ : Human marker expressed or chromosome present

- : Human marker not expressed or chromosome absent

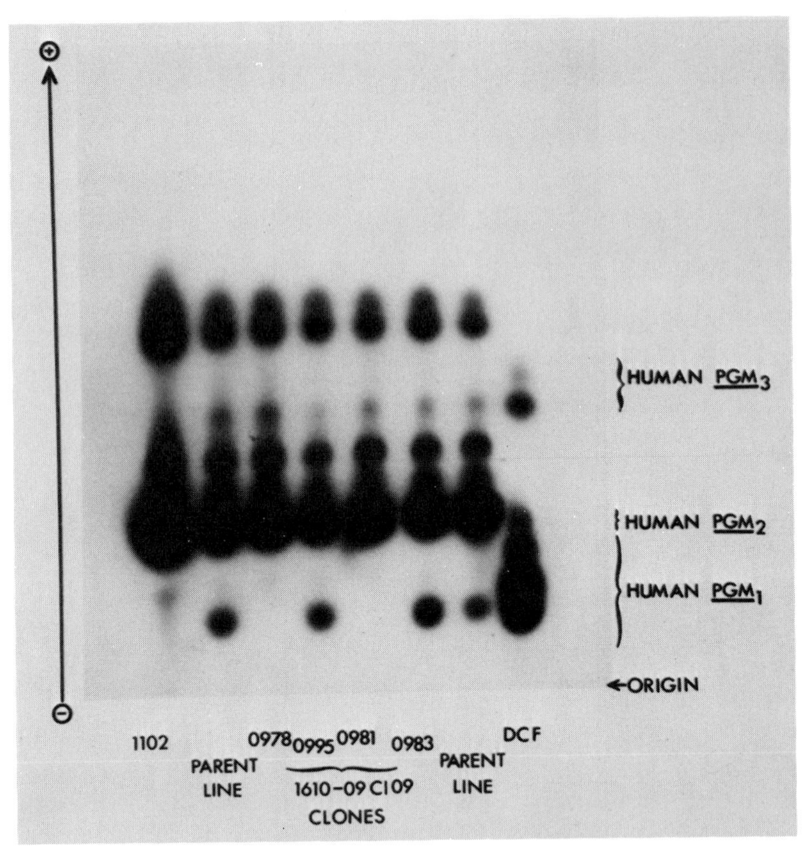

Fig. 1. Phosphoglucomutase isozyme patterns of somatic cell hybrids. 1102: Chinese hamster cell line; parent line: somatic cell hybrid line 1610; 0978, 0995, 0981 and 0983: clones of parent line, 0995 and 0983 express human PGM_1 locus; DCF: human fibroblasts. Parent line and all clones express human PGM_3 locus. (After Douglas et al., 1973b).

---

chromosome 1 consisted of a single FH component with electrophoretic mobility identical to that of the hamster band, together with a second FH isozyme that migrated slightly more anodally than the hamster isozyme (Fig. 2). This more anodally migrating isozyme is considered to be a human-hamster FH heteropolymer. If FH of both human and hamster origin has a tetrameric structure, as does porcine FH, then the observed FH heteropolymer as judged by its electrophoretic mobility presumably consists of one human subunit and three hamster subunits. As the overall level of FH activity in the lysates examined was low, the other human-hamster FH heteropolymers

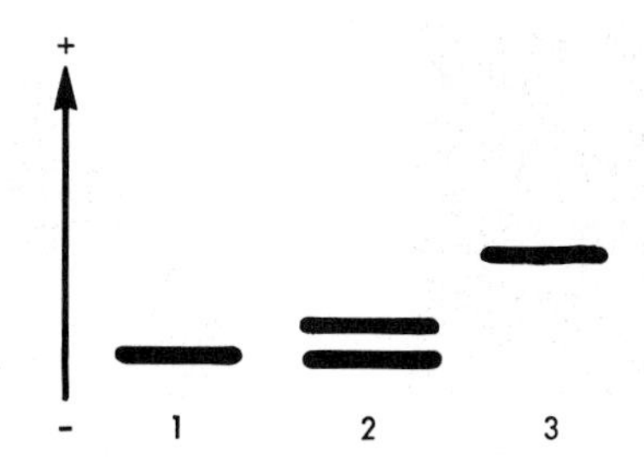

Fig. 2. Diagrammatic representation of fumarate hydratase (FH) isozymes in somatic cell hybrid lysates. Channel 1: Chinese hamster fibroblasts; Channel 2: somatic cell hybrid line expressing human FH; Channel 3: human fibroblasts.

which are theoretically possible were not observed. The nature of the FH isozyme pattern observed in the somatic cell hybrid lines expressing human FH suggests that all FH monomers in humans and hamsters are related to a single gene locus in each species.

A CHROMOSOME 4 MARKER

Phosphoglucomutase (E.C.2.7.5.1) is the enzyme which catalyzes the reversible interconversion of glucose-1-phosphate and glucose-6-phosphate. Family studies have indicated that the phosphoglucomutase (PGM) isozymes observed after electrophoretic examination of human tissues are determined by three separate and distinct structural gene loci, designated as PGM_1, PGM_2, and PGM_3, which are not closely linked (Spencer et al., 1964, Hopkinson and Harris, 1965, 1966, 1968 and Parrington et al., 1968). The PGM_1 gene locus is now known to be located on chromosome 1, while recent evidence (Jongsma et al., 1973 and Pearson et al., 1973) indicates that the PGM_3 gene locus may be on chromosome 6 in man.

When Chinese hamster-human somatic cell hybrids are analyzed electrophoretically for PGM, isozyme *e*, the most cathodal member of the isozyme set associated with the common allele, PGM_2^1, at the PGM_2 locus in man, migrates to approximately the same position in starch gels as the most cathodal hamster PGM component. This cathodal hamster component stains very intensely and thus, under these conditions of analysis, it is not possible to determine whether or not human PGM_2 components are present in these somatic cell hybrids (Fig. 3). However, as the isozymes related to the PGM_2 locus can be distinguished from those related to the PGM_1 and PGM_3 loci in man on the basis of their molecular weight (McAlpine et

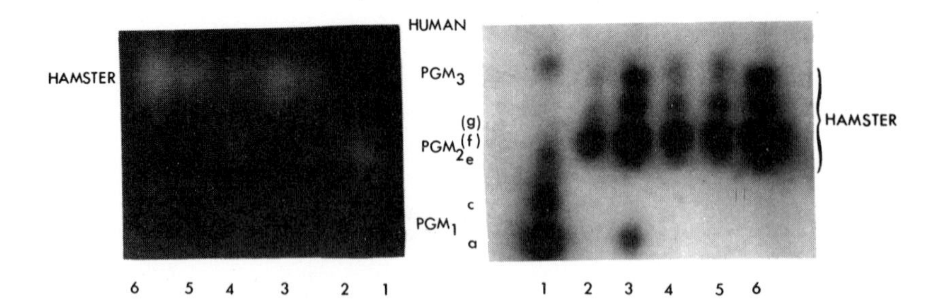

Fig. 3. Mirror images of a starch gel stained for phosphopentomutase (PPM) activity (left gel) and phosphoglucomutase (PGM) activity (right gel). Anode is at the top. Channel 1: human fibroblasts; Channels 2 and 4: somatic cell hybrid lines with weak human PPM activity; Channels 3 and 5: somatic cell hybrid lines not expressing human PPM activity; Channel 6: hamster fibroblasts. The hamster fibroblasts have an additional set of PGM isozymes which migrate more anodally than the isozymes shown.

al., 1970a), thermal stability (McAlpine et al., 1970b), relative activities in different tissues (McAlpine et al., 1970c), and substrate specificity (Quick et al., 1972) the possibility that one of these characteristics could be used to distinguish the human PGM_2 components from the hamster PGM isozymes was considered. Of these characteristics, substrate specificity was investigated first as the technique involved appeared to be the most expedient for analyzing large numbers of somatic cell hybrids.

The isozymes associated with each of the three *PGM* loci in man have catalytic activity when glucose-1-phosphate is used as substrate but only the components related to the PGM_2 locus have catalytic activity toward ribose-1-phosphate (Quick et al., 1972). Thus, in addition to having phosphoglucomutase activity, the human PGM_2 isozymes also have phosphopentomutase (PPM) activity, as evidenced by their ability to effect the catalytic interconversion of ribose-1-phosphate and ribose-5-phosphate. When lysates of somatic cell hybrids are examined by starch gel electrophoresis for their phosphopentomutase activity, the human and hamster isozymes can be distinguished quite readily on the basis of their electrophoretic mobility. Usually only one band of PPM activity corresponding to isozyme *e*, the most cathodally migrating member of the isozyme set associated with the PGM_2^1 allele in man, is observed in human fibroblasts. The hamster pattern consistently shows a single isozyme migrating more anodally than the human PPM isozyme, and on occasion, a second isozyme, migrating more cathodally than the usual hamster PPM isozyme, may be observed. Somatic cell hybrids which express the human *PGM*$_2$ gene locus show a single human PPM isozyme together with one or both of the hamster PPM isozymes (Fig. 3). This pattern resembles that expected from a superimposition of the human PPM isozyme pattern on the hamster PPM pattern. The absence of a heteropolymer with PPM activity in somatic cell hybrids which express the human *PGM*$_2$ gene locus is consistent with the previous conclusion of Hopkinson and Harris (1966) that each of the PGM isozymes in man consists of a single polypeptide chain.

Table 2 summarizes the analysis of PPM (PGM_2) isozyme patterns in sixteen independently derived somatic cell hybrid lines together with the human chromosome complements of these cell lines. 4 and X are the only human chromosomes present in every cell line which expresses the human *PGM*$_2$ gene locus. As the *PGM*$_2$ gene locus was not expressed when chromosome 4 was absent although X was present, and family studies have indicated that this gene locus is on an autosome (Hopkinson and Harris, 1965 and 1966); these data allow the *PGM*$_2$ gene locus in man to be assigned to chromosome 4. Human PPM was **not detected** in one cell line in which chromosome 4 was present in only 22% of the cells examined, presumably because the amount of human PPM enzyme synthesized by this cell line was not sufficient to be visualized in the gels. Table 3 presents the correlation of the expression of the human *PGM*$_2$ gene locus with the presence of chromosome 4 in these sixteen somatic cell lines, together with the results of the analysis of seven secondary clones and eleven tertiary clones derived from one of these cell lines. In one of the cell lines which expressed the human *PGM*$_2$ gene locus, a presumably spontaneous chromosome break had occurred at (4)(q26) resulting in the deletion of the portion of the long arm of chromosome 4 distal to 4q26 (Fig. 4). Therefore, as the *PGM*$_2$ gene locus can be excluded from the (4)(q26→qter) region, it must be elsewhere on the chromosome, i.e. on the short arm of chromosome 4 or on the long arm proximal to the q26 region.

A CHROMOSOME 6 MARKER

The use of the substrate glucose-1-phosphate allows the isozymes related to the *PGM*$_3$ gene locus in man to be identified in PGM isozyme patterns of somatic cell hybrids (Fig. 1). Under these conditions of analysis usually only the most cathodally migrating member of the isozyme sets associated with either the *PGM*$_3^1$ or with the *PGM*$_3^2$ allele in man can be

TABLE 2

ASSIGNMENT OF STRUCTURAL GENE FOR PHOSPHOGLUCOMUTASE$_2$ (PGM_2) TO CHROMOSOME 4 IN MAN

Hybrid Cell Line	Human PGM_2	1	2	3	4	5	6	7	8	9	10	11	12	13	14	15	16	17	18	19	20	21	22	X	Y
1610-09	+	+	+	+	+	+	+	+	+	+	+	+	+	-	+	(+)	(+)	-	-	+	+	+	+	+	+
4105	+	-	-	-	+[1]	-	-	-	-	-	-	-	+*	+*	+	+	-	-	+	-	-	-	-	+*	-
4109	+	-	-	+	+	+	-	+	+	+	+	+	+	-	-	+	+	+	-	-	+	-	+	+	-
4110	+	-	-	+	+	-	+	-	-	-	+	-	-	+	+	+	+	-	-	+	+	+	+	+	-
4113	+	+	+	+	+	+	+	+	-	+	-	+	+	+	+	-	+	+	+	+	+	+	+	+	-
4117	+	-	-	+	+	+	-	+	-	-	+	-	+	+	+	-	-	-	+	-	+	-	+	+	-
4119	+	-	-	+*	+	-	-	-	+	-	-	-	-	+	-	-	-	+	-	+	-	+	+	+	-
4122	+	+	+	+	+	+	+	+	+	+	+	+	+	+	+	-	+	+	+	+	-	-	+	+	-
1710	-	-	-	-	(+)	-	(+)	-	-	-	+	+	-	+	-	+	+	(+)	(+)	-	(+)	+	-	+	-
1705	-	-	-	-	-	+	-	-	-	-	-	-	+	-	-	-	-	-	+	-	-	-	-	+	-
1781	-	+	-	+	-	-	-	-	-	-	-	+	+	-	+	+	(+)	-	-	+	-	-	+	+	-
2911	-	-	-	+	-	+	+	+	+	+	-	+	-	-	+	-	+	+	-	+	-	+	+	+	-
4106	-	-	-	+	-	+	-	-	-	+	+	-	-	+	+	+	+	+	-	+	+	-	+	+	-
4111	-	+	-	+	-	+	+	+	+	-	-	+	+	+	+	-	+	+	+	+	+	+	+	+	-
4116	-	+	-	+	-	+	-	-	+	+	-	+	+	-	-	+	+	+	-	-	+	+	+	+	-
4120	-	-	-	-	-	-	-	-	-	-	-	-	-	-	-	-	-	-	-	-	-	-	-	+*	-

\+ Enzyme present or chromosome present in>25% of cells analyzed

\- Enzyme absent or chromosome absent or present in <10% of cells analyzed

(+) Chromosome present in 10-25% of cells analyzed

1 (4) (q26→qter) region deleted

* Chromosomes involved in structural rearrangements

TABLE 3

CORRELATION OF THE EXPRESSION OF HUMAN PGM_2 GENE LOCUS WITH THE PRESENCE OF HUMAN CHROMOSOME 4

Source of lysate	Expression of human PGM_2 gene locus	Human chromosome 4	Number
Hybrid Cell Lines	+	+*	8
	-	(+)	1
	-	-	7
Secondary Clones	+	+	2
	-	-	5
Tertiary Clones	+	+	11
	-	-	-

* : (4) (q26→qter) region deleted in one cell line
\+ : human gene locus expressed or human chromosome present
\- : human gene locus not expressed or human chromosome absent
(+) : human chromosome present in a low proportion of cells analysed

visualized in the gels. A human PGM_3 isozyme was detected in four somatic cell hybrid lines and its presence in these lines correlated positively only with the presence of human chromosome 6. Seven of the cell lines examined did not express the human PGM_3 gene locus and had lost chromosome 6 from man (Table 4). These data, therefore, confirm the previously tentative assignment of the PGM_3 gene locus in man to chromosome 6 (Jongsma et al., 1973, Pearson et al., 1973). The isozyme patterns observed indicate that the PGM_3 isozymes in man are single polypeptides.

A CHROMOSOME 10 MARKER

Glutamic oxaloacetic transaminase (GOT), also referred to as aspartate aminotransferase (E.C.2.6.1.1.), from human

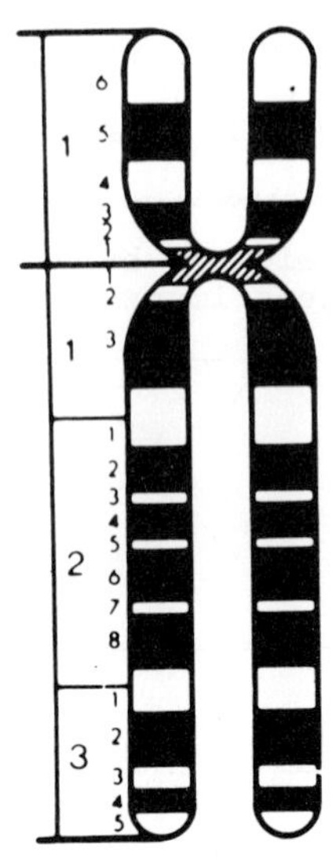

Fig. 4. Diagrammatic representation of banding patterns of human chromosome 4. (After Paris Conference 1971).

tissues exists in two forms. The more cathodally migrating components are the mitochondrial form of this enzyme while the more anodally migrating isozymes represent the cytoplasmic form of GOT (Nisselbaum, 1965 and Nisselbaum and Bodansky 1965). Family studies indicate that the cytoplasmic GOT isozymes are related to a single gene locus which has been designated as *sGOT* (Chen and Giblett, 1971) or as GOT_1.

Analysis of GOT$_1$ isozyme patterns in seventeen of our somatic cell hybrid lines obtained after electrophoretic examination using Cellogel indicated that when the human GOT_1 gene locus was expressed, a single GOT heteropolymer, presumably consisting of one human and one hamster subunit, was present in addition to a hamster GOT homopolymer and, when it could be visualized, a human GOT homopolymer. These isozyme patterns indicating that GOT$_1$ has a dimeric molecular structure, are consistent with those observed in hemolysates from individuals who are heterozygous at the GOT_1 gene locus (Chen and Giblett, 1971). Comparison of the human chromosome complements of the seven cell lines expressing the human GOT_1 gene locus with the seven cell lines in which no evidence of human GOT$_1$ was observed indicated that the expression of the human GOT_1 locus could be correlated positively only with the presence of human chromosome 10 (Table 5).

Using Cellogel, the electrophoretic analysis of

TABLE 4

CORRELATION OF THE EXPRESSION OF THE HUMAN *PGM_3* GENE LOCUS AND THE PRESENCE OF HUMAN CHROMOSOME 6

Expression of human *PGM_3* gene locus	Presence of human chromosome 6	Number of independent hybrid cell lines
+	+	4
-	-	6

+ : gene locus expressed or human chromosome present

- : gene locus not expressed or human chromosome absent

hexokinase (HK) in human and hamster fibroblasts revealed two sets of isozymes in each species. The more cathodal components, which had relatively more enzyme activity are presumed to be high Km forms of this enzyme while the more anodal components were very difficult to visualize in the gels and may be low Km forms of hexokinase. HK isozyme patterns were analyzed in ten of the seventeen cell lines which were examined for *GOT_1*. Five of the cell lines which expressed the human *GOT_1* gene locus also expressed this human high Km form of HK and in these cell lines chromosome 10 was the only human chromosome which was consistently present. All of the remaining five cell lines which did not express the human *HK* or *GOT_1* gene loci had lost chromosome 10 (Table 5). Thus, the concordant segregation of the *HK* and *GOT_1* human gene loci indicates that these two loci are syntenic, i.e. on the same chromosome, and the positive correlation of their expression with the presence of chromosome 10 from man confirm the previously reported (Creagan et al., 1973, Shows, 1973) chromosome assignment of these two loci using human-mouse somatic cell hybrids.

A CHROMOSOME 21 MARKER

Two different forms of superoxide dismutase (SOD) or

TABLE 5

CORRELATION OF THE EXPRESSION OF THE GOT_1 AND *HK* HUMAN GENE LOCI WITH THE PRESENCE OF HUMAN CHROMOSOME 10

Expression of human gene loci for		Human chromosome 10	Number of hybrid cell lines
GOT_1	*HK*		
+	+	+	5
+	NA	+	2
-	-	-	5
-	NA	-	5

+ : human gene locus expressed or human chromosome present

- : human gene locus not expressed or human chromosome absent

NA: not analyzed

indophenol oxidase as it was previously called, occur in human tissues and are related to separate and distinct gene loci (Brewer, 1967). The more cathodally migrating SOD isozymes have a tetrameric structure and are related to the *SOD B* (formerly *SOD 2* and *IPO B*) gene locus while the more anodal components exist as dimers and are related to the *SOD A* (*SOD 1*, *IPO A*) gene locus.

Electrophoretic examination of lysates of eighteen independently derived somatic cell hybrid lines revealed that a human-hamster SOD dimer, indicating the expression of the human *SOD A* gene locus, was present in eight of these cell lines. Correlation of the expression of the human *SOD A* with the chromosome complements of these cell lines indicated that human dimeric *SOD* gene locus was expressed only in those cell

TABLE 6

CORRELATION OF THE EXPRESSION OF THE HUMAN *SOD A* GENE LOCUS WITH THE PRESENCE OF HUMAN CHROMOSOME 21

Expression of human *SOD A* gene locus	Human chromosome 21	Number of independent hybrid cell lines
+	+	8
-	-	10

\+ : human gene locus expressed or human chromosome present

\- : human gene locus not expressed or human chromosome absent

lines which had retained chromosome 21 and was not expressed when chromosome 21 had been lost (Table 6). These data provide additional confirmation of the assignment (Tan et al., 1973, Creagan et al., 1973) of the gene locus for the dimeric form of *SOD* to chromosome 21 in man.

SUMMARY

The application of isozyme analysis of somatic cell hybrids to gene assignments and to determining the molecular structure of enzymes has been described. The expression of both human fumarate hydratase (FH) and guanylate kinase (GuK) in somatic cell hybrids was found to be dependent upon the presence of chromosome 1 from man. FH appears to be a tetramer composed of four identical subunits while GuK may exist as a monomer. Additonal evidence for the monomeric structure of the phosphoglucomutase$_1$ (PGM_1) and peptidase C (pep C) isozymes and for the dimeric nature of 6-phosphogluconate dehydrogenase (PGD) was obtained. The isozymes related to the PGM_2 gene locus in man were found to be monomeric and the PGM_2 locus has been assigned to the (4) (pter→q26) region of chromosome 4. Further evidence for the dimeric structure of the cytoplasmic form of glutamic oxaloacetic transaminase was obtained. The tentative assignments of the PGM_3 gene locus to chromosome 6, the gene loci for the cytoplasmic form of GOT (*GOT_1*), and a high Km form of hexokinase expressed in fibroblasts to chromosome 10, and the gene locus for the dimeric form of superoxide dismutase (*SOD B*) to chromosome 21 have been confirmed.

ACKNOWLEDGEMENTS

The authors gratefully acknowledge the expert technical assistance provided by Ms. E. Hosea, Ms. L. Komarnicki, Ms. H. Maledy, Miss V. Niewczas-Late and Miss A. Vust. This research was supported by the Medical Research Council of Canada (Grant MA-4061 to J.L.H.) and by The Children's Hospital of Winnipeg Research Foundation Inc. T.M. holds a postdoctoral fellowship from the Medical Research Council of Canada. P.J.M. is supported by The Children's Hospital of Winnipeg Research Foundation Inc.

REFERENCES

Brewer, G. 1967. Achromatic regions of tetrazolium stained starch gels: Inherited electrophoretic variation. *Amer. J. Hum. Genet.* 19: 674-680.

Burgerhout, W., H. van Someren ,and D. Bootsma 1973. Cytological mapping of the genes assigned to the human A1 chromosome by use of radiation - induced chromosome breakage in a human-Chinese hamster hybrid cell line. *Humangenetik* 20: 159-162.

Burgerhout, W. G., A. P. M. Jongsma ,and P. Meera Khan 1974. Regional assignments of seven enzyme loci on chromosome 1 of man. *Cytogenet. Cell Genet.* 13: 73-75.

Chen, S.-H. and E. R. Giblett 1971. Genetic variation of soluble glutamic-oxaloacetic transaminase in man. *Amer. J. Hum. Genet.* 23: 419-424.

Creagan, R., J. Tischfield, F. A. McMorris, S. Chen, M. Hirschi, T.-R. Chen, F. Ricciuti ,and F. H. Ruddle 1973. Assignment of the genes for human peptidase A to chromosome 18 and cytoplasmic glutamic oxaloacetate transaminase to chromosome 10 using somatic-cell hybrids. *Cytogenet. Cell Genet.* 12: 187-198.

Creagan, R., J. Tischfield, F. Ricciuti, and F. H. Ruddle 1973. Chromosome assignments of genes in man using mouse-human somatic cell hybrids: mitochondrial superoxide dismutase (indophenoloxidase-B tetrameric) to chromosome 6. *Humanagenetik* 20: 203-210.

Douglas, G. R., P. A. Gee ,and J. L. Hamerton 1973a. Chromosome identification in Chinese hamster/human somatic cell hybrids. Nobel Symposium 23: 170-176.

Douglas, G. R., P. J. McAlpine, and J. L. Hamerton 1973b. Regional localization of loci for human PGM_1 and 6*PGD* on human chromosome one by use of hybrids of Chinese hamster-human cells. *Proc. Natl. Acad. Sci. U.S.A.* 70:2737-2740.

Gee, P. A., M. Ray, T. Mohandas, G. R. Douglas, H. A. Palser, B. J. Richardson, and J. L. Hamerton 1974. Characteristics of an HPRT deficient Chinese hamster cell line. *Cytogenet. Cell Genet.* (in press).

Hamerton, J. L., G. R. Douglas, P. A. Gee, and B. J. Richardson 1973. The association of glucose phosphate isomerase expression with human chromosome 19 using somatic cell hybrids. *Cytogenet. Cell Genet.* 12: 128-135.

Hopkinson, D. A. and H. Harris 1965. Evidence for a second 'structural' locus determining human phosphoglucomutase. *Nature* 208: 410-412.

Hopkinson, D. A. and H. Harris 1966. Rare phosphoglucomutase phenotypes. *Ann. Hum. Genet.*, London. 30: 167-181.

Hopkinson, D. A. and H. Harris 1968. A third phosphoglucomutase locus in man. *Ann. Hum. Genet.*, Lond. 31: 359-367.

Jongsma, A., H. Van Someren, A. Westerveld, A. Hagemeijer, and P. Pearson 1973. Localization of genes on human chromosomes using human-Chinese hamster somatic cell hybrids. Assignment of PGM_3 to chromosome C6 and regional mapping of the PGD, PGM_1 and Pep-C genes on chromosome A1. *Humangenetik* 20: 195-202.

Kanarek, L., E. Marler, R. A. Bradshaw, R. E. Fellows, and R. L. Hill 1964. The subunits of fumarase. *J. Biol. Chem.* 239: 4207-4211.

McAlpine, P. J., D. A. Hopkinson, and H. Harris 1970a. Molecular size estimates of the human phosphoglucomutase isozymes by gel filtration chromatography. *Ann. Hum. Genet.*, Lond. 34: 177-185.

McAlpine, P. J., D. A. Hopkinson, and H. Harris 1970b. Thermostability studies on the isozymes of human phosphoglucomutase. *Ann. Hum. Genet.*, Lond. 34: 61-71.

McAlpine, P. J., D. A. Hopkinson, and H. Harris 1970c. The relative activities attributable to the three phosphoglucomutase loci (PGM_1, PGM_2, PGM_3) in human tissues. *Ann. Hum. Genet.*, Lond. 34: 169-175.

McAlpine, P. J., L. Kormanicki, and H. Maledy 1974. Detection of human gene products in somatic cell hybrids using electrophoretic methods. (In preparation).

Meera Khan, P., B. A. Doppert, A. Hagemeijer and A. Westerveld 1974. The human loci for phosphopyruvate hydratase and guanylate kinase are syntenic with the PGD-PGM_1 linkage group in man-Chinese hamster somatic cell hybrids. *Cytogenet. Cell Genet.* 13: 130-131.

Monn, E. and R. O. Christiansen 1972. Guanylate kinase in man-multiple molecular forms. *Human Heredity* 22: 18-27.

Nisselbaum, J. S. 1965. Erythrocyte glutamic-oxaloacetic transaminase. *Fed. Proc.* 24: 356.

Nisselbaum, J. S. and O. Bodansky 1965. Glutamic-oxaloacetic transaminase in reticulocytes and erythrocytes. *Science* 149: 195-197.

Paris Conference 1971. Standardization in Human Cytogenetics. *Birth Defects: Original Article* Series VIII No. 7. The National Foundation, New York.

Parr, C. W. 1966. Erythrocyte phosphogluconate dehydrogenase polymorphism. *Nature* 210: 487-489.

Parrington, J. M., G. Cruickshank, D. A. Hopkinson, E. R. Robson, and H. Harris 1968. Linkage relationships between the three phosphoglucomutase loci PGM_1, PGM_2 and PGM_3. *Ann. Hum. Genet.*, Lond. 32: 27-34.

Pearson, P. 1972. The identification of chromosomes in hybrid cells. *Bull. Europ. Soc. Hum. Genet.* (Nov. 1972) p. 54.

Penner, P. E. and L. H. Cohen 1971. Fumarase: Demonstration, separation and hybridization of different subunit types. *J. Biol. Chem.* 246: 4261-4265.

Povey, S., G. Corney, W. H. P. Lewis, E. B. Robson, J. M. Parrington and H. Harris 1972. The genetics of peptidase C in man. *Ann. Hum. Genet.*, Lond. 35: 455-465.

Quick, C. B., R. A. Fisher, and H. Harris 1972. Differentiation of the PGM_2 locus isozymes from those of PGM_1 and PGM_3 in terms of phosphopentomutase activity. *Ann. Hum. Genet.*, Lond. 35: 445-454.

Robson, E. R., P. J. L. Cook, G. Corney, D. A. Hopkinson, J. Noades, and T. E. Cleghorn 1973. Linkage data on *Rh*, PGM_1, *PGD*, *Peptidase C* and *Fy* from family studies. *Ann. Hum. Genet.*, Lond. 36: 393-399.

Santachiara Benerecetti, S. A. 1970. Studies in African pygmies. III. Peptidase C polymorphism in Babinga pygmies; a frequent erythrocytic enzyme deficiency. *Amer. J. Hum. Genet.* 22: 228-231.

Shows, T. 1974. Synteny of human genes for glutamic oxaloacetic transaminase and hexokinase in somatic cell hybrids. *Cytogenet. Cell Genet.* 13: 143-145.

Spencer, N. A., D. A. Hopkinson, and H. Harris 1964. Phosphoglucomutase polymorphism in man. *Nature* 204: 742-745.

Tan, Y. H., J. Tischfield, and F. H. Ruddle 1973. The linkage of genes for the human interferon-induced antiviral protein and indolphenol oxidase-B traits to chromosome G-21. *J. Exp. Med.* 137: 317-330.

Van Cong, N., C. Billardon, J.-Y. Picard, J. Feingold, and J. Frezél 1971. Liason probable (linkage) entre les locus

PGM_1 et peptidase C chez l'homme. *C. R. Acad. Sc.* Paris 272: 485-487.

Van Someren, H., H. Beyersbergen van Henegouwen, and J. de Wit 1974. Evidence for synteny between the human loci for fumarate hydratase, UDP glucose pyrophosphorylase, 6-phosphogluconate dehydrogenase, phosphoglucomutase, and peptidase-C in man-Chinese hamster somatic cell hybrids. *Cytogenet. Cell Genet.* 13: 150-152.

Westerveld, A. and P. Meera Khan 1972. Evidence for linkage between human loci for 6-phosphogluconate dehydrogenase and $phosphoglucomutase_1$ in man-Chinese hamster somatic cell hybrids. *Nature* 236: 30-32.

Woodfin, B. M. 1975. A new purification procedure for pig heart fumarase. *Isozymes I. Molecular Structure,* C. L. Markert, editor. Academic Press, New York. pp. 797-806.

ISOZYME VARIANTS AS MARKERS OF POPULATION MOVEMENT IN MAN

R.L. KIRK
Department of Human Biology
The John Curtin School of Medical Research
Canberra ACT Australia 2601

ABSTRACT. Examples have been selected from three isozyme systems in which genetically controlled variants can be used to demonstrate migration of human populations.
(1) Variants of superoxide dismutase (SOD) occur with frequencies of 1 per 1,000 or less except in a few exceptional populations. Variants are widespread in Finland where the variant gene frequency is about 1 per cent, and reaches 5 per cent in the north of Sweden. The gene is polymorphic on the Island of Westray off Northeast Scotland, and probably identical variants have been described in Europe and England and among white Australians. The same variant has been found also in some villages south of the Caspian Sea in Iran. The distribution of this SOD variant suggests that it was distributed across Europe by Vikings and has been further disseminated by their descendants.
(2) A unique LDH variant 'Calcutta-1' is widespread in India and occurs with a frequency ranging from under 1 to 4 percent. It occurs in all caste groups and in tribal populations and may well have been introduced into India by the early Aryan invaders.
(3) The distribution of several phosphoglucomutase (PGM) alleles in Southeast Asia and the Western Pacific is consistent with hypotheses about the colonization of Micronesia and Melanesia based on archaeological and linguistic evidence.

The examples given indicate that specific isozyme variants can be used to complement other information on population movements.

Wherever Man stops on his wanderings around the world the genes which he leaves behind reflect not only the hospitality, sometimes enforced, that he has received but also the homeland from where he came. For example, in Holland one can trace the families of present-day patients with Huntington's chorea to the towns and villages along the Rhine: undoubtedly some sailor of Scottish extraction plied more than material goods along the waterway. Similarly, de Beer in a scholarly essay (de Beer, 1965) showed that one could use the pattern of genes in the south of France and along the Rhone to trace the influence of the Roman conquest on the living populations. More recently a book on *Genetic Variation in Britain* (Roberts and Sunderland, 1973) in a series of papers has explored some of the historical and proto-historical reasons for the local variation in gene frequencies across the British Isles.

During the past decade the development of techniques for readily detecting genetically controlled electrophoretic variants of isozymes in human tissues has made available a vastly increased number of specific markers which can be used in such studies. The present essay gives three examples from separate systems and for separate parts of the world.

1. *Superoxide Dismutase*

Brewer (1967) detected the first genetic variant of superoxide dismutase (tetrazolium oxidase: indophenol oxidase), and he described several examples of heterozygotes for this variant in a single family in the United States. During the intervening 7 years relatively few other examples of SOD variants have been reported despite intensive surveys of populations in nearly all parts of the world. Shinoda (1970) found three examples of SOD heterozygotes in a Japanese family, and the variant was said to be similar to that of Brewer. Ritter and Wendt (1971) in a survey of 5,100 persons in Germany found 6 heterozygous phenotypes and their paper implied they were all similar. At this stage it appeared therefore that the 'oxidase' activity detectable in the lysates from human red cells was genetically unusually stable, but that rare variant families occurred in different parts of the world. At this point, Welch and Mears (1972) found 12 heterozygotes in just under 400 persons sampled on the Island of Westray, in the Orkneys off the northeast coast of Scotland, and they speculated that since the Island of Westray had been colonized from Scandinavia at an earlier period other examples of this variant might be found in Scandinavia itself. This prediction has been amply fulfilled. Gunhild Beckman (1973) surveyed more than 4,400 persons in north Sweden and found 48 heterozygotes in the counties of Norrbotten and Vasterbotten, close to the border with Finland. Later that year Beckman et al. (1973) reported on further studies in the north of Sweden and in Finland. She found the highest frequency of the variant gene $\underline{SOD}^2$ was 2.5 per cent in Tornedalen, between Sweden and Finland, while the frequency for north Sweden in general was 0.4 per cent and in Finland around 1.0 per cent. She reported in the same paper the first two homozygotes in man for a variant oxidase allele. It appears therefore that the SOD[2] gene is widely distributed in northern Scandinavia (Beckman and Pakarinen, 1973). In a more recent paper Harris et al. (1974) have reported 7 SOD variants in 11,000 persons examined in the United Kingdom. These variants are the same as those found in Scandinavia (Harris: personal communication).

Our own interest was directed to this system some two years ago when two separate oxidase variants were detected in our laboratories on the same day. These two samples were from white Australians and were the first such variants we had observed from more than 10,000 persons examined. Since then we have found one other example in a white Australian, and several examples from persons around the southern shores of the Caspian sea.

Recently, through the kindness of Dr. Welch and Dr. Gunhild Beckman we have been able to compare their variants with our own. We find the Swedish, Orkney, Caspian and 2 of the 3 white Australian samples to be identical in band mobility and relative band intensities. The other white Australian sample has not been adequately studied, but unfortunately the propositus is now dead. It is however, quite distinctly different from the other variants mentioned above.

The historical significance of the distribution of the SOD[2] gene is our main focus of interest here. The links between the Orkney Islands and Scandinavia have been referred to already, and the historical records attest to the occupancy of the Orkneys by Scandinvaians over a considerable period of time. Recording this expansionist phase of the Nordic peoples, Treece (1962) writes that in the 9th century A.D. the Vikings were moving through Europe having been driven from their own Scandinavian lands by a shift in the feeding grounds of the herrings and also by the system of splitting up land between all the sons after the death of the father. In a polygamous society such a practice left too little land for survival and they set forth to find new homes. Their dreams and stories abound with references to the vast rich place their heroes find - 'Miklagard' or 'The great city'. "Some of them" continues Treece "sought this Miklagard in England, Scotland, Ireland, France, some in Iceland and even Greenland. Still others, taking the overland route southward down the rivers of Europe came at last to Constantinople and there indeed they found their gold and their palaces so commonly that they even confused this city with Asgard, the home of the gods. The Saracens saw them appear on the Caspian Sea and gave them the name

'Russians'."

There is evidence that they married local girls and settled and there is at least one contemporary account of these Nordic 'Russians' or 'Varangians' as they were termed by the Greeks. Ibn Fadhlan an Arab missionary wrote in 922 A.D. "I have seen the Rusiya when they came hither on their trading voyages and had encamped by the river Atil. . . They are tall as date palms, blond and ruddy so that they do not need to wear either a qurtag nor a kaftan" . . . "They come from their own country, moor their barks on the strand of the Atil, which is a great river, and build on its bank great houses out of wood. In a house like this ten or twenty people more or less, live together. Each of them has a resting bench wheron he sits and with them are the fair maidens (slave girls) who are destined for sale to the merchants and they may have intercourse with their maiden while their comrades look on." (Quoted by Coon, 1948). There is dispute about the origin of the term Rusiya, some authorities claiming that it is derived from the Finnish *Ruotsi*, others that the Rusiya came from Roslagen in the area around the modern Swedish city of Uppsala (Wilson, 1970). The river Atil mentioned in Ibn Fahdil's account, however, undoubtedly is the Volga, and the city of Atil stands near the point where the great waterway used by the Scandinavian traders enters the Caspian sea. Whether the Rusiya were Vikings or Finns it is easy to see that the $\underline{SOD}^2$ gene found in the Caspian littoral today spread from a focus in the far north of Europe at least 1,000 years ago. Similarly the Viking adventures across the North Sea and down the Vistula, and Oder into central Europe accounts for the widespread dispersal of this gene in other European groups, a dispersal which is illustrated schematically in Figure 1.

The two isolated examples in Australia present a more difficult problem. In neither case have we been able to trace the genealogies backwards in time. Our only clue rests on the family names: in one case Spence, and in the other Randall both of which are names which occur in Scottish records as far back as the 13th century. Though not confined to Scotland it is not improbable that the Viking presence made its way into the genetic constitution of the Spences and Randalls, as well as into many other families in Scotland at a time when the Vikings were exploring, conquering, or settling around the British Isles.

2. *Lactate Dehydrogenase*

Moving south from the shores of the Caspian Sea following

SPREAD OF SCANDINAVIAN SOD^2 GENE

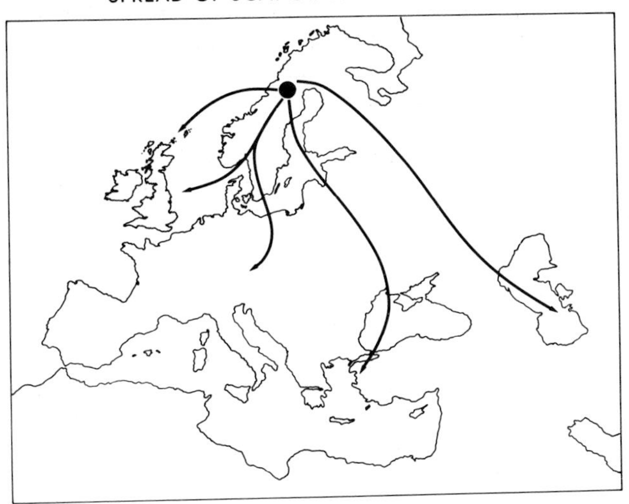

Figure 1

the caravan routes through to Afghanistan and finally across the passes of the Himalayas into the fertile plains of northern India one is confronted with a subcontinent teeming with people whose variety and origins, whose social structure and range of environmental adaptations arouse the greatest interest in the human population geneticist.

Who were the pre-Dravidians and their Dravidian-speaking successors? How great an impact on the populations of various parts of the sub-continent was made in the biological sense by the Aryan invaders or in later times by their Muslim successors? What effect on the genetic structure of populations has resulted from the system of Varnas and endogamous castes, and what are the differences, in genetic terms, between the areas of high frequency of cousin marriage in the south and effectively zero cousin marriage in the north? As a tool to help probe for answers to some of these questions we have been using an isozyme variation which we discovered some five years ago in Calcutta. It is a heterozygote state for two alleles at the A locus of lactate dehydrogenase and which we called 'Calcutta-1' (Das et al., 1970).

The first genetic variant of LDH in man was discovered in an itinerant Yoruba peddlar in West Africa (Boyer et al., 1963) and variants at both the A and B loci were described by a number of workers, the highest combined frequency approaching 1 per cent in Black Americans (Kraus and Neely, 1964). Variants in most populations are reported sporadically: for example in Caucasians, in tests of over 4,000 randomly sampled persons in the United States, England, Iceland, and Sweden only 7 persons with A variants and 1 person with a B variant were detected (Vessell, 1965; Mourant et al., 1968). Exception to this low frequency was the discovery of 6 persons with an A sub-unit variation in 406 members of the Murapin clan of the Enga, in the Western Highlands of Papua New Guinea (Blake et al., 1969). This variant however, is very localized, and we have now examined many thousands of persons from many parts of New Guinea without finding another example of an LDH variant.

The situation in India is very different. OUr first report listed 10 'Calcutta-1' variants in 614 Bengalis living in Calcutta, a variant frequency of 1.63 per cent. Since then our Indian collaborators and ourselves have tested populations from many parts of India. The results are summarized in Figure 2.

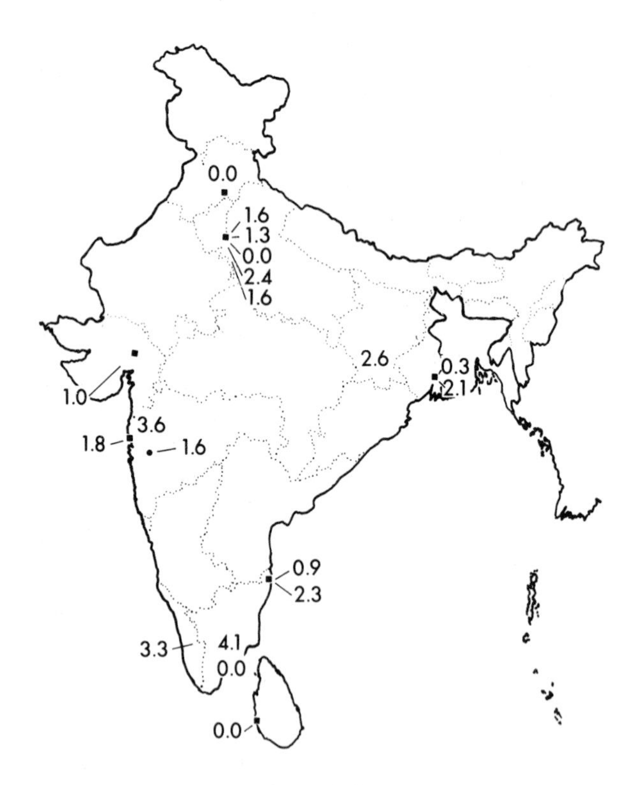

Figure 2

In a few of the series no variants were detected, but in the great majority 'Calcutta-1' was present with frequencies ranging from under 1 percent to a maximum of 4 per cent.

What are the implications of these figures for the student of population movements in India? Our second series of studies in Bengal (Das et al., 1972) suggested that persons from lowly castes had a higher frequency of 'Calcutta-1' than did persons from high castes. We made the hypothesis therefore that the gene $\underline{LDH}^{Cal-1}$ had arisen as a mutant allele in tribal populations predating the Aryan invasions, and that the incorporation of tribal elements into Hindu society had introduced the gene into low castes, and gradually with the passage of time it had permeated even into high caste Brahmin groups. Such a pattern of gene flow would be contrary of course to what happened for another specific marker, HbS, which achieves very high frequencies in many tribal populations in India, but which occurs in Hindu populations only rarely (Saha and Banerjee, 1974). However, we or our Indian colleagues have looked at several tribal groups and some populations like the Kaora in Bengal which are probably derived from tribal people. LDH 'Calcutta-1' is present in these groups but not in greater frequency than in many other Indian populations. By contrast one of the highest frequencies has been observed among the Parsis in Bombay.

The Parsis are an endogamous group living now mainly in Maharastra State, of which Bombay is the capital. They came first to India some 1200 years ago to escape the Muslim persecution of their Zoroastrian faith, and they have a remarkable collection of genetic specialties in the living population, included among these being homozygotes for the Rh allele $\underline{r}^{Y}$, a high incidence of G6PD deficiency, and some cases of HbS (Undevia et al., 1969). Since the Zoroastrians came from Persia, maybe'Calcutta-1' came in at an earlier date with the Aryans, or with the later, almost legendary arrival of Alexander the Great. Since the genetic impact of the latter on Indian populations is likely to have been restricted to the northwest it seems unlikely that Alexander or his followers could have led to such a widespread distribution but maybe Persia or central Asia could have been the focus from which the gene spread, firstly into the north and then as the priests took their beliefs and their genes further south, to the rest of the country.

There is another group of migrants from Persia living in Maharastra. They are the Iranis who arrived within the last two hundred years. We have sampled a mere 48 without finding a single 'Calcutta-1' variant. However, in testing samples in our own series from the Caspian littoral we have found one example of 'Calcutta-1', but only 1, in more than 1,000 persons tested (unpublished observations). This suggests that the $\underline{LDH}^{Cal-1}$ gene may be present outside India in the Middle-East but the search for its center of distribution has so far not been completely successful.

3. Phosphoglucomutase

For my third example I want now to turn attention to an area for which there is no written history until very recent times, and yet despite this deficiency presents problems of great interest to the student of human population migration, integration and assimilation. The Western Pacific is peopled by many groups with differing languages and customs. Broadly they can be classified as Melanesians, inhabiting New Guinea, the Solomon Islands and places across to Fiji; Micronesians who inhabit a multitude of islands in the Marianas, Carolines, Marshalls and Gilberts, straddling the equator and covering an area of immense size as ocean, but very small as land; and finally Polynesians who represent the outer fringes of an even greater dispersal of peoples across the remaining parts of the Pacific Ocean to the east and north and southwest to include New Zealand.

In the absence of written history we can attempt to plot its course only by reference to the artifacts recovered from archaeological sites together with reconstructions based on linguistic differentiation and the study of the human biology of living populations. Howells (1973) recently has summarized the evidence which suggests that Melanesia was settled first by a population whose origins remain a mystery. These people, speaking Papuan languages, inhabited the area of present day New Guinea and Australia and possibly areas to the north in Indonesia, and Malaysia as long as 50,000 years ago. About 6,000 years ago, when Australia was already separated from New Guinea by rising seas at the end of the Pleistocene, there began a period of drastic change in Melanesia. People speaking Austronesian languages moved in from the west into the Solomons, New Hebrides and New Caledonia as far as Fiji, and touching along the coastal fringes of New Guinea. A similar thrust went more directly eastward through the Carolines to the Marshall, Gilbert, and Ellice Islands, but leaving permanent settlements only on the High Islands. By about 4,000 years ago some permanent settlements were occurring on atolls in eastern Micronesia, and during the next few hundred years some of these atoll-dwelling Micronesians were moving into eastern Melanesia and reaching as far as Fiji. These people were reef-fishers, with only limited agricultural interests. Differentiation in Fiji continued and by 3,000 years ago colonists from Fiji were moving into Tonga and to parts of the Solomons. During the next 1,000 years these movements of Austronesian speakers continued, reaching New Guinea and New Britain. In Micronesia the atoll-dwellers from eastern Micronesia started colonizing the Marshalls and Gilbert Islands. From 2,000 years ago onward the westward movement of the atoll-dwellers continued, and further Melanesian immigrants arrived by chance, although the Islands of Yap and Palau were not affected. A further movement to the east brought about the occupation of Polynesia. In turn the Polynesians sent people away in many directions, colonizing some of the areas back in the Western Pacific where Melanesian influence was not too strongly established.

The broad movements outlined in these hypotheses of Howells' should be reflected in the patterns of genes in the Western Pacific area, though the task of collecting and analyzing the material is immense. Our own laboratories, together with a limited number of others, have been able to assemble some results which do have a bearing on the interpretation given above. Among the enzyme systems studied, one which is proving of very great interest is phosphoglucomutase (PGM).

The isozymes of PGM are controlled by genes at 3 loci, but the gene products from only 2 of these loci are detectable in lysates of human red cells. Most of what is known therefore about the distribution of isozyme variants for PGM in human populations refers to the loci $\underline{PGM}_1$ and $\underline{PGM}_2$. The common alleles $\underline{PGM}_1^1$ and $\underline{PGM}_1^2$ at the first locus are polymorphic in all human populations, but in addition a number of rarer alleles have been described and two of these are of interest in the present context because they are of widespread occurrence in many parts of the Western Pacific and S.E. Asia. One of these alleles, $\underline{PGM}_1^3$ occurs at many localities in India, Malaysia, New Guinea, Western Carolines and Fiji and it has been reported also for Chinese and Japanese. It has not been detected however in Australia. Most of the frequencies for $\underline{PGM}_1^3$ are low, but in parts of New Guinea the frequency for this gene is around 10 per cent. The other of the two rarer alleles, $\underline{PGM}_1^7$ is completely absent in New Guinea and Fiji and also in Australia, but is present in the Western Carolines with frequencies up to 8 per cent, and has been reported sporadically in S.E. Asia, Japan and India(Figure 3).

Locus $\underline{PGM}_2$ is monomorphic in nearly all human populations

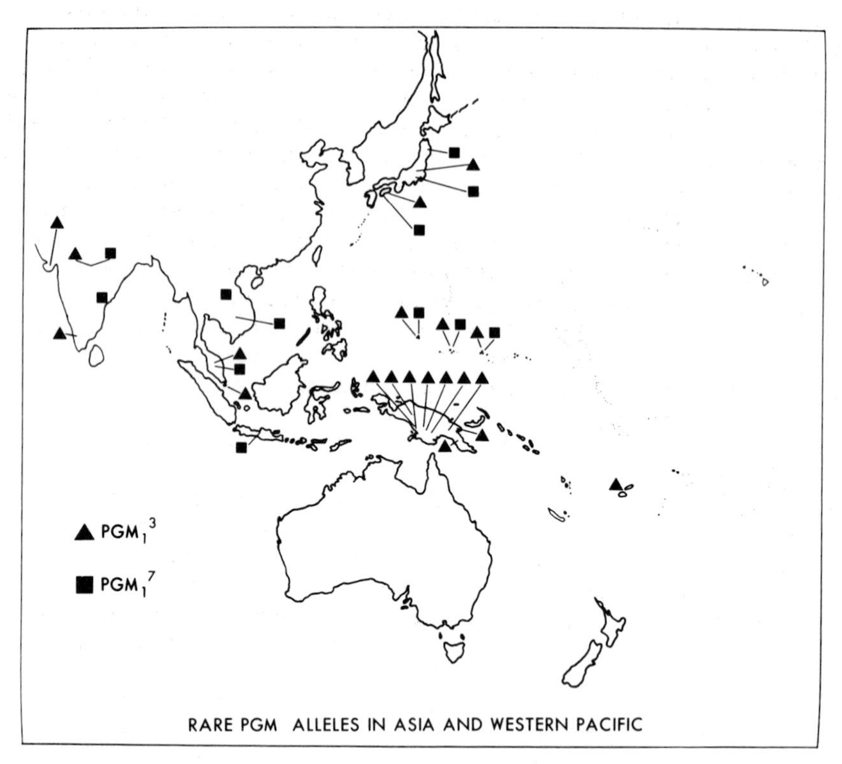

Figure 3

except Black Africans. Of special interest therefore is the occurrence of a widespread polymorphism for two rare alleles $\underline{PGM_2^9}$ and $\underline{PGM_2^{10}}$. These are found with varying frequency throughout the Island of New Guinea, with a tendency for a higher frequency of the alleles in highland populations but also with an appreciable frequency of both the $\underline{PGM_2^9}$ and $\underline{PGM_2^{10}}$ alleles in some of the coastal populations studied. Neither of these alleles occurs outside New Guinea, with the exception of a single person in N.E. Arnhem Land, Australia.

It seems probable, when reviewing this evidence for the pattern of distribution of rare PGM alleles, that both $\underline{PGM_1^3}$ and $\underline{PGM_1^7}$ were carried into Micronesia from S.E. Asia, with a separate wave moving along the New Guinea coast and out as far as Fiji with $\underline{PGM_1^3}$, but lacking $\underline{PGM_1^7}$. The locus 2 variants represent an older Melanesian population in New Guinea, but one not so old that it shared its genes with the Aboriginal populations of Australia. More recent arrivals in coastal areas of New Guinea have mixed with the older peoples in peripheral areas, confining the Papuan speakers mainly to Highland areas. This is reflected in the higher frequencies of $\underline{PGM}_2^9$ and $\underline{PGM}_2^{10}$ in the Highlands, the lower values around the coast representing dilution due to new migrants lacking these two older Melanesian genes.

The phosphoglucomutase alleles therefore support in broad terms Howells' model for the population movements through what is a region of great complexity in terms of human origins. As further study of the isozyme patterns in these and other parts of the world proceeds we can expect to find additional support or even contradictions for theories based on different types of study, and from these arrive at a better understanding of our own evolution.

REFERENCES

Beckman, G. 1973. Population studies in northern Sweden. *Hereditas* 73: 305-310.

Beckman, G., L. Beckman and L.O. Nilsson 1973. A rare homozygous phenotype of superoxide dismutase, SOD 2. *Hereditas* 75: 138-139.

Beckman, G. and A. Pakarinen 1973. Superoxide dismutase. *Human Heredity* 23: 346-351.

Blake, N.M., R.L. Kirk, E. Pryke, and P. Sinnett 1969. Lactate dehydrogenase electrophoretic variant in New Guinea Highland Population. *Science* 163: 701.

Boyer, S.H., D.C. Fainer and E.J. Watson-Williams 1963. Lactic dehydrogenase variant from human blood: Evidence for molecular subunits. *Science* 141: 642-643.

Brewer, G.J. 1967. Achromatic regions of tetrazolium stained starch gels: Inherited electrophoretic variation. *Amer. J.Hum. Genet.* 19: 674-680.

Coon, C.S. 1948. *A reader in General Anthropology*. Henry Holt, New York.

Das, S.R., B.N. Mukherjee, S.K. Das, R. Ananthakrishnan, N.M. Blake and R.L. Kirk 1970. LDH variants in India. *Humangenetik* 9: 107-109.

Das, S.R., B.N. Mukherjee and S.K. Das 1972. Caste and age variations of the incidence of LDH variants in the Bengali Hindus. *Humangenetik* 14: 151-154.

de Beer, G. 1965. Genetics and Prehistory. *The Rede Lecture*, Cambridge University Press.

Harris, H., D.A. Hopkinson and E.B. Robson 1974. The incidence of rare alleles determining electrophoretic variants: Data on 43 enzyme loci in man.

Ann.Hum.Genet., Lond. 37: 237-253.
Howells, W.W. 1973. *The Pacific Islanders.* Weidenfeld and Nicolson, London.
Kraus, A.P. and C.L. Neely 1964. Human erythrocyte lactate dehydrogenase: Four genetically determined variants. *Science* 145: 595-597.
Mourant, A.E., L. Beckman, G. Beckman, L.O. Nilsson and D. Tills 1968. Erythrocyte lactate dehydrogenase variants in the Icelandic and Swedish populations. *Acta Genet.* (Basel) 18: 553.
Ritter, H. and G.G. Wendt 1972. Indophenol oxidase variability. *Humangenetik* 14: 72.
Roberts, D.F. and E. Sunderland 1973. *Genetic variation in Britain.* Symposia of the Society for the Study of Human Biology. Vol. XXI Taylor & Francis Ltd., London.
Saha, N. and B. Banerjee 1974. Haemoglobinopathies in the Indian subcontinent: A review of Literature. *Acta Genet.Med.Gemell.* (in press)
Shinoda, T. 1970. Inherited variation in tetrazolium oxidase in human red cells. *Japan.J. Human Genet.* 15: 144-152.
Treece, H. 1962. *The Crusades.* Mentor Books, New York.
Undevia, J.V. 1969. Population genetics of the Parsis: Comparison of genetical characteristics of the present Parsi population with its ancestral and affiliated groups. Ph.D. thesis, Bombay University.
Vessell, E.S. 1965. Genetic control of isozyme patterns in human tissues. *Progress in Medical Genetics IV.* Eds. A.G. Steinberg and A.B. Bearn. Grune and Stratton, New York.
Welch, S.G. and G.W. Mears 1972. Genetic variants of human indophenol oxidase in the Westray Island of the Orkneys. *Human Heredity* 22: 38-41.
Wilson, D. 1970. *The Vikings and their Origins.* Scandinavia in the First Millennium. McGraw-Hill Book Company, New York.

EVOLUTION OF VERTEBRATE HEMOGLOBIN AMINO ACID SEQUENCES

MORRIS GOODMAN[1], G. WILLIAM MOORE[1], and GENJI MATSUDA[2]

Department of Anatomy
Wayne State University School of Medicine
Detroit, Michigan 48201[1], and
Department of Biochemistry
Nagasaki University School of Medicine
Nagasaki, Japan[2]

ABSTRACT. The maximum parsimony method was applied to the 55 best sequenced globins to refine their sequence alignments and depict their phylogeny. Several insertions and many deletions as well as 1636 nucleotide replacements occurred in descent from the ancestral metaphyte-metazoan globin of 160 amino acid residues in a 175 position archetype alignment. Not long after, the myoglobin-hemoglobin gene duplication in primordial vertebrates homopoly-(probably tetra-) meric hemoglobin emerged. A β-α gene coding apparently for ^{B}4-like hemoglobin duplicated in primitive gnathostomes; and by basal amniote times, sophisticated $\alpha_2\beta_2$ type heterotetramers had evolved. Very fast nucleotide replacement rates in globin genes in earlier vertebrates became many times slower in amniote α and β lineages, averaging 8 to 35 times less at residue positions with the cooperative functions of heme and $\alpha_1\beta_2$ contacts and Bohr effect related salt bridge formation. In earlier genes for monomers which lacked cooperative behavior and especially in duplicated genes which were silent during part of their history, neutral mutations readily accumulated. This eventually facilitated the discovery by positive directional selection of specializations in α and β chains for functionally superior tetramers. Stringent selection maintained the improvements, drastically limiting types and numbers of subsequent mutations.

Previous investigations (Goodman and Moore, 1973; Goodman et al., 1974) have given us some inkling of the respective roles played by gene duplication, mutation, drift, and natural selection in the evolutionary development of tetrameric hemoglobins. In this report we explore in further detail the processes and events which characterized the emergence and descent of this protein in the jawed vertebrates. The method of maximum parsimony (Moore et al., 1973; Goodman et al., 1974) is again used to reconstruct the evolutionary history of hemoglobin genes from the amino acid sequences of different globin chains. The reconstruction is based on 55 contemporary globins (Figure 1) which encompass plant and animal kingdoms and include, in addition to the 47 chains previously used, 8 newly completed sequences - α and β hemoglobin chains of the old world monkey *Cercopithecus aethiops* (Matsuda et al., 1973a), and the new world monkey *Cebus apella* (Matsuda et al., 1973b), α chain of another new world monkey, *Ateles geoffroyi* (Matsuda et al., 1973c), β chain of mouse (Popp, 1973), α chain of a prototherian mammal, the echidna (Whittaker et al., 1973), and a muscle globin, classed functionally as a myoglobin, from the mollusc *Aplysia limacina* (Tentori et al., 1973). These 55 globin chains represent the most satisfactory globin amino acid sequence data in that they have either been completely (46 chains) or about two thirds to nine tenths (9 chains) sequenced by rigorous chemical procedures. Other sequences of globins reported in the literature have been inferred over large portions of their chain lengths from amino acid compositions of peptide fragments by homology with related, known sequences.

MAXIMUM PARSIMONY APPROACH

An evolutionary tree constructed by the parsimony principle describes the descent of related protein amino acid sequences by the fewest number of mutational changes. Such a tree maximizes the number of identities due to common inheritance and minimizes those due to parallel or back mutations. The empirical findings from previous investigations (e.g. Goodman et., 1974; Goodman 1973, 1974), as well as from the present investigation indicate that the trees constructed by this parsimony principle from protein sequences are in general agreement with phylogenies derived from paleontological evidence on the organisms represented by the sequences.

Our maximum parsimony method (Moore et al., 1973a; Goodman et al., 1974) resembles the ancestral sequence technique of Dayhoff (1972), but is more precise in that it determines the codon rather than just amino acid ancestors which yield the fewest mutations over a tree and calculates tree lengths as numbers of nucleotide rather than amino acid replacements. For a data set of aligned contemporary protein amino acid sequences we use several computer algorithms, called TPLNG, TPITR, and PTANC, to execute the maximum parsimony calculations. Each algorithm starts by translating these contemporary amino acid sequences into codon sequences, including all necessary alternative synonymous codons among the 61 specifying the 20 amino acids[1]. TPLNG then calculates the minimum number of nucleotide replacements over the entire tree at each aligned position and sums these lengths for the total maximum parsimony length of the give tree. (The lengths position by position and the total length summed for all positions are given in the computer print out.) TPITR not only calculates the length of the initial tree, but examines in a branch swapping procedure all one-step nearest neighbor changes in tree topology. The nature of such one-step changes is described in Moore et al. (1973b). In a tree of N exterior points, 2(N-3) alternative topologies are examined and the topology which reduced the tree length the most is the start for the next cycle of nearest neighbor branch swapping. This iterative search procedure stops when the tree length can not be lowered by further swapping of adjacent branches. We then resume the search by submitting dendrograms with more extensive changes in topology and persist in searching (examining hundreds of topologies) until we are satisfied that a sufficiently wide range of phylogenetic possibilities have been examined. Finally on the tree of lowest length found by this procedure, algorithm TPANC prints out for each aligned position the maximum parsimony codons at the evolutionary forks (interior points) and terminal buds (exterior points) of the tree and then combines the results for all aligned positions to yield the mutational length of each individual link of the tree (i.e. the number of differing nucleotides between each pair of connected adjacent points). Where alternative configurations of ancestral and descendant codons are equally parsimonious, the algorithm chooses the configuration which distributes the mutations to regions of the tree least well represented by sequence data, since nucleotide change tends to be grossly underestimated in such

[1]TPLNG, TPITR, and TPANC carry out the maximum parsimony calculations at all aligned amino acid residue positions in the data set, whereas our earlier algorithms, PSLNG, PSITR, and PSANC, which were used in the previous investigation (Goodman and Moore 1973; Goodman et al., 1974) carried out the calculations only at those residue positions occupied by amino acids in all sequences of the data set.

relatively empty regions (Barnabas et al., 1972; Goodman et al., 1974). Even with this choice of codons (we call them the A solution maximum parsimony codons), mutational lengths will be more underestimated when a branch descends without going through any evolutionary forks, or only a few forks, than when it goes through many, since in the latter case any successive nucleotide replacements at single nucleotide positions occurring at the forks are still detectable.

With another computer procedure, called MNAUG (Goodman et al., 1974), we more fully estimate evolutionary change in the regions of the parsimony tree which are sparse in ancestors. Using only the empirical distribution of nucleotide replacements on the parsimony tree, MNAUG obtains from the tree's paths with the most intervening ancestors, minimum estimates of the numbers of mutations to add to the paths with fewer intervening ancestors. These augmented mutational lengths depict amounts of evolutionary change between the more remotely separated globins which are comparable to the amounts Jukes and Holmquist (1972) estimated using their REHC (random evolutionary hits per codon) method (Holmquist et al., 1972).

In the evolutionary tree of globins in Figure 1 the mutational link lengths which were augmented by the MNAUG algorithm are shown as italicized numbers. This figure also employs an ordinate scale which attempts to portray the time spans during which the globin lineages were descending. In placing the evolutionary forks on this ordinate time scale, we followed the views of Young (1962) and Romer (1966) on the fossil history of the vertebrates, as Dickerson (1971) also did in his treatment of evolutionary rates of proteins. Consequently, in Figure 1 the vertebrate (lamprey-gnathostome) ancestor is placed at 500 million years ago, the teleost-tetrapod ancestor at 400 million years ago, the tetrapod (frog-mammalian) ancestor at 340 million years ago, the amniote (chicken-mammal) ancestor at 300 million years ago, and the eutherian mammalian ancestor at 90 million years ago.

RECONSTRUCTING THE GLOBIN EVOLUTIONARY TREE AND REFINING THE ALIGNMENT OF THE GLOBIN CHAINS

It was too expensive in computer time to run the TPITR algorithm on the full data set of 55 globin sequences. Therefore, the sequences comprising regions of less remotely related branches in the globin evolutionary tree, such as the region of amniote α sequences, were first treated as

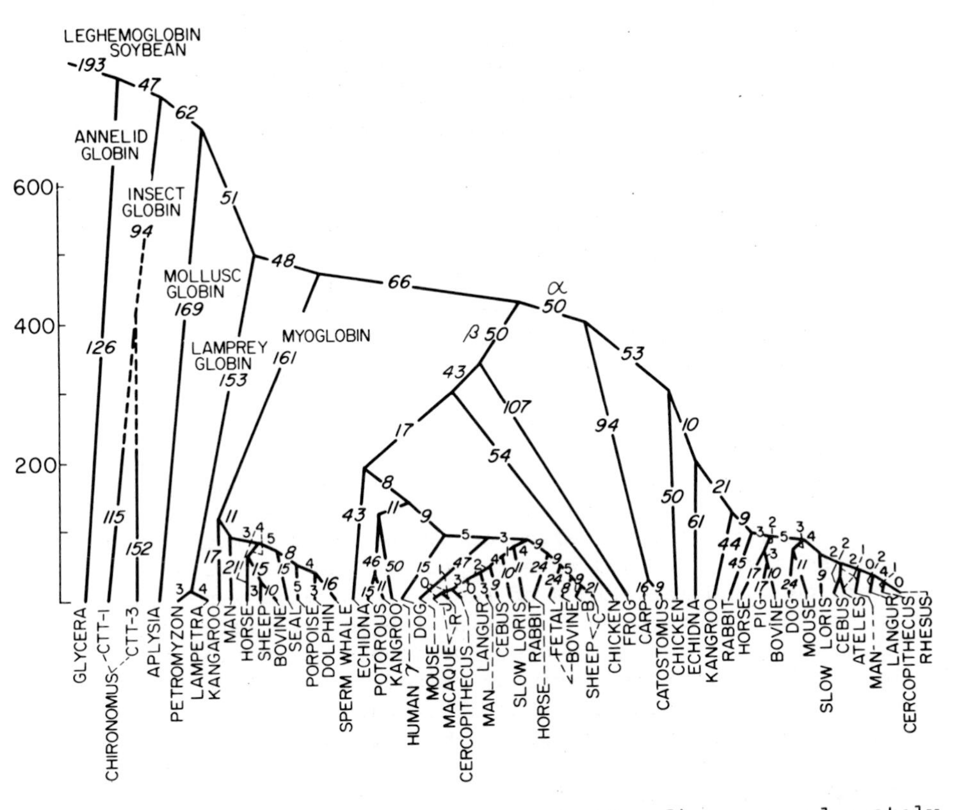

Figure 1. Maximum parsimony tree of fifty-five more adequately sequenced globins. The references to most of these sequences are in Goodman et al. (1974). New references cited in this paper are for α and β hemoglobin chains of *Cercopithecus* (Matsuda et al., 1973a), *Cebus* (Matsuda et al., 1973b), slow loris (Matsuda et al., 1973d), langur (Matsuda et al., 1973e), α hemoglobin chain of *Ateles* (Matsuda et al., 1973c), and echidna (Whittaker et al., 1973), β hemoglobin chain of mouse (Popp, 1973), and a "myoglobin" from *Aplysia* (Tentori et al., 1973). The mutational link lengths which were augmented by the MNAUG algorithm are shown as italicized numbers. Originally observed and corresponding augmented link lengths are as follows: 0=0, 1=1, 2=2, 3=3, 4=4, 5=5, 6=*8*, 7=*9*, 8=*10*, 9=*11*, 10=*15*, 11=*16*, 12=*17*, 15=*21*, 18=*24*, 20=*43*, 21=*44*, 22=*45*, 23=*46*, 24=*47*, 25=*48*, 27=*50*, 28=*51*, 30=*53*, 31=*54*, 35=*61*, 36=*62*, 40=*66*, 49=*94*, 54=*107*, 59=*115*, 65=*126*, 70=*152*, 71=*153*, 79=*161*, 83=*169*, 93=*193*. The ordinate scale is in millions of years. The

Figure 1 legend, continued:
placing of branch points on this scale was based on fossil evidence on the ancestral separation of the organisms from which the globins came. Dashed lines are used to designate the ancestral separation of the two *Chironomus* globins, CTT-1 and CTT-3, since the time of this gene duplication can only wildly be guessed at from the magnitude of the mutational distance between CTT-1 and CTT-3.

---

separate data sets. This permitted us to use TPITR to search for the least length branching topologies of the less remotely related sequences. After establishing regional topologies by this means, we then searched for the maximum parsimony arrangement of the major evolutionary forks in the composite globin tree, eventually examining by the TPLNG algorithm many different arrangements for the following 10 anciently separated branches, amniote and teleost α branches, amniote and frog β branches, therian myoglobin branch, lamprey, mollusc, insect, annelid, and plant globin branches.

At the onset we realized that the outcome of this search could be significantly affected by how the individual globin sequences were aligned with respect to one another, i.e. by where gaps and insertions of residue positions were placed in the different chains. To find the most satisfactory alignments of these sequences, we developed, again on the basis of the maximum parsimony method, a new algorithm called ALIGN. Like our other algorithms for the parsimony method, the input data required by ALIGN are the file of amino acid sequences and a file describing an order of branching among the sequences, i.e. a dendrogram file. In any one computer run, ALIGN can compare the respective alignments of the sequences at two adjacent points on the evolutionary tree described by the dendrogram. The gaps that can appear in one sequence with respect to the other are slided back and forth and the tree's mutational lengths are calculated by the maximum parsimony method for the different positions of these gaps. The computer print out then shows where the gaps should be placed for the particular tree to have its lowest parsimony length. Limits are placed on the stretch of sequence positions (no more than 50 at a time) over which a gap (a deleted amino acid or contiguously deleted amino acids) can be slided in the ALIGN procedure. We took advantage of this restriction in the algorithm to utilize information (e.g. that of Perutz et al., 1965; Perutz 1970a, b; and Perutz and TenEyck 1972) on the structural and functional roles of the different amino acid residues in the globin chains to fix individual alignment positions among the 55 contemporary sequences.

Initially we did not use this whole collection of 55 globins in the individual ALIGN runs, but rather used overlapping subsets of the collection, representing each time only branches of the globin tree with a well understood order of phylogenetic relationship. After the mutual alignments of all sequences were checked and in some cases refined by this procedure, we resumed the search for the lowest length tree topology of the major globin branches. Two arrangements were clearly more promising than any others. In one arrangement, the stem line descending from the metazoan ancestor to the vertebrate hemoglobin β-α ancestor gave off the following branches: the first to the insect *Chironomus* globins, the second to the mollusc *Aplysia* "myoglobin", the third to the lamprey globins and the fourth to a common ancestor of the annelid *Glycera* globin and the mammalian myoglobins. In the second arrangement, the first branch off the descending stem line was to the annelid globin, the second to the insect globins, the third to the mollusc globin, the fourth to lamprey globins, and the fifth to mammalian myoglobins.

For each of these two branching arrangements we further refined the alignments of the different sequences after many ALIGN runs on the full set of 55 globins. The most satisfactory alignments were found with the dendrogram for the second branching arrangement, the one shown in Figure 1 in which all vertebrate globins have a common ancestor separating them from invertebrate and plant globins, because the alignments found for this arrangement conformed closely to knowledge on the structural and functional properties of the different globin chains, whereas the alignments found with the other dendrogram did not conform so well. In the more satisfactory alignments, the ones found with the dendrogram specifying the tree shown in Figure 1 in which the *Glycera* sequence descends as a separate branch from the earliest metazoan globin ancestor, the lamprey chains have a gap over 9 amino acid positions starting in the GH interhelical region and ending in the H helical region, whereas in the alignments found when the *Glycera* sequence joins the mammalian myoglobins, the lamprey chains have the 9 amino acid gap start in the G helix and end in the GH interhelical region. The former alignment agrees very closely with that proposed by Li and Riggs (1970) from the x-ray crystallographic evidence of Hendrickson & Love (1971). Furthermore with the dendrogram yielding the more satisfactory alignments the *Glycera* sequence has a gap of seven amino acid positions exactly where the D helix forms in other globins, which agrees with the findings from x-ray crystallography (Padlan and Love 1968) that the D helix is deleted in the *Glycera* sequence (Imamura et al., 1972). In contrast, the alignment found for the *Glycera* sequence when it was placed with the mammalian myoglobin branch had, instead of one gap over the D helical region, two gaps, one in the CD interhelical region and another in the middle of the E helix. Disallowing these two gaps and using the single gap over the D helix, but otherwise keeping the sequence alignments found with this dendrogram which joined the annelid sequence to the mammalian myoglobins, added 5 mutations to the globin tree raising its length to 1638 nucleotide replacements. In contrast with the other dendrogram, the one specifying the globin evolutionary tree shown in Figure 1, and with the sequence alignments found with it, the tree's length was 1636. With these alignments no other dendrogram tried was as parsimonious. Twenty two alternative topologies for the ten major globin branches were examined and these trees had mutational lengths ranging from 1640 to 1657.

One hundred seventy five amino acid residue positions are required to account for the sequence alignments used to construct the globin evolutionary tree in Figure 1. The reconstructed earliest metazoan globin ancestor, which also represents the metazoan - metaphyte ancestor, has 160 amino acids. Using the helical-interhelical notation of Perutz (1969) for vertebrate hemoglobin chains and the 175 position archetype alignment, this earliest ancestral sequence encompasses the following positions: NA1(9)-A13(24), A14(27)-E12(80), E13(82)-EF7(96), EF8(98)-F8(112), and F9(116)-HC9(175). Insertions in the descent of the ancestral lamprey lineage (at residue positions 1-8 and 97), the annelid lineage (at positions 25-26 and 113-115) and the teleost α lineage (position 81) account for the remaining 15 positions in the archetype alignment. Notable features of the reconstructed metaphyte-metazoan globin ancestor are the presence of six amino acids at positions 99-104 between the end of the EF interhelical region and the beginning of the F helix and also of six amino acids at the C-terminal end at positions HC4(170)-HC9(175). The former accounts for the greater length of the F helix found in *Glycera* globin (Padlan and Love 1968) and the latter accounts for the six amino acid extension of the C-terminal end of vertebrate myoglobins. These stretches of amino acids occur, therefore, in these contemporary sequences, not as insertions, but as retentions of the primitive alignment. Although this primitive globin ancestor has more amino acids than any of the particular contemporary chains from which its nature was inferred, globin sequences with relatively long chain lengths are still common among plant leghemoglobins were chain lengths up to 183 amino acid residues occur (Broughton and Dilworth 1973). Deletions appear to far out weigh insertions in the descent of the globin lineages from their early ancestors. There are 6 deletions in the descent of soybean leghemoglobin (at D7(68)-E3(71), EF5(94)-EF7(96), F7(111), G4(125), G11(132), and HC2(168)-HC9(175); 6 in descent of *Glycera* globin (at NA2(10), CD2(55), D2(63)-E1(69), EF5(94)-EF7(96), FG4(120), and H21(166)-HC4(170); 11 in descent of the *Chironomus* lineage (at NA2(10), B3(33)-B6(36), CD7(60), E13(82)-E16(85), EF4(93), F7(111), G4(125), GH2(142)-GH3(143), H5(150), H21 (166)-HC2(168), and HC6(172)-HC9(175), with further deletions at NA1(9) and FG3(119) in CTT-3 and at H4(149) in CTT-1; 6 in descent of Aplysia "myoglobin" (at NA1(9), CD7(60), E20(89) -EF2(91), G4(125), H5(150)-H7(152), and HC4(170)-HC9(175); 1 deletion in descent from the invertebrate-vertebrate globin ancestor to the basal vertebrate globin ancestor (at 99-104 between the EF interhelical and the F helical regions) with 2 more deletions in the descent of the lamprey lineage (at GH4(144)-H7(152) and HC3(169)-HC9(175)) and further deletions at G10(131) in *Lampetra* globin and at H18(163) in *Petromyzon* globin; 1 deletion in descent of the ancestral mammalian myoglobin lineage (at NA2(10); 1 deletion in descent from the vertebrate myoglobin-hemoglobin ancestor to the β-α hemoglobin ancestor (at HC4(170)-HC9(175)); 1 deletion in descent from the β-α ancestor to the tetrapod β ancestor (at A16(29)-AB1(30)) followed by deletions at NA1(9)-A3(14) in frog β and at NA2(10) in bovid β as well as at NA1(9)-A2(13) in sheep Cβ; and 2 deletions in descent from the β-α ancestor to the teleost-tetrapod α ancestor (at NA2(10) and D2(63)-D6 (67)) followed by deletions at E6(74) in the teleost α branch and at CD5(58) in the branch to the amniote α ancestor.

BROAD PATTERNS OF EVOLUTIONARY CHANGE IN THE DESCENT OF VERTEBRATE HEMOGLOBIN AMINO ACID SEQUENCES

The gene duplication which separated the ancestor of vertebrate myoglobins from hemoglobin genes occurred, from the evidence depicted in Figure 1., in the basal vertebrates. A descending hemoglobin gene in primitive jawed fish ancestral to teleosts and tetrapods then duplicated to give rise to separate α and β loci. The phylogeny for the genes at these loci described in Figure 1 parallels the phylogeny of the vertebrates. Thus in the descent of the α stem line from the β-α ancestor to the eutherian mammalian ancestor, the

the teleost divergence is followed first by the avian separation, next by the monotreme, and then by the marsupial. Similarly, in the β stem line, the amphibian divergence is followed first by the avian and then by the monotreme separation, after which the β-γ gene duplication occurs with a marsupial branch emerging from the γ locus as the β genes descend to the eutherian β ancestor. During the past 400 million years from the teleost-tetrapod ancestor to the present, α genes evolved at the average rate of 20 nucleotide replacements per 100 codons per 10^8 years. However, during the initial 100 million years from the teleost-tetrapod ancestor to the amniote ancestor, the rate was 38 nucleotide replacements per 100 codons per 10^8 years, a much faster rate than during the subsequent 300 million years from the amniote ancestor to the present in which α genes accumulated on the average 15 nucleotide replacements per 100 codons per 10^8 years. A similar pattern of rate changes is presented by β genes. During the past 340 million years from the tetrapod ancestor to the present, β genes evolved at the average rate of 23 nucleotide replacements per 100 codons per 10^8 years. But again, initially there was more rapid evolution than later, for during the first 40 million years from the tetrapod to amniote ancestor, there were 74 nucleotide replacements per 100 codons per 10^8 years, whereas during the subsequent 340 million years from the amniote ancestor to the present, the β genes accumulated on the average only 14 nucleotide replacements per 100 codons per 10^8 years.

If the myoglobin-hemoglobin gene duplication indeed occurred in the basal vertebrates, as our present evidence indicates, then globin sequences evolved at extremely fast rates in these early vertebrates. The globin lineage descending from the basal vertebrate globin ancestor of about 500 million years ago through the myoglobin-hemoglobin and β-α ancestor to the teleost tetrapod α ancestor of 400 million years ago evolved at the rate of 111 nucleotide replacements per 100 codons per 10^8 years, about 3 times faster than during the period from the teleost-tetrapod to amniote α ancestor and 7 to 8 times faster than during the period of amniote phylogeny. When early globins are traced from this 500 million year old ancestor to the tetrapod β ancestor of 340 million years ago, the rate of nucleotide replacements is not quite as fast. It comes out to 69 per 100 codons per 10^8 years. Apparently after the β-α gene duplication the β stem line evolved at a slower rate than the α stem line. This suggests as previously discussed (Goodman and Moore 1973; Goodman et al., 1974), that the new locus produced by the duplication was the α locus and that at first it was phenotypically relatively inactive, thus free to accumulate mutations, whereas the active gene at the older β-type locus continued to function under the full constraints of natural selection. Moreover if multi-chained β-type hemoglobin had already evolved by the time of the β-α gene duplication, then the subsequent emergence from the α locus of an active gene for a specialized non-β chain would permit natural selection to replace β_4-type hemoglobins by $\alpha_2\beta_2$ type hemoglobins. In turn this would have intensified evolutionary pressures at the β locus and thus account for the rapid rate of β globin evolution observed between the tetrapod and amniote ancestors. It appears that during this period improved tetrameric hemoglobins of modern type were selected. Moreover, once having discovered the α and β genes responsible for the improvements, natural selection must have markedly increased their resistance to further changes, because rates of hemoglobin evolution drastically slowed after the beginning of the amniote radiation at 300 million years ago. Conversely the extremely rapid rates of globin evolution observed in the early vertebrates in the period when the myoglobin-hemoglobin divergence first began suggests that the unsophisticated monomeric hemoglobin of that time still had numerous residue positions where amino acid substitutions could readily be tolerated. The pattern of change in early and later vertebrate hemoglobin sequences is explored in more detail in the following section.

PATTERNS OF MUTATIONAL CHANGE AT RESIDUE POSITIONS WITH DEFINED FUNCTIONAL PROPERTIES IN α AND β HEMOGLOBIN SEQUENCES

The residue postions examined are listed in Table 1 according to their present day functions and are designated by both their helical and archetype alignment locations. Table 2 shows the codon changes found on the parsimony tree at these positions on going from the vertebrate myoglobin-hemoglobin ancestor to the β-α ancestor and then separately to the amniote α and β ancestors. Table 3 then compares, for these lines of descent, the average mutational lengths of different functional types of residue positions. In this region of the tree between the early vertebrates and the basal amniotes, the most consistently conservative residue positions in the reconstructed hemoglobin sequences are the positions which function in both present day α and β chains as heme contacts. Residue postions which, if the three

TABLE 1
Residue Positions with Defined Functional Roles in α and β Chains

Heme Contacts in Both α and β Chains: B13(43)[a], C4(50)[a], C7(53)[b], CD1(54)[a], CD3(56), CD4(57)[a], E7(75), E11(79)[a], F4(108), F7(111), F8(112), FG3(119)[b], FG5(121)[a,c], G4(125)[d], G5(126)[a], G8(129)[a], H15(160)[a], H19(164)[a],

Heme Contacts Just in α Chains: H12(157)[a]; Heme Contacts Just in β Chains: E10(78), E14(83), E15(84)[a],

Non-Heme $\alpha_1 \beta_2$ Contacts in Both α and β Chains: C2(48), C3(49), C5(51)[e], C6(52), FG4(120), G1(122), G2(123), G3(124), HC2(168)[a]

Non-Heme $\alpha_1 \beta_2$ Contact Just In α Chains: CD2(55); Like-Like Chain Contact Just in α Chains: G6(127)

Like-Like Chain Contacts in Both α and β Chains: NA1(9)[e,h], H10(155)[g], HC3(169)[d,i]

$\alpha_1 \beta_1$ Contacts in Both α and β Chains: B12(42)[f], B15(45), B16(46), C1(47)[f], G10(131), G14(135), G18(139), GH2(142), GH5(145), H2(147), H5(150)[g], H6(151), H9(154)[g]

$\alpha_1 \beta_1$ Contacts Just in α Chains: B11(41), G11(132)[a], G13(134); $\alpha_1 \beta_1$ Contacts Just in β Chains: D2(63), D6(67), G17(138), H1(146), H3(148)

Just α Chains Interacts with Bohr Effect Site: A4(15)

Bohr Effect Just in β Chains: FG1(117)

2,3-DPG Binding Just in β Chains: EF6(95), H21(166)

Remaining Interior Positions (Stabilization of Tertiary Structure): A8(19), A11(22), A12(23), A15(28), B6(36), B9(39), B10(40), B14(44), D5(66)[j], E4(72), E8(76), E12(80), E18(87), E19(88), F1(105), G12(133), G16(137), H8(153), H11(156)

The Functional roles of these positions have been identified from Perutz et al., (1965), Perutz (1969, 1970a, 1970b), and Perutz and TenEyck (1972).

TABLE 1 Footnotes

[a] Also an interior position in both α and'β chains.
[b] Also an $\alpha_1 \beta_2$ contact just in α chains.
[c] Also an $\alpha_1 \beta_2$ contact in both α and β chains.
[d] Also an $\alpha_1 \beta_2$ contact just in β chains.
[e] Also Bohr effect site just in α chains.
[f] Also just in β chains interacts with Bohr effect site.
[g] Also just in α chains interacts with Bohr effect site.
[h] Also 2, 3-DPG binding just in β chains.
[i] Also Bohr effect site in both α and β chains.
[j] Deleted in α chains.

---

dimensional structures of these hypothetical ancestral hemoglobin chains could be reconstructed, would be located at the interior of the molecules where they would have the general function of stabilizing tertiary structure, are also conservative. In contrast, throughout this evolutionary period, the residue postions which function in both α and β chains of modern hemoglobin as interchain contacts, evolve at rather fast rates. Indeed, in the earliest period, that between the myoglobin-hemoglobin and β-α ancestors, the subset of interchain contact positions, which in both α and β chains are $\alpha_1\beta_1$ contacts, evolve at a rate faster than the average of all positions, and in the next period, that between the β-α and amniote ancestors, the subset of interchain contact positions, which function in both α and β chains of modern hemoglobin as $\alpha_1\beta_2$ contacts, evolve at a faster than average rate in the descending α line and also in the descending β line. Moreover in the early α line, the residue positions which acquire functions specific for α chains, change at a faster rate than any other set of positions; as likewise, in the early β line do the residue positions which acquire functions specific for β chains. It can also be noted in Table 3 that after the β-α gene duplication that the residue positions categorized functionally as common to β and α chains diverge more from the ancestral state in the α line than in the β line.

How are we to interpret these patterns of mutational change? Firstly, we can assume that the primitive heme binding protein which was ancestral to later myoglobins and hemoglobins existed in the primordial vertebrates largely as single-chained molecules. These molecules must have lacked the properties of cooperativity, such as heme-

TABLE 2

The codon changes in descent from the myoglobin-hemoglobin ancestor to the amniote α and β ancestors at residue positions showing functions in present day tetrameric hemoglobins.

A) Common to both β and α chains

	Heme Contact					Non-Heme $\alpha_1\beta_2$ Contact					
	B13 (43)	C7 (53)	CD3 (56)	G4 (125)	H15 (160)	C3 (49)	C5 (51)	C6 (52)	FG4 (120)	G2 (123)	G3 (124)
Myo-Hb ancestor	CUU	UAU	AAU	UAU	UUU	GAG	CAG	GAG	GAG	GCU	CAG
β-α ancestor	CUU	UAU	AAU	AAU	GUU	GAG	CAG	AAG	CAG	CCU	GAG
Tel-Tet α ancestor	AUG	UAU	CAU	AAU	GUU	GAG	AAG	ACG	CGG	CCU	GAG
Amniote α ancestor	AUG	UAU	CAU	AAU	GUU	ACG	AAG	ACG	CGG	CCU	GUG
Tetrapod β ancestor	CUU	UAU	AAU	AAU	GUU	UGG	CAG	AGG	CAU	CCU	GAG
Amniote β ancestor	CUU	UUU	AGU	AAU	GUU	UGG	CAG	AGG	CAU	CCU	GAG

	$\alpha_1\beta_1$ Contact									Like-Like Chain Contact	
	B15 (45)	B16 (46)	C1 (47)	G10 (131)	G14 (135)	GH2 (142)	H2 (147)	H5 (150)	H9 (154)	NA1 (9)	HC3 (169)
Myo-Hb ancestor	AAA	GCU	UAU	GAU	GAU	GCG	GCU	CAG	GAG	GGG	AAG
β-α ancestor	AUA	GUU	UAU	AAU	GUU	CCG	CCU	CAG	GAG	GUG	AAG
Tel-Tet α ancestor	AUA	GUU	UAU	AAU	GUU	CCG	CCU	CAU	GAU	GUG	AGG
Amniote α ancestor	AUA	GGU	UUU	CAU	GUU	CCG	CCU	CAU	GAU	GUG	AGG
Tetrapod β ancestor	AUA	GUU	UAU	AUU	AAU	CCG	CCU	CAG	GAG	GUG	CAU
Amniote β ancestor	AUA	GUU	UAU	AAU	AUU	UCG	CCU	CAG	CAG	GUG	CAU

TABLE 2 continued:

The codon changes in descent from the myoglobin-hemoglobin ancestor to tha **amniote** α and β ancestors at residue positions showing functions in present day tetrameric hemoglobins.

B) Specific either to β or α chains

	Heme Contact				$\alpha_1\beta_2$ Contact	Like-Like Chain Contact	α_1 β_1 Contact					Bohr Effect
	H12α (157)	E10β (78)	E14β (83)	E15β (84)	CD2α (55)	G6α (127)	B11α (41)	G11α (132)	G13α (134)	D6β (67)	H1β (146)	FG1β (117)
Myo-Hb ancestor	UUU	AGG	GCG	AUU	GCU	GAG	GUA	GUU	AUA	AAG	GGU	AAU
β-α ancestor	UUU	AAG	GCG	AUU	GCU	AAG	GUA	GUU	AUA	AAG	GCU	AAU
Tel-Tet α ancestor	UUU	AAG	GCG	AUU	GCU	AAG	GAA	UGU	AUA	-	GCU	AAU
Amniote α ancestor	CUU	AAG	GCG	AUU	CCU	AAG	GAA	UGU	CUA	-	GCU	CAU
Tetrapod β ancestor	UUU	AAG	GCG	AUU	GCU	AAG	GCA	GUU	AUA	AAG	ACU	AAU
Amniote β ancestor	GUU	AAG	UCG	UUU	GCU	AAG	GCA	GUU	AUA	AUG	ACU	GAU

	Bohr Effect Related	2,3-DPG Binding		C) Remaining Interior Positions							
	A4α (15)	EF6β (95)	H21β (166)	A11 (22)	B9 (39)	B14 (44)	E12 (80)	F1 (105)	G12 (133)	H8 (153)	H11 (156)
Myo-Hb ancestor	GAG	GAG	GCG	GUU	GUU	UUU	CUU	UUG	UUU	UUG	CUU
β α ancestor	GAG	GAG	GCG	GUU	GCU	UUU	CUU	UUG	UUU	UUG	CUU
Tel-Tet α ancestor	GUA	GAG	GCG	GUU	GCU	UUU	GUU	UUG	UUU	UUG	UUU
Amniote α ancestor	GAU	GAG	GCG	GUU	GCU	UUU	GCU	UUG	UUU	UUG	UUU
Tetrapod β ancestor	GAG	AAG	GAG	GUU	GCU	CUU	CUU	UUU	UUU	UUG	CUU
Amniote β ancestor	GAG	AAG	CAG	CUU	GCU	CUU	CUU	UUU	CUU	UGG	CUU

A residue position may have in present day hemoglobin chains, in addition to the function of the group in which it is placed, other functions. These can be identified from Table 1.

TABLE 3

Effect of Functions on Mutational Lengths in Descent from

Type of Residue Positions	Myo-Hb Ancestor to β-α Ancestor		β-α Ancestor to Amniote α Ancestor		β-α Ancestor to Amniote β Ancestor	
	No. of Positions	Average length	No. of Positions	Average length	No. of Positions	Average length
Heme Contacts in both β and α chains	18	0.11	18	0.17	18	0.11
Non-heme $\alpha_1 \beta_2$ contacts in both β and α chains	9	0.33	9	0.67	9	0.44
$\alpha_1 \beta_1$ contacts in both β and α chains	13	0.46	13	0.38	13	0.23
Like-Like chain contacts in both β and α chains	3	0.33	3	0.33	3	0.67
Total interchain contacts common to β and α chains	25	0.40	25	0.48	25	0.36
Function just in α chains	7	0.14	7	1.00	7	0.29
Function just in β chains	11	0.18	9	0.11	11	0.73
Remaining interior positions	19	0.05	18	0.17	19	0.26
Remaining exterior positions	68	0.35	64	0.48	66	0.32

TABLE 3 continued

Type of Residue Positions	No. of Positions	Average length	No. of Positions	Average length	No. of Positions	Average length
All Positions	148	0.27	141	0.40	146	0.32

The mutational length for each residue position for each period of descent is the sum of mutations observed at that residue postion on the parsimony globin tree for that period of descent. Positions were grouped together on the basis of their functional roles in present day hemoglobin chains, as identified from Perutz et al., (1965), Perutz (1969, 1970a, 1970b), and Perutz and Ten Eyck (1972), and the mutational lengths were then averaged for the positions in each group.

heme interactions, which regulate the oxygen affinities of modern tetrameric hemoglobins in physiologically advantageous ways. We can also assume that tetrameric or at least multi-chained hemoglobins had emerged before the β-α gene duplication and that the β-α ancestor formed such a polymer, for if α and β chains had already diverged while hemoglobin was still a monomeric protein there would be little reason for the residue positions which became responsible for the contacts between subunits when the tetrameric hemoglobin did evolve to be mostly the same in aligned α and β chains. Moreover, homotetramers are readily formed by β chains, although not by α (Benesch et al., 1973), demonstrating that the formation of a globin tetramer does not require α chains to be present with β chains, as well as suggesting that the chain type in the original β-α globin homotetramer was more β-like than α-like in properties. Secondly, we know that in the normal heterotetrameric state of the hemoglobins in higher organisms, the α_1 β_2 contact positions have a more advanced functional role than the α_1 β_1 contact positions; the α_1 β_2 contacts are more intimately associated with co-operativity than are the α_1 β_1 contacts (Perutz 1969, 1970a, b; Perutz and TenEyck 1972). Bearing these considerations in mind, we can interpret the patterns of mutational change described from the data in Table 3 as follows.

After the myoglobin-hemoglobin gene duplication in primordial vertebrates, but before the β-α gene duplication in primitive gnathostomes, a multichained hemoglobin evolved from its single chained ancestor. At the beginning of this evolutionary transformation, the heme binding residue positions of the ancestral hemoglobin and the positions inside the folded molecule where tertiary structure is stabilized were subjected to stronger selective constraints, due to their ancient well established functions, than the remaining positions at the protein surface. There many amino acid substitutions occurred without encountering any especially adverse selection. Indeed, certain of these surface substitutions were positively selected for, because they produced multichained hemoglobins (probably β-α_4 type homotetramers) which imparted some new physiological advantage to their bearers. The positions at which these particular substitutions occurred could henceforth be considered inter-chain contact positions. Later, in most cases, after the β-α gene duplication, they would become α_1 β_1 and α_1 β_2 contact positions. Functionally these first homotetramers were probably more β_4 like than α_2 β_2 like, although they might have exhibited a Bohr effect and cooperative oxygen binding in a rudimentary way. In any event, the existence of these first crude homotetramers and the selective pressures on the primitive jawed vertebrates for a more active mode of life set the stage for the evolution of sophisticated hemoglobin heterotetramers with fully devoloped allosteric properties.[2] The β-α gene duplication spurred this process. The reason for the rapid rates of evolution at the α and β positions with the common $\alpha_1\ \beta_2$ contact function and at those α positions with functions specific for α chains and those β positions with functions specific for β chains is that natural selection was discovering specializations in α and β chains which produced improved tetrameric hemoglobins. By the time of the basal amniotes, fully perfected $\alpha_2\ \beta_2$ type hemoglobins had replaced the older models. From then on natural selection acted strongly against any further amino acid substitutions at the different residue positions involved in heme and $\alpha_1\ \beta_2$ contacts and in salt bridge formation related to the Bohr effect.

It can be seen in Table 4 for amniote α globins that the average mutational lengths of these residue positions with strong functional roles is only about one sixth the average of all positions in α chains. This same markedly slower rate of substitution at positions with strong functions is also evident in amniote β globins (Table 5). In contrast the exterior positions in α chains which appear to be without functions have an average length almost twice as large as the average of all α positions - similarly in β chains. The $\alpha_1\ \beta_1$ contact positions evolve at an average speed comparable to that of all positions, thus are not nearly as conservative as the $\alpha_1\ \beta_2$ contact positions. This correlates with the lesser functional importance of the former compared to the latter. The mutational length of the three positions (NA1, EF6, and H21) in β chains thought to be involved in binding 2,3-diphosphoglycerate can be related to the functional specializations which distinguish fetal and adult

---

2. Dr. M.F. Perutz, in reply to a letter from one of us (M.G.) pointed out that "it is generally believed that large fast-moving creatures could not supply their muscles with oxygen at a sufficiently high rate without having a respiratory pigment with a sigmoid oxygen equilibrium curve and a Bohr effect. On the basis of this argument the evolution of a tetrameric hemoglobin with two different chains would seem to be a precondition".

TABLE 4

Effect of Functions on Mutational Lengths in Descent of Amniote α Globins

Type of Residue Positions	Number of Positions	Average Length
Heme Contacts	19	0.37
Non-Heme α_1 β_2 Contacts	10	0.20
Salt Bridges, Associated Bohr effect, α_1-α_2 Contact	7	0.14
Non-Salt Bridge α_1 β_1 Contacts	14	2.00
Remaining Interior Positions	19	1.47
Remaining Exterior Positions	72	2.36
(Subset of "non-functional" exterior positions from clinical evidence)	(18)	(2.89)
All Positions	141	1.67

The mutational length for each residue position is the sum of mutations observed at that residue position in the amniote α region of the parsimony globin tree. Positions were grouped together on the basis of their functional roles, as identified from Perutz et al., (1965), Perutz (1969, 1970a, 1970b), and Perutz and TenEyck (1972), and the mutational lengths were then averaged for the positions in each group. The "nonfunctional" exterior positions from clinical evidence were the general positions from Figure 7 in Huisman and Schroeder (1971) except that F1 as an interior position and HC3 as a position with multiple functions were not included in our "nonfunctional" group.

hemoglobins in higher primates. The major contribution to the length at these 2, 3-DPG binding positions are mutations at NA1(val/GUG → Gly/GGG) and H21 (His/CAU → Ser/UCU or AGU) which occurred in the γ line. They occurred when the γ locus, which had been produced in primitive therian mammals by a β gene duplication, was accumulating nucleotide replacements extensively (Figure 1). A few potentially beneficial mutations (which might have been neutral when they occurred if the γ locus had been inactive for part of its history) eventually were discovered by natural selection and expressed

TABLE 5

Effect of Functions on Mutational Lengths in Descent of Amniote β Globins

Type of Residue Positions	Number of Positions	Average Length
Heme Contacts	21	0.43
Non-Heme α_1 β_2 Contacts	10	0.40
Salt Bridges, Associated Bohr Effect	4	0.00
2, 3-DPG Binding	3	2.00
Non-Salt Bridge α_1 β_1 Contacts	16	2.81
Remaining Interior Positions	21	1.57
Remaining Exterior Positions	71	3.31
(Subset of "non-functional" exterior positions from clinical evidence)	(27)	(4.11)
All Positions	146	2.27

The mutational length for each residue position is the sum of mutations observed at that residue position in the amniote β region of the parsimony globin tree. Positions were grouped together on the basis of their functional roles, as identified from Perutz et al., (1965), Perutz (1969, 1970a, 1970b), and Perutz and TenEyck (1972), and the mutational lengths were then averaged for the positions in each group. The "nonfunctional" exterior positions from clinical evidence were the general positions from Figure 8 in Huisman and Schroeder (1971).

through the γ chain of higher primate fetal hemoglobin.

To appreciate how extremely conservative α and β chain evolution became at most residue positions with strong functions, we need to realize that the average mutational lengths in Figures 4 and 5 represent vastly greater stretches of time than the lengths in Figure 3. Each average length is simply the sum of the lengths of the different positions in a particular group of aligned positions divided by the number of the positions. Each position length which went into the averages in Table 4 is the sum of mutations (actually observed, not augmented) in the amniote α region of the tree in Figure 1, and when related to the number of α lineages tracing back to the amniote ancestor at 300 million years ago, represents about 1.5 billion years of evolution. Similarly each position length used to calculate the averages in Table 5 is the sum of mutations in the amniote β region of the tree and represents about 1.7 billion years of evolution. However the lengths in Table 3 are only first for the link between the vertebrate myoglobin-hemoglobin and β-α ancestor and then for α globins only for the one line of descent from the β-α ancestor to the amniote α ancestor and similarly for β globins only for the one line from the β-α ancestor to the amniote β ancestor. Thus even the sum of the two lines and the initial link probably represents only about 300 million years of evolution. When these time dimensions are considered it becomes apparent that the heme contact positions in the amniotes were evolving about 8 times faster in the earlier vertebrates and that the α_1 β_2 contact, salt bridge, and Bohr effect positions were evolving about 35 times faster.

FURTHER EVIDENCE FROM MUTATIONS ON THE PERVASIVE ROLE OF NATURAL SELECTION

When the different mutations between ancestor and descendant codon sequences on the globin parsimony tree of Figure 1 are counted (Table 6), transversions are twice as common as transitions at the second nucleotide position of the codons, but transitions of guanine (G) and adenine (A) are in great excess over any other kinds of base replacements at the first nucleotide position of the codons. As previously discussed (Goodman and Moore 1973; Goodman et al., 1974), this reflects the fact that non-conservative amino acid interchanges encounter much more adverse selection than conservative interchanges. The structure of the genetic code is such that the amino acids specified by G beginning codons are usually more similar in physical chemical properties to the amino acids specified by A beginning codon (and vice versa) than they are to those specified by either C (cytosine) or U (uracil) beginning codons. Thus transversions of G or A beginning codons are more likely to encounter adverse selection than are transitions of these codons. This is not the case at the second nucleotide position of the codons where transversions are as likely as transitions to result in amino acid interchanges in which the exchanged amino acids have similar selective properties.

The excess of transitions for G and A beginning codons

TABLE 6

Comparison Of Transitions to Transversions at First and Second Nucleotide Positions of the Codons in Globin Phylogency

	1st Position	2nd Position
G → A Transitions	183	42
A → G Transitions	116	85
Total	299	127
G → C Transversions	105	49
G → U Transversions	80	14
A → C Transversions	61	114
A → U Transversions	29	55
Total	275	232
C → U Transitions	21	45
U → C Transitions	22	62
Total	43	107
C → A Transversions	45	76
C → G Transversions	43	77
U → A Transversions	22	31
U → G Transversions	29	25
Total	139	209

observed in Table 6 for the total span of globin descent from the metaphyte-metazoan ancestor to the present also occurs with high regularity in most local regions of the globin tree. One of the few exceptions is in the early α lineage in its descent from the β-α ancestor to the amniote α ancestor, for here transitions of G and A are half as frequent as transversions (6 to 13). In contrast in the early β lineage in its descent from the β-α ancestor to amniote β ancestor, the usual excess of transitions over transversions in G and A beginning codons occurs (9 to 9). This again suggests that after the β-α duplication the β locus continued to function under the full view of natural selection, whereas the α locus was much less restrained and, therefore, accumulated mutations in a more random fashion.

When natural selection finally screened these α genes with their multiple changes, it chose a radically restructured one which coded for the α chain type of a functionally superior hemoglobin tetramer.

ACKNOWLEDGMENTS

We thank Mr. Walter Farris and Miss Elaine Krobock for assistance. This work was supported by grant GB-36157 from the National Science Foundation Systematic Biology Program and by computer time provided by the Wayne State University Computing Center.

REFERENCES

Barnabas, J., M. Goodman, and G.W. Moore 1972. Descent of mammalian alpha globin chain sequences investigated by the maximum parsimony method. *J. Molec. Biol.* 69: 249-278.

Benesch, R., R.E. Benesch, and S. Yung 1973. The solubility of hemoglobin β_4, the mutant subunits of sickle cell hemoglobin. *Biochem. Biophys. Res. Commun.* 55:261-265.

Broughton, W.J. and M.J. Dilworth 1973. Amino acid composition and relationships of lupin and serradella leghaemoglobins. *Biochem. Biophys. Acta* 317: 226-276.

Dayhoff, M.O. 1972. *Atlas of Protein Sequence and Structure,* Vol. 5: National Biomedical Research Foundation, Silver Spring, Md.

Dickerson, R.E. 1971. The structure of cytochrome C and the rates of molecular evolution. *J. Molec. Evolution* 1: 26-45.

Goodman, M. 1973. The chronicle of primate phylogeny contained in proteins. *Symp. Zool. Soc. Lond.* No. 33: 339-375.

Goodman, M. 1974. Biochemical evidence on hominid phylogeny. *Ann. Rev. Anthropology* 3: 203-228.

Goodman, M., and G.W. Moore 1973. Phylogeny of hemoglobin. *Systematic Zoology* 22: 508-532.

Goodman, M., G.W. Moore, J. Barnabas, and G. Matsuda 1974. The phylogeny of human globin genes investigated by the maximum parsimony method. *J. Molec. Evolution* 3: 1-48.

Hendrickson, W.A., and W.E. Love 1971. Structure of lamprey haemoglobin. *Nature* 232: 187-203.

Holmquist, R., C.R. Cantor, and T.H. Jukes 1972. Improved procedures for comparing homologous sequences in molecules of proteins and nucleic acids. *J. Molec. Biol.* 64: 145-161.

Huisman, T.H.J. and W.A. Schroeder 1971. *New Aspects of the Structure, Function, and Synthesis of Hemoglobins.* CRC Press, Cleveland.

Inamura, T., T.O. Baldwin, and A. Riggs 1972. The amino acid sequence of the monomeric hemoglobin component from the bloodworm, *Glycera dibranchiata. J. Biol. Chem.* 247: 2785-2797.

Jukes, T.H. and R. Holmquist 1972. Estimation of evolutionary changes in certain homologous polypeptide chains. *J. Molec. Biol.* 64: 163-179.

Li, S.L. and A. Riggs 1970. The amino acid sequence of hemoglobin V from the lamprey, *Petromyzon marinus. J. Biol. Chem.* 245: 6149-6169.

Matsuda, G., T. Maita, B. Watanabe, A. Araya, K. Morokuma, M. Goodman, and W. Prychodko 1973a. The amino acid sequences of the α and β polypeptide chains of adult hemoglobin of the savannah monkey *(Cercopithecus aethiops). Hoppe-Seyler's Z. Physiol. Chem.* 354: 1153-1155.

Matsuda, G. T. Maita, B. Watanabe, A. Araya, K. Morokuma, Y. Ota, M. Goodman, J. Barnabas, and W. Prychodko 1973b. The amino acid sequences of the α and β polypeptide chains of adult hemoglobin of the capuchin monkey *(Cebus apella). Hoppe-Seyler's Z. Physiol.Chem.* 354: 1513-1516.

Matsuda, G., T. Maita, Y. Suzuyama, M. Setoguchi, Y. Ota, A. Araya, M. Goodman, J. Barnabas, and W. Prychodko 1973c. Studies on the primary structures of α and β polypeptide chains of adult hemoglobin of the spider monkey *(Ateles geoffroyi). Hoppe-Seyler's Z. Physiol. Chem.* 354: 1517-1520.

Matsuda, G., T. Maita, B. Watanabe, H. Ota, A. Araya, M. Goodman, and W. Prychodko 1973d. The primary structure of the alpha and beta polypeptide chains of adult hemoglobin of the slow loris *(Nycticebus coucang). Int. J. Peptide Protein Res.* 5: 419-422.

Matsuda, G., T. Maita, Y. Nakashima, J. Barnabas, P.K. Ranjekar, and N.S. Gandhi 1973e. The primary structures of the α and β polypeptide chains of adult hemoglobin of the Hanuman langur (*Presbytis entellus*). *Int. J. Peptide Protein Res.* 5: 423-426.

Moore, G.W., J. Barnabas, and M. Goodman 1973a. A method for constructing maximum parsimony ancestral amino acid sequences on a given network. *J. Theor. Biol.* 38: 459-485.

Moore, G.W., M. Goodman, and J. Barnabas 1973b. An iterative approach from the standpoint of the additive hypothesis to the dendrogram problem posed by molecular data sets. *J. Theor. Biol.* 38: 423-457.

Padlan, E.A. and W.E. Love 1968. Structure of the haemoglobin of the marine annelid worm, *Glycera dibranchiata*, at 5.5A° resolution. *Nature* 220: 376-378.

Perutz, M.F. 1969. The haemoglobin molecule. *Proc. Roy. Soc. B.* 173: 113-140.

Perutz, M.F. 1970a. Sterochemistry of cooperative effects in haemoglobin. Haem-haem interaction and the problem of allostery. *Nature* 228: 726-734.

Perutz, M.F. 1970b. The Bohr effect and combination with organic phosphates. *Nature* 228: 734-739.

Perutz, M.F., J.C. Kendrew, and H.C. Watson 1965. Structure and function of haemoglobin. II. Some relations between polypeptide chain configuration and amino acid sequence. *J. Molec. Biol.* 13: 669-678.

Perutz, M.F. and L.F. TenEyck 1972. Stereochemistry of cooperative effects in hemoglobin. *Cold Spring Harbor Symp. Quant. Biol.* 36: 295-310.

Popp, R.A. 1973. Sequence of amino acids in the β chain of single hemoglobins from C57 Bl, SWR, and NB mice. *Biochem. Biophys. Acta* 303: 52-60.

Romer, A.S. 1966. *Vertebrate Paleontology*. University of Chicago Press, Chicago.

Tentori, L., G. Vivaldi, S. Carta, M. Marinucci, A. Massa, E. Antonini, and M. Brumori 1973. The amino acid sequence of myoglobin from the mollusc *Aplysia limacina*. *Int. J. Peptide Protein Res.* 5: 187-200.

Whittaker, R.G., W.K. Fisher, and E.O.P. Thompson 1973. Studies on monotreme proteins. II. Amino acid sequence of the α-chain in haemoglobin from the echidna, *Tachyglossus aculeatus aculeatus*. *Aust. J. Biol. Sci.* 26: 877-888.

Young, J.Z. 1962. *The Life of the Vertebrates*. 2nd ed. University Press, Oxford.

EVOLUTION OF THE CARBONIC ANHYDRASE ISOZYMES

RICHARD E. TASHIAN[1], MORRIS GOODMAN[2], ROBERT J. TANIS[1], ROBERT E. FERRELL[1], and WILLIAM R. A. OSBORNE[1]
Department of Human Genetics[1]
University of Michigan Medical School
Ann Arbor, Michigan 48104

and

Department of Anatomy[2]
Wayne State University School of Medicine
Detroit, Michigan 48207

ABSTRACT. Portions of the primary sequences for the carbonic anhydrase isozymes (CA I and CA II) of man, five higher primates (chimpanzee, orangutan, vervet, rhesus macaque, and and baboon), two ruminants (ox and sheep), and a lagomorph (rabbit) have been compared. A maximum parsimony statistical method was used to determine the number of nucleotide substitutions occurring during the descent of CA I and CA II in these mammalian species. Over the past 90 million years, the rate of incorporation of mutational change appears to have been about twice as fast in the high-activity CA II isozymes as in the low-activity CA I isozymes. However, during the past 35 million years the evolutionary rate of the CA II enzymes of the higher primates seems to have decreased markedly. Over the same time period, the evolutionary rate of CA I does not appear to have changed significantly, with the exception of the human and chimpanzee enzymes where the rate has decelerated considerably. Preliminary studies on the carbonic anhydrases of 12 species of mammals and three species of birds suggest that the duplication which produced the carbonic anhydrase isozymes may have occurred early in mammalian evolution.

INTRODUCTION

The zinc metalloenzyme carbonic anhydrase (EC 4.2.1.1) appears to be present in almost all organisms, and is found in many different tissues of plants and animals. It is believed to be involved in a variety of physiological functions such as photosynthesis, calcification, acid-base balance, secretory processes, and gas exchange where the specific catalytic role of the enzyme is the interconversion of CO_2 and HCO_3^-. Although carbonic anhydrase can occur as an aggregate, the basic molecular weight is approximately 30,000 with the exception of the oyster (Nielsen and Frieden, 1972) and the shark (Maynard and Coleman 1971) enzymes where a basic molecular weight of 38,000 has been reported. For excellent comprehensive reviews of the physiological and chemical aspects of carbonic anhydrase see Maren (1967), Lindskog et al. (1971), and Carter (1972).

Two genetically distinct isozymes of carbonic anhydrase, designated CA I (or CA B) and CA II (or CA C) are known to occur in mammals (cf. Tashian 1969; Tashian et al., 1972). As compared to the CA I isozymes, the evolutionarily homologous CA II isozymes of mammals generally show higher specific CO_2 hydrase activities, higher sulfonamide binding affinities, and possibly higher thermostabilities (Osborne and Tashian, 1974a). As yet only one main form of carbonic anhydrase has been isolated from such nonmammalian sources as parsley (Tobin, 1970), bacteria (Adler et al., 1972), shark (Maynard and Coleman, 1971), tuna (Shimizu and Matsuura, 1962), and chicken (Bernstein and Schraer, 1972), and the enzymes from these organisms exhibit specific CO_2 hydrase activities similar to those of the so-called "high activity" CA II forms of mammals. Thus, it has been suggested that after the duplication of the ancestral high-activity carbonic anhydrase gene, its functional properties have been retained in the evolutionarily homologous high-activity isozymes in comparison to the homologous low-activity isozymes which have undergone changes in these properties (cf. Tashian et al., 1972).

The primary structures of both human isozymes have now been completed (Andersson et al., 1972; Henderson et al., 1973; Lin and Deutsch, 1973 and 1974). A 60% identical homology of the amino acid sequence of these two isozymes clearly confirms their descent from a common ancestral gene. In addition, the three-dimensional structures of human CA II and CA I have now been completed at high resolution (Kannan et al., 1971; Liljas et al., 1972; Notstrand et al., 1975). As might be expected, the tertiary structures of the two carbonic anhydrases are very similar (Notstrand et al., 1975). These similarities notwithstanding, there are, as mentioned above, a number of differences in kinetic and physical properties between these enzymes suggesting that each isozyme is associated with a different physiological role.

In this report, the evolutionary origins of the mammalian carbonic anhydrase isozymes will be examined, and their respective rates of evolution in different species of mammals will be compared. This information not only compares the rates of evolution between the two isozymes but will also indicate whether these rates changed or remained constant during their descent.

MATERIALS AND METHODS

For comparative purposes, actual or inferred sequences were used from the following sources: human CA I (Andersson et al., 1972); human CA II (Henderson et al., 1973); chimpanzee (*Pan troglodytes*) CA I, orangutan (*Pongo pygmaeus*) CA I, vervet (*Cercopithecus aethiops*) CA I and CA II, rhesus macaque (*Macaca mulatta*) CA I and CA II, baboon (*Papio cynecephalus*) CA I (Tashian and Stroup, 1970; Ferrell, unpublished orangutan data; Tanis, unpublished rhesus data); rabbit (*Orycytolagus cuniculus*) CA I (Ferrell, unpublished data); sheep (*Ovis aries*) CA II (Tanis and Tashian, 1974); ox (*Bos taurus*) CA II (Foveau et al., 1974), and European elk (*Alces alces*) CA II (Carlsson et al., 1973).

Construction of the Phylogenetic Tree. Since complete sequences are presently available for only human CA I and CA II, and sheep CA II, our strategy was to compare the maximum number of positions common to a majority of the available sequences. Table 1 shows the sequences which were used to construct the evolutionary tree of mammalian carbonic anhydrases. One hundred and forty-five residue positions were shared by human CA I with human, ox and sheep CA II, encompassing 136 positions of rhesus and vervet CA II; 126 positions of vervet, rhesus, and orangutan CA I; 117 positions of baboon, vervet, and chimpanzee CA I; and 78 positions of rabbit CA I. Thus, except for the rabbit data which represents about 30% of the molecule, the residues examined cover from 45 to 56% of the 260 possible positions that could have been compared.

The phylogenetic tree (Fig. 1) was constructed by the maximum parsimony method (Moore et al., 1973; Goodman et al., 1974). The numbers on the branches are the minimum amounts of nucleotide change required to account for the descent of the different carbonic anhydrases. These amounts are represented in terms of the number of nucleotide replacements (i.e., base changes) per 100 codons. The 12 known amino acid sequences in the data set were first translated into codon sequences, including all necessary alternative synonymous codons among the 61 codons specifying the 20 amino acids. Those ancestral descendant configurations which yielded the fewest mutations over the tree were then calculated. Where alternative configurations were equally parsimonious at a particular residue position, the computer program selected the configuration which distributed the mutations to those regions of the tree with the smallest amount of sequence data. This is because a nucleotide change tends to be grossly

TABLE 1

Partial Amino Acid Sequences of Mammalian Carbonic Anhydrase Isozymes (CA I and CA II). Numbering Based on Human CA I Sequence (Andersson et al., 1973). One Letter Amino Acid Notation Is That of Dayhoff (1972).

```
             2                                                                            80
Human I      SPDWGYDDKNGPEQWSKLYPIANGNNQSPVDIKTSETKHDTSLKPISVSYNPATAKEIINVGHSFHVNFEDNNDRSVLK
Chimp I      SPEWGYDDKNGPEQWSKLYPIANGNNQSPVDIKTSETKHDTSLKPISVSYNPATAK                  SVLK
Orang I      SPEWGYDDKNGPEQWSKLYPTANGNNQSPVDIKTSETKHDTSLKPISVSYNPATAK                  SVLK
Vervet I     SPEWGYDDKNGPEQWSKLYPTANGNNQSPVDIKTSETKHDTSLKPISVSYNPATAK                  SVLK
Rhesus I     SPDWGYDDKNGPEQWSKLYPTANGNNQSPVDIKTSETKHDTSLKPISVSYNPATAK                  SVLK
Baboon I     SPDWGYDDKNGPEQWSKLYPTANGNNQSPVDIKTSETKHDTSIKPISVSYNPATAK                  SVLK
Rabbit I                      LYPTIA           TSEVK DTSLKP

             2                                                                            80
Human II     SHHWGYGKHNGPEHWHKDFPIAKGERQSPVDIDTHTAKYDPSLKPISVSYDQATSLRILNNGHAFNVEFDDSZBKAVLK
Vervet II    SHHWGYGKHNGPEHWHKDFPIAKGERQSPVDIDTHTAKYDPSLKPISVSYDQATSLRILNNGHAFNVEFDDSZBKAVLK
Rhesus II    SHHWGYGKHNGPEHWHKDFPIAKGERQSPVDIDTHTAKYDPSLKPISVSYDQATSLRILNNGHAFNVEFDDSZBKAVIK
Sheep II     SHHWGYGEHNGPEHWHKDFPIADGERQSPVDIDTKAVVPDPALKPLALLYEQAASRRMVNNGHSFNVEFDDSQDKAVLK
Ox II        SHHWGYGKHBGPBHWHKDFPIANGERQSPVNIDTKAVVQDPALKPLALVYGEATSRRMVNNGHSFNVEYDDSQDKAVLK

             90     98  150                        178 227                             259
Human I      LFQFHEHWG  VGEANPKLQKVLDALQAIKTKGKR   SLLSNVEGDNAVPMQHNNRPTQPLKGRTVRASF
Chimp I                 VGEABPKLQKVLDALQAIKTKGKR   SLLSNVEGDNAVPMZHNNRPTQPLKGRTVRASF
Orang I      LFQFHFHWG  VGEABPKLQKVLDALQAIKTKGKR   SLLSNVEGDNAVPMEHNNRPTQPLKGRTVRASF
Vervet I                VGEABPKLQKVLDALHAIKTKGKR   SLLSNVEGDNPVPMZHNNRPTQPLKGRTVRASF
Rhesus I     LRQFHFHWG  VGEADPKLQKVLDALHAIKTKGKR   SLLSNVEGSNPVPMZRNNRPTQPLKGRTVRASF
Baboon I                LGEABPKLQKVLDALHAIKTKGKR   SLLSNVEGSNPVPMZRNNRPTQPLKGRTVRASF
Rabbit I     SRQRHFHWG  VGEADPKLQKVLDALHAVKTKGKR   SLLSNAEGEAAVPMLHNNRPP     GRTVKASF
```

TABLE 1 (continued)

	90–98	150–178	227–259
Human II	LIQFHFHWG	VGSAKPGLQKVVDVLDSIKTKGKS	KLNFDGEGEPEELMVDNWRPAQPLKNRQIKASF
Vervet II		VGSAKPGLQKVVDVLDSIKTKGKS	KLNFDGEGEPEELMVDNWRPAQPLKNRQIKASF
Rhesus II		VGSAKPGIQKVVDVIDSIKTKGKS	KLNFDGEGEPEELMVDNWRPAQPLKNRQIKASF
Sheep II	LVQFHFHWG	VGDANPALQKVLKVLKSIKTKGKS	SLNFNAEGEPELIMIANWRPAQPLKNRQVRVFP
Ox II	LVQFHFHWG	VGDANPALQKVIDALDSIKTKGKS	TLNFNAEGEPELIMLANWRPAQPLKNRQVRGFP

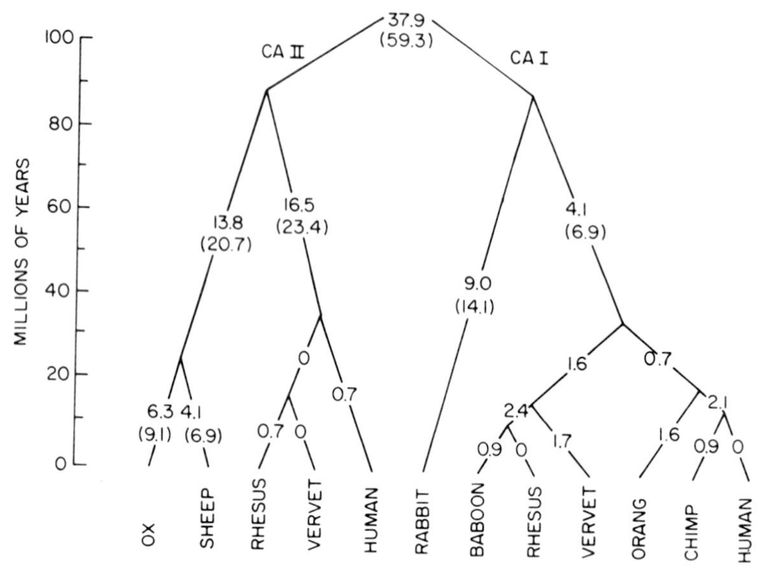

Fig. 1. Maximum parsimony phylogenetic tree of the carbonic anhydrase isozymes (CA I and CA II) based on comparisons of the partial sequences from nine species of mammals. The link length numbers are the number of nucleotide replacements per 100 codons; the link lengths in parentheses were calculated by the augmented maximum parsimony program.

underestimated in such relatively empty regions (Goodman et al., 1974). By this type of analysis, many different branching topologies could be examined in the search for the tree of minimum nucleotide change. No other topology in those examined was more parsimonious than the tree presented in Figure 1.

In addition, an augmented computer procedure, called MNAUG (Goodman et al., 1974), was used to obtain a better estimate of evolutionary change in the regions of the parsimony tree with the smallest number of ancestors.

C-terminal sequence determinations. In addition to the 12 C-terminal sequences listed in Table 1, five additional C-terminal sequences were obtained from pig CA I, horse CA I and CA II, deer CA II (Ashworth et al., 1972), and the European elk CA II (Carlsson et al., 1973). The sequences from erythrocyte carbonic anhydrases of dog, cat, pigeon, duck, and chicken were determined in our laboratory. In each case, a single major form of carbonic anhydrase was isolated from hemolysates of these animals by affinity chromatography using sulfonamide-coupled Sephadex columns by the method of

Osborne and Tashian (1974). The C-terminal sequences were determined on purified, heat-denatured, carbonic anhydrase using DFP-treated carboxypeptidase. Mixed carboxypeptidase A and B digestions were carried out for varying time periods and the released residues quantitated using routine amino acid analyzer techniques.

RESULTS AND DISCUSSION

COMPARISON OF ACTIVE SITE RESIDUES

Because the three-dimensional structures of human CA I and CA II have now been determined at high resolution, it has been possible to locate the positions of a number of residues located in the active sites of both enzymes. In view of the activity and binding differences between the two enzymes, it was of interest to compare some of the active site residues from both isozymes in different mammalian species to see which have remained constant and which have changed during the evolution of the isozymes.

In Table 2, 21 residues located in the active site cavities of the homologous forms of both isozymes are compared in several mammalian species. As might be expected, the three histidyl residues forming ligands to the zinc (positions 94, 96, and 119) are the same for all enzymes. In addition, His-64, Gln-92, His-107, Glu-117, Leu-141, Val-143, Tyr-193, Leu-198, Thr-199, His-200, Pro-201, Pro-202, Val-207, and Trp-209 are common to both isozymes. Those residues which are different, and might account for the different activities and binding properties of the two isozymes, are at positions: 62 (Val in CA I, Asn in CA II); 67 (His in CA I, Asn in CA II); 106 (Glu in CA I, Glu or Gln in CA II); 131 (Leu in CA I, Phe in CA II); and 200 (His in CA I, Thr or Asn in CA II).

COMPARISONS OF AROMATIC CLUSTERS

Two pronounced aromatic clusters which are probably responsible for the maintenance of the stability of the active site cavities have been described on the basis of the three-dimensional structures of human CA I and CA II. One of these, cluster I, is located in the amino terminal region near the active site, and the other, cluster II, is located in the interior of the molecule below the active site. In Table 3, the residues which make up these clusters are compared from the carbonic anhydrases of ten different mammals. In cluster I, the only difference occurs at position 20 where the residue is Tyr for the CA I enzymes and Phe for the CA II enzymes.

TABLE 2

COMPARISONS OF RESIDUES LOCATED IN THE ACTIVE SITE REGIONS OF MAMMALIAN CARBONIC ANHYDRASES. RESIDUES LIGANDED TO ZINC DESIGNATED BY A 'Z'

CA Isozyme		Residue Number*																				
		62	64	67	92	94	96	106	107	117	119	131	141	143	193	198	199	200	201	202	207	209
						Z	Z				Z											
Human	I	Val	His	His	Gln	His	His	Glu	His	Glu	His	Leu	Leu	Val	Tyr	Leu	Thr	His	Pro	Pro	Val	Trp
Orang	I	-	-	-	Gln	His	His	-	-	-	-	-	-	-	Tyr	-	-	His	Pro	Pro	-	-
Vervet	I	-	-	-	-	-	-	-	-	-	-	-	-	-	-	Leu	Thr	-	-	-	-	-
Rhesus	I	-	-	-	Gln	His	His	Glu	His	-	-	-	-	-	Tyr	Leu	Thr	-	Pro	Pro	-	-
Rabbit	I	-	-	-	Gln	His	His	-	-	-	-	-	-	-	-	-	-	-	-	-	-	-
						Z	Z				Z											
Human	II	Asn	His	Asn	Gln	His	His	Gln	His	Glu	His	Phe	Leu	Val	Tyr	Leu	Thr	Thr	Pro	Pro	Val	Trp
Vervet	II	Asn	His	Asn	-	-	-	-	-	-	-	-	-	-	-	-	-	-	-	-	-	-
Rhesus	II	Asn	His	Asn	-	-	-	-	-	-	-	-	-	-	-	-	-	-	-	-	-	-
Sheep	II	Asn	His	Asn	Gln	His	His	Glu	His	Glu	His	Phe	Leu	Val	Tyr	Leu	Thr	Asn	Pro	Pro	Val	Trp
Ox	II	Asn	His	Asn	Glx	His	His	Glx	His	Glu	His	Phe	Leu	Val	Tyr	Leu	Thr	Asx	Pro	Pro	Val	Trp
Elk	II	-	-	-	Glx	His	His	-	-	-	-	-	-	-	-	-	-	-	-	-	-	-

*Based on human CA I sequence.

TABLE 3

COMPARISON OF RESIDUES LOCATED IN TWO AROMATIC CLUSTERS OF MAMMALIAN CARBONIC ANHYDRASES

	CA Isozyme	Aromatic Cluster I Residue Number*					Aromatic Cluster II Residue Number							
		5	7	16	20	64	66	70	93	95	98	176	179	226
Man	I	Trp	Tyr	Trp	Tyr	His	Phe	Phe	Phe	Phe	Trp	Phe	Phe	Phe
Chimpanzee	I	Trp	Tyr	Trp	Tyr	-	-	-	-	-	-	-	-	-
Orangutan	I	Trp	Tyr	Trp	Tyr	-	-	-	Phe	Phe	Trp	Phe	Phe	Phe
Vervet	I	Trp	Tyr	Trp	Tyr	-	-	-	-	-	-	-	-	-
Rhesus	I	Trp	Tyr	Trp	Tyr	-	-	-	Phe	Phe	Trp	Phe	Phe	Phe
Baboon	I	Trp	Tyr	Trp	Tyr	-	-	-	-	-	-	-	-	-
Rabbit	I	Trp	Tyr	Trp	Tyr	-	-	-	Phe	Phe	Trp	Phe	Phe	Phe
Human	II	Trp	Tyr	Trp	Phe	His	Phe	Phe	Phe	Phe	Trp	Phe	Phe	Phe
Vervet	II	Trp	Tyr	Trp	Phe	His	Phe	Phe	-	-	-	-	-	-
Rhesus	II	Trp	Tyr	Trp	Phe	His	Phe	Phe	-	-	-	-	-	-
Sheep	II	Trp	Tyr	Trp	Phe	His	Phe	Phe	Phe	Phe	Trp	Phe	Phe	Phe
Ox	II	Trp	Tyr	Trp	Phe	His	Phe	Tyr	Phe	Phe	Trp	Phe	Phe	Phe
Elk	II	-	-	-	-	-	-	-	Phe	Phe	Trp	-	-	-

*Based on human CA I sequence.

In cluster II, the phenylalanyl residues at all six positions have remained unchanged except for Tyr at position 70 in the ox enzyme.

COMPARATIVE RATES OF EVOLUTION

In Figure 1, an ordinate scale of millions of years is used for the period of eutherian (placental mammals) phylogeny with the times for some of the branch points in the tree chosen from paleontological evidence on the divergence times of the eutherian lineages. Thus, the eutherian ancestor from which ungulate, primate, and lagomorph lineages diverged is placed at 90 million years ago (McKenna 1969), and the bovid ancestor at 25 million years ago (Gentry, 1970), the catarrhine (Old World monkeys, apes, and man) ancestor at 35 million years ago (Uzzell and Pilbeam, 1971; Simons, 1967), and the human-chimpanzee ancestor at 15 million years ago (Pilbeam, 1968). Using these times and the amounts of nucleotide change shown on the branches of the tree, the rates of evolution for carbonic anhydrases I and II can be compared during the period of eutherian phylogeny (Table 4). CA I shows an average evolutionary rate during the past 90 million years of one base change per 100 codons every 11.5 million years. About the same average rate (one base change/100 codons/10.0 million years) is maintained in the catarrhine lineages during the past 35 million years. However, in the human and chimpanzee lineages, the rate appears to have decreased to a base change per 100 codons every 33 million years. This is based on the assumption that the 14 million year old *Ramapithecus* specimen from Kenya (Pilbeam, 1968) is indeed an antecedent of the genus *Homo* which had already separated in descent from the most recent common ancestor of chimpanzees and humans. If the most recent common ancestor of chimpanzee and human still existed within the last 10 million years, then the apparent deceleration in the rate of evolution of human CA I would not be as marked.

With respect to the CA II isozymes, an exceptionally pronounced deceleration in evolutionary rate seems to have occurred in the catarrhine primates. This isozyme shows an average rate during the past 90 million years of one nucleotide change/100 codons every 5.1 million years which is over twice the rate of CA I. The faster rate (one nucleotide change/100 codons/4.8 million years) was maintained in the bovid lineages. In the catarrhine lineages, however, the rate of evolution in the descent of the CA II isozymes decreased drastically to only one base change/100 codons every 75 million years.

Using the augmented program (cf. Table 4), the average

TABLE 4

COMPARATIVE RATES (BASE CHANGES/100 CODONS) OF DESCENT FOR CARBONIC ANHYDRASE ISOZYMES FROM VARIOUS COMMON ANCESTORS*

	Eutherian ancestor (90×10^6 yrs)				Catarrhine ancestor (35×10^6 yrs)		Bovid ancestor (25×10^6 yrs)		Human-chimpanzee ancestor (15×10^6 yrs)
	CA I		CA II		CA I	CA II	CA II		CA I
	A**	B**	A	B	A	A	A	B	A
Human	6.9	9.7	17.2	24.1	2.8	0.7			0
Chimpanzee	7.8	10.6	-	-	3.7	-			0.9
Orangutan	6.4	9.2	-	-	2.3	-			
Vervet	7.4	10.2	16.5	23.4	3.3	0			
Rhesus	8.1	10.9	17.2	24.1	4.0	0.7			
Baboon	9.0	11.8	-	-	4.9	-			
Rabbit	9.0	14.1	-	-					
Sheep	-	-	17.9	27.0			4.1	6.9	
Ox	-	-	20.1	29.3			6.3	9.2	
Total	54.6	76.5	88.9	127.9	21.0	1.4	10.4	16.1	0.9
Average time ($\times 10^6$ yrs) for one base change/100 codons	11.5	8.2	5.1	3.5	10.0	75.0	4.8	3.1	33.0

*Link lengths summed from Fig. 1.
**A = Original link lengths; B = augmented link lengths.

rates show an increase for both isozymes during their descent from a common eutherian ancestor, and in the descent of ox and sheep from a common bovid ancestor. Since no change in the rates of CA I and CA II were found with the augmented program in the descent of these enzymes from a common catarrhine ancestor, the decrease in the evolutionary rates for the primate CA II isozymes is even more marked.

Obviously, these kinds of trends should be viewed with circumspection until more data are available; nevertheless, they do present an interesting challenge to the "molecular clock" concept which holds that homologous proteins evolve at approximately equal rates in different lines of descent (cf. Dickerson, 1971; Ohta and Kimura, 1971). A similar deceleration in primate globins has been shown by the maximum parsimony method (cf. Goodman et al., 1974).

If we assume that the mutation rates for both CA I and CA II have remained essentially the same, it might indicate that selective forces were responsible for the observed differences in evolutionary rates. For example, CA II appears to have been evolving at about twice the rate of CA I over the past 90 million years; however, this is strikingly reversed in the higher primates where the evolution of CA II may have been seven times slower that CA I. This would suggest that it has been physiologically important to preserve the primary structure of CA II in the higher primates.

DUPLICATION OF THE CARBONIC ANHYDRASE GENE

We would have a better understanding of the evolutionary history of the carbonic anhydrase isozymes if we could determine the period in evolutionary time where the proposed gene duplication occurred. One approach would be to identify the the oldest group of organisms in which both isozymes are known to occur. As previously mentioned, only one form of carbonic anhydrase has as yet been isolated from non-mammalian sources. If only one form is present in birds and reptiles as well as in all evolutionarily older organisms, and two forms are present in such primitive mammals as marsupials, then one could speculate that the duplication took place sometime before the separation of marsupials from the ancestors of other mammals (about 120 million years ago) and the point of divergence of the reptile-bird line from the line leading to mammals (about 300 million years ago). Since it is critical to establish whether or not one high-activity form of carbonic anhydrase is characteristic for all birds, we undertook a preliminary study to examine the red cell carbonic anhydrases of pigeon, duck, and chicken utilizing an affinity chromatography method which is capable of separately isolating both isozymes from mammalian hemolysates based on their different sulfonamide binding affinities (Osborne and Tashian, 1974). Both forms are bound to a sulfonamide-coupled CM-Sephadex column. The CA I forms are then selectively eluted with 0.4 M KI, and the more firmly bound CA II forms are eluted with 0.2 M KCN. Our results indicate, on the basis of differential sulfonamide binding affinities, that pigeon, chicken and duck possess a high-activity form apparently homologous to the CA II of mammals.

In addition to these binding properties, a comparison of the C-terminal residues might also be helpful in indentifying the homologies of the bird carbonic anhydrases. Table 5 lists the C-terminal CA I and CA II sequences from a number of mammals, along with the sequences from the single main form of carbonic anhydrase purified from dog, cat, pigeon, duck and chicken erythrocytes. The CA I forms characteristically show Phe or Ser as the penultimate residue, and Phe as the C-terminal residue. On the other hand, the CA II molecules show Ser or Phe as the residue third from the end, Phe, Pro, or Leu as the penultimate residue, and Lys or Arg as the C-terminal residue.

As shown in Table 5, only one form of carbonic anhydrase was found in the red cells of the three species of birds that were examined. Since all of the bird carbonic anhydrases showed both high CO_2 hydrase activities and high sulfonamide binding affinities (Tashian, unpublished data), it would indicate that they are functionally more homologous to the high-activity forms of mammals. However, until more sequence data are available, and more species of birds are examined, we cannot be certain that birds characteristically possess only one form of carbonic anhydrase.

The evidence now available would suggest that the gene duplication which gave rise to the two mammalian anhydrase isozymes took place early in mammalian evolution, perhaps 100-150 million years ago. As yet we have no clear-cut evidence that both isozymes are present in a marsupial. The one species that we have looked at most carefully, the red kangaroo (*Macropus rufus*) showed only one form (high-activity) in the red cell (Tashian, unpublished data). If two isozymes of carbonic anhydrase are not found in marsupials, then the duplication may have occurred after the separation of the placental mammals. It is of interest that the two carbonic anhydrase isozyme loci appear to be closely linked in mammals (Carter, 1973; DeSimone et al., 1973), a fact which might indicate that the duplication is relatively recent.

TABLE 5

COMPARISON OF C-TERMINAL SEQUENCES OF MAMMALIAN AND AVIAN RED CELL CARBONIC ANHYDRASES

Source	CA Isozyme	C-terminal sequence
Human	I	-Gly-Arg-Thr-Val-Arg-Ala-Ser-Phe-COOH
Chimpanzee	I	-Gly-Arg-Thr-Val-Arg-Ala-Ser-Phe-COOH
Orangutan	I	-Gly-Arg-Thr-Val-Arg-Ala-Ser-Phe-COOH
Vervet	I	-Gly-Arg-Thr-Val-Arg-Ala-Ser-Phe-COOH
Rhesus	I	-Gly-Arg-Thr-Val-Arg-Ala-Ser-Phe-COOH
Baboon	I	-Gly-Arg-Thr-Val-Arg-Ala-Ser-Phe-COOH
Rabbit	I	-Gly-Arg-Thr-Val-Lys-Ala-Ser-Phe-COOH
Pig	I	-Lys-Ala-Ser-Phe-COOH
Horse	I	-Val-Arg-Ala-Phe-Phe-COOH
Dog*	I	-Gly-Arg-Ile-Val-Lys-Ala-Ser-Phe-COOH
Human	II	-Asn-Arg-Gln-Ile-Lys-Ala-Ser-Phe-Lys-COOH
Vervet	II	-Asn-Arg-Gln-Ile-Lys-Ala-Ser-Phe-Lys-COOH
Rhesus	II	-Asn-Arg-Gln-Ile-Lys-Ala-Ser-Phe-Lys-COOH
Horse	II	-Ile-Arg-Ala-Ser-Phe-Lys-COOH
Sheep*	II	-Asn-Arg-Gln-Val-Arg-Val-Phe-Pro-Lys-COOH
Ox*	II	-Asn-Arg-Glu-Val-Arg-Gly-Phe-Pro-Lys-COOH
Elk*	II	-Leu-Arg-COOH
Deer*	II	-Pro-Arg-COOH
Cat*	II	-His-Phe-Ser-Lys-COOH
Duck*	?	-Ala-Ser-Phe-Ser-COOH
Pigeon*	?	-Val-Arg-Ala-Ser-Phe-Ser-COOH
Chicken*	?	-Arg-Ala-Ser-Phe-Ser-COOH

*CA present as only one form in the red cell.

ACKNOWLEDGEMENTS

This work was supported by grant GM-15419 from the U. S. Public Health Service, and grant GB-36157 from the National Science Foundation.

We wish to thank Dr. G. William Moore for his assistance in preparing the maximum parsimony computer program used in this study. We are also especially thankful to Dr. K. K. Kannan and Dr. Louis E. Henderson for the valuable discussions we had with them concerning the primary and tertiary structures of the human carbonic anhydrase molecules. The excellent research assistance of Mrs. Ya-shiou L. Yu is gratefully acknowledged.

REFERENCES

Adler, L., J. Brundell, S. O. Falkbring, and P. O. Nyman 1972. Carbonic anhydrase from *Neisseria sicca*, strain 6021. I. Bacterial growth and purification of the enzyme. *Biochim. Biophys. Acta* 284: 298-310.

Andersson, B., P. O. Nyman, and L. Strid 1972. Amino acid sequence of human erythrocyte carbonic anhydrase B. *Biochem. Biophys. Res. Commun.* 48: 670-677.

Ashworth, R. B., J. M. Brewer, and R. L. Stanford, Jr. 1971. Composition and carboxyterminal amino acid sequences of some mammalian erythrocyte carbonic anhydrases. *Biochem. Biophys. Res. Commun.* 44: 667-674.

Bernstein, R. S. and R. Schraer 1972. Purification and properties of avian carbonic anhydrase from the erythrocytes of *Gallus domesticus*. *J. Biol. Chem.* 247: 1306-1322.

Carlsson, U., U. Hannestad, and S. Lindskog 1973. Purification and some properties of erythrocyte carbonic anhydrase from the European moose. *Biochim. Biophys. Acta* 327: 515-527.

Carter, M. J. 1972. Carbonic anhydrase: isozymes, properties distribution, and functional significance. *Biol. Rev.* 47: 465-513.

Carter, N. D. 1972. Carbonic anhydrase isozymes in *Cavia porcellus*, *Cavia aperea* and their hybrids. *Comp. Biochem. Physiol.* 43B: 743-747.

Dayhoff, M. O. 1972. *Atlas of Protein Sequence and Structure*. Vol. 5, National Biomedical Research Foundation, Silver Spring, Maryland.

DeSimone, J., M. Linde, and R. E. Tashian 1973. Evidence for linkage of carbonic anhydrase genes in the pig-tailed macaque, *Macaca nemestrina*. *Nature New Bio.* 242: 55-56.

Dickerson, R. 1971. The structure of cytochrome c and the rates of molecular evolution. *J. Mol. Evolution* 1: 26-45.

Foveau, D., M. Sciaky, and G. Laurent 1974. Nouvelles donneés sur la parenté chimique des anhydrases carboniques érythrocytaires B and C humaines et CI bovine. *Compt. Rend. Acad. Sc.* 278, Serie D: 959-962.

Gentry, A. W. 1970. The bovidae (mammalia) of the Fort Ternan fossil fauna. In (L.S.B. Leakey and R.J.G. Savage, eds.), *Fossil Vertebrates of Africa*, Vol. 2, Academic Press, London.

Goodman, M., G. W. Moore, and J. Barnabas 1974. The phylogeny of human globin genes investigated by the maximum parsimony method. *J. Mol. Evolution* 3: 1-48.

Henderson, L. E., D. Henriksson, and P. O. Nyman 1973. Amino acid sequence of human erythrocyte carbonic anhydrase C. *Biochem. Biophys. Res. Commun.* 52: 1388-1394.

Kannan, K. K., A. Liljas, I. Waara, P.-C. Bergstén, S. Lövgren, C. Strandberg, U. Bengtsson, U. Carlbom, K. Fridborg, L. Järup, and M. Petef. 1971. Crystal structure of human erythrocyte carbonic anhydrase C. VI. The three-dimensional structure at high resolution in relation to other mammalian carbonic anhydrases. *Cold Spring Harbor Symp. Quant. Biol.* 36: 221-231.

Liljas, A., K. K. Kannan, P.-C. Bergstén, I. Waara, K. Fridborg, B. Strandberg, U. Carlbom, L. Järup, S. Lövgren, and M. Petef.1972. Crystal structure of human carbonic anhydrase C. *Nature New Biol.* 235: 131-137.

Lin, K.-T. D. and H. F. Deutsch 1973. Human carbonic anhydrases. XI. The complete primary structure of carbonic anhydrase B. *J. Biol. Chem.* 248: 1885-1893.

Lin, K.-T. D. and H. F. Deutsch 1974. Human carbonic anhydrases. XII. The complete primary structure of the C isozyme. *J. Biol. Chem.* 249: 2329-2337.

Lindskog, S., L. E. Henderson, K. K. Kannan, A. Liljas, P. O. Nyman, and B. Strandberg 1971. Carbonic anhydrase. p. 587. In: (P. D. Boyer, ed.). *The Enzymes*, Vol. 5, Academic Press, New York.

McKenna, M. C. 1969. The origin and early differentiation of therian mammals. *Ann. N. Y. Acad. Sci.* 167: 217-240.

Maren, T. 1967. Carbonic anhydrase: chemistry, physiology, and inhibition. *Physiol. Rev.* 47: 595-781.

Maynard, James R., and J. E. Coleman 1971. Elasmobranch carbonic anhydrase. Purification and properties of the enzyme from two species of shark. *J. Biol. Chem.* 246: 4455-4464.

Moore, G. W., J. Barnabas, and M. Goodman 1973. A method for constructing maximum parsimony ancestral amino acid sequences on a given network. *J. Theroret. Biol.* 38: 459-485.

Nielsen, S. A. and E. Frieden 1972. Some chemical and kinetic properties of oyster carbonic anhydrase. *Comp. Biochem. Physiol.* 41B: 875-889.

Notstrand, B., I. Vaara, and K.K. Kannan 1975. Structural relationship of human carbonic anhydrase isozymes B and C. In: *I. Isozymes: Molecular Structure*, C. L. Markert, editor, Academic Press, New York.

Ohta, T. and M. Kimura 1971. On the constancy of the evolutionary rate of cistrons. *J. Mol. Evolution* 1: 18-25.

Osborne, W. R. A. and R. E. Tashian 1974a. Thermal inactivation studies of normal and variant human erythrocyte carbonic anhydrases by using a sulphonamide binding assay. *Biochem. J.* 141: 219-225.

Osborne, W. R. A. and R. E. Tashian 1974b. The large scale purification of carbonic anhydrase isozymes by affinity chromatography. *Analyt. Biochem.* (in press).

Pilbeam, P. 1968. The earliest homonids. *Nature* 219: 1335-1338.

Simons, E. 1967. The earliest apes. *Sci. Am.* 217: 28-35.

Shimizu, C. and F. Matsuura 1962. Some physico-chemical properties of crystalline carbonic anhydrase from erythrocytes of ox, blue-white dolphin and yellow-fin tuna. *Bull. Japan. Soc. Sci. Fish.* 28: 924-929.

Tanis, R. J. and R. E. Tashian 1974. Amino acid sequence of sheep carbonic anhydrase C. *Biochim. Biophys. Acta* (in press).

Tashian, R. E. 1969. The esterases and carbonic anhydrases of human erythrocytes. p. 307. In: (J.J. Yunis, ed.) *Biochemical Methods in Red Cell Genetics*, Academic Press, New York.

Tashian, R. E., R. J. Tanis, and R. E. Ferrell 1972. Comparative aspects of the primary structures and activities of mammalian carbonic anhydrases. *Alfred Benzon Symposium* 4: 353-362.

Tashian, R. E. and S. R. Stroup 1970. Variation in the primary structure of carbonic anhydrase B in man, great apes, and Old World monkeys. *Biochem. Biophys. Res. Commun.* 41: 1457-1462.

Tobin, A. J., 1970. Carbonic anhydrase from parsley leaves. *J. Biol. Chem.* 245: 2656-2666.

Uzzell, T., and P. Pilbeam 1971. Phyletic divergence dates of homonoid primates: A comparison of fossil and molecular data. *Evolution* 25: 615-635.

BLOOD PROTEIN POLYMORPHISMS AND POPULATION STRUCTURE OF THE JAPANESE MACAQUE, *MACACA FUSCATA FUSCATA*

KEN NOZAWA, TAKAYOSHI SHOTAKE AND YOSHIKO OKURA

Department of Variation Research
Primate Research Institute
Kyoto University
Inuyama, Aichi-ken 484, JAPAN

ABSTRACT. Genetic variability in individual troops of the Japanese macaque was quantified by two measures, that is, the proportion of polymorphic loci and the proportion of heterozygous loci per individual. The former averaged 10.8% and the latter 1.7%, and these values were remarkably lower than those estimated for other animal populations. Observations of the distribution patterns of genetic variations among the macaque troops indicated that an individual troop could not be regarded as a genetic isolate because of exchange of individuals with neighboring troops. Assuming the neutrality of segregating alleles and the two-dimensional stepping-stone model of population structure, the genetic migration rate between troops was estimated to average less than 5% per generation. Analyses of correlation between geographic and genetic distances between troops revealed that the gene constitutions of two troops apart more than 100 km could be regarded as practically independent of each other. These results suggest that the population structure of the Japanese macaque species has a tendency to split into a number of local subpopulations in which the effect of random genetic drift is prevailing.

INTRODUCTION

Ecological and sociological information on the Japanese macaque, *Macaca fuscata fuscata*, have been abundantly accumulated, mainly by Japanese field biologists, for more than 20 years. The whole population of this species, of which the census number is estimated as 20,000 - 70,000, is composed of several hundred troops. Each macaque troop contains one or more adult males, adult females (the number of which is generally more than that of adult males), and young and infants of both sexes accompanying the adult females. The troop is a social unit integrated by a rank system among the adult members, but this does not necessarily mean that a macaque troop is a genetic isolate, because some solitary males from neighboring troops come into contact with the adult females in the breeding season. Thus, the problem is whether an individual troop is a closed system from the viewpoint of population genetics. If the troop is not a closed system, what is the rate of genetic migration between adjacent troops, and how widely dispersed are the genes from a troop?

The present study is designed to obtain information on the above points by using the frequencies of genes controlling some enzymatic and non-enzymatic blood-protein polymorphisms. The use of blood-protein polymorphisms as genetic markers for clarifying population dynamics is justified because of the theory that these genes are considered to be nearly neutral to natural selection (Kimura, 1968; King and Jukes, 1969), and because, even if not exactly neutral, the over-all variations of frequencies of a group of such polymorphic genes can be considered to give us the most reliable data for the solution of our problems if a sufficient number of loci are studied.

MATERIALS AND METHODS

A total of 812 blood samples were collected from 18 troops of the Japanese macaque during a period from April, 1971 through March, 1973. The names of these troops which are indicated by the names of localities and the numbers of samples are as follows: Fukushima (F, 31), Yugawara-T (YT, 61), Ihama (I, 40), Ryozenyama (R, 63), Mikata-I (MI, 14), Mikata-II (MII, 36), Takahama (T, 49), Arashiyama-A (AA, 148), Awajishima-I (AwI, 49), Awajishima-II (AwII, 37), Kochi (K, 33), Shodoshima-I (SI, 55), Shodoshima-K (SK, 23), Takasakiyama-A (TsA, 71), Takasakiyama-B (TsB, 18), Takasakiyama-C (TsC,44), Kamae (Km, 30), and Koshima (Ko, 10). The locations of these troops are shown in the maps of Figs. 1 - 3. The samples from the troops YT, I, R, AA, AwI, AwII, and K were collected directly from the macaque individuals at the respective locations, and all or a part of the samples from the other troops were collected in the zoological gardens or institutes to which the macaques, captured in their natural nomadic ranges, had been transferred for public exhibition or for experimental use.

A total of 21 genetic loci controlling the 19 kinds of blood proteins were examined by electrophoresis. The survey techniques adopted are as follows. Serum prealbumin (PA, Gahne, 1966), serum albumin (Alb, Ishimoto, 1972a), serum transferrin (Tf, Ishimoto, 1972a), serum haptoglobin (Hp, Ishimoto, 1972a), serum slow α_2-macroglobulin (examined on the plate for Tf), serum ceruloplasmin (Cp, Imlah, 1964), serum amylase (Amy, Ogita, 1966), serum thyroxin-binding prealbumin (TBPA, Tanabe et al., in prep.), hemoglobin (Hb, Ishimoto, 1972a), cell phosphohexose isomerase (PHI, Ishimoto, 1972b), cell 6-phosphogluconate dehydrogenase (PGD, Ishimoto, 1972b),

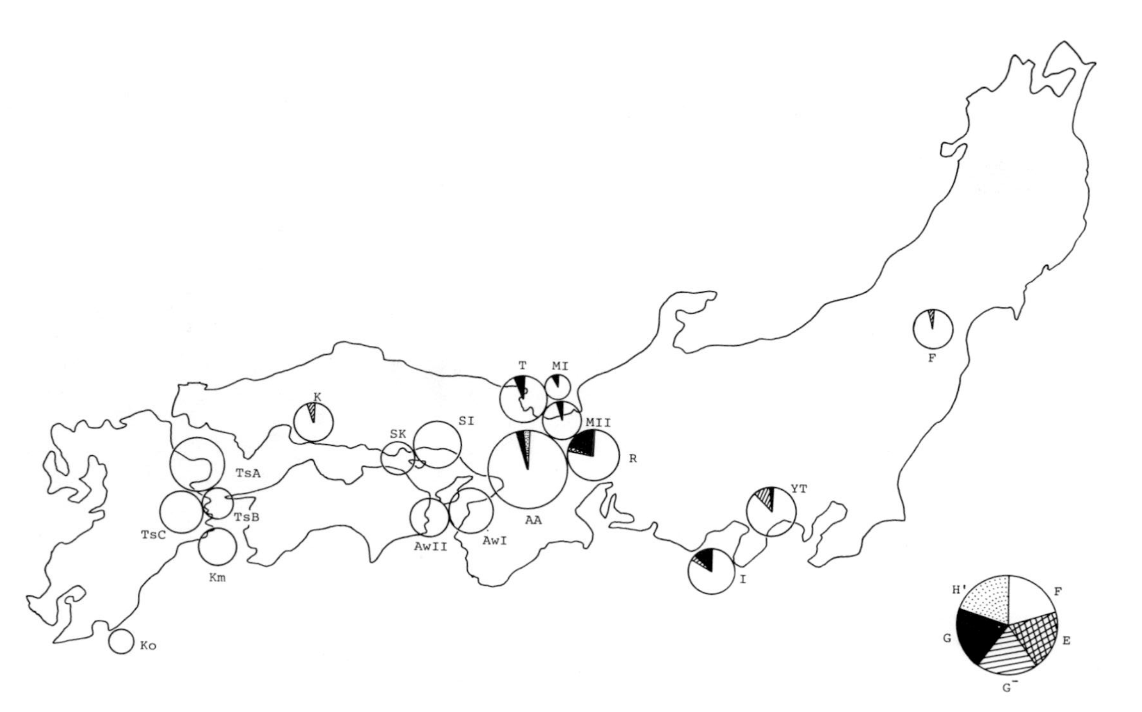

Figure 1. Distribution of transferrin (Tf) variants.

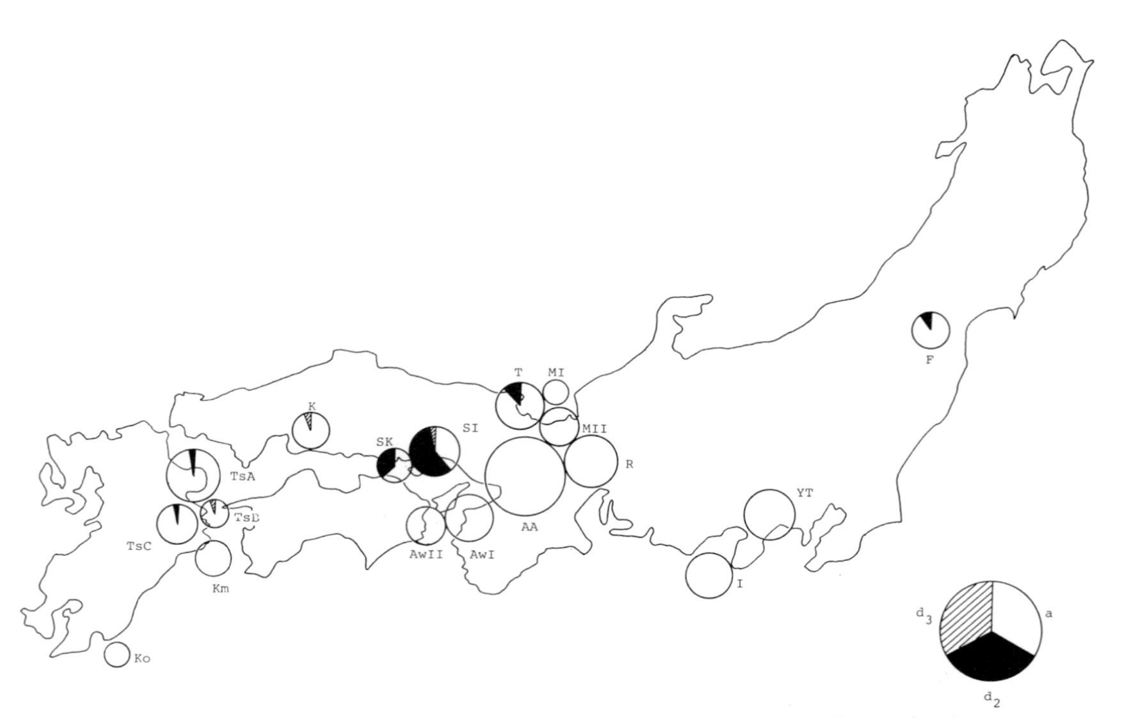

Figure 2. Distribution of carbonic anyhdrase-I (CA I) variants.

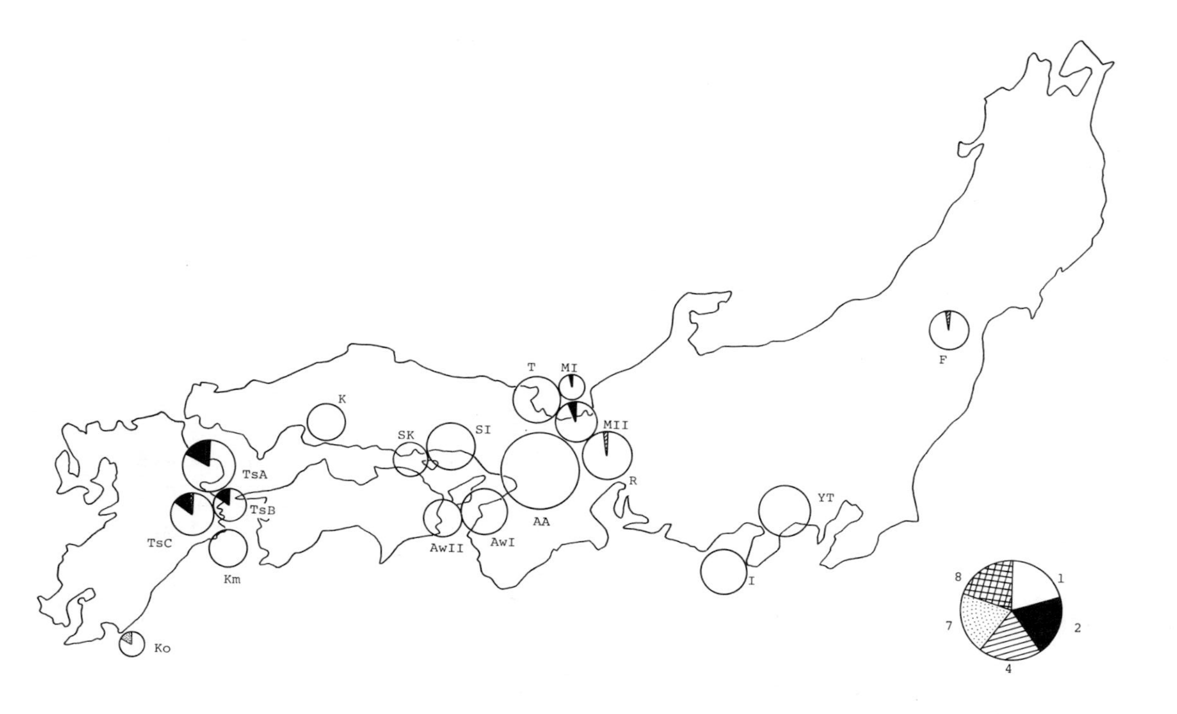

Figure 3. Distribution of phosphohexose isomerase (PHI) variants.

cell phosphoglucomutase I and II (PGM-I and PGM-II, Ishimoto, 1972b), cell adenosine deaminase (ADA, Spencer et al., 1968) cell NADH diaphorase (Dia, Ishimoto, 1972b), cell carbonic anhydrase-I (CA-I, Tashian et al., 1971), cell acid phosphatase (AP, Ishimoto, 1972b), cell glucose-6-phosphate dehydrogenase (G6PD, Shaw and Prasad, 1970), cell malate dehydrogenase (MDH, Shotake and Nozawa, in press), and cell lactate dehydrogenase A and B (LDH-A and LDH-B, Shotake, in press). Sampling of a protein locus was at random, having no relation to whether the existence of polymorphism at this locus was likely or not.

Allele frequencies at each locus in individual troops were calculated in accordance with the mode of inheritance of the variation already established, or by assuming codominance in those cases where equivalent electrophoretic bands could be observed and complete recessiveness in those cases where a band was missing. The genetic variability within troops was quantified by measuring the proportion of polymorphic loci (the criterion of polymorphism was the frequency of the commonest allele < 0.99), and the expected proportion of heterozygous loci per individual (Lewontin, 1967). Investigations of the population structure were accomplished by two methods of analysis of the genetic variability between troops. The first was to describe the observed distributions of allele frequencies among troops on the map of Japan and compare them with the mode of distribution expected mathematically from a model of population structure. The second procedure was to analyze the relationship between geographic distance and genetic distance between troops. The geographic distance between the j-th and k-th troops was expressed by the straight-line distance (km) between them, and the genetic distance ($\bar{D}_{jk}$) was measured by the following formula:

$$\bar{D}_{jk} = \frac{1}{l} \sum_{m=1}^{l} \sqrt{\sum_{i=1}^{n} (q_{imj} - q_{imk})^2}$$

where q_{imk} and q_{imk} were the frequencies of the i-th allele ($i=1, 2, \ldots, n$) at the m-th locus ($m=1, 2, \ldots, l$) in the j-th and k-th troops, respectively.

RESULTS AND DISCUSSION

GENETIC VARIABILITY WITHIN TROOPS

Of the 21 genetic loci examined 10 loci, that is, PA, Alb, Hp, Amy, TBPA, PGD, ADA, Dia, AP, and G6PD, were observed to lack any variation. Electrophoretic variations observed in the Tf, Hb, PHI, PGM-I, PGM-II, CA-I, MDH, LDH-A and LDH-B loci are illustrated in Fig. 5 and the postulated genotypes for the respective variations are shown in the figure. In the other two kinds of protein, α_2 and Cp, the normal band was missing in a few samples. Table 1 gives the numbers of different alleles counted at each locus in the 18 macaque troops. Table 2 gives the results of measurements of genetic variability within each troop. The proportion of polymorphic loci were in a range 0 ∿ 23.8%, the average being 10.8%. The proportion of heterozygous loci per individual expected from the estimated gene frequencies was in a range 0 ∿ 3.5%, the average being 1.7%.

Selander et al. (1970) tabulated the results of quantification of genetic variability in populations of several animal species made by different researchers. The table showed that the proportion of polymorphic loci was mostly distributed in a range from 25 to 40% and that the proportion of heterozygous loci per individual was from 5 to 15% irrespective of taxonomic position. Compared with the above values the genetic variability in the troops of the Japanese macaque is clearly much lower. Another remarkable feature of the genetic variation of the Japanese macaques is that the variants are not distributed uniformly in the whole species but occur only in some limited areas, the Tf^G allele in the central region of Japan (Fig. 1), CA^{d2} allele in Shodoshima Island (Fig. 2), and PHI^2 allele in northeastern Kyushu (Fig. 3) being typical examples. But, it should be recognized that the occurrence of variants is not limited to a single troop and that neighboring troops maintain common genetic variants. Such data deserve further consideration in connection with the population structure of this species.

MIGRATION RATE BETWEEN ADJACENT TROOPS

When we consider the migration problem between subpopulations of a species, it is fundamentally important to have a clear image about the breeding structure of that species. A convenient way to achieve this is to select the most suitable model from among the three models of population structure proposed by the mathematical geneticists, that is, the continuous distribution model or isolation by distance model (Wright, 1940, 1943), the island model (Wright, 1940), and the stepping-stone model (Kimura and Weiss, 1964). Considering the troop organization and the behavior of solitary males in the breeding season mentioned above, the population structure of the Japanese macaque would most suitably be approximated by the two-dimensional stepping-stone model (Kimura and Weiss, 1964). This model assumes that the whole population consists of an

Table 1. Genetic variations in the 18 troops of Japanese macaque (* Results of examination by Dr. Y. Tanabe, Gifu Univ., ** Including the results of examination by Dr. G. Ishimoto, Mie Univ.)

Troop	Number of alleles counted in variable loci																										
	Tf					Hb		PHI					PGM·I**			PGM·II		CA·I			MDH			LDH·A		LDH·B	
	E	F	G^-	G	H'	S	X	1	2	4	7	8	1	2	3	1	2	a	d_2	d_3	1	2	3	1	2	1	2
F		60	2			62		61		1			62			60	2	58	4		60		2	62		62	
YT		106	15	1		114	8	122					109		13	122		122			112	10		122		122	
I		66	5	9		70	10	80					77		3	80		80			80			80		80	
R	4	99		23		126		125		1			126			126		126			122	4		126		126	
MI		26		2		28		27	1				28			28		28			28			28		28	
MII		68		4		72		68	4				72			72		72			72			72		72	
T		91		7		98		98					94	4		98		87	11		82		16	98		98	
AA		283		6	7	296		296					296			296		296			296			296		296	
AwI		98				98		98					98			98		98			98			98		98	
AwII		74				74		74					74			74		74			74			74		74	
K		62		4		66		66					66			66		63		3	66			66		58	8
SI		110				110		110					110			110		51	57	2	110			110		110	
SK		46				46		46					46			46		34	12		46			46		46	
TsA		142				142		118	24				142			142		139	3		142			139	3	142	
TsB		36				36		30	6				36			36		35		1	36			35	1	36	
TsC		88				88		75	12			1	88			88		87	1		88			88		88	
Km		60				60		60					60			60		60			60			60		60	
Ko		20				20		17			3		20			20		20			20			20		20	

Normal band of α_2 was missing in one individual each of R and AA troops, and Cp band was missing in one of SI troop. Loci without variations : PA, Alb, Hp, Amy, TBPA*, PGD**, ADA, Dia**, Acp**, G6PD.

Table 2. Genetic variability of Japanese macaque

Troop	No. of polymorphic loci	Proportion of polymorphic loci	Proportion of heterozygous loci per individual
F	5	0.2380	0.0162
YT	4	0.1904	0.0330
I	3	0.1428	0.0283
R	3	0.1428	0.0307
MI	2	0.0952	0.0096
MII	2	0.0952	0.0100
T	4	0.1904	0.0329
AA	2	0.0952	0.0112
AwI	0	0	0
AwII	0	0	0
K	3	0.1428	0.0197
SI	2	0.0952	0.0352
SK	1	0.0476	0.0223
TsA	3	0.1428	0.0173
TsB	3	0.1428	0.0183
TsC	2	0.0952	0.0132
Km	1	0.0476	0.0031
Ko	1	0.0476	0.0121
Average		0.1084	0.0174

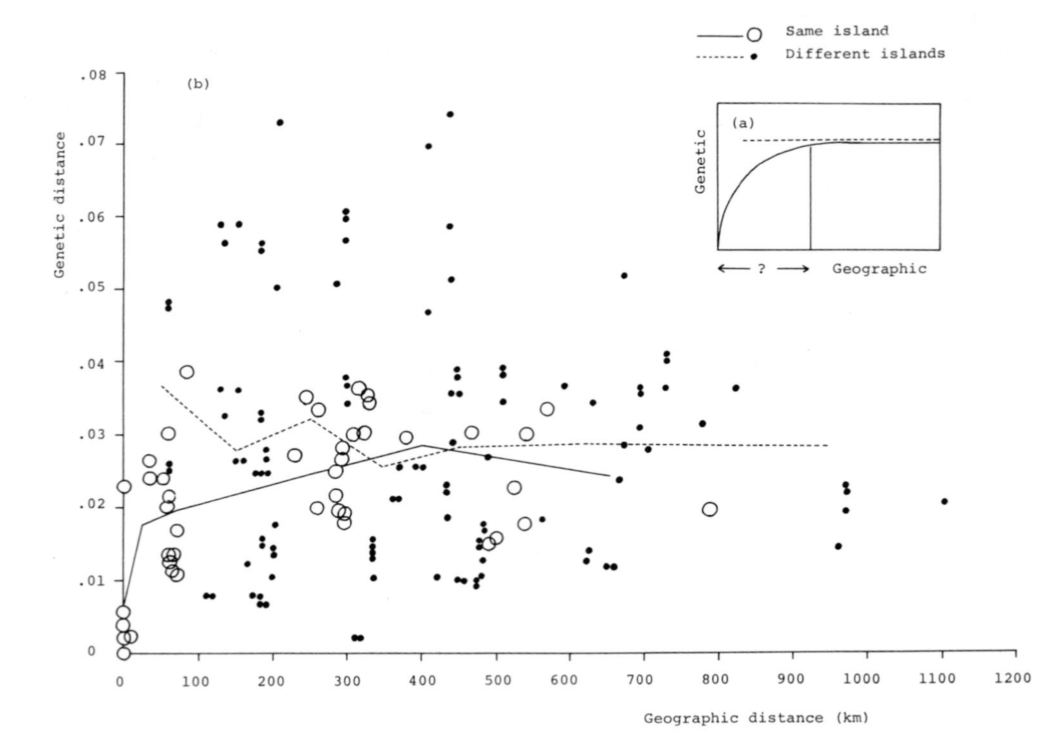

Figure 4. Relationship between geographic distance and genetic distance.

array of subpopulations with a gridlike arrangement, that the effective size of each subpopulation is constantly N, and that each subpopulation exchanges individuals with four adjacent subpopulations at the rate of m each generation. Kimura and Maruyama (1971) revealed by computer simulation that in this model, if a pair of selectively neutral alleles was segregating, marked local differentiation of gene frequencies could occur when $mN < 1$, and the whole population behaved as if it were panmictic and the gene frequencies became uniform over the entire distribution range when $mN \geq 4$. The situation of the Japanese macaques would undoubtedly be considered to fit into the case of $mN < 1$.

Population census of the Japanese macaque carried out by Takeshita (1964) shows that the average troop size is about 66. Assuming a possible overestimation, let us take a value of 60 as the average census number of a troop. The genetically effective troop size (N) of the Japanese macaques is estimated by Nozawa (1972) as about one-third of its census number. Then, the average of N is 20. This means that the average migration rate of a troop should be less than 5% per generation.

DISTANCE OF GENE DISPERSION FROM A TROOP

From the results of the above analyses, it is obvious that a macaque troop is not a genetic isolate but an open system, although the average rate of gene exchange between troops has been estimated as a low value. But, the above estimation has been obtained from a straight application of the mathematical model of population structure. The actual breeding population of the Japanese macaques, of course, deviates from such an ideal state: neither the population size nor the migration rate can be considered as constant, the condition of the natural environment is variable so that the arrangement of the troops is not exactly gridlike but irregular, and the exchange of individuals of a troop should not necessarily be restricted to its adjoining troops. Thus it is desirable to obtain quantitative information concerning the amount of migration of individuals from one troop to another without using any mathematical models.

A matrix presented in Table 3 gives the results of calculations of the geographic distance (km) under the diagonal and of the genetic distance above the diagonal between every pair of the troops examined. Values without underline show the distance values between troops located on same island and underlined values between troops located on different islands. Correlation coefficients between the geographic and genetic

Table 3. Geographic distance (km) under the diagonal, and genetic distance above the diagonal, between troops of the Japanese macaque. Underlined values show the distances between troops on different islands.

	F	YT	I	R	MI	MII	T	AA	AwI	AwII	K	SI	SK	TsA	TsB	TsC	Km	Ko
F	0	.030	.029	.030	.014	.015	.022	.017	.011	.011	.019	.051	.028	.021	.022	.019	.014	.020
YT	322	0	.020	.034	.026	.028	.035	.029	.025	.025	.032	.073	.050	.039	.040	.035	.027	.035
I	380	60	0	.027	.019	.021	.036	.024	.021	.021	.029	.069	.046	.035	.036	.031	.023	.031
R	468	246	232	0	.021	.023	.038	.016	.024	.024	.034	.072	.049	.038	.038	.034	.026	.033
MI	492	294	290	62	0	.002	.024	.010	.007	.007	.018	.055	.032	.016	.017	.012	.009	.013
MII	496	294	284	56	10	0	.026	.011	.007	.007	.018	.055	.032	.014	.014	.010	.009	.012
T	526	328	316	86	36	36	0	.030	.026	.026	.033	.058	.035	.037	.038	.035	.028	.036
AA	540	308	284	72	72	64	62	0	.008	.008	.020	.056	.033	.022	.023	.018	.010	.018
AwI	652	394	364	182	180	174	156	114	0	0	.015	.048	.025	.014	.014	.010	.002	.010
AwII	652	394	364	182	180	174	156	114	0	0	.015	.048	.025	.014	.014	.010	.002	.010
K	788	570	542	332	298	296	262	262	186	186	0	.058	.036	.027	.026	.024	.017	.025
SI	674	440	410	208	186	184	154	136	60	60	132	0	.022	.059	.060	.056	.050	.058
SK	674	440	410	208	186	184	154	136	60	60	132	0	0	.036	.037	.034	.027	.035
TsA	972	732	696	508	484	480	448	436	336	336	192	300	300	0	.002	.004	.013	.013
TsB	972	732	696	508	484	480	448	436	336	336	192	300	300	0	0	.005	.013	.014
TsC	972	732	696	508	484	480	448	436	336	336	192	300	300	0	0	0	.012	.010
Km	962	706	666	492	474	470	442	422	314	314	204	288	288	62	62	62	0	.012
Ko	1104	826	780	634	626	622	594	564	452	452	370	440	440	200	200	200	166	0

Table 4. Correlation coefficient (r) between geographic (km) and genetic distances.

	N	r
Whole Japan	153	+0.0911
Different islands	109	-0.0246
Same island	44	+0.3613**
Same island, distance > 100 km	25	-0.1624
" , distance < 100 km	19	+0.5901**
" , distance > 50 km	36	+0.2020
" , distance < 50 km	8	+0.7304*

* $p < 0.05$, ** $p < 0.01$

distances were calculated (Table 4). The correlation coefficient over the whole country was regarded as zero, and between troops located on different islands also zero. But, a significant correlation was observed between troops on the same island; and furthermore, between troops separated by less than 100 km on the same island the correlation was highly significant. The correlation coefficient between troops separated by more than 50 km on the same island was regarded statistically as zero.

Now, let us assume that the Japanese macaques can not migrate across the sea. Then, the genetic distance between troops located on two different islands would be independent of the geographic distance between them, and the genetic distance between troops located on the same island would be considered to increase from a small value to the same level as that between troops on different islands. It is of interest to to evaluate the geographic distance on the same island over which the genetic distance is at the same level as the genetic distance between troops on different islands (Fig. 4a). Fig. 4b shows the results of these observations. Considering a broad range of distribution of genetic distance values at about the same geographic distance, especially between troops located on different islands, we can see from the figure that in order for the genetic variation maintained in a troop to exert any effective influence on other troops, these troops should be located in a circle with a radius of no more than 100 km from the former troop. In other words, the gene constitutions of two troops separated by more than 100 km can be regarded as practically independent of each other.

CONCLUSION

The above analyses suggest that the population structure of the Japanese macaque species as a whole has a remarkable tendency to split into a number of local subpopulations, although we do not know whether such a local subpopulation is equivalent to an individual troop. In any event, it is certain that the Japanese macaque population has a structure capable of being influenced by random genetic drift. Such circumstances are considered responsible for the low genetic variability within each troop and the marked genetic differentiation between troops.

ACKNOWLEDGEMENTS

The authors wish to thank Professors G. Ishimoto, Mie University, and Y. Tanabe, Gifu University, for permitting the use of their unpublished data of electrophoresis examina-

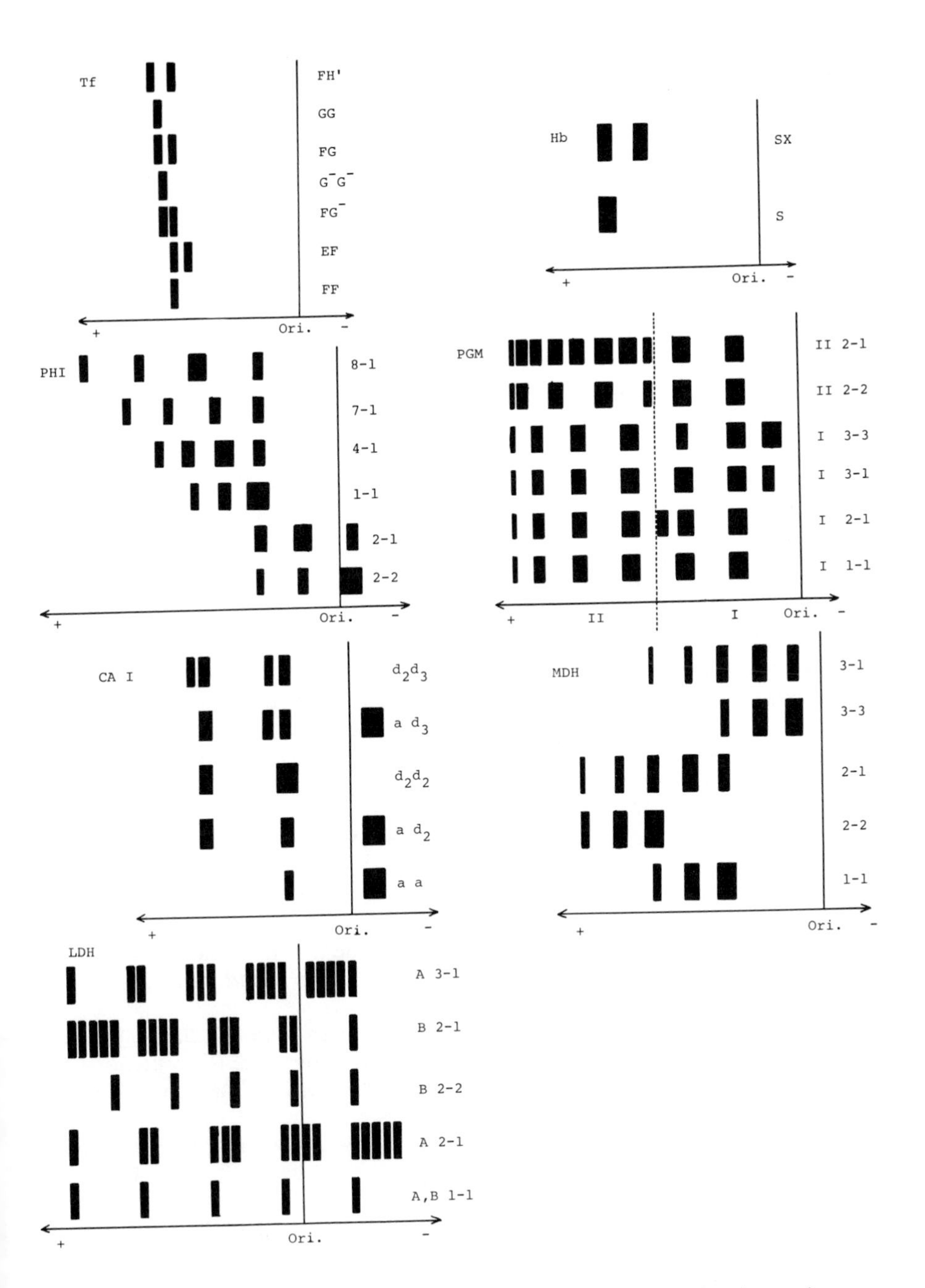

Fig. 5. Electrophoretic patterns of protein variations in *Macaca fuscata*.

tions for the present work. The authors are also indebted to many field biologists working with the Japanese macaque for their valuable assistance in collecting blood samples from natural troops of this species.

REFERENCES

Gahne, B. 1966. Studies on the inheritance of electrophoretic forms of transferrins, albumins, prealbumins and plasma esterases of horses. *Genetics* 53: 681-694.

Imlah, P. 1964. Inherited variants in serum ceruloplasmin of the pig. *Nature* 203: 658-659.

Ishimoto, G. 1972a. Blood protein variations in Asian macaques. I. Serum proteins and hemoglobin. *J. Anthrop. Soc. Nippon.* 80: 250-274.

Ishimoto, G. 1972b. Blood protein variations in Asian macaques. II. Red cell enzymes. *J. Anthrop. Soc. Nippon.* 80: 337-350.

Kimura, M. 1968. Evolutionary rate at the molecular level. *Nature* 217: 624-626.

Kimura, M. and T. Maruyama 1971. Pattern of neutral polymorphism in a geographically structured population. *Genet. Res.* 18: 125-131.

Kimura, M. and G. H. Weiss 1964. The stepping stone model of population structure and the decrease of genetic correlation with distance. *Genetics* 49: 561-576.

King, J. L. and T. H. Jukes 1969. Non-Darwinian evolution. *Science* 164: 788-798.

Lewontin, R. C. 1967. An estimate of average heterozygosity in man. *Amer. J. Human Genet.* 19: 681-685.

Nozawa, K. 1972. Population genetics of Japanese monkeys. I. Estimation of the effective troop size. *Primates* 13: 381-393.

Ogita, Z. 1966. Genetico-biochemical studies on the salivary and pancreatic amylase isozymes in human. *Med. J. Osaka Univ.* 16: 271-286.

Selander, R. K., S. Y. Yang, R. C. Lewontin, and W. E. Johnson 1970. Genetic variation in the horseshoe crab (*Limulus polyphemus*), a phylogenetic "relic". *Evolution* 24: 402-414.

Shaw, C. R. and R. Prasad 1970. Starch gel electrophoresis of enzymes--a compilation of recipes. *Biochem. Genet.* 4: 297-320.

Shotake, T. (in press). Genetic polymorphisms in blood proteins in the troops of Japanese monkeys, *Macaca fuscata*. II. Erythrocyte lactate dehydrogenase polymorphism in *Macaca fuscata*. *Primates*.

Shotake, T. and K. Nozawa (in press). Genetic polymorphisms in blood proteins in the troops of Japanese monkeys, *Macaca fuscata*. I. Cytoplasmic malate dehydrogenase polymorphism in *Macaca fuscata* and other non-human primates. *Primates*.

Spencer, N., D. A. Hopkinson, and H. Harris 1968. Adenosine deaminase polymorphism in man. *Ann. Human Genet.* 32: 9-14.

Takeshita, M. 1964. Distribution and population of the wild Japanese monkeys. *Yaen* 19: 6-13; 20·21: 12-21.

Tanabe, Y., M. Ogawa and K. Nozawa (in preparation). Polymorphisms of thyroxin-ginding prealbumin (TBPA) in primate species. *Jap. J. Genet.*

Tashian, R. E., M. Goodman, V. E. Headings, J. DeSimone, and R. H. Ward 1971. Genetic variation and evolution in the red cell carbonic anhydrase isozymes of macaque monkeys. *Biochem. Genet.* 5: 183-200.

Wright, S. 1940. Breeding structure of populations in relation to speciation. *Amer. Nat.* 74: 232-248.

Wright, S. 1943. Isolation by distance. *Genetics* 28: 114-138.

RELATIVE SIGNIFICANCE OF ISOZYMES, CONTROL GENES, AND CHROMOSOMES IN GRADE EVOLUTION

ANN L. KOEN
C. S. Mott Center for Human Growth and Development
Department of Gynecology and Obstetrics
Wayne State University School of Medicine
Detroit, Michigan 48201

ABSTRACT. Some of the results of isozymic analysis of primates are not adequately explained by theories of cladogenetic evolution. A combination of the cladogenetic theories, with some modifications, and with incorporation of recent chromosome data led to the formulation of a theory of grade or anagenetic evolution. Control genes in the form of centromeric heterochromatin may have been agglutinating at the centromeres of certain chromosomes throughout primate evolution. Extrapolation of this theory may help to explain such results as progressive, directional changes in isozyme ratios and appearance of isozymes in one species which properly belong to other species that separate later from the primate line.

INTRODUCTION

There are two aspects of evolution: (1) phylogenetic branching and subsequent radiation of species and (2) progressive or "upward" evolution. The two have been designated as cladogenesis and anagenesis (Rensch, 1959, 1971; Goodman, 1963, 1965; Dobzhansky, 1972). Cladogenesis has been much more extensively studied, probably because it is more easily defined and more amenable to direct investigation. Detailed phylogenetic trees have been contructed on the basis of data obtained by modern analytical techniques of protein sequencing (Fitch and Margoliash, 1967; Fitch, 1971; Barnabas et al., 1971), DNA hybridization (Kohne et al., 1971), and immunological methods (Goodman, 1971).

Several mechanisms have been proposed for the separation of species; e.g., geographic or niche separation of groups (Mayr, 1966); duplication of genes or whole chromosomes (Ohno et al., 1968; Nei, 1969; Klose et al., 1969; Watts and Watts, 1968a); reduced heterozygote fitness (Bazykin, 1969); the "founder event" (Carson, 1970); and the "master-slave" concept of Edström (1968).

Anagenesis, on the other hand, has been more elusive to definition, especially in molecular terms; consequently it has been implicitly accepted by some and ignored by most. We know from paleontological and anatomical evidence that advancing animal grades are distinguished by increasing organization of cells, organs, and organisms. At the molecular level, however, its traces have not been obvious. It is already clear from chromosome mapping data that functionally related genes are not spatially related on the chromosomes (Nichols and Ruddle, 1973; Creagan and Ruddle, 1974). Attempts to find directional patterns in base substitutions of DNA have for the most part been disappointing (Zuckerkandl et al., 1971; Bak et al., 1972), although Fitch (1969) has presented evidence that non-random substitution in DNA may occur.

Britten and Kohne (1968) reported the occurrence of saltatory replications of DNA during evolution, which led to the theory of gene control in higher organisms proposed by Britten and Davidson (1969). They suggested that integrator or receptor genes might be translocated within the genome and by assuming control of new batteries of structural genes, give rise to new tissues or new organs. They considered that in higher organizational grades, evolution might proceed via changes in regulatory systems rather than in structural genes.

In the order Primates, different anagenetic stages are represented by still living species. Because their cladogenetic relationships are well-defined from paleontological as well as molecular evidence, they are good subjects for investigation of grade evolution. As observed in the Primates, anagenesis is evidenced not by the appearance of new tissues or organs, but by modification of existing ones. For instance, the neocortex already existed in a primitive form in the hedgehog, but expanded and differentiated in the Primates. (Diamond and Hall, 1969). Similarly, in the Primates, the placenta underwent progressive modification from the more primitive epitheliochorial type in lower primates to the hemochorial placenta of higher primates. Thus, while the appearance of new tissues or organs may accompany large steps in grade evolution, as from one order to another, it is not a necessary prerequisite for smaller steps, as from family to family.

In the same context, the striking increase in DNA content of the genome through evolutionary grades on a broad scale from virus to mammal is well correlated with anagenetic evolution, as pointed out by Britten and Davidson (1969). On the narrower scale of primate evolution, however, there is no observable gradation of DNA content from lowest to highest primate (Manfredi Romanini and De Stefano, 1974). Until more definitive criteria are found, therefore, we are dependent upon the rather vague specification of increasing organization as the *sine qua non* of grade evolution.

Evolutionary events, both cladogenetic and anagenetic, occur side by side in members of breeding groups, probably frequently in the same individuals. It is likely then that the two overlap, especially that an anagenetic event may be mirrored in cladistic relationships. Changes in control genes, for instance, must express themselves either qualitatively or quantitatively through structural genes. When these effects are observed in structural genes, they may or may not appear to fit into theories of cladistic evolution. Among the many isozyme variations observed, there are some that do not seem to fall within the scope of any cladistic theory. In some of the recent chromosome data may lie a clue to a supra-genic organization which could explain the isozyme data and yield a possible mechanism of anagenesis.

QUALITATIVE CHANGES IN ISOZYMES

It is well recognized that gene duplication has occurred at many structural gene loci during evolution. Phosphoglucomutase (PGM) is one, since at least PGM_1, $_{-2}$ and $_{-3}$ are known to be encoded at different loci. However, gene duplication within the primate line after separation from other mammals has been thought to be rare. One such event may have been the duplication of the PGM_1 locus. The double bands of PGM_1 always appear together, and one never varies alone. We have extracted the two bands individually from preparative gels and re-run them in regular gels, and have found no conversion from one to the other, such as might be expected if they were conformational isomers or polymers.

We have observed, in several species of *Tupaia*, *Nycticebus coucang* and *Symphalangus syndactylus*, what we presume to be homologous PGM_1 bands present as single bands. This could be interpreted as gene duplication, the duplicate genes remaining very closely linked. This does not explain, however, why the various mutations occurring throughout the Primates, theoretically independent of each other, should always affect both bands. Mutation of a common subunit is not a satisfactory explanation, either, because it provides no explanation for the failure of subunits <u>not</u> held in common to vary independently.

Another observation, also in *N. coucang*, was that of an apparent homozygote for the more anodal (fast-band) form of cytoplasmic aconitase; this individual also had a low level of the slow-band activity (Koen and Goodman, 1969a) (Figure 1). The homozygous slow-band animals had none of the fast band, and heterozygotes had equal amounts of each. In this series, most animals were homozygous for the slow band, a few were heterozygous, and only one was an apparent fast-band homozygote. The fast-band therefore appeared to be the new form of the

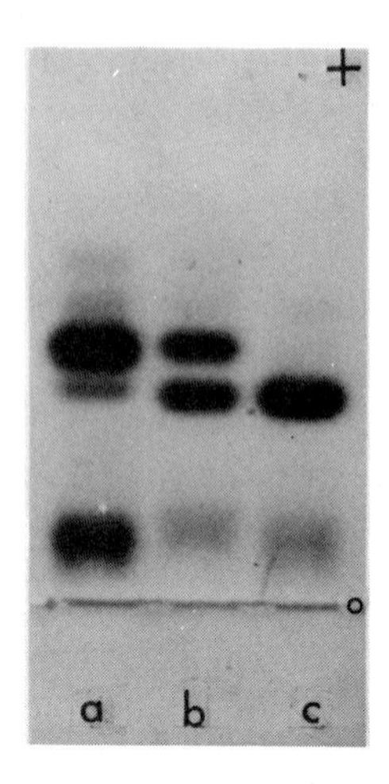

Fig. 1. *N. coucang* aconitase. (a) fast-band homozygote; (b) heterozygote; (c) slow-band homozygote. Anode is at the top.

enzyme; the slow-band activity in the fast-band animal was interpreted according to a duplicated gene theory. It could also be interpreted according to Edström's "master-slave" hypothesis, in which the "slave" genes represent "memories" of previous meiotic partners. Their expression is controlled by the "master" gene. If in *N. coucang*, the "master" gene in the fast-band homozygote were similar enough in sequence to the slow-band gene, then some slow "slaves" at that locus could have slipped by without being silenced.

In *N. coucang* also we observed a clear case of duplication of the LDH B gene (Koen and Goodman, 1969b) (Figure 2). In several animals, a minor LDH B_4 isozyme which corresponded to the common primate LDH B_4 (CPB) was present, in addition to the common *N. coucang* LDH B_4 (CNB). One of these animals was also heterozygous at the LDH B locus, having therefore six LDH 1 bands. We interpreted this to mean that the CNB was the result of a primary gene duplication of the CPB. Further investigation revealed that some *N. coucang* also possessed an LDH A_4 isozyme electrophoretically the same as that of *Tupaia*, *Lemur* and the cercopithecoids, while some *Tupaia* possessed an LDH A_4 of the same mobility as that of the *Lorisidae*, *Tarsidae* and *Cebidae*, in addition to their own (Figure 2). In a mother and fetus of *N. coucang*, the mother possessed the CPB as well as the CNB, while the fetus had only the CNB. The mother's CPB could have been present as a single allele, and the fetus could have received the chromosome lacking that allele. Many of the *N. coucang* examined, like the fetus, did not have the CPB; it

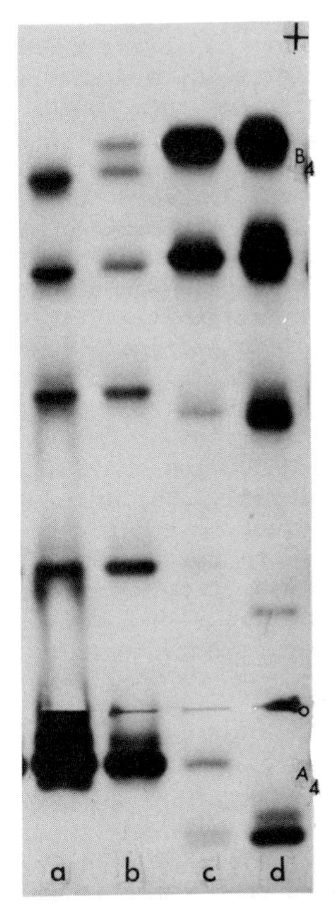

Fig. 2. LDH: CPB in *N. coucang*, NCB in *Tupaia*. (a) *N. coucang* without CPB; (b) *N. coucang* with CPB; (c) *Tupaia* with CNB; (d) *Tupaia* without CNB. Anode is at the top. From anode to cathode, the composition of the bands is: B_4, B_3A_1, B_2A_2, B_1A_3, A_4.

could therefore be a "memory" in the process of elimination from the species. The lorisoids separated from the rest of the primates 65-80 million years ago (Goodman et al., 1971a). Even this long afterwards, therefore, on the verge of elimination from the species, the CPB is still in the same form that it was in the beginning. This does not support the theory that "silent" genes may vary more freely than those assigned a useful function.

A somewhat different interpretation is required to explain the presence in *Tupaia* of an LDH A_4 similar to that of species which separated later from the primate line. If the common ancestor of *Tupaia* and the lorisoids were heterozygous at that locus, and each of the two lines fixed a different allele, then the reappearance of the more cathodal form in species separating after the lorisoids must be ascribed to an independent back mutation, since they would have fixed the more anodal form. While back mutation has been hypothesized, intra-species variations have not been found to mimic inter-species variations. However, the lorisoid form in *Tupaia* cannot be a "memory" of a previous meiotic partner, since the lorisoids separated subsequently to the tupaids. Indeed, *Tupaia* is thought to be a very primitive mammal (McKenna, 1969), and its inclusion in the order Primates is still debated.

Another instance of an isozyme appearing before its time was observed in the glucosephosphate isomerase (GPI) bands of an individual *Presbytis senex* (Figure 3). In the Primates, the GPI bands become progressively more cathodal as the evolutionary scale is ascended, with human GPI the most cathodal of all (Goodman et al., 1971a). The extra GPI band of the *P. senex* individual was electrophoretically indistinguishable from that of man. As with the *Tupaia* A_4, this cannot represent a "memory" of a previous meiotic partner, because humans had not yet appeared on the scene when the line leading to *Presbytis* branched away from the rest of the primates.

QUANTITATIVE CHANGES IN ISOZYMES

Of the quantitative effects, the most clearcut is that of the increasing LDH B/LDH A ratios across a span of primate evolution (Koen and Goodman, 1969c) (Figure 4). The increasing ratios were apparently unrelated to diet, since the liver was the only organ in which they remained constant in all primates tested (Table 1). These changes were independent of the molecular structure of the isozymes. From amino acid sequencing data on hemoglobins, the ceboids diverged from the cercopithecoids just before the hominoids branched away (Goodman et al., 1971b). The ceboids and cercopithecoids remained at the same grade level while the hominoids advanced. The relative concentrations of the LDH isozymes reflect this evolutionary relationship.

Progressive quantitative changes also occurred during grade evolution in at least some of the enzymes which exist in both cytoplasmic and mitochondrial forms. In erythrocytes, ratios of cytoplasmic to mitochondrial forms of aconitase, isocitrate dehydrogenase, and malate dehydrogenase (NAD-linked) increased from lower to higher primates (Goodman et al., 1971a).

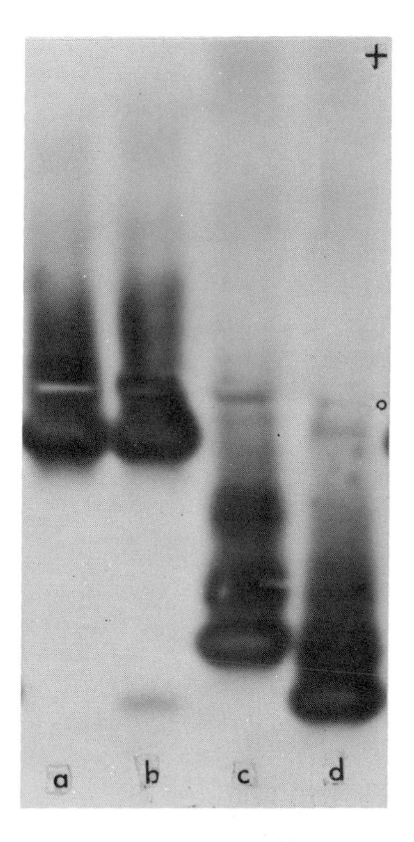

Fig. 3. GPI in primates. (a) *P. senex* without human GPI band; (b) *P. senex* with human GPI band; (c) *Pan*; (d) *Homo*.

QUALITATIVE AND QUANTITATIVE CHANGES IN ISOZYMES

The appearance of new isozymes of an enzyme may be the result of a new gene duplication, but a gradual increase in prominence thereafter must be due to control gene activity. Such is the case with what we have called the secondary LDH bands, to distinguish them from the familiar sub-bands of the five main isozymes (Koen and Goodman, 1970)(Figure 4). These bands may vary electrophoretically without effect on the main bands. They are absent in mice and *Prosimii*, appearing in the infraorder *Anthropoidea* and increasing in prominence through the hominoids. In humans, the ratio of activities of the main:secondary LDH bands on lactate is about 10:1 by densitometric analysis. With α-hydroxy butyrate as substrate the ratio of their activities is also approximately 10:1. However, when β-hydroxy butyrate is used as substrate, the

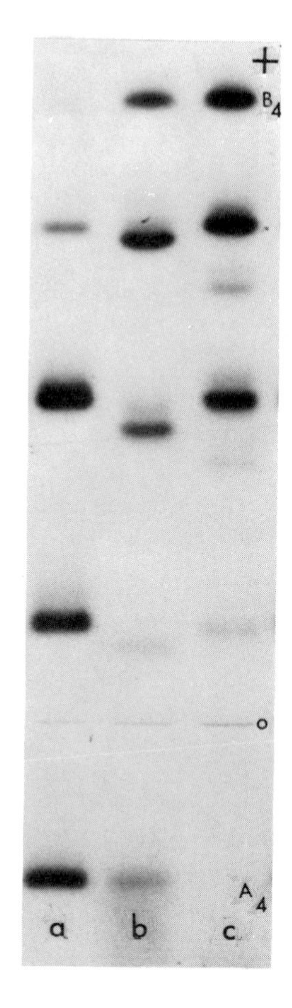

Fig. 4. Relative concentrations of LDH isozymes in primate erythrocytes: (a) *Tupaia*; (b) *C.Niger*; (c) *Homo*. Note secondary bands between LDH 2 and 3 and between 3 and 4 in *Homo*.

ratio is much closer to 1:1. These bands may therefore represent the formation of a new isozyme of LDH with progressive modification of substrate specificity.

CHROMOSOME DATA

The "founder event" mechanism of evolution proposed by Carson (1970) depends on chromosome translocation or inversion. These changes in chromosome structure are being traced through primate evolution by means of chromosome banding stains

TABLE 1

ERYTHROCYTE LACTATE DEHYDROGENASE B/A SUBUNIT RATIOS AMONG PRIMATES

PROSIMII	
Tupaiidae	0.63 (0.15)[a]
Lemuridae	0.66 (0.34)
Lorisidae	0.83 (0.25)
Tarsiidae	0.96 (0.02)
ANTHROPOIDEA	
Cebidae	1.81 (0.35)
Cercopithecidae	1.80 (0.39)
Pongidae	2.13 (0.35)
Hominidae	2.92 (0.38)

[a]standard deviation in parentheses

(deGrouchy et al., 1972). There can be little doubt that in many cases, speciation may have occurred in just such a way.

Although in the Primates there is no corresponding increase in nuclear DNA with advancing grade evolution, there is nevertheless corroborative evidence for translocation of control genes as postulated by Britten and Davidson. This has come from studies of chromosome constitution using new methods which stain heterochromatin differentially, both in the centromeres and in the chromosome arms, and from methods of *in situ* DNA-RNA hybridization. It has been shown that centromeric heterochromatin is evenly distributed among the chromosomes in the mouse (Hsu et al., 1971) and in the guinea pig (Bianchi and Ayres, 1971). In lower primates it is also fairly evenly distributed, and in man there are three autosomes with large blocks of centromeric heterochromatin (chromosomes 1, 9 and 16) while the Y chromosome is heterochromatic in the long arms (Figure 5). By means of *in situ* hybridization, Jones (1970) localized some satellite DNA to the centromeric heterochromatin. Subsequently, Jones (1974) demonstrated that one of three fractions of human heterochromatin, that of human chromosome 9, hybridized to about 20 sites in the chromosomes of *Pan*, *Gorilla*, and *Pongo*, many more than in man himself.

Accepting rodents as having branched off from the primates after the primates diverged from the rest of the mammals (McKenna, 1969), the data indicate that an agglutination of heterochromatin may have been taking place from more diffuse sites into more compact areas, possibly with elimination of some of it.

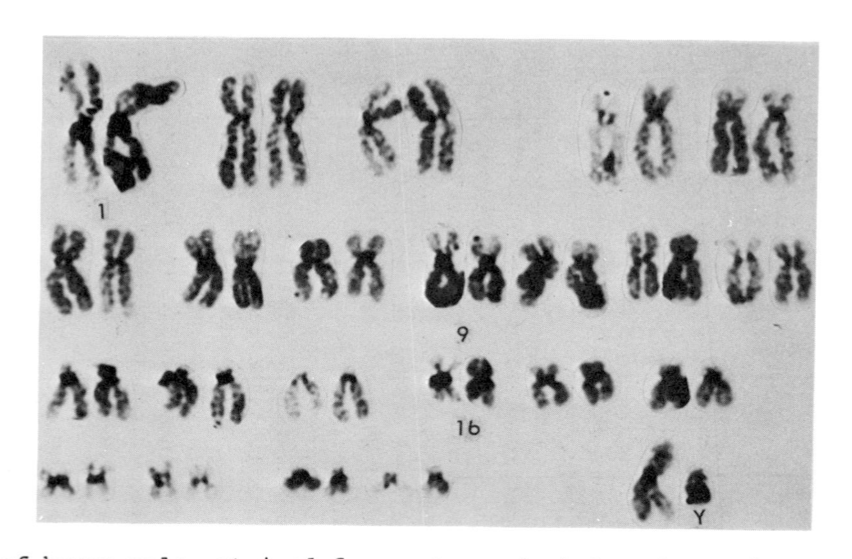

Fig. 5. Karyotype of human male, stained for centromeric heterochromatin. Note large blocks of heterochromatin in chromosomes 1, 9, 16 and Y.

Different chromosome banding methods and different conditions are required to demonstrate the heterochromatin of the centromeres and that of the arms (Dutrillaux et al., 1972). These are therefore assumed to be of different types, either in their DNA content or in the type of protein bound to them.

DISCUSSION

Yunis and Yasmineh (1971) suggested that heterochromatin areas play a structural role in chromosomes, serving primarily to protect the vital cistrons that code for 18S and 28S ribosomal RNA. Brown (1966), however, believed that evolutionary processes may quickly result in rearrangement of heterochromatin between closely related species. The work of Jones (1974) established that, at least with regard to human fraction III (the satellite DNA of chromosome 9), the mechanism has been rearrangement and possibly some elimination from the genome in passing from one grate to another, rather than old heterochromatin being replaced with new.

Heterochromatin has long been known to influence structural genes in their function. Hannah (1951) demonstrated that in *Drosophila*, some structural genes were depressed by translocation into heterochromatin and reactivated by translocation back to euchromatin. Brown (1966) noted that this was not a generalized depressive effect of heterochromatin, since other genes when translocated from euchromatin to heterochromatin were activated.

If regulatory genes located in centromeric heterochromatin have been condensing into certain areas of the genome, this could represent Britten and Davidson's control gene translocation. By moving into closer contact with one another, integrator genes, for example, might achieve more efficient interrelationships with each other and more exact control over their structural genes.

The impetus for this continuing orderly process of agglutination could lie in the nature of centromeric heterochromatin itself. It is more adhesive than euchromatin, especially during certain stages of the cell cycle, probably because of the type of protein that it manufactures (DuPraw, 1966). Heterochromatic regions of chromosomes have been observed to adhere together non-specifically when they are near each other, even when they are on non-homologous chromosomes (Maguire, 1972). Thus, it may be a natural process for the adhesive material to agglutinate gradually through generations of meiotic processes. Heterozygosity of centromeric heterochromatin is relatively common in humans, usually without observable phenotypic effect (Koen, 1974). The polymorphism may result from homologous unequal crossing-over. It is usually found in the chromosomes which contain large or moderately large blocks of heterochromatin.

Our data on isozymes require certain elements in a theory of evolution, specifically that there be: (1) some rationale for isozymes properly belonging to one species appearing in another as a minor component, even when there is no known continuity in their line of descent; (2) a central control mechanism allowing for progressive adjustment of quantitative isozyme ratios independently of their structure; (3) an explanation for anomalous concentrations of enzymes produced by apparently simple alleles.

The basic difference between the "master-slave" gene control theory of Edström and the more elaborate theory of Britten and Davidson is that in the former the structural genes are duplicated sequences and the control gene is single, while in the latter there are multiple control genes and the structural genes are single. Our data would be most satisfactorily explained if there were multiple copies--but not necessarily exact copies--of structural genes in combination with a control gene complex. Elements of other cladistic theories of evolution, with some modification and incorporation of chromsosme data, can be synthesized into a model yielding a possible mechanism of anagenetic evolution. A sequence of evolutionary events could be visualized as follows.

A burst of replicative activity might occur in certain portions of the genome of one or a few individuals (the saltatory replications of Britten and Kohne or the gene duplications of Ohno); some of the replications would be of control genes and some of structural genes. The individuals possessing the extra DNA would be relatively or completely incompatible reproductively with the parent population, and a new species would arise in one or a few generations. This new DNA, being heterochromatic in nature and therefore adhesive, would begin to rearrange itself within the genome, thereby causing more non-homology of chromosomes and still more rapid speciation.

In the replication process, all of the new sequences would not necessarily be exact copies of the original (Britten and Davidson, 1969), but might differ from each other by one or more base substitutions each, much in the manner suggested by Smith (1970), except that the substituted genes would be present all together simultaneously instead of occurring in a temporal sequence at the same locus. The duplicated sequences would be arranged in a tandem fashion at first, as described by Edström (1968). However, with the passage of generations, either they or the control genes might be translocated to different parts of the genome, the "slave" genes thus becoming separate loci instead of units of the same locus. However, their expression or repression would still be dependent upon the "master" genes which would now be distantly located control genes, communicating with the structural genes through a messenger, probably intra-nuclear RNA.

If, as suggested by Eström, the "master" gene is periodically discarded and one of the "slaves" becomes the "master", this could allow the expression of another previously repressed gene and repress another previously expressed gene, thus giving the outward impression of a new mutation. The appearance of a small amount of aconitase slow-band in a fast-band homozygote and of a small amount of the common primate LDH B_4 in *N. coucang* could be interpreted this way, except that the repression of the old gene was not complete. Alternatively, the new "master" gene could allow the expression of a previously repressed gene without repressing the old one. This would be recognized as gene duplication. The transition of PGM_1 from single to double bands could be an example of this. The PGM_1-2 of the tantulus monkey and the chimpanzee are electrophoretically the same as the PGM_1-1 of the macaque and man, indicating continuity of these bands through the Primates. The other forms of PGM_1 are each different in the different species. If these forms have all been present together in the primate genome from the beginning, with one or another form (or both) active in different species, then it would not be necessary to postulate simultaneous but independent mutations affecting duplicate genes each time a new species arose.

Watts and Watts (1968b) have expressed the view that nucleotide substitutions, even though neutral at the protein level, might still exert selective pressure at the DNA level, because of the difficulty of incorporating many mutant genes into the genome. Except during the replicative periods, the mechanism outlined above would theoretically create little genetic load at the DNA level, since the DNA, although expressed at different times, would have been incorporated into the genome long before. The pressure of natural selection would be exerted on the proteins as they are exposed to it. If they are accepted, their expression relative to other genes would be determined according to the central control genes without regard to their particular structure. Evidence for this lies in the relative concentrations of the LDH isozymes, which are similar at similar grade levels, regardless of structure.

In the meantime, point mutations, neutral or adaptive, would be taking place in the structural genes; their relationship to the control genes would remain the same, except in cases where the mutation was such as to change the response to control gene influence. In this case, selection would operate. When a sufficient number of these structural gene mutations had accumulated to result in reproductive incompatibility, a new species might arise, but it would remain at the same grade level as before, until a rearrangement of the control gene-structural gene relationship occurred.

Similarly, if the reproductive incompatibility is a result of a chromosomal rearrangement of euchromatin, a "founder event" may occur. But again the new species would remain at the same grade level as before, until the relationship of control gene to structural gene is disrupted.

This model of evolution contains implications for clinical enzymology. If there are indeed many copies of a gene which are silenced, then enyzme deficiencies due to mutation of a structural gene might be treatable by inducing the control gene to activate another structural gene of the set which might be capable of producing an active enzyme. This gene might be a "memory" of a chimpanzee or orangutan enzyme form or another as yet unexpressed form of the gene.

REFERENCES

Bak, A. L., J. F. Atkins, and S. A. Meyers 1972. Evolution of base compositions in microorganisms. *Science* 179: 1391-1393.

Barnabas, J., M. Goodman, and G. W. Moore 1971. Evolution of hemoglobin in primates and other therian mammals. *Comp. Biochem. Physiol.* 39B: 455-482.

Bazykin, A. D. 1969. Hypothetical mechanism of speciation. *Evolution* 23: 685-687.

Bianchi, N. O. and J. Ayres 1971. Polymorphic patterns of heterochromatin distribution in guinea pig chromosomes. *Chromosoma* 34: 254-260.

Britten, R. J. and E. H. Davidson 1969. Gene regulation for higher cells: a theory. *Science* 165: 349-357.

Britten, R. J. and D. E. Kohne 1968. Repeated sequences in DNA. *Science* 161: 529-540.

Brown, S. W. 1966. Heterochromatin. *Science* 151: 417-425.

Carson, H. L. 1970. Chromosome tracers of the origin of species. *Science* 168: 1414-1418.

Creagan, R. and F. H. Ruddle 1974. Mapping of human chromosomes. In: Proc. Intl. Cong. Anthr. Eth. Sci.: *Comparative Karyology of Primates*. Eds. A. B. Chiarelli and A. L. Koen. Mouton, The Hague. (In press).

DeGrouchy, J., C. Turleau, M. Roubin, and M. Klein 1972. Evolutions caryotypiques de l'homme et du chimpanzee: etude comparative des topographies de bandes apres denaturation menagee. *Ann. de Genet.* 15: 79-84.

Diamond, I. T. and W. C. Hall 1969. Evolution of neocortex. *Science* 164: 251-262.

Dobzhansky, Th. 1972. Darwinian evolution and the problem of extra-terrestrial life. In: *Perspectives in Biology and Medicine* 15: 157-175.

DuPraw, E. J. 1966. Evidence for a "folded fibre" organization in human chromosomes. *Nature* 209: 577-581.

Dutrillaux, B., C. Finas, J. DeGrouchy, and J. Lejeune 1972. Comparison of banding patterns of human chromosomes obtained with heating, fluorescence, and proteolytic digestion. *Cytogenet.* 11: 113-116.

Edström, J. E. 1968. Masters, slaves and evolution. *Nature* 220: 1196-1198.

Fitch, W. M. 1971. Toward defining the course of evolution: minimum change for a specific tree topology. *Syst. Zool.* 20: 406-416.

Fitch, W. M. 1969. Evidence suggesting a non-random character to nucleotide replacements in naturally occurring mutations. *J. Mol. Biol.* 26: 499-507.

Fitch, W. M. and E. Margoliash 1967. The construction of phylogenetic trees. *Science* 155: 279-284.

Goodman, M. 1963. Man's place in the phylogeny of the primates as reflected in serum proteins. P. 204-234. In: *Classification and Human Evolution*. Ed. S. L. Washburn. Viking Fund Publ. In Anthr. Chicago.

Goodman, M. 1965. The specificity of proteins and the process of primate evolution. P. 70-86. In: *XII Colloquium on Protides of the Biological Fluids*. Ed. H. Peeters. Elsevier, Amsterdam.

Goodman, M. 1971. Molecular evolution at the population level in higher primates. P. 81-87. *Proc 3rd Intl Cong. Primat.* Vol. 2. Karger, Basel.

Goodman, M. J. Barnabas, G. Matsuda, and G. W. Moore 1971b. Molecular evolution in the descent of man. *Nature* 238: 604-613.

Goodman, M., A. L. Koen, J. Barnabas, and G. W. Moore 1971a. Evolving primate genes and proteins. P. 155-208. In: *Comparative Genetics in Monkeys, Apes and Man*. Ed. A. B. Chiarelli. Academic Press, New York.

Hannah, A. 1951. Localization and function of heterochromatin in *Drosophila melanogaster*. *Adv. Gen.* 4: 87-125.

Hsu, T. C., J. E. K. Cooper, M. L. Mace, Jr., and B. A. Brinkely 1971. Arrangement of centromeres in mouse cells. *Chromosoma* 34: 73-87.

Jones, K. W. 1970. Chromosomal and nuclear location of mouse satellite DNA in individual cells. *Nature* 225: 912-915.

Jones, K. W. 1974. Related satellite DNA in man and higher

primates. In: Proc Intl. Cong. Anthr. Eth. Sci.: *Comparative Karyology of Primates*. Eds. A. B. Chiarelli and A. L. Koen. Mouton, The Hague. (In press).

Klose, J., U. Wolf, H. Hitzeroth, and H. Ritter 1969. Polyploidization in the fish family *Cyprinidae*, Order Cyriniformes. II. Duplication of the gene loci coding for lactate dehydrogenase and 6-phosphogluconate dehydrogenase in various species of *Cyprinidae*. *Humangenetik* 7: 245-250.

Koen, A. L. 1974. Observations on centromeric heterochromatin. In: Proc. Intl Cong. Anthr. Eth. Sci.: *Comparative Karyology of Primates*. Eds. A. B. Chiarelli and A. L. Koen. Mouton, The Hague. (In press).

Koen, A. L. and M. Goodman 1969a. Aconitate hydratase isozymes: subcellular location, tissue distribution and possible subunit structure. *Biochim. Biophys. Acta* 191: 698-701.

Koen, A. L. and M. Goodman 1969b. Evidence that genes at two loci code for lactate dehydrogenase subunit B in *Nycticebus coucang*. *Biochem. Genet.* 3: 595-601.

Koen, A. L. and M. Goodman 1969c. Lactate dehydrogenase isozymes: qualitative and quantitative changes during primate evolution. *Biochem. Genet.* 3: 457-474.

Koen, A. L. and M. Goodman 1970. Enzyme studies on primates: evolutionary trends and physiological significance (a summary of findings). *6571st Aeromed. Res. Tech. Rep.* ARL-TR-70-8.

Kohne, D. E., J. A. Chiscon, and B. H. Hoyer 1971. Evolution of primate DNA sequences. *Carnegie Inst. Yr. Book* 69: 488-495.

Maguire, M. P. 1972. Role of heterochromatin in homologous chromosome pairing: evaluation of evidence. *Science* 176: 543-544.

Manfredi Romanini, M. G. and G. F. DeStefano 1974. Critical approach to the interpretive problem of the DNA nuclear variations in primates. Proc. Intl. Cong. Anthr. Eth. Sci.: *Comparative Karyology of Primates*. (In press).

Mayr, E. 1966. *Animal Species and Evolution*. Harvard Univ. Press, Cambridge.

McKenna, M. C. 1969. The origin and early differentiation of therian mammals. *Ann. N. Y. Acad. Sci.* 167: 217-239.

Nei, M. 1969. Gene duplication and nucleotide substitution in evolution. *Nature* 221: 40-42.

Nichols, E. A. and F. H. Ruddle 1973. A review of enzyme polymorphisms, linkage and electrophoretic conditions for mouse and somatic cell hybrids in starch gels. *J. Histochem. Cytochem.* 21: 1066-1081.

Ohno, S., U. Wolf, and N. B. Atkins 1968. Evolution from fish to mammals by gene duplication. *Hereditas* 59: 169-187.

Rensch, B. 1959. *Evolution Above the Species Level*. Methuen Press, London.

Rensch, B. 1971. *Biophilosophy*. Columbia Univ. Press, New York.

Smith, J. M. 1970. Natural selection and the concept of a protein space. *Nature* 225: 563-564.

Watts, R. L. and D. C. Watts 1968a. Gene duplication and the evolution of enzymes. *Nature* 217: 1125-1130.

Watts, R. L. and D. C. Watts 1968b. The implications for molecular evolution of possible mechanisms of primary gene duplication. *J. Theor. Biol.* 20: 227-244.

Yunis, J. J. and W. G. Yasmineh 1971. Heterochromatin, satellite DNA, and cell function. *Science* 174: 1200-1209.

Zuckerkandl, E., J. Derancourt, and H. Vogel 1971. Mutational trends and random processes in the evolution of informational macromolecules. *J. Mol. Biol.* 59: 473-490.

ISOZYMES IN PLANT POPULATION GENETICS

R.W. ALLARD,[1] A.L. KAHLER,[1] and M.T. CLEGG[2]
[1]Department of Genetics, University of California
Davis, California 95616 and
[2]Division of Biological and Medical Sciences
Brown University, Providence, Rhode Island 02912

ABSTRACT. Studies of population structure in plants, utilizing isozyme markers, have revealed a complexity of genetic organization which belies the assumptions of simple population models. Thus, when precise quantitative data have been obtained, mating systems have been found to vary from population to population within a species and also to differ among closely related species with similar ecological preferences. Mating systems are highly flexible and capable of rapid shifts in time as well as in space. Studies of selection indicate that it takes a diversity of forms. Viability differences tend to be larger than differences in fecundity, indicating that much of the reproductive potential of populations is associated with differential mortality. Estimates of selection at the multilocus level show little correspondence to single-locus estimates supporting the view that epistatic forces are important in molding the genetic organization of populations. Natural populations exhibit an astonishing degree of genetic organization, including dramatic correlations in allelic state over loci within single populations as well as markedly non-random distributions of genotypes in space, correlated with environment. Finally, studies of experimental plant populations have shown that selection can organize the genetic materials into a highly non-random state in a few generations, indicating that evolution of multilocus organization proceeds at a rapid pace.

INTRODUCTION

It has been less than a decade since the technique of gel electrophoresis was introduced into population genetics. Yet this technique has so greatly increased the resolving power of experiments in population genetics that knowledge of population structure and of the magnitude of the various forces involved in molding the genetic architecture of populations has grown enormously in this short period. The technique has four main advantages: first, it makes large numbers of single-gene characters available for study; second, it reveals individual alleles at each locus; third, most banding patterns are inherited codominantly so that there is a one-to-one correspondence between phenotype and genotype; and fourth, the technique is adaptable to a wide range of plant and animal species. This discussion will illustrate the descriptive power of the electrophoretic tool in three aspects of plant population genetics: in studies of mating systems; in the estimation of selection intensities; and in studies of organizations at the multilocus level.

MATING SYSTEMS

The mating system is a natural point to begin an analysis of the genetic organization of populations. Systems of mating determine the patterns in which gametes are brought together and consequently they exert a primary control on genotypic frequency distributions. The mating system also affects the recombinational potential of populations thereby moderating or accelerating the rate at which new gene combinations are produced and existing gene combinations are lost (Allard, in press).

Allelic isozymes (allozymes) have provided a powerful tool for determining how gametes carrying particular gene markers are brought together in both natural and experimental plant populations. The basic technique has involved collecting seeds from adult plants, germinating the seeds in the laboratory and determining the array of genotypes in the progeny of single adults. Given the observed progeny distributions, the genotype of each maternal plant as well as the composition of the pollen pool can be inferred, and from such data parameters entering models of mating systems can be estimated (see Brown and Allard (1970) for a treatment of the statistical problems associated with estimating mating system parameters.) The mating system model usually adopted is known as the _mixed mating_ model. This model assumes that progeny from a particular parent can arise through self-pollination, with probability s_i (for the ith maternal genotype), or by the random draw of pollen from an external pollen pool with probability t_i (= 1 - s_i). It is further assumed that the gene frequencies in the pollen pool are independent of maternal type and uniformly distributed over the population. Questions which immediately present themselves regarding mating systems are: 1) Does $t_i = t$ (a constant over genotypes) i.e., are outcrossing rates the same for different genotypes? 2) Are gene frequencies the same in the adult population and in the pollen pool it produces? 3) Do the parameters which describe the mating system of a population change through time (i.e. does the mating system undergo evolutionary change?) 4) Do mating system parameters differ among local populations within a species? Until the advent of electrophoretic techniques few attempts had been made to answer these questions. The answers remain incomplete in plants and quantitative data on parallel questions in animals is for practical purposes non-existent.

Taking the questions in order, the most extensive data on the issue of whether outcrossing rates are the same among the different genotypes in a population comes from studies of morphological polymorphisms. The generalization which arises from such studies is that outcrossing rates are often not the same for different morphs in a population. For example, Jain and Allard (1960) found wide differences in t for different morphs in a barley population and Harding and Tucker (1964) obtained similar results in the lima bean. More recently, several investigators have exploited the advantages of the codominant inheritance of allelic isozymes in estimating mating system parameters. Allard et al. (1972) determined outcrossing rates over four isozyme loci for ten separate generations in an experimental barley population and found their estimates of t to be homogeneous over loci. Similarly, Brown and Allard (1970) found no heterogeneity over estimates of t obtained using nine and seven isozyme markers respectively, in two experimental corn populations. Marshall and Allard (1970) have investigated the question whether t varies for different genotypes within loci in two natural populations of the slender wild oat, *Avena barbata*. These investigators found that estimates of t are homogeneous both within and between the three and four isozyme loci monitored in the two populations. The evidence at hand thus suggests that outcrossing rates often depend upon genotype for morphological variants but not when the markers are enzyme loci. However the sample of enzyme markers studied is still small and clearly more studies are required to establish that the apparent difference between morphological and enzyme variants is real.

Our second question is whether gene frequencies are the same in the effective pollen pool (the pool from which pollen involved in outcrossing events is drawn) as in the population of adult plants which produce pollen in the pool. Most investigations of mating systems in plants have simply equated gene frequencies in the pollen pool to those among adult plants, thus tacitly assuming no fertility differences (i.e. differences among genotypes in average number of gametes contributed to the pollen pool), and also no selection in gametophytic stages of the life cycle. However isozyme loci, because of their codominant inheritance, provide sufficient degrees of freedom to estimate the relevant mating system

TABLE I

Generation	Allele					
	p_{1m}	p_{1f}	p_{2m}	p_{2f}	p_{3m}	p_{3f}
8	.41	.48	.46	.39	.13	.13
	(.11)	(.02)	(.11)	(.01)	(.07)	(.01)
19	.61	.65	.29	.22	.10	.13
	(.08)	(.01)	(.08)	(.01)	(.05)	(.01)
28	.48	.66	.45	.31	.07	.03
	(.08)	(.01)	(.08)	(.01)	(.04)	(.005)

Pollen pool and adult gene frequencies denoted p_{im} and p_{if} respectively, for the ith allele at Esterase A locus in barley Composite Cross V. Standard errors are given in parentheses. (From Kahler et al., in preparation.)

parameters. Brown et al. (1974) in an investigation of a natural population of soft chess (*Bromus mollis*) estimated pollen pool frequencies cojointly with the outcrossing rate and found no significant differences between pollen pool and adult gene frequency at an alcohol dehydrogenase locus. Kahler et al. (in preparation) have analyzed an average of 9 individuals per family in 1,100 families in each of three different generations in Composite Cross V, an experimental barley population. A sample of the data is given in Table I. Before discussing these results an important feature of this experiment needs emphasis: the adult plants from the three generations involved were all grown in three isolated plots in the same year at Davis, California. Thus three stages in the evolutionary history of this population were grown in the same temporal and spatial environment which means that observed differences over generations reflect actual evolutionary changes in the population. Turning to Table I, at locus EA, it is apparent that pollen pool (p_{im}) and adult (p_{if}) gene frequencies agree rather closely in generations 8 and 19, but a significant discrepancy exists between p_{1m} and p_{1f} in generation 28 ($P < .03$). Significant differences between pollen pool and adult gene frequencies were also found in generation 28 for a second locus (EC), which reinforces the impression that the differences are real and not artifacts of sampling. Evidently, the gametes marked by these loci are contributed to the gene pool differentially or they are being selected in the gametophytic stage. Additional evidence on fertility differences among adult genotypes (Clegg et al., in preparation) points to gametophytic selection as a major factor influencing pollen pool frequencies.

The fact that the composition of the pollen pool relative

to the composition of the adult population changed through time in Composite Cross V goes part way towards answering question three. Additional evidence comes from estimates of t in Composite Cross V which were made cojointly with the pollen pool estimates. The average values of t over loci per generation are $\bar{t}_8 = 0.0057 \pm 0.0010$, $\bar{t}_{19} = 0.0088 \pm 0.0009$ and $\bar{t}_{28} = 0.0124 \pm 0.0011$ (Kahler et al., in preparation). Clearly there has been a steady increase in outcrossing rate during the evolution of this population. Thus the mating structure of a population, together with its effects on genotypic frequency distributions and recombinational potential, are subject to evolutionary processes. It is especially interesting that outcrossing has increased in this population because, if attention is limited to considerations of mating system alone, genes predisposing an individual to selfing should be favored since they will be transmitted to a greater proportion of the progeny of an individual (Fisher, 1941). The general theory associated with evaluating the advantages of recombination in evolution is very complex; however Karlin (1973) asserts that in infinite populations recombination may be favored if there is negative disequilibrium ($D < 0$, where D is the gametic disequilibrium parameter to be discussed in a later section) between interacting genes (more specifically, where the relative viabilities are supermultiplicative). Such a theory could apply in the present instance because the population was initiated by intercrossing a relatively small number (30) of barley varieties and population size was subsequently very large.

One answer to the fourth question is given by comparisons of outcrossing rates between local populations within species. When actual measurements are made on local populations variations in amount of outcrossing are found to be large from population to population within a species (Table II). Also the amount of outcrossing is often correlated with identifiable environmental variables such as moisture availability. Thus Marshall and Allard (1970) report that a population of *Avena barbata* occurring in a damp site has an outcrossing rate which is five fold greater than another population which occupies a dry harsh site ($t = 0.075$ versus $t = 0.014$). Variation in outcrossing over short distances ($\simeq 15$ meters) is indicated in a study of *Bromus mollis* in Australia, and here too the damper enivronments support populations with higher values of t (Brown et al., in press). Further evidence of microgeographical variation in mating structure comes from *A. barbata* where t varies from 0.001 to 0.031 over six sites along a 180 meter transect in California (Hamrick and Allard, 1972). These data also suggest that t is higher

TABLE II

Species	$\bar{t}$ range	number of marker loci used
Avena barbata	0.1-7.5	5
Avena fatua	0.1-1.6	6
Bromus mollis	7.0-14.0	6
Lolium multiflorum	68.0-100.0	2
Collinsia sparsiflora	0.0-83.0	4
Hordeum vulgare	0.0-8.5	4
Festuca microstachys	0.0-7.0	3

Outcrossing rates (%) determined using isozyme marker loci in several plant species.

in moist environments. There has been a tendency to dichotomize species of plants into "selfers" or "outcrossers," but even the meager data given in Table II show that a wide spectrum of mating systems is often represented within a single species, and also among species with similar ecological preferences. Interestingly, comparisons among populations and species with different levels of outcrossing reveal no clear relationship between mating system and amount of genetic variability. As an example, *Avena barbata* outcrosses at a substantially higher rate than *A. fatua* (about 3x higher on the average) but *A. fatua* is much more variable genetically than *A. barbata*. Evidently mating system is only one of the factors affecting genetic variability in populations, and, among the other factors which might be involved, selection appears to be a major candidate.

ESTIMATION OF SELECTION INTENSITIES

Once having characterized the mating system of a population, we know the expected genotypic frequency distribution immediately after mating; however, all the diverse phenomena which fall under the rubric selection will transform this frequency distribution into new distributions. Enzyme variants have provided useful genetic markers for studying the direction and intensity of selection on the chromosomal segments which they mark. The basic methodology is to take successive samples from a population, and then determine the set of scalars which map the genotypic frequency distribution in one stage into the distribution observed in the succeeding stage, taking proper account of the mating cycle if it occurs between the two sampling points. The set of scalars which determine the transformation from one frequency distribution to another are called relative fitness values (w_i); they are determined uniquely by defining the w_i relative to some standard genotype, say w_1, so that $w_i = w_i^1/w_1$.

Annual plant species have a number of advantages for the estimation of selection intensities in natural populations. Chief among these advantages is the fact that such populations lack a complex age structure; hence demographic changes are not confounded with selective effects. Other advantages are ease of collection and sedentary habit which allows precise identification of populational units and accurate description of ecological relationships. Consequently it is not surprising that isozyme markers have been used to measure selective intensities in natural populations of several plant species. The species investigated include *Avena barbata* (Marshall and Allard, 1970; Hamrick and Allard, 1972; Clegg and Allard, 1973), *A. fatua* (Clegg and Allard, 1974), *Bromus mollis* (Brown et al., in press) and *Lolium multiflorum* (Pahlen, 1969). Two generalizations emerge from the results; first, heterozygous genotypes are usually favored and second, the extent of heterozygous advantages is often correlated with the harshness of the habitat. Thus Brown et al. (in press) find two to four times as many heterozygotes as expected at five enzyme loci in *B. mollis* subpopulations occupying dry sites, while a slight deficit of heterozygotes occurs in adjacent subpopulations in damp areas. Similarly, Marshall and Allard (1970) found substantial heterozygous excess in a population of *A. barbata* occurring in a harsh dry site; but a second population in a moist environment exhibited only very mild heterozygous advantage. In an attempt to analyze further the heterozygote advantage often observed in natural populations of *A. barbata*, Clegg and Allard (1973) partitioned selection estimates into viability and fecundity components. Their results, illustrated in Table III, show that heterozygotes tend to be favored in viability rather than fecundity.

Isozymes have also featured in studies of the dynamics of selection in experimental populations. Allard et al. (1972) found highly significant directional changes in frequencies of four esterase loci in a study which spanned 25 generations in the aforementioned barley Composite Cross V (CCV) population. There was also a consistent excess of heterozygotes in this population. More recently Clegg et al. (in preparation) have partitioned the selective values in CCV into viability and fecundity components. They found intense viability and fecundity differences among genotypes. However, the genotypes favored in viability were usually less fertile, leading to a kind of balancing effect, so that total observed selection over the whole life cycle was smaller than that esti-

TABLE III

	Genotype		
	11	12	22
viability	1.00	1.82	1.03
fecundity	1.00	0.87	0.83
total life cycle	1.00	1.59	0.84

Estimates of viability and fecundity components of selection averaged over three loci (E_4, E_9, and E_{10}) in *Avena barbata*. (From Clegg and Allard, 1973.)

mated for viability or fecundity components alone. Thus enzyme markers have revealed intense selection in natural and in experimental plant populations. However, a caveat must be entered at this point; the relative fitness values measured for the various marker loci cannot be attributed solely to the effects of the different allozymes on the fitness of their carriers. To measure selection at a single locus the rest of the background genotype must be either isogenic, or perfectly randomized with respect to the marker locus. Such conditions must rarely obtain in finite populations. Rather, the typical condition is expected to vary from one of mild to very strong correlation in allelic state over loci depending on population size, extent of linkage, mating systems and the pattern of selective forces acting over the whole genome. Hence the extent to which the entire population genotype is organized into correlated systems of genes is a fundamental issue in contemporary population genetics and one which has recently been approached experimentally using electrophoretically detectable markers.

MULTILOCUS ORGANIZATION

A natural focus for studies of multilocus organization has been the gametic frequency distribution taken over two or more loci. The reasons for this convergence are straightforward; the departure from a random state can be measured from this distribution in terms of marginal gene frequencies and various disequilibrium functions. In the case of two diallelic loci a single disequilibrium function variously called linkage disequilibrium, gametic phase disequilibrium, gametic phase unbalance etc., suffices to parameterize departures from randomness. More generally, a two-locus measure of disequilibrium is defined as,

$$D_{ij} = g_{ij} - p_i q_j \qquad i = 1, 2, \ldots k; \; j = 1, 2, \ldots \ell$$

where g_{ij} is the relative frequency of gametes carrying the

i^{th} and j^{th} alleles at the first and second loci respectively, and

$$p_i = \sum_{j=1}^{\ell} g_{ij} \quad \text{and} \quad q_j = \sum_{i=1}^{k} g_{ij}$$

are marginal frequencies of the ith and j^{th} alleles at the first and second loci. Analogous disequilibrium functions have been defined for second-and higher order-interaction terms when the joint distribution of three or more loci is considered (e.g. Bennett, 1954; Slatkin, 1972).

Theoretical expressions for the dynamics of two-locus disequilibrium for neutral alleles have been long known for random mating populations,

$$D_{ij}^{(t)} = \left(\frac{1+\lambda}{2}\right)^t D_{ij}^{(0)},$$

where t is time in generations and λ is a linkage parameter [$\lambda \varepsilon (0,1)$]. More recently, Weir and Cockerham (1973) have given general expressions for the dynamics of disequilibrium for neutral alleles under the mixed mating model. Providing $\lambda < 1$ and $s < 1$, D_{ij} approaches zero at the geometric rate

$$1 - 1/2\{\frac{1+\lambda+s}{2} + [\ (\frac{1+\lambda+s}{2})^2 - 2s\lambda]^{1/2}\}.$$

Thus, under neutrality, the genic material of mixed mating populations is expected to become randomized, albeit at a slower rate depending jointly upon λ and s, than under random mating. In opposite vein, disequilibrium ($D \neq 0$) is more likely to develop in inbreeding than in random mating populations because much less intense selection is required to hold together favorably interacting combinations of alleles at different loci. Inbreeding species of plants are therefore favorable materials in which to study gametic phase disequilibrium and both natural and experimental populations have been studied (Allard, in press).

Investigations of multilocus structure in natural plant populations have revealed strikingly non-random patterns of organization in each of the predominantly selfing species which has been studied. The best studied case is that of *A. barbata* in which strong indications of multilocus organization were obtained for both macro- and microgeogeographical patterns of allozyme differentiation (Clegg and Allard, 1972; Hamrick and Allard, 1972). In a detailed study of a very small area (8m x 18m) Allard et al. (1972) determined joint five-locus frequencies over three esterase, a peroxidase, and a phosphatase loci and found that genotypic frequencies were highly correlated with environmental features. In particular, one five-locus genotype shown earlier to be characteristic of hot dry climatic regions (Clegg and Allard, 1972; Hamrick and Allard, 1972) was found to be in great excess in topographically raised portions of the study area characterized by shallow rocky soils. Conversely, a shallow depression which served as a natural drainage area exhibited a great excess of an exactly complementary five-locus genotype known to be characteristic of cool damp habitats (Hamrick and Allard, 1972). These results bring out the intense correlation of genotype with microenvironment and the highly non-random organization of the genetic material of the population in both space and also among loci.

The rate at which multilocus organization can develop has been studied in two experimental populations of barley, Composite Cross V and Composite Cross II, which were synthesized by crossing 30 and 28 barley varieties, respectively, in all possible combinations. This method of synthesis is expected to produce random combinations of alleles at different loci, i.e. $D = 0$, is expected in the initial generations of the population. In a study of the dynamics of disequilibrium in CCV, Weir et al. (1972, and in press) found that disequilibrium for pairs of esterase loci increased in this population at rates much greater than expected under neutrality. In a subsequent study in which all four loci were considered simultaneously, Clegg et al. (1972) found that certain gametic types increased much more rapidly in frequency than predicted from their marginal allelic frequencies. They also found that the four-locus gametic types which achieved high frequency tended to be perfectly complementary in allelic composition. Thus the genetic resources of CCV very rapidly achieved a highly non-random state. Further, the same four-locus gametic types become predominant in CCII, even though this population had different parents. Clearly, selection played the major role in determining the genetic structure of these two populations. Thus, a major problem is to determine the pattern of selection at the multilocus level.

Efforts to measure selection for multilocus genotypes are rare because the sampling problems associated with even the ten genotypes produced by two diallelic loci demand very large scale experiments. Nevertheless, Weir et al. (1972) measured fitness values for locus pairs in ten separate generations of CCV. These experiments involved assaying 68,230 plants for their genotype at four esterase loci. The results indicate that single-locus selection estimates bear little relationship to two-locus estimates and hence, that complex epistatic selective forces are operating on this population.

In an effort to increase the dimensionality of selection experiments Clegg et al. (in preparation) estimated fitness values for joint three-locus genotypes in CCV. Samples were taken from the population at both the adult and the seed stages, permitting viability estimates from the seed to adult transition. Four features of the results are particularly noteworthy: first, viability differences among the eight homozygous genotypes were very large; second, two of the three homozygous genotypes with the largest fitness values in generation 28 were the genotypes which contribute the gametes shown to be in great excess in the earlier investigation (Clegg, et al., 1972); third, heterozygous genotypes, especially multiple heterozygous genotypes, were strongly favored in viability; finally, fecundity was not very different for different multilocus genotypes and if anything, genotypes which were highest in viability were lower in fecundity. Thus, as in the one-locus case discussed earlier, selection differs in direction and intensity in different stages of the life cycle so that total observed selection is smaller over the whole life cycle than for specific stages.

ACKNOWLEDGEMENTS

This work was supported in part by grants from the National Institutes of Health GM 10476 and the National Science Foundation GB 39898X.

REFERENCES

Allard, R. W. The mating system and microevolution. XIII *Intern. Genetics Congress* (in press).

Allard, R. W., G. R. Babbel, M. T. Clegg, and A. L. Kahler 1972. Evidence for coadaption in *Avena barbata*. *Proc. Natl. Acad. Sci. USA* 69: 3043-3048.

Allard, R. W., A. L. Kahler, and B. S. Weir 1972. The effect of selection on esterase allozymes in a barley population. *Genetics* 72: 489-503.

Bennett, J. H. 1954. On the theory of random mating. *Ann. Eugen.* 18: 311-317.

Brown, A. H. D., and R. W. Allard 1970. Estimation of the mating system in open-pollinated maize populations using isozyme polymorphisms. *Genetics* 66: 133-145.

Brown, A. H. D., D. R. Marshall, and A. Albrecht. The maintenance of alcohol dehydrogenase polymorphism in *Bromus mollis* L. *Aust. J. Biol. Sci.* (in press).

Clegg, M. T., and R. W. Allard 1973. Viability versus fecundity selection in the slender wild oat, *Avena barbata*.

Science 181: 667-668.

Clegg, M. T., and R. W. Allard. Genetics of inter- and intra population structure in *Avena fatua* (in preparation).

Clegg, M. T., R. W. Allard, and A. L. Kahler 1972. Is the gene the unit of selection? Evidence from two experimental plant populations. *Proc. Natl. Acad. Sci. USA* 69: 2474-2478.

Clegg, M. T., A. L. Kahler, and R. W. Allard. The estimation of two life cycle components of selection in an experimental plant population (in preparation).

Fisher, R. A. 1941. Average excess and average effect of a gene substitution. *Ann. Eugen.* 11: 53-63.

Hamrick, J. L., and R. W. Allard 1972. Microgeographical variation in allozyme frequencies in *Avena barbata*. *Proc. Natl. Acad. Sci. USA* 69: 2100-2104.

Harding, J., and C. L. Tucker 1964. Quantitative studies on mating systems. I. Evidence for the non-randomness of outcrossing. *Heredity* 19: 369-381.

Jain, S. K., and R. W. Allard 1960. Population studies in predominantly self-pollinated species. I. Evidence for heterozygote advantage in a closed population of barley. *Proc. Natl. Acad. Sci. USA* 46: 1371-1377.

Kahler, A. L., M. T. Clegg, and R. W. Allard. Evolutionary changes in the mating system of an experimental plant population (in preparation).

Karlin, S. 1973. Sex and infinity: a mathematical analysis of the advantages and disadvantages of genetic recombination in *The Mathematical Theory of the Dynamics of Biological Populations* (eds., M. S. Bartlett and R. W. Hiorns), Academic Press, London.

Marshall, D. R., and R. W. Allard 1970. Maintenance of isozyme polymorphisms in natural populations of *Avena barbata*. *Genetics* 66: 393-399.

Pahlen, A. The genetics of isozyme polymorphism in natural populations of *Lolium multiflorum* Lam. Unpublished Ph.D. thesis. University of California, Davis.

Slatkin, M. 1972. On treating the chromosome as a unit of selection. *Genetics* 72: 157-168.

Weir, B. S., R. W. Allard, and A. L. Kahler 1972. Analysis of complex allozyme polymorphisms in a barley population. *Genetics* 72: 505-523.

Weir, B. S., R. W. Allard, amd A. L. Kahler. Further analysis of complex allozyme polymorphisms in a barley population (in preparation).

Weir, B. S., and C. C. Cockerham 1973. Mixed self and random mating at two loci. *Genet. Res., Camb.* 21: 247-262.

ALLOZYME VARIATION IN *E. COLI* OF DIVERSE NATURAL ORIGINS

ROGER MILKMAN
Department of Zoology, The University of Iowa
Iowa City, Iowa 52242

ABSTRACT. The results of an electrophoretic analysis of allozyme (allelic isozyme) variation at 5 loci in *E. coli* from diverse natural sources are incompatible with the "neutral" hypothesis, that electrophoretic mobility variation is due largely to the random genetic drift of many adaptively equivalent alleles. The effective number of mobility classes is small, and they are distributed discontinuously over a considerable range. Neither the relationship of charge change to amino acid substitution, nor a sharp reduction in the number of possible equivalent alleles, nor any of the likely spatiotemporal population structures of the species *E. coli* can provide a way out of this conclusion.

Genotypic variation of the *coli* within a host, at a given time and over periods of several months is substantial. Evidence of frequent recombination is presented, and other preliminary observations are discussed, together with the prospects of further investigations of protein variation, recombination, infectivity, and factors influencing selection in *E. coli*.

I have recently reported (Milkman, 1973) the results of an electrophoretic study of five loci in 829 clones of *E. coli* derived from 156 fecal samples of diverse natural origin. Table I provides an updated summary of these results. The main conclusion of this study is that the electrophoretic variation observed must be attributed to natural selection rather than to random genetic drift. There is good reason to generalize this conclusion to higher organisms, as well. In addition, several conclusions were drawn from this study with respect to the role of heterosis in maintaining genic polymorphism, the occurrence of recombination in natural populations, and the genetic structure of the species *E. coli*.

In the present paper I should like to consider three aspects of the argument for selection and against the "neutral hypothesis". These aspects are the ones that have been probed extensively in order to determine whether the argument against the neutral hypothesis can be escaped. Methods of processing and analysis have been described (Milkman, 1973); acquisition of samples will be discussed later. Then I shall present some detailed data from which certain of the additional conclusions were drawn.

The "neutral hypothesis" attributes almost all genic polymorphism to random genetic drift. In the absence of a clear quantitative relationship between phenotypic difference and genotypic difference, proponents of the neutral hypothesis consider that a very small part of genetic variation is sufficient to account for the adaptive variation -- in space and time -- in higher phenotypes. The rest is "evolutionary noise".

Since the major measure of genic polymorphism is currently electrophoretic analysis, the controversy has centered on the role of selection in mobility variation. Thus some relationship between allelic variation and mobility variation must be described. There are two such relationships, and between them they seem to cover the reasonable possibilities. Both models, applied to the observations on *E. coli* lead to the rejection of the neutral hypothesis. I shall consider each of them.

First, Ohta and Kimura (1973) have recently proposed a means of estimating the effective number of mobility classes in a population. They assume that a given protein can undergo only unit changes of charge in the usual pH range of isozyme electrophoresis, 8.0-9.0. This assumption excludes the possibility of fractional charge changes due either to conformational changes or to incomplete dissociation of side groups. The presumably infrequent substitution of a charged amino acid residue for one of opposite charge, resulting in a two-unit change, is also ignored for simplicity. Nei and Chakraborty (1973) and King (1973) have also considered the implications of limiting mobility variants to integer classes.

Ohta and Kimura conclude that the steady state relationship, reached after some $4N_e$ generations at population size N_e, will be $m_e = (1+8N_e v)^{\frac{1}{2}}$, where m_e is the effective number of mobility classes, analogous to the effective number of alleles, n_e, (The use of the letter m is my own emendation). N_e is effective population size, and v is the rate of neutral mutations (or "nearly neutral", i.e., having too small an effect to prevail over random genetic drift). The number of possible essentially equivalent alleles must be large in comparison to the actual number of alleles present. The more familiar equation dealing with the alleles themselves is $n_e = 1+4N_e v$. Sometimes v is called u.

The widespread distribution of *E. coli* among mammals and the existence of mammals for some 70 million years suggests that *E. coli* has been numerous for at the very least 40 billion generations, permitting us to use 10^{10} for N_e (a tiny fraction of all the *coli* in the world, but enough for present purposes). A per-locus mutation rate of 10^{-5} - 10^{-6} suggests that 10^{-8}

is a reasonable minimum estimate of v, the neutral mutation rate per locus. The model of Ohta and Kimura thus would lead us to expect in the present study an effective number of mobility classes between 25 and 30, since $m_e = (1 + 8 \times 10^{10} \times 10^{-8})^{\frac{1}{2}}$. But in fact m_e is between 1 and 2 in each of the 5 cases. Furthermore, the best comparison is probably between $(m_e - 1)$ expected and $(m_e - 1)$ observed, since 1 allele per locus is

TABLE I

FREQUENCIES OF MOBILITY CLASSES, IN ORDER OF INCREASING MOBILITY

	ADH	AP	G-6-PDH	6-PGDH	MDH
	17	9*	1	1	14
	1	815	793	10	1
	611	5**	10	7	2
	3		25	16	39
	2			6	2
	142			21	769
	1			65	2
				34	
				22	
				591	
	(null:52)			26	
				30	
m_e	1.54	1.00-1.03	1.09	1.92	1.16

* Two bands, + and slow

** Two bands, + and fast (4 cases) or slow and fast (1 case). Allelic variation uncertain: see M. Schlesinger, this symposium.

the minimum. So the model's prediction is two orders of magnitude off the mark. In addition, it should be noted that the mobility classes are observed to be distributed discontinuously, and there is generally daylight between them (Milkman, 1973). Clearly, the neutral hypothesis cannot be sustained by the use of this model.

Frankly, I consider this model to be a gross oversimplification. Titration curves for hemoglobin and other proteins are never flat between pH 8 and pH 9, nor between pH 4 and pH 10, for that matter. So partial dissociation must be in force throughout the range of pH commonly used in electrophoresis. Furthermore, the interactions of ionizable groups are well known, and they depend upon distance in space as well as along a molecular chain. The dissociation constants of fumaric and maleic acids, which are cis-trans isomers, constitute one simple example. The pk's of fumaric acid are 3.0 and 4.4, while those of maleic acid are 1.8 and 6.1. In any event, the frequently-quoted paper of Henning and Yanofsky (1963) on the electrophoretic analysis of proteins stemming from apparently single mutations seems conclusive: "As was observed with some of the other proteins (compare A33 and wild type [,] and A88 and A3), a fraction of a single charge difference seems to distinguish the A58 and A46 proteins...". It is surprising that this paper has actually been cited (Ohta and Kimura, 1973) in support of the unit-charge-change view.

We now proceed to a second model, in which an essential continuum of possible mobilities exists. In this case one is interested, once again, in the frequency distribution of the mobility classes. In Figure 1, frequency distributions are displayed for four loci. The most important feature -- and it holds for all 5 loci -- is that the commonest mobility class contains over 70% of all the individuals. This mobility class might include many different allozymes (different with respect to amino acid sequence) that could not be distinguished electrophoretically. But if this were so, one would expect to find almost all the remaining 30% grouped into adjacent classes of equal width. A random distribution would require that. Less important, but true, is the requirement that the distribution be reasonably symmetrical.

In fact, the distribution is strikingly different. First of all, classes adjacent to the commonest class are generally empty. This cannot be passed over. And equally striking is the great mobility range over which the classes are distributed. Finally, we see no monotonic decline in frequency with distance from the common class. The first two observations clearly disqualify this combination of neutral hypothesis and model. The third alone would constitute a serious contradiction.

I see no "middle ground" relationship between alleles and mobility classes that has conceptual merit, and that might therefore be explored. Accordingly I consider that the refutation of the neutral hypothesis cannot be avoided by the choice of a particular charge-variation model for protein molecules.

I have avoided consideration of the effects of conformational changes alone on mobility, on the grounds that conformational changes sufficient to alter mobility by themselves would doubtless alter net charge as well.

A second suggestion for a way out is the possibility of "limited neutralism". Is it possible that there are very few possible acceptable allelic states, and that they are all equally acceptable (neutral)? This possibility could obviously be tailored to fit the small effective number of mobility classes in *E. coli* , but not the distribution of the classes. Furthermore, the required reduction in the number of acceptable alternatives, k, would drive the estimate of n_e down to an unacceptable level in higher organisms. This would be due both to the direct effect on k and to the consequent lowering of v, which is the rate of mutations among acceptable states only. Indeed, a lower v might be hailed as the basis for explaining the allele frequencies observed in vast populations of flies, but it would run afoul of smaller populations, for they have a great deal of genetic variation, too. Thus a severe limitation on the possible number of neutral mobility classes would not reconcile the *E. coli* data to the predictions of the neutral hypothesis, nor would they fit the collective observations on higher organisms at all.

Finally, it has been suggested that the genetic structure of the species *E. coli* might possess certain properties that would make these observations consistent with the neutral hypothesis. But again, no reasonable genetic structure has such properties. For example, an essentially panmictic species, with enough gene flow and recombination over the billions of generations to make the species one great population serves as the model by which the neutral hypothesis is refuted. In contrast, a set of essentially isolated populations would be expected to differ strikingly in the alleles at high frequencies, whereas the commonest allele at each locus is the commonest in every locality sampled in this study. The absence of genotypic dependence upon the host's phyletic position in studies to date must also be taken as evidence for a good deal of communication among the myriad *E. coli* populations.

While an essentially panmictic nature of *E. coli* would be consistent with the data, and the hypothesis of considerable isolation over long periods of time would not, a third view has been occasionally proposed, though casually. This view is that human beings have altered the environment to such an extent that a special strain of *E. coli* has evolved to meet the challenge, first in man, and then, by successful competition, has eliminated the *coli* living in the vast and diverse number of mammals in the world. It is hard to take this view seriously, in light of what we know about recombination and selection in bacteria, and in view especially of the rapid succession of diverse genotypes described later and illustrated in Table 6.

E. coli was identified (Milkman, 1972) as a "no-excuse" organism on which to test the neutral hypothesis because of a favorable set of properties, including its vast distribution and short generation time. It can also recombine with other genera of bacteria. The likelihood that *E. coli* recently

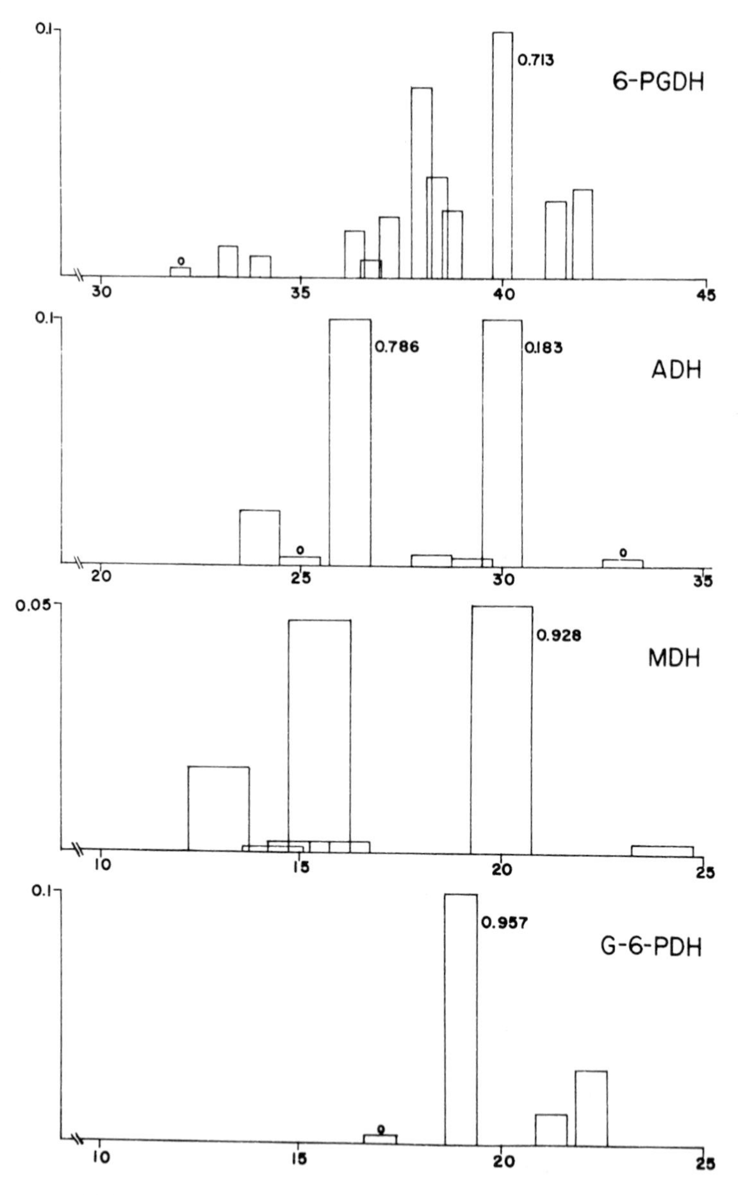

Fig. 1. Relative frequencies and spacing of mobility classes. Ordinate, relative frequency. Abscissa, distance in mm from origin (top row) after standard run. High-frequency classes are labeled since they extend beyond the scale. Low frequency classes (below 0.002) are exaggerated in height for clarity and labeled "o".

passed through a population bottleneck (thus nullifying the expectations of large values of n_e and m_e) is far too small to be taken seriously by any biologist; it is probably even less likely than the possibility of a recent Caribbean glacier, suggested by Ohta and Kimura (1973).

It has so far been impossible to associate particular alleles with any particular locale, environmental factor, or phyletic position of the host. Needless to say, the observations to date are merely first indications. In any event, Table II shows that the commonest mobility class overall at each locus is commonest in each of several categories. And Table III shows that the mobility classes at the 6-PGD locus are widely distributed.

TABLE II

FREQUENCY OF COMMONEST CLASS

Source	ADH	G-6PDH	6-PGDH	MDH
Wild Animals	0.75	1.00	0.67	0.75
Domestic, New Guinea	0.72	0.94	0.67	0.95
Man and Domestic, Iowa	0.69	1.00	0.63	0.77
Short Captivity, Zoo	0.86	0.97	0.75	0.93
Longer Captivity, Zoo	0.76	0.93	0.76	0.87
Lab Stock (mice)	0.97	0.99	1.00	1.00

TABLE III

DISTRIBUTION OF THE 6-PGDH MOBILITY CLASSES

Source	S_9	S_8	S_7	S_6	S_5	S_4	S_3	S_2	S_1	+	F_1	F_2
Human, Iowa	+	+	-	+	+	+	+	-	+	+	+	+
Primates, Zoo	-	+	+	+	+	+	+	+	+	+	-	+
Other animals, Zoo	-	+	+	+	-	+	+	+	+	+	+	+
New Guinea	-	+	-	+	-	+	+	+	+	+	+	+
Other locales	-	-	-	-	-	+	+	-	+	+	-	-

ADDITIONAL CONCLUSIONS

Obviously, since *E. coli* is essentially haploid, heterosis cannot account for the genic polymorphism observed in this study. So whatever the role of heterosis in eukaryotic diploids, it is not the universal mechanism in all living things. Also, it should be noted that each *E. coli* clone studied exhibited only one band for each allozyme, with 14 exceptions, all at the alkaline phosphatase locus (See Table I). These cases can be explained by a mechanism of the type described by Schlesinger (1975). The variation in mobility of these pairs of bands does suggest allelic variation. It is of interest that seven +/S strains are clones from one fecal sample (from a New Guinea steer) and three +/F strains are clones from another sample (a young woman from Iowa). These and the other three strains showing two bands (from a wild-caught gorilla, a zoo elephant, and a deer) were all tested for heterogeneity by subcloning. All subclones showed two bands.

This study apparently provides the first evidence for frequent recombination in natural populations of *E. coli* . In a number of instances, up to 20 clones were taken from a single fecal sample. When individual clones from a given sample differed genetically at more than one locus, several combinations of alleles were generally observed. The likeliest explanation is recombination within each host, and two features of the data argue against an alternative view, that exceedingly rare recombination over a long period of time has resulted in a vast array of widely distributed genotypic combinations. First, the number of alleles at each locus is relatively low in an individual, but the proportion of possible combinations of these few alleles is high. And second, two animals living close to one another may have quite different *coli* populations, one with numerous genotypes and one with few. Some representative genotypes are illustrated in Table IV. These observations are of course minimal, and explicit experiments are now underway to characterize recombination in natural populations of *E. coli*.

Because very little is presently known about natural genetic variation in *E. coli*, some additional observations are presented of a rather anecdotal nature. In Table V, the number of alleles at each of 4 loci (alkaline phosphatase is excluded) are listed for those animals from which 10-20 clones were examined. Perhaps some conclusions can eventually be drawn about feeding habits, communication, and other characteristics of the hosts, when we know more. In Table VI are listed the changes in successive samples, several months apart, from individual animals at the Woodland Park Zoo, in Seattle. Beyond the fact that changes do occur, little can be concluded at present.

TABLE IV

COMBINATIONS OF ALLELES SEEN

Strain	Loci	Combinations	Strain	Loci	Combinations
200 Pig #1	A,G,6	+++	216 Pig #3	A,G,6	+++
		F+S			F++
		++S			FF+
		F++			+F+
		++M			++M
		F+M			F+M
		FF+			FFM
211 Sheep #1	A,6,M	SEP	212 Sheep #2	A,6,M	+++ only
		+EP			
		S++			
213 Goat #1	A,G,6	+++			
		++T			
		+FT			
		F++			
		FFE			

A number of interesting problems in the biology of *E. coli* can be approached by simple electrophoretic analysis. Samples from distant places can be placed directly in "Whirlpak" bags and shipped in double-walled containers (nested cardboard canisters) with generally good results and the approval of the U.S. Postal Service. The abundance of natural markers makes possible studies of recombination within particular hosts, the possible dependence of infectivity (or competitive ability in a particular host) on the source of the bacteria, and of course the dependence of *coli* genotype on the host's phyletic position, niche, and habitat. The sequencing of allozymes in *E. coli* should prove an exciting landmark in the study of the evolutionary diversity of proteins, following comparisons among species and genera that are currently being made (Li and Hoch, 1974). Meanwhile cellulose acetate electrophoresis, employing multiple application (Milkman, 1973), provides a convenient approach to the study of biological variation where large and diverse samples are required.

TABLE V

NUMBER OF ALLELES PER LOCUS IN A SET OF 10 OR MORE CLONES

	Locus				
	ADH	G-6PDH	6-PGDH	MDH	Sets (N)
	1	1	1	1	12
	1	1	2	1	6
	2	1	3	1	4
	1	1	3	1	3
	2	1	2	1	2
	1	1	2	2	2
	2	1	2	2	2
	3	2	3	2	2
	2	2	3,5,7	1	3
	2	1	2,3	2	2
	3	1	2,4	1	2
	2	2	2,4	2	2
	1	1	1	2	1
	2	1	1	1	1
	1	2	2	1	1
	3	2	5	1	1
	4	1	2	1	1
AV.	1.6	1.2	2.3	1.3	Total 47

TABLE VI

GENOTYPES OF *E. COLI* IN SUCCESSIVE SAMPLES*

Host	Sample #	Clones (N)	ADH	G-6PDH	6-PGDH	MDH
Leopard "Lita"	2	10	+	+	+H	+
	3	10	+E	+	+GHJ	+A
Leopard "Liz"	1	1	+	+	+	+
	2	10	+E	+	+GH	+
	3	10	+CDE	+C	+BJ	+D
Cougar "Bonnie"	1	1	+	+	J	+
	2	10	+	+	+J	+D
	3	10	+E	+C	+J	+D
Lion "Rex"	1	1	+	+	+	+
	2	10	+	+	+	+
	3	10	E	+C	+G	+
Giraffe "Duchess"	1	10	+	+	+	+
	2	10	+E	+	+	+
	3	10	+	+	+	+
Giraffe "Princess"	1	1	+	+	+	+
	2	10	+	+	+	+
	3	10	+	+	+H	+
Giraffe "Contessa"	1	1	+	+	+	+
	2	10	+	+	+FG	+
	3	10	+	+	+	+
Elephant "Bamboo"	1	1	E	+	D	+
	2	10	+E	+C	+I	+
	3	10	+	+	+H	+D
Hippo "Gertrude"	1	1	+	+	+	+
	2	10	+E	+	+DJ	+D
	3	3	+	+	+	+
Gorilla "Nina"	1	3	+E	+	+F	+
	2	2	+	+	G	+

TABLE VI, continued.

Host	Sample #	Clones (N)	ADH	G-6PDH	6-PGDH	MDH
Gorilla	1	3	+	+	+F	+
"Caboose"	2	10	+	+	F	+
Celebes Black	1	10	+ACE	+	+G	+
Ape "B"	2	10	+E	+	+DE	+
Celebes Black	1	3	+	+	+G	+
Ape "E"	2	9	+	+	+	+

* For ease of visual recognition, mobility classes are coded by letters (commonest class is "+") in order of increasing mobility. Refer to Table I for frequencies.

ACKNOWLEDGEMENTS

The work was supported in part by NIH Grant GM-18967-01-02.

REFERENCES

Crow, James F. 1972. Darwinian and non-Darwinian evolution. In: *Proceedings of the Sixth Berkeley Symposium on Mathematical Statistics and Probability. Volume V: Darwinian, Neo-Darwinian, and Non-Darwinian Evolution* (Ed. L.M. Le Cam, J. Neyman, and E.L. Scott). Berkeley, California pp. 1-22.

Henning, Ulf and C. Yanofsky 1963. An electrophoretic study of mutationally altered A proteins of the tryptophan synthetase of *Escherichia coli*. *J. Molec. Biol.* 6:16-21.

King, Jack L. 1973. The probability of electrophoretic identity of proteins as a function of amino acid divergence. *J. Molec. Evol.* 2:317-322.

Li, Shoe-Lung and S.O. Hoch 1974. Amino-terminal sequence of tryptophan synthetase α chain of *Bacillus subtilis*. *J. Bact.* 118:187-191.

Milkman, Roger 1972. How much room is left for non-Darwinian evolution? In: *Evolution of Genetic Systems* (Ed. H.H. Smith). Gordon and Breach, New York, pp. 217-229.

Milkman, Roger 1973. Electrophoretic variation in from natural sources. *Science.* 182:1024-1026.

Nei, Masatoshi and R. Chakraborty 1973. Genetic distance and electrophoretic identity of proteins between taxa. *J. Molec. Evol.* 2: 323-328.

Ohta, Tomoko and M. Kimura 1973. A model of mutation appropriate to estimate the number of electrophoretically detectable alleles in a finite population. *Genet. Res., Camb.* 22:201-204.

Schlesinger, M. J. 1975. Differences in the structure, function, and formation of two isozymes of *Escherichia coli* alkaline phosphatase. *Isozymes I. Molecular Structure,* C. L. Markert, editor, Acadmic Press, New York. pp. 333-342.

ISOZYME POLYMORPHISMS IN THE STUDY OF EVOLUTION IN THE *PHASEOLUS VULGARIS* - *P. COCCINEUS* COMPLEX OF MEXICO

J.R. WALL AND SARAH W. WALL
Department of Biology, George Mason University
4400 University Drive, Fairfax, Virginia 22030

ABSTRACT. Alcohol dehydrogenase and anodal and cathodal esterase seed polymorphisms are present in wild, cultivated, and feral populations of the Mexican *Phaseolus vulgaris-P. coccineus* complex. Three zones of anodal ADH mobility are found. Allozymes (allelic isozymes) of the Adh-3 locus are controlled by two codominant alleles in *P. vulgaris*. Individuals heterozygous for these alleles usually produce one band of intermediate mobility, suggesting that these allozymes are dimers. The Adh-1 zone is monomorphic in *P. vulgaris* but polymorphic in *P. coccineus*. It is suggested that Adh-1 isozymes are controlled by a second locus, and that Adh-2 isozymes are interaction products formed by the association of monomeric products coded by alleles at the Adh-1 and Adh-3 loci. Anodal esterase allozymes of the Est-1 locus are found consistently only in *P. vulgaris*. Faster and slower variants are controlled by codominant alleles. Cathodal esterases are found in all populations of the *Phaseolus* complex. Codominant alleles control two cathodal esterases at the Est-2 locus, and two other bands are found. Adh-3 and Est-1 allozyme data for several wild and cultivated populations of *P. vulgaris* are indicative of outcrossing well in excess of previous estimates, suggesting a more flexible breeding system for *P. vulgaris*.

INTRODUCTION

Studies of electrophoretic variants of enzymes and proteins have had numerous applications in evolutionary biology in recent years, among which are investigations of the genetic structure of populations (Prakash et al., 1969; Brown and Allard, 1969), estimates of genic heterozygosity (Lewontin and Hubby, 1966), degree of genetic similarity between species (Webster et al., 1972; Gottlieb, 1973), distribution of geographical variation in allelic frequencies (Johnson et al., 1969; Marshall and Allard, 1970), and evidence of cryptic chromosomal structural differences between species (Wall and Whitaker, 1971). In this paper we report on the initial application of starch gel electrophoresis to evolutionary problems of the *Phaseolus vulgaris* - *P. coccineus* complex of

Mexico.

The genus *Phaseolus* includes about 150 wild species of annuals and perennials found mainly in the warmer mesic regions of the New and Old Worlds. The New World species are largely concentrated in Chiapas in southern Mexico, and in Guatemala, although they have at least discontinuous distribution as far as north central Mexico and northern Argentina. Several species also occur sporadically in the United States and perhaps elsewhere in the New World. The wild progenitor of the common bean, *P. vulgaris*, has been unknown until recent investigations showed it growing in numerous restricted sites in Mexico and Central America (Gentry, 1969). The cultivated and wild forms are now considered to be conspecific. The center of origin of cultivated *P. vulgaris* was considered by Vavilov (1935) to be Central America. However, his conclusions were based upon the knowledge that Central America was the center of diversity of cultivated *P. vulgaris*, not that the wild progenitor occurred there. It now appears that there have been multiple domestications of cultivated forms from the wild *P. vulgaris* in the New World (Gentry, 1969).

P. vulgaris cultivars are by far the most numerous and the most economically important beans in Latin America. However, *P. coccineus* is a species which has produced a number of cultivars having special importance in cool habitats at elevations above 6,000 feet. Cultivars of these two species are frequently grown in close proximity to each other and thereby have sympatric distribution (Fig. 1). The wild forms are allopatric even though they occur in the same geographical area, with allopatry conferred by altidutinal separation of the species. Wild *P. coccineus* is restricted to elevations above 6,000 feet, whereas wild *P. vulgaris* is limited to elevations below 5,800 feet (Wall, 1970). Introgression (Anderson, 1949) of genes from *P. coccineus* into *P. vulgaris* is reported by Freytag (1955) on the basis of morphological studies to be an important contributing factor to the variability of the cultivated Mexican black beans, which are cultivars of *P. vulgaris*. Introgression has also been cited (Gentry, personal communication) to explain the origin of *P. coccineus* ssp. *darwinianus*. It is a widely held belief that this subspecies is in reality a stabilized introgressive type between *P. coccineus* and *P. vulgaris*.

In order to gain evidence on introgression and on phylogenetic relationships within the cultivated-wild Mexican bean complex, we initiated this investigation. This paper reports: (1) electrophoretic patterns of cathodal and anodal esterases and of anodal alcohol dehydrogenases; (2) genetic control of these esterases and alcohol dehydrogenases; (3) evidence of

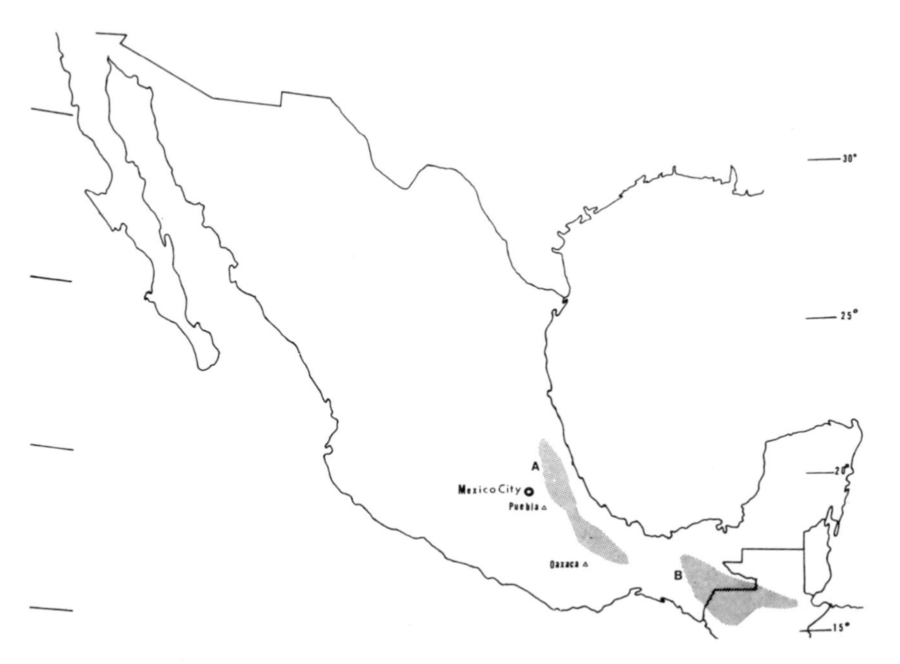

Fig. 1. The two most extensive areas of sympatric distribution of cultivated *P. vulgaris* and *P. coccineus* in Mexico and Guatemala (Gentry, personal communication).

relatively frequent outcrossing in certain populations of *P. vulgaris*; and (4) evidence of introgression from *P. vulgaris* into cultivated *P. coccineus* via *P. vulgaris* cultivars.

MATERIALS AND METHODS

Thirty-eight collections of mature seed of wild and cultivated *P. vulgaris* and *P. coccineus* were made in the Mexican states of Oaxaca, Morelos, Puebla, and Veracruz (Table 1, Fig. 2). Seed of the cultivated species were obtained at farm sites and at markets. Crossing experiments have not yet been conducted. However, seed from several collection sites have been grown and allowed to self in the greenhouse; and seed from some of these selfed plants have been analyzed electrophoretically to determine the mode of inheritance of

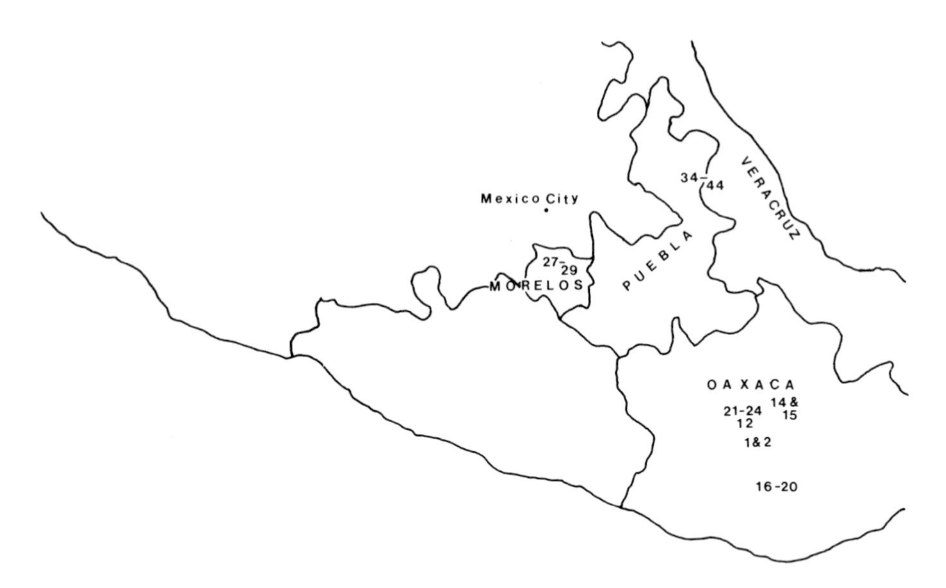

Fig. 2. Sites of the *P. vulgaris* and *P. coccineus* populations screened for ADH, Est-1, and Est-2 phenotypes. Population Nos. 18, 35, 40, 42, and 43 have not been screened. All sites occur between 17° and 20° north latitude.

---

anodal alcohol dehydrogenase (ADH, E.C. 1.1.1.1) isozymes at the Adh-3 locus and of anodal esterase (Est, E.C. 3.1.1.-) isozymes at the Est-1 locus of *P. vulgaris*. The inheritance of cathodal esterases at the Est-2 locus has been previously determined (Wall, unpublished).

Horizontal starch gel electrophoresis was used to screen seed of twenty-three of the original populations for Adh and Est patterns. Samples for electrophoresis were prepared as follows: Whole mature seed were soaked 9-15 hrs in distilled H_2O at room temperature. Seed coats were then removed from the whole seed or from portions of seed. Further preparations were carried out at 4°C. The seed portions (or whole seed) were ground with cold gel buffer in mortars and pestles. The resulting crude extract was allowed to stand 15-30 min and then absorbed onto 5 mm X 7 mm filter paper (Beckman No. 319329) wicks, which were inserted into cuts made 3.0 cm and 12.5 cm from the cathodal end of a 13.0% starch gel. This arrangement allowed accomodaton of up to 40 unknown samples,

20 inserted in each cut, and resulted in gel slices of 3.0 cm, 9.5 cm, and 7.0 cm in length for subsequent staining. The middle section of each gel was long enough (9.5 cm) so that Est-1 ioszymes migrating anodally from the first cut and Est-2 isozymes migrating cathodally from the second cut did not run into each other. Isozyme patterns of wild and cultivated Mexican populations were compared to the patterns of *P. coccineus* cv. Scarlet Runner and *P. vulgaris* cv. Black Turtle Soup, commercial U.S. cultivars, of which one sample of each was included on each gel. Seed ADH and Est patterns of *P. vulgaris* cvs. Blue Lake, Kentucky Wonder, and Top Crop were also determined.

The EBT continuous triple buffer system, pH 8.6, of Boyer et al. (1963) was used for gel and electrode chamber buffers. Electrophoresis was allowed to proceed at 4°C anodally at 300 v for 15 min. Sample wicks were then removed and electrophoresis continued at 350 v for a total of 5.5 hrs. Gels were then sliced once horizontally. The three portions of the top horizontal slice were stained for esterase visualization. The most cathodal portion of each horizontal bottom slice was discarded and the two remaining portions were stained for alcohol dehydrogenase.

Esterase activity was visualized by incubating the gel slices at 37°C for 15-30 min in a staining solution consisting of 100 ml of buffer, pH 7.0, 0.05 M in tris, 0.05 M in maleic acid, and 0.05 NaOH, to which was added 25 mg each of alpha- and beta-naphthyl acetate, dissolved together in 2.5 ml acetone, and 50 mg Fast Garnet GBC salt. Alcohol dehydrogenase activity was visualized by incubating gel slices at 37°C for about 3 hrs in a staining solution (modified from Shaw and Prasad, 1970) consisting of 15 ml of 0.5 M tris-HCl buffer, pH 7.1, 5 ml ethanol, and 80 ml distilled H_2O, to which was added 15 mg beta-nicotinamide adenine dinucleotide (NAD), 15 mg nitro blue tetrazolium (NBT), and 2.5 mg phenazine methosulfate (PMS). "Nothing dehydrogenase" (NDH) and "formazan oxidase" (FO) interference was eliminated by reduction of NAD concentration to 15 mg/100 ml staining solution and by addition of 10 ppm $MnCl_2$ (Hare, 1973). Interference from background staining was eliminated by avoidance of exposure to light.

RESULTS AND DISCUSSION

SEED ADH ISOZYME PATTERNS

Electrophoretic analysis shows three zones of ADH mobility. The Adh-3 zone in *P. vulgaris* is known from the recovery of two-and three-banded phenotypes from segregation of seed Adh-3 isozymes of selfed plants of wild *P. vulgaris* to be controlled by two codominant alleles (Fig. 3, e-i; Table 2). Parental genotypes of the plants are concluded to be $Adh\text{-}3^F/Adh\text{-}3^S$. The occurrence of a band intermediate in mobility between the two isozyme variants is evidence of the dimeric structure of the isozymes controlled by the Adh-3 locus in *P. vulgaris*. Phenotypic segregation conforms well to a 1:2:1 ratio of Fast:Fast/Slow:Slow for pooled data, as shown by the χ^2 value for pooled data ($0.80 > P > 0.70$) and by the heterogeneity χ^2 ($0.50 > P > 0.30$) (Table 2).

Codominant alleles coding for plant alcohol dehydrogenase monomers and dimeric structure of the allozymes have been reported in *Zea mays* by Schwartz and Endo (1966) and by Scandalios (1967), in *Triticum dicoccum* by Hart (1969), in *Carthamus* species by Efron et al. (1973), and in *Helianthus annuus* by Torres (1974). Monomers of equal activity which are produced in equal amounts (by virtue of equal allelic activity) and which associate at random would be indicated by a staining reaction roughly twice as intense at the site of heterodimer migration as at the site of homodimer migration, while monomers of unequal activity or which are coded for by alleles of unequal activity would be indicated by a staining reaction in which relative intensity is reduced at the site of heterodimer migration and at one site of homodimer migration and increased at the other site of homodimer migration (Schwartz, 1971). These phenomena occur in the wild *P. vulgaris* seed heterozygous at the Adh-3 locus and in the heterozygous progeny of selfed plants from such seed. Densitometric readings of gels were not made. However, staining reactions obviously approaching 1:2:1 in intensity were visually observed in heterozygote phenotypes, indicating equal monomeric activity as well as equal allelic activity. Also observed were heterozygotes in which one homodimer site stained more intensely than either the heterodimer or the other homodimer site, as well as heterozygotes in which, although random subunit association did not occur (Fig. 3, h-i), one homodimer site stained more intensely than the other, indicating in each case either unequal allelic activity or unequal monomeric activity. These phenomena have also been reported for the ADH allozymes in the papers already cited by Scandalios (1967), by Hart (1969), by Efron et al (1973), who attribute them to unequal monomeric activity and suggest that they may have important evolutionary consequences, and by Torres (1974). Schwartz (1971) presents evidence that in maize they are due to unequal allelic activity.

TABLE I

ADH AND EST ISOZYMES OF WILD AND CULTIVATED MEXICAN *P. VULGARIS* AND *P. COCCINEUS* AND OF U.S. CULTIVARS

				Adh-3*			Adh-2†	Adh-1†	Est-1			Est-2 (Bands)					
Pop.	Collection Site	Elev.	Species Description	F	F/S	S	M or P	M or P	F	F/S	S	1	2	3	4	1/2	2/3
69-1	Near Cuilapam, Oax.	5100'	*P. cocc.*cv. 'Frijol Ayocote'	0	0	12	M	M	-	-	-	0	12	0	0	0	0
69-2	De Ixtlan de Juarez, Oax.	5100'	*P. cocc.*cv. 'Frijol Bayo'	0	0	9	M	M	-	-	-	0	10	0	0	0	0
69-12	Telixtlahuaca, Oax.	5500'	*P. vulg.*,wild	0	0	19	M	M	0	0	19	No data--------------					
69-14	Oaxaca, Oax. 6 mi. N	5500'	*P. vulg.*,wild	0	12	12	P	M	0	0	24	0	24	0	0	0	0
69-15	Oaxaca, Oax. 6 mi. N	5500'	*P. vulg.*,wild	7	8	17	P	M	1	8	24	0	34	0	0	0	0
69-16	Near San Andres, Oax.	7600'	*P. cocc.*cv. 'Frijol Tavalle'	0	0	19	M	M	-	-	-	0	19	0	0	0	0
69-17	Near San Andres, Oax.	7600'	*P. cocc.*cv. 'Frijol Taxene'	0	0	19	M	M	-	-	-	0	19	0	0	0	0
69-19	Near Miahuatlan, Oax.	6900'	*P. vulg.*cv. 'Frijol Virgen'	0	0	19	M	M	3	2	14	0	2	16	0	0	1
69-20	Near Miahuatlan, Oax.	7000'	*P. cocc.*,wild	0	0	16	M	M	-	-	-	6	2	0	0	4	0
69-21	Nacaltepec, Oax.	6500'	*P. cocc.* cv. 'Ayocote'	0	?	?	P	P	-	-	-	0	38	0	0	0	0
69-22	Nacaltepec, Oax.	6500'	*P. cocc.*,mixed feral types	0	0	40	M	M	-	-	-	0	39	0	0	0	0

TABLE I, continued

Pop.	Collection Site	Elev.	Species Description	Adh-3*			Adh-2†	Adh-1†	Est-1			Est-2 (Bands)					
				F	F/S	S	M or P	M or P	F	F/S	S	1	2	3	4	1/2	2/3
69-23	Nacaltepec, Oax.	6500'	*P. vulg.*,cult., mixed seed types	0	0	19	M	M	10	1	8	0	1	18	0	0	0
69-24	Nacaltepec, Oax.	6500'	*P. cocc.*,feral, mixed seed types	0	?	?	P	P	-	-	-	3	14	10	0	0	0
69-27	Near Tepoztlan, Mor.	5800'	*P. vulg.*,wild	0	0	19	M	M	0	0	19	0	0	19	0	0	0
69-28	Cuernavaca, Mor. 5 mi. N	6400'	*P. cocc.*, wild	0	0	17	M	M	-	-	-	0	17	0	0	0	0
69-29	Near Tepoztlan, Mor.	5450'	*P. vulg.*, wild	0	0	16	M	M	0	0	16	0	0	16	0	0	0
69-34	Hueyapan, Pueb.	6000'	*P. cocc.* cv. 'Frijol Xaxana'	0	0	19	M	M	0	0	16	1	18	0	0	0	0
69-36	Teziutlan, Pueb.	6500'	*P. cocc.* ssp. *darwinianus*, cv. 'Frijol Bayo Gordo'	0	0	26	M	M	3	0	31	0	7	21	7	0	0
69-37	Teziutlan, Pueb.	6500'	*P. cocc.* ssp. *darwinianus* cv. 'Frijol Acalete'	0	0	43	M	M	-	-	-	0	51	0	0	0	0
69-38	Tlatlauquitepec, Pueb.	5900'	*P. vulg.* cv. 'Frijol Enredador'	0	0	38	M	M	21	8	7	8	0	30	0	0	0
69-39	Tlatlauquitepec, Pueb.	5900'	*P. vulg.*-*P. cocc.* mixture, cv. 'Frijol Coco Huaque'	0	0	19	P	P	-	-	-	No data---------------					
69-41	Altotonga, Vera.	6300'	*P. cocc.*, feral, mixed seed types	0	0	16	M	M	0	0	16	8	6	0	0	2	0

TABLE I, continued

Pop.	Collection Site	Elev.	Species Description	Adh-3* F	Adh-3* F/S	Adh-3* S	Adh-2† M or P	Adh-1† M or P	Est-1 F	Est-1 F/S	Est-1 S	Est-2 (Bands) 1	Est-2 (Bands) 2	Est-2 (Bands) 3	Est-2 (Bands) 4	Est-2 (Bands) 1/2	Est-2 (Bands) 2/3
69-44	Jalacingo, Vera.	5900'	*P. cocc.* ssp. *darwinianus* 'Frijol Acalete'	0	0	18	M	M	-	-	-	No data---------------					
69-15-1, selfed			*P. vulg.* wild, selfed	0	0	6	M	M	0	0	6	0	6	0	0	0	0
69-15-2, selfed			*P. vulg*, wild,selfed	7	18	11	P	M	0	0	39	0	39	0	0	0	0
69-15-3, selfed			*P. vulg.*, wild, selfed	0	0	5	M	M	0	0	6	0	5	0	0	0	0
69-15-4, selfed			*P. vulg.*, wild, selfed	0	0	6	M	M	0	0	6	0	6	0	0	0	0
69-15-5, selfed			*P. vulg.*,wild, selfed	0	0	6	M	M	0	0	6	0	6	0	0	0	0
69-15-6, selfed			*P. vulg.*, wild,selfed	0	0	6	M	M	0	0	6	0	6	0	0	0	0
69-15-7, selfed			*P. vulg.*, wild,selfed	8	26	14	P	M	8	9	15	0	46	0	0	0	0
69-15-8, selfed			*P. vulg.*,wild, selfed	0	0	6	M	M	0	0	6	0	6	0	0	0	0
69-15-9, selfed			*P. vulg.*,wild,selfed	0	0	6	M	M	0	0	6	0	6	0	0	0	0
69-15-10, selfed			*P. vulg.*,wild,selfed	11	10	7	P	M	8	12	7	0	28	0	0	0	0
69-15-11, selfed			*P. vulg.*,wild,selfed	0	0	6	M	M	0	0	6	0	6	0	0	0	0
69-15-12, selfed			*P. vulg.*,wild,selfed	15	22	12	P	M	0	0	47	0	50	0	0	0	0
			*P. vulg.*cv. Blue Lake	0	0	16	M	M	0	0	16	0	0	16	0	0	0
			*P. vulg.*cv. Kentucky Wonder	0	0	8	M	M	0	0	8	0	0	8	0	0	0
			P. vulg. cv. Top Crop	0	0	22	M	M	0	0	38	0	0	38	0	0	0
			P. cocc. cv. Scarlet Runner	0	0	35	P	P	-	-	-	0	38	0	0	0	0

*Only Adh-3 bands shown in Fig.3,a-k,are included in this Table. †Adh-2 and Adh-1 are scored M(monomorphic) or P (polymorphic).

TABLE II

SEGREGATION OF SEED ADH-3 ALLOZYMES OF FOUR SELFED $ADH\text{-}3^F/ADH\text{-}3^S$ PLANTS OF WILD *P. VULGARIS* AND GOODNESS OF FIT TO A 1:2:1 RATIO

Plant No.	Seed Adh-3 phenotypes of progeny (F_2)				df	χ^2	P
	Fast	Fast/Slow	Slow	Total			
69-15-2 selfed	7	18	11	36	2	0.889	0.70 > P > 0.50
69-15-7 selfed	8	26	14	48	2	1.833	0.50 > P > 0.30
69-15-10 selfed	11	10	7	28	2	3.429	0.20 > P > 0.10
69-15-12 selfed	15	22	12	49	2	0.867	0.70 > P > 0.50
	41	76	44	161	8	7.018	0.70 > P > 0.50
				Pooled	2	0.615	0.80 > P > 0.70
				Heterogeneity	6	6.403	0.50 > P > 0.30

Two other bands are assigned to the Adh-3 zone. These bands are both slower than the Adh-3 allozymes of *P. vulgaris*, and they occurred in 8-9% of the *P. coccineus* ssp. *darwinianus* seed analyzed (Fig. 3, l-m). Other *darwinianus* seed have the phenotype typical of most wild and cultivated Mexican *P. coccineus* (Fig. 3, c). It is postulated that the slower of these two bands is controlled by a third allele of the Adh-3 locus. Presently we have no evidence on the origin of the bands of the Adh-1 and Adh-2 zones in either *P. vulgaris* or *P. coccineus*. However, in both species the relative positions of Adh-2 bands with respect to Adh-3 and Adh-1 bands suggest that Adh-1 may be another genetic locus and that the Adh-2 bands could be the result of subunit association between monomeric gene products of the presumed Adh-1 and Adh-3 loci, each consisting of three alleles. Figure 4 presents a model which would account for the bands of Fig. 3 based on a two locus, three alleles per locus, system.

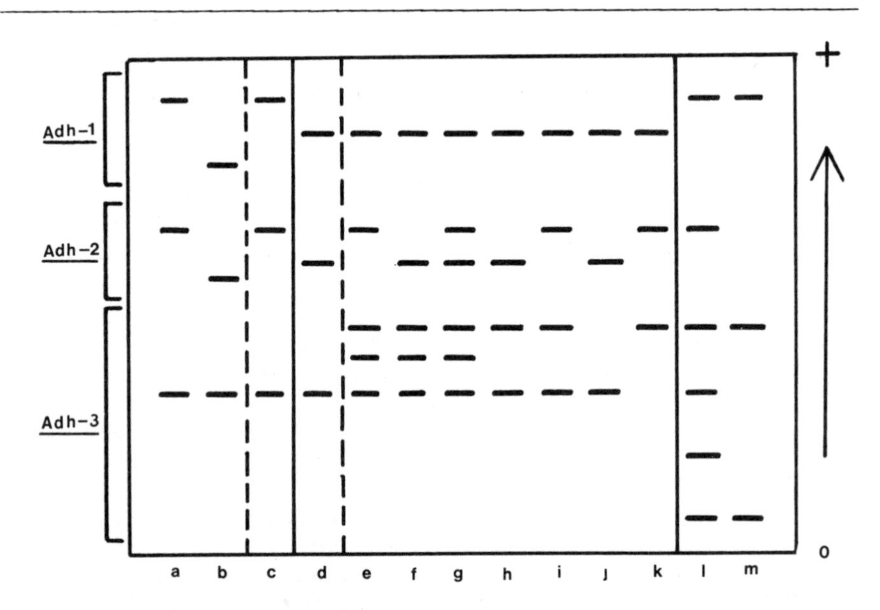

Fig. 3. *Phaseolus* seed ADH phenotypes showing three zones of electrophoretic mobility. (a-b) *P. cocc.* cv. Scarlet Runner; (c) wild and cult. Mexican *P. cocc.*, including ssp. *darwinianus* cv. 'Acalete' (pops. 69-37, 69-44); (d) *P. vulg.* cvs. Black Turtle Soup, Blue Lake, Kentucky Wonder, and Top Crop; (e-k) wild Mexican *P. vulg.* and wild Mexican *P. vulg.* selfed; (j) wild and cult. Mexican *P. vulg.*; *P. cocc.* ssp. *darwinianus* cv. 'Bayo Gordo' (pop. 69-36); (l-m) 9 seed (6 of l, 3 of m) of *P.*

Fig. 3, continued.
cocc. ssp. *darwinianus* cv. 'Acalete'. Adh-3 genotypes of wild *P. vulg.* and wild *P. vulg.* selfed: Adh-3^F/Adh-3^F (k); Adh-3^F/Adh-3^S (e-i); Adh-3^S/Adh-3^S (j). Evidence of subunit association between fast and slow monomers coded by codominant alleles at the Adh-3 locus in *P. vulg.* is seen in e-g, but not in h-i.

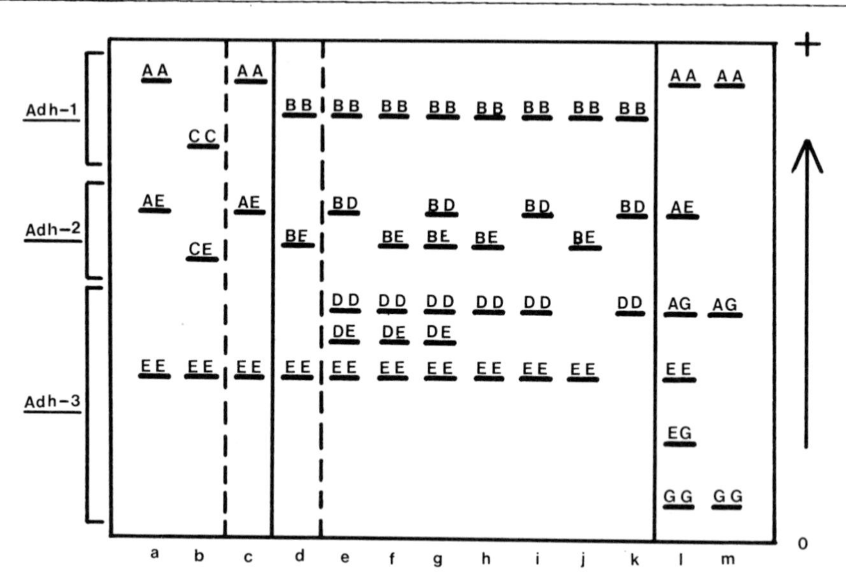

Fig. 4. Proposed dimeric structures of ADH isozymes from Fig. 3 based on a two locus gene product interaction model with three codominant alleles at each locus. Monomers A, B, and C are coded by alleles at the Adh-1 locus; monomers D, E, and G are coded by alleles at the Adh-3 locus. Bands AA, BB, CC, DD, EE, and GG are autodimers; bands AE, AG, BD, BE, CE, DE, and EG are allodimers. Autodimer DD (e-i, k) migrates to the same position as allodimer AG (l-m); allodimers AE (a, c, l) and BD (e, g, i, k) migrate to the same position. If the six alleles are designated A, B, C, D, E, and G to correspond to the monomers for which they code, the following genotypes result: Adh-1^A/Adh-1^A, Adh-3^E/Adh-3^E (a,c); Adh-1^C/Adh-1^C, Adh-3^E/Adh-3^E (b); Adh-1^B/Adh-1^B, Adh-3^E/Adh-3^E (d, j); Adh-1^B/Adh-1^B, Adh-3^D/Adh-3^E (e-i: Adh-3^D and Adh-3^E are designated Adh-3^F and Adh-3^S in Tables 1 and 2, Fig. 3, and in text); Adh-1^B/Adh-1^B, Adh-3^D/Adh-3^D (k); Adh-1^A/Adh-1^A, Adh-3^E/Adh-3^G (l); Adh-1^A/Adh-1^A, Adh-3^G/Adh-3^G (m).

All isozyme data, with the exception of ADH data for the *P. coccineus* ssp. *darwinianus* seed having the phenotypes shown in Fig. 3, l-m, are summarized in Table 1. The Adh-1 and Adh-2 zones are scored simply as monomorphic or polymorphic for each population. All known *P. vulgaris* material analyzed was monomorphic in the Adh-1 zone for a single band intermediate in mobility between the two *P. coccineus* Adh-1 bands (Fig. 3, a-b, d). Population 69-39, which is polymorphic for Adh-1 (and for Adh-2), consists of a mixture of *coccineus* and *vulgaris* seed types. Wild populations 69-14 and 69-15 and four 69-15 plants selfed were polymorphic for both the Adh-3 and the Adh-2 zones. All other *P. vulgaris* populations were monomorphic for both these zones.

Most (twelve populations) wild and cultivated Mexican *P. coccineus* were monomorphic in the Adh-1, Adh-2, and Adh-3 zones, and conformed to the phenotype shown in Fig. 3, c, which is also one of the common phenotypes of *P. coccineus* cv. Scarlet Runner. The remaining *P. coccineus* populations (including Scarlet Runner) and population 69-39, the *vulgaris-coccineus* mixture, were polymorphic in the Adh-2 zone; and all but one of these (69-37) were polymorphic in the Adh-1 zone. Two *P. coccineus* populations (69-21 and 69-24) appear to be polymorphic in the Adh-3 zone, with a high frequency of Adh-3 heterozygote phenotypes. The suspicion that these are genetic heterozygotes is reinforced by the occurrence of double bands in the Adh-2 zone. Population 69-24 is a feral *coccineus*, exhibiting a mixture of highly variable seed types. But population 69-21 is a cultivated *P. coccineus* having large, uniform *coccineus*-type seed size and color. In addition to both Adh-1 *P. coccineus* variants, this population also contains the *P. vulgaris* Adh-1 and probably the *vulgaris* Adh-2 variant. On the basis of these considerations, population 69-21 may be an introgressant between *P. vulgaris* and *P. coccineus*.

SEED EST ISOZYME PATTERNS

The Est-1 (anodal, alpha-naphthyl acetate) zone in *P. vulgaris* is known from segregation of Est-1 isozymes of selfed plants of wild *P. vulgaris* to be controlled by codominant alleles (Fig. 5, c-e; Table 3). Parental genotypes are concluded to be $Est\text{-}1^F/Est\text{-}1^S$. The pooled data of Table 3 do not fit a 1:2:1 ratio for Fast:Fast/Slow:Slow phenotypic segregation (0.05 P 0.025). However, since segregation of the progeny of one plant does conform to a 1:2:1 ratio (0.90 P 0.80), it would seem that more data are needed either to confirm or to reject the 1:2:1 ratio for the other plant. *P. coccineus*, with the exceptions of the cultivated populations

69-34 and 69-36 and the feral population 69-41, did not react with alpha-naphthyl acetate.

There is one zone of cathodal esterase activity, found in all *P. vulgaris* and *P. coccineus* populations, which exhibits activity only with alpha-naphthyl acetate. Six phenotypes have been identified from wild and cultivated populations (Fig. 6). Phenotypes b, c, and f of Fig. 6 are known from interspecific crosses (*P. vulgaris* cv. Black Turtle Soup x *P. coccineus* cv. Scarlet Runner) to be determined by codominant alleles (Wall, unpublished). The symbols assigned to the alleles controlling bands 2 and 3 are $\underline{Est\text{-}2^2}$ and $\underline{Est\text{-}2^3}$, respectively; their heterozygote is designated $\underline{Est\text{-}2^2/Est\text{-}2^3}$. Bands 1 and 4 have not been subjected to genetic analysis. However, their zymogram patterns strongly suggest that they are controlled by two additional alleles of the Est-2 locus. Cathodal esterases 1, 2, and 3 are all found in *P. coccineus* and *P. vulgaris*. Band 4 is unique to *P. coccineus* cultivar 69-36.

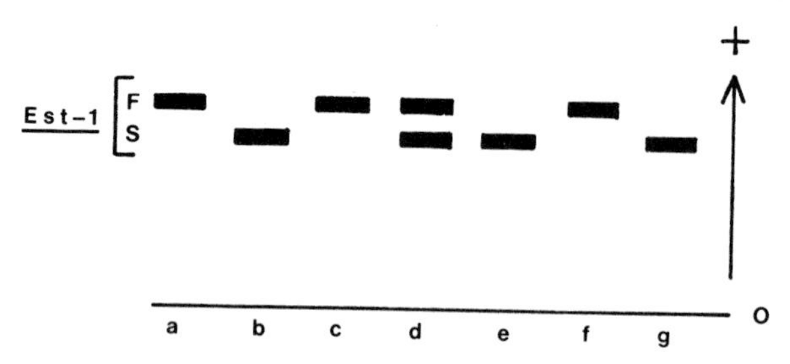

Fig. 5. *P. vulgaris* seed Est-1 phenotypes. (a) *P. vulgaris* cv. Black Turtle Soup; (b) *P. vulgaris* cvs. Kentucky Wonder, Blue Lake, and Top Crop, and wild Mexican populations; (c-e) wild and cultivated populations and a wild population selfed; (f-g) a cultivated Mexican population.

INTROGRESSION

Our findings indicate that, in the material we have analyzed, introgression is occurring, and that it is resulting primarily in the transmission of *P. vulgaris* germ plasm into *P. coccineus* cultivars via cultivated *P. vulgaris*. We base this assertion on the following: (1) *P. coccineus* cultivars 69-21 and 69-36 and the feral 69-24 carry the wild and cultivated *P. vulgaris* Adh-1 variant. (2) Populations 69-21 and

TABLE III

SEGREGATION OF SEED EST-1 ALLOZYMES OF TWO SELFED $EST\text{-}1^F/EST\text{-}1^S$ PLANTS OF WILD *P. VULGARIS* AND GOODNESS OF FIT TO A 1:2:1 RATIO

Plant No.	Fast	Fast/ Slow	Slow	Total	df	χ^2	P
69-15-7 selfed	8	9	15	32	2	9.188	0.025>P>0.01
69-15-10 selfed	8	12	7	27	2	0.407	0.90 >P>0.80
	16	21	22	59	4	9.595	0.05 >P>0.025
				Pooled	2	6.119	0.05 >P>0.025
				Heterogeneity	2	3.476	0.20 >P>0.10

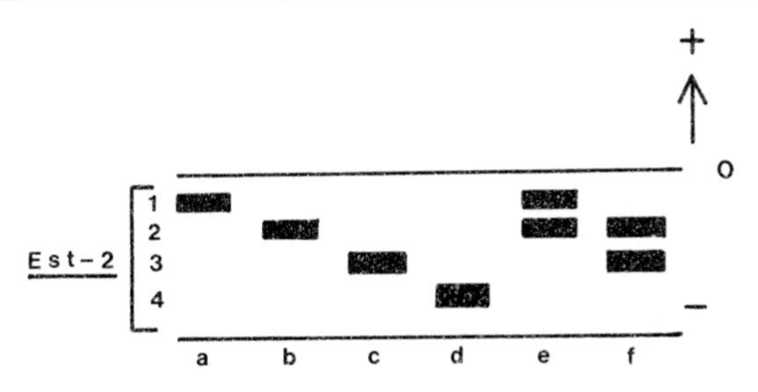

Fig. 6. *Phaseolus* seed Est-2 phenotypes. (b) *P. coccineus* cv. Scarlet Runner, wild and cultivated Mexican *P. coccineus*, wild *P. vulgaris*, and wild *P. vulgaris* selfed; (c) *P. vulgaris* cvs. Black Turtle Soup, Kentucky Wonder, Blue Lake, and Top Crop, and wild Mexican *P. vulgaris*; (a,b) cultivated Mexican *P. coccineus*; (a,c) cultivated Mexican *P. vulgaris*; (b,c) cultivated Mexican *P. vulgaris*; (a,b,e) feral Mexican *P. coccineus* and perhaps wild *P. coccineus*; (a-c) feral Mexican *P. coccineus*; (b-d) *P. coccineus* ssp. *darwinianus*; (b,c,f) cultivated Mexican *P. vulgaris*.

69-24 show a high degree of heterozygosity at the Adh-3 locus, a situation which could arise by crossing to $Adh\text{-}3^F/Adh\text{-}3^F$ *P. vulgaris* cultivars (although such cultivars were not found). (3) Est-1 alleles of wild and cultivated *P. vulgaris* occur in *P. coccineus* cultivars 69-34 and 69-36, and in feral

69-41. (4) The Est-2[3] allele of wild and cultivated *P. vulgaris* occurs in the *P. coccineus* cultivar 69-36 and in feral 69-24. (5) One instance of presumed introgression from *P. coccineus* to *P. vulgaris* is found: The Est-2[1] variant of wild and cultivated *P. coccineus* occurs in *P. vulgaris* cultivar 69-38.

BREEDING SYSTEM OF *P. VULGARIS*

The genetic structure and evolutionary potential of plant populations are largely determined by their breeding systems (Brown and Allard, 1970; Stebbins, 1957). The prevailing view has been that the breeding system of *P. vulgaris* is a highly self-fertilizing system adapted to the establishment of a series of inbreeding homozygous genotypes. Kristofferson (1921) reported outcrossing of 0-1.42% in a wide range of *P. vulgaris* cultivars. However, plants cultivated in habitats far removed from their centers of origin may exhibit breeding systems very different from those prevailing in their centers of origin. For example, the cultivated tomato is highly self-fertilizing throughout most of its cultivated range but exhibits high rates of outbreeding in its native region of South America (Rick, 1950). Limited data from this study suggest that *P. vulgaris* has a more flexible breeding system than has been previously appreciated.

Isozymes are uniquely suited for estimating outcrossing, for it is a relatively simple matter to identify heterozygotes by direct inspection of zymograms (Brown and Allard, 1970). Outcrossing estimates can be made most accurately by determining the proportion of heterozygotes among the progeny of homozygotes by the method of Allard and Workman (1963). Estimates made in this manner require that all seed assayed be collected from single plants in order that maternal genotypes may be identified. Populations used in this study were not collected in this manner because an estimate of outcrossing was not an object at the time the collections were made. Cultivars were obtained as samples of recently harvested crops; wild populations were bulk-collected in the field with extensive random sampling of each colony.

Even though statistically reliable outbreeding estimates cannot be obtained from these populations, the high frequency of heterozygotes in several polymorphic populations clearly indicate outcrossing in excess of previous reports of less than 2% (Table 4). The highest rate of outcrossing appears to occur in 69-14, a wild population exhibiting 50% Fast/Slow and 50% Slow phenotypes at the Adh-3 locus. Heterozygotes

TABLE IV

EVIDENCE OF OUTCROSSING IN NATURAL AND CULTIVATED MEXICAN *P. VULGARIS* POPULATIONS BASED ON SEED ADH-3 AND EST-1 PHENOTYPES

Population	Collection Site	Altitude	Description	Phenotypes Adh-3 F	Adh-3 F/S	Adh-3 S	Est-1 F	Est-1 F/S	Est-1 S
69-14	Near Oaxaca, Oaxaca	5500'	Wild colony	0	12	12	0	0	24
69-15	Near Oaxaca, Oaxaca	5500'	Wild colony	7	8	17	1	8	24
69-19	Near Miahuatlan, Oaxaca	6900'	Local cultivar, variable seed type	0	0	19	3	2	14
69-38	Tlatlauquitepec, Puebla	5900'	Local cultivar, variable seed type	0	0	38	21	8	7

can only be obtained from the outcrossing of maternal homozygotes or from the self-pollination of heterozygotes in the population. A few of the heterozygotes of 69-14 may be segregants from selfed heterozygotes, but most of them probably result from outcrossing of female Slow to male Fast phenotypes, since no Fast phenotypes were recovered. Selfed heterozygotes would be expected to produce 25% Fast phenotypes.

ACKNOWLEDGEMENTS

Supported in part by Grant No. 4925 from the Penrose Fund of the American Philosophical Society.

REFERENCES

Allard, R.W. and P.L. Workman 1963. Population studies in predominantly self-pollinating species. IV. Seasonal fluctuations in estimated values of genetic parameters in lima bean populations. *Evolution* 17:470-480.

Anderson, E. 1949. *Introgressive Hybridization*. John Wiley and Sons, New York.

Boyer, S.H., D.C. Fainer, and M.L. Naughton 1963. Myoglobin: Inherited structural variation in man. *Science* 140:1228-1231.

Brown, A.H.D. and R.W. Allard 1969. Inheritance of isozyme differences among the inbred parents of a reciprocal recurrent selection population of maize. *Crop Sci.* 9: 72-75.

Brown, A.H.D. and R.W. Allard 1970. Estimation of the mating system in open-pollinated maize populations using isozyme polymorphisms. *Genetics* 66:133-145.

Efron, Y., M. Peleg, and A. Ashri 1973. Alcohol dehydrogenase allozymes in the safflower genus *Carthamus* L. *Biochem. Genet.* 9:299-308.

Freytag, G. 1955. Variation of the common bean *(Phaseolus vulgaris L.)* in Central America. *Ph.D. thesis*, Washington University, St. Louis.

Gentry, H.S. 1969. Origin of the common bean, *Phaseolus vulgaris*. *Econ. Bot.* 23:55-69.

Gottlieb, L.D. 1973. Enzyme differentiation and phylogeny in *Clarkia franciscana*, *C. rubicunda*, and *C. amoena*. *Evolution* 27:205-214.

Hare, R.C. 1973. Reducing interference by "tetrazolium oxidase" and "nothing dehydrogenase" when staining for dehydrogenase isozymes in polyacrylamide gels. *Isozyme Bull.* 6:8-9.

Hart, G.E. 1969. Genetic control of alcohol dehydrogenase isozymes in *Triticum dicoccum*. *Biochem. Genet.* 3:617-625.

Johnson, F.M., H.E. Schaffer, J.E. Gillaspy, and E.S. Rockwood 1969. Isozyme genotype-environment relationships in natural populations of the harvester ant, *Pogonmyrmex barbatus*. from Texas. *Biochem. Genet.* 3:429-450.

Kristofferson, K.B. 1921. Spontaneous crossing in the garden bean, *Phaseolus vulgaris*. *Hereditas* 2:395-400.

Lewontin, R.C. and J.L. Hubby 1966. A molecular approach to the study of genic heterozygosity in natural populations. II. Amount of variation and degree of heterozygosity in natural populations of *Drosophila pseudoobscura*. *Genetics* 54:595-609.

Marshall, D.R. and R.W. Allard 1970. Isozyme polymorphisms in natural populations of *Avena fatua* and *A. barbata*. *Heredity* 25:373-382.

Prakash, S., R.C. Lewontin, and J.L. Hubby 1969. A molecular approach to the study of genic heterozygosity in natural populations. IV. Patterns of genic variation in central, marginal and isolated populations of *Drosophila pseudoobscura*. *Genetics* 61:841-858.

Rick, C.M. 1950. Pollination relationships of *Lycopersicon esculentum* in native and foreign regions. *Evolution* 4:110-122.

Scandalios, J.G. 1967. Genetic control of alcohol dehydrogenase isozymes in maize. *Biochem. Genet.* 1:1-8.

Schwartz, D. 1971. Genetic control of alcohol dehydrogenase-A competition model for regulation of gene action. *Genetics* 67:411-425.

Schwartz, D. and T. Endo 1966. Alcohol dehydrogenase polymorphism in maize-Simple and compound loci. *Genetics* 53:709-715.

Shaw, C.R. and R. Prasad. 1970. Starch gel electrophoresis-A compilation of recipes. *Biochem. Genet.* 4:297-320.

Stebbins, G.L. 1957. Self-fertilization and population variability in the higher plants. *Amer. Nat.* 91:337-354.

Torres, A.M. 1974. Sunflower alcohol dehydrogenase: ADH_1 genetics and dissociation-recombination. *Biochem. Genet.* 11:17-24.

Vavilov, N.I. 1935. Origin, variation, immunity, and breeding of cultivated plants. K.S. Chester trans. *Chron. Bot.* 1951.

Wall, J.R. 1970. Genetic variation in wild beans (*Phaseolus vulgaris-P. coccineus* complex) and squashes of Mexico. *Year Book of the American Philosophical Society:*345-347.

Wall, J.R. and T.W. Whitaker 1971. Genetic control of leucine aminopeptidase and esterase isozymes in the interspecific cross *Cucurbita ecuadorensis* x *C. maxima*. *Biochem. Genet.* 5:223-229.

Webster, T.P., R.K. Selander, and S.Y. Yang 1972. Genetic variability and similarity in the *Anolis* lizards of Bimini. *Evolution* 26:523-535.

EVOLUTION OF ISOZYMES FORMING
STORAGE POLYGLUCANS IN ALGAE

JEROME F. FREDRICK
The Research Laboratories
Dodge Chemical Company
Bronx, New York 10469

ABSTRACT. Three groups of isozymes involved in the biosynthesis of amylose, amylopectin, phytoglycogen, and glycogen are traced throughout phylogenesis. In algae, the changes in polyacrylamide gel patterns of these isozymes indicate evolutionary trends. It is possible to follow the biphyletic origin from the blue-green algae of the red algae and the green algae. The isozyme patterns lend supporting evidence for the "transition" category occupied by the hot-springs alga, *Cyanidium caldarium*. Evidence from immunochemical studies, amino acid sequence analyses, and polyacrylamide gel electrophoresis, indicates a close similarity between phosphorylase and synthetase isozymes, and between phosphorylase and branching enzymes. The branching enzymes themselves, show two isozymic types, one capable of branching amylose, and the other capable of inserting further alpha-1,6-glucosyl linkages into amylopectins, forming phytoglycogens and other highly branched glucose polymers. The evidence points to the origins of all three groups of isozymes involved in polyglucan formation in algae from a single bifunctional primordial enzyme capable of synthesizing both alpha-1,4 and alpha-1,6-glucosyl bonds.

INTRODUCTION

Klein and Cronquist ('67) have stressed the almost universal use of carbohydrate as <u>the</u> primary respirable substrate for plants, and that because of energy requirements, the development of <u>storage</u> forms of carbohydrate must have occurred quite early in evolutionary history. As the available organic molecules became depleted in the primordial soup (probably as a result of some primitive cells which had already developed glycolytic pathways), the processes of evolution became selective for the cells which had developed a potential for photosynthesis via some type of chlorophyll derived from an available primitive porphyrin in the melange. This step, which Calvin ('69) has termed, <u>reflexive catalysis</u>, was probably autocatalytic and self-perpetuating. It conferred upon those cells the distinct evolutionary advantage of using available solar energy for the synthesis of their own nutritional polysaccharide <u>stores</u>.

Among the Thallophytes, the use of polyglucans as storage polysaccharides is fairly widespread (Table No. 1). The presence of phytoglycogens, amylopectins and amyloses in the algae in particular, presents a wide spectrum of alpha-1,4 and alpha-1,6-glucosyl bonded polymers. It would seem logical, therefore, to seek suggestive evolutionary trends in those enzymes responsible for the biosynthesis of these glucose polymers.

Two main types of bonding between glucose residues are present in algal storage polyglucans (Figure 1). The alpha-1,4 type, synthesized by phosphorylase (alpha-1,4-glucan: orthophosphate glucosyltransferase, E.C. 2.4.1.1) from glucose-1-phosphate, and by synthetase (ADP-glucose: alpha-1,4-glucan alpha-4-glucosyltransferase, E.C. 2.4.1.11) from various nucleotide sugars, results in the formation of linear polymers such as amylose. Branched polyglucans, such as glycogen and amylopectin, are the result of the insertion of alpha-1,6-glucosyl linkages into these linear chains, and is mediated by the branching enzyme (alpha-1,4-glucan: alpha-1,4-glucan 6-glucosyltransferase, E.C. 2.4.1.18).

Multiple molecular forms (isozymes) of all three groups of enzymes were first reported by Fredrick ('62) in the Cyanophyte, *Oscillatoria princeps*. Since then, these isozymes have been found in other Cyanophytes (Fredrick '71), in Rhodophytes and Chlorophytes (Fredrick '64). Reports indicate that multiple molecular forms of phosphorylase (other than the b and a conversion types) are also present in higher plants (DeFekete '68, Tsai and Nelson '68, Slabnik and Frydman '70, Shivram et al '71), animal cells (Yunis and Arimura '66) and organs (Davis et al '67). The synthetases too, appear to be present as isozymes in algae (Preiss et al '67) and other plants (Frydman and Cardini '65). At least two forms of branching enzyme were reported by Lavintman ('66) in her sweet corn preparations, one having the classical "Q" enzyme action (or able to insert branch points into amylose so that an amylopectin results), and the other having the property of being able to further branch amylopectins to form phytoglycogens (true branching enzyme, or b.e.).

The three groups of isozymes show some interesting and suggestive properties in common. For example, the phosphorylases and the synthetases show common immunological cross-reactions (Schliselfeld and Krebs '67, Schliselfeld '73), as do also the phosphorylases and branching enzymes of blue-green algae (Fredrick '61). In addition, the phosphorylated sites of both phosphorylase (Nolan et al '64) and of glycogen synthetase (Larner and Sanger '65) show identical amino acid sequences when their polypeptides are analyzed. Phosphorylase and synthetase also appear to have similar reactions for converting the inactive forms of the enzymes to the active forms (Larner

TABLE 1
STORAGE POLYGLUCANS IN THALLOPHYTES

Organism	Polyglucan	Linkages	Reference
Cyanophyceae: *Oscillatoria*	Phytoglycogen	α-1,4, α-1,6 (highly branched)	Fredrick ('68a)
Rhodophyceae: *Serraticardia*	Floridean starch	α-1,4, α-1,6 (moderately branched)	Nagashima et al ('71)
Chlorophyceae: *Chlorella*	True starch	α-1,4, α-1,6 (amylopectins) α-1,4 (amyloses)	Fredrick ('68b)
Transition: *Cyanidium*	Phytoglycogen	α-1,4, α-1,6 (identical with *Oscillatoria*)	Fredrick ('68b)
Fungi: *Saccharomyces*	Glycogen	α-1,4, α-1,6 (highly branched)	Cabib et al ('73)
Hericium	Amylose	α-1,4 (linear)	McCracken & Dodd ('71)
Slime Mold: *Dictyostelium*	Glycogen	α-1,4, α-1,6 (highly branched)	Wright et al ('73)

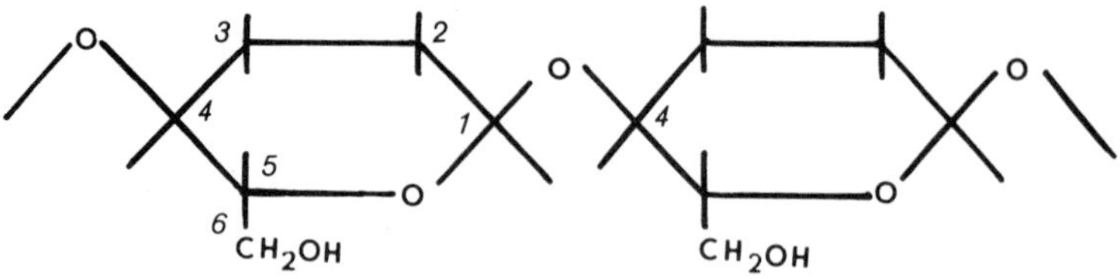

Amylose

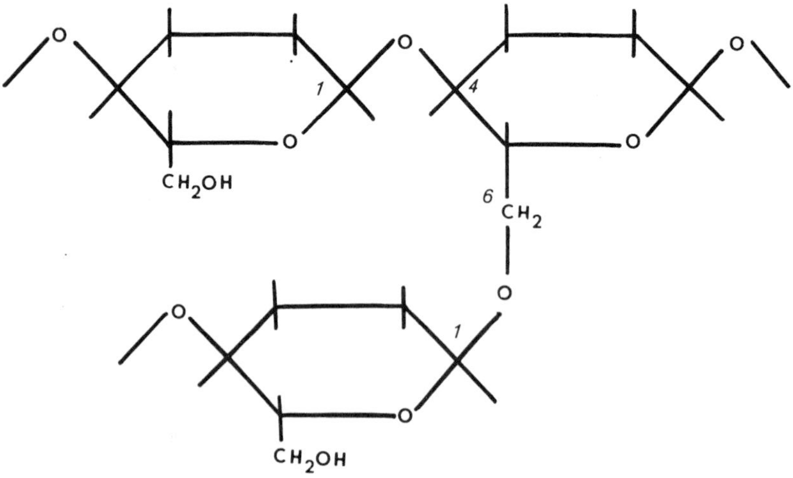

Amylopectin

Fig. 1. The glucosyl bonding in storage glucans. The alpha-1,4 type of bonding forms linear amylose chains. The alpha-1,6 type is characteristic of branched glucans such as glycogens and amylopectins.

'66). All-in-all, there is much data indicative of a common origin for these three groups of enzymes.

Isozyme Patterns in Algae

The use of two-dimensional (orthogonal) polyacrylamide gel electrophoresis has made possible the isolation and study of patterns of the three groups of isozymes in various phylogenetic groups of algae. These gel patterns (Figure 2), show a suggestive trend insofar as these isozymes are concerned. When these patterns are viewed in conjunction with the particular polyglucan formed by the alga, a definitive progression can be seen. The storage polyglucans of prokaryotic algae are very highly branched and resemble animal glycogen. Those of eukaryotic algae, except for the Floridean "starch" of the red algae, are very much akin to the true starches found in higher plants. These starches consist of mixtures of amylose (the unbranched, linear component) and amylopectins (moderately branched glucans).

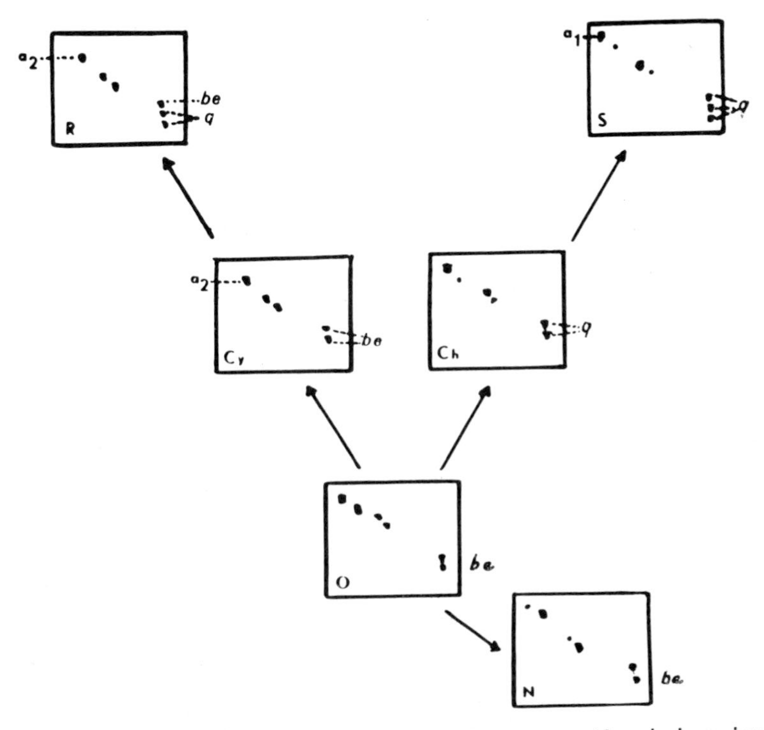

Fig. 2. Orthogonal gel patterns of glucan-synthesizing isozymes of various algae. R, *Rhodymenia pertusa*; Cy, *Cyanidium caldarium*; O, *Oscillatoria princeps*; N, *Nostoc muscorum*; Ch, *Chlorella pyrenoidosa*; S, *Spirogyra setiformis*. a_1, primer-requiring phosphorylase isozyme; a_2, non-primer isozyme; q, classical branching enzyme; b.e., true branching isozyme. Published with permission of the New York Academy of Sciences (after Fredrick '73b).

Of interest, is the important position occupied in such a

synthetic scheme by the enigmatic alga, *Cyanidium caldarium*. The polyglucan of this alga is in every respect similar to that of the Cyanophyte, *Oscillatoria princeps* (Fredrick '68). Klein and Cronquist ('67) had proposed this alga as a "transition" form between the blue-greens and the red algae (or the green algae). Indeed, although this hot-springs alga had been variously classified as a blue-green (Copeland '36), a red (Hirose '50) and as a green alga (Allen '59), and even as a "blue-green Chlorella" (Allen '54), the biochemical data accumulated thus far, suggest that it may be a transition form. The data presented in Table No. 2, extracted from a paper by Brown and Richardson ('68), on the pigments of algae, shows the intermediate position of *Cyanidium*, and particularly the close relationship of this alga with the blue-green algae. The recent evidence for the classification of this alga as a eukaryote, such as the presence of mitochondria and an endoplasmic reticulum (Margulis '68, Seckbach '72), establishes it as a primitive type of eukaryotic cell, capable of fulfilling the criteria for a "transition" form between the prokaryotic blue-greens and the eukaryotic red algae.

TABLE 2
ABSORPTION MAXIMA OF PIGMENTS IN VARIOUS ALGAE*
(nm)

Alga	Chlorophyll a				Chlorophyll b		Phycocyanin
CYANIDIUM	437	-	-	679	-	-	624
NOSTOC	438	-	-	678	-	-	629
PORPHYRIDIUM	436	-	-	680	-	-	626
CHLORELLA	438	538	628	680	470	654	628

*Adapted from Brown and Richardson ('68)

If one studies the polyglucan-synthesizing isozyme patterns of *Cyanidium caldarium*, its role in this evolutionary progression becomes obvious. For example, although Klein and Cronquist ('67) had proposed it as a transition form either from the blue-greens to the red algae, or from the blue-greens to the green algae, it appears that this alga is excluded from that part of the biphyletic pathway leading to the Chlorophytes (Figure 3). Both of its branching isozymes are of the type capable of introducing further alpha-1,6-glucosyl linkages into moderately branched glucans as amylopectins, while the branching isozymes in a morphologically similar form such as the Chlorophyte, *Chlorella pyrenoidosa*, are of the classical "Q" type, or able to branch amylose to amylopectin (Fredrick '68).

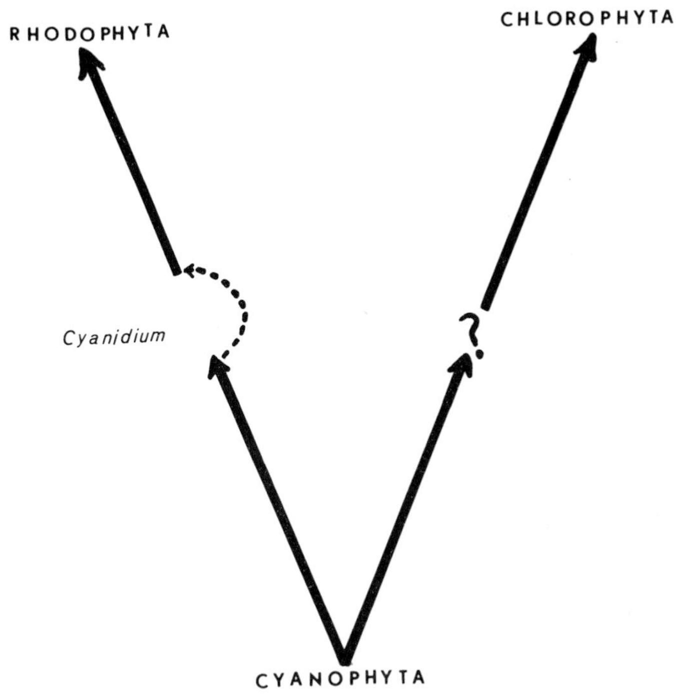

Fig. 3. Position proposed for *Cyanidium caldarium* in the bi-phyletic scheme leading from the Cyanophytes to the Rhodophytes and to the Chlorophytes.

The dual line of evolution from the Cyanophyceae to the Rhodophyceae on the one hand, and from the Cyanophyceae to the Chlorophyceae on the other, has been further clarified recently through the use of the method of analogy: homology "goodness of fit", proposed by Klein ('70) for comparing the chemical characteristics in these three groups of algae. The same method, applied to *Cyanidium* vis-a-vis the red algae and the green algae, reinforces the correctness of the position assigned to *Cyanidium caldarium* as a transition form from the blue-green algae to the red algae, rather than to the green algae (Table No. 3).

Fredrick ('71a) has shown that the "Q" type of branching isozymes of the red alga, *Rhodymenia pertusa* are charge isomers

TABLE 3

ANALOGY: HOMOLOGY "GOODNESS OF FIT" COMPARISONS*

Chemical Character	Chloro:Rhodo	Cyanid:Chloro	Cyanid:Rhodo	Cyano:Rhodo
Chlorophyll *a*	+	+	+	+
Chlorophyll *b*	-	-	-	+
Biliproteins	-	-	-	+
Carotenoids	+	+	+	+/-
Carbohydrates				
Trehalose	-	-	+	+
Storage polyglucan	-	-	+	+
Isozymes	+/-	+/-	+	+
Sterols	+	?	?	+
Tocopherols	+	?	?	+

*After Klein ('70)

\+ goodness of fit

\- no apparent relationship

? data lacking

of the one true branching isozyme (b.e.) also found in this alga. The difference between them seems to be only in the substitution of amino acid residues. However, while *Rhodymenia* contains both types of branching isozymes, the green algae studied, *Spirogyra setiformis* and *Chlorella pyrenoidosa*, have only the "Q" type of isozymes. This, when viewed alone, might be suggestive of a possible derivation of the Chlorophytes from a Rhodophycean ancestor.

It seems probable that the Cyanophyceae served as the progenitor of both the red and the green algae. On the basis of the three groups of isozymes involved in storage polyglucan formation, it appears likely that these prokaryotic algae gave rise to the Rhodophyceae, an eukaryotic group, via the transitional form as the extant alga, *Cyanidium caldarium*. This alga appears to combine the morphological properties of eukaryotic algae with the enzymology of the blue-green prokaryotes.

The line of evolution from the Cyanophycean ancestor to the Chlorophyceae does not seem to be as clear as that leading to the Rhodophyceae. Although there is a continuity of the two phosphorylase isozymes and the two synthetase isozymes of the blue-greens in both of the green algae, *Chlorella* and *Spirogyra*, there is no occurrence of any true branching isozyme, b.e. (see Figure 2). Rather, all of the branching isozymes present in these algae are of the "Q" type. In addition, the a_2 phosphorylase isozyme appears to be diminished in absolute quantity in the Chlorophytes (Figure 2). This, of course, does not rule out the possibility of the occurrence of a transitional form between the blue-green and the green algae similar to *Cyanidium* which serves the function between the blue-greens and the reds (Figure 3).

Continuity in Other Life Forms

While the line of evolution seems to be quite suggestive in the Thallophytes, there is also a persistence of these isozymes in higher life forms. Despite the obvious large evolutionary gap between such diverse life forms as the primitive *Oscillatoria* and the rabbit, both appear to be dependent upon pyridoxal for the biosynthesis of active catalytic phosphorylases (Fredrick '71b, Fischer et al '66). One of the phosphorylase isozymes in *Oscillatoria princeps* (Fredrick '71c) and in *Cyanidium caldarium* (Fredrick '73a) has been shown to be a glycoprotein, which did not require a "primer" maltosaccharide in order to initiate synthesis of linear glucans. A similar phosphorylase glycoprotein had been reported in rabbit skeletal muscle by Feigin et al ('51), and more recently, in a tuber (Slabnik and Frydman '70, Frydman and Slabnik '73). Similar, primer-less synthesis of glucans may also be present in some of the synthetase isozymes (Fox et al '73).

The branching isozymes too, appear to show continuity throughout organic evolution. The branching isozyme of *Oscillatoria princeps* appears to transfer maltoglucosides of the same chain length as the branching enzymes present in animal liver (Verhue and Hers '66) and muscle (Brown and Brown '66).

A Consideration of the Possible Primitive Enzyme

In general, the biosynthesis of highly branched storage polyglucans appears to be associated with the more primitive plant forms (Schoch '47, Fox et al '73). In higher plants, the synthesis of moderately branched polyglucans is usually encountered. Here, less-branched sugars, such as the amylopectins, are intimately associated with the more typical higher plant form of linear, unbranched polyglucan, amylose.

It seems possible that the helical structure of amylose (Rundle et al '44, Fredrick '55, Zugenmaier and Sarko '73, Blackwell et al '69) may be an important factor in the conservation of space within the cell. Such a conservation of space was undoubtedly needed as the result of the continued specialization of intracellular structures: the formation of plastids, etc. Because of the absence of cellular organelles (i.e., photosynthetic plastids) in prokaryotes, there was no similar imposition of space-limiting requirements in the cytoplasm of these cells. Hence, more highly branched and ramified structures could be accommodated in the Cyanophycean cell than in the Chlorophycean cell. It does not seem surprising therefore, that the more highly branched polyglucans are found in the more primitive algal forms. Frey-Wyssling ('69), has shown, using the dimensions calculated by Fredrick ('55) for an amylose molecule having six glucosyl residues per helix turn unit, that the amylose molecule would fit correctly into the lamellae of the plastids of higher plants. The primitive types of plastids present in *Cyanidium caldarium* (Seckbach '72) and in Rhodophyceae (Gibbs '70) would still accommodate large branching polyglucan molecules (and hence, the branched storage sugars found in these algae), whereas the more advanced Chlorophytes with their specialized but restrictive plastids, could not.

The intraenzymic similarities in those isozymes involved in polyglucan synthesis, the suggestive trends of these isozymes in algae, their continuity in higher life-forms, and the possible imposition of space limitations insofar as the ultimate structure of the polyglucan they form, when going from the prokaryotic primitive cellular condition to the more advanced, eukaryotic condition, are all indicative of the possible derivation of these extant isozymes from a single ancestral catalytic protein. Yourno et al. ('70) recently showed that it was possible to cause the fusion of two adjacent genes by the elimination of the so-called "punctuation signal" or nonsense codon, and thereby obtain an enzymatically bifunctional polypeptide combining the catalytic activities of the originally separate two enzymes. A recent paper by Fredrick ('73b) shows that the alpha-1,4 and the alpha-1,6-glucosyl bond synthesizing ability of phosphorylase and branching enzyme could have both been contained in one primordial enzyme protein, and that the insertion of a nonsense codon during transcription of this primitive polycistron could give rise to two separate proteins wherein each contained the catalytic activity for the synthesis of only one type of glucosyl bond (Figure 4). Further differ-

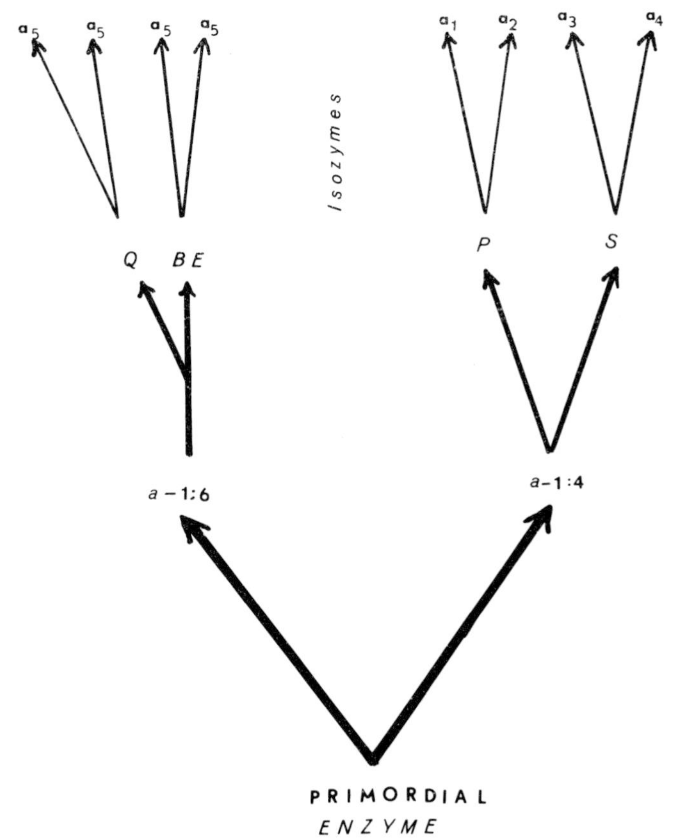

Fig. 4. Possible evolution of the original polyglucan-forming enzyme leading to alpha-1,4-synthesizing and alpha-1,6-synthesizing enzymes. These, in turn, underwent further finer differentiation into the Q and BE branching isozymes (a_5), and

(Fig. 4 legend continued) the P and S linear glucan-forming isozymes. In turn, the phosphorylases (P) differentiated into the extant a_1 and a_2 isozymes, while the synthetases (S) gave rise to the a_3 and a_4 isozymes.

---

entiation probably occurred once this bifurcation was established, so that the synthetases and phosphorylases were derived from the alpha-1,4 synthesizer, while the "Q" and b.e. isozymes were derived from the alpha-1,6-glucosyl bond forming protein. In essence, this would be akin to the reverse process of gene fusion described by Yourno et al ('70).

ACKNOWLEDGMENT

This work was supported by a research grant from the Dodge Institute for Advanced Studies, Cambridge, Mass. 02140. The assistance of Alan M. Fredrick in the preparation of this manuscript is gratefully acknowledged.

LITERATURE CITED

Allen, M.B. 1954. Studies with a blue-green Chlorella. *Proc. Int. Bot. Cong.* 17: 41-42.

Allen, M.B. 1959. Studies on Cyanidium caldarium, an anomalously pigmented Chlorophyte. *Arch. für Mikrobiol.* 32: 270-277.

Blackwell, J., A. Sarko, and R.H. Marchessault. 1969. Chain conformation in β-amylose. *J. Mol. Biol.* 42: 379-383.

Brown, D.H., and B.I. Brown. 1966. Action of muscle branching enzyme on polysaccharide enlarged from UDP (^{14}C) glucose. *Biochim. Biophys. Acta* 130: 263-266.

Brown, T.E., and F.L. Richardson. 1968. Effect of growth environment on the physiology of algae: light intensity. *J. Phycol.* 4: 38-54.

Cabib, E., L.B. Rothman-Denes, and K. Huang. 1973. The regulation of glycogen synthesis in yeast. *Annals N.Y. Acad. Sci.* 210: 192-206.

Calvin, M. 1969. *Chemical Evolution.* Oxford U. Press, New York City.: 147-148.

Copeland, J.J. 1936. Yellowstone thermal myxophyceae. *Annals*, N.Y. Acad. Sci. 36: 1-229.

Davis, C.H., L.H. Schliselfeld, D.P. Wolf, C.A. Leavitt, and E.G. Krebs. 1967. Interrelationship among glycogen phosphorylase isozymes. *J. Biol. Chem.* 242: 4824-4833.

DeFekete, M.A.R. 1968. Die Rolle der Phosphorylase in stoffwechsel der Starke in den Plastiden. *Planta* 79: 208-221.

Feigin, I., Fredrick, J.F., and A. Wolf. 1951. Mechanism of holophosphorylase action. *Fed. Proc.* 10: 182.

Fischer, E.H., and E.G. Krebs. 1966. Relationship of structure to function of muscle phosphorylase. *Fed. Proc.* 25: 1511-1520.

Fox, J., L.D. Kennedy, J.S. Hawker, J.L. Ozbun, E. Greenberg, C. Lammel, and J. Preiss. 1973. De novo synthesis of bacterial glycogen and plant starch by ADPG-alpha-glucan 4-glucosyltransferase. *Annals N.Y. Acad. Sci.* 210: 90-102.

Fredrick, J.F. 1955. Proposed molecular structure for straight and branched polymers of glucose. *Physiol. Plant.* 8: 288-290.

Fredrick, J.F., and F.J. Mulligan. 1955. Mechanism of action of branching enzyme from Oscillatoria and the structure of branched polysaccharides. *Physiol. Plant.* 8: 74-83.

Fredrick, J.F. 1961. Immunochemical studies of the phosphorylases of Cyanophyceae. *Phyton* 16: 21-26.

Fredrick, J.F. 1962. Multiple molecular forms of 4-glucosyltransferase in Oscillatoria princeps. *Phytochem.* 1: 153-157.

Fredrick, J.F. 1964. Polyacrylamide gel studies of the isozymes involved in polyglucoside synthesis in algae. *Phyton* 21: 85-89.

Fredrick, J.F. 1968a. Properties of branching enzymes. *Phytochem.* 7: 931-936.

Fredrick, J.F. 1968b. The polyglucoside and enzymes of Cyanidium caldarium. *Phytochem.* 7: 1573-1576.

Fredrick, J.F. 1971a. Polyglucan branching isoenzymes of algae. *Physiol. Plant.* 24: 55-58.

Fredrick, J.F. 1971b. Requirements for pyridoxal in the biosynthesis of phosphorylases in the algae. *Phytochem.* 10: 1025-1029.

Fredrick, J.F. 1971c. Storage polyglucan-synthesizing isozyme patterns in the Cyanophyceae. *Phytochem.* 10: 395-398.

Fredrick, J.F. 1971d. De novo synthesis of polyglucans by a phosphorylase isoenzyme in algae. *Physiol. Plant.* 25: 32-35.

Fredrick, J.F. 1973a. Further studies of primer-independent phosphorylase isozymes in algae. *Phytochem.* 12: 1933-1936.

Fredrick, J.F. 1973b. A primordial bifunctional polyglucan-synthesizing enzyme. *Annals N. Y. Acad. Sci.* 210: 254-264.

Frey-Wyssling, A. 1969. On the ultrastructure of the starch granule. *Am. J. Bot.* 56: 696-701.

Frydman, R.B., and C.E. Cardini. 1965. Studies on ADPG: alpha-1,4-glucan alpha-4-glucosyltransferase of sweet corn endosperm. *Biochim. Biophys. Acta* 96: 294-303.

Frydman, R.B., and E. Slabnik. 1973. Role of phosphorylase in starch biosynthesis. *Annals N.Y. Acad. Sci.* 210: 153-168.

Gibbs, S.P. 1970. The comparative ultrastructure of the algal chloroplast. *Annals N.Y. Acad. Sci.* 175: 454-473.

Hirose, H. 1950. Studies on a thermal alga, Cyanidium caldarium. *Bot. Mag. (Tokyo)* 63: 745-746.

Klein, R.M., and A. Cronquist. 1967. A consideration of the evolutionary and taxonomic significance of some biochemical, micromorphological and physiological characters in the thallophytes. *Quart. Rev. Biol.* 42: 105-296.

Klein, R.M. 1970. Relationships between blue-green and red algae. *Annals N.Y. Acad. Sci.* 175: 623-633.

Larner, J., and F. Sanger. 1965. The amino acid sequence of the phosphorylated site of muscle UDPG-alpha-1,4-glucan alpha-4-glucosyltransferase. *J. Mol. Biol.* 11: 491-500.

Larner, J. 1966. Hormonal and non-hormonal control of glycogen metabolism. *Annals N.Y. Acad. Sci.* 29: 192-209.

Lavintman, N. 1966. Formation of branched glucans in sweet corn. *Arch. Biochem. Biophys.* 116: 1-8.

Margulis, L. 1968. Evolutionary criteria in thallophytes; a radical alternative. *Science* 161: 1020-1022.

McCracken, D.A., and J.L. Dodd. 1971. Molecular structure of starch-type polysaccharides from Hericium ramosum and Hericium coralloides. *Science* 174: 419.

Nagashima, H., S. Nakamura, K. Nisizawa, and T. Hori. 1971. Enzymic synthesis of floridean starch in a red alga, Serraticardia maxima. *Plant and Cell Physiol.* 12: 243-253.

Nolan, C., W.B. Novoa, E.G. Krebs, and E.H. Fischer. 1964. Further studies on the site phosphorylated in the phosphorylase b to a reaction. *Biochem.* 3: 542-551.

Preiss, J., H.P. Ghosh, and J. Wittkop. 1967. Regulation of the biosynthesis of starch in spinach leaf chloroplasts. In *Biochemistry of Chloroplasts* (ed. T.W. Goodwin), Acad. Press, New York City.: 131-153.

Rundle, R.E., J.F. Foster, and R.R. Baldwin. 1944. On the nature of the starch-iodine complex. *J. Am. Chem. Soc.* 66: 2116-2120.

Schliselfeld, L., and E.G. Krebs. 1967. Immunological comparison of phosphorylase isozymes. *A.C.S. Abstracts*, 154th Meeting, Chicago, Ill.: C-197.

Schliselfeld, L. 1973. Comparative studies of phosphorylase isozymes from the rabbit. *Annals N.Y. Acad. Sci.* 210: 181-191.

Schoch, T.J. 1947. A decade of starch research. *Bakers Digest* 21: 1-7.

Seckbach, J. 1972. On the fine structure of the acidophilic hot-spring alga, Cyanidium caldarium: a taxonomic approach. *Microbios* 5: 133-142.

Shivram, K.N., H. Stegmann, R. Siepmann, and H. Boser. 1971. Appearance and disappearance of phosphorylases in potatoes. *Zeit. Naturforsch.* 266: 69-70.

Slabnik, E., and R.B. Frydman. 1970. A phosphorylase involved in starch biosynthesis. *Biochem. Biophys. Res. Commun.* 38: 709-714

Tsai, C.Y., and O.E. Nelson. 1968. Phosphorylases I and II of maize endosperm. *Plant Physiol.* 43: 103-112.

Verhue, W., and H.G. Hers. 1966. A study of the reaction catalyzed by the liver branching enzyme. *Biochem. J.* 99: 222-227.

Wright, B.E., Rosness, P., T.H.D. Jones, and R. Marshall. 1973. Glycogen metabolism during differentiation in *Dictyostelium discoideum*. *Annals N.Y. Acad. Sci.* 210: 51-62.

Yourno, J., T. Kohno, and J.R. Roth. 1970. Enzyme evolution: generation of a bifunctional enzyme by fusion of adjacent genes. *Nature* 228: 820-824.

Yunis, A.A., and G.K. Arimura. 1966. Enzymes of glycogen metabolism in white blood cells. *Biochim. Biophys. Acta* 118: 335-343.

Zugenmaier, P., and A. Sarko. 1973. Packing analysis of carbohydrates and polysaccharides. II. β-amylose. *Biopolymers* 12: 435-444.

HOMOLOGY AND ANALOGY OF DEHYDROGENASES IN FUNGAL PHYLOGENY

H. B. LEJOHN
Department of Microbiology
University of Manitoba
Winnipeg,
Manitoba, R3T 2N2 CANADA

ABSTRACT. The distribution of the analogous NAD- and NADP-linked glutamic dehydrogenases was determined for more than eighty species of fungi, most of them belonging to the Phycomycetes, organisms that lack the NADP-linked analogue of these enzymes. The higher fungi, Ascomycetes, Deuteromycetes, and Basidiomycetes contain both types of glutamic dehydrogenase analogues.

Over 60 species of Oomycetes were analyzed electrophoretically and the isozyme patterns for NAD-linked glutamic, D(-)lactate, and malate dehydrogenases as well as the NADP-linked isocitric dehydrogenases determined. The results showed that, by and large, the electrophoretic properties of isozymes within the same genus are uniform, but are significantly different when intergeneric properties are compared. In a few cases, organisms that have been placed in different genera showed remarkable enzyme correlations and this hinted at a closer relationship than hitherto suspected using purely taxonomic criteria.

Besides electrophoretic mobility, glutamic dehydrogenase isozymes of the Oömycetes were compared on the basis of their response to 6 allosteric activators, $NADP^+$, P-enol-pyruvate, GTP, ATP, UTP, and acetyl-CoA. A correlation was observed to exist at this level of enzyme control because isozymes from members of the Peronosporales were more sensitive to the effectors than isozymes from the Saprolegniales and Leptomitales. A similar correlation was observed for the allosteric inhibition of the D(-) lactate dehydrogenases by GTP. The isozymes of the Peronosporales were the most sensitive.

Using the parameter of enzyme kinetic mechanism, it was observed that NAD-linked glutamic dehydrogenases of the Phycomycetes could be separated into two groups, one with a kinetic mechanism in which the order of addition of substrates and release of products is typical of NAD-linked glutamic dehydrogenases of the higher fungi and bacteria, and the other in which the order is slightly altered in a way that is typical of the NADP-linked enzyme of other microorganisms. The same was found to be true for NAD- and NADP-linked isocitric dehydrogenases of both the lower and higher forms of the fungi. The reaction pattern for the NAD-linked isocitric dehydrogenase is synonymous with that for the NAD-linked glutamic dehydrogenase, and the pattern for NADP-isocitric dehydrogenase is synonymous with the NADP-linked glutamic dehydrogenase. Such findings have provided us with basic information about the possible ancestral relations between two extant genes, the NAD- and NADP-linked dehydrogenases, products that may not be erstwhile twins as one is led to believe.

INTRODUCTION

Extremely useful and precise information about the evolutionary relationships between species has been obtained from studies that involve the determination of the primary structures of homologous proteins (Smith, 1972). But less tedious comparisons of characteristic features of protein homologues such as allosteric modes of enzyme regulation (Jensen et al., 1967; Jensen, 1970; LéJohn, 1971a), behavior in an electric field (Markert and Møller, 1959; Watts, 1970; Browman et al., 1967; Clare, 1963), molecular weight and subunit composition (Hutter and DeMoss, 1967; Rutter, 1965) and kinetic mechanism of enzyme action (Stevenson and LéJohn, 1971; Watts, 1971) must certainly reflect a specific but undefined amino acid sequence and these could prove to be very informative in elucidating broad relationships of species homology. This discourse is confined to the fungi where, in addition to the above, information about cell wall structure and uniqueness in biosynthetic pathways (Bartnicki-Garcia, 1970) has been used very effectively to simplify an otherwise complex prolem of systematics and phylogeny. However, important questions do remain unanswered about the ancestral roots of some of the lower fungi and our studies on enzyme regulations and the kinetic mechanisms operative in the catalytic reactions of analogous and homologous enzymes were initiated to try and explain some of the anomalies that exist.

This paper summarizes studies of the kinetic (LéJohn, et al., 1969; LéJohn and Stevenson, 1970), regulatory (LéJohn, et al., 1969; LéJohn and Jackson, 1968; LéJohn et al., 1970), and physical (Wang and LéJohn, 1974a) properties of a variety of NAD-linked glutamic dehydrogenases that are found exclusively in the lower fungi belonging to the obsolete class of Phycomycetes. Similar studies on D(-)lactate dehydrogenases that are prevalent in the Phycomycetes, and less detailed analyses of NAD-linked malate and isocitric dehydrogenases as well as NADP-linked isocitric dehydrogenases of the fungal organisms provide ancillary information for any major conclusions reached about their phylogenetic relationships. Some eighty species of the lower fungi were compared in this manner, and a handful of the higher fungal forms belonging to the Ascomycetes, Deuteromycetes, and Basidiomycetes were analyzed in a general way.

MATERIALS AND METHODS

Full details about the various techniques used in the numerous aspects of these studies have been reported in earlier communications (see Wang and LéJohn, 1974 a,b,c; LéJohn, 1971a; LéJohn et al., 1970; LéJohn, 1971b).

RESULTS

HOMOLOGUES AND ANALOGUES OF GLUTAMIC DEHYDROGENASES

More than eighty species of the lower fungi, Myxomycetes, Chytridiomycetes, Hyphochytridiomycetes, Oömycetes, and Zygomycetes have been examined and found to possess only an NAD-linked glutamic dehydrogenase. The higher fungi, Deuteromycetes, Ascomycetes, and Basidiomycetes, seem to produce two distinct forms of the enzyme, one NAD-linked, the other NADP-linked (Table 1). The unique distribution of the two coenzyme-specific forms of these catalysts provided an opportunity to determine what mechanisms of enzyme regulation of the glutamic dehydrogenases might be operative in these fungi.

The NAD-linked glutamic dehydrogenases of the Chytridiomycetes, Hyphochytridiomycetes, Oömycetes and Zygomycetes can be placed into three groups based on their regulatory properties (Table 2). Both the Chytridiomycetes and Zygomycetes contain species whose glutamic dehydrogenases are unregulated. This is type I enzyme. No Zygomycetes glutamic dehydrogenase has yet been found that displays any form of allosteric control. Many of the Chytridiomycetes glutamic dehydrogenases are allosterically regulated and have been described as type II in Table 2. Activators of the enzyme are Ca^{++}, Mn^{++}, and AMP, with citrate, ATP, GTP, and fructose-1, 6-diphosphate as inhibitors. The effects of these modulators are magnified in the reductive amination reaction and reduced or weak in the oxidative deamination process. This nonequilibrium unidirectional effect has been described (LéJohn,1968) and shown to be widespread in the glutamic dehydrogenases of the fungi (LéJohn, 1971a).

A type III glutamic dehydrogenase is common in members of the Oömycetes and a modified form of it is present in the Hyphochytridiomycetes. This is the most interesting type of these enzymes because although NAD^+ is the substrate, $NADP^+$

Table 1. Distribution of Glutamic Dehydrogenases in Fungi.

Organism	Order	Class	Coenzyme Specificity	
			NAD-linked	NADP-linked
Dictyostelium discoideum	Acrasiales	Myxomycetes	+	-
Polysphondylium violaceum	Acrasiales	Myxomycetes	+	-
Entophlyctis sp.	Chytridiales	Chytridiomycetes	+	-
Rhizophydium sphaerocarpum	Chytridiales	Chytridiomycetes	+	-
Allomyces arbuscula	Chytridiales	Chytridiomycetes	+	-
Blastocladiella emersonii	Blastocladiales	Chytridiomycetes	+	-
Rhizidiomyces apophysatus	Hyphochytriales	Hyphochytridiomycetes	+	-
Hyphochytrium catenoides	Hyphochytriales	Hyphochytridiomycetes	+	-
Achlya sp. (1969)	Saprolegniales	Oomycetes	+	-
A. ambisexualis (male)	Saprolegniales	Oomycetes	+	-
A. ambisexualis (female)	Saprolegniales	Oomycetes	+	-
A. americana	Saprolegniales	Oomycetes	+	-
A. dubia	Saprolegniales	Oomycetes	+	-
A. colorata	Saprolegniales	Oomycetes	+	-
A. bisexualis (male)	Saprolegniales	Oomycetes	+	-
A. bisexualis (female)	Saprolegniales	Oomycetes	+	-
A. crenulata	Saprolegniales	Oomycetes	+	-
A. intricata	Saprolegniales	Oomycetes	+	-
A. flagellata	Saprolegniales	Oomycetes	+	-
A. racemosa	Saprolegniales	Oomycetes	+	-
A. radiosa	Saprolegniales	Oomycetes	+	-
A. heterosexualis (male)	Saprolegniales	Oomycetes	+	-
A. heterosexualis (female)	Saprolegniales	Oomycetes	+	-
Isoachlya itoana	Saprolegniales	Oomycetes	+	-
I. eccentrica	Saprolegniales	Oomycetes	+	-
I. unispora	Saprolegniales	Oomycetes	+	-
I. intermedia	Saprolegniales	Oomycetes	+	-
I. toruloides	Saprolegniales	Oomycetes	+	-
Protoachlya paradoxa	Saprolegniales	Oomycetes	+	-
Aphanomyces laevis	Saprolegniales	Oomycetes	+	-
A. cladogamus	Saprolegniales	Oomycetes	+	-
A. euteiches	Saprolegniales	Oomycetes	+	-
A. stellatus	Saprolegniales	Oomycetes	+	-
Thraustotheca clavata	Saprolegniales	Oomycetes	+	-
Aphanopsis terrestris	Saprolegniales	Oomycetes	+	-
Leptolegnia caudata	Saprolegniales	Oomycetes	+	-
Saprolegnia asterophora	Saprolegniales	Oomycetes	+	-

Table 1. Continued

Organism	Order	Class	Coenzyme Specificity NAD-linked	NADP-linked
Saprolegnia parasitica	Saprolegniales	Oomycetes	+	-
S. turfosa	Saprolegniales	Oomycetes	+	-
S. lapponica	Saprolegniales	Oomycetes	+	-
S. ferax	Saprolegniales	Oomycetes	+	-
S. litoralis	Saprolegniales	Oomycetes	+	-
S. mixta	Saprolegniales	Oomycetes	+	-
S. terrestris	Saprolegniales	Oomycetes	+	-
S. monoica	Saprolegniales	Oomycetes	+	-
S. delica	Saprolegniales	Oomycetes	+	-
S. megasperma	Saprolegniales	Oomycetes	+	-
S. furcata	Saprolegniales	Oomycetes	+	-
Sapromyces androgynus	Leptomitales	Oomycetes	+	-
S. elongatus	Leptomitales	Oomycetes	+	-
Apodachlya sp.	Leptomitales	Oomycetes	+	-
A. brachynema	Leptomitales	Oomycetes	+	-
Leptomitus sp.	Leptomitales	Oomycetes	+	-
Pythium ultimum	Peronosporales	Oomycetes	+	-
P. debaryanum	Peronosporales	Oomycetes	+	-
P. heterothallicum (male)	Peronosporales	Oomycetes	+	-
P. heterothallicum (female)	Peronosporales	Oomycetes	+	-
P. butleri	Peronosporales	Oomycetes	+	-
P. splendens	Peronosporales	Oomycetes	+	-
P. intermedium (plus)	Peronosporales	Oomycetes	+	-
P. intermedium (minus)	Peronosporales	Oomycetes	+	-
P. undulatum	Peronosporales	Oomycetes	+	-
P. sylvaticum (male)	Peronosporales	Oomycetes	+	-
P. sylvaticum (female)	Peronosporales	Oomycetes	+	-
P. catenulatum	Peronosporales	Oomycetes	+	-
Phytophthora cinnamoni	Peronosporales	Oomycetes	+	-
P. palmivora	Peronosporales	Oomycetes	+	-
Absidia glauca	Mucorales	Zygomycetes	+	-
Cunninghamella blaskesleeana	Mucorales	Zygomycetes	+	-
Mucor hiemalis	Mucorales	Zygomycetes	+	-
Rhizopus stolonifer	Mucorales	Zygomycetes	+	-
Zygorhynchus mollieri	Mucorales	Zygomycetes	+	-

Table 1. Continued

Organism	Order	Class	Coenzyme Specificity NAD-linked	NADP-linked
<u>Phycomyces blakesleeanus</u>	Mucorales	Zygomycetes	+	-
<u>Piricularia oryzae</u>	Moniliales	Deuteromycetes	+	+
<u>Fusarium oxysporum</u>	Moniliales	Deuteromycetes	+	+
<u>Candida utilis</u>	Moniliales	Deuteromycetes	+	+
<u>Hypomyces rosellus</u>	Hypocreales	Ascomycetes	+	+
<u>Saccharomyces cerevisiae</u>	Endomycetales	Ascomycetes	+	+
<u>Hansenula subpelliculosa</u>	Endomycetales	Ascomycetes	+	+
<u>Sordaria fimicola</u> (plus)	Sphaeriales	Ascomycetes	+	+
<u>S. fimicola</u> (minus)	Sphaeriales	Ascomycetes	+	+
<u>Neurospora crassa</u>	Sphaeriales	Ascomycetes	+	+
<u>Coprinus lagopus</u>	Agaricales	Basidiomycetes	+	+
<u>Schizophyllum commune</u>	Agaricales	Basidiomycetes	+	+

The above represents data taken from (ref. 19 and 27)

is an allosteric activator. In addition, GTP and ATP which strongly inhibit the type II enzyme form, activate the type III form. Similarly, the bivalent metals, Ca^{++}, and Mn^{++}, activators of type II enzyme inhibit type III enzyme strongly. The type III glutamic dehydrogenase has additional modifiers such as acyl CoA derivatives and P-enolpyruvate (Table 2).

TABLE 2

DIFFERENT TYPES OF NAD-LINKED GLUTAMIC DEHYDROGENASE FOUND IN THE PHYCOMYCETES

Fungal class	Enzyme type	Activators	Inhibitors
Chytridiomycetes and Zygomycetes	Type I	None	None
Chytridiomycetes and Zygomycetes	Type II	Ca^{++}, Mn^{++}, AMP	citrate, ATP, FDP, GTP
Hyphochytridio- mycetes	Type IIIA	$NADP^+$, NADPH, AMP, short chain acylCoA derivatives, PEP	citrate, GTP, ATP, Ca^{++} Mn^{++}, long chain acyl-CoA derivatives
Oömycetes	Type IIIB	$NADP^+$, NADPH, ATP, short chain acylCoa derivatives, PEP	AMP citrate, Ca^{++}, Mn^{++}, long chain acyl-CoA derivatives

PEP, P-enolpyruvate; short chain acylCoA derivatives refer to fatty coenzyme A esters with less than 8 carbons in the hydrocarbon side chain; long chain acylCoA derivatives refer to fatty coenzyme A esters with more than 8 carbons in the hydrocarbon side chain; FDP, fructose-1, 6-diphosphate. Type IIIA is a hypothetical functional hybrid of Type II and Type IIIB. Type IIIB is referred to as Type III throughout the text.

Because a large number of different allosteric modulators have been found to interact with types II and III glutamic dehydrogenases with some of the modulators having diametrically opposite effects on the two enzyme types, it was of interest to make an in-depth study of this catalyst from as many species as possible from the various classes of the lower fungi. To date, progress has been made on type III enzyme from 50 species of the Oömycetes. This type of enzyme in the Oömycetes has been examined for its electrophoretic properties as well as its catalytic responses to 6 allosteric effectors. The type II and III enzymes have been studied to compare their kinetic mechanisms of catalysis.

Of the 50 Oömycetes examined, 33 species belonged to the Saprolegniales, 8 species to the Peronosporales, and 2 species to the Leptomitales. Some of the Saprolegniales and Peronosporales used had two mating types and this feature was useful because it served to emphasize any significant differences that were observed.

The electrophoretic mobilities of the type III glutamic dehydrogenase in species of the genus *Achlya* (R_e of between 2.8 to 4.2 cm) were very similar to those of species of the genus *Saprolegnia*. A distinguishing feature was the existence of two isozymes of type III glutamic dehydrogenase in two species of Saprolegnia (*S. litoralis* and *S. monoica*) while only one form of the enzyme was found in all of the other Saprolegniales. In contrast, 4 of the 8 species of the Peronosporales examined had two type III isozymes (Fig. 1). Otherwise, they were similar in their electrophoretic property to the homologous enzymes from the Saprolegniales. The type III enzyme from the Leptomitales was quite different from those of the other two families.

ENZYME REGULATION

Analysis of the relative reactivity of the various type III enzymes to the several allosteric modifiers gave results that are summarized in Table 3. The most reactive type III enzyme was found in the Peronosporales where all 8 species contained glutamic dehydrogenases that were activated by P-enolpyruvate, $NADP^+$, GTP, ATP, UTP, and short chain acyl CoA derivatives. But for a few cases, none of the type III glutamic dehydrogenases in the Saprolegniales were activated by all six modulators. In those cases where interaction between all modulators and enzyme was observed, the responses were marginally small. $NADP^+$ and P-enolpyruvate remained in every case the two most effective allosteric activators.

Studies of inhibition of enzyme activity were restricted to the use of citrate. All members of the Peronosporales contained a type III glutamic dehydrogenase whose activity was inhibited by citrate (Table 4). Some of the species (*P. butleri*, *P. heterothallicum* ♂ and ♀) whose type III enzymes were strongly inhibited by citrate contained two isozymes implying that both isozymes were affected. It would be interesting to isolate the two isozymes in pure form and examine them for their reactivity towards the allosteric

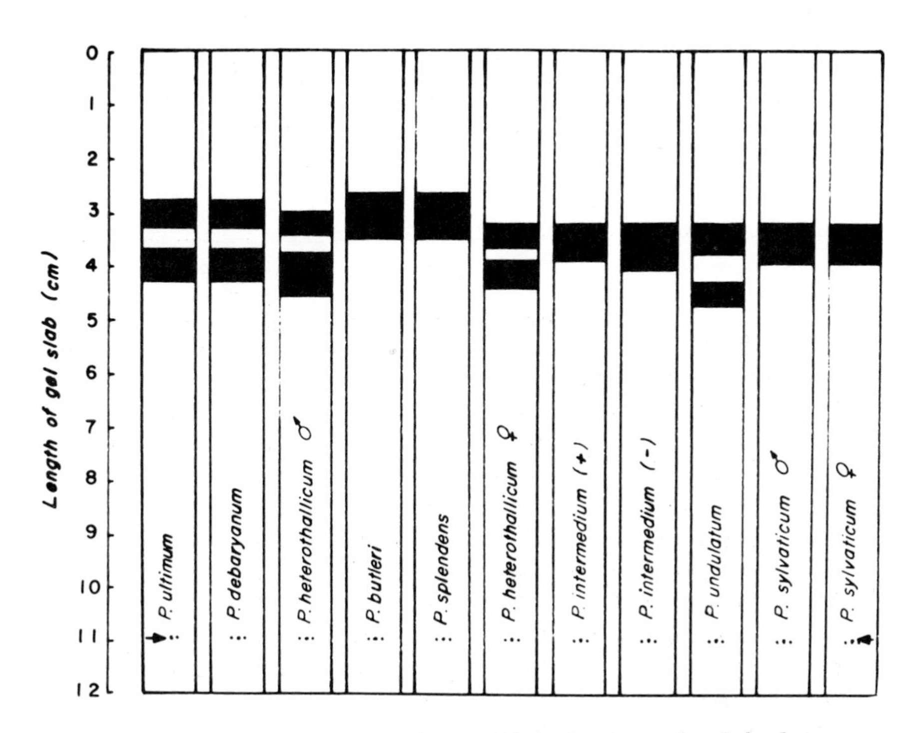

Fig. 1. Electropherogram of NAD-linked glutamic dehydrogenase isozymes of eleven species of *Pythium*. Electrophoresis was for 2 hr at 10°C in a buffer composed of 5.52 g barbital diethyl-barbituric acid, 1 g Tris-base in 1 liter of H_2O, pH 8. Arrows indicate position of tracking dye. Two hundred and fifty micrograms of protein sample applied to each slot of gel slab. Twenty milliliters of buffer was added to the gel solution.

---

activators and inhibitors. No attempt has yet been made to determine whether those species that have two type III isozymes produce them at all stages of the growth cycle and under different nutritional conditions.

TABLE 3

Activation of NAD-dependent glutamic dehydrogenase isozymes of Oömycetes by $NADP^+$, P-enolpyruvate, GTP, ATP, UTP and acetyl-CoA. The ratio v_a/v_o represents the increase in enzyme activity over control where no activator is present. The assay was carried out spectrophotometrically in a 3-ml reaction solution consisting of 30 mM NH_4^+, 10 mM α-ketoglutarate, 0.167 mM NADH, 33.3 mM Tris-acetate buffer, pH 8 with or without one of the following activators 1 mM $NADP^+$; 0.5 mM P-enolpyruvate; 0.33 mM GTP: 1 mM ATP: 1 mM UTP; and 0.33 mM acetyl-CoA. Reaction rate was determined from the change in absorbance at 340 nm. v_a is the rate in the presence of one activator and v_o is the rate in the absence of an activator.

Enzyme source	Modulator effect, v_a/v_o					
	$NADP^+$	PEP	GTP	ATP	UTP	acetyl-CoA
Pythium heterothallicum ♀	11.0	11.0	10.0	1.7	3.9	1.8
P. heterothallicum ♂	4.8	3.5	3.0	1.3	2.0	1.5
P. ultimum	8.8	4.2	6.2	2.0	2.7	1.5
P. splendens	8.0	4.0	3.3	3.1	1.4	1.4
P. undulatum	5.1	3.4	2.9	1.5	2.5	1.3
P. debaryanum	3.7	2.3	2.8	1.5	1.4	1.6
P. intermedium (-)	2.8	1.5	1.5	1.3	1.4	1.4
P. intermedium (+)	2.5	1.6	1.9	1.4	1.6	1.4
P. sylvaticum ♀	1.8	1.4	1.6	1.3	1.5	1.3
P. sylvaticum ♂	1.7	1.4	1.7	1.3	1.5	1.3
P. butleri	2.1	1.5	1.8	1.5	1.5	1.5
Achlya dubia	4.0	2.0	1.5	-	-	-
A. americana	3.8	3.7	1.5	-	1.4	-
A. ambisexualis ♂	3.6	3.6	-	-	-	-
A. ambisexualis ♀	3.4	3.4	1.3	1.4	1.4	-
A. colorata	3.0	1.3	1.3	1.3	1.3	1.3
A. bisexualis ♂	5.0	1.3	-	-	-	-
A. bisexualis ♀	2.3	1.2	-	-	-	-
A. crenulata	6.0	2.0	1.4	1.4	1.4	-
A. intricata	3.0	-	-	-	-	-
A. flagellata	3.5	-	-	-	-	-

Table 3. Continued

Enzyme source	Modulator effect, v_a/v_o					
	NADP^{+}	PEP	GTP	ATP	UTP	acetyl-CoA
Achlya racemosa	2.6	1.4	1.3	1.4	1.4	1.4
A. radiosa	4.0	-	-	-	-	-
A. heterosexualis ♂	1.8	1.3	-	-	-	-
A. heterosexualis ♀	3.7	1.7	-	-	-	-
Saprolegnia asterophora	2.0	1.7	1.7	1.5	1.5	1.3
S. lapponica	3.3	-	-	-	-	-
S. ferax	3.1	1.4	1.2	-	1.3	1.2
S. mixta	2.0	-	-	-	-	-
S. terrestris	3.0	-	-	-	-	-
S. monoica	3.5	1.3	-	-	-	-
S. megasperma	2.6	1.3	-	-	-	-
S. furcata	3.0	-	-	-	-	-
S. litoralis	-	-	-	-	-	-
S. delica	-	-	-	-	-	-
Aphanomyces laevis	3.0	1.2	-	-	-	-
A. cladogamus	1.4	1.2	-	-	-	-
Isoachlya eccentrica	3.5	-	-	-	-	-
I. unispora	2.6	1.2	-	-	-	-
I. intermedia	3.2	1.4	-	-	-	1.2
I. toruloides	3.4	1.3	1.2	-	-	-
Thraustotheca clavata	2.4	-	-	-	-	-
Aphanopsis terrestris	3.2	1.3	1.3	-	-	-
Protoachlya paradoxa	2.1	1.2	1.2	-	-	-

TABLE 4

Inhibition of NAD-dependent glutamic dehydrogenase isozymes of Pythium spp. by citrate. Assay conditions as described elsewhere (18). Citrate was kept constant at 10 mM and substrates and the modulator, $NADP^+$ provided at saturating concentrations. The K_m values of all isozymes are similar.

Organism	% Inhibition
Pythium ultimum	40
P. debaryanum	40
P. heterothallicum (male)	64
P. heterothallicum (female)	67
P. butleri	82
P. splendens	31
P. intermedium (plus)	65
P. intermedium (minus)	80
P. undulatum	63
P. sylvaticum (male)	87
P. sylvaticum (female)	85

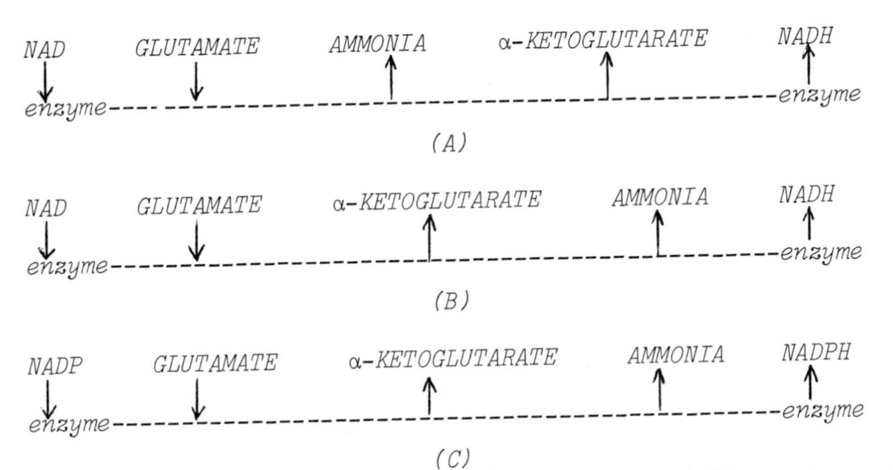

Fig. 2. Line diagrams of the sequential ordered binary-ternary kinetic mechanism of action of (A) Type II; (B) Type III, NAD-linked glutamic dehydrogenases of the Phycomycetes; and (C) NADP-linked glutamic dehydrogenase of the higher fungi and bacteria.

KINETIC MECHANISM OF ENZYME ACTION

A wider comparison was made of these glutamic dehydrogenases by a study of their kinetic mechanism of enzyme action. NAD-specific and NADP-specific glutamic dehydrogenases from various microorganisms have been shown to catalyze the reversible oxidation-reduction reactions in the ordered sequences shown in Fig. 2 (Stevenson and LéJohn, 1971) where the order of addition or release of ammonia and α-ketoglutarate are reversed in the two schemes. The first mechanism has been called by us, "NAD-type" and the second mechanism "NADP-type" (Stevenson and LéJohn, 1971).

Because of the allosteric effects of $NADP^+$ on the NAD-linked glutamic dehydrogenases of the Oömycetes, it seemed an ideal system to use to investigate what physiological or evolutionary reasons might underlie the differences in kinetic mechanism that have been observed between the NAD- and the NADP-linked glutamic dehydrogenases of microorganisms.

Product inhibition studies patterned after Cleland (Cleland, 1963) were carried out to elucidate the kinetic mechanism of the NAD-linked glutamic dehydrogenase purified from *Pythium debaryanum* and *Achlya* sp (1969), representatives of the Peronosporales and Saprolegniales respectively. The enzyme from both species gave identical results as summarized in Table 5. This contrasts with the product inhibition data obtained for *Blastocladiella emersonii* and *Allomyces arbuscula*

NAD-linked glutamic dehydrogenases (also summarized in Table 5). The data for *Blastocladiella* have been reported (LéJohn, et al., 1969) but that for *Allomyces* is unpublished (F. S. Yen, and H. B. LéJohn, unpublished results).

Blastocladiella and *Allomyces* belong to the Chytridiomycetes, and the regulatory properties of their glutamic dehydrogenases place them in type II category. They show the NAD-type kinetic mechanism that has already been elucidated for *Neurospora crassa* (Ascomycetes) NAD-linked glutamic dehydrogenase (Stevenson and LéJohn, 1971; Stachow, 1965) and the autotrophic sulfur-oxidizing bacterium, *Thiobacillus novellus* (LéJohn et al., 1968). Both of these organisms, it so happens, also contain NADP-linked glutamic dehydrogenases whose kinetic mechanisms have been determined by similar product inhibition studies and shown to be typically NADP-type. Consequently, we see that this rather diverse group of lower fungi, the Phycomycetes, displays a rather remarkable uniformity in possessing only the NAD-linked variety of glutamic dehydrogenase, but has a subtle dichotomy that segregates the species into two major groups based on such a surprising property as kinetic mechanism. Therefore, the types II and III glutamic dehydrogenases should be regarded as analogous rather than homologous.

Engel and Dalziel (Engel and Dalziel, 1970) have questioned the conclusions of Fahien and Strmecki (Fahien and Strmecki, 1969) and Frieden (Frieden, 1959) that the ox liver glutamic dehydrogenase that is non-specific for either coenzyme has an ordered reaction mechanism. They proposed instead a random mechanism based largely on initial reaction rate studies. Cross and Fisher (Cross and Fisher, 1970) have shown that in the presence of NADPH, the ox liver enzyme catalyzes an obligatory order of reaction. This is interesting in the light of our findings that an NAD-linked glutamic dehydrogenase mimics the reaction order of an NADP-type catalyst because $NADP^+$ can bind to the enzyme, alter its reaction rate, but the coenzyme itself is not catalyzed in the process. The existence of an NADP(H) site on the enzyme, seemingly, gives it a kinetic predestination.

LACTATE DEHYDROGENASE ISOZYMES

Data on phylogenetic interrelations of organisms based on those enzyme properties that have been discussed for glutamic dehydrogenase are likely to lead to spurious correlations unless they are supported by alternate parameters or by corresponding studies of other enzyme systems. Glutamic dehydrogenase is involved in aerobic metabolism and, therefore,

Table 5. Predicted and Observed Product Inhibition Patterns for Elucidating the Kinetic Mechanisms of Enzyme Action for Types II and III NAD-linked Glutamic Dehydrogenases of the Phycomycetes. The Type II enzyme was purified from _Blastocladiella emersonii_ (see ref. 16), and the Type III enzyme was isolated from _Pythium debaryanum_ and _Achlya_ sp. (1969) (see ref. 25).

Substrate/Inhibitor	Inhibition Patterns			
	Type II Enzyme		Type III Enzyme	
	Predicted	Observed	Predicted	Observed
NAD^+/α-Ketoglutarate	UC	UC	UC	NC[a]
NAD^+/ammonia	NC	NC	NC[a]	UC
NAD^+/NADH	C	C	C	C
Glutamate/α-Ketoglutarate	UC	CP[b]	NC	NC
Glutamate/ammonia	NC	NC	NC	NC
Glutamate/NADH	NC	NC	NC	NC

The abbreviations used are: UC, uncompetitive; NC, noncompetitive; C, competitive; CP, cooperative phenomenon. [a]Indicates that the order of addition or release of ammonia and α-ketoglutarate are reversed in this system when compared to Type II enzyme. [b]The term cooperative phenomenon is used to described an allosteric effect which distorts the uncompetitive pattern that underlies the interaction as discussed in ref. 16.

another enzyme involved in anaerobic metabolism that is also allosterically regulated was selected. This enzyme, NAD-linked D(-) lactate dehydrogenase, shares the same distribution pattern as the types II and III NAD-linked glutamic dehydrogenases of the lower fungi. There are also two classes of the D(-)lactate dehydrogenase, one that is inhibited by ATP and GTP, the latter being an allosteric modulator, is designated as class I. The second type inhibited by ATP but not by GTP is designated as class II. The inhibition by ATP is explicable on simple competition terms without recourse to allostery as was found necessary for GTP inhibition (LéJohn, 1971b). GTP has a distinct binding site and it influences the activity of the enzyme in a unique manner; cooperative effects are shown when GTP is present and the limiting substrate is either NADH or lactate.

The class I D(-)lactate dehydrogenase appears to be confined to the Oömycetes and possibly the Acrasiales (true slime molds). This corresponds with the distribution of the type III glutamic dehydrogenase.

An electrophoretic study of the physical properties of the D(-)lactate dehydrogenase from 50 Oömycetes species showed that all members of the Saprolegniales listed in Table 1 except *Leptolegnia caudata*, had very similar enzymes. Under the conditions of electrophoretic analysis, the enzymes had an R_e of about 7 cm. *Apodachlya brachynema* (Leptomitales) D(-) lactate dehydrogenase had similar electrophoretic characteristic as *Achlya* and *Saprolegnia* lactate dehydrogenases, but the second member of the Leptomitales tested, *Sapromyces elongatus*, had an uniquely different enzyme. This enzyme had an R_e value of 1 cm (Fig. 3) and was present to the extent of about 5% of the total cell protein. This is about 50 times more than the cellular concentration of the other species. Because of this significant difference in physical property, a detailed kinetic study of the kinetic mechanism has been carried out and the data is to be published.

D(-)lactate dehydrogenases of the Peronosporales (Fig. 4) proved to be interesting from several points of view. (a) At least 2 of the 8 species studied contained two isozymes of the lactate dehydrogenase; one of them was the sexual mating type of *Pythium intermedium*. (b) The enzymes from the male and female mating types of *P. heterothallicum* were electrophoretically different. (c) Except for *P. undulatum*, the lactate dehydrogenases of the Peronosporales were different, electrophoretically, from the enzyme of the Saprolegniales. (d) There was a greater variation in the electrophoretic mobilities of the lactate dehydrogenases of the Peronosporales than shown by the Saprolegniales whose electrophoretic patterns

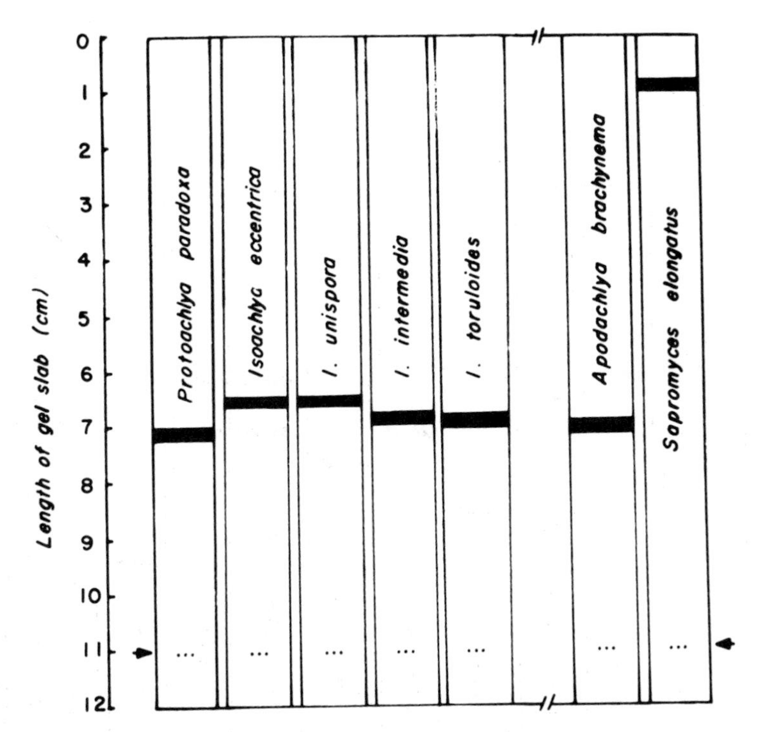

Fig. 3. Electropherogram of NAD-linked D(-)lactate dehydrogenase isozymes of some members of the Leptomitales and Saprolegniales. Electrophoresis was for 4 hr at 10°C in a buffer of the same composition as described in Fig. 1 legend. Arrows indicate position of tracking dye. Protein sample per slot as for Fig. 1 legend.

---

are not all presented here.

In general, the lactate dehydrogenases of the three major families of the Oömycetes tested were uniformly different from each other. This is in sharp contrast to the gross similarity in electrophoretic property of the glutamic dehydrogenases of the same three families discussed earlier.

ENZYME REGULATION

All of the lactate dehydrogenases of the Oömycetes were tested for their sensitivity to GTP inhibition under conditions where the concentration of NADH was varied from less than K_m

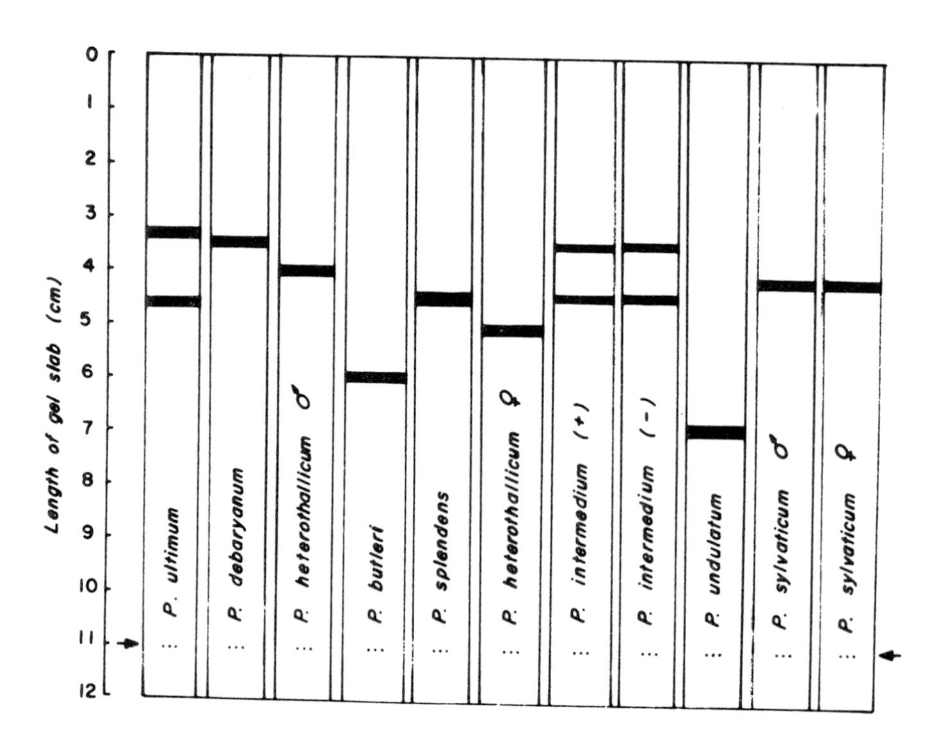

Fig. 4. Electropherogram of NAD-linked D(-)lactate dehydrogenase isozymes of eleven species of *Pythium*. All other conditions as described for the legend to Fig. 3.

---

value to several times greater than K_m. This was essential because the response of the enzyme to the inhibitor is a cooperative process. A condensed version of the results is shown in Table 6. The lactate dehydrogneases of the Peronosporales were the most sensitive to GTP inhibition. Except for *P. splendens*, most of the enzymes were inhibited by as much as 85 to 100 percent at low NADH concentrations. This meant that those species with two lactate dehydrogenase isozymes must have both enzymes being susceptible to the inhibitor.

Lactate dehydrogenases of the Saprolegniales were less

TABLE 6

GTP INHIBITION OF NAD-DEPENDENT D(-) LACTATE DEHYDROGENASE ISOZYMES OF OÖMYCETES

Organism		% Inhibition of Control Activity low conc. NADH	high conc. NADH
Peronosporales			
Pythium ultimum		87	35
P. debaryanum		92	68
P. heterothallicum	(male)	89	53
P. heterothallicum	(female)	100	100
P. butleri		93	95
P. splendens		55	32
P. intermedium	(plus)	100	99
P. intermedium	(minus)	97	95
P. undulatum		88	77
P. sylvaticum	(male)	100	90
P. sylvaticum	(female)	100	96
Saprolegniales			
Achlya ambisexualis	(male)	60	57
A. ambisexualis	(female)	33	25
A. dubia		50	25
A. colorata		25	0
A. bisexualis	(male)	25	17
A. bisexualis	(female)	67	50
A. flagellata		100	33
A. racemosa		50	33
Saprolegnia mixta		100	20
S. terrestris		90	60
S. monoica		33	12
S. furcata		50	20
Aphanomyces cladogamus		60	4
A. euteiches		33	0
Aphanopsis terrestris		33	17
Isoachlya unispora		71	57
Protoachlya brachynema		60	17
Leptomitales			
Apodachlya brachynema		41	22
Sapromyces elongatus		33	20

low NADH means less than K_m value; high NADH means five times K_m value.

sensitive to GTP inhibition and there was a greater disparity in the inhibition observed for isoyzmes from different mating types of the same species. For example, *A. bisexualis* *A. ambisexualis*, male and female mating types, had enzymes that responded differently to the same amount of GTP. The significance of these observations is not immediately apparent since the isozymes have the same electrophoretic mobilities and relative K_m values for substrates.

Sapromyces elongatus contained an unique D(-)lactate dehydrogenase (see Fig. 3) and GTP inhibition of this enzyme, while not especially strong, was comparable to the inhibition of the enzyme from the other Leptomitales, *Apodachlya brachynema*. This observation was proof enough that the electrophoretic mobility and allosteric property may not necessarily bear any correlation.

KINETIC MECHANISM OF ENZYME ACTION

The kinetic mechanism of enzyme action has been determined for the lactate dehydrogenases of several members of the Oömycetes but this discussion will be confined to the enzyme from *P. debaryanum*, *Saprolegnia ferax*, and *Sapromyces elongatus*. Data on the kinetics of *Pythium* lactate dehydrogenase have been reported (LéJohn, 1971b) and completed studies of the other two species are to be published.

The kinetic mechanism for the enzyme from all of the three sources was found to be obligatory ordered binary-binary processes with the coenzymes binding first to the enzyme followed by either pyruvate or lactate. GTP inhibition studies have shown the existence of two binding sites for NADH and four for lactate. Pyruvate and NAD^+ did not seem to have more than one binding site for each. All three enzymes therefore have the same kinetic mechanism of enzyme action although they differ greatly in their physical properties as deduced from the electrophoretic studies. The physiological reasons behind GTP inhibition have been discussed elsewhere (LéJohn, 1971b) and will not be reiterated here.

MALATE AND ISOCITRIC DEHYDROGENASE ISOZYMES

The distribution and variety of isozymes of NAD-linked malate and NADP-linked isocitric dehydrogenases of 50 species of the Oömycetes was also determined by slab gel electrophoresis. The results were not very informative. Two isocitric dehydrogenase isozymes were detected in the same species in only 2 of 50 cases; one was *Aphanomyces cladogamus*, and the other *Aphanopsis terrestris* (Fig. 5). Analysis of the lactate and glutamic dehydrogenases by electrophoresis and regulation

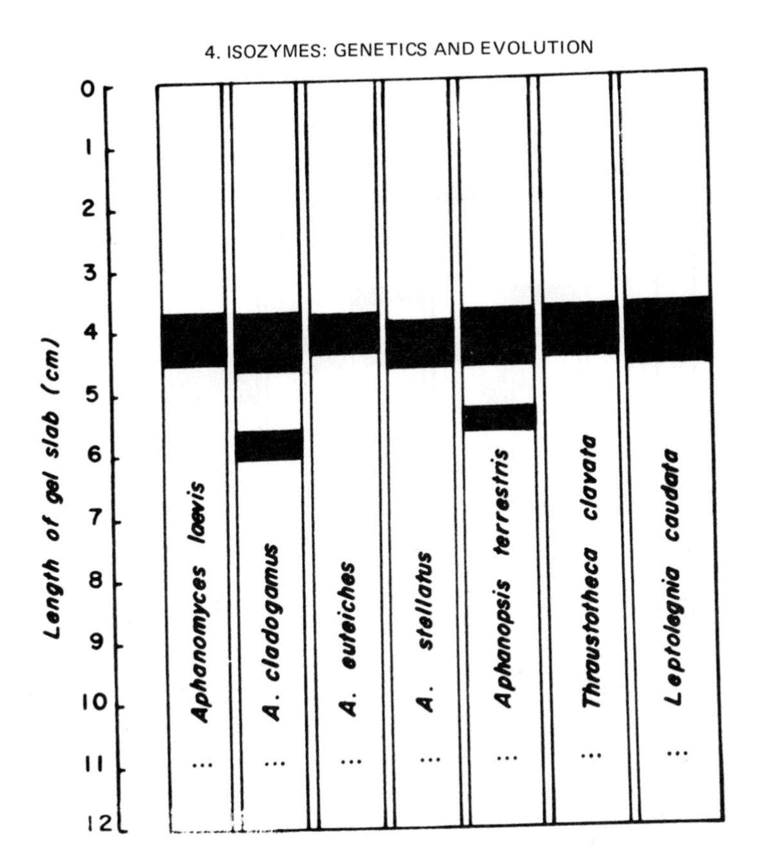

Fig. 5. Electropherogram of NADP-linked isocitric dehydrogenase isozymes of some Saprolegniales. Electrophoresis was for 4 hr at 10°C in a buffer composed of 34.55 g Tris-base, 5.25 g glycine, pH 9.3, and H_2O to 1 liter. Twenty milliliters of buffer was added to the gel solution. Two hundred and fifty micrograms of protein sample applied to each slot of gel slab.

---

had vaguely suggested the possibility that these two species in different genera may be closely linked. In contrast, 38 of the 50 Oömycetes showed evidence of multiple isozymic species of malate dehydrogenases (Fig. 6). Only representative electropherograms are shown here. In the few cases where there were only single enzyme activity bands, these were rather broad implying that there may well be more than one enzyme type but that they were not properly resolved by the electrophoretic procedures used. As many as 5 electrophoretically distinct isozymes were found in *Aphanomyces cladogamus*, 3 each in *Saprolegnia delica* and *Aphanomyces stellatus*. In the other cases, there were two malate dehydrogenase isozymes

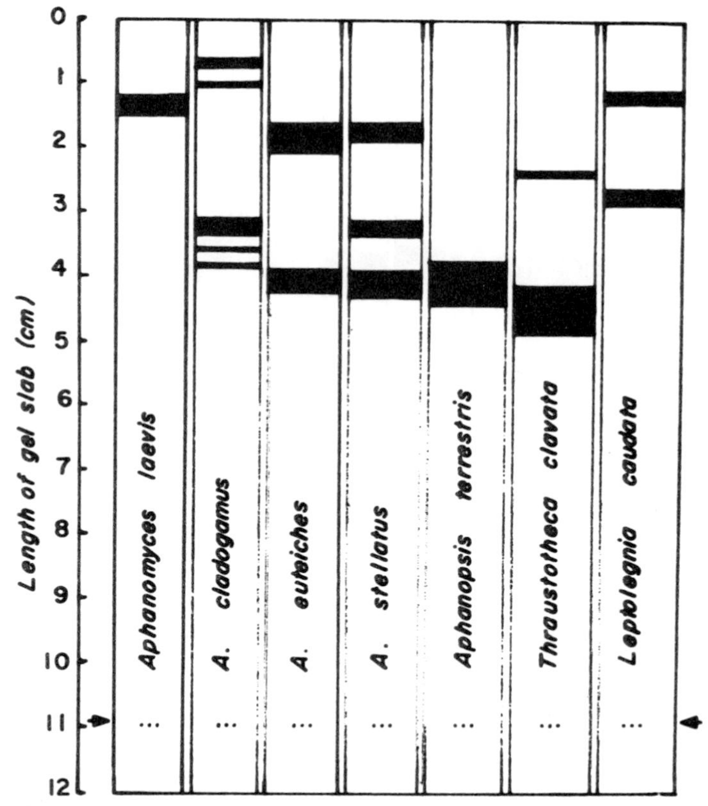

Fig. 6. Electropherogram of NAD-linked malate dehydrogenase isozymes of some Saprolegniales. Electrophoresis under same conditions as described in the legend to Fig. 5. Arrows indicate position of tracking dye.

---

per species and even these showed broad activity bands which could be masking multiple isozyme species as well. No special electrophoretic patterns could be seen that could distinguish the various genera of the Oömycetes as was done in the cases of the lactate and glutamic dehydrogenases. Consequently, the use of electrophoretic mobility of enzymes as a criterion to relate these species was rather disappointing.

However, a feature that may have considerable evolutionary significance was detected in our studies of the kinetic mechanisms of action of the isocitric dehydrogenases in particular. In these studies, work was extended to other fungal groups that contain the NAD-linked variety of isocitric dehydrogenase as well as the NADP-linked form. We have observed to date that the NADP-linked isocitric dehydrogenases of the Oömycetes (unpublished data of LéJohn and Smaluck), *Saccharomyces cerevisiae*, *Neurospora crassa*, *Blastocladiella emersonii*, and *Allomyces arbuscula* (unpublished data of H. B. L.) have an ordered sequential kinetic mechanism analogous to the NADP-type glutamic dehydrogenase mechanism. This means that substitution of CO_2 for NH_4^+ in Fig. 2 will describe the mechanism. Of course, isocitrate and glutamate interchange. The NAD-linked isocitric dehydrogenases of *Blastocladiella*, yeast, and *Neurospora* show the NAD-type kinetic mechanism deduced for glutamic dehydrogenase with isocitrate and CO_2 making the corresponding replacements for glutamate and ammonia. This latter mechanism was deduced after the allosteric effects that exist in the reactions of these NAD-linked isocitric dehydrogenases had been neutralized with the pertinent effectors, citrate, AMP or ADP. An extensive analysis is under way to see if these analogous NAD- and NADP-linked isocitric dehydrogenases and their metabolically related dehydrogenase analogues sustain the kinetic patterns being uncovered and if they can be extended to mammalian and other organismic forms as well. If they do, it may prove to be another useful parameter to use in examining the possibility of parallel evolution of two dehydrogenase genes, one NAD-type, the other NADP-type previously considered by me to be erstwhile twins (LéJohn, 1971a) but with no strong evidence for that speculation.

SUMMARY

The above results have provided some fundamental support for the idea that changes in enzyme structure may be important factors in natural selection. In time, it may not even be necessary to perform tedious analyses of the primary sequences of enzymes to obtain meaningful phylogenetic and phyletic correlations. Refinement of data through studies of the chemical, catalytic, regulatory, and basic physical properties of enzymes should provide the framework for our understanding of the physiological meanings behind molecular and evolutionary changes in proteins. The established classical procedures of studying phylogeny, taxonomy, and evolution would still be necessary to give us the basic definitions from which to start applying these newer biochemical techniques.

ACKNOWLEDGEMENTS

This work was supported by grants from the National Research Council of Canada and the Fisheries Research Board of the University of Manitoba. I thank Roselynn Stevenson for reading an abbreviated version of this paper at the Third International Isozyme Conference at Yale, April, 1974. I thank the publishers of the Canadian J. Microbiology for allowing me to use Figs. 1, 4, 5, and 6, and Tables 3, 4, and 6 in this report. I also thank the publishers of Nature for allowing me to reproduce Table 2.

REFERENCES

Bartnicki-Garcia, S. 1970. "Cell wall composition and other biochemical markers in fungal phylogeny" In: *Phytochemical Phylogeny;* ed. Harborne, J. B., pp. 80-103. Academic Press, New York.

Browman, J. E., R. R. Brubaker, H. Fischer, and P. E. Carson 1967. Characterization of eubacteria by starch gel electrophoresis of glucose-6-phosphate dehydrogenase and phosphogluconate dehydrogenase. *J. Bacteriol.* 94: 544-551.

Clare, B. G., and G. A. Zentmyer 1966. Starch gel electrophoresis of proteins from species of *Phytophthora*. *Phytophathology* 56 : 1334-1335.

Cleland, W. W. 1963. The kinetics of enzyme-catalyzed reactions with two or more substrates of products I, II and III. *Biochim. Biophys. Acta* 67: 104-137; 173-187; 188-196.

Cross, D. G. and H. F. Fisher 1970. The mechanism of glutamate dehydrogenase reaction: III. The binding of ligands at multiple subsites and resulting kinetic effects. *J. Biol. Chem.* 245: 2612-2621.

Engel, P. C. and K. Dalziel 1970. Kinetic studies of glutamate dehydrogenase: The reductive amination of 2-oxoglutarate *Biochem. J.* 118: 409-419.

Fahien, L.A. and M. Strmecki 1969. Studies of gluconeogenic mitochondrial enzymes: III. The conversion of α-ketoglutarate by bovine liver mitochondrial glutamate dehydrogenate and glutamate-oxäloacetate transaminase. *Arch. Biochem. Biophys.* 130: 468-477.

Freiden, C. 1959. Glutamic dehydrogenase III. The order of substrate addition in the enzymatic reaction. *J. Biol. Chem.* 234: 2891-2896.

Hutter, R. and J. A. DeMoss 1967. Organization of the tryptophan pathway: a phylogenetic study of the fungi. *J. Bacteriol.* 94: 1896-1907.

Jensen, R. A., D. S. Nasser, and E. W. Nester 1967. Comparative control of a branch point enzyme in microorganisms. *J. Bacteriol.* 94: 1582-1593.

Jensen, R. A. 1970. Taxonomic implications of temperature dependence of the allosteric inhibition of 3-deoxy-D-arabinoheptulosonate 7-phosphate synthetase in bacillus. *J. Bacteriol.* 102: 489-497.

Jordan, E. M. and S. Raymond 1967. Multiple analyses on a

single gel electrophoresis preparation. *Nature* 216: 78-80.

LéJohn, H. B. 1968. Unidirectional inhibition of glutamate dehydrogenase by metabolites. *J. Biol. Chem.* 243: 5126-5131.

LéJohn, H. B., I. Suzuki, and J. A. Wright 1968, Glutamate dehydrogenases of *Thiobacillus novellus:* kinetic properties and a possible control mechanism. *J. Biol. Chem.* 243: 118-128.

LéJohn, H. B. and S. Jackson 1968. Allosteric interactions of a regulatory nicotinamide adenine dinucleotide-specific glutamic dehydrogenase from *Blastocladiella*. *J. Biol. Chem.* 243: 3447-3457.

LéJohn, H. B., S. G. Jackson, G. R. Klassen, and R. V. Sawula 1969. Regulation of mitochondrial glutamic dehydrogenase by divalent metals, nucleotides, and α-ketoglutarate: correlations between the molecular and kinetic mechanisms, and the physiological implication. *J. Biol. Chem.* 244: 5346-5356.

LéJohn, H. B., and R. M. Stevenson 1970. Multiple regulatory processes in nicotinamide adenine dinucleotide-specific glutamic dehydrogenases: catabolite repression: nicotinamide adenine dinucleotide phosphate, reduced nicotinamide adenine dinucleotide phosphate, and phosphoenolpyruvate as activators: allosteric inhibition by substrates. *J. Biol. Chem.* 245: 3890-3900.

LéJohn, H. B., R. M. Stevenson, and R. Meuser 1970. Multivalent regulation of mitochondrial glutamic dehydrogenase from fungi: effects of adenylates, guanylates, and acyl coenzyme A derivatives. *J. Biol. Chem.* 245: 5569-5576.

LéJohn, H. B. 1971a. Enzyme regulation, lysine pathways and cell wall structures as indicators of major lines of evolution in fungi. *Nature* 231: 164-168.

LéJohn, H. B. 1971b. D(-)lactic dehydrogenases in fungi: kinetics and allosteric inhibition by GTP. *J. Biol. Chem.* 246: 2116-2126.

Lowry, O. H., N. J. Rosebrough, A. L. Farr, and R. J. Randall 1951. Protein measurement with the Folin phenol reagent. *J. Biol. Chem.* 193: 265-275.

Markert, C. L. and F. Møller 1959. Multiple forms of enzymes: tissue, ontogenetic, and species specific patterns. *Proc. Nat. Acad. Sci.* U. S. A. 45: 753-763.

Rutter, W. J. 1965. "Enzymatic homology and analogy in phylogeny" In: *Evolving genes and Proteins* (ed. Bryson, V., and Vogel, H. J.) pp. 279-291. Academic Press, New York.

Stachow, C. S. 1965. "Glutamate dehydrogenase: kinetic analysis and enzyme regulation" Doctoral thesis, University of Manitoba.

Stevenson, R. M., and H. B. LéJohn,1971. Glutamic dehydrogen-

ases of Oömycetes: kinetic mechanism and possible evolutionary history. *J. Biol. Chem.* 246: 2127-2135.
Smith, E. L. 1972. "Evolution of enzymes" In: *The Enzymes* (third edition), ed. Boyer, P. D. vol. 1, 267-341. Academic Press, New York.
Wang, H. S. and H. B. LéJohn 1974. Analogy and homology of dehydrogenases of Oömycetes. I. Regulation of glutamic dehydrogenases and isozyme patterns. *Can. J. Microbiol.* 20: 567-574.
Wang, H. S. and H. B. LéJohn 1974. Analogy and homology of dehydrogenases of Oömycetes. II. Regulation by GTP of D(-)lactic dehydrogensaes and isozyme patterns. *Can. J. Microbiol.* 20: 575-580.
Wang, H. S. and H. B. LéJohn 1974. Analogy and homology of dehydrogenases of Oömycetes. III. Isozyme patterns of malic and isocitric dehydrogenases. *Can. J. Microbiol.* 20: 581-586.
Watts, R. L. 1970. "Proteins and plant phylogeny" In: *Phytochemical Phylogeny*, ed. Harborne, J. B. pp. 145-178. Academic Press, New York.
Watts, D.C. 1971. "Evolution of phosphagen kinases" In: *Molecular Evolution: Biochemical Evolution and the Origin of Life*. vol. 2. ed. Schoffeniels, E. pp. 150-173. North-Holland Publishing Co. Amsterdam.

BIOCHEMICAL GENETICS OF *ARABIDOPSIS* ACID PHOSPHATASES: POLYMORPHISM, TISSUE EXPRESSION, AND GENETICS OF AP_1, AP_2, AND AP_3 LOCI

M. JACOBS AND F. SCHWIND
Laboratory of Plant Genetics, Vrije Universiteit Brussel,
B - 1640 Sint-Genesius-Rode, Belgium

ABSTRACT. Following gel electrophoresis, three distinct groups of acid phosphatase enzymes (AP_1, AP_2, and AP_3) have been identified in leaf extracts of *Arabidopsis, thaliana*. The AP_1 group includes three electrophoretic variants - a fast (F), an intermediate (I), and a slow (S) type. The genetic analysis shows that the AP_1 variants are under the control of three codominant alleles at the AP_1 locus. Each allele specifies a series of four isozymes and isozymes belonging to the I series are found to correspond in migration rate with some bands of the F series and of the S series. The distribution of staining intensity between the bands in heterozygotes may indicate a dimeric structure of the AP_1 enzymes. The use of electrophoretic and activity mutants as well as biochemical data leads to the conclusion that the multiple molecular forms within a series are all products of a single gene and that the AP_1 isozymes are due to variation in charge rather than in size. The AP_2 group comprises two bands which migrate rapidly towards the anode. Three electrophoretic variants with a slow, an intermediate, and a fast migration rate have been identified to be under the control of the AP_2 locus with three codominant alleles, AP_2^F, AP_2^I, and AP_2^S. There is no evidence for the formation of hybrid enzymes in heterozygotes. The AP_3 set is represented by the most anodal band. A slow variant has been discovered and in heterozygotes between a slow and a fast type a three band pattern is obtained which is consistent with the hypothesis that the AP_3 isozymes exist as dimers. The AP_3 locus comprises two allelic forms AP_3^F and AP_3^S. Organ specific differences in acid phosphatase pattern are mainly quantitative, except in seeds where other isozymes than those specified by the

AP_1 gene seem to be operative during this stage of development.

INTRODUCTION

There are several plant species within which varying numbers of molecular forms of acid phosphatases (orthophosphoric monoesterphosphohydrolase, E.C. 3.1.3.2.) have been described (Dorn,1965; Brown and Allard, 1969; Efron, 1970; Endo et al., 1971; Te Niejenhuis, 1971; Cherry and Ory, 1973; Baker and Takeo, 1973). Despite numerous biochemical investigations into substrate specificity (Ikawa et al., 1964), action of inhibitors, isolation procedures, and kinetic studies of these enzymes (Alvarez, 1961; Verjee, 1969; Yoshida and Tamiya, 1971; Horgen and Horgen, 1974) there are very few explanations for the basis of this heterogeneity. The precise physiological roles of these enzymes remains largely unknown. They have been implicated in various aspects of cell metabolism such as transport of sugars across membranes (Sauter, 1972), differentiation of plastid structure (Mlondianowski, 1972), senescence in fruits (De Leo and Sacher, 1970), and in catabolic processes in the seed (Rychter et al., 1972). They have been localized both in the cell wall and in the cytoplasm with varied subcellular distribution: vacuoles, chloroplasts, plant lysosomes, and along the plasmalemma. (Bieleski, 1973; Hanower and Baxoxowska, 1973; Klis et al., 1974).

This paper describes a genetical approach to the control of acid phosphatase isozymes in *Arabidopsis thaliana*. This species is a very convenient tool for combined genetic and biochemical analysis of isozymes. The main advantages of this self-fertile crucifer are a short life cycle, the possibility of aseptic culture on conditioned media, clearcut physiological stages of development, a low chromosome number (n=5), and the existence of numerous geographical races with distinct developmental features. Natural populations display a high range of polymorphism for various enzymes (Jacobs, unpublished).

This report deals with the variation observed in acid phosphatase patterns, the changes occuring at defined stages of development together with the study of the genetic system which controls this polymorphism. A preliminary account of some of these results has been published (Jacobs and Schwind, 1973).

METHODS

Preparation of extracts. Various geographical races of *Arabidopsis thaliana* were used in these experiments. The tested plants were grown in a greenhouse or in controlled growth chamber (24° ± 1° C, 10,000 lux, continuous illumination). Plant material was homogenized (1g plant/1 ml buffer), in 0.1M Hepes buffer, pH 7.4 containing 5 x 10^{-4}M dithiothreitol and 10^{-3}M EDTA. After centrifugation for 15 min at 17000 x g. at 4° C, the supernatants, were used for electrophoresis ahd enzyme assays. For analysis of individual plants, the crude extract of 2-4 rosette leaves, obtained by squashing them in a drop of buffer, was directly absorbed into the filter paper sections (Whatman 3MM) and inserted in the starch gels. By the latter method it is possible to determine the genotype of one plant and still to keep it for further experiments. All operations were performed at 4°, according to the method of Schwartz (1960) in a sodium borate buffer pH 8.3. Electrophoresis was carried out at 4° C using 10v/cm during 4-4,5 hr. Acid phosphatase bands were located by flooding the gel slices with a solution of 0.05 M acetate buffer, pH 4.8, containing 1mg/ml sodium *d*-naphtylphosphate and 1mg/ml Fast Garnet GBC, for 15-30 min at room temperature. Areas of phosphatase activity appeared as brown-purple bands (Allen et al., 1963).

Polyacrylamide gel electrophoresis. Vertical gel slab electrophoresis of the phosphatases was performed in a 10, 8, 4 1/2 % gradient polyacrylamide gel, using a tris-citrate gel buffer pH 9.0 and a Tris-borate electrode buffer 0.065 M, pH 9.0. Electrophoresis was performed at a constant current of 30 mA per cell during 1.30 hr. After separation the gels were washed in 0.1 M acetate buffer pH 4.8 and stained at room temperature for 30-60 min with the same solution used for starch gels.

Isoelectric gel focusing. Isoelectric focusing was performed in a thin layer of acrylamide following the procedure of Awdeh et al., (1968). Ampholine carrier ampholytes were of pH range 3-10.

Assay for enzyme activity. The reaction mixture contained p-nitrophenylphosphate 0.005 M as substrate, 0.15 ml plant extract and 2 ml acetate buffer 0.05 M, pH 4.8. This mixture was incubated at 25° C, for periods up to 10 min, and 2 ml of NaOH 1 M was added to stop the enzyme reaction. The increase in absorbance due to the product p-nitrophenol was measured spectrophotometrically at 400 nm. One enzyme unit is defined as the amount of enzyme that liberated 1 μM p-nitrophenol.

Gel filtration. Column of 30 cm x 1.5 (Pharmacia K_{15}) was packed to a height of 25 cm with P_{150} (Biogel) or G 200 (Sephadex) suspended in a 0.01 M Tris - HCl buffer at pH 7.4. Sample volumes of 1-2 ml were applied to the column and were eluted at 4°C with the same buffer at a flow rate of 20 ml/hr. 2.5 ml fractions were collected and assayed for enzyme activity at 400 nm and protein at 280 nm.

Neuraminidase treatment. Leaf extracts were incubated at room temperature in 0-1 M acetate buffer, pH 4.8, with 500 μg/ml of neuraminidase *Clostridium perfringens* type VI (0.22 units /mg, Sigma) for 30 min. 1 hr. and 1.30 min. All aliquots were subjected immediately to electrophoresis.

RESULTS AND DISCUSSION

POLYMORPHISM

Three electrophoretically distinct groups of acid phosphatase (AP_1, AP_2, and AP_3) have been found in *Arabidopsis*. The characteristics of the AP_1 set have been principally identified after starch gel electrophoresis. AP_2 and AP_3 sets have been clearly resolved by the use of flat bed gel acrylamide electrophoresis. These groups differ in their relative migration rates towards the anode and in their levels of activity.

The AP_1 pattern shows four bands (numbered from 1 to 4 starting with the least anodal) and contains about 70% of the activity of acid phosphatase present in the leaves. The intensity of the AP_1 bands is graded so that band 3 has the highest activity, bands 1 and 2 a lower comparable activity, and band 4 is only slightly active. Three types of electrophoretic variants of acid phosphatase were discovered among 115 geographical races, originating from the Laibach collection of natural races (Robbelen, 1965) as well as from various botanical gardens. We have designated these types as S (slow), I (intermediate), and F(fast) in order of electrophoretic mobility with respect to the anode (Fig. 1). Each electrophoretic variant contains a set of four bands which in comparison to each other are displaced in a parallel fashion from S to I and from I to F towards the anode. The value of this displacement is twice the distance between two successive bands so that the patterns overlap between S and I types and I and F types.

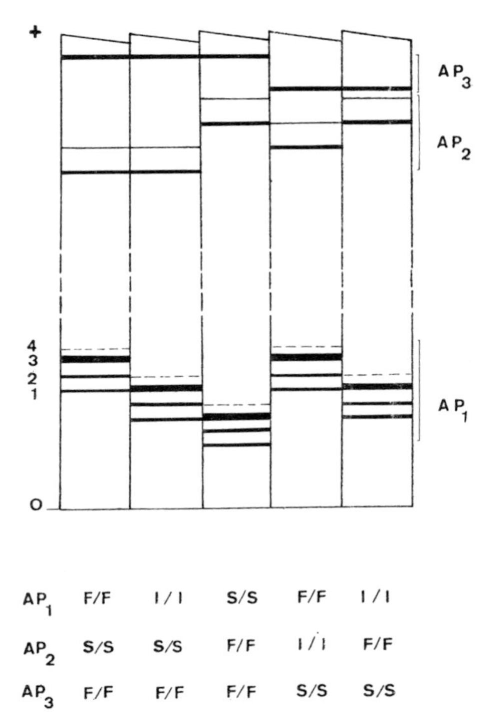

Fig. 1. Schematic representation of the three acid phosphatase systems AP_1, AP_2, and AP_3 in *Arabidopsis thaliana*. AP_1 bands are numbered from 1 to 4 in regard to their migration rate towards the anode. In each system, the F/F, I/I, or S/S genotypes indicate respectively a fast, intermediate or slow variant. + indicates the anode, 0 the origin.

A second acid phosphatase group (AP_2) has been identified on polyacrylamide gel. This group which migrates more rapidly towards the anode than the AP_1 bands consists of two isozymes and three distinct phenotypes have been recognized: a slow migrating type (S), an intermediate type (I) represented in only two races and a fast type (F) (Fig. 1). No correlations could be found between the variation observed within the AP_1 and AP_2 systems.

A third acid phosphatase group is represented by the most anodal band. This form is the only one among the acid phosphatase isozymes which does not show activity when stained in a buffer pH 6.5 or above. One slow variant has been identified in a few geographical races (Fig. 1).

These three groups of acid phosphatases can also be distinguished after thin-layer acrylamide isoelectric focusing using pH 3-10 carrier ampholytes. The AP_1 group is comprised of 6 bands with high isoelectric points in the range pH 6.8-5.9. The second distinct group (AP_2) is composed of 3-4 isozymes in the range pH 5.3-4.8 and finally a strongly acid band with an isoelectric point of pH 4.3 is observed.

Table 1 shows the distribution of the AP_1, AP_2, and AP_3 phenotypes among 115 different geographical races spreading from Sweden to Portugal and from Siberia to England.

TABLE 1
Distribution of acid phosphatases phenotypes of *Arabidopsis* among 115 geographical races (%)

Isozyme System	Acid Phosphatase Phenotypes		
	S	I	F
AP_1	3.5	73.9	22.6
AP_2	64.6	1.8	33.6
AP_3	95.6	-	4.4

GENETICS OF THE ACID PHOSPHATASE VARIANTS

To determine the mode of inheritance of the phosphatase variants of these systems, crosses were made between a few races representative of the defined types.

In the case of the AP_1 group, the pattern of the three possible heterozygous hybrids (S/I, I/F, S/F) shows both group of isozymes each of which comes from the homozygous parents. On a zymogram of a cross between S and I type (Fig. 2), 6 bands are present instead of the expected eight parental bands which can be explained by overlapping between bands 3^S and 1^I and between bands 4^S and 2^I. A hybrid between F type and S type displays 8 bands on the gel as there is no overlapping between the two parental sets (Fig. 3).

Hybrid enzymes with electrophoretic mobility differing from those of the parental bands do not appear in heterozygotes. It seems unlikely that the isozymes are composed of monomers which could randomly associate to generate the active enzymes. However, the distribution of activity between the bands of the heterozygote does not correspond to that expected from

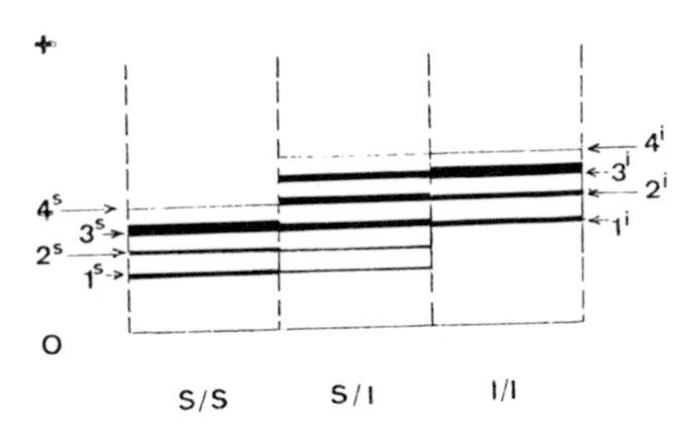

Fig. 2. Schematic picture of a AP_1 S/I heterozygote in comparison with the S/S and I/I homozygotes. The hybrid has six isozymes. The bands of the parental types are numbered from 1^S to 4^S for S/S and from 1^I to 4^I for I/I (see Table 2). + indicates the anode, 0 the origin.

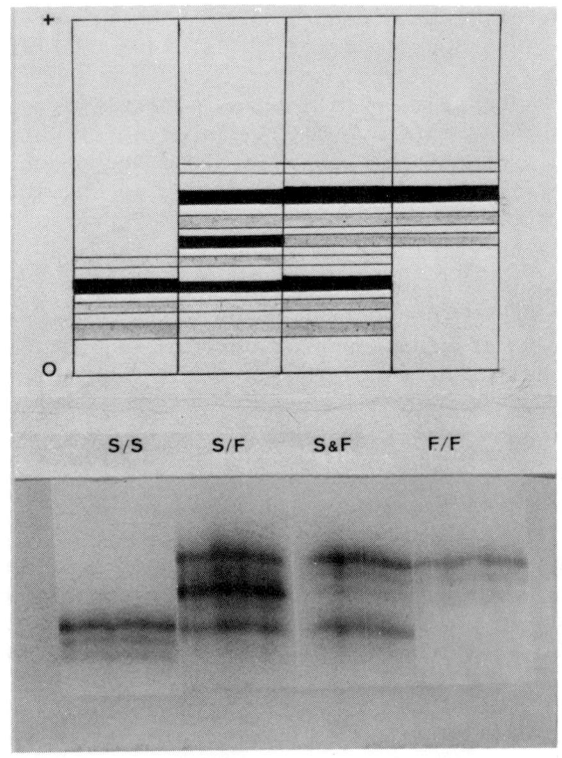

Fig. 3. Picture and schematic representation of a AP_1 S/F hybrid in comparison with a 1:1 mixture of extracts from

Legend for Fig. 3, continued: S/S and F/F types and their respective parents S/S and F/F. The activity levels of the acid phosphatase isozymes are indicated by the relative density of the dots; a high enzyme activity is represented by a solid bar. + indicates the anode, 0 the origin.

---

the addition of the parental bands. This pattern can be obtained with mixtures in a 1:1 ratio of the F and S types (see Fig. 3).

The particular distribution of activity in heterozygotes can be considered as indication of a dimeric constitution for the enzyme, with a restriction in the dimerization process to subunits belonging to the same corresponding bands in the three migration types S, I, and F. Table 2 gives an example of a possible subunit composition of AP_1 isozymes in the case of two heterozygotes S/I and S/F.

The intensity distribution of the bands in heterozygotes suggests that each subunit retains its characteristic activity in the heterodimer. Thus here the combinations 3^S3^I or 3^S3^F give the densest bands, the bands 1^S1^F, 1^S1^I, 2^S2^F, and 2^S2^I are less dense and the bands 4^I4^S, 4^S4^F are the least intense.

In the F_2 generation, all three possible types (the two parental and their corresponding heterozygote) were found in progenies in proportions showing a close agreement with the expected 1: 2: 1 ratio. The data presented in Table 3 are consistent with the hypothesis that the AP_1 variants are controlled by three alleles at one locus (AP_1^S, AP_1^I, AP_1^F) acting without dominance.

The AP_2 system is characterized by the following properties. There is an absence of correlation between the migration rates of AP_1 and AP_2 enzymes. Fast AP_1 isozymes as well as slow and intermediate types can be associated with a S or F AP_2 type. This indicates that AP_1 and AP_2 bands are under the control of non-allelic genes. The heterozygous condition is characterized by the presence of both parental bands (fig. 4). There is no evidence for the formation of hybrid enzyme types. Crosses between races representative of the two main types S and F lead to the conclusion that the two bands involved in this system are allelic and co-dominant. Segregations for AP_2 phenotypes are presented in Table 4.

Polymorphism of AP_3 is caused by two alleles S and F of the AP_3 locus which determine the primary structure of the polypeptides S and F. Presumably, these aggregate to form enzymatically active dimers immediately after they have been

TABLE 2
Proposed subunit composition in AP_1 homozygotes and heterozygotes

Bands	Genotypes I/I	S/I	S/S	S/F	F/F
4^F				4^F4^F	4^F4^F
3^F				3^F3^F	3^F3^F
4^I-2^F	4^I4^I	4^I4^I		$2^F2^F+4^F4^S$	2^F2^F
3^I-1^F	3^I3^I	$3^I3^I+4^I4^I$		$1^F1^F+3^F3^S$	1^F1^F
2^I-4^S	2^I2^I	$2^I2^I+4^S4^S+3^S3^I$	4^S4^S	$4^S4^S+2^F2^S$	
1^I-3^S	1^I1^I	$1^I1^I+3^S3^S$	3^S3^S	$3^S3^S+1^F1^S$	
2^S		$2^S1^S+1^I1^S$	2^S2^S	2^S2^S	
1^S		1^S1^S	1^S1^S	1^S1^S	

TABLE 3
Segregation observed in the progeny of crosses made to determine the inheritance of AP_1 variation

Parent Female		Male	AP_1 patterns of offspring F	I	S	Hetero-zygote	Total	Probability of X^2 value
F	X	F	14	0	0	0	14	-
F	X	I	39	32	0	74	145	0.70>P>0.60
I	X	F	27	30	0	53	110	0.60>P>0.50
I	X	I	0	22	0	0	22	-
I	X	S	0	16	15	36	67	0.90>P>0.80
S	X	F	125	0	130	233	488	0.70>P>0.60
S	X	I	0	24	25	59	108	0.90>P>0.80

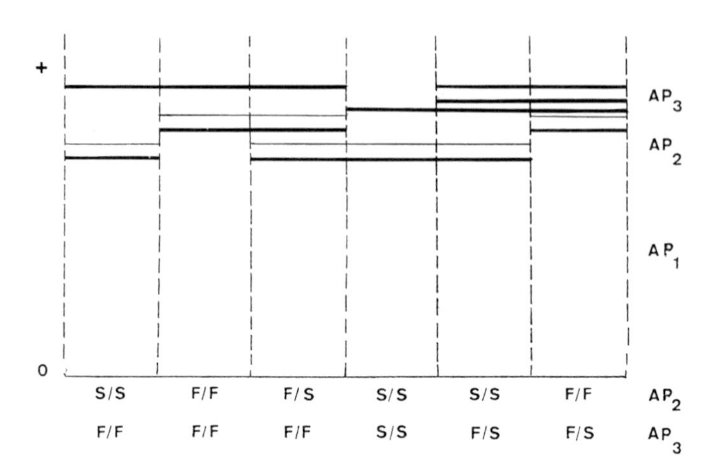

Fig. 4. Schematic representation of various combinations of AP_2 and AP_3 genotypes after acrylamide gel electrophoresis. Notice that F/S AP_2 heterozygotes show both parental bands and the 3-band pattern of the F/S AP_3 hybrid due to the formation of a hybrid enzyme. + indicates the anode, 0 the origin.

TABLE 4
Segregation observed in the progeny of crosses made to determine the inheritance of AP_2 variation

Parent Female		Male	AP_2 patterns of offspring F	S	Heterozygote	Total	Probability of X^2 value
F	X	S	33	45	79	157	$0.50>P>0.30$
S	X	F	21	33	62	116	$0.30>P>0.20$

synthesized. Heterozygotes display a three band pattern composed of the parental homodimer bands and a heterodimer hybrid band (Fig. 4). Phenotypes resulting from the crosses between the variants are exactly predictable on the basis that the alleles are codominantly expressed (Table 5).

Linkage data between the AP_1, AP_2, and AP_3 loci have been gathered (Tables 6, 7, 8). In the three cases, comparison of the observed class and group frequencies and those expected on a zero linkage basis shows that close linkage between the loci is ruled out. AP_1 - AP_2 loci and AP_1 - AP_3 loci seem to segregate independently.

TABLE 5
Segregation observed in the progeny of crosses made to determine the inheritance of AP_3 variation

Female		Male	AP_3 patterns of offspring F	S	Heterozygote	Total	Probability of X^2 value
F	X	S	26	33	60	119	0.80>P>0.70

A possibility of linkage exists between AP_2 and AP_3 loci as a lack of group II genotypes (39 individuals) was found compared to group I plus group III (61 individuals). Test crosses are now under study.

ORGAN-SPECIFIC PATTERNS

Zymograms for the AP_1 isozymes have been compared for various organs. Activity in the AP_1 group is present in all tissues, in the same 4-band pattern, except for the seeds. However the four isozymes are represented in varying concentrations in roots; bands 1 and 2 are very weak, band 3 is particulary intense in flower homogenates. The acid phosphatase isozyme pattern of extracts from seeds appears very distinct and comprises 8-9 anodal bands and two cathodal bands.

Seeds hydrated for 24 hr show a similar pattern to the one observed for dry seed extracts. A progressive reduction of the number and activity of these seed bands is observed 40-50 hr after sowing on moistened filter paper. These progressive changes lead finally to a typical leaf pattern with the emergence of the cotyledons. This emergence is accompanied by an increase in activity of a defined band which corresponds to band 3 of the leaf isozyme set (Fig. 5). The seed enzymes produced in the presence of AP_1^F, AP_1^I, or AP_1^S- alleles exhibit electrophoretic mobilities which do not parallel those exhibited by the leaf extracts of the same genotypes. This lack of parallelism may suggest absence of genetic relationships between the two isozyme systems. There is however a correspondence between leaf bands and some of the seed bands.

The changing pattern during development may be interpreted as evidence for differential timing of gene expression correlated with physiological changes during germination as observed in other plants (Presley et al., 1965; Johnson et al., 1973). In tissue such as *Arabidopsis* seeds, different

TABLE 6

F_2 linkage data between AP_1 and AP_2 loci

Cross	Group	Genotypic classes	No[a]	Frequency observed		Zero linkage probability
				class	group	
$AP_1^I/AP_1^F, AP_2^S/AP_2^F$		$1^I/1^I, 2^S/2^S$	13	0.131		
self-pollinated		$1^I/1^I, 2^F/2^F$	6	0.061	0.263	
	I	$1^F/1^F, 2^S/2^S$	6	0.061		
		$1^F/1^F, 2^F/2^F$	1	0.010		
						P>0.80
		$1^I/1^I, 2^S/2^F$	16	0.161		
		$1^F/1^F, 2^S/2^F$	15	0.151	0.515	
	II	$1^I/1^F, 2^S/2^S$	12	0.120		
		$1^I/1^F, 2^F/2^F$	8	0.081		
	III	$1^I/1^F, 2^S/2^F$	22	0.222	0.222	
		total	99			

[a] Number of plants in each class

TABLE 7

F_2 linkage data between AP_1 and AP_3 loci

Cross	Group	Genotypic classes	No[a]	Frequency observed		Zero linkage probability
				class	group	
$AP_1^S/AP_1^I, AP_3^S/AP_3^F$	I	$1^S/1^S, 3^S/3^S$	5	0.0520		
self-pollinated		$1^S/1^S, 3^F/3^F$	6	0.0625	0.260	
		$1^I/1^I, 3^S/3^S$	6	0.0625		
		$1^I/1^I, 3^F/3^F$	8	0.0830		P>0.80
	II	$1^S/1^S, 3^S/3^F$	11	0.115		
		$1^I/1^I, 3^S/3^F$	9	0.094	0.469	
		$1^S/1^I, 3^S/3^S$	18	0.188		
		$1^S/1^I, 3^F/3^F$	7	0.073		
	III	$1^S/1^I, 3^S/3^F$	26	0.271	0.271	
		total	96			

[a] Number of plants in each class

TABLE 8

F_2 linkage data between AP_2 and AP_3 loci

Cross	Group	Genotypic classes	No[a]	Frequency observed		Zero linkage probability
				class	group	
$AP_2^S/AP_2^F, AP_3^S/AP_3^F$	I	$2^S/2^S, 3^S/3^S$	9	0.090		
self-pollinated		$2^S/2^S, 3^F/3^F$	7	0.070	0.280	
		$2^F/2^F, 3^S/3^S$	8	0.080		
		$2^F/2^F, 3^F/3^F$	4	0.040		0.10>P>0.05
	II	$2^S/2^S, 3^S/3^F$	11	0.110		
		$2^F/2^F, 3^S/3^F$	8	0.080	0.390	
		$2^S/2^F, 3^S/3^S$	10	0.100		
		$2^S/2^F, 3^F/3^F$	10	0.100		
	III	$2^S/2^F, 3^S/3^F$	33	0.330	0.330	
		total	100			

[a] Number of plants in each class

isozymes might be present in different parts of the seed, for example cotyledons versus embryo.

In relation to the occurence of cathodal bands in the dry seeds and in the early phase of germination, it is interesting to mention that faint cathodal bands showing phosphatase activity are obtained from leaf extracts if *Arabidopsis* plants are grown in submerged culture with a mineral medium including 2% glucose. Organic phosphates in the medium such as β-glycerophosphate are substrates for the acid phosphatases.

AP_1 GENE - AP_1 ISOZYMES RELATIONSHIPS

The problem of why in homozygous conditions the AP_1 gene controls a set of four bands in most of the plant organs remains to be answered. To obtain more critical evidence that a single gene was involved in the formation of different isozymes, we first tried to induce a mutation in this gene. Seeds of the race Wilna (which gives a AP_1^I pattern) were treated by ethyl methanesulfonate (80 mM-6 hr) and analyzed in the M_2 generation. In one out of 1500 we found a shift in the electrophoretic pattern from I to F and the occurrence of F/I hybrids. The mutant showed all the morphological and physiological characteristics of the race Wilna used in this experiment which excludes a possible contamination. The spacing between the isozyme bands in the set is not altered. Another type of mutant (mutant 96) discovered among previously isolated *Arabidopsis* morphological mutants was characterized by lack of AP_1 activity (Fig. 6). This mutation did not affect the isozymes controlled by the AP_2 and AP_3 loci (Fig. 6a). Thus a single mutation at the AP_1 locus eliminates all the bands simultaneously as has been reported for *E. coli* with alkaline phosphatase (Signer et al., 1961). This may serve as additional indication for the assumption that the same gene controls the synthesis of the AP_1 isozymes and it is unlikely that the three electrophoretic types S, I, and F are under the control of different closely linked non-allelic genes.

One possible explanation for this situation, the existence of aggregates, does not seem to fit the experimental data obtained by calculating the retardation coefficients according to Smithies (1962) using 10 and 15% gel concentrations. The enzyme forms inside the same group have the same "retardation coefficients" (43%). This was confirmed by gel filtration on Bio-gel P 150. The four bands elute in the same fractions suggesting that the enzyme was homogeneous with respect to molecular weight.

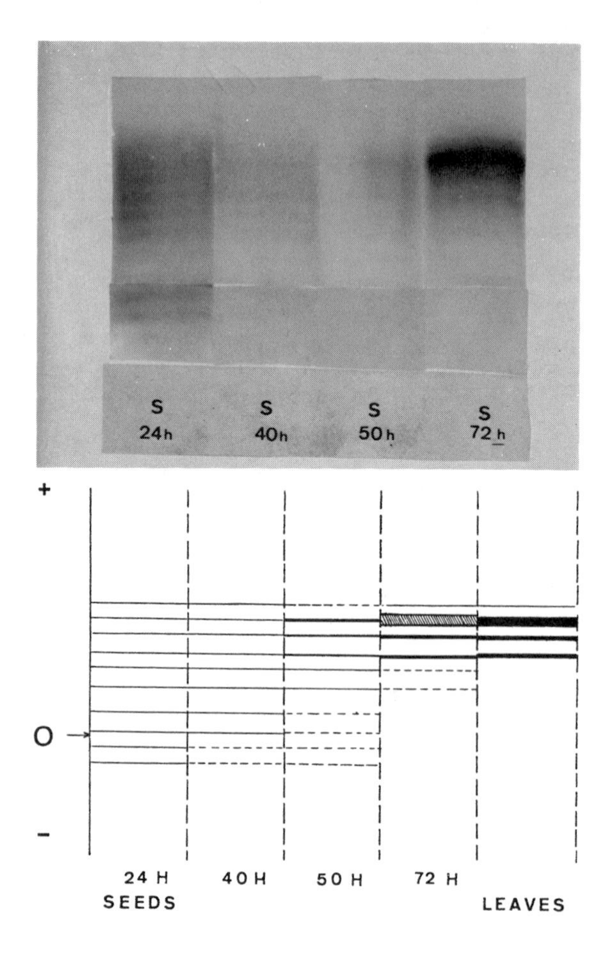

Fig. 5. Picture and drawing showing the acid phosphatase[1] pattern in seeds during germination. Starch gel electrophoresis was performed 24, 40, 50, and 72 hr. after sowing + indicates the anode, 0 the origin.

Another possibility may be that different bands in a set differ by the amount of charge groups conjugated to the protein e.g. polysaccharide residues. Prostatic acid phosphatase has been shown to be a glycoprotein with differing numbers of sialic acid residues (Smith and Whitby, 1971).

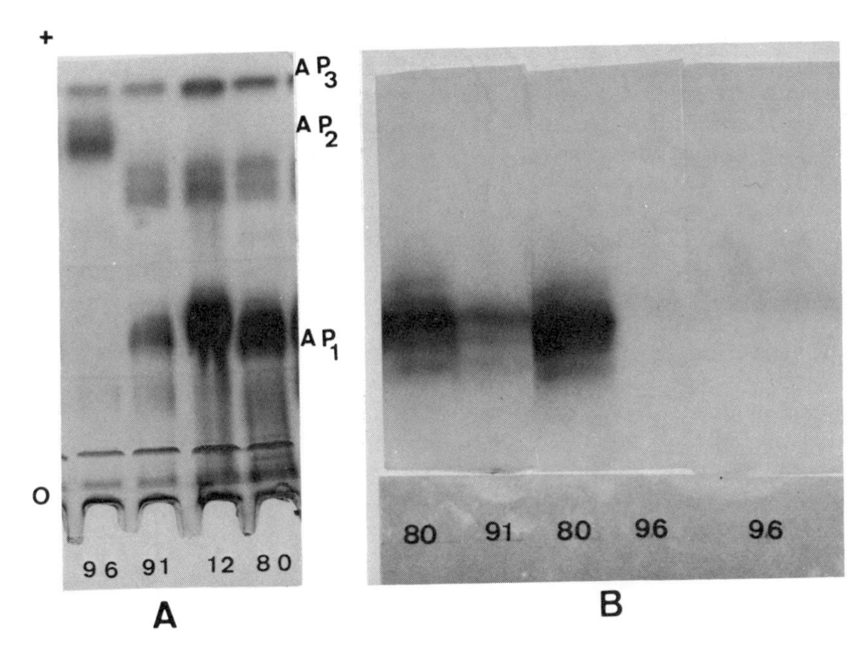

Fig. 6. Zymogram showing different levels of AP_1 activity in a few *Arabidopsis* races. 12 = F variant with a normal activity level, 80 = I variant with a normal activity level, 91 = I variant with a low activity level, 96 = I variant with a "null-activity" level. The gels have been stained for the same period of time.

A. AP_1, AP_2, and AP_3 patterns after acrylamide gel electrophoresis. Notice the unchanged activity of AP_2 and AP_3 sets in the AP_1 "null-activity" mutant (96)

B. AP_1 patterns after starch gel electrophoresis of races 80, 91, and 96.

+ indicates the anode, 0 the origin.

However, *Arabidopsis* extract incubated with neuraminidase does not show any conversion of acidic forms of acid phosphatase to more basic forms. Therefore the association of various numbers of sialic acid residues with *Arabidopsis* acid phosphatase does not account for the heterogeneity of the AP_1 enzyme.

We must also mention that when a defined band such as band 3 was cut out after electrophoresis on starch gel and re-run under the same conditions it again migrates as a single band and with its characteristic migration rate. Similar situations where a single allele determines the appearance of several electrophoretically distinct isozymes have been reported for red cell acid phosphatase (Fisher and Harris (1971) and esterase E_4 in maize (Harris, 1968).

As a point mutation in the gene which determines the structure of acid phosphatase 1 modifies the entire isozyme pattern, the bands may be formed as a result of epigenetic modifications. Secondary modifications of the enzyme molecule after synthesis could arise for example from cleavage of peptide bonds by proteases as reported by Gazith et al. (1968) for yeast hexokinase and by Gardner et al. (1971) for phytochrome. The existence of the same enzyme in various conformations has also been suggested as a possible explanation for several enzymes determined by a single allele (Kitto et al., 1968; Pahlich, 1972).

CONCLUSION

The results obtained suggest that the electrophoretic patterns of AP_1 isozymes are due to variation in charge rather than in size and that the bands within a set are not due to differences in the amino acid sequence of the originally synthesized enzyme. The different groups within an AP_1 series may differ in charge distribution with no appreciable differences in size.

AP_3 isozymes consist of two polypeptide chains which associate to form dimers as observed in the 3-bands pattern of the heterozygote $AP_3^S/AP_3^F/AP_1$ isozymes probably also behave as dimers on the basis of the relative distribution of intensity among the bands of the three hybrid types, AP_1^S/AP_1^I, AP_1^S/AP_1^F, AP_1^I/AP_1^F. In potato, three electrophoretically distinct molecular forms of acid phosphatase have the same molecular weight and all present a dimeric structure (Kubicz, 1973).

The partial purification of the different AP_1 isozymes should permit a detailed analysis of the physico-chemical differences between the isozymes. The knowledge of their intracellular localization might allow more evidence to be obtained about the physiological significance of the developmental changes in the molecular forms of acid phosphatase. In this perspective the null-activity mutant represents a critical tool to evaluate the role of the AP_1 enzymes in the development of the plant as indicated by Bell et al. (1972) for *Drosophila*.

ACKNOWLEDGEMENTS

The competent technical assistance of Mr. H. Van Hemelrijck is gratefully acknowledged. Part of this work was financially supported by the "Fonds voor Kollektief Fundamenteel Onderzoek" under the contract No. 10239. For critical reading of the manuscript we thank Dr. B. Harford.

REFERENCES

Allen, S.L., M.S Misch, and B.M. Morrison 1963. Genetic control of an acid phosphatase in *Tetrahymena:* formation of a hybrid enzyme. *Genetics* 48: 1635-1658.

Alvarez, E.F. 1961. The kinetics and mechanism of the hydrolysis of phosphoric acid esters by potato acid phosphatase. *Biochim. Biophys. Acta* 59: 663-72.

Awdeh, Z.L., A.R. Williamson, and B.A. Askonas 1968. Isoelectric focusing in polyacrylamide gel and its application to immunoglobulins. *Nature* 219: 66-67.

Baker, J.E. and T. Takeo 1973. Acid phosphatase in plant tissues I. Changes in activity and multiple forms in tea leaves and tomato fruit during maturation and senescence. *Plant and Cell Physiol.* 14: 459-71.

Bell, J.B., R.J. MacIntyre, and A.P. Olivieri 1972. Induction of null-activity mutants for the acid phosphatase -1 gene in *Drosophila melanogaster*. *Biochem. Genet.* 6: 205-216.

Bieleski, R.L. 1973. Phosphate pools, phosphate transport and phosphate avaibility. *Ann. Rev. Plant. Physiol.* 24: 225-52.

Brown, A.H.D. and R.H. Allard 1969. Inheritance of isozyme differences among the inbred parents of a reciprocal recurrent selection population of maize. *Crop. Sci.* 9: 72-75.

Cherry, J.P. and R.L. Ory 1973. Electrophoretic characterization of six selected enzymes of peanut cultivars. *Phytochemistry* 12: 283-9.

DeLeo, P. and J.A. Sacher 1970. Senescence: association of synthesis of acid phosphatase with banana ripening. *Plant Physiol.* 46: 208-11.

Dorn, G. 1965. Genetic analysis of the phosphatase in *Aspergillus nidulans*. *Genet. Res.* 6: 13-26.

Efron, Y. 1970. Tissue specific variation in the isozyme pattern of the AP_1 acid phosphatase in maize. *Genetics* 65: 575-583.

Endo, Toru, B.B. Shahi, and Chiang Pai 1971. Genetic convergence of the specific acid phosphatase zymograms in *Oryza sativa*. *Japan. J. Genetics* 46: 147-52.

Fisher, R.A. and H. Harris 1971. Studies on the separate isozymes of red cell acid phsophatase phenotypes A and B. II Comparison of kinetics and stabilities of the isozymes. *Ann. Hum. Genet.* 34: 439-448.

Gardner, G., C.S Pike, H.V. Rice, and W.R. Briggs 1971. "Disaggregation" Of phytochrome *in vitro*. A consequence of proteolysis. *Plant Physiol.* 48: 686-93.

Hanower, P. and J. Bazozowska 1973. Influence d'un choc hydrique sur l'activité de la phosphatase acid chez le Cotonnier (Gossypium). *Physiol. Veģ.* 11: 385-94.

Harris, J.W. 1968. Isozymes of the E_4 esterase in maize. *Genetics* 60: s-186 - s-187.

Horgen, I.A. and P.A Horgen 1974. Purification and properties of acid phosphatase I from the cellular slime mold *Polysphondylium pallidum*. *Canad. J. Biochem.* 52: 126-136.

Ikawa, T., K.Nisizawa,and T. Miwa 1964. Specificities of several acid phosphatases from plant sources. *Nature* 203: 939-40.

Jacobs, M. and F. Schwind 1973. Genetic control of acid phosphatase in *Arabidopsis thaliana*. *Plant Science Letters* 1: 95-104.

Johnson, C.B., B.R. Holloway, H. Smith, and D. Grierson 1973. Isoenzymes of acid phosphatase in germinating peas. *Planta* 115: 1-10.

Klis, F.M., R. Dalhuizen, and K. Sol 1974. Wall bound enzymes in callus of *Convolvulus Arvensis*. *Phytochemistry* 13: 55-57.

Kubicz A. 1973. Acid phosphatase III from potato tubers, molecular weight and subunit structure. *Acta Biochim. Polonica* 20: 223-229.

Lerner, H.R., A.H. Mayer, and E. Harel 1972. Evidence for conformational changes in grape catecholoxidase. *Phytochemistry* 11: 2415-2421.

Mlodianowski, F. 1972. The occurrence of acid phosphatase in the thylakoids of developing plastids in lupin cotyledons. *Z. Pflanzenphysiol.* 66: 362-5.

Pahlich, E. 1972. Sind die multiplen formen der glutamatdehydrogenase aus Erbsenkeimlingen Conformer? *Planta* 104: 78-88.

Presley, H. and L. Fowden 1965. Acid phosphatase and isocitritase production during seed germination. *Phytochemistry* 4: 169-76.

Robbelen, G. 1965. The Laibach Standard Collection of natural races. *Arabidopsis Inform. Serv.* 2: 36-47.

Rychter, A., K. Szkutnicka, and S. Lewak 1972. Acid phosphatase and the low temperature requirement of apple seed stratification. *Physiol. Veg.* 10: 671-6.

Sauter, J.J. 1972. Respiratory and phosphatase activities in contact cells of wood rays and their possible role in sugar secretion. *Z. Pflanzenphysiol.* 67: 135-42.

Schwartz, D. 1960. Electrophoretic and immunochemical studies with endosperm proteins of maize mutants. *Genetics* 45: 1419-24.

Signer, E., A. Torriani, and C. Levinthal 1961. Gene expression in intergeneric merozygotes. *Cold Spring Harbor Symp. on quantitative biology.* 26: 31.

Smith, J.K. and I.G. Whitby 1968. The heterogeneity of prostatic acid phosphatase. *Biochem. Biophys. Acta* 151: 607-18.

Smithies, O. 1962. Molecular size and starch gel electrophoresis. *Arch. Biochem. Biophys. Supp.* I: 125-131.

Te Nijenhuis, B. 1971. Estimation of the proportion of inbred seeds in Brussels sprouts hybrid seeds by acid phosphatase isozyme analysis. *Euphytica* 20: 498-507.

Verjee, Z.H.M. 1969. Isolation of three acid phosphatases from wheat germ. *J. Biochem.* 9: 439-44.

Yoshida, H. and N. Tamiya 1971. Acid phosphatases from *Fusarium moniliforme*. *J. Biochem.* 69: 525-34.

GENETIC STUDIES OF BACTERIAL ISOZYMES

JAMES N. BAPTIST, DIANE THOMAS,
MARY ANN BUTLER, and THOMAS S. MATNEY
The University of Texas System Cancer Center
M. D. Anderson Hospital and Tumor Institute and
The University of Texas Health Science Center
Graduate School of Biomedical Sciences
Texas Medical Center
Houston, Texas 77025

ABSTRACT. Two methods, based on data from zone electrophoresis and specific enzyme stains, were developed to identify electrophoretically distinguishable gene-products and to map the corresponding structural genes in enteric bacteria. The first method employed induced mutations which resulted in electrophoretically detectable enzyme variants (Shaw et al., 1973). Such variants are mapped by linkage as unselected markers following Hfr X F^- crosses. Their precise position is determined by linkage with known markers in P1 transduction. These methods have positioned the structural gene for malate dehydrogenase (*mdh*) at approximately 62.3 min. The second method is based on the construction of *Escherichia coli/Salmonella typhimurium* merodiploids. These are formed by introducing F'-episomes carrying various regions of the *E.coli* chromosome into *S.typhimurium*. Each interspecies F-merodiploid is tested for a number of isozymes whose separate species-forms are identifiable by starch gel electrophoresis. The malate dehydrogenase gene *(mdh)* was found to reside on F' 141. F' 152 carried the gene for phosphoglucomutase *(pgm)*, and the banding patterns in the hybrid were interpreted to indicate the possibility of non-identical subunits in the active form.

INTRODUCTION

At present, more than 460 genes have been located on the genetic map of *Escherichia coli* K12, and the map no longer contains any large blank spaces (Taylor and Trotter, 1972). However, the genes coding for many of the enzymes which produce clear activity bands on gels after zone electrophoresis are not yet represented on the K12 map. A possible reason for this situation is that many of these enzymes are essential in the metabolism of the cell and therefore defective mutations would be lethal. However, the fact that metabolically important enzymes are involved makes it all the more interesting to position their corresponding structural genes on the genomic map. For example, it would be interesting to know if the structural genes coding for portions of the citric acid cycle or the glycolytic pathway are clustered and thus perhaps subject to coordinate regulation. The gene map for *Salmonella typhimurium* (Sanderson, 1972) is similar to that of *E.coli* (Taylor and Trotter, 1972), and it is reasonable to assume that these common gene sequences are maintained by selective forces.

Lew and Roth (1971) suggested the use of F' episomes (autonomous F factors which carry segments of the bacterial chromosome) to determine the quaternary structure of enzymes and to learn the approximate map locations of their structural genes. They transferred an F' from *E.coli* into a strain of *S.typhimurium* to produce a partially diploid bacterium with two different genes for 6-phosphogluconate dehydrogenase (*gnd*). Since the enzyme types produced by those two genes were different in electrophoretic migration, they were distinguishable in the hybrid strain. In addition, a single hybrid enzyme band was found approximately midway between the parental types indicating that 6-phosphogluconate dehydrogenase contains two polypeptide subunits in these organisms. Following their suggestion, we have developed a set of inter-species merodiploid derivatives in which most of the *E.coli* genome has been introduced as F-merogenotes into *S.typhimurium*.

MATERIALS AND METHODS

The bacterial derivatives employed in this study are listed in Table 1 in the order of their appearance in the paper. All F-primes used in this study were derived in *E.coli* K12. The unnumbered F' *his*$^+$ also carries the *gnd*$^+$ locus (Lew and Roth, 1971) and was isolated in these laboratories (Matney et al., 1964). F' *his*$^+$ is essentially the same as F' 131 as shown in Figure 1. The remaining F' derivatives were described by Low (1972). They were obtained through the courtesy of Dr. Barbara J. Bachmann, Curator of the *E.coli* Genetic Stock Center at Yale University. The genomic segments carried by this set of F-merogenotes are displayed in Figure 1. They were transferred into derivatives of *S.typhimurium* carrying nutritional mutations which corresponded to wild-type alleles carried by the F'. Thus, rare F-ductants appearing in these restricted crosses could be selected as previously described (Goldschmidt et al., 1970). The larger F' particles are frequently lost during growth. Cells carrying F' particles may be constantly selected by growing the derivative on minimal

TABLE 1

RELEVANT GENETIC MARKERS IN DERIVATIVES OF *ESCHERICHIA COLI* K12 AND *SALMONELLA TYPHIMURIUM* LT-2

UTH No.*	Electrophoretic Markers	Nutritional Markers
6402	F⁻ *gnd-6390·C***	F⁻ *hisG, argG, metB, leu-6*
6490	F' *gnd⁺·C/gnd-6390·C*	F' *his⁺/hisG; argG, metB, leu-6*
4349	F' *gnd⁺·C/(Δgnd)·C*	F' *his⁺/hisG-EΔ*
4349/4905	F' *gnd⁺·C/gnd⁺·S*	F' *his⁺/hisG-EΔ*
4905	F⁻ *gnd⁺·S*	F⁻ *hisG-EΔ*
6653	F' 152 *pgm⁺·C/pgm⁺·S*	F' 152 *nadA⁺/nadA; met, tyr, try*
6581	F' 129 -- */pgm⁺·S*	F' 129 *his⁺/his; try*
6677	F' 140 -- */pgm⁺·S*	F' 140 *argG⁺/argE***; tyr, ser*
6700	F' 254 -- */pgm⁺·S*	F' 254 *lac⁺/Δlac; his, try*
475	F⁻ *pgm⁺·C*	F⁻ *argG, metB, leu-6, hisG*

*UTH designates The University of Texas/Houston stock culture collection. Symbols indicate the mating types, F⁻ and F' (presence of F-merogenote whose mutational content preceeds /); the nutritional requirements for histidine (*his*), arginine (*arg*), methionine (*met*), leucine (*leu*), nicotinic acid (*nad*), tyrosine (*tyr*), serine (*ser*), tryptophan (*try*); the inability to utilize lactose (*lac*); and structural loci for the enzymes 6-phosphogluconate dehydrogenase (*gnd*) and phosphoglucomutase (*pgm*); Δ signifies genetic deletion.

***C* = *E.coli*; *S* = *Salmonella typhimurium*.

****argE* in *Salmonella* corresponds to *argG* in *E.coli* (Sanderson, 1972).

media which is devoid of the supplement that corresponds to the wild-type allele carried on the F'. Thus, if a daughter cell fails to inherit the F', it will no longer be able to grow without the supplement. *S.typhimurium* derivatives carrying specific F' particles were tested genetically as F' donors by back crossing against nutritional mutants of *E.coli* K12 strains, which were F-duced with low efficiency due to restriction, and against restrictionless *E.coli* C (nutritional

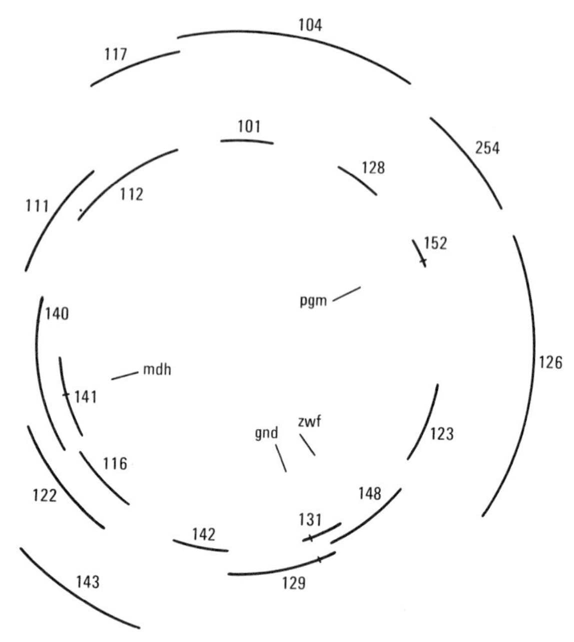

Fig. 1. A set of F' factors. The approximate portion of the *E.coli* chromosome believed present on each is indicated by an arc where the entire chromosome would be a complete circle (Low, 1972). Enzymes found in this study on each F' factor are indicated. F' factors 140 and 148 have internal deletions which are not indicated.

---

mutants which could be F-duced with high efficiency).

As noted in Table 1, the electrophoretic marker allele is followed by *C* to denote *E.coli* origin and by *S* to designate *S.typhimurium*. Thus, UTH 6653 is a *S.typhimurium* strain that carries the F' 152 derived in *E.coli*. This interspecies hybrid carries the wild-type *E.coli* allele for phosphoglucomutase in the F' (*pgm*$^{+}$· *C*) and the *Salmonella* counterpart in the chromosome (*pgm*$^{+}$· *S*). In this case the nutritional markers in the right half of Table 1 indicate that the entry of F' 152 which carried the *nadA*$^{+}$ allele into the *Salmonella* bacterium conferred upon it the ability to grow on a minimal agar without nicotinic acid, provided of course that the medium was still supplemented with methionine, tyrosine, and tryptophan.

The mutant with the electrophoretically altered 6-phosphogluconate dehydrogenase, *gnd-6390*, was isolated from a population of UTH 475 cells which survived treatment by the chemical mutagen, N-methyl-N'-nitro-N-nitrosoguanidine (MNNG) and formed colonies on nutrient agar plates. The UTH 6390 isolate also carried a mutation, *phe-6390*, that invoked a requirement for phenylalanine. The two new mutations had been induced concomitantly by MNNG since all of the cells in the original clone contained them. The procedures for the induction and detection of such electrophoretic variants has been described in detail (Shaw et al., 1973). A mutant of UTH 6390, resistant to T6 virulent bacteriophage, was isolated and numbered UTH 6402. The techniques for starch gel electrophoresis have been described in detail by Baptist et al. (1969).

RESULTS

Both of the techniques described here rely on the use of zone electrophoresis of enzymes to identify allelic differences. The first method relies on electrophoretic variants which are induced and isolated in the laboratory as described by Shaw et al. (1973). We have isolated mutants which produce electrophoretic variants for the *E.coli* enzymes: malate dehydrogenase (*mdh*), glyceraldehyde-3-phosphate dehydrogenase (*gpd*), esterase (*est*), and 6-phosphogluconate dehydrogenase (*gnd*). The gene for 6-phosphogluconate dehydrogenase had already been mapped at 39 min. in *E.coli* (Garrick-Silversmith and Hartman, 1970), but we were able to confirm the position with an electrophoretic variant. The mapping of *mdh* is now complete. Bacteriophage P1 transduction experiments placed the gene within 1.3 min. of *argG* (61 min.), and Hfr X F^- crosses have shown the gene sequence to be *argG-mdh-metB*. Therefore, the map position is approximately 62.3 min. Thus, we were able to predict that mdh^+ should appear on F' 140 and F' 141. Transfer of the F' 141 episome into *Salmonella* did produce a strain with three electrophoretically separable bands of enzyme activity as expected. However, F' 140 apparently has undergone an internal deletion so that while still carrying $argG^+$ it no longer carries mdh^+. The mapping of the other two enzyme variants, glyceraldehyde-3-phosphate dehydrogenase and esterase is in progress.

Figure 2 shows the staining patterns for the 6-phosphogluconate dehydrogenase isozyme types. The first channel shows the slower migrating mutant form, *gnd-6390·C*. When this mutant was infected with a F' carrying the wild-type allele, $gnd^+ \cdot C$, the resulting diploid yielded the three banded pattern shown in channel 2. The slowest band matches that of the mutant. The fastest band matches that of the wild-type shown in channel 3. The middle band is in the position expected for a hybrid, containing one wild-type monomer and one mutant monomer. When the same F' was transferred into *S.typhimurium*,

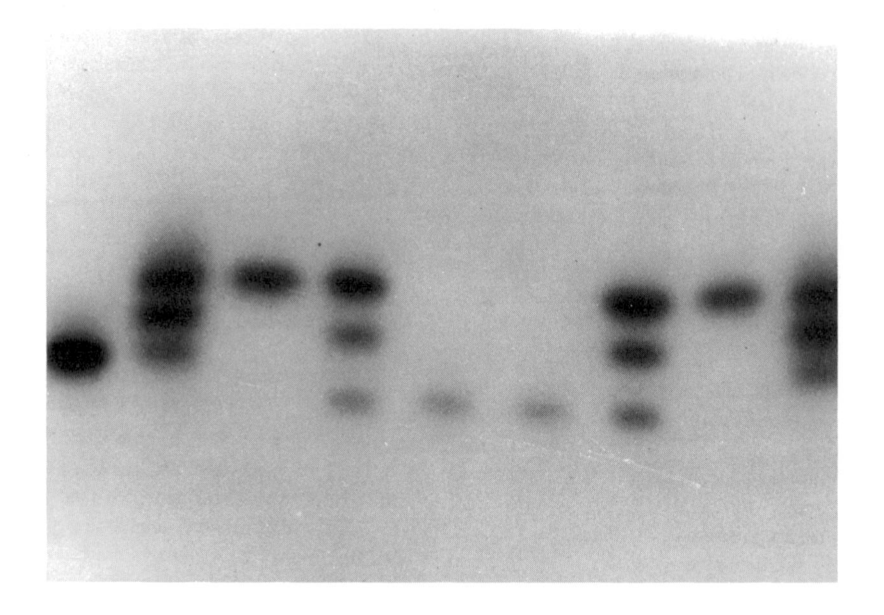

Fig. 2. A starch gel stained for 6-phosphogluconate dehydrogenase. The extracts used, starting from left, are: (1) 6402, (2) 6490, F' his^+, (3) 4349, (4) 4349/4905, F' his^+, (5) 4905, (6) 4905, (7) 4349/4905, F' his^+, (8) 4349, and (9) 6490, F' his^+. The anode is at the top.

the 3-banded pattern in channel 4 resulted. Again the top band conformed to the position of $gnd^+ \cdot C$, the bottom band to $gnd^+ \cdot S$ (as evidenced by channels 5 and 6) while the central band represented the interspecies hybrid form of the enzyme. Channels 7, 8, and 9 are repeats of channels 4, 3, and 2 respectively.

Phosphoglucomutase is another enzyme which had already been mapped in *E.coli* at 15 min. (Adhya and Schwartz, 1971). Therefore, we expected it to occur on F' 152 (Low, 1972). Figure 3 shows the result obtained when F' 152 was introduced into a strain of *S.typhimurium*. The first channel contained extract from the partially diploid derivative and produced five bands of activity. The next three channels contained extracts from partially diploid derivatives whose F-merogenote did not carry the $pgm^+ \cdot C$ gene. They produced only the *Salmonella*-type of enzyme, and it appeared at the same position noted in haploid controls in other experiments. The fifth channel contained extract from haploid *E.coli*. The position

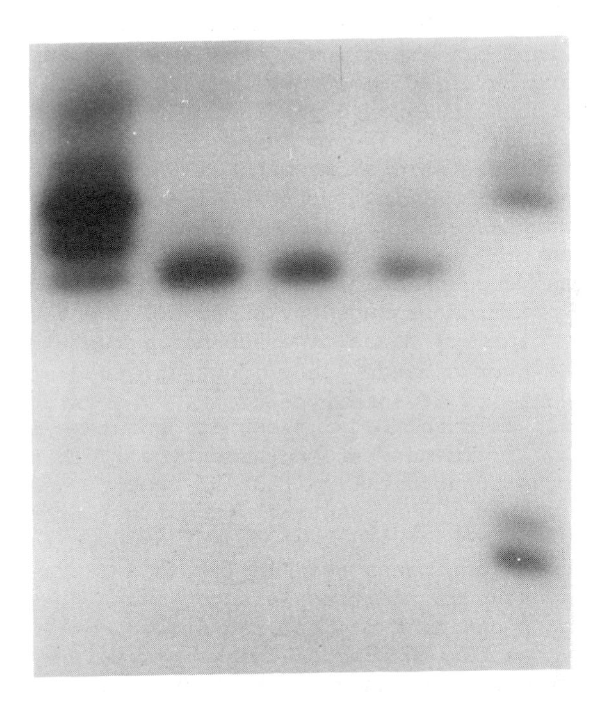

Fig. 3. A starch gel stained for phosphoglucomutase. The extracts used, starting from the left, are: (1) 6653, F' 152, (2) 6581, (3) 6677, (4) 6700, and (5) 475. The anode is at the top.

of the fast migrating band corresponded to the most intense band produced by the diploid. The slow bands which also appear in channel 5 are usually not seen. Their cause is unknown but could be due to other enzymes resulting from the lack of specificity in the staining procedure. The interspecies diploid has always produced at least four bands: a light band corresponding to the *Salmonella* form of the enzyme, the most intense band conforming to the *E.coli* enzyme, a hybrid band between the two, and a band that is slightly faster moving than the *E.coli*-type band. The fifth most rapidly migrating band shown in Figure 3, channel 1, is not always visible.

DISCUSSION

These experiments show that zone electrophoresis is a useful tool for gene mapping in bacteria. Although this technique is more complex than most of the methods used to distinguish bacterial gene markers, our experience is that it is quite practical. More of the technical problems and delays encountered in these experiments have involved the genetic work than have involved electrophoresis or the gel stain methods. We plan to map *gpd* and *est* by the same methods. The merodiploid strains should provide us with a convenient method to find approximate chromosome map positions and subunit structures for the genes coding for a considerable number of enzymes. Examples are: catalase, esterase, glutamic and isocitric dehydrogenases, adenylate kinase, triosephosphate isomerase, and superoxide dismutase.

The gel patterns produced by phosphoglucomutase are not sufficiently clear to reveal the subunit structure. However, the fact that hybrid bands occur in strain 6653 strongly suggests that subunits of some type occur. The data of Joshi and Handler (1964) on pure *E.coli* phosphoglucomutase indicate that the subunits, if present, are not identical. These arguments would support a structure like that of rabbit muscle phosphoglucomutase with two non-identical subunits (Duckworth and Sanwal, 1972). Such a structure cannot account for the number and positions of the bands seen in Figure 3 (Strain 6653) unless we invoke some additional factor such as a secondary modification of at least one enzyme form.

Another possible outcome of this work would be the development of experimental techniques to study regulation of biosynthesis of the enzyme studied. For example, in Figure 2, the diploid strain (channels 4 and 7) is synthesizing both *E.coli* and *S.typhimurium* forms of 6-phosphogluconate dehydrogenase. However, the pigment band produced by the *E.coli* enzyme is much stronger than that produced by the *Salmonella* enzyme. Similarly, in Figure 3, the partially diploid strain, 6653, produces a stronger band representing *E.coli*-type enzyme activity than the band of *Salmonella*-type enzyme. Although these differences may reflect some extraneous factor such as a greater resistance of the *E.coli* enzymes to extraction or electrophoresis conditions, they could also reflect real differences in the rates of synthesis of the enzymes. In the latter case, these experiments may lead to new experimental methods to study the control of protein biosynthesis.

ACKNOWLEDGEMENTS

This study was supported in part by the U. S. Atomic Energy Commission Contract No. AT-(40-1)-4024 and Public Health Service Research Grant GM 15597.

We gratefully acknowledge the technical assistance of Beulah S. Harriell. We are grateful to Dr. Barbara J. Bachmann, Curator of the *E.coli* Genetic Stock Center at Yale University, for supplying the F-merogenotes used in this study.

REFERENCES

Adhya, S. and M. Schwartz 1971. Phosphoglucomutase mutants of *Escherichia coli* K-12. *J. Bacteriol.* 108: 621-626.

Baptist, J. N., C. R. Shaw, and M. Mandel 1969. Zone electrophoresis of enzymes in bacterial taxonomy. *J. Bacteriol.* 99: 180-188.

Duckworth, H. W. and B. D. Sanwal 1972. Subunit composition of rabbit muscle phosphoglucomutase. *Biochem.* 11: 3182-3188.

Garrick-Silversmith, L. and P. E. Hartman 1970. Histidine-requiring mutants of *Escherichia coli* K-12. *Genetics* 66: 231-244.

Goldschmidt, E. P., M. S. Cater, T. S. Matney, M. A. Butler, and A. Greene 1970. Genetic analysis of the histidine operon in *Escherichia coli* K12. *Genetics* 66: 219-229.

Joshi, J. G. and P. Handler 1964. Phosphoglucomutase I. Purification and properties of phosphoglucomutase from *Escherichia coli*. *J. Biol. Chem.* 239: 2741-2751.

Lew, K. K. and J. R. Roth 1971. Genetic approaches to determination of enzyme quaternary structure. *Biochem.* 10: 204-207.

Low, K. B. 1972. *Escherichia coli* K-12 F-prime factors, old and new. *Bacteriol. Rev.* 36: 587-607.

Matney, T. S., E. P. Goldschmidt, N. S. Erwin, and R. A. Scroggs 1964. A preliminary map of genomic sites for F-attachment in *Escherichia coli* K12. *Biochem. Biophys. Res. Commun.* 17: 278-281.

Sanderson, K. E. 1972. Linkage map of *Salmonella typhimurium*, edition IV. *Bacteriol. Rev.* 36: 558-586.

Shaw, C. R., J. N. Baptist, D. A. Wright, and T. S. Matney 1973. Isolation of induced mutations in *E.coli* affecting the electrophoretic mobility of enzymes. *Mutation Res.* 18: 247-250.

Taylor, L. and C. Trotter 1972. Linkage map of *Escherichia coli* strain K-12. *Bacteriol. Rev.* 36: 504-524.

EVOLUTION OF THE LACTATE DEHYDROGENASE ISOZYMES OF FISHES

G. S. WHITT[1], J. B. SHAKLEE[1,2], and C. L. MARKERT[2]

[1]Department of Zoology
University of Illinois
Urbana, Illinois 61801

[2]Department of Biology
Yale University
New Haven, Connecticut 06520

ABSTRACT. The evolution of the L-lactate dehydrogenase (LDH) structural genes as well as the evolution of their regulation has been elucidated by investigating the number, electrophoretic properties, tissue specific expression, and immunochemical relatedness of the LDH isozymes in fishes. The most primitive vertebrates probably possessed a single LDH locus similar to the LDH A locus of today. This single locus genotype is reflected in the LDH phenotype of some contemporary Agnatha. A duplication of this ancestral locus, via polyploidization, has apparently given rise to the LDH A and B loci, each of which exhibits extensive homology in structure, function, and regulation throughout the vertebrates today. All of the Chondrichthyes examined possess both LDH subunits. A third LDH locus, LDH C, which arose from a duplication of the LDH B locus, is first observed in the primitive bony fishes, the chondrosteans. Many of the more primitive bony fishes (including the chondrosteans, holosteans, and some primitive teleost orders) exhibit a generalized function of the C gene, with the C_4 isozyme being synthesized in many tissues and possessing different relative electrophoretic mobilities from species to species. In the more advanced orders of teleosts the C gene is restricted sharply in function and in net charge. In most orders of advanced teleosts the C gene encodes a retinal specific isozyme with a large net negative charge, whereas in some families in the orders Cypriniformes and Gadiformes the C gene codes for a liver predominant isozyme with a net positive charge. The LDH C gene in the teleosts may be ancestral to the third LDH gene of birds and mammals (characteristically active in the primary spermatocyte). These studies demonstrate that the lactate dehydrogenase isozymes provide an excellent model system for studying the origin and evolution of the structure and regulation of homologous genes.

During the past decade considerable insight has been gained into the evolution of protein structure and function. However investigations of structure and function have usually not been accompanied by analysis of evolutionary changes in the regulation of the genes encoding these proteins.

The complete process of gene evolution can probably best be examined through the study of multilocus isozyme systems (Markert and Møller, 1959), particularly those which exhibit dramatic differential gene regulation during embryogenesis and cellular differentiation (Markert and Ursprung, 1971). The isozymes of lactate dehydrogenase (LDH) (E.C. 1.1.1.27) constitute such a system.

The current understanding of the evolution of vertebrate LDH is based on research from many different laboratories (Markert and Appella, 1963; Wilson et al, 1964; Zinkham et al, 1969; Goldberg, 1971; Ohno, 1973; Shaklee et al, 1973; Taylor et al, 1973; Whitt et al, 1973).

Early in the evolution of the vertebrates there was probably only a single LDH gene. Since nearly all contemporary vertebrates possess at least two LDH genes, the ancestral LDH gene must have been duplicated early in vertebrate evolution, probably by polyploidization of the genome as proposed by Ohno (1970; 1973). Thus the genes now designated A and B (Shaw and Barto, 1963) would have arisen from a common ancestral gene. The homology of these two LDH genes is supported by several kinds of data concerning the properties of the A_4 and B_4 isozymes: (1) the A and B polypeptides readily associate to form functional tetramers both in vivo and in vitro even when the homopolymers are from the most distant classes of vertebrates (Markert, 1963; 1965; 1968); (2) the dodecapeptides at the enzymatically active site have nearly identical amino acid sequences (Taylor et al, 1973); and (3) these polypeptides, particularly in the lower vertebrates, possess some of the same antigenic determinants (Holmes and Markert, 1969; Horowitz and Whitt, 1972; Holmes and Scopes, 1974).

Despite the common ancestry of these genes they have diverged markedly, as shown by the differences in amino acid composition of the isozymes (Markert, 1963; Pesce et al, 1967), by antigenic properties (Markert and Appella, 1963; Wilson et al, 1964), and by kinetic properties (Pesce et al, 1967).

Based on immunochemical data there is a high degree of homology among the A subunits and among the B subunits of the vertebrates (Wilson et al, 1964; Horowitz and Whitt, 1972; Holmes and Scopes, 1974). In addition to this conservation of structure, the distinctive kinetic properties of the A_4 and B_4 isozymes have also been maintained (Pesce et al, 1967). In most vertebrates the A_4 isozyme predominates in tissues producing or using large amounts of lactic acid, particularly if oxygen is in limited supply, and the B_4 isozyme is most abundant in the more aerobic tissues (Cahn et al, 1962; Markert and Ursprung, 1962). These observations suggest that the A and B subunits diverged substantially during early vertebrate evolution to occupy specific metabolic niches which are common to most vertebrates today. Such specialization of function and regulation undoubtedly increased metabolic efficiency but probably also diminished evolutionary flexibility. Major evolutionary changes in the A and B genes may have been resisted because the products of these genes fulfilled such essential roles in metabolism.

One obvious way by which such restraints on further enzyme evolution may be circumvented is by duplication of one of the LDH genes. Such a duplication has occurred at least once during the evolution of the vertebrates to give rise to a third LDH gene, the LDH C locus (Zinkham et al, 1969; Goldberg, 1971). The highly specialized LDH C gene in mammals and birds, which appears to have arisen from the B gene by a tandem duplication (Zinkham et al, 1969), is restricted in function to the primary spermatocyte.

A third LDH gene is also found in the fishes, which exhibit an extremely rich isozyme repertory (Markert and Faulhaber, 1965; Whitt et al, 1973). It is in these lower vertebrates that the evolution of the LDH genes may be most profitably investigated.

A typical teleost LDH isozyme pattern is displayed in Fig. 1. In addition to the A_4 and B_4 isozymes found in all vertebrates, there is also a retinal specific isozyme (or eye band). This unique homopolymer in the retina of many teleosts has been referred to as LDH E_4 and LDH C_4. In this presentation we shall refer to it as LDH C_4. At a pH of 7.0 this isozyme has a very large net negative charge. It is synthesized primarily in the photoreceptor cells (Whitt and Booth, 1970) and first appears at the time of retinal differentiation (Whitt et al, 1973; Miller and Whitt, 1974). This retinal specific isozyme is much more closely related to the B_4 isozyme than to the A_4 isozyme in its kinetic (Whitt, 1970), immunological (Whitt, 1969; Holmes and Markert, 1969), and physical properties (Whitt, 1970). This retinal isozyme has been demonstrated, by genetic analyses, to be under the control of a third LDH locus (Whitt et al, 1971; Vrijenhoek. 1972).

Earlier surveys of many orders of fishes revealed that none of the more primitive, non-teleostean fishes possessed an eye specific isozyme, and even some of the primitive orders of teleosts lacked this isozyme (Horowitz and Whitt, 1972).

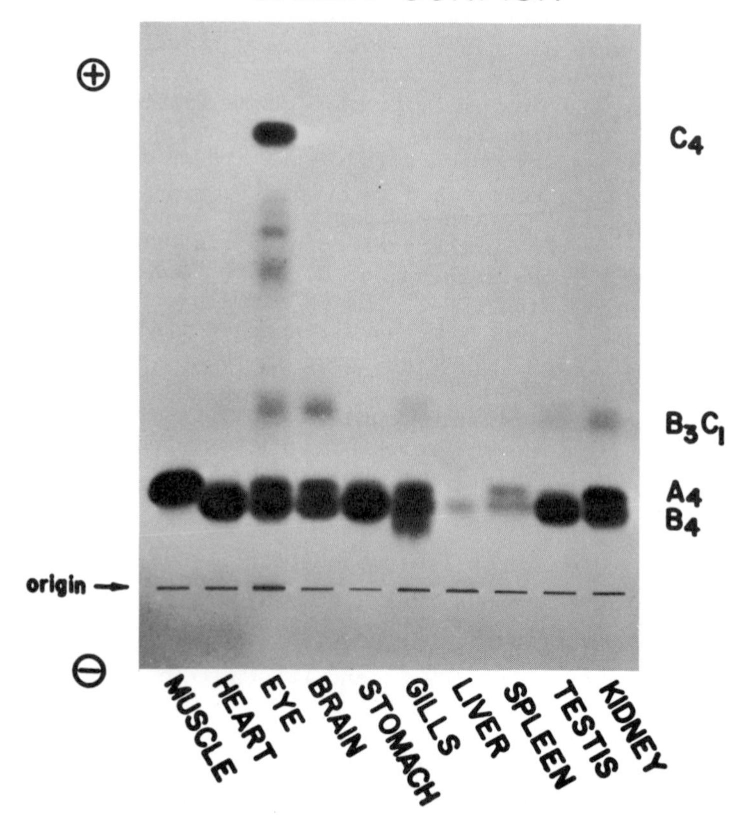

Fig. 1. The lactate dehydrogenase isozymes of tissues in the green sunfish *(Lepomis cyanellus)*. The LDH C_4 is predominantly synthesized in the retina of the eye.

However, its widespread distribution in many orders of teleosts led us to propose that the duplication of the B gene, giving rise to this third LDH locus, occurred prior to the adaptive radiation of the teleosts (Horowitz and Whitt, 1972; Whitt et al, 1973). The results of these earlier studies prompted the investigations leading to the present report. Our objectives were to determine more precisely the number of times that duplications of the LDH gene(s) have taken place and to identify the evolutionary levels at which they have occurred. We were particularly interested in the evolution of the regulation of this third LDH gene, which exhibits such a specialized function in the more advanced teleosts.

The agnathans, the most primitive class of vertebrates, are of particular interest because they exhibit two distinct types of LDH patterns.

Lampreys (order Petromyzontiformes) possess an isozyme pattern which is identical for all their tissues. Note the pattern exhibited by the American brook lamprey as shown in Fig. 2. Only one isozyme is present (the faint more anodal band is only a sub-band arising from the same single LDH gene). Early studies by Wilson et al (1964) led these investigators to conclude that lamprey tissues contain only the A_4 LDH, "as judged by immunological, electrophoretic, and catalytic criteria." The results of our studies support this interpretation. We have also been able to hybridize in vitro the lamprey LDH with other isozymes (Markert, 1961) and it behaves as a homotetramer. This isozyme is also similar to the A_4 isozyme of other vertebrates with respect to its behavior during affinity chromatography (Nadal-Ginard and Markert, 1974). These results also suggest that the ancestral LDH gene may have been similar to the LDH A gene of today. Horowitz and Whitt (1972) noted that antibodies made against teleost (weakfish) A_4 and B_4 isozymes react with the lamprey isozyme to the same extent. These data indicate that the lamprey isozyme has antigenic determinants in common with those of both the A_4 and the B_4 isozymes of contemporary teleosts.

The hagfish (order Myxiniformes), another agnathan, on the other hand,clearly possesses an isozyme pattern consistent with the presence of two LDH genes. The single isozyme in the lamprey may reflect the most primitive genetic arrangement in ancestral vertebrates, and the hagfish, the isozyme pattern arising after gene duplication. Alternatively, the ancestors of the lamprey may have possessed two LDH genes, with the B gene then lost during lamprey evolution. In any event, the results of investigations into the isozymes of the agnathans provide examples of two major steps in the evolution of the LDH genes, namely the existence of primitive vertebrates possessing only one or two LDH loci.

The second group we have examined consists of the cartilaginous fishes in the class Chondrichthyes. The isozyme pattern of the dogfish shark is shown in Fig. 3. As in other vertebrates the A_4 isozyme is predominant in white skeletal muscle,

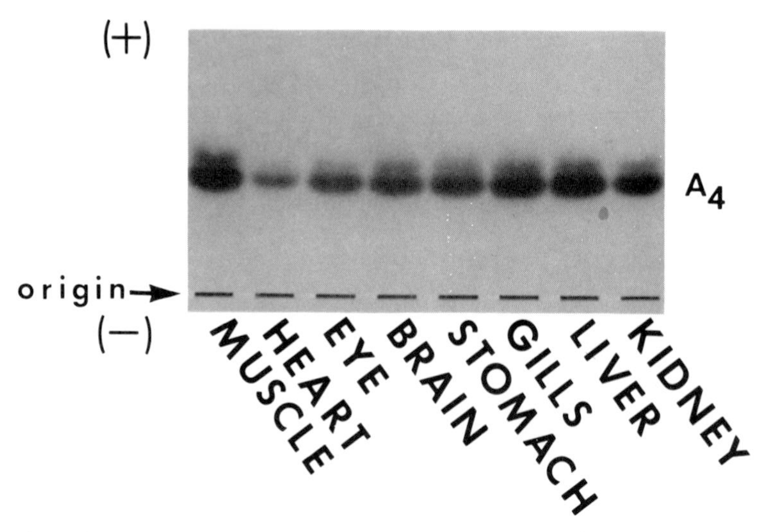

Fig. 2. The lactate dehydrogenase isozyme of tissues in the American brook lamprey *(Lampetra lamottei)*. Only the A_4 isozyme, and its sub-band, is expressed in all the tissues examined.

---

and the B_4 isozyme is predominant in heart muscle. This four-isozyme pattern is typical of many sharks and can be reproduced by in vitro hybridization of the two homopolymers. Only the A and B LDH genes have been found in the Chondrichthyes; there is no evidence in these vertebrates of a third LDH gene.

The third major group we have investigated is the bony fishes (class Osteichthyes). In the major line of bony fishes (subclass Actinopterygii), which leads directly to the teleosts, we have detected the presence of at last three LDH loci in nearly all fish investigated.

The most primitive vertebrate in which we have unequivocally detected three LDH genes is the sturgeon (Acipenseriformes) (Fig. 4). As in other vertebrates the A_4 isozyme predominates

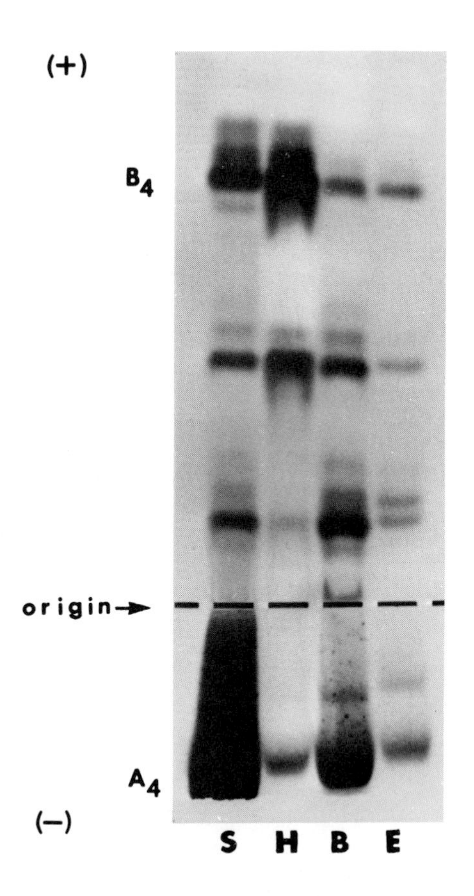

Fig. 3. The lactate dehydrogenase isozymes of tissues in the smooth dogfish *(Mustelus canis)*. S = skeletal muscle; H = heart muscle; B = brain; E = eye.

---

in white skeletal muscle while the B_4 isozyme predominates in heart muscle. The C_4 isozyme has a cathodal migration in electrophoresis and appears to be present to some extent in nearly all tissues. Moreover, it is important to note that in each tissue the relative C_4 isozyme activity tends to

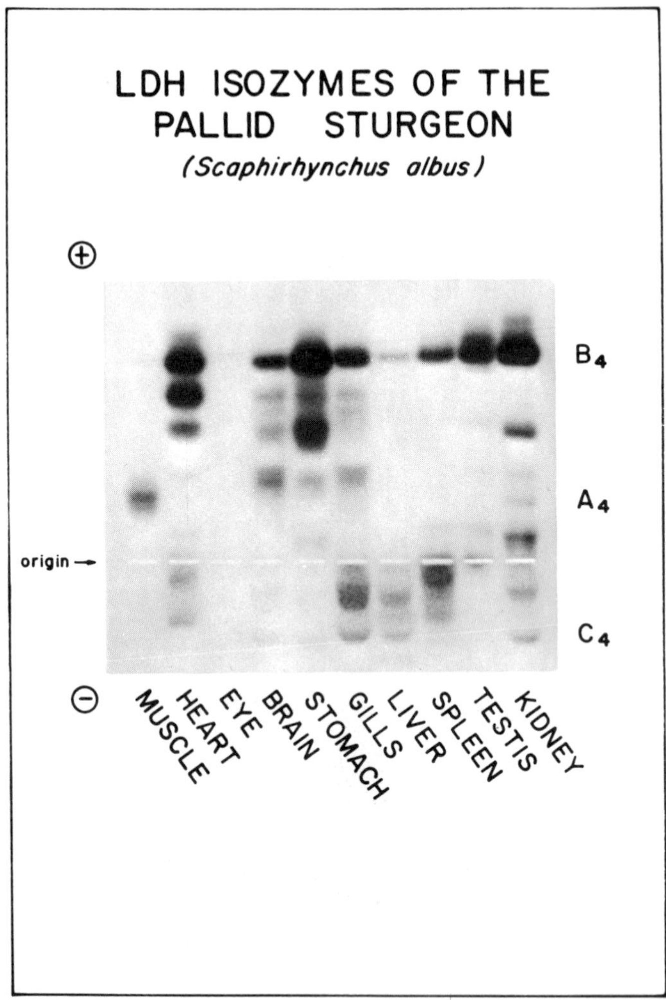

Fig. 4. The lactate dehydrogenase isozymes of tissues in the pallid sturgeon *(Scaphirhynchus albus)*.

---

approximate that of the B_4 isozyme rather than that of the A_4. The isozyme pattern of the paddlefish (also in the order Acipenseriformes) is similar, but in this species the C_4 isozyme possesses a more intermediate electrophoretic mobility and its presence is restricted to certain tissues (kidney, liver, gills). In both species the expression of the C gene is quite parallel to that of the B gene from which it was derived and noticeably different from that of the more distantly related A gene.

The isozyme pattern of the bowfin (Amiiformes) is shown in Fig. 5. Again the C_4 isozyme is synthesized in many tissues, but tends to be abundant precisely in those tissues in which the B_4 isozyme predominates. In this species the C_4 isozyme has an anodal electrophoretic mobility.

To summarize the data from the chondrosteans and holosteans, the expression of the A and B genes of these fishes is like that of the A and B genes of more advanced fishes and indeed of all other vertebrates. In contrast, the function of the C gene is not as tissue specific as that in more advanced fishes. Furthermore, the C gene tends to be expressed in the same tissues in parallel with the B gene.

Examination of the LDHs of the primitive orders of teleosts (e.g., Elopiformes, Osteoglossiformes, and Anguilliformes) reveals that the function of the C gene is still fairly generalized. The isozyme pattern of the purplemouth moray (in the order Anguilliformes) (Fig. 6) is characteristic of other primitive teleosts. We do see that although the C_4 isozyme is present in most tissues it occurs abundantly in only a few tissues. In fact, in the Anguilliformes and other primitive teleost orders, the synthesis of the C subunit tends to predominate in tissues such as kidney, spleen, gills, and stomach, depending upon the species examined. In addition, in these more primitive groups of teleosts the relative electrophoretic mobility of the C_4 isozyme varies considerably from species to species in contrast to the more stable relative mobilities of the A_4 and B_4 isozymes.

The above results are quite consistent with the conclusion that the C gene arose from the B gene. Immediately after the B gene duplicated the two genes were probably regulated in perfect concert and expressed to the same extent in the same tissues. However, with the accumulation of mutations in both the LDH structural genes and the regulatory genes, divergence in function and expression would be expected to occur. In fact we do see some divergence in immunological properties, net charge, and in tissue specificity of these isozymes. However, in these primitive bony fishes the C gene, and its protein product, occupy an intermediate evolutionary level and have not yet acquired the sharply restricted specificity of tissue expression and physiological function, evident in the advanced teleosts.

When we examined the more advanced teleosts, we found a dramatic specialization in both the _structure_ and _regulation_ of the C gene in accord with the evolutionary history of the

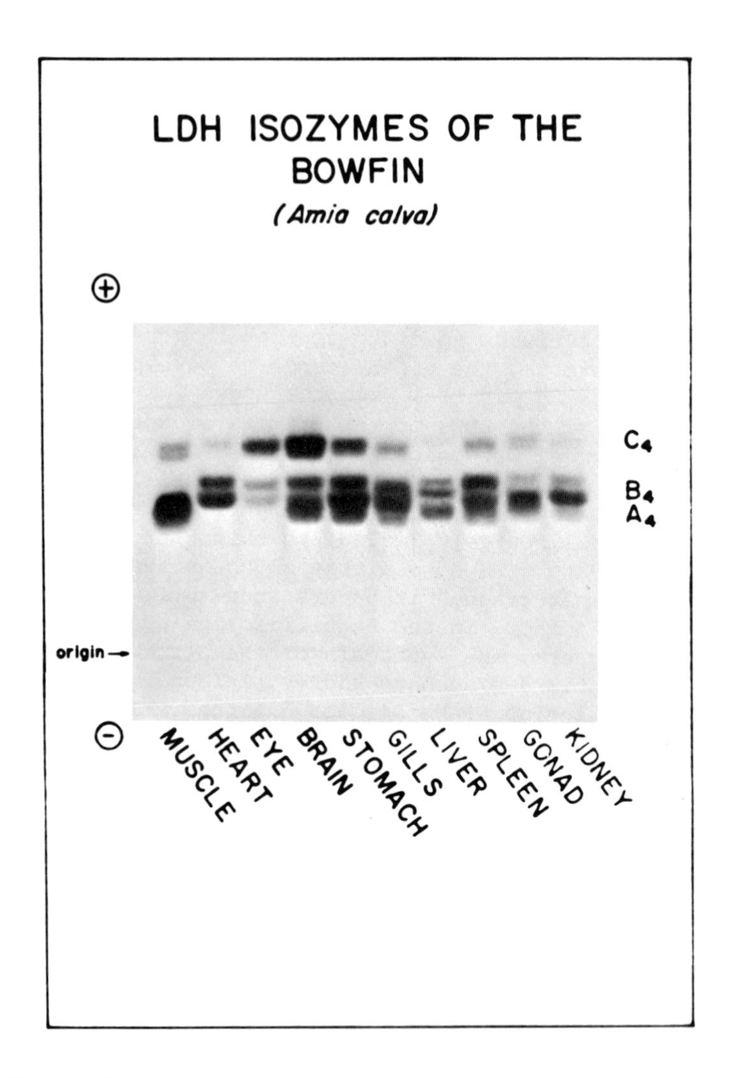

Fig. 5. The lactate dehydrogenase isozymes of tissues in the bowfin *(Amia calva)*.

order or family of each fish.

By far the majority of the teleost orders examined (42 families in nine orders) possess the highly anodal retinal specific LDH isozyme (Fig. 1). However, some families of fishes in a couple of advanced orders of teleosts exhibit a quite different pattern of function of the LDH C gene, particular-

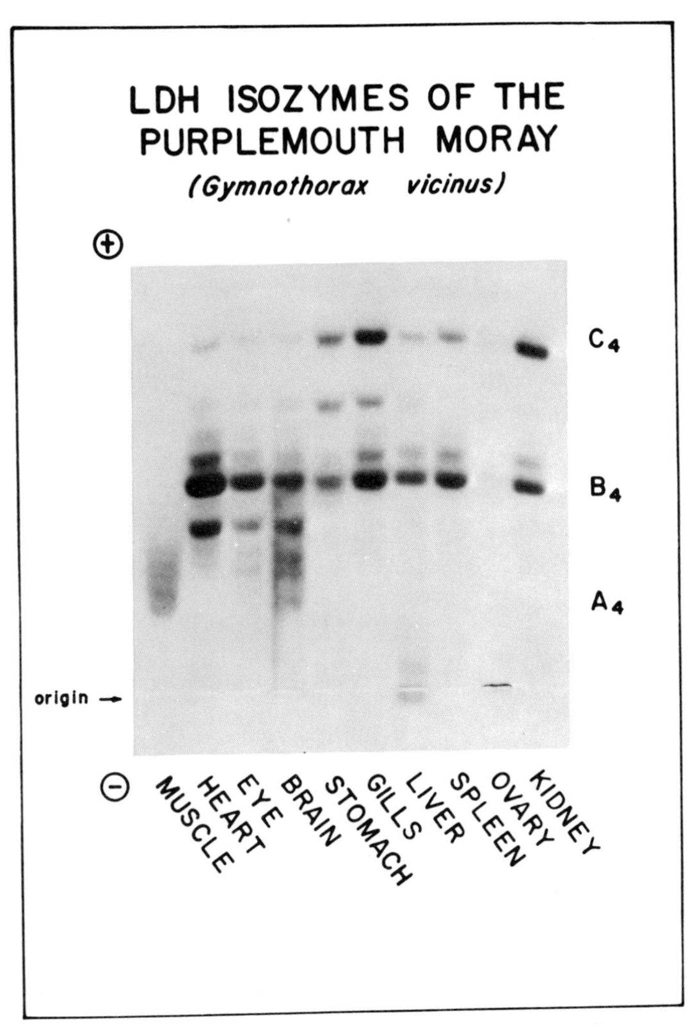

Fig. 6. The lactate dehydrogenase isozymes of tissues in the purplemouth moray *(Gymnothorax vicinus)*.

ly in the Cypriniformes and Gadiformes. For example, no anodal eye band is present in the isozyme pattern of the Atlantic cod (Gadiformes) (Fig. 7). The important characteristic to note is that a third homopolymeric isozyme (the C_4 isozyme--see Shaklee et al, 1973) is synthesized mainly in the liver and

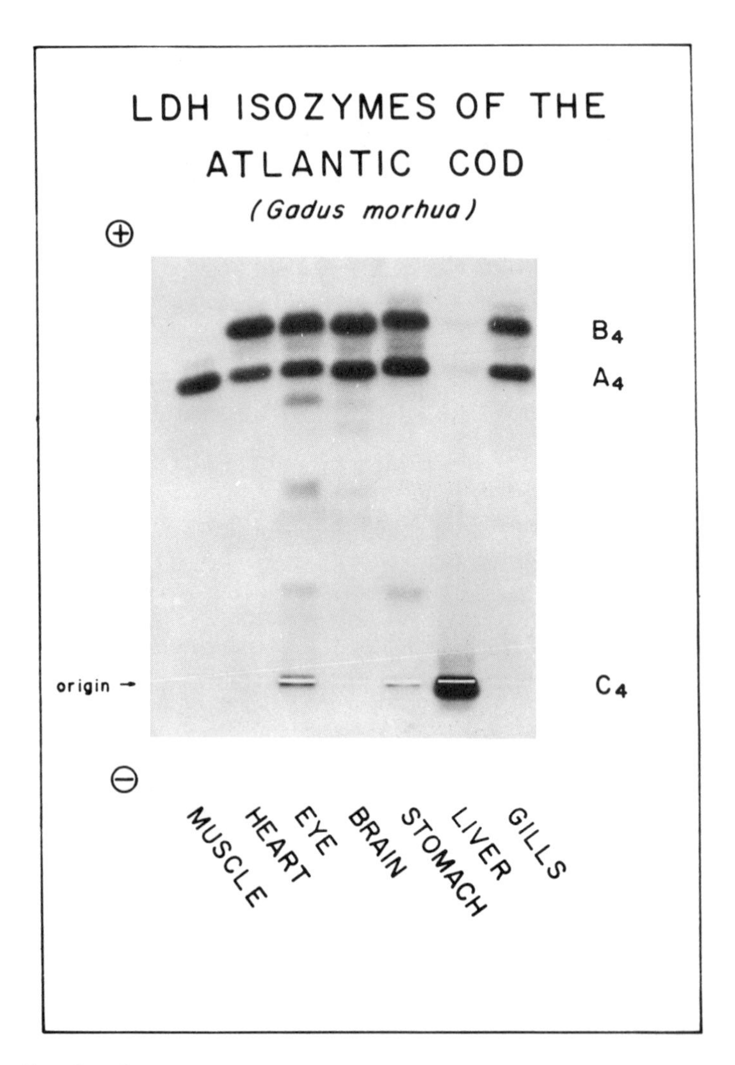

Fig. 7. The lactate dehydrogenase isozymes of tissues in the Atlantic cod *(Gadus morhua)*.

possesses a cathodal mobility. A similar pattern is also observed for many cypriniform fishes. As shown in Fig. 8, the function of the C gene in one such cyprinid, the hornyhead chub, is again restricted to the liver, and the C_4 isozyme again possesses a cathodal mobility.

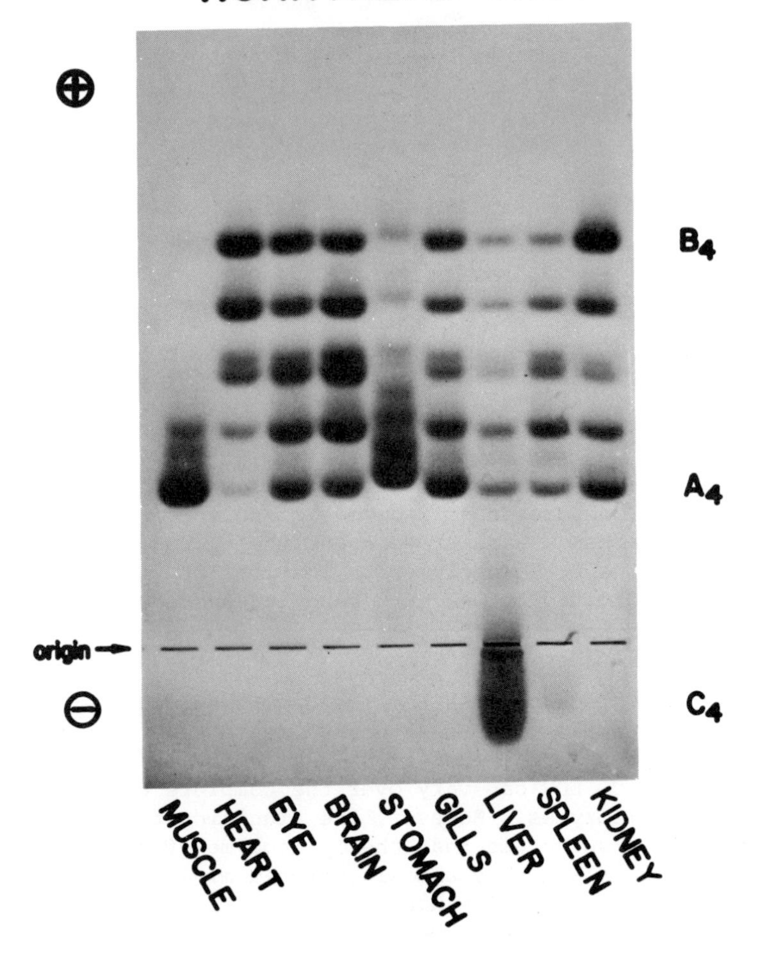

Fig. 8. The lactate dehydrogenase isozymes of tissues in the hornyhead chub *(Nocomis biguttatus)*.

In can be concluded that the advanced teleosts possess two main patterns of expression of the C gene--an anodal isozyme in the retina or a cathodal isozyme in the liver. Although this general correlation of net charge with tissue distribution holds for most species, a few interesting exceptions do

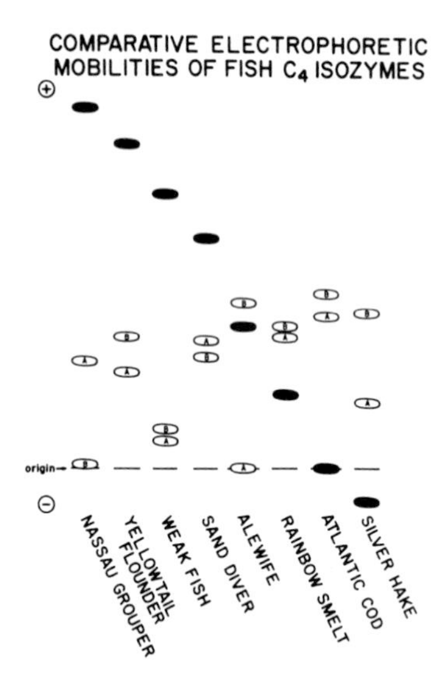

Fig. 9. Relative electrophoretic mobilities of the homopolymeric lactate dehydrogenase isozymes of selected species of teleosts. The bands with A represent the A_4 homopolymers, those with B the B_4 homopolymers. The mobility of the C_4 isozyme is indicated by the dark band in each species.

---

exist.

During our study of fish LDHs we have observed that a few species possess C_4 isozymes with intermediate net charges and/or tissue expressions. In fact, as shown in Fig. 9, the net charge of the C_4 LDH can vary over the complete electrophoretic range of LDH isozymes. However, the significant generalization from this survey is that the C_4 isozyme of teleost fishes exists either as a highly anodal eye-predominant isozyme or as a cathodal liver-predominant isozyme.

Despite the differences between the retinal and liver isozymes in both net charge and in tissue specificity, our investigations indicate that the retinal specific and liver specific isozymes of teleosts are probably encoded by the same basic genetic locus, which has undergone two evolutionary paths of specialization in the teleosts.

One of the most persuasive lines of evidence supporting this thesis is that all of the many species of teleosts we have examined possess either a characteristic eye band <u>or</u> a characteristic liver band--never both. This mutual exclusivity of isozyme patterns is what one would expect for a single C locus that had diverged in tissue expression during the evolution of different groups of fishes.

The other approach that we have taken involved an immunochemical assessment of the structural relatedness of the liver isozyme, the eye isozyme, and the B_4 homopolymer. It has been previously shown that the retinal isozyme and the liver isozyme are both much more closely related to the B_4 isozyme than to the A_4 isozyme (Whitt, 1969; Kepes and Whitt, 1972; Sensabaugh and Kaplan, 1972). Additional immunochemical studies were carried out to determine whether the liver and retinal LDH isozymes are encoded in two loci (arising from two separate gene duplication events at the B locus) or in one locus (arising from a single gene duplication event at the B locus) (Shaklee et al, 1973; Whitt et al, 1973). In Fig. 10 are shown the retinal LDH isozymes of the weakfish *(Cynoscion regalis)* after treatment with different antisera. To the right is shown the effect of the anti-B_4 antiserum (against alewife) which reacts with both B_4 and C_4 isozymes. But most significant are the results on the left hand side of the zymogram. Antibodies were prepared against the purified liver-predominant C_4 isozyme of the cod. At low concentrations, the anti-liver-specific LDH antiserum cross reacts with the retinal specific LDH isozyme but not with the B_4 isozyme, and it is only at higher antiserum concentrations that the B_4 isozyme is precipitated. The same kind of results have been obtained for other species of fishes (Shaklee et al, 1973; Whitt et al, 1973).

Although these data are not highly quantitative, they do clearly indicate that these two tissue specific isozymes, exhibiting very different net charges and tissue distributions, are more closely related to each other than either is to the B_4 isozyme. These results clearly support the hypothesis of a single duplication of the B locus followed by a divergence of the duplicated gene in both its structure and regulation so that it evolved to be expressed either in the retina as in most teleosts or in the liver as in a few teleosts.

A general outline of the evolutionary origin and divergence of vertebrate LDHs, as observed in fishes, is presented in Fig. 11. This figure depicts the evolutionary history of the LDH genes: originally one ancestral LDH gene gave rise by duplication to the A and B loci, and then more recently a second duplication occurred, this time involving a duplication of the B gene to give rise to the C gene.

When one considers the class Osteichthyes, the only fishes to contain all three LDH genes (A, B, and C), it is clear that the more primitive groups exhibit a generalized function of the C gene as well as a more variable net charge on the C_4

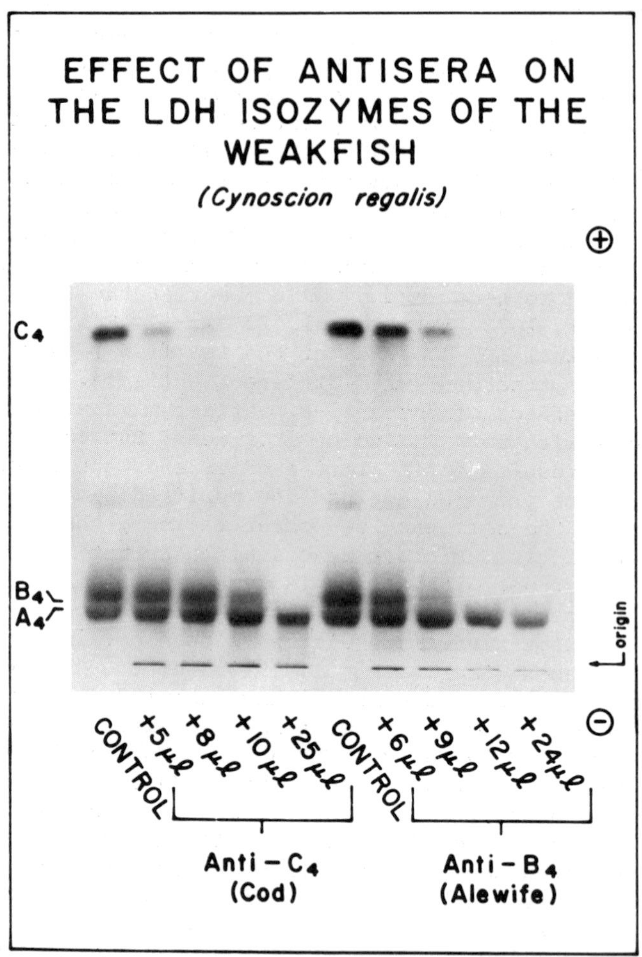

Fig. 10. The effect of antisera on the retinal lactate dehydrogenase isozymes of the weakfish *(Cynoscion regalis)*. Channels 2 through 5 show the effects of antisera to the cod liver predominant lactate dehydrogenase isozyme (anti-C_4) upon the retinal specific LDH C_4 isozyme and the LDH B_4 of the weakfish. The antisera to cod liver specific LDH preferentially precipitates the retinal specific LDH at concentrations having no apparent effect on the LDH B_4 isozyme. These data are consistent with the hypothesis that the retinal specific LDH is more closely related to the liver specific LDH than either is

Fig. 10 legend continued
to the LDH B_4 isozyme, indicating that both these isozyme types are encoded in the same locus, LDH C, and are thus both referred to as LDH C_4.

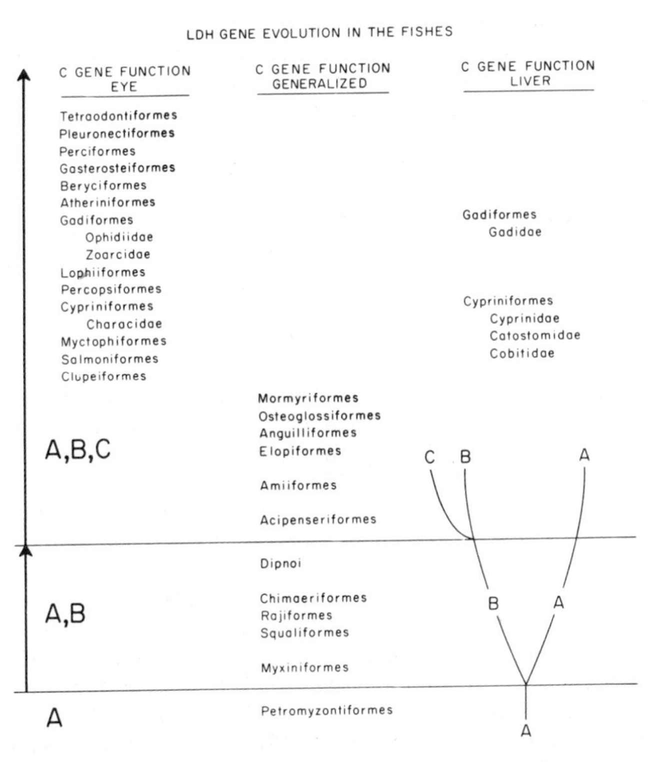

Fig. 11. Schematic portrayal of the evolution of lactate dehydrogenase genes in the fishes. Although not included in this diagram, additional, more recent duplications of the LDH genes within isolated orders (Salmoniformes and Cypriniformes) have occurred through polyploidization and these have been discussed in more detail elsewhere (Bailey and Wilson, 1968; Holmes and Markert, 1969; Ohno, 1970).

isozyme. In the more advanced teleosts the C gene has become sharply restricted in its tissue expression and in its function. In most orders, it functions predominantly in the eye, but in some families in a few orders it functions predominantly in the liver.

Some of these results present an apparent paradox. In both the orders Cypriniformes and Gadiformes, some families are characterized by a liver band, others by an eye band. Either a rearrangement of evolutionary relationships is in order or we must conclude that the expression of the C gene has undergone convergent evolution in these orders--convergent in tissue specificity and in net charge. Obviously more detailed immunochemical assessments and eventually amino acid sequence analyses will be necessary to resolve these issues, as well as to determine whether the LDH C gene of fishes is ancestral to the LDH C gene of birds and mammals.

ACKNOWLEDGMENTS

This research was supported by NSF grants GB 16425 and GB 43995 to G. S. W. and GB 5440X to C. L. M. The expert assistance of Mike Burns is gratefully acknowledged.

REFERENCES

Bailey, G. S. and A. C. Wilson 1968. Homologies between isoenzymes of fishes and those of higher vertebrates. Evidence for multiple H_4 lactate dehydrogenases in trout. *J. Biol. Chem.* 243: 5843-5853.

Cahn, R. D., N. O. Kaplan, L. Levine, and E. Zwilling 1962. Nature and development of lactate dehydrogenases. *Science* 136: 962-969.

Goldberg, E. 1971. Immunochemical specificity of lactate dehydrogenase-X. *Proc. Natl. Acad. Sci. USA* 68: 349-352.

Holmes, R. S. and C. L. Markert 1969. Immunochemical homologies among subunits of trout lactate dehydrogenase isozymes. *Proc. Natl. Acad. Sci. USA* 64: 205-210.

Holmes, R. S. and R. K. Scopes 1974. Immunochemical homologies among vertebrate lactate-dehydrogenase isozymes. *Eur. J. Biochem.* 43: 167-177.

Horowitz, J. J. and G. S. Whitt 1972. Evolution of a nervous system specific lactate dehydrogenase isozyme in fish. *J. Exp. Zool.* 180: 13-32.

Kepes, K. L. and G. S. Whitt 1972. Specific lactate dehydrogenase gene function in the differentiated liver of cyprinid fish. *Genetics* 71 (no. 3/part 2): S29.

Markert, C. L. 1961. Nucleocytoplasmic interactions during development. In *Proc. First International Conference on Congenital Malformations*. J. B. Lipincott Co., Philadelphia. pp. 158-165.

Markert, C. L. 1963. Lactate dehydrogenase isozymes: Dissociation and recombination of subunits. *Science* 140:

1329-1330.

Markert, C. L. 1963. Epigenetic control of specific protein synthesis in differentiating cells. In *Cytodifferentiation and Macromolecular Synthesis. Twenty-first Symposium of Society for Study of Development and Growth* (M. Locke, ed.). Academic Press, N.Y. pp. 65-84.

Markert, C. L. 1965. Mechanisms of cellular differentiation. pp. 230-238. In *Ideas in Modern Biology* (J. Moore, ed.). Natural History Press, Garden City, N.Y. 563 pp.

Markert, C. L. 1968. The molecular basis for isozymes. *Ann. N.Y. Acad. Sc.* 151: 14-40.

Markert, C. L. and E. Appella 1963. Immunochemical properties of lactate dehydrogenase isozymes. *Ann. N.Y. Acad. Sc.* 103: 915-929.

Markert, C. L. and I. Faulhaber 1965. Lactate dehydrogenase isozyme patterns of fish. *J. Exp. Zool.* 159: 319-332.

Markert, C. L. and R. S. Holmes 1969. Lactate dehydrogenase isozymes of the flatfish, Pleuronectiformes: Kinetic, molecular, and immunochemical analysis. *J. Exp. Zool.* 171: 85-104.

Markert, C. L. and F. Møller 1959. Multiple forms of enzymes: Tissue, ontogenetic, and species specific patterns. *Proc. Natl. Acad. Sci. USA* 45: 753-763.

Markert, C. L. and H. Urpsrung 1962. The ontogeny of isozyme patterns of lactate dehydrogenase in the mouse. *Devel. Biol.* 5: 363-381.

Markert, C. L. and H. Ursprung 1971. *Developmental Genetics.* In *Foundations of Developmental Biology Series* (C. L. Markert, ed.). Prentice-Hall, Inc., Englewood Cliffs, N.J. 214 pp.

Miller, E. T. and G. S. Whitt 1975. Lactate dehydrogenase isozyme synthesis and cellular differentiation in the teleost retina. *Isozymes III. Developmental Biology,* C. L. Markert, editor, Academic Press, N.Y. pp. 359-374.

Nadal-Ginard, B. and C. L. Markert 1975. Use of affinity chromatography for purification of lactate dehydrogenase and for assessing the homology and function of the A and B subunits. *Isozymes II. Physiological Function,* C. L. Markert, editor, Academic Press, N.Y. pp. 45-67.

Ohno, S. 1970. *Evolution by Gene Duplication.* Springer-Verlag, New York. 160 pp.

Ohno, S. 1973. Ancient linkage groups and frozen accidents. *Nature* 244: 259-262.

Pesce, A., T. P. Fondy, F. Stolzenbach, F. Castillo, and N. O. Kaplan 1967. The comparative enzymology of lactic dehydrogenases. III. Properties of the H_4 and M_4 enzymes from a number of vertebrates. *J. Biol. Chem.* 242:2151-2167.

Sensabaugh, G. F., Jr. and N. O. Kaplan 1972. A lactate dehydrogenase specific to the liver of gadoid fish.

J. Biol Chem. 217: 585-593.
Shaklee, J. B., K. L. Kepes, and G. S. Whitt 1973. Specialized lactate dehydrogenase isozymes: The molecular and genetic basis for the unique eye and liver LDHs of teleost fishes. *J. Exp. Zool.* 185: 217-240.
Shaw, C. and E. Barto 1963. Genetic evidence for the subunit structure of lactate dehydrogenase isozymes. *Proc. Natl. Acad. Sci. USA* 50: 211-214.
Taylor, S. S., S. S. Oxley, W. S. Allison, and N. O. Kaplan 1973. Aminoacid sequence of dogfish M_4 lactate dehydrogenase. *Proc. Natl. Acad. Sci. USA* 70: 1790-1794.
Vrijenhoek, R. C. 1972. Genetic relationships of unisexual hybrid fishes to their progenitors using lactate dehydrogenase isozymes as gene markers (Poeciliopsis, Poeciliidae). *Am. Nat.* 106: 754-766.
Whitt, G. S. 1969. Homology of lactate dehydrogenase genes: E gene function in the teleost nervous system. *Science* 166: 1156-1158.
Whitt, G. S. 1970. Developmental genetics of the lactate dehydrogenase isozymes of fish. *J. Exp. Zool.* 175: 1-35.
Whitt, G. S. and G. M. Booth 1970. Localization of lactate dehydrogenase activity in the cells of the fish *(Xiphophorus helleri)* eye. *J. Exp. Zool.* 174: 215-224.
Whitt, G. S., W. F. Childers, and T. E. Wheat 1971. The inheritance of tissue-specific lactate dehydrogenase isozymes in interspecific bass *(Micropterus)* hybrids. *Biochem. Genet.* 5: 257-273.
Whitt, G. S., E. T. Miller, and J. B. Shaklee 1973. Developmental and biochemical genetics of lactate dehydrogenase isozymes in fishes, pp. 243-276. In *Genetics and Mutagenesis of Fish* (J. H. Schröder). Springer-Verlag, Berlin. 356 pp.
Wilson, A. C., N. O. Kaplan, L. Levine, A. Pesce, M. Reichlin, and W. S. Allison 1964. Evolution of lactic dehydrogenase. *Fed. Proc.* 23: 1258-1266.
Zinkham, W. H., H. Isensee, and J. H. Renwick 1969. Linkage of lactate dehydrogenase B and C loci in pigeons. *Science* 164: 185-187.

GENE DUPLICATION IN SALMONID FISH: EVOLUTION OF A LACTATE DEHYDROGENASE WITH AN ALTERED FUNCTION

GEORGE S. BAILEY and LIM SOO THYE
Department of Biochemistry
University of Otago
Dunedin, NEW ZEALAND

ABSTRACT: As a result of gene duplication salmonids produce two polypeptides, B and B', for B_4-type lactate dehydrogenase and appear to produce two other peptides, A and A', for A_4-type lactate dehydrogenase. The tissue distribution of lactate dehydrogenase isozymes has been examined in these fish and appears to be unusual relative to higher vertebrates. Most salmon and trout tissues exhibit varying levels of B_4 and B'_4, rather than B_4 and A_4.

Salmon B_4, B'_4 and A_4 - A'_4 have now been purified to high specific activity and examined chemically, immunologically, and catalytically. The B_4 and B'_4 proteins are similar in amino acid composition and closely related immunologically. Quantitative immunological studies suggest that B_4 and B'_4 have diverged in structure for approximately 80-100 million years. Both proteins are easily distinguishable from the A_4 - A'_4 preparation in composition and fail to show any immunological relationship with A_4 - A'_4, either qualitatively or quantitatively, by our methods.

The B'_4 isozyme is somewhat intermediate between B_4 and A_4 - A'_4 in Michaelis constants, substrate optima, and susceptibility to substrate inhibition. In particular the B'_4 isozyme is intermediate to B_4 and A_4 - A'_4 lactate product inhibition, a catalytic parameter suggested to be of major regulatory importance to A_4-type isozymes. These findings suggest that salmon B'_4 is diverging from B_4 catalytically, perhaps to substitute in some tissues for the absence of a balanced B_4 - A_4 isozyme complement.

INTRODUCTION

Considerable experimental evidence has accumulated to indicate that gene duplication has been very extensive in salmonid fish, and possibly occurred through an ancient tetraploidy event (for a review of the relevant experimental evidence and literature see Lim et al., 1974). However, while an evolutionary model for the origin of the many duplicated gene products in salmonids thus may be offered, there is as yet little experimental information on the possible functional significance or usefulness to present-day salmonids of extra viable gene copies for many proteins of differing function. The question arises in salmonids as to what number, if any, of the many duplicated gene products have evolved new functions, how many have failed to diverge in function, and how many "extra" genes have been eliminated?

Recent studies on the biochemistry and genetics of salmonid cytoplasmic malate dehydrogenase (MDH) by Bailey et al. (1970) demonstrated the presence of two loci coding for the B_2 form of MDH, both of which appear identical and undifferentiated in structure, regulation and function; two identical loci probably exist for A_2 MDH as well.

This provides direct evidence that at least some genes in salmonids have survived in duplicated form, but have failed at the molecular level to diverge in function in any obvious way. In direct contrast, we present biochemical evidence here that the duplicated genes for B_4-type lactate dehydrogenase (LDH) have diverged in time to produce two genes, Ldh-<u>B</u> and Ldh-<u>B</u>', coding for polypeptides B and B'. These polypeptides are shown to be closely related in structure, but appear to have diverged considerably in functional behavior since the initial duplication event.

EXPERIMENTAL PROCEDURE

MATERIALS

Chemicals. Sheep red cells in Alsever's solution were obtained from the Microbiology Department, Otago University; commercial immunological reagents were purchased from Difco Labs. All other chemicals were of reagent grade quality.

Biological specimens. Quinnat salmon *(Oncorhynchus tshawytscha)*, brook trout *(Salvelinus fontinalis)*, brown trout *(Salmo trutta)*, and rainbow trout *(Salmo gairdnerii)* were obtained from Otago and Canterbury streams, and stored at -20°. For tissue distribution studies, trout were sacrificed immediately prior to use.

METHODS

Enzyme and protein assays; electrophoresis. LDH activity was measured by following the rate of NADH oxidation as reported elsewhere (Bailey and Wilson, 1968; Lim et al., 1974).

Protein concentrations were routinely estimated by absorbance at 280 nm, assuming that $E_{1cm}^{0.1\%} = 1.0$. It has been shown for crystalline preparations of A_4 and B_4 LDH from ten species representing birds, mammals, amphibians, bony fish, and cartilaginous fish, that for LDH at 280 nm, $E_{1cm}^{0.1\%} = 1.35 \pm 0.06$

(Pesce et al., 1967), and this value was used for computing the final specific activity of LDH in our preparations.

Methods for electrophoresis are described in detail elsewhere (Lim et al., 1974).

Purification of quinnat salmon LDH isozymes. Salmon B_4 LDH was purified from pooled salmon hearts to a final specific activity of 970 units of enzyme per A_{280}, or 1310 units per mg LDH. Salmon B'_4 from pooled livers was purified to a specific activity of 905 units per A_{280}, or 1220 units per mg LDH. Salmon A_4-type LDH, which appears to consist of a mixture of A_4, A_3A', and $A_2A'_2$ isozymes, was purified from skeletal muscle to a specific activity of 939 units per A_{280}, 1270 units per mg LDH. These procedures, which will be presented in detail elsewhere (Lim et al., 1974) are summarized in Table 1.

Preparation of antisera and other immunological methods are reported in detail elsewhere (Lim et al., 1974).

RESULTS

ISOZYME DISTRIBUTION

The distribution of lactate dehydrogenase isozymes was investigated in various tissues of brown trout (Fig. 1), quinnat salmon, and rainbow trout, and found to be similar in all three species as judged qualitatively by starch gel electrophoresis. Striated muscle tissues are seen to exhibit a series of isozymes apparently resulting from the random combination into tetrameric isozymes of two different polypeptides, A and A' (Fig. 1-E). Although a total of five isozymes is possible, namely A_4, A_3A', $A_2A'_2$, AA'_3 and A'_4, some salmonids show only three or four isozymes, apparently due either to unequal rates of synthesis of A and A' or to differential inherent catalytic efficiencies of the A and A'subunits. For simplicity we refer to this entire series of isozymes as salmon A_4 - A'_4 LDH in this communication.

Most tissues do not contain significant levels of the A_4-A'_4 series, but instead exhibit a second series of isozymes composed of B and B' subunits. These subunits, which have been shown in their tetrameric forms to be immunologically related to the extensively studies B_4 LDH isozyme of higher vertebrates (Bailey and Wilson, 1968) and of certain other teleosts (Massaro and Markert, 1968), are seen to differ from each other in tissue distribution (Fig. 1), presumably due to differential rates of synthesis and/or breakdown (Fritz et al., 1969). In some tissues such as spleen (Fig. 1-M) and liver (Fig. 1-B), the B'_4 homotetramer predominates; in other tissues such as heart (Fig. 1-A) the B_4 homotetramer is predominant. Eye

TABLE I

PURIFICATION OF QUINNAT SALMON LACTATE DEHYDROGENASE ISOZYMES

Purification step	Total enzyme I.U. x 10^{-4}	Total protein A_{280} x 10^{-3}	Specific activity I.U./A_{280}	Total purification	Total yield %
B_4 Isozyme					
Extraction	12.7	11	11		
Heat	9.9	4.1	24	2.1	78
$(NH_4)_2SO_4$	6.4	2.7	235	20	50
Sephadex	6.2	1.9	321	28	49
First DEAE	4.3	0.55	775	67	34
Second DEAE	1.2	0.14	898	78	9.6
Third DEAE	0.63	0.065	970	84	4.9
B'_4 Isozyme					
Extraction	23.7	413	0.57		
Heat	18.3	195	0.94	1.6	77
$(NH_4)_2SO_4$	17.9	28	6.5	11	76
DEAE	8.1	0.30	285	499	36
Sephadex	7.2	0.14	515	904	31
Hydroxyl-apatite	2.3	0.024	905	1590	9.6
A_4 - A'_4 Isozyme					
Extraction	25.6	115	2.18		
$(NH_4)_2SO_4$	25.2	59	4.33	1.9	98
DEAE-Sephadex	23.2	5.6	41.2	19.1	91
1st DEAE	18.2	0.60	300	138	71
Sephadex G 200	14.9	0.29	539	248	58
2nd DEAE	8.6	0.11	763	350	34
Hydroxyl-apatite	3.2	0.034	939	442	13

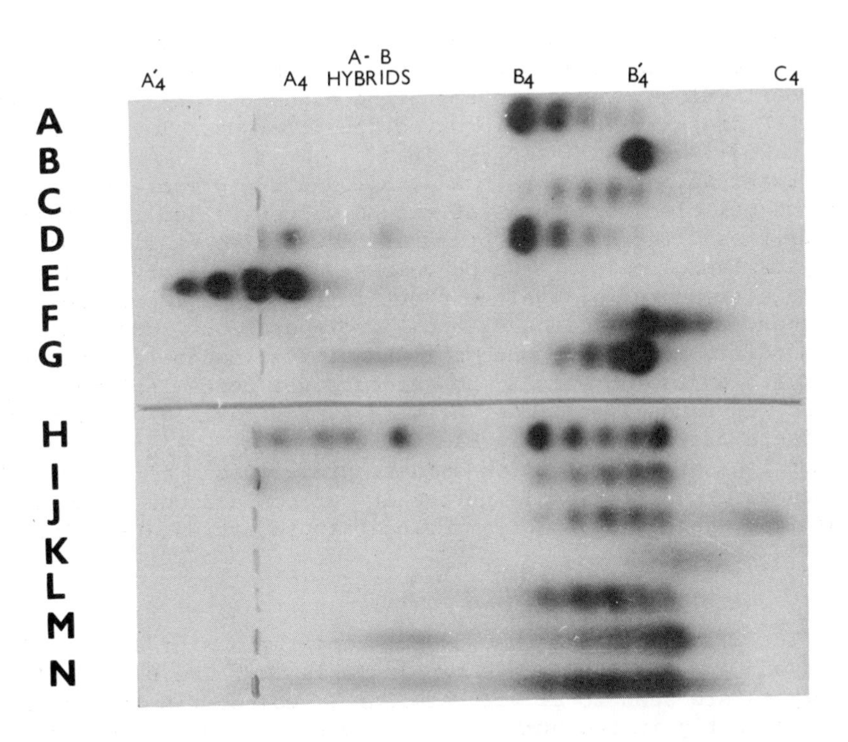

Fig. 1. Electrophoresis of crude extracts of various tissues from brown trout. A - Heart. B - Liver. C - Brain. D - Lateral line tissue. E - Epaxial striated muscle. F - Pyloric Caeca. G - Gall bladder. H - Gill filaments. I - Swim bladder. J - Eye. K - Fore-gut. L - Oesophagus. M - Spleen. N - Kidney. Electrophoresis was at pH 7.8 for 18 hours at 4°. Staining was for 1.0 hour at 25°.

extracts exhibit approximately equal levels of B_4 and B'_4 isozymes (Fig. 1-J) and also exhibit a fifth isozyme type, the C_4 lactate dehydrogenase unique to retinal and certain neural tissues of teleost fish (see Shaklee et al., 1974). Similar patterns of isozyme distribution have been reported for brook trout by Morrison (1970).

The relative amount of A_4 - A'_4 isozymes compared to B_4-

B'_4 - C_4 in various tissues has also been assessed quantitatively (Lim et al., 1974). In agreement with starch gel patterns, only gill, lateral line, and striated muscle showed appreciable percentages of the A_4 - A'_4 series; all other tissues contained 98-100% B_4 - B'_4 (plus C_4 in eye).

PURIFICATION OF SALMON LDH ISOZYMES

Salmon lactate dehydrogenase isozymes were purified as summarized in Table I. Taking into account assay conditions and definitions, these are among the highest specific activities for LDH preparations yet reported (see, for example, Pesce et al., 1967).

Examination of the A_4 - A'_4, B_4, and B'_4 preparations by starch gel electrophoresis at two pH values failed to reveal any impurities. The A_4 - A'_4 preparation was also examined on polyacrylamide gels at two pH values and by SDS-polyacrylamide gel electrophoresis, again without detection of impurities. Further, antisera to the B_4 and B'_4 preparations were tested by immunodiffusion and by much more sensitive immunoelectrophoretic analysis at two pH values; no evidence for contaminating proteins was found (see Lim et al., 1974, for details). Although antisera are not yet available to test the A_4 - A'_4 preparation immunologically, we feel that these preparations of salmon LDH isozymes are sufficiently pure that our comparisons of their chemical, catalytic, and immunological properties are valid.

AMINO ACID ANALYSIS

The amino acid analyses of our preparations of salmon B_4, B'_4, and A_4 - A'_4 LDH are shown in Table II. The B_4 and B'_4 isozymes are seen to be very similar in composition but differ significantly in the number of residues of alanine and isoleucine and possibly in valine, methionine, and tyrosine as well. Both proteins differ from the A_4 - A'_4 system in several residues, notably histidine, arginine, aspartic acid, glycine, alanine, and methionine.

Lactate dehydrogenases appear to be quite variable in composition among distantly related species (Pesce et al., 1967). The most consistent observation is that B_4 isozymes contain about half as much histidine (7.0 residues/polypeptide, S. D. $\pm$ 0.6) as A_4 isozymes (13.3 residues/polypeptide, S. D. $\pm$ 3.5). In this respect, B_4 and B'_4 LDH from salmon are typical of B_4 isozymes from other vertebrates, and salmon A_4 - A'_4 are typical A_4 isozymes.

TABLE II

AMINO ACID COMPOSITION OF SALMON B_4, B'_4 AND A_4 - A'_4

Amino acid[a]	B_4	B'_4	A_4 - A'_4
Lys	26.0 ± 0.7	25.7 ± 0.6	30.0 ± 0.8
His	8.7 ± 0.8	8.6 ± 0.3	17.6 ± 0.6
Arg	11.2 ± 0.2	11.4 ± 0.2	8.2 ± 0.7
Asp	37.7 ± 0.7	35.9 ± 0.6	28.9 ± 0.8
Thr[b]	17.3 ± 0.6	18.2 ± 0.3	17.4 ± 0.6
Ser[b]	28.3 ± 0.9	25.5 ± 0.9	27.8 ± 0.9
Glu	28.1 ± 0.9	30.4 ± 0.7	32.2 ± 0.9
Pro	12.1 ± 0.8	11.5 ± 0.5	9.9 ± 0.5
Gly	27.9 ± 0.6	26.9 ± 0.5	33.8 ± 1.1
Ala	21.1 ± 0.6	17.7 ± 0.9	11.4 ± 0.3
Val	40.1 ± 1.2	35.6 ± 0.8	36.8 ± 2.5
Met	8.8 ± 0.6	10.8 ± 0.2	6.3 ± 0.3
Iso	12.9 ± 0.6	16.9 ± 0.5	14.6 ± 0.5
Leu	33.6 ± 2.9	33.8 ± 0.7	36.9 ± 2.4
Tyr	4.9 ± 0.5	5.9 ± 0.2	4.7 ± 0.1
Phe	7.4 ± 0.3	7.9 ± 0.3	5.9 ± 0.4
Trp[c]	4.9 ± 0.4	4.8 ± 0.4	5.6 ± 0.4

[a]Mean ± standard error from duplicate hydrolysates of 24, 48 and 72 hours. Values reported as residues of amino acid per 36,000 grams of protein.

[b]Serine and threonine values extrapolated to zero time of hydrolysis.

[c]Tryptophan determined by direct analysis of hydrolysate containing 4% thioglycollic acid and assuming an 80% recovery.

IMMUNOLOGICAL STUDIES

The microcomplement fixation (MC'F) technique is known to be sensitive to even very small sequence differences between related proteins and capable of providing an approximate quantitative measure of structural relatedness between homologous proteins (Sarich, 1969; Prager and Wilson, 1971). Antisera against purified salmon B_4 and B'_4 LDH's were tested by MC'F against salmon B_4, B'_4, and A_4 - A'_4 (Fig. 2). The ratio of antiserum concentrations required for homologous and cross-reacting proteins to show equal fixation is called the index of dissimilarity. The quantity which appears to vary linearly with percent sequence difference between related proteins is the immunological distance (Sarich, 1969; Prager and Wilson, 1971), where:

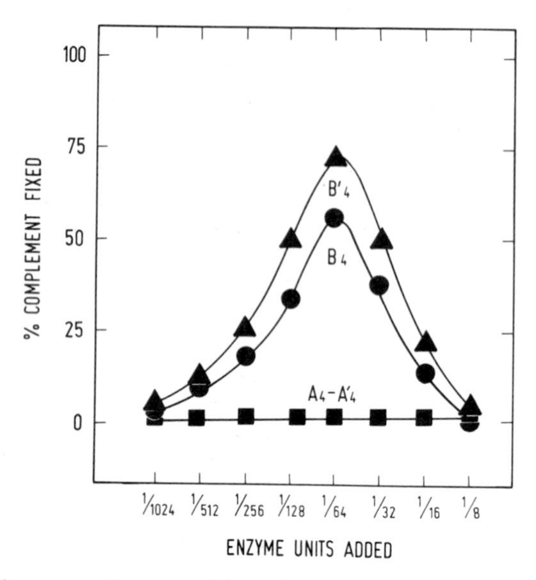

Fig. 2. Microcomplement fixation. Antiserum 102B1 (to purified B'_4 lactate dehydrogenase) was tested against purified B_4, B'_4 and A_4 - A'_4 lactate dehydrogenases. The reaction mixture contained 1 ml of a 1:7000 dilution of the antiserum. No cross-reaction with A_4 - A'_4 could be observed even at antiserum concentrations up to the anticomplementary limit of 1/500 dilutions.

---

immunological distance = 100 log (index dissimilarity)

Similar immunological distances between salmon B_4 and B'_4 LDH were obtained using either anti-B_4 or anti-B'_4 antiserum (15 and 18, respectively). Immunological distances of this magnitude have been reported for chicken versus duck B_4 LDH, and for halibut versus salmon A_4 LDH (Wilson et al., 1964). Neither of our antisera could be made to cross-react with salmon A_4 - A'_4 LDH, even at antiserum concentrations up to the anti-complementary limit of 1/500 dilution (Fig. 2).

Immunological cross-reactions were also assessed quantitatively by the enzyme precipitation/inhibition test. As shown in Fig. 3, anti-B_4 antiserum is capable of completely precipitating or inhibiting the enzymatic activity of B_4 LDH, but even high levels of antiserum were incapable of reacting with A_4 - A'_4. Similar results were obtained with brook trout LDH isozymes.

Immunological cross reaction can also be monitored qualitatively by incubation of isozyme mixtures with antiserum prior to electrophoresis and staining (Holmes and Markert,

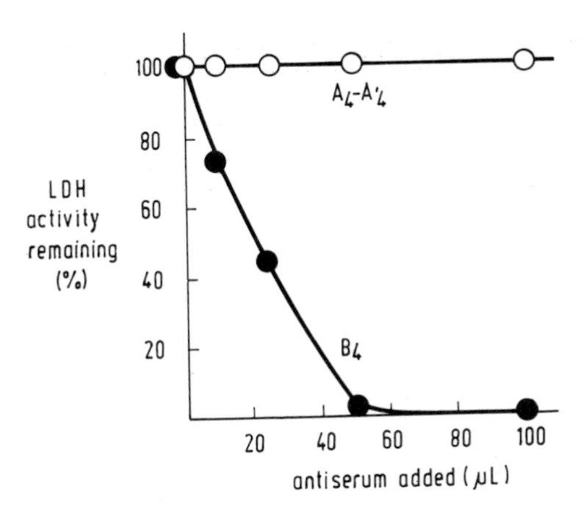

Fig. 3. Enzyme precipitation/inhibition of salmon lactate dehydrogenases by anti-salmon B_4 antiserum. Antiserum 201B1 (to purified salmon B_4) was added to a mixture containing 2.5 enzyme units of B_4 or A_4 - A'_4 in 0.5 ml final volume, incubated 18 hours at 4°, centrifuged (100,000 x g, 30 min) and the supernatant assayed for remaining LDH activity spectrophotometrically and by starch gel analysis (Holmes and Markert, 1969). All dilutions were made in microcomplement fixation buffer containing 0.1% albumin.

1969). Using this technique we were unable to detect any cross-reaction of our anti-salmon B'_4 or anti-salmon B_4 antisera with any A_4 - A'_4 LDH from salmon, brook trout, or brown trout. Nor could cross reaction be observed qualitatively by immunodiffusion, or quantitatively by the much more sensitive microcomplement fixation or enzyme precipitation/inhibition techniques.

CATALYTIC PROPERTIES

Purified salmon B_4, B'_4, and A_4 - A'_4 lactate dehydrogenases were examined for possible functional differences by comparison of several *in vitro* catalytic parameters. The response of these isozymes to variations in substrate and coenzyme levels, as well as their susceptibilities to product inhibition by pyruvate and lactate are summarized in Table III.

The B_4 isozyme is seen to have a greater affinity for each substrate and lower substrate optima than the A_4 - A'_4 series, and is more susceptible to inhibition by high levels of pyruvate. These same properties are characteristic of B_4 and A_4 lactate dehydrogenases of higher vertebrates (Wilson et al.,

TABLE III

CATALYTIC PROPERTIES OF SALMON LACTATE DEHYDROGENASE ISOZYMES

Property	Isozyme		
	B_4	B'_4	A_4 - A'_4
Substrate Response			
Pyruvate			
Km (mM) [a]	0.037 ± 0.003	0.075 ± 0.002	0.65 ± 0.08
Optimum (mM)	0.10	0.50	1.0
Inhibition (5 mM pyruvate)	40%	34%	11%
Lactate [b]			
Km (mM)	5.8 ± 0.6	11 ± 1	30 ± 3
Optimum (mM)	40	50	150
NADH [c]			
Km (mM)	0.026 ± 0.001	0.024 ± 0.001	0.031 ± 0.001
Optimum (mM)	0.14	0.14	0.10
Product Inhibition			
Pyruvate Ki (mM)[d]	0.075 ± 0.005	0.119 ± 0.005	0.187 ± 0.005
Lactate Ki (mM) [e]	29 ± 2	53 ± 4	103 ± 6

[a]Apparent Km and optimum for pyruvate were obtained at 0.14 mM NADH, 25°, pH 7.5, 0.1 M potassium phosphate. Values were computed as described by Lim et al., 1974, and listed together with computed standard deviations.

[b]Apparent Km and optimum for lactate were obtained at 2.0 mM NAD, 25°, pH 9.0, 0.01 M Tris-HCl.

[c]Apparent Km and optimum for NADH were measured using 0.33 mM sodium pyruvate, 25°, pH 7.5, 0.1 M potassium phosphate.

[d]Ki values for pyruvate were determined at pH 7.5, 0.1 M potassium phosphate, 25°, 2 mM NAD, computed as described elsewhere (Lim et al., 1974).

[e]Ki values for lactate were determined as above, with NAD at 0.14 mM.

1963; Wilson and Kaplan, 1964; Stambaugh and Post, 1966; Everse and Kaplan, 1973). Note that the B'_4 isozyme has properties somewhat intermediate between B_4 and A_4 - A'_4. Of particular importance for A_4-type LDH's is the resistance of these isozymes to lactate product inhibition as reflected in their Ki for lactate (Stambaugh and Post 1966; Everse and Kaplan, 1973). The B'_4 isozyme is seen to be nearly intermediate between salmon B_4 and A_4 - A'_4 in its resistance to lactate product inhibition.

DISCUSSION

STRUCTURAL AND EVOLUTIONARY RELATIONSHIPS

The results presented here on purified B_4 and B'_4 lactate dehydrogenase from quinnat salmon demonstrate that these two proteins are closely related, but not identical, in amino acid composition and in immunological behavior. Salmon B_4 and B'_4 are seen to be about as different immunologically as are the B_4 lactate dehydrogenases of chicken and duck, or the A_4 isozymes of halibut and salmon. Unfortunately the fossil record does not allow these lineage splits to be fixed chromologically with great precision; current estimates suggest that lineages leading to ducks (order Anseriformes) and to chickens (order Galliformes) split approximately 80-100 million years ago (Romer, 1945). A similar date can be given for the splitting of halibuts (superorder Acanthopterygii) and salmon (superorder Protacanthopterygii) (Greenwood et al., 1966). Thus if a duplication event gave rise to B_4 and B'_4 LDH in salmon, it must have occurred 80-100 million years ago; this event may coincide with the origin of the salmonid lineage.

FUNCTIONAL RELATIONSHIPS

As suggested above, several lines of evidence presented here point to the evolution of an altered and biologically essential function for one of the duplicated B_4-type lactate dehydrogenases in salmon, the B'_4 isozyme. First, salmon lactate dehydrogenases appear to be unusual in tissue distribution. The A_4 - A'_4 isozymes were observed in significant amounts in only three tissue types: striated muscle, lateral line, and gill tissue. Although very little quantitative information is available on lactate dehydrogenase isozyme ratios in various higher vertebrate tissues, most cells previously examined appear to require the presence of significant levels of both A and B subunits in the form of enzymatically active tetramers to meet a balanced catalytic requirement for lactate dehydrogenase (Markert and Møller; 1959, Cahn et al., 1962; Wilson et al., 1963; Vessell, 1965; Markert, 1965). In salmonids, it appears to be the B' subunit, rather than A, which occurs concomitantly with B in varying amounts in most tissues.

In addition, by a number of catalytic criteria, such as Michaelis constants for pyruvate and lactate, substrate inhibition by pyruvate and lactate, and substrate optima, the B'_4 isozyme is somewhat intermediate between B_4 and A_4 - A'_4 from salmon. For example, the B'_4 isozyme shows resistance to product inhibition by lactate with an inhibition constant (Ki) nearly midway between that for the B and A_4 - A'_4 isozymes. It has been previously suggested (Stambaugh and Post, 1966; Everse and Kaplan, 1973) that resistance to lactate product inhibition is the crucial catalytic property which accounts for the physiological role of A_4 lactate dehydrogenases. It seems clear that the primary catalytic requirement for the A_4 isozyme is that it be able to catalyze the conversion of pyruvate to lactate under relatively anaerobic conditions, when lactate (product) levels and NADH/NAD ratios are high, Thus the *in vitro* catalytic properties of B'_4, together with the unusual tissue distribution of isozymes, suggest that the B'_4 isozyme in salmonids is diverging from B_4 in functional behavior, perhaps to satisfy the normal physiological requirement for balanced levels of B_4 and A_4 lactate dehydrogenases in certain cell types. For reasons which are unclear, these fish appear unable to synthesize or maintain A-type polypeptides in most cell types; the B' subunit may be able to fulfill this need. At the same time, the B'_4 isozyme appears to be similar enough to B_4 in certain of its catalytic properties, such as Km for lactate, to serve as an effective catalyst for lactate scavenging in such tissues as salmon liver, which has only the B'_4 isozyme. In this sense the B'_4 isozyme in salmon may be a very unusual generalized lactate dehydrogenase, capable of functioning aerobically as well as anaerobically.

Finally, the *in vitro* catalytic properties of salmon B_4 lactate dehydrogenase reported here are similar to those of higher vertebrate B_4 isozymes (Stambaugh and Post, 1966; Everse and Kaplan, 1973). For an extensive discussion of the metabolic regulation of B_4 isozymes via *in vivo* ternary complex formation, and the relationship of these complexes to catalytic properties measured by conventional *in vitro* methods, see Everse and Kaplan, 1973.

ACKNOWLEDGEMENTS

This work was supported in part by research grants from the University Grants Committee, the University of Otago Research Committee, and the Medical Research Council of New Zealand. Part of this work was performed while one of us (S. T. L.) was supported by pre-doctoral fellowships from the Medical Research Coundil of New Zealand and from the Lee Foundation. We are grateful for the cooperation of the Otago Acclimitisation Society and the Ministry of Agriculture and Fisheries for their support in obtaining specimens for this study.

REFERENCES

Bailey, G. S. and A. C. Wilson 1968. Homologies between isozymes of fishes and those of higher vertebrates: evidence for multiple H_4 lactate dehydrogenases in trout. *J. Biol. Chem.* 243: 5843-5853.

Bailey, G. S., A. C. Wilson, J. E. Halver and C. L. Johnson 1970. Multiple forms of supernatant malate dehydrogenase in salmonid fishes. *J. Biol. Chem.* 245: 5927-5940.

Cahn, R. D., N. O. Kaplan,L. Levine, and E. Zwilling 1962. Nature and development of lactic dehydrogenases. *Science* 136: 952-969.

Everse, J. and N. O. Kaplan 1973. Lactate dehydrogenases: structure and function. *Advan. Enzymol.* 37: 61-133

Fritz, P. J., E. S. Vesell, E. L. White, and K. M. Pruitt 1969. The roles of synthesis and degradation in determining tissue concentrations of lactate dehydrogenase-5. *Proc. Nat. Acad. Sci.* 62: 558-565.

Greenwood, H. P., D. E. Rosen, S. H. Weitzman and G. S. Myers 1966. Phyletic studies of teleostean fishes, with a provisional classification of living forms. *Bulletin of the American Museum of Natural History* 131: 341-455.

Holmes, R. S. and C. L. Markert 1969. Immunochemical homologies among subunits of trout lactate dehydrogenase isozymes. *Proc. Nat. Acad. Sci. U.S.A.* 64: 205-210.

Lim, T. S., R. M. Kay, and G. S. Bailey 1974. Lactate dehydrogenase isozymes of salmonid fish: evidence for structural and functional divergence of duplicated H_4 lactate dehydrogenases. Submitted for publication.

Markert, C. L. 1965. Developmental genetics. In: *The Harvey Lectures* (Series 59). Academic Press, New York, pp. 187-218.

Markert, C. L. and F. Møller 1959. Multiple forms of enzymes: Tissue, ontogenetic, and species specific patterns. *Proc. Nat. Acad Sci. U. S. A.* 45: 753-763.

Massaro, E. J. and C. L. Markert 1968. Isozyme patterns of salmonid fishes: evidence for multiple cistrons for lactate dehydrogenase polypeptides. *J. Exp. Zool.* 168: 223-238.

Morrison, W. J. 1970. Nonrandom segregation of two lactate dehydrogenase subunit loci in trout. *Trans. Am. Fish Soc.* 99: 193-206.

Pesce, A., T. P. Fondy, F. Stolzenbach, F. Castillo, and N. O. Kaplan 1967. The comparative enzymology of lactate dehydrogenases. II. Properties of the H_4 and M_4 enzymes from a number of vertebrates. *J. Biol. Chem.* 242: 2151-2167.

Prager, E. M. and A. C. Wilson 1971. The dependence of immunological cross-reactivity upon sequence resemblance among lysozymes. I. Micro-complement fixation studies. *J. Biol. Chem.* 246: 5978-5989.

Romer, A. S. 1945. *Vertebrate Paleontology*. University of Chicago Press Chicago Illinois, p. 267.

Sarich, V. M. 1969. Pinniped origins and the rate of evolution of carnivore albumins. *Syst. Zool.* 18: 286-295.

Shaklee, J. B., K. L. Kepes, and G. S. Whitt 1973. Specialized lactate dehydrogenase isozymes: the molecular and genetic basis for the unique eye and liver LDH's of teleost fishes. *J. Exp. Zool.* 185: 217-240.

Stambaugh, R. and D. Post 1966. Substrate and product inhibition of rabbit muscle lactic dehydrogenase heart (H_4) and muscle (M_4) isozymes. *J. Biol. Chem.* 241: 1462-1467.

Vesell, E. S. 1965. Genetic control of isozyme patterns in human tissues. In: *Progress in Medical Genetics* (H. G. Strinberg and A. G. Bearn, eds.) Greene and Stratton: New York, N. Y. pp. 128-175.

Wilson, A. C., R. D. Cahn and N. O. Kaplan 1963. Functions of the two forms of lactic dehydrogenase in the breast muscles of birds. *Nature* 197: 331-334.

Wilson, A. C. and N. O. Kaplan 1964. Enzyme structure and its relation to taxonomy. In: *Taxonomic Biochemistry and Serology* (C. A. Leone, ed.) Ronald Press Co.

Wilson, A. C., N. O. Kaplan, L. Levine, A. Pesce, M. Reichlin, and W. S. Allison 1964. Evolution of lactic dehydrogenases. *Federation Proc.* 23: 1258-1266.

GENE DUPLICATION WITHIN THE FAMILY SALMONIDAE:
II. DETECTION AND DETERMINATION OF THE GENETIC CONTROL OF DUPLICATE LOCI THROUGH INHERITANCE STUDIES AND THE EXAMINATION OF POPULATIONS

FRED W. ALLENDORF, FRED M. UTTER, AND BERNARD P. MAY
Northwest Fisheries Center
National Marine Fisheries Service
National Oceanic and Atmospheric Administration
2725 Montlake Boulevard East
Seattle, Washington 98112

ABSTRACT. We examine the interpretation of electrophoretic data reflecting duplicated gene loci in salmonid fishes. Genetic models are considered to explain electrophoretic patterns of proteins reflecting duplicated genes where common alleles give rise to proteins of identical electrophoretic mobilities. It is shown that in the absence of breeding data it is impossible to distinguish between a model of tetrasomic inheritance and one of disomic inheritance where alleles segregating at two loci occur at the same frequency. However, it is shown that disomic inheritance can be verified by the examination of population phenotypic distributions under certain conditions. Family data are presented verifying disomic inheritance of malate dehydrogenase in rainbow trout (*Salmo gairdneri*) and aspartate aminotransferase in chum salmon (*Oncorhynchus keta)*. Examination of 19 salmonid biochemical systems indicate that 8 systems clearly reflect gene duplication, 8 do not, and 3 systems lack variation, precluding detection of gene duplication; genetic variants at all of the duplicated loci appear to segregate disomically. No significant evidence for genetic linkage was found among 10 jointly segregating pairs of loci in rainbow trout and one pair of loci in chum salmon; one comparison in rainbow trout (IDH-AGPDH) approached significance (P=.06).

INTRODUCTION

Measuring the patterns and amounts of genetic variation in natural populations of many species is presently a major thrust of experimental population genetics. This examination has only been possible since the development of the electrophoretic separation of proteins as a means of examining the gene products of many individual loci. Since 1968, our group has conducted an extensive survey of genetic variation in populations of fish and marine invertebrates. Our efforts have been concentrated somewhat on the salmonid fishes because of our interest in the evolutionary implications of their extensive gene duplication and their usefulness in inheritance studies. (The ripe sex products from both sexes can be artificially removed and stored under refrigeration while the parental types are determined electrophoretically prior to selecting the desired matings.)

The efforts of one author (Allendorf) have been concentrated on an intensive survey of genetic variation in populations of *Salmo* species with particular emphasis on the rainbow trout (*S. gairdneri*). The extensive gene duplication found in salmonids often confounds the genetic interpretation of the observed variation. Because of the critical importance of understanding this genetic basis we are involved in a series of inheritance studies directed toward the identification of the genetic control of polymorphic loci in salmonids.

The initial electrophoretic studies with salmonids revealed additional loci coding for lactate dehydrogenase (LDH) in comparison with other vertebrate species (Morrison and Wright, 1966). Subsequent studies of salmonids have revealed similar gene duplications for a number of other enzymes--e.g. malate dehydrogenase (Bailey et al., 1970) and isocitrate dehydrogenase (Wolf et al., 1970). Ohno has postulated that the salmonids are descended from a recent tetraploid ancestor based upon this gene duplication, relative DNA contents, and chromosomal characteristics (Ohno et al., 1969). Although Ohno's theory is sound and based upon valid evidence, some subsequent attempts to reinforce this theory have not been sound (see Allendorf and Utter, 1973). Upon reading the relevant literature one is left with the impression that virtually all enzymes in salmonids are coded for by duplicate loci and that many of these loci are segregating tetrasomically. Our studies are in conflict with this impression. We have found no evidence for the duplication of many protein loci and we have found no indication of tetrasomic inheritance even for those loci which have been reported to be inherited tetrasomically (this in no way rules out the existence of tetrasomic loci in salmonids but rather points out that they have not yet been detected).

The purpose of this present paper is to examine the interpretation of duplicate loci detected electrophoretically and to review our genetic investigations of rainbow trout in view of the accepted level of gene duplication and patterns of inheritance of salmonid loci. The evolutionary question to be considered in the course of these studies is the fate of duplicated loci. We feel that the salmonids represent an outstanding opportunity to examine this question experimentally.

DETERMINATION OF GENETIC CONTROL OF DUPLICATED LOCI

The multiple loci coding for many salmonid proteins often make the interpretation of genetic control rather difficult. The disagreements which can result are illustrated by the reported number of loci coding for LDH in salmonids; values all the way from five to eight have been reported in the literature (Massaro and Markert, 1968).

We will examine a hypothetical case of a single enzyme and outline some of the difficulties in determining the exact nature of genetic control of that enzyme. We will restrict our example to the case of a maximum of two disomic loci (or a single tetrasomic locus) with two alleles (A and A'). Table 1 outlines possible genetic models controlling these loci, while Figure 1 presents typical electrophoretic phenotypes for a dimeric enzyme in this hypothetical case. Two basic questions are to be examined: (1) Is there any evidence of gene duplication? and (2) If there has been gene duplication, is there a single tetrasomic locus or two disomic loci?

PHENOTYPIC DISTRIBUTIONS IN POPULATIONS

Predictions of the expected phenotypic distributions for each of the models can be made assuming Hardy-Weinberg proportions--see Table 2. Certain conclusions can be drawn pertaining to the nature of genetic control involved by comparing the observed distributions within populations with these theoretical distributions.

Models A, C, and G all predict a single-banded phenotype to be found in all individuals of a population because of the lack of polymorphism. Therefore, no conclusions of genetic control can be made if this distribution is found in a population. However, model D which also lacks polymorphism, predicts a fixed multi-banded phenotype to be found in all individuals. If this "fixed heterozygote" situation is found in a population it is best explained by two monomorphic, disomic loci fixed for alleles with differing electrophoretic mobilities.

Model B is also distinguishable in that there are no asymmetrical banded phenotypes and the single banded alter-

TABLE 1
List of the possible genetic models to be considered that could be controlling a protein system examined electrophoretically. (P = frequency of allele A; Q = frequency of allele A'.)

Genetic Model	Disomic Loci Mono-morphic	Poly-morphic	Tetrasomic Loci Mono-morphic	Poly-morphic	Allele Frequencies
A	1	0	0	0	$p=1$
B	0	1	0	0	p
C	2	0	0	0	$p_1=p_2=1$
D	2	0	0	0	$p_1=1, p_2=0$
E	1	1	0	0	$p_1=1, p_2$
F	0	2	0	0	p_1, p_2
G	0	0	1	0	$P=1$
H	0	0	0	1	P

nate homozygote type is found at a much higher frequency than under any of the other models.

Models E, F, and H are not so easily distinguished. In fact, Models F (two polymorphic disomic loci) and H (a single polymorphic tetrasomic locus) predict identical phenotypic distributions when $q_1=q_2$ for model F. Therefore, the existence of tetrasomic inheritance can never be conclusively demonstrated by examining the phenotypic distribution in a population since a disomic model can always fit the data equally well. When there is a comparatively high amount of variation model E is distinguishable from models F and H by virtue of the absence of phenotypes four and five. Accordingly, when the amount of variation is low these models are not easily distinguished. It is interesting to examine the predicted phenotypic distributions of these three models as shown in Figure 2. The expected Hardy-Weinberg proportions of a population are represented by the family of values between these two curves in the case where the correct model cannot be distinguished. Similarly, if Model F is shown to be the correct one (through inheritance studies as outlined

TABLE 2

Expected Hardy-Weinberg proportions for each of the eight genetic models considered as outlined in Table 1.

Duplicated Phenotype	Non-Dupl Phenotype	Expected Phenotypic Distributions Genetic Model: A	B	C	D	E	F	G	H
A_4	A_2	1	p^2	1	0	p_2^2	$p_1^2 p_2^2$	1	P
A_3A'	---	0	0	0	0	$2p_2q_2$	$2p_1^2 p_2 q_2 + 2p_1 p_2^2 q_1$	0	$4P^3Q$
$A_2A'_2$	AA'	0	2pq	0	1	q_2^2	$p_1^2 q_2^2 + 4p_1 p_2 q_1 q_2 + p_2^2 q_1^2$	0	$6P^2Q^2$
AA'_3	---	0	0	0	0	0	$2p_1 q_1 q_2^2 + 2p_2 q_1^2 q_2$	0	$4PQ^3$
A'_4	A'_2	0	q^2	0	0	0	$q_1^2 q_2^2$	0	Q^4

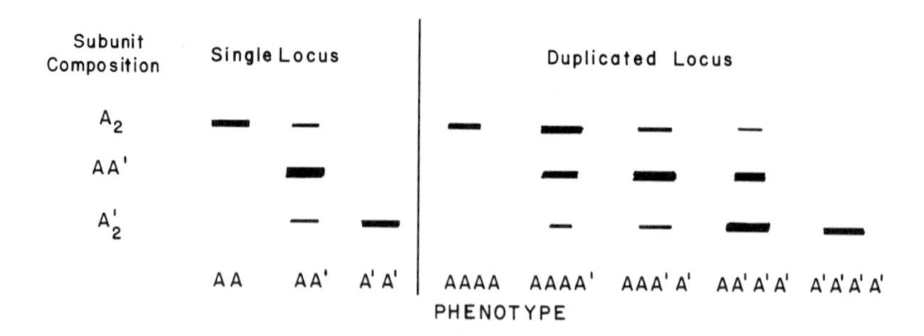

Figure 1. Diagrammatic representation of typical electrophoretic phenotypes for a dimeric enzyme controlled by either a single locus or a duplicated locus.

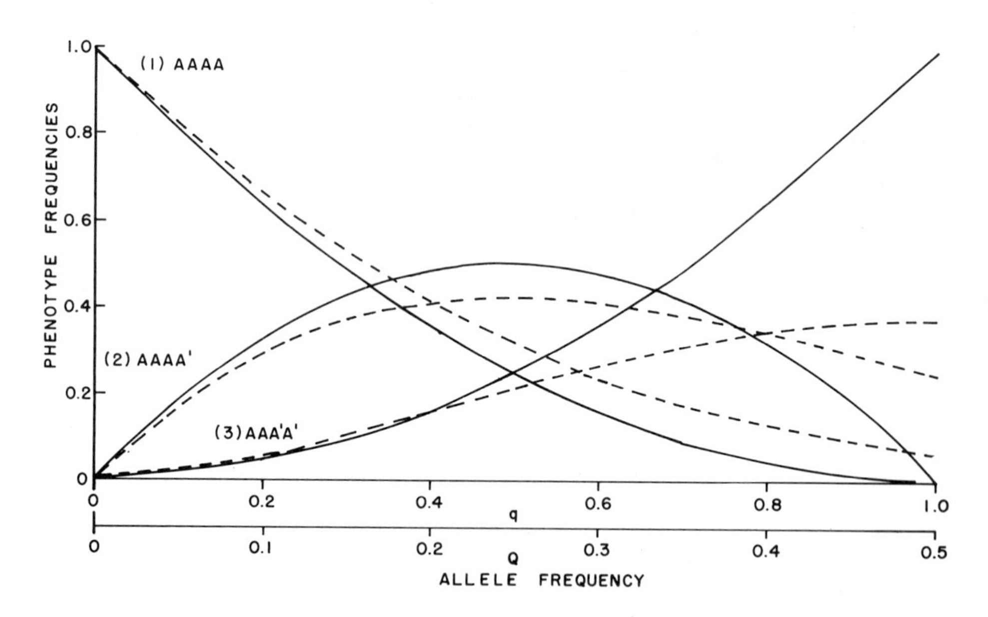

Fig. 2. Graphical representation of expected phenotypic distributions for Model E (solid line) and Models F and H (dotted line) when $q_1=q_2$ under Model F. Upon examining a population, the overall allele frequency (Q) of the A' allele can be determined by dividing the total number of A' alleles observed by 4N (where N=number of individuals examined). This frequency is the allele frequency for a tetrasomic locus

Legend for Fig. 2, continued:
(Q) and is also equal to both q_1 and q_2 ($Q=q_1=q_2$) under Model F when $q_1=q_2$. This frequency is related to q_2 from Model E by the equation $q_2=2Q$. This graph demonstrates the differences in expected Hardy-Weinberg proportions for a duplicated locus for these three cases; (1) a single polymorphic locus, (2) both loci polymorphic and (3) a single tetrasomic locus. When $q_1 \neq q_2$ for case (2) the expected Hardy-Weinberg proportions lie between these curves.

---

in the next section), the expected Hardy-Weinberg proportions are represented by this same family since gene frequencies cannot be assigned to individual loci.

The major conclusions that can then be drawn from this section are that (1) tetrasomic inheritance cannot be conclusively demonstrated only by the phenotypic distribution of a population and (2) under certain conditions, disomic inheritance can be demonstrated in this way.

INHERITANCE EXPERIMENTS

The mode of inheritance of duplicated loci (either disomic or tetrasomic can be positively verified through inheritance studies. Individuals with single doses of a variant allele do not provide useful information since their gametes are expected to segregate 1:1 (AA : AA') under both a tetrasomic and disomic model. However, individuals with a double dose of a variant allele do provide sufficient information to distinguish between these models.

Examination of the gametic output from an individual typed as AAA'A' will provide the desired information. The phenotype AAA'A' can represent any one of three genotypes.

Genotype I (Disomic) $A_1A_1A_2'A_2'$: A/A ; A'/A'

Genotype II (Disomic) $A_1A_1'A_2A_2'$: A/A' ; A/A'

Genotype III (Tetrasomic) AAA'A': A/A/A'/A'

These genotypes and their expected gametic ratios are presented in Table 3. It can be seen that each genotype predicts a unique gametic distribution. In the disomic case, the proportion of genotypes I and II reflect the gene frequencies p_1 and p_2. The gametic ratios for genotypes II and III are variable because of possible linkage in the disomic model and possible double reduction division in the tetrasomic model. Therefore, to verify disomic inheritance, parental types corresponding to both genotypes I and II should be observed. In addition, it should be noted that examining the gametic ratios from genotype II will also provide a measure of possible linkage between the two loci.

MATERIALS AND METHODS

The results presented here are based on the examination of thousands of rainbow trout from numerous populations as well as from a continuing series of experimental matings. We have examined both non-anadromous and anadromous (steelhead) populations of *S. gairdneri*, and have found no reason for treating these two forms as being distinct. For purposes of this paper, we will use the common name rainbow trout to represent both the anadromous and non-anadromous forms.

Our methods have been documented elsewhere (Utter et al., 1973a). Three buffer systems were used: (1) a discontinuous system described by Ridgway et al., (1970), (2) a continuous tris-borate-EDTA system (pH 8.6) described by Markert and Faulhaber (1965), and (3) a continuous phosphate system (pH 6.5) described by Wolf et al., (1970).

The following proteins were examined in the course of these studies: AAT (aspartate aminotransferase); ADH (alcohol dehydrogenase); AGPDH (alphaglycerophosphate dehydrogenase); DIA (diaphorase); EST (esterase); IDH (isocitrate dehydrogenase); LDH (lactate dehydrogenase); MDH (malate dehydrogenase); ME (malic enzyme); 6PGDH (6-phosphogluconate dehydrogenase); PGM (phosphoglucomutase); SDH (sorbitol dehydrogenase); TFN (transferrin); TO (tetrazolium oxidase); and XDH (xanthine dehydrogenase). All of these proteins migrated anodally with the exception of ADH.

The mating experiments presented for MDH-B and AAT were carried out in the 1973-1974 spawning season. The progeny from these matings are still being examined. A more detailed report of the inheritance of these loci will be presented elsewhere.

TABLE 3

Comparison of disomic verus tetrasomic inheritance in an individual carrying two doses of a variant allele.

PARENTAL GENOTYPE

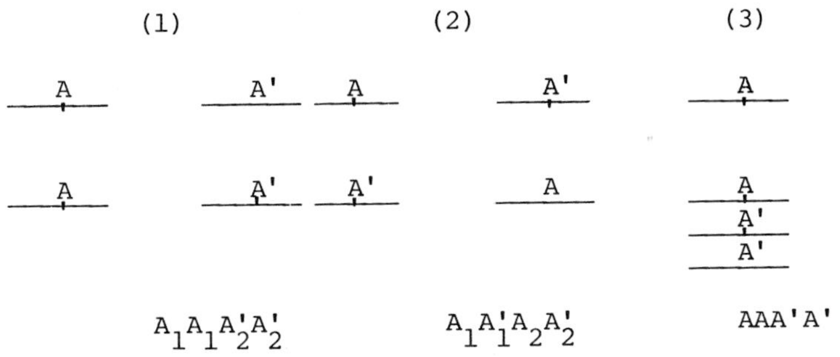

(1)		(2)	(3)
A / A	A' A / A' A'	A' / A	A / A / A' / A'
	$A_1A_1A_2'A_2'$	$A_1A_1'A_2A_2'$	AAA'A'

COMPARATIVE SEGREGATION RATIOS OF GAMETES

Gamete Genotype	Parental Genotype (I)	(II)*	(III)**
AA	0	1	1
AA'	1	2	4
A'A'	0	1	1

* Assuming no linkage.

** Assuming chromosome segregation.

RAINBOW TROUT PROTEIN LOCI

DIA, 6PGDH, XDH. These enzymes were found to be represented by a single invariant band in all individuals examined. Since this result is in agreement with models A, C, and G, no conclusions can be made as to whether they are coded for a single locus or duplicated loci.

ME. Two forms of this enzyme were observed--one which was predominant in muscle extracts and one which was predominant in liver extracts. Both of these forms were uniformly seen as a single band. As explained above, the number of loci coding for each form could therefore not be determined.

ADH. The great majority of individuals examined also displayed a single invariant band for this enzyme. A few variant types with a more cathodal allele have been seen however. In these variant individuals, there is no indication of any asymmetrical banding intensities as would be expected for a duplicated locus. For this reason ADH is assumed to be coded for by a single disomic locus.

PGM, TO, TFN. These enzymes have all been previously reported by our laboratory to be represented by single disomic polymorphic loci in rainbow trout (Utter and Hodgins, 1972). A report of possible gene duplication of TO (Cederbaum and Yoshida, 1972) was found to be in error (Utter et al., 1973b).

LDH. Our previously reported results (Utter et al., 1973a) indicate five loci coding for LDH in salmonids. This enzyme is an example of fixed heterozygosity in that all five loci have common alleles with differing mobilities. The duplicated LDH-B loci present an especially interesting evolutionary history. In addition to evolving common alleles of different mobilities, these loci have also evolved differential tissue specificity. Only the B_2 locus is expressed in liver tissue while the B_1 locus is strongly predominant in heart tissue.

Although LDH is the classic example of duplicate loci in salmonids, not all salmonid LDH loci display gene duplication. The eye form of LDH is present in many families of fish and is presumed to be the result of a single evolutionary event (Horowitz and Whitt, 1972). This indicates that the LDH eye form was present in the salmonid lineage before the presumed polyploid event. However, genetic evidence has been presented which shows the salmonid eye LDH to be coded by a single locus (Morrison and Wright, 1966; Wright and Atherton, 1970).

AGPDH. Previous publications from our group have reported this enzyme to be represented by a single polymorphic disomic locus (Utter and Hodgins, 1972). Examination of this enzyme with buffer system (3) clearly revealed the presence of an additional locus not detectable with buffers (1) or (2). The gene products of this second locus do interact with the products of the first locus in that the appropriate heterodimers are formed. This additional locus is fixed for an allele with a different mobility than both alleles seen at the first locus, thereby creating a fixed heterozygote effect.

EST. A single polymorphic disomic locus predominantly expressed in liver extracts has been observed. Inheritance studies have confirmed the genetic basis of this variation but there are indications of possible ontogenetic and environmental effects on the expression of this locus. There is no indication of duplication of this locus.

SDH. This enzyme has been reported to be tetrasomically inherited in rainbow trout on the basis of the observed phenotypic distribution in a single population (Engel et al., 1970). As shown in a prior section, tetrasomic inheritance cannot be verified in this manner. We have found multiple banded phenotypes in all rainbow trout examined in accordance with model D, indicating this enzyme to be coded for by two disomic loci with common alleles of different mobilities. We have seen genetic variation at low frequency for this enzyme but have not found it in the populations used for the mating experiments. However, we do have preliminary inheritance data for a similar variant found in cutthroat trout (*S. clarki*) which verifies this interpretation.

IDH. The supernatant form of this enzyme (IDH-s) has also been reported to be inherited tetrasomically in rainbow trout on the basis of the phenotypic distribution in a population (Wolf et al., 1970). In an inheritance study previously reported (Allendorf and Utter, 1973), we have shown this variation to be controlled by two disomic loci.

The mitochondrial form of this enzyme (IDH-m), as best seen in muscle extracts, is represented by three nonvariant bands indicating the presence of two monomorphic disomic loci with common alleles of different mobilities.

MDH-B. Bailey et al., (1970) presented an excellent biochemical and genetic analysis of salmonid MDH demonstrating the existence of duplicate loci coding for the B form. However, since all of their matings involved individuals with at most a single dose of the variant allele they could not distinguish between disomic and tetrasomic inheritance. We have recently made the necessary crosses to answer this

TABLE 4

Inheritance of duplicate loci. In each case only the results from one family are presented here. Additional families (some with parental types corresponding to genotype 1 in Table 3) further confirm these conclusions.

(a) Rainbow Trout MDH-B

PROGENY PHENOTYPES: OBSERVED, (DISOMIC), /TETRASOMIC/

PARENTAL PHENOTYPES		BBBB	BBBB'	BBB'B'	X^2	d.f.	P
BBBB	BBB'B'	80	165	70			
		(78.8)	(157.5)	(78.8)	1.36	2	> 0.5
		/52.5/	/210.0/	/52.5/	33.36	2	< 0.0001

(b) Chum Salmon AAT

PROGENY PHENOTYPES: OBSERVED, (DISOMIC), /TETRASOMIC/

PARENTAL PHENOTYPES		AAAA	AAAA'	AAA'A'	X^2	d.f.	P
AAAA	AAA'A'	44	85	43			
		(43.0)	(86.0)	(43.0)	0.03	2	> 0.9
		/28.7/	/114.7/	/28.7/	27.80	2	< 0.0001

TABLE 4, continued

Conclusions for both cases:

(1) Inherited disomically

(2) Parental genotypes - (a) $B_1B_1B_2B_2$ X $B_1B'_1B_2B'_2$

(b) $A_1A_1A_2A_2$X $A_1A'_1A_2A'_2$

(3) No indication of linkage

---

question. Table 4 outlines the results from one of these crosses. There are clearly two disomic loci both of which are polymorphic in the population we examined. Additional results from other families confirmed this conclusion. In addition, Table 4 examines possible linkage between the two loci by comparing the frequencies of the two linkage classes (BBBB + BBB'B' : BBBB'). There is no indication of linkage on this basis.

MDH-A. Bailey et al., (1970) presented some evidence for the duplication of this locus in brown trout (*S. trutta*) and subsequent evidence has supported this conclusion (Bailey, personal communication). We have seen a variant of this form at a low frequency in some rainbow trout populations. Based on the relative intensities of bands in the variant phenotypes it appears that this locus is not duplicated in rainbow trout. Additional evidence in the form of inheritance data is desirable before it can be firmly concluded that this enzyme is coded for by different number of loci in these two closely related species.

AAT. The common phenotype found in rainbow trout for this enzyme is a single distinct band. There is a low frequency variant found in some populations. The intensities of bands found in the variant phenotypes are typical of a duplicated locus. Inheritance studies have not been carried out for this enzyme with rainbow trout. However, inheritance studies have been done for this enzyme with the chum salmon (*Oncorhynchus keta*). The results from one family from these experiments are shown in Table 4. Comparison with the expected disomic and tetrasomic ratios indicate that chum salmon AAT is controlled by two polymorphic disomic loci.

As with rainbow trout MDH-B loci, there is no indication of linkage between these duplicated loci. One cannot conclude on this basis that a similar mode of genetic control exists in rainbow trout. However, these data directly apply to the question of genetic control of duplicated loci in salmonids.

JOINT SEGREGATION OF LOCI

Whenever possible in the course of our inheritance experiments, we have tested for linkage between segregating loci by testing for any aberrant 2-way joint segregation in progeny from double heterozygote individuals. The results of these tests are summarized in Table 5. There is no definite indication of linkage. In the case of IDH - AGPDH, the deviation from random segregation did approach significant proportions ($P > .06$). Unfortunately, it has not been possible to examine further the joint segregation of these loci.

Of particular interest is the joint segregation of duplicated loci. Aberrant joint segregation of duplicated LDH loci in both brook trout (*Salvelinus fontinalis*) and rainbow trout have been reported (Morrison, 1970; Davisson et al., 1973). Recently, Aspinwall (1974) has reported limited data indicating aberrant joint segregation of duplicated MDH loci in pink salmon (*O. gorbuscha*). Our results with MDH-B in rainbow trout and AAT in chum salmon do not indicate any abnormal joint segregation of these duplicated loci in these species.

SUMMARY

Table 6 presents a summary of the rainbow trout loci we have examined. Contrary to the statements of some authors, a significant portion of loci in this salmonid species show no evidence of gene duplication. If we accept the tetraploid origin of salmonids, these loci then seem to represent evidence of Haldane's (1933) original suggestion that if a gene is duplicated one of the two resulting genes may become nonfunctional because of the fixation of a deleterious mutation. One other possible fate of a duplicated gene is the evolution of differential function as demonstrated by the duplicated LDH-B loci which have evolved differential tissue specificity.

We have found no evidence of tetrasomic inheritance in salmonids. We have also shown that previous reports of tetrasomic inheritance in salmonids are not sound. These results do not, of course, rule out the existence of tetrasomic loci in salmonids, but rather are meant to show that there is

TABLE 5
Joint segregation ratios of rainbow trout (and chum salmon) loci.

Loci		Linkage Class I	Linkage Class II	Chi-square*
MDH-B_1	IDH-3	62	55	.42
		34	23	2.12
TO	IDH-3	69	50	3.03
		52	46	.37
MDH-B_1	TO	40	53	1.82
		71	76	.17
		89	79	.60
MDH-B_1	AGPDH-1	73	72	.01
LDH-B_2	IDH-3	66	59	.39
LDH-B_2	TO	53	57	.15
		86	74	.90
		100	108	.31
LDH-B_2	MDH-B_1	53	44	.84
		48	52	.16
		121	127	.15
TO	AGPDH-1	70	79	.54
IDH-3	AGPDH-1	49	70	3.71
MDH-B_1	MDH-B_2	150	165	.71
**AAT-1	AAT-2	85	87	.02

* - d.f. = 1
** - Chum salmon

TABLE 6
Summary of rainbow trout loci examined.

Protein	Duplicated?	Loci	Mode of inheritance (if duplicated)
AAT	Yes	AAT-1 AAT-2	(Disomic in chum salmon)
ADH	No	ADH	
AGPDH	Yes	AGPDH-1 AGPDH-2	Disomic
DIA	?	DIA	
EST	No	EST	
IDH(m)	Yes	IDH-1 IDH-2	Disomic
IDH(s)	Yes	IDH-3 IDH-4	Disomic
LDH-A	Yes	LDH-A_1 LDH-A_2	Disomic
LDH-B	Yes	LDH-B_1 LDH-B_2	Disomic
LDH-C	No	LDH-C	
MDH-A	No	MDH-A	
MDH-B	Yes	MDH-B_1 MDH-B_2	Disomic
ME	No	ME-1 ME-2	
6PGDH	?	6PGDH	
PGM	No	PGM	

TABLE 6, continued:

Protein	Duplicated?	Loci	Mode of inheritance (if duplicated)
SDH	Yes	SDH-1 SDH-2	Disomic
TFN	No	TFN	
TO	No	TO	
XDH	?	XDH	

at present no evidence for tetrasomic inheritance in salmonids.

REFERENCES

Allendorf, F.W. and F.M. Utter 1973. Gene duplication within the family Salmonidae: Disomic inheritance of two loci reported to be tetrasomic in rainbow trout. *Genetics* 74: 647-654.

Aspinwall, N. 1974. Genetic analysis of duplicate malate dehydrogenase loci in the pink salmon, *Oncorhynchus gorbuscha*. *Genetics* 76: 65-72.

Bailey, G.S., A.C. Wilson, J.E. Halver, and C.L. Johnson 1970. Multiple forms of supernatant malate dehydrogenase in salmonid fishes. *J. Biol. Chem.* 245: 5927-5940.

Cederbaum, S.D. and A. Yoshida 1972. Tetrazolium oxidase polymorphism in rainbow trout. *Genetics* 72: 363-367.

Davisson, M.T., J.E. Wright, and L.M. Atherton 1973. Cytogenetic analysis of pseudolinkage of LDH loci in the teleost genus *Salvelinus*. *Genetics* 73: 645-658.

Engle, W., J. Op't Hof, and U. Wolf 1970. Genduplikation durch polyploide Evolution: die Isoeyzyme der Sorbit dehydrogenase bei herings-und lachsartigen Fischen (Isospondyli). *Humangenetik* 9: 157-163.

Haldane, J.B.S. 1933. The part played by recurrent mutation in evolution. *Amer. Nat.* 67: 5-19.

Horowitz, J.J. and G.S. Whitt 1972. Evolution of a nervous system specific lactate dehydrogenase isozyme in fish. *J. Exp. Zool.* 180: 13-32.

Markert, C.L. and I. Faulhaber 1965. Lactate dehydrogenase isozyme patterns of fish. *J. Exp. Zool.* 159: 319-332.

Massaro, E.J. and C.L. Markert 1968. Isozyme patterns of salmonid fishes: evidence for multiple cistrons for lactate dehydrogenase. *J. Exp. Zool.* 168: 223-238.

Morrison, W.J. 1970. Nonrandom segregation of two lactate dehydrogenase subunit loci in trout. *Trans. Amer. Fish. Soc.* 99: 193-206.

Morrison, W.J. and J.E. Wright 1966. Genetic analysis of three lactate dehydrogenase isozyme systems in trout: Evidence for linkage of genes coding subunits A and B. *J. Exp. Zool.* 163: 259-270.

Ohno, S.M., J. Muramato, J. Klein, and N.B Atkin 1969. Diploidtetraploid relationship in Clupeoid and Salmonid fish. *Chromosomes Today* 2: 139-147.

Ridgway, G.J., S.W. Sherburne, and R.D. Lewis 1970. Polymorphism in the esterases of Atlantic herring. *Trans. Amer. Fish. Soc.* 99: 147-151.

Utter, F.M., F.W. Allendorf, and H.O. Hodgins 1973a. Genetic variability and relationships in Pacific salmon and related trout based on protein variations. *Syst. Zool.* 22: 257-270.

Utter, F.M., F.W. Allendorf, H.O. Hodgins, and A.G. Johnson 1973b. Response to tetrazolium oxidase polymorphism in rainbow trout. *Genetics* 73: 159.

Utter, F.M. and H.O. Hodgins 1972. Biochemical genetic variation at six loci in four stocks of rainbow trout. *Tran. Amer. Fish. Soc.* 101: 494-502.

Wolf, U., W. Engel, and J. Faust 1970. Zum mechanismus der Diploidisierung in der Wirbeltier evolution: Koexistanz von tetrasomen und disomen Genloci der isocitrat-Dehydrogenasen bei: der Regenbagen forelle (*Salmo irideus*). *Humangenetik* 9: 150-156.

Wright, J.E. and L. M. Atherton 1970. Polymorphism for LDH and transferrin loci in brook trout populations. *Trans. Amer. Fish. Soc.* 99: 179-192.

GENETICS OF MULTIPLE SUPERNATANT AND MITOCHONDRIAL MALATE DEHYDROGENASE ISOZYMES IN RAINBOW TROUT *(Salmo gairdneri)*

J. W. CLAYTON,[1] D. N. TRETIAK,[1]
B. N. BILLECK,[1] and P. IHSSEN[2]
Canada Department of the Environment
Fisheries and Marine Service
[1]Freshwater Institute, Winnipeg, Manitoba R3T 2N6
[2]Ontario Ministry of Natural Resources
Research Branch, Maple, Ontario

ABSTRACT. Rainbow trout *(Salmo gairdneri)* were previously shown to produce two kinds of supernatant malate dehydrogenase (MDH) subunits, A and B. Furthermore, it was adduced by analogy with genetic studies in another salmonid species that the "B" MDH locus was duplicated and carried two alleles (Bailey et al., 1970). The electrophoretic nature of rainbow trout MDH has been reinvestigated using a pH 6.1 amine-citrate buffer for electrophoresis. This buffer system resolves MDH isozymes that are indicative of four alleles for the "B" supernatant MDH subunit in several unrelated populations. These electrophoretic procedures also resolve several mitochondrial MDH isozymes, one of which is genetically polymorphic and appears to be specified by four genes.

Evidence from both genetic experiments and population surveys indicates that one "B" supernatant MDH locus is represented by all four known alleles and the other by at least three alleles. The data are all consistent with the view that salmonid fishes have passed through an event of gene duplication in the course of evolution and have subsequently achieved a double diploid state.

INTRODUCTION

Malate dehydrogenase (MDH) (L-malate; NAD oxidoreductase, EC 1.1.1.37) occurs in two forms, supernatant (or cytoplasmic), and mitochondrial in all animal species. In many fish species the supernatant MDH genes, at least, seem to be duplicated and three kinds of supernatant MDH isozyme AA, AB, and BB are seen in every individual (Numachi 1970; Wheat and Whitt, 1971, and Clayton et al., 1971). In two of these reports (Numachi, 1970; Clayton et al., 1971) evidence of genetic polymorphism of the B supernatant MDH was also presented. In salmonid fishes not only is there evidence for genetic polymorphism of B supernatant MDH but there is also evidence for a redoubling of the genes for B supernatant MDH (Bailey et al., 1970, Numachi et al., 1972). We report here a further investigation of MDH in rainbow trout that has disclosed a total of four alleles for B supernatant MDH as well as genetic polymorphism of some of the mitochondrial MDH isozymes.

MATERIALS AND METHODS

The fish utilized in this study came from a variety of sources. The lots designated Idaho No. 1 and No. 2 originated as "eyed" eggs from Mr. A. Dunn, Cariboo Trout Ranch, Soda Springs, Idaho. The fish designated Tunkwa Lake originated as fertilized eggs taken from a native population of rainbow trout at Tunkwa Lake, British Columbia (50°55' N, 120°50' W). The Idaho and Tunkwa fish were hatched and reared at the Freshwater Institute hatchery at Rockwood, Manitoba. The Maple, Ontario, stock are fish that have been maintained for many generations in the Ontario Ministry of Natural Resources fish hatchery at Maple, Ontario. The Nottawasaga River stock were hatched and reared at Maple from eggs taken from fish that normally spawn in the Nottawasaga River, Ontario. This river is a tributary of Georgian Bay, Lake Huron. These fish have become naturalized to a steelhead-like life pattern and are thought to be derived from various introductions of fish from California made in the latter part of the nineteenth century. All the specific matings for genetic studies were made at Maple and the progeny were also reared there.

Starch gel electrophoresis was conducted with the apparatus described by Tsuyuki et al., 1966. The general procedures for extract preparation, electrophoresis and isozyme staining were those of Clayton et al., 1971. The pH 6.1 N-(3-aminopropyl)-morpholine-citrate electrophoresis buffer used routinely here was described by Clayton and Tretiak 1972. The responses of isozymes to heat denaturation and to coenzyme analogues were measured according to the technique developed by Clayton et al., 1973.

RESULTS AND DISCUSSION

The abundant isozymes found in fishes have often presented some problems in nomenclature and also in the determination of homology of isozymes in interspecies comparisons. It has already been shown that in walleye *(Stizostedion vitreum vitreum)* and sauger *(S. canadense)* the least anodal supernatant MDH isozyme predominates in liver (Clayton et al., 1973) The least anodal supernatant isozyme also predominates in the livers of various bass *(Micropterus)* species (Wheat et al.,

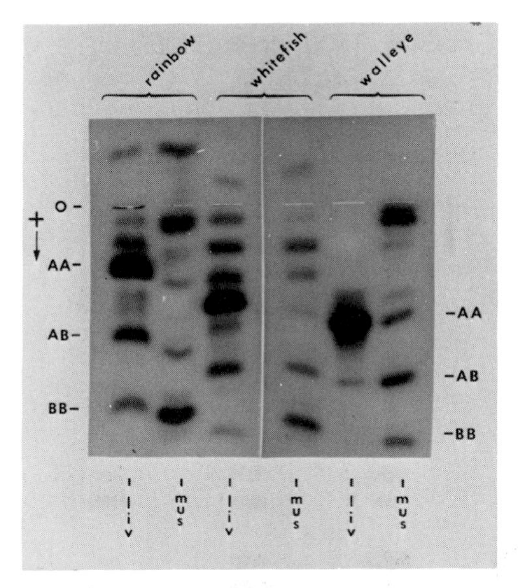

Fig. 1. Comparison of MDH isozymes in liver (liv) and white skeletal muscle (mus) tissue of the BB homozygous rainbow trout, lake whitefish, and walleye.

1971). The bass, walleye, and sauger are all spiny-rayed fishes distantly related to the salmonids and possible homology in tissue specificity of isozyme synthesis is uncertain. However, as Fig. 1 shows, the least anodal supernatant isozyme does indeed predominate in the liver of rainbow trout and lake whitefish (*Coregonus clupeaformis*), both of the family *Salmonidae*, as well as in the liver of the spiny-rayed walleye. In the nomenclature developed by Bailey et al., 1971, this is the AA supernatant MDH isozyme.

The model for the genetic and molecular structure of salmonid supernatant MDH isozymes developed by Bailey et al., 1970, is shown in Fig. 2. The supernatant isozymes are designated AA, AB, and BB and we show in Fig. 2 the five phenotypes generated by the alleles $\underline{b}^2$ and $\underline{b}^3$. The genetics of this system will be discussed in detail later. At this point we wish to draw attention to the three kinds of heterozygotes that are produced by this reduplicated genetic system that is made up of two loci represented by two alleles, $\underline{b}^2$ and $\underline{b}^3$. The ratios of intensity of the BB supernatant isozymes in the heterozygotes vary through the series 9:6:1 to 1:2:1 to 1:6:9 as shown in Fig. 2 for B phenotypes 2223, 2233, and 2333. The AB isozymes also vary in intensity through a parallel series 3:1, 1:1, 1:3 for the same phenotypes directly reflecting the $\underline{b}$ gene dosage. This much was described by Bailey et al. in 1970. We have also included in Fig. 2. three other isozymes XX, YY,

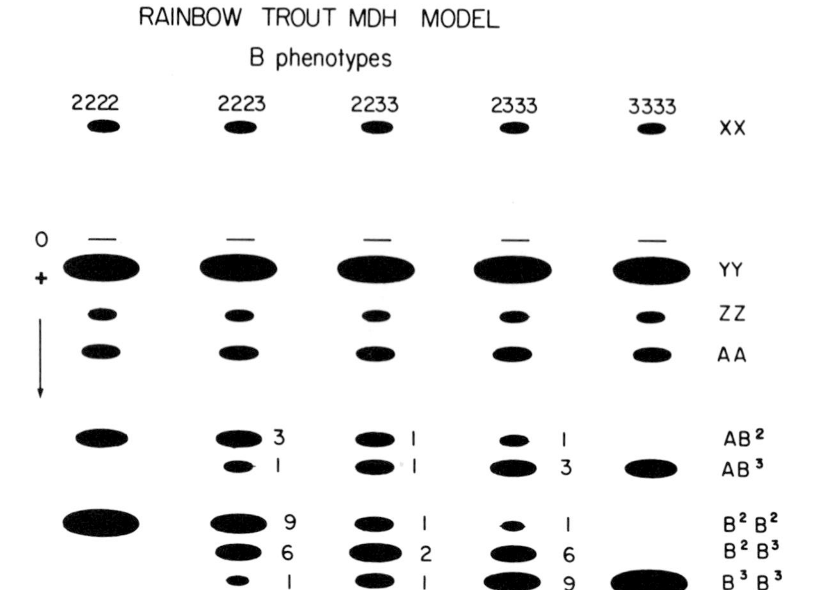

Fig. 2. Diagram (after Bailey et al., 1970) of MDH isozymes in rainbow trout generated by two alleles ($\underline{b}^2$ and $\underline{b}^3$) at duplicated B supernatant MDH loci. Isozymes AA, AB, BB, supernatant MDH; isozymes XX, YY, ZZ, mitochondrial MDH.

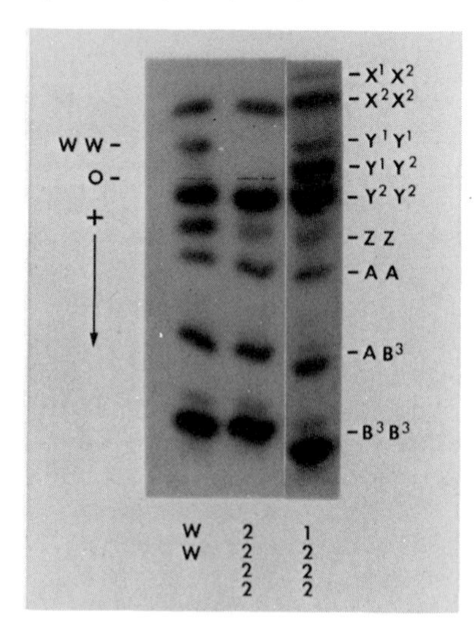

Fig. 3. Zymograms of Y mitochondrial MDH phenotypes in rainbow trout.

and ZZ. As will be shown later these are the major mitochondrial MDH isozymes in rainbow trout that are resolved by our electrophoretic procedures.

Zymograms (electropherograms) representing the three presently known mitochondrial MDH phenotypes of rainbow trout are shown in Fig. 3. In this and all subsequent figures of l.c.Z O is the origin or sample slot, the arrow indicates electrophoretic migration toward the anode, numbers below each channel such as 2222 indicate the phenotype designation and letters at the sides such as X^2X^2, AA, AB^1 etc. indicate the subunit composition of individual isozymes.

The similarity of the asymmetric ratio of isozyme intensity of the Y^1Y^1, Y^1Y^2, Y^2Y^2 isozymes in the 1222 phenotype in Fig. 3 to the B^2B^2, B^2B^3, B^3B^3 isozymes in the 2333 B supernatant MDH phenotype (Fig. 2) is very obvious and is suggestive of a similar genetic basis that may consist of two different alleles present in a gene dosage ratio of 1 : 3. The mitochondrial MDH phenotype 1222 also displays (Fig. 3) another isozyme, X^1X^2, that is not seen in the presumed homozygote 2222. The presence of this X^1X^2 isozyme is difficult to explain and to some extent it disrupts the analogy between the supernatant and mitochondrial MDH phenotypes. Thus, although there are some marked similarities in the appearance of the isozyme phenotypes, the detailed genetic and molecular basis of these mitochondrial and supernatant MDH isozymes could be quite different. The third phenotype WW displays an additional isozyme of rather low enzymatic activity. The WW phenotype is shown here for completeness of the record but because it is of uncertain relationship to the other phenotypes it will not be considered further.

The responses of the isozymes of phenotypes 1222 and 2222 to heat denaturation are shown in Fig. 4. The ZZ isozyme has been omitted to reduce crowding in Part a, Fig. 4. The corresponding responses to the nicotinamide-adenine dinucleotide (NAD) analogues, 3-acetylpyridine-adenine dinucleotide (APAD), and nicotinamide-hypoxanthine dinucleotide (NHD) are shown in Fig. 5. As in our previous study (Clayton et al., '73) it was necessary to choose one isozyme as a 'standard' in each electrophoresis channel to compensate for variable amounts of total protein, time of staining , etc. In the present work the B^3B^3 isozyme was taken as the 'standard' and all other isozymes were assigned absorbances in relation to the B^3B^3 isozyme set to an absorbance of 1.0. Taken together the data of Figs. 3, 4, and 5 show that the isozymes designated XX, YY, and ZZ are typical mitochondrial MDH isozymes with marked sensitivity toward heat denaturation and low reactivity with the NHD analogue of NAD. The typical enhancement of mitochondrial MDH activity elicited by APAD (Kitto '66) is displayed by the X^2X^2 isozymes and to a lesser extent by the

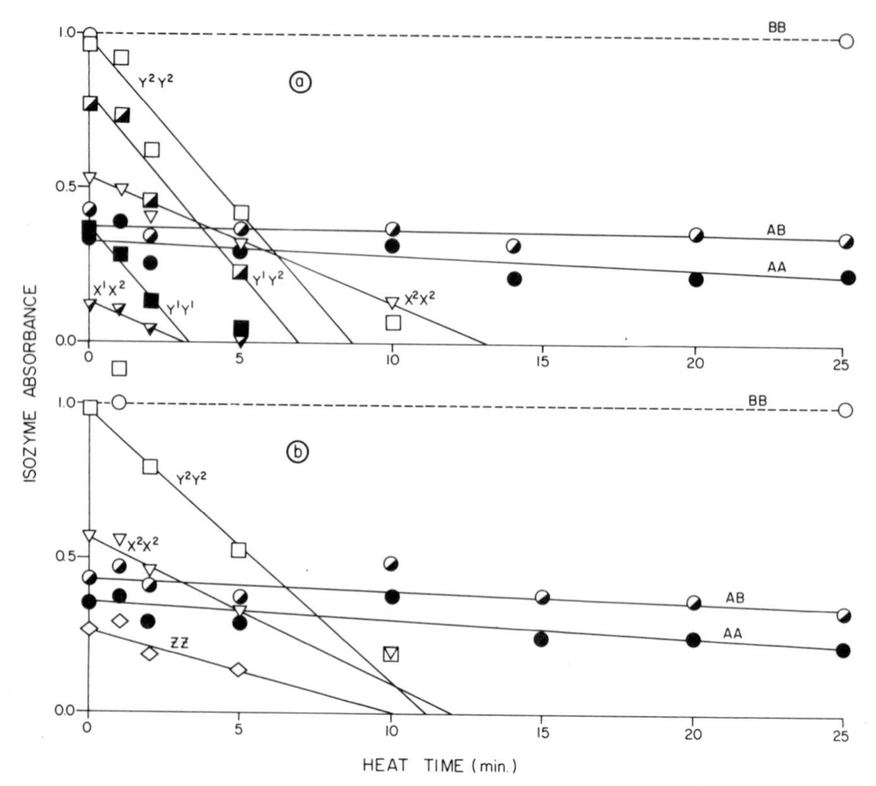

Fig. 4. Absorbance of rainbow trout MDH isozymes heated at 40° C for various times before specific staining. Part a, Y mitochondrial MDH 1222 heterozygote; part b, Y mitochondrial MDH 2222 homozygote.

Y^2Y^2 isozymes (Fig. 5). The heat denaturation data of Fig. 4 also indicate that there is a slight but consistent difference in the heat sensitivity of the X^1X^2, X^2X^2 isozymes relative to the Y^1Y^1, Y^1Y^2, Y^2Y^2 isozymes. This adds support to our interpretation of the 1222 phenotype as a genetic variant of the 2222 phenotype.

The segregation of these mitochondrial MDH phenotypes in specific family groups of rainbow trout is shown in Table I. The segregation ratios are consistent with the interpretation that phenotype 2222 is specified by a homozygous 2222 genotype and phenotype 1222 by a heterozygous 1222 genotype. The

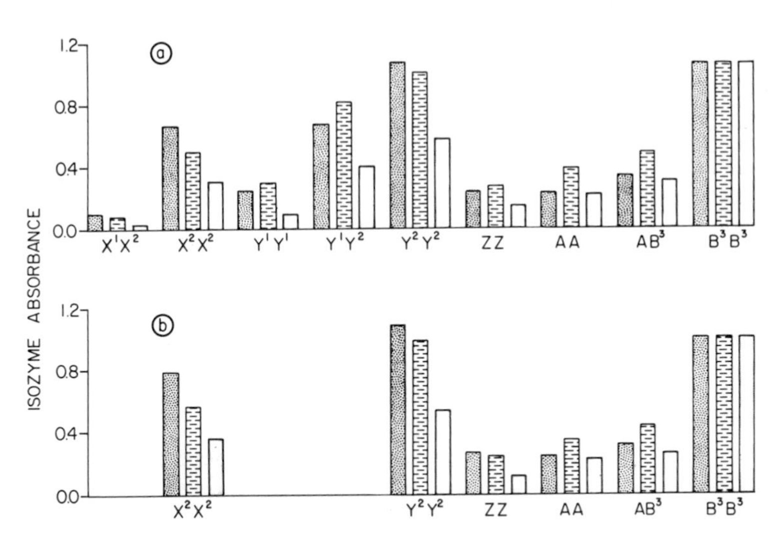

Fig. 5. Absorbance of rainbow trout MDH isozymes when stained with the natural coenzyme nicotinamide-adenine dinucleotide (NAD) and the coenzyme analogues acetylpyridine-adenine dinucleotide (APAD) and nicotinamide-hypoxanthine dinucleotide (NHD). For each isozyme center bar ≡ NAD; left bar ≡ APAD and right bar ≡ NHD. Part a, Y mitochondrial MDH 1222 heterozygote; Part b, Y mitochondrial MDH 2222 homozygote.

TABLE I

No. of families	No. of fish and phenotype		Probable parentage
	1222	2222	
14	-	165	2222 x 2222
5	32	32	2222 x 1222
			Known parentage
2	11	13	2222 x 1222
2	-	23	2222 x 2222

Segregation of 'Y' mitochondrial MDH phenotypes in rainbow trout.

2222 mitochondrial MDH phenotype is by far the most common one in several rainbow trout populations (Table II). This population and segregation evidence as well as the asymmetric intensity distribution of the YY isozymes in the heterozygote all seem to be consistent with a two-locus, two-allele system

TABLE II

	No. of fish and phenotype		
Population	2222	1222	WW
Idaho #1	64	5	7
Idaho #2	61	7	10
Maple, Ontario	69	26	6
Nottawasaga River, Ontario	52	0	0
Tunkwa Lake, B.C.	57	1	2

Occurrence of mitochondrial MDH phenotypes in rainbow trout populations.

analogous to the model proposed by Bailey et al., 1970 for supernatant MDH. However, the presence of the X^1X^2 isozyme in the heterozygous phenotype is not accommodated by such a model and it is possible that the genetic variation observed here is not confined to a structural gene at a Y mitochondrial MDH locus.

In addition to the polymorphism of mitochondrial MDH we have observed a remarkable array of supernatant MDH phenotypes in rainbow trout and a sample of these is shown in Fig. 6. The electropherograms in Fig. 6 represent the five possible phenotypes produced by alleles $\underline{b}^2$ and $\underline{b}^3$ as well as some of the phenotypes produced by the allelic combinations $\underline{b}^1$, $\underline{b}^3$

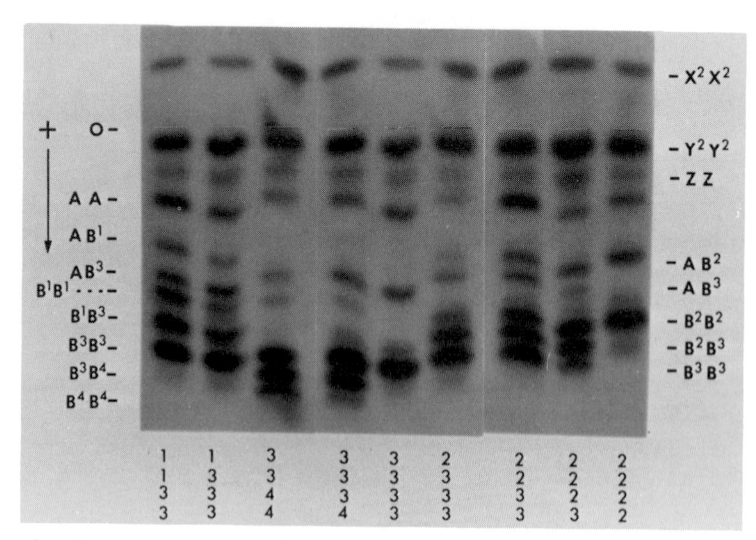

Fig. 6. Zymograms of various B supernatant MDH phenotypes in rainbow trout.

TABLE III

Population	Gene frequency $\underline{b}^1$	$\underline{b}^2$	$\underline{b}^3$	$\underline{b}^4$	No. of fish with n alleles n = 1	2	3	Total
Idaho #1	0.008	0.107	0.885	-	63	31	2	96
Idaho #2	0.021	0.116	0.845	0.018	53	40	4	97
Maple, Ontario	-	0.104	0.896	-	62	39	-	101
Nottawasaga R. Ontario	0.055	0.015	0.887	0.043	65	29	6	100
Tunkwa Lake, B.C.	0.021	-	0.829	0.15	23	35	2	60

Distribution of b supernatant MDH alleles in rainbow trout populations.

TABLE IV

No. of families	No. of fish and phenotype 2233	2333	3333	3334	Probable parentage
1	2	6	4	-	2333 x 2333
6	-	39	38	-	2333 x 3333
1	-	-	5	7	3334 x 3333
8	-	-	90	-	3333 x 3333
					Known parentage
4	-	-	96	-	3333 x 3333

Segregation of B supernatant MDH phenotypes in rainbow trout.

and $\underline{b}^4$, $\underline{b}^3$. The phenotypes in Fig. 6 provide clear evidence for four kinds of B supernatant MDH subunits in rainbow trout. These phenotypes seem to be similar to those reported by Bailey et al., 1970 (Fig. 2) except for the increased number of subunits. If all four alleles are present in both loci a theoretical total of 35 phenotypes will be generated. A diagram analogous to Fig. 2 was prepared representing these 35 phenotypes and a number of rainbow trout in several populations were analyzed for B supernatant MDH phenotype and classified according to the theoretical phenotypes.

Results of the population survey of supernatant MDH are presented in Table III in terms of gene frequency. This gene frequency presentation is necessary because of the complex and unwieldy nature of the 35 phenotype model. Bailey et al. 1970 found, according to their terminology, a common B' subunit and a variant B subunit in rainbow trout MDH phenotypes. It seems probable, on the basis of the distribution of the alleles in Table III, that the $\underline{b}$ and $\underline{b}'$ alleles of Bailey et al., 1970 correspond to our $\underline{b}^2$ and $\underline{b}^3$ respectively. The data presented in Table 3 indicate a considerable fraction of the fish are heterozygous and also that a few fish in each population carry three different alleles. The presence of the latter three allele individuals is strong evidence for two B MDH loci and it also lends support to the suggestion that both loci may contain all four alleles. The fish that carry only one allele were with a single exception 3333 homozygotes; the exceptional individual was a solitary 2222 homozygote in the Idaho No. 2 population.

The segregation of some of these phenotypes in family groups of rainbow trout is presented in Table IV. Only a limited number of the parental fish was available for phenotype analysis and thus in most families the parental phenotypes were inferred from the segregation ratios of the progeny. The data of Table IV closely parallel the results obtained by Bailey et al., 1970 in a similar study with spring salmon *(Oncorhynchus tshawytscha)*. It seems clear that these simple segregation ratios reflect the presence of three alleles and two loci. The 1 : 2 : 1 ratio of 2233, 2333 and 3333 progeny in one family indicates that in both 2333 parental fish, allele 2 was in the same locus.

A more complex set of B supernatant MDH inheritance data, from three families related through a common male parent, is presented in Table V. There is little doubt that the male parent was of the 1334 B MDH phenotype. This has to be so because the 3333 phenotype is by far the most frequent in all populations (Table III) and it would have been an extraordinary coincidence to find two females of phenotype 1334 to

TABLE V

Family	No. of fish and phenotype									Probable parentage (Genotype)
	1 1 3 3	1 1 3 4	1 2 3 3	1 2 3 4	1 3 3 3	1 3 3 4	2 3 3 4	3 3 3 3	3 3 3 4	
1 x 2	2	1	1	3	1	2	2	1	3	$\frac{\underline{b}^2,\underline{b}^3}{\underline{b}^1,\underline{b}^3}$ X $\frac{\underline{b}^1,\underline{b}^4}{\underline{b}^3,\underline{b}^3}$
2 x 2					5				7	
3 x 2					5			1	5	$\frac{\underline{b}^3,\underline{b}^3}{\underline{b}^3,\underline{b}^3}$ X $\frac{\underline{b}^1,\underline{b}^4}{\underline{b}^3,\underline{b}^3}$

Segregation of 'B'supernatant MDH phenotypes from multiple heterozygous matings.

TABLE VI

Genotype	1,2 / 1,3	1,3 / 1,3	2,4 / 1,3	3,4 / 1,3	1,4 / 1,3
No. obs.	0	2	3	2	1
Genotype	1,2 / 3,3	1,3 / 3,3	2,4 / 3,3	3,4 / 3,3	3,3 / 3,3
No. obs.	1	1	2	3	1

Genetic analysis of B MDH in a multiple heterozygous family.

satisfy the occurrence of equal numbers of 1333 and 3334 progeny in families 2 x 2 and 3 x 2. The solitary 3333 fishes found in families 3 x 2 and 1 x 2 could have been strays from other families and these two fish will not be considered further. The male parent of phenotype 1334 can be assigned the genotype $\underline{b}^1, \underline{b}^4/\underline{b}^3, \underline{b}^3$ since this is the only genotype that could yield equal numbers of 1333 and 3334 progeny in matings with 3333 homozygotes.

An analysis of family 1 x 2 can be undertaken on the assumption that the male parent has the $\underline{b}^1, \underline{b}^4/\underline{b}^3, \underline{b}^3$ genotype. A quick inspection of the progeny phenotypes in family 1 x 2 shows that alleles (subunits) 1 and 3 are represented more frequently than 2 and 4. The progeny can then be reclassified according to genotype with one locus either 1,3 or 3,3 as is shown in Table VI. The arrangement of the progeny in this manner leads to the assignment of the parental genotypes as $\underline{b}^2,\underline{b}^3/\underline{b}^1,\underline{b}^3$ x $\underline{b}^1,\underline{b}^4/\underline{b}^3,\underline{b}^3$ since this is the only combination of genotypes that will account for the majority of the progeny. One class of progeny of genotype 1,2/1,3 predicted by this model was not found but this is probably not significant in such a small group of specimens where the number expected in each class is only two. On the other hand two progeny were found that were not predicted by the model. One of these, the 3333, could be a stray from some other family but the 1134 fish cannot be a stray since the 1 and 4 B MDH subunits are not found together in any other family. Also, other isozyme systems were examined in this fish and again all the evidence indicates that it definitely belongs in the 1 x 2 family.

The zymograms from this family are shown in Fig. 7. These electropherograms were made with the usual pH 6.1 amine-citrate buffer but with apparatus equivalent to a macro adaption of the starch gel strip technique of Tsuyuki et al., 1966. The time of electrophoresis was about 20 hours and it

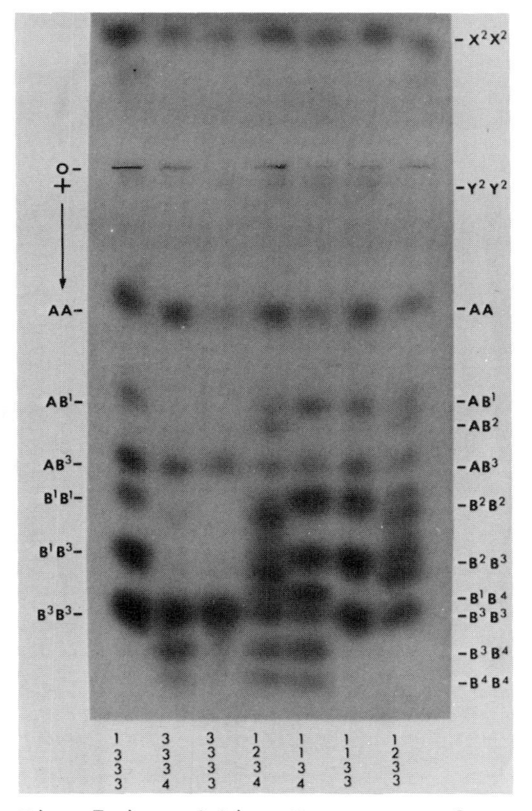

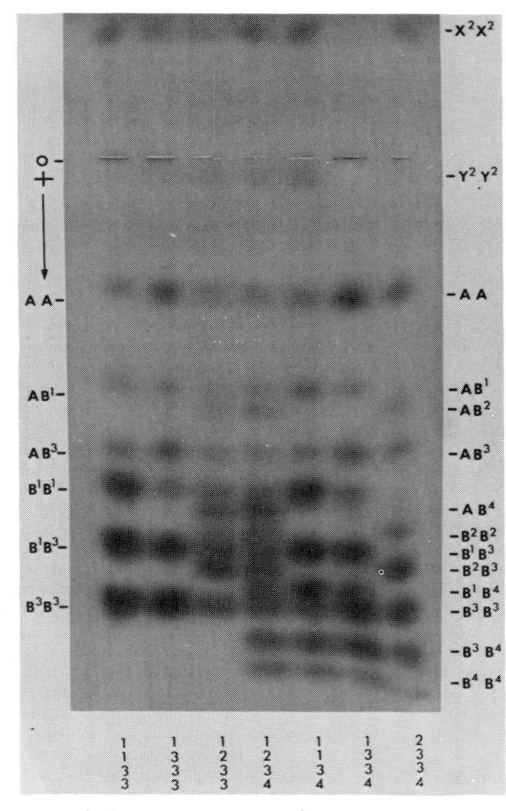

Fig. 7 (a and b). Zymograms of progeny from one rainbow trout mating (1 x 2) heterozygous for all four B supernatant MDH alleles.

seems significant that much of the mitochondrial MDH is evidently denatured in the process. These electropherograms do however provide resolution of many of the supernatant isozymes and the complexity of the multiple heterozygous phenotypes is evident. These figures also show clearly that the 1134 phenotype is definitely unique and is not a misclassified 1334 or any other for that matter.

The incidence of such unexpected progeny at similar frequencies seems to have been observed in two other investigations that were also concerned with four allele systems (Allendorf and Utter 1973; Ohno et al 1969). It was suggested by Ohno et al 1969 that the occurrence of unexpected, and in their case, completely new phenotypes of 6-phosphogluconate dehydrogenase in Japanese quail, *(Coturnix c. japonica)*, was due to a process of intragenetic recombination. A similar process could explain our results and those of Allendorf and Utter 1973, who studied liver isocitrate dehydrogenase in rainbow trout. It may be significant that all three of these studies that seem to have discovered a moderate frequency of intragenetic recombination are concerned with four allele systems. On the other hand, conventional genetic studies that have been designed to detect intragenetic recombination (Smith et al 1970) employ quite different genetic systems and show clearly that such recombination events are extremely rare.

This problem with one of the progeny from mating 1 x 2, Table VI, does not interfere with the interpretation of the remainder of our genetic data as a clear confirmation of the two-locus model for B supernatant MDH in salmonids originally proposed by Bailey et al 1970. The present data also show very definitely that both B MDH loci are discrete and that the genetics of B MDH are accurately described by a duplicated diploid model. The results obtained here may also be interpreted in terms of the overall genetic model for salmonids proposed several years ago (Klose et al 1968; Ohno et al 1968). In outline this model suggests, on the basis of biochemical and cytological evidence, that the major salmonid fishes have gone through a process of genome duplication followed by diploidization so that they now have a genetic constitution equivalent to a double diploid. It has been further proposed that such a genetic constitution should leave one of the duplicated loci relatively free to accumulate various mutations (multiple alleles) while the other locus of the duplicated pair is maintained relatively conservatively by natural selection (Ohno et al 1968; Ohno 1970). Our observation of four b MDH alleles and one a allele in a number of populations of rainbow trout seems to fulfill this prediction.

This observation also is in general agreement with another principle proposed by Ohno (Ohno et al, 1969;Ohno, 1969) that "polymorphism generates more polymorphism" and perhaps additional b MDH alleles will be discovered in rainbow trout. In any case the rainbow trout seems already to be well endowed with many potential phenotypes and genotypes for supernatant MDH on the basis of present information. Thus, if as seems probable all four b alleles are maintained at both loci then 35 phenotypes and 100 genotypes are available for this isozyme system alone. These estimates are for the structural genes only and would be increased if each locus can be regulated independently.

If the evolution of vertebrates has proceeded by genome duplication then in species such as rainbow trout many loci should be duplicated. In this context it would seem that our discovery of genetic polymorphism of mitochondrial MDH isozymes, and in particular the asymmetric enzymatic activity of isozymes in the heterozygote that may be interpreted as evidence of a 1 : 3 gene dosage ratio, probably adds yet another gene locus to the extensive number now known to be duplicated in salmonids. The proposal that this group of fishes has experienced an event of genome duplication in the course of evolution is thus further supported by our observations regarding mitochondrial MDH as well as supernatant MDH.

REFERENCES

Allendorf, F. W. and F. M. Utter 1973. Gene duplication within the family *Salmonidae:* Disomic inheritance of two loci reported to be tetrasomic in rainbow trout. *Genetics* 74: 647-654.

Bailey, G. S., A. C. Wilson, J. E. Halver, and C. L. Johnson 1970. Multiple forms of supernatant malate dehydrogenase in salmonid fishes. Biochemical, immunological and genetic studies. *J. Biol. Chem.* 245: 5927-5940.

Clayton, J. W., R. E. K. Harris, and D. N. Tretiak 1973. Identification of supernatant and mitochondrial isozymes of malate dehydrogenase on electropherograms applied to the taxonomic discrimination of walleye *(Stizostedion vitreum vitreum)*, sauger *(S. canadense)* and suspected interspecific hybrid fishes. *J. Fish. Res. Board Can.* 30: 927-938.

Clayton, J. W. and D. N. Tretiak 1972. Amine-citrate buffers for pH control in starch gel electrophoresis. *J. Fish. Res. Board Can.* 29: 1169-1172.

Clayton, J. W., D. N. Tretiak, and A. H. Kooyman 1971. Genetics of multiple malate dehydrogenase isozymes in skeletal muscle of walleye *(Stizostedion vitreum vitreum)*. *J. Fish. Res. Board Can.* 28: 1005-1008.

Kitto, G. B. 1966. The comparative enzymology of malate dehydrongenases. Ph.D. Thesis. Brandeis Univ., Waltham, Mass. 381 pp. (Univ. Microfilms, Ann Arbor, Michigan (Diss. Abstr. 27B: 2624)).

Klose, J., U. Wolf, H. Hitzeroth, H. Ritter, and S. Ohno 1968. Duplication of the LDH gene loci by polyploidization in the fish order *Clupeiformes*. *Humangenetik* 5: 190-196.

Numachi, K. 1970. Polymorphism of malate dehydrogenase and genetic structure of juvenile population in saury *Cololabis saira*. *Bull. Japan. Soc. Sci. Fish.* 36: 1235-1241.

Numachi, K., Y. Matsumiya, and R. Sato 1972. Duplicate genetic loci and variant forms of malate dehydrogenase in chum salmon and rainbow trout. *Bull. Japan. Soc. Sci. Fish.* 38: 699-706.

Ohno, S. 1970. *Evolution by Gene Duplication*. Springer-Verlag, Berlin.

Ohno, S. 1969. The spontaneous mutation rate revisited and the possible principle of polymorphism generating more polymorphism. *Can. J. Genet. Cytol.* 11: 451-467.

Ohno, S., J. Stenius, L. Christian, and G. Schipmann 1969. *De novo* mutation-like events observed at the 6PGD locus of the Japanese quail and the principle of polymorphism breeding more polymorphism. *Biochem. Genet.* 3: 417-428.

Ohno, S., U. Wolf, and N. B. Atkin. Evolution from fish to mammals by gene duplication. *Hereditas* 59: 169-187.

Smith, P. D., V. G. Finnerty, and A. G. Chovnick 1970. Gene conversion in *Drosophila:* Non-reciprocal events at the maroon-like locus. *Nature* 228: 442-444.

Tsuyuki, H., E. Roberts, R. H. Kerr, and A. P. Ronald 1966. Micro starch gel electrophoresis. *J. Fish Res. Board Can.* 23: 929-933.

Wheat, T. E., W. F. Childers, E. T. Miller, and G. S. Whitt 1971. Genetic and in vitro molecular hybridization of malate dehydrongenase isozymes in interspecific bass *(Micropterus)* hybrids. *Anim. Blood Grps. Biochem. Genet.* 2: 3-14.

Wheat, T. E. and G. S. Whitt 1971. In vivo and in vitro molecular hybridization of malate dehydrogenase isozymes. *Experientia* 27: 647-648.

DIPLOID-TETRAPLOID RELATIONSHIPS IN TELEOSTEAN FISHES

WOLFGANG ENGEL, JÖRG SCHMIDTKE and ULRICH WOLF
Institut für Humangenetik und Anthropologie
der Universität
D-7800 Freiburg
Albertstr.11

ABSTRACT. Gene duplication in man is assumed to be due to both tandem duplication and polyploidization. The polyploidization event in human evolution has been traced back to the fish ancestor of mammals. The ancient occurrence of polyploidization in recent fish species is now well documented. Members of the orders *Salmoniformes* and *Cypriniformes* have been demonstrated to be in a diploid-tetraploid relationship to each other, with respect to the number of chromosomes, the DNA content per cell, and the number of specific gene loci (using isozymes as genetic markers). The occurrence of tetrasomic inheritance has been suggested for several enzyme loci in tetraploid species, but it seems more likely that the genes for all isozyme systems studied so far are inherited disomically. In a few instances no gene duplication could be evidenced in the tetraploids on the basis of genetic isozyme polymorphisms. It is concluded that the phylogenetically tetraploid species have undergone a process of diploidization. This view has been substantiated by the finding that the tetraploids have regulated their protein content, enzyme activities, and cell size down to the level of the diploids. Preliminary data from our laboratory indicate that this is not due to a loss of ribosomal cistrons in the tetraploids, a mechanism which has been suggested by others. There seems to be a clear-cut 1 : 2 ratio in the number of ribosomal genes between diploids and tetraploids.

INTRODUCTION

Ohno (1970) suggested that big leaps in evolution were not accomplished simply by allelic mutations at already existing gene loci, but rather via duplication of genes which made possible the accumulation of mutations unmolested by selection. On principle, gene duplication can be regionally confined, or involve, through polyploidization, the entire genome. If the first means is employed, duplicated genes should be arranged in tandem on the same chromosome, while after polyploidization the genes should be located on different chromosomes, and therefore be unlinked.

In the human genome there are a number of examples for tandem duplications which presumably have occurred during vertebrate evolution, e.g. the genes for amylase, the blood group system Rhesus, the hemoglobin β-cluster, and the genes for the heavy chains of the immunoglobulins (see Bender, 1972). Tandem duplications can occur at all stages of eukaryotic evolution, as is known for the haptoglobin α-gene locus in man (Smithies et al., 1961; Black and Dixon, 1968), or the triose phosphate isomerase gene locus in the hominoid species (Rubinson et al., 1973). Polyploidy as an evolutionary mechanism is restricted to lower vertebrates. In higher vertebrates it may be even incompatible with life, as in humans. But there is some indication that the human genome is of tetraploid origin. Quinacrine stained chromosomes can be arranged in sets of four based on their banding patterns (Comings, 1972). Hemoglobin α-and β-genes as well as the genes for light and heavy chains of immunoglobulins are unlinked. Genes for isozymic forms of various enzymes, e.g. glutamic pyruvic transaminase, peptidase, phosphoglucomutase, and lactate dehydrogenase are likewise unlinked (see Bender, 1972). The gene for LDH-A is located on chromosome 11 (Boone et al., 1972) and the gene for LDH B on the short arm of chromosome 12 (Mayeda et al., 1974). The banding of these two chromosomes suggests that they are ancestrally homologous. It may therefore be expected that the LDH A gene is located on the short arm of chromosome 11.

Tandem duplication and polyploidy complement each other as mechanisms of achieving the DNA increase which has been a prerequisite for evolution; however, polyploidization no doubt was the more powerful mechanism in this connection (Ohno, 1970). Since polyploidy is incompatible with a well established chromosomal sex determining mechanism (Muller, 1925), the polyploidization event in mammalian evolution has to be traced back to a fish ancestor 300 million years ago. During this period of time originally tetrasomic gene loci must have been diploidized so that all mammals of present time have become diploid organisms.

In some examples of lower vertebrates, a more recent occurrence of polyploidization can be demonstrated. They may serve as models to elucidate the evolutionary process following polyploidization, i.e. diploidization. It is our conception that the history of the human genome was governed by similar rules.

In fishes of the orders *Salmoniformes* and *Cypriniformes*, analysis of the genome with respect to the DNA content per cell and the number of chromosomes revealed a diploid-tetraploid relationship (Ohno et al., 1968; Hinegardner and Rosen 1972). The analysis of individual gene loci, using various genetically determined isozyme polymorphisms as markers, however, does not always confirm this diploid-tetraploid relationship (Figure 1). The results of these investigations can be classified into three groups:

1. Duplicated genes are demonstrable and behave as alleles; they have been suggested to follow a tetrasomic mode of inheritance.
2. Duplicated genes are demonstrable and have evolved divergently; they follow a disomic mode of inheritance
3. No gene duplication is demonstrable.

Another independent measure of the polyploidization hypothesis involves the genes for ribosomal RNA. The number of ribosomal cistrons should be duplicated after polyploidization. During the diploidization process, however, this number may have been reduced again towards the original diploid level, as was proposed by Pedersen (1971). To test this possibility, we have evaluated the number of ribosomal cistrons in the diploid and tetraploid group of cyprinid species, as will be explained below.

RESULTS AND DISCUSSION

TETRASOMIC INHERITANCE SUGGESTED IN TETRAPLOID SALMONOID FISHES

A newly arisen tetraploid species has four homologous chromosomes, and only quadrivalents are found during meiosis. Such a situation exists in the recent tetraploid amphibian species *Odontophrynus americanus* (Beçak et al., 1966). In this species the first example of tetrasomic inheritance in vertebrates, namely for the albumin gene locus was demonstrated (Beçak et al., 1968). The tetraploid species *Salmo gairdneri*, *Salmo trutta* and *Salmo salar* of the order *Salmoniformes* exhibit bivalents and multivalents in one and the same meiotic figure, indicating the persistence of large homologies of the duplicated genome in these species (Ohno et al., 1965; Nygren et al., 1971). This finding has been supported by indications for tetrasomic inheritance of several enzyme systems (Wolf et al., 1970; Engel et al., 1970; Stegeman and Goldberg, 1972). Stegeman and Goldberg (1972) investigated the polymorphism of hexose-6-phosphate dehydrogenase in brook trout; they concluded that in one population this enzyme was inherited tetrasomically while it followed a disomic mode of inheritance in another one. No breeding experiments were performed.

We have suggested a tetrasomic mode of inheritance for the gene locus of the supernatant form of NADP-dependent isocitrate dehydrogenase (S-NADP-IDH; E.C. 1.1.1.42) in *Salmo gairdneri* (Wolf et al., 1970) and for the gene locus of sorbitol dehydrogenase in *Salmo gairdneri* as well as in *Coregonus lavaretus*,

NUMBER OF GENE LOCI CODING FOR VARIOUS ISOZYME SYSTEMS IN SPECIES OF THE ORDERS <u>ISOSPONDYLI</u> AND <u>OSTARIOPHYSI</u>. COMPARISON BETWEEN SPECIES ON THE DIPLOID AND TETRAPLOID LEVEL.

Enzyme	Number of gene loci			
	Isospondyli		Ostariophysi	
Genome:	2n	4n	2n	4n
S-form NADP-IDH*	1[1]	2[1]	2[2]	2[2]
SDH*	1[3]	2[3]	1[2]	1[5]
S-form AAT	1,2[6]	2[6]	1[6]	1,2[6]
M-form NADP-IDH	1[1]	2[1]	1[2]	2[2]
LDH**	2[7]	4[7,16]	2[8]	2,3[8,9]
6-PGD	1[3]	1[3]	1[8,10]	2[8,10,11]
α-GPDH	3[4]	3[4]	1[12]	2[12]
PGI	1,2[13]	2,3[13]	2[13]	3,4[13]
S-form NAD-MDH	1[14]	2[15]	1,2[12]	2,3[12]
M-form NAD-MDH	1[14]	2[12]	1,2[12]	2[12]

*Tetrasomic inheritance suggested in some species

(1) Wolf et al. (1970); (2) Engel et al. (1971); (3) Engel et al. (1970); (4) Engel et al. (1971a); (5) Lin et al. (1969); (6) Schmidtke and Engel (1972); (7) Klose et al. (1968); (8) Klose et al. (1969); (9) Engel et al. (1973); (10) Schmidtke and Engel (1974); (11) Bender and Ohno (1968); (12) own unpublished results; (13) Schmidtke et al. (1974); (14) Ohno (1974); (15) Bailey et al. (1970); (16) Massaro and Markert (1968).

Figure 1

another ancient tetraploid species of the order *Salmoniformes* (Engel et al., 1970). In a random sample of 135 specimens of *S.gairdneri* 9 different phenotypes for S-NADP-IDH were observed (Figure 2, a-i) and seemed best to be interpreted under the assumption of a tetrasomic mode of inheritance of this gene locus (Wolf et al., 1970). In order to substantiate tetrasomic inheritance for the S-NADP-IDH gene locus in rainbow trout breeding experiments were performed (Ropers et al., 1973). The results from the majority of informative matings support a genetic model of two disomic gene loci, A and B rather than a tetrasomic gene locus A for this enzyme. This interpretation is based on the assumption of an identical electrophoretic position of the A and B isozymes. Population studies and breeding experiments of the S-NADP-IDH in rainbow trout by Allendorf and Utter (1973) confirm our results. In Figure 3 two types of the 20 matings examined are shown.

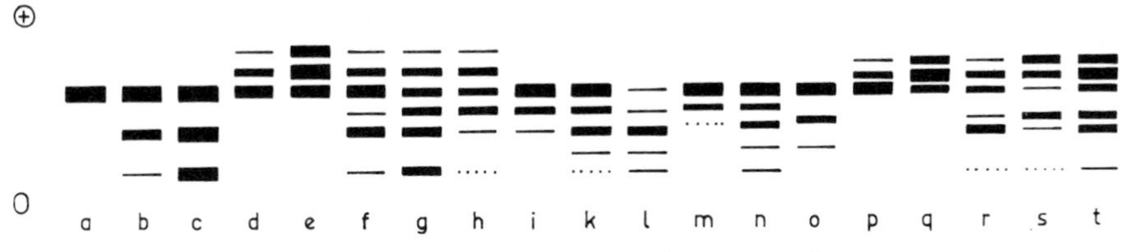

Fig. 2. S-NADP-IDH phenotypes of *Salmo gairdneri* observed in our laboratory.

Following the nomenclature of a tetrasomic mode of inheritance, the parental isozymic pattern of mating type I is designated as ♀ $AAAA_1$ x ♂ AAA_1A_1, of mating type II as ♀ AAA_1A_1 x ♂ AAA_1A_3. $AAAA_1$ may produce the two types of gametes AA and AA_1 in the ratio 1:1, while AAA_1A_1 may form the three types of gametes AA, AA_1 and A_1A_1 in the ratio 1:4:1. The offspring of mating type I should exhibit the phenotypes AAAA : $AAAA_1$: AAA_1A_1 : $AA_1A_1A_1$ in the 1:5:5:1 ratio, but only $AAAA_1$ and AAA_1A_1 in the ratio of approximately 1:1 were found. These results can be explained best by the assumption of two separate gene loci A and B, for S-NADP-IDH in rainbow trout. According to this model, the maternal parent in mating I has the phenotype $AABB_1$ and would have produced AB- or AB_1-gametes, and the paternal parent (phenotype AAB_1B_1) only AB_1-gametes. However, very strict preferential pairing between chromosomes carrying identical alleles would also fit the assumption of tetrasomic inheritance in mating type I. In mating type II, again following the nomenclature of a tetrasomic mode of inheritance, the female should produce the same gametes and in the same ratio as the male of mating type I. The male of mating type II with the phenotype AAA_1A_3 may form the gametes AA, AA_1, AA_3 and A_1A_3 in the ratio 1:2:2:1. Under the assumption of a strict tetrasomic inheritance the offspring of the mating type II should be AAAA, $AAAA_3$, $AAAA_1$, AAA_1A_1, AAA_1A_3, $AA_1A_1A_3$, $AA_1A_1A_1$ and $A_1A_1A_1A_3$ in the ratio of 1:2:6:9:9:6:2:1. From 60 offspring investigated, 13 $AAAA_1$, 12 AAA_1A_1, 20 AAA_1A_3 and 15 $AA_1A_1A_3$ phenotypes were identified. These results also argue against strict tetrasomic inheritance, but can be explained by a disomic model with the two loci A and B unlinked; A_1 should be allelic to B, while A_3 should be allelic to A. The phenotypes of mating II should therefore be written as ♀ AAB_1B_1 x ♂ AA_1BB_1 with the offspring $AABB_1$, AA_1BB_1, AAB_1B_1, and $AA_1B_1B_1$ in the ratio 1:1:1:1; indeed considering the small sample size, this ratio was approximately found.

At present no definite example of tetrasomic inheritance in salmonoid fishes is known. Since Robertsonian fusion is generally observed in tetraploid salmonoid fishes, and since meiotic figures from the same species can contain varying numbers of multivalents (Ohno et al., 1965), the existence of a pure tetrasomic mode of inheritance for a definite protein might never be demonstrated. We might have approached the limits of the usage of isozymes as genetic markers to clarify this problem.

In the fish family *Cyprinidae*, order *Cypriniformes*, based on the DNA content per cell and the number of chromosomes, a diploid-tetraploid relationship due to ancestral polyploidization has likewise been assumed (Ohno et al., 1968; Wolf et al., 1969; Hinegardner and Rosen, 1972). Also a number of individual genes (Figure 1) have been demonstrated to be duplicated in the tetraploids. However, no multivalents have been observed in meiosis nor an indication for a tetrasomic mode of inheritance has been found in any of the tetraploids. Therefore, it has been suggested that the tetraploids of the order *Cypriniformes* have undergone polyploidization earlier than those of the order *Salmoniformes*, i.e. carp-like tetraploids are more diploidized than trouts (Engel et al., 1971).

DISOMIC MODE OF INHERITANCE FOR DUPLICATED GENE LOCI IN FISHES OF THE ORDERS CYPRINIFORMES AND SALMONIFORMES

The diploidization process in which the ancient tetraploids were involved subsequent to polyploidization, suggests that tetrasomic gene loci become functionally diverse and subject to a disomic mode of inheritance. Indeed, the great majority of isozyme systems tested (Figure 1) could be shown to be duplicated and disomic. As an example the gene loci for glucosephosphate isomerase (PGI, E.C. 5.3.1.9) shall be mentioned here. In *Barbus tetrazona* which is considered to be a diploid cyprinid fish, two PGI-gene loci exist, while in the tetraploid *Barbus barbus* 3 PGI-loci can be deduced from the electrophoretic pattern (Figure 4). In the order *Clupeiformes*, *Clupea harengus* as a diploid species is endowed with a single PGI-locus, while the tetraploid *Salmo trutta (Salmoniformes)* is endowed with two PGI loci (Figure 5). Beyond the results presented in Figure 1, it has been noted on many occasions that tetraploid salmonoid and cyprinid fishes have more gene loci for a specific enzyme than other groups of vertebrates (for review see Ohno, 1974).

NO GENE DUPLICATION IS DEMONSTRABLE IN TETRAPLOID FISHES OF THE ORDERS SALMONIFORMES AND CYPRINIFORMES

As is shown in Figure 1, in a few instances tetraploids of both the orders *Cypriniformes* and *Salmoniformes* have the

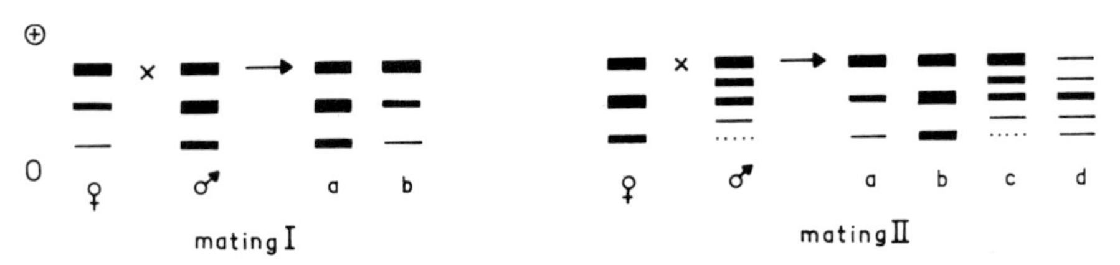

Fig. 3. Parental and offspring S-NADP-IDH patterns of two matings of *Salmo gairdneri*.

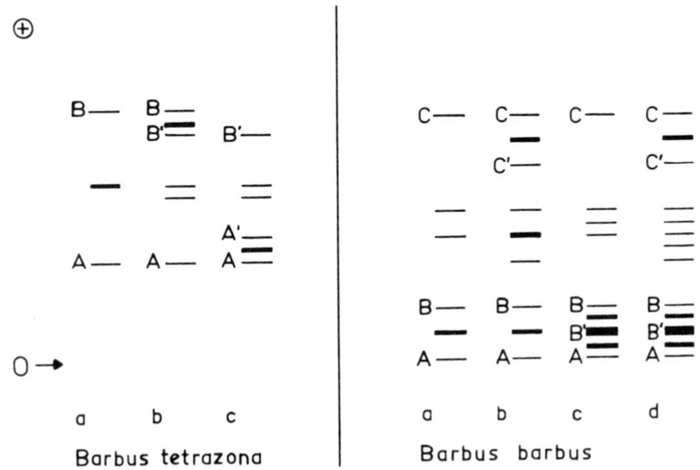

Fig. 4. Schematic representation of the PGI isozymes after starch gel electrophoresis in the diploid *Barbus tetrazona* and in the tetraploid *Barbus barbus*. Phenotypes in *Barbus tetrazona:* (a)AABB, (b) AABB', (c) AA'B'. Phenotypes in *Barbus barbus:* (a) AABBCC, (b) AABBCC', (c) AABB'CC (d) AABB'CC'.

same number of gene loci coding for a specific protein as do the respective diploids. Disregarding the estimation that only about 25% of all mutations can be demonstrated by electrophoresis, duplicated genes in the fish that became tetraploids during phylogeny may not have mutated and thus cannot be visualized electrophoretically. Such a situation was suggested in *Salmo gairdneri* for the gene loci of S-MDH (Bailey et al., 1970) and of S-NADP-IDH (Ropers et al., 1973), and in the tetraploid natural hybrid *Carassius carassius* x *Carassius auratus* for the gene loci of the supernatant form of aspartate aminotransferase (S-AAT; E.C. 2.6.1.1) (Schmidtke and Engel, 1972). If, on the other hand, a regionally confined gene duplication for an enzyme has occurred later during evolution

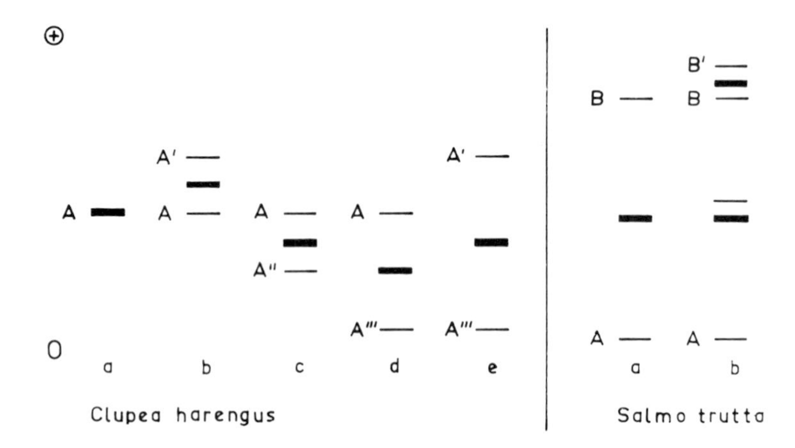

Fig. 5. Schematic representation of the PGI isozymes after starch gel electrophoresis in the diploid *Clupea harengus* and in the tetraploid *Salmo trutta*. Phenotypes in *Clupea harengus:* (a) AA, (b) AA', (c) AA'', (d) AA''', (e) A'A'''. Phenotypes in *Salmo trutta:* (a) AABB, (b) AABB'.

of the diploids, i.e., after separation from the tetraploid group, a diploid-tetraploid relationship for the enzyme in question cannot be demonstrated. The DNA variability within the diploid group of cyprinid fishes was interpreted to be a result of such regionally confined duplications (Muramoto et al., 1968; Wolf et al., 1969), including the 6-phosphogluconate dehydrogenase gene loci (Klose et al., 1969), thus resulting in a similar number of gene loci for this enzyme in diploids and tetraploids. Meanwhile we have shown, however, that all diploids are endowed with one 6-PGD gene locus, while only the tetraploids have two loci. In contrast, *Osmerus esperlanus*, a diploid species of the order *Salmoniformes* may even offer two examples for the individual duplication of a single gene, namely for PGI and S-AAT (Schmidtke and Engel, 1974a). Moreover, a duplicated gene may not be functionally essential for the organism and could therefore be eliminated. We recently observed a variant in the phylogenetic tetraploid carp which may be due to a defective mutation of an originally duplicated but dispensable gene coding for lactate dehydrogenase (LDH; E.C. 1.1.1.27) (Engel et al., 1973).

According to Klose et al. (1969) the LDH pattern demonstrable in most organs of the carp consists of 14 isozymes and can be interpreted under the assumption of the three different gene loci A, B^1 and B^2. Additional LDH loci are demonstrable in the liver. In a wild carp population as well as in a bred population three electrophoretic types of LDH were demonstrable

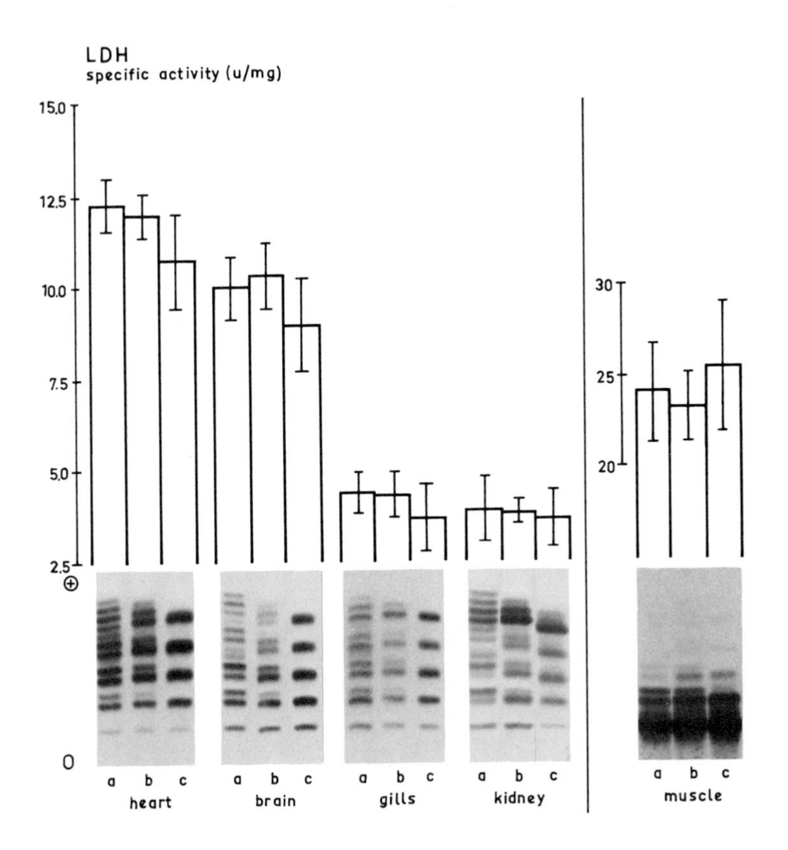

Fig. 6. LDH isozyme pattern of *Cyprinus carpio* from various organs. Phenotypes (a) $AAB^2B^2B^1B^1$, (b) $AAB^2B^2B^1_+B^1_-$, (c) $AAB^2B^2B^1_-B^1_-$, and mean specific LDH activities and standard deviation.

(Figure 6). According to our interpretation the three phenotypes can be designated as: (a) $AB^2B^1_+B^1_+$, (b) $AB^2B^1_+B^1_-$ and (c) $AB^2B^1_-B^1_-$. The frequencies of these three types in both carp populations are in good agreement with the Hardy-Weinberg equilibrium indicating that the respective types have no selective disadvantage (Engel et al., 1973). The measurement of LDH activity in different organs of the three types showed the following: the wild type and the assumed hemizygote for B^1 exhibit no significant activity difference, while the deficient homozygote has a significantly decreased activity in the heart only (Figure 6). From these results it can be inferred that the partial or total lack of synthesis of B^1 -subunits in types $AB^2B^1_+B^1_-$ and $AB^2B^1_-B^1_-$ can be compensated for by an increase in the production rate of the other LDH subunits.

This example might reflect an evolutionary process which started in the production of isozyme genes and is now on the way of eliminating the newly created gene which is of no use to the organism. The fact that vertebrates including mammals, despite their tetraploid origin have fewer genes for definite proteins or enzymes than the tetraploid fishes might then be due in part to similar processes as described for the LDH B^1 gene locus in the carp.

REGULATORY PHENOMENA INVOLVED IN THE PROCESS OF DIPLOIDIZATION AND THE NUMBER OF RIBOSOMAL GENES

Unless some interacting control system is established, gene duplication must imply a doubling of the amount of product to be synthesized, as has been evidenced in various organisms throughout the plant and animal kingdoms (for review see Farber, 1973; Lucchesi and Rawls, 1973). Accordingly, polyploidization is followed by a proportional increase of synthetic activity and cell size, as is known for many experimentally induced polyploids. The diploid-tetraploid relationship in cyprinid fish should therefore be reflected also on the level of cell size, total soluble protein, and enzyme activities per cell. To determine these parameters, we used erythrocytes from the diploid species *Tinca tinca, Abramis brama,* and from the tetraploid species *Barbus barbus, Carassius auratus* and *Cyprinus carpio.* Contrary to our assumption, cell size and total soluble protein per cell declined with increasing genome size. Likewise mean enzymatic activity per cell for LDH, PGD and GPI per active gene locus is reduced in the tetraploids as compared to the diploids (Figure 7). Thus it can be stated that contrary to newly polyploidized species, in the ancient tetraploid species studied here the synthetic output per cell is reduced and apparently regulated down to the level of the diploids (Schmidtke and Engel, 1974b). Similar results have been obtained in diploid and tetraploid *Odontophrynus* species (Beçak and Pueyo, 1970).

Pedersen suggested (1971) that the adjustment of the synthetic activity of tetraploids to the diploid level might be due to a reduction of ribosomal cistrons in the tetraploids. In contrast to this assumption we determined the mean number of ribosomal genes per haploid genome in diploid and tetraploid cyprinid fish and found them to be 125 and 262 respectively. RNA-DNA hybridization experiments were carried out essentially as described by Birnstiel et al. (1972). DNA was

GENE-DOSIS RELATIONSHIP IN ERYTHROCYTES OF CYPRINID FISH SPECIES

Species	DNA content per cell (% human leucocytes)	cell volume per % DNA	protein per % DNA	number of gene loci expressed ( ) and mean enzymatic activity (mU x 10^{-7}/cell/gene locus) LDH	6-PGD	PGI
Tinca tinca	30	6.79	2.35	(2) 24.45	(1) 1.50	(1) 9.30
Abramis brama	36	5.06	1.75	(2) 12.95	(1) 1.40	(1) 5.90
Barbus barbus	49	3.89	1.44	(3) 9.00	(2) 0.50	(1) 16.40
Cyprinus carpio	52	3.49	1.24	(3) 9.87	(2) 0.35	(2) 4.85
Carassius auratus	53	3.41	1.22	(3) 7.43	(2) 0.25	(2) 2.70

Figure 7

purified from erythrocytes of diploid and tetraploid species, while labelled r-RNA was prepared from HeLa S 3 suspension cultures incubated with ^{3}H-uridine. Thus, a clear-cut diploid-tetraploid relationship with respect to the number of ribosomal genes exists in cyprinid fishes, strongly supporting the view that tetraploidization has occurred in an ancestor of some of the cyprinid fish species of today. However, the regulatory mechanism responsible for the process of diploidization during polyploid evolution is still unknown.

ACKNOWLEDGEMENTS

We thank Drs. K. Bross and H. Dittes for cooperation with the RNA-DNA hydridization experiments. Thanks are also due to Mrs. Marianne Niemetscheck for skillful photographic assistance. This work was supported by the Deutsche Forschungsgemeinschaft.

REFERENCES

Allendorf, F. W. and F. M. Utter 1973. Gene duplication within the family *Salmonidae*. Disomic inheritance of two loci reported to be tetrasomic in rainbow trout. *Genetics* 74: 647-654.

Bailey, G. S., A. C. Wilson, J. E. Halver, and C. L. Johnson 1970. Multiple forms of supernatant malate dehydrogenase in salmonid fishes. *J. Biol. Chem.* 245: 5927-5940.

Beçak, M. L., W. Beçak, and M. N. Rabello 1966. Cytological evidence of constant tetraploidy in the bisexual South American frog. *Odontophrynus americanus*. *Chromosoma* 19: 188-193.

Beçak, W., A. R. Schwantes, and M. L. B. Schwantes 1968. Polymorphism of albumin-like proteins in the South American tetraploid frog *Odontophrynus americanus*.

(*Salientia: Ceratophrydidae*). *J. Exp. Zool.*168: 473-476.
Beçak, W. and M. T. Pueyo 1970. Gene regulation in the polyploid amphibian *Odontophrynus americanus*. *Exptl. Cell. Res.* 63: 448-451.
Bender, K. and S. Ohno 1968. Duplication of the autosomally inherited 6-phosphogluconate dehydrogenase gene locus in tetraploid species of Cyprinid fish. *Biochem. Genet.* 2 : 101-107.
Bender, K. 1972. Zur Bedeutung der Chromosomenkarte des Menschen für die Ermittlung genetischer Mechanismen. *Habilitationsschrift*, Universität Freiburg.
Birnstiel, M. L., B. H. Sells, and I. F. Purdom 1972. Kinetic complexity of RNA molecules. *J. Mol. Biol.* 63: 21-39.
Black, J. A. and G. H. Dixon 1968. Amino-acid sequence of alpha chain of human haptoglobins. *Nature* 218: 736-741.
Boone, C., T. R. Chen, and F. Ruddle 1972. Assignment of three human genes to chromosomes (LDH-A to 11, TK to 17, and IDH to 20) and evidence for translocation between human and mouse chromosomes in somatic cell hybrids. *Proc. Nat. Acad. Sci.* USA 69: 510-514.
Comings, D. E. 1972. Evidence for ancient tetraploidy and conservation of linkage groups in mammalian chromosomes. *Nature* 238: 455-457.
Engel, W., J. Op't Hof, and U. Wolf 1970. Genduplikation durch polyploide Evolution: die Isoenzyme der Sorbitdehydrogenase bei herings- und lachsartigen Fischen (*Isospondyli*). *Humangenetik* 9: 157-163.
Engel, W., J. Faust, and U. Wolf 1971. Isoenzyme polymorphism of the sorbitol dehydrogenase and the NADP-dependent isocitrate dehydrogenases in the fish family Cyprinidae.*Anim. Blood Grps. Biochem. Genet.* 2: 127-133.
Engel, W., J. Schmidtke, and U. Wolf 1971 a. Genetic variation of α-glycerophosphate dehydrogenase isoenzymes in Clupeoid and Salmonoid fish. *Experientia* 27: 1489-1491.
Engel, W., J. Schmidtke, W. Vogel, and U. Wolf 1973. Genetic polymorphism of lactate dehydrogenase isoenzymes in the carp (*Cyprinus carpio*) apparently due to a "null allele". *Biochem. Genet.* 8: 281-289.
Farber, R. A. 1973. Gene dosage and the expression of electrophoretic patterns in heteroploid mouse cell lines. *Genetics* 74: 521-531.
Hinegardner, R. and D. E. Rosen 1972. Cellular DNA content and the evolution of teleostean fishes. *Am. Nat.* 106: 621-644.
Klose, J., U. Wolf, H. Hitzeroth, H. Ritter, N. B. Atkin, and S. Ohno 1968. Duplication of the LDH gene loci by polyploidization in the fish order *Clupeiformes*. *Humangenetik* 5: 190-196.
Klose, J., U. Wolf, H. Hitzeroth, H. Ritter, and S. Ohno 1969.

Polyploidization in the fish family Cyprinidae, order *Cypriniformes*. II. Duplication of the gene loci coding for lactate dehydrogenase (E.C.: 1.1.1.27) and 6-phosphogluconate dehydrogenase (E.C.: 1.1.1.44) in various species of *Cyprinidae*. *Humangenetik* 7: 245-250.

Lin, C.-C., G. Schipman, W. A. Kittrell, and S. Ohno 1969. The predominance of heterozygotes found in wild goldfish of Lake Erie at the gene locus for sorbitol dehydrogenase. *Biochem. Genet.* 3: 603-607.

Lucchesi, J. C. and J. M. Rawls 1973. Regulation of gene function: A comparison of enzyme activity levels in relation to gene dosage in diploids and triploids of *Drosophila melanogaster*. *Biochem. Genet.* 9: 41-51.

Massaro, E. J. and C. L. Markert 1968. Isozyme patterns of salmonid fishes: Evidence for multiple cistrons for lactate dehydrogenase polypeptides. *J. Exp. Zool.* 168: 223-238.

Mayeda, K., L. Weiss, R. Lindahl, and M. Dully 1974. Localization of the human lactate dehydrogenase B gene on the short arm of chromosome 12. *Am. J. Human Genet.* 26: 59-64.

Muller, H. J. 1925. Why polyploidy is rarer in animals than in plants. *Amer. Natural.* 59: 346-353.

Muramoto, J., S. Ohno, and N. B. Atkin 1968. On the diploid state of the fish order *Ostariophysi*. *Chromosoma* 24: 59-66.

Nygren, A., B. Nilsson, and M. Jahnke 1971. Cytological studies in *Salmo trutta* and *Salmo alpinus*. *Hereditas* 67: 259-268.

Ohno, S., C. Stenius, E. Faisst, and M. T. Zenzes 1965. Post-zygotic chromosomal rearrangements in rainbow trout (*Salmo irideus* Gibbons). *Cytogenetics* 4: 117-129.

Ohno, S., U. Wolf, and N. B. Atkin 1968. Evolution from fish to mammals by gene duplication. *Hereditas* 59: 169-187.

Ohno, S. 1970. *Evolution by Gene Duplication*. Springer-Verlag, Berlin, Heidelberg, New York.

Ohno, S., 1974. *Cytogenetics of chordates, protochordates, cyclostomes, and fishes*. Gebr. Bornträger Verlag. Stuttgart.

Pedersen, R. A. 1971. DNA content, ribosomal gene multiplicity and cell size in fish. *J. Exp. Zool.* 177: 65-78.

Ropers, H.-H., W. Engel, and U. Wolf 1973. Inheritance of the S-form of NADP-dependent isocitrate dehydrogenase polymorphism in rainbow trout. pp. 319-327 In: (J. H. Schröder, ed.) *Genetics and Mutagenesis of Fish*, Springer-Verlag, Berlin, Heidelberg, New York.

Rubinson, H., M. C. Meienhofer, and J. C. Dreyfus 1973. A new isozyme of triose phosphate isomerase specific to hominoids. *J. Molec. Evol.* 2: 243-250.

Schmidtke, J. and W. Engel 1972. Duplication of the gene loci coding for the supernatant aspartate aminotransferase by polyploidization in the fish family Cyprinidae. *Experientia* 28: 976-978.

Schmidtke, J., G. Dunkhase, and W. Engel 1974. Genetic variation of phosphoglucose isomerase isoenzymes in fish of the orders *Ostariophysi* and *Isospondyli*. *Comp. Biochem. Physiol.* (In Press).

Schmidtke, J. and W. Engel 1974a. On the problem of regional gene duplication in diploid fish of the orders *Ostariophysi* and *Isospondyli*. *Humangenetik* 21: 39-45.

Schmidtke, J. and W. Engel 1974b. Gene action in fish of tetraploid origin. I. Cellular and biochemical parameters in cyprinid fish. *Biochem. Genet.* (In Press).

Smithies, O., G. E. Cornell, and G. H. Dixon 1961. Chromosomal rearrangements and the evolution of haptoglobin genes. *Nature* 196: 232-236.

Stegeman, J. J. and E. Goldberg 1972. Inheritance of hexose-6-phosphate dehydrogenase polymorphism in brook trout. *Biochem. Genet.* 7: 279-288.

Wolf, U., H. Ritter, N. B. Atkin, and S. Ohno 1969. Polyploidization in the fish family Cyprinidae, order *Cypriniformes*. I. DNA-content and chromosome sets in various species of Cyprinidae. *Humangenetik* 7: 240-244.

Wolf, U., W. Engel, und J. Faust 1970. Zum Mechanismus der Diploidisierung in der Wirbeltierevolution: Koexistenz von tetrasomen und disomen Genloci der Isocitrat-Dehydrogenasen bei der Regenbogenforelle (Salmo irideus). *Humangenetik* 9: 150-156.

GENE DOSAGE IN DIPLOID AND TRIPLOID UNISEXUAL FISHES (*Poeciliopsis*, POECILIIDAE)

R. C. VRIJENHOEK[1]
Department of Biology
Southern Methodist University
Dallas, Texas 75275

ABSTRACT. Isozyme patterns of diploid and triploid all-female "species" of fishes of hybrid origin show codominant expression of allelic isozymes characteristic of their parental species. The sexual progenitors of these hybrids, *P. monacha* and *P. lucida*, have allelic isozymes (allozymes) of distinctly different electrophoretic mobilities for two gene loci. Isozyme patterns of the monomeric muscle protein (MP 1) in the diploid unisexual-hybrid "species" *P. monacha-lucida*, reveal expression of both parental allozymes (allelic isozymes) with approximately equal staining intensities. *P. monacha-lucida* has a three banded alcohol dehydrogenase (ADH) pattern composed of the two parental homodimers and a heterodimer of intermediate mobility. Slight deviations from the expected 1:2:1 ratio of isozymes might be due to partial repression of the maternal ADH allele.

ADH and MP 1 isozyme patterns of the triploid unisexual-hybrid "species," *P. 2 monacha-lucida*, clearly reveal the effects of having a double genomic dose from *P. monacha* and a single genomic dose from *P. lucida*. Those of *P. monacha*-2 *lucida* clearly reveal the effects of having a single genomic dose from *P. monacha* and a double genomic dose from *P. lucida*. The relationships of allelic expression and genomic dosage are discussed in reference to the possible roles of unisexuality, hybridization, and polyploidy in the process of gene duplication.

INTRODUCTION

Electrophoresis of proteins has provided a useful tool for determining the origins and genetic relationships of diploid and triploid unisexual "species" of vertebrates (Uzzell and Goldblatt, 1967; Neaves and Gerald, 1969; Balsano et al, 1972; Vrijenhoek, 1972a). Although the intent of previous studies was to use isozymes as a convenient tool for supporting the hybrid origin hypothesis for these all-female "species," they also revealed the additive effects of allelic dosage on isozyme patterns. Now that the hybrid origins and genomic

[1]Present address: Department of Zoology, Rutgers University, New Brunswick, N.J. 08903.

constitutions are well established for diploid and triploid unisexual fishes in the genus *Poeciliopsis* (Poeciliidae), attention can be focused on the relationships between allelic expression and genomic dosage.

Morphological and biochemical evidence indicate that an original mating between the diploid sexual species, *Poeciliopsis monacha* (2n=48) and *Poeciliopsis lucida* (2n=48), produced the diploid (2n=48) unisexual "species," *P. monacha-lucida* (Schultz, 1969; Vrijenhoek, 1972a). Experimental support for the hybrid origin hypothesis is provided by the recent laboratory synthesis of this unisexual "species" by matings of *P. monacha* x *P. lucida* (Schultz, 1973). All-female populations of *P. monacha-lucida* are sustained in nature by hybridogenesis, a unique reproductive mode that transmits only the 24 maternal, *monacha*, chromosomes to the ova (Schultz, 1966, 1969; Cimino, 1972a). Upon fertilization of the haploid *monacha* ova by sperm from *P. lucida*, diploid *monacha-lucida* F_1 hybrids, expressing traits from both parents, are re-established.

Occasional failures of the maturation divisions during oogenesis of *P. monacha-lucida* apparently have resulted in in the production of diploid ova (2n=48), which upon fertilization by sperm from *P. monacha* or *P. lucida* gave rise to the two triploid (3n=72) all-female "species," *P.* 2 *monacha-lucida* and *P. monacha*-2 *lucida* (Schultz, 1967, 1969; Cimino, 1972a). The "species" designations of the diploid and triploid unisexuals express both their hybrid constitution and genomic dosages (Schultz, 1969). The triploids are gynogenetic; sperm from a sexual host species is necessary for activation of cleavage, but no paternal chromatin enters the zygotic nucleus of the triploid eggs (Schultz, 1967; Cimino, 1972b).

Morphological comparisons among the three unisexual-hybrid "species" have revealed an array of character states that is distributed between those of their progenitors, clearly related to the additive effects of genomic dosage (Schultz, 1969). For meristic characters such as dentition and vertebral numbers, *P. monacha-lucida* is approximately intermediate between its progenitors, whereas the two triploids, *P.* 2 *monacha-lucida* and *P. monacha*-2 *lucida*, more closely resemble the progenitor having contributed a double genomic dose.

A more direct approach for determining the relationships between genomic dosage and allelic expression is by examination of allelic isozymes. Polymorphisms have been found at a locus encoding the eye-tissue specific isozyme of lactate dehydrogenase (LDH E_4) in both *P. monacha* and *P. lucida*

(Vrijenhoek, 1972a). Depending upon the parental combinations, *P. monacha-lucida* unisexuals can have four genotypes: E'E, EE, EE", and E'E". Five distinct LDH E_4 isozymes resulting from assembly of two distinct polypeptide subunits into enzymatically active tetrameric molecules are discernible in E'E" and E'E heterozygotes. Due to the close mobilities of the parental isozymes, EE" heterozygotes *P. monacha-lucida* have only a single wide LDH E band, with a peak of activity midway between the parental E_4 and E''_4 homotetramers. This wide band presumably results from the overlapping of E_4, $E_3E''_1$, $E_2E''_2$, $E_1E''_3$, and E''_4 isozymes. The triploid "species," *P.* 2 *monacha-lucida* and *P. monacha*-2 *lucida*, also have a single wide LDH E_4 isozyme of similar mobility to that in EE" heterozygous *P. monacha-lucida*. Nevertheless, in each case the peak of enzymatic activity in that single band is not midway between the parental isozymes, but is highly skewed, and thus suggests the presence of a double dose of the E allele in *P.* 2 *monacha-lucida* and of the E" allele in *P. monacha*-2 *lucida*. Their genotypes are presumed to be EEE" and EE"E", respectively.

The lack of clear resolution among the LDH E_4 isozymes in the triploids has led to a search for other isozymes that would allow more detailed examinations into the relationships of allelic expression and genomic dosage in *Poeciliopsis*. This paper reports on allelic expression at an alcohol dehydrogenase (ADH) locus specific to liver tissue and a muscle protein (MP) locus in diploid and triploid "species" of this genus. Allelic expression and genomic dosage relationships of other polyploid organisms are discussed with special reference to the potential roles of hybridization, unisexuality, and polyploidization in the process of gene duplication.

MATERIALS AND METHODS

Live fishes in this study were either from wild collections made by the author or from strains maintained by R. J. Schultz at the University of Connecticut and originally collected from the Rio Fuerte in the state of Sinaloa, Mexico. All wild specimens were carefully examined to insure that tissue specific isozyme patterns of the fish were not altered by parasitism (Vrijenhoek, 1974). Individual tissues were removed from fish killed by cold temperature shock. Tissues were ground in two volumes of cold distilled water and the extracts centrifuged in a Beckman Microfuge for five minutes at 4°C prior to electrophoresis.

Isozymes were separated by vertical starch gel electrophoresis (Buchler Instruments) in 12% Electrostarch (Otto Hiller

Inc.) by imposition of 200 V for 16 hours at 4°C. The two buffer systems employed were modified after Whitt et al (1973): 1) 0.75 M tris-0.25 M citrate stock buffer pH 6.8, diluted 1:9 for gels and 1:5 for the electrode vessels, provided the best resolution of muscle proteins: 2) 0.9 M tris-0.02 M Na_4EDTA-5.0M borate stock buffer pH 8.0, diluted 1:9 for gels and used undiluted for electrode vessels, provided the best resolution of alcohol dehydrogenase. ADH was stained as by Shaw and Prasad (1970). Muscle proteins were stained with 5% Ponceau red S in a 7.5% solution of trichloroacetic acid; gels were destained in a mixture of methanol, water, and acetic acid (5:5:1). The presumption that staining intensities of the tetrazolium dye for ADH and Ponceau red S for muscle myogens is directly proportional to the concentrations of these proteins was supported by comparisons among a series of samples, diluted 1:1, 1:2, 1:3, and 1:4 parts of tissue to distilled water (w/v).

RESULTS

The enzyme alcohol dehydrogenase (ADH) is encoded at a single gene locus and expressed predominantly in liver tissues of *Poeciliopsis* as a major, cathodally migrating isozyme on starch gels. Associated with this major band is a weakly expressed satellite band of slightly slower mobility. The satellite band is apparently not a separate gene product since its mobility and concentration are directly related to those of the major band. Allozymes (allelic isozymes) of two distinct cathodal mobilities were observed in the sexual species: *P. lucida* has a major band of fast mobility that migrates approximately 5 cm from the origin, whereas *P. monacha* has a major band migrating approximately one-half this distance (Fig. 1). The quantity of enzyme included in the satellite bands also differs between the species. *P. monacha* has a more highly concentrated band than *P. lucida*, in which it is barely visible. The homozygous genotypes assigned these allozymes are $ADH^{b/b}$ in *P. monacha* and $ADH^{c/c}$ in *P. lucida;* the $ADH^{a/a}$ genotype is assigned to the very slow cathodally migrating allozyme in *P. latidens* (Vrijenhoek, 1972b). Recent collections from wild populations revealed no polymorphism at this locus within a sample of 57 *P. monacha* from three geographically disjunct populations (Vrijenhoek, unpublished). No polymorphism was observed among six inbred strains of *P. lucida* derived independently from four disjunct populations.

The diploid unisexual *P. monacha-lucida* is heterozygous, $ADH^{b/c}$ having three isozymes: a fast band equivalent to that encoded by the ADH^c allele in *P. lucida*, a slow band equivalent to that encoded by the ADH^b allele in *P. monacha*, plus

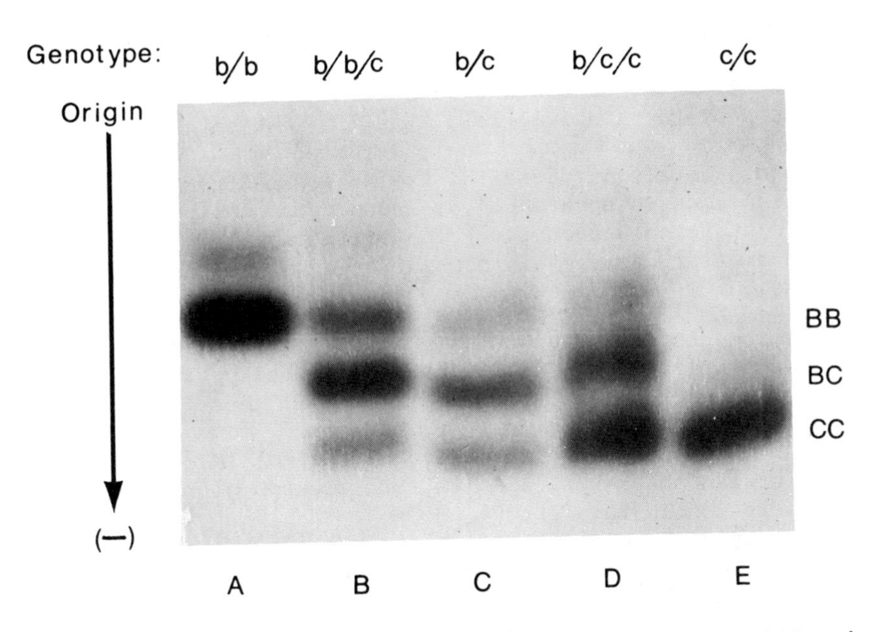

Fig. 1. Alcohol dehydrogenase (ADH) patterns in *Poeciliopsis*: A) *P. monacha;* B) *P.* 2 *monacha-lucida;* C) *P. monacha-lucida;* D) *P. monacha*-2 *lucida;* E) *P. lucida.*

---

a band of intermediate mobility (Fig. 1). The three-banded heterozygous pattern is consistent with previous indications that enzymatically active ADH of vertebrates is a dimeric molecule (Hitzeroth et al, 1968; Smith et al, 1973). The dimeric molecules BB, BC, and CC, in *P. monacha-lucida* are not distributed in the precise 1:2:1 ratio of concentrations expected for random assembly of dimers from equal amounts of equally active B and C polypeptide subunits. Consistently, in each of 23 individuals, the paternally derived CC isozyme appears in slightly higher concentration than the maternal BB isozyme. Satellite bands of the BB, BC, and CC isozymes were not observed in the diploid unisexuals.

Although the maternal and paternal allelic isozymes of ADH are not equally expressed in *P. monacha-lucida*, the isozyme patterns of the triploid unisexuals reveal distinct dosage effects (Fig. 1). The genomic dosages of the triploids, *P.* 2 *monacha-lucida* and *P. monacha*-2 *lucida* would indicate $ADH^{b/b/c}$ and $ADH^{b/c/c}$ genotypes, respectively. The concentrations of isozymes in *P.* 2 *monacha-lucida* resembles the 4 BB:4 BC:1 CC ratio predicted by expansion of the binomial equation $(2B+C)^2$ that presumes two doses of the ADH^b allele and one dose of the ADH^c allele. The isozyme pattern in *P. monacha-2 lucida* closely resembles the 1 BB:4 BC:4 CC ratio expected from the presence of one dose of the ADH^b allele and two doses of the ADH^c allele. A single weak satellite of the BB isozyme was visible in *P.2 monacha-lucida*. In both triploids the slight deviation from the expected ratio of BB: BC:CC isozymes is apparently due to a decrease in those isozymes containing B subunits. These findings are consistent with a similar decrease in concentration of the maternal BB isozyme in *P. monacha-lucida*.

Extracts of soluble proteins from muscle tissues (MP) reveal three distinct regions of electrophoretic migration in the anodal direction (Fig.2). The genetically variable region (MP 1) is encoded by a single gene locus. Those in the mid-anodal region (MP 2) do not vary between *P. monacha* and *P. lucida*. Those in the lower third of the gel (MP 3) appear to vary between the sexual species, but are too poorly resolved for inclusion in this study.

The rapidly migrating MP 1 protein A of *P. monacha* and the slow MP 1 protein B of *P. lucida* are assigned the homozygous genotypes MP $1^{a/a}$ and MP $1^{b/b}$, respectively. Populations of *P. monacha* from the Rio Fuerte are monomorphic at this locus (Vrijenhoek, unpublished). It is not known whether *P. lucida* are monomorphic since individuals from only two inbred strains were available for this portion of the study.

The unisexual-hybrids show codominant expression of both progenitor alleles at the MP 1 locus. *P. monacha-lucida* is heterozygous (MP $1^{a/b}$) having two bands of apparently equal staining intensity that correspond to the MP 1 A and MP 1 B bands in the progenitors (Fig. 2). Due to the lack of hybrid bands of intermediate mobility, MP 1 is presumed to be a monomer. Dosage effects are clearly observed in the triploid hybrids. *P.2 monacha-lucida* has a darkly staining MP 1 A band and an MP 1 B band of lower concentration. Conversely, *P. monacha-2 lucida* has an MP 1 A band of lower concentration than the intensely staining MP 1 B band. Triploid genotypes are presumed to be MP $1^{a/a/b}$ and MP $1^{a/b/b/}$, respectively.

DISCUSSION

Diploid and triploid unisexual "species" of *Poeciliopsis* have isozyme patterns that display heterozygous expression of allelic isozymes derived from the sexual species *P. monacha* and *P. lucida*. Allelic expression for loci encoding the enzyme alcohol dehydrogenase (ADH), and a soluble protein in muscle tissues (MP 1) is directly related to genomic dosage in the unisexuals. These results reaffirm previous morphological (Schultz, 1969) and biochemical (Vrijenhoek, 1972a) evidence for the hybrid origin and genomic constitution of the all-

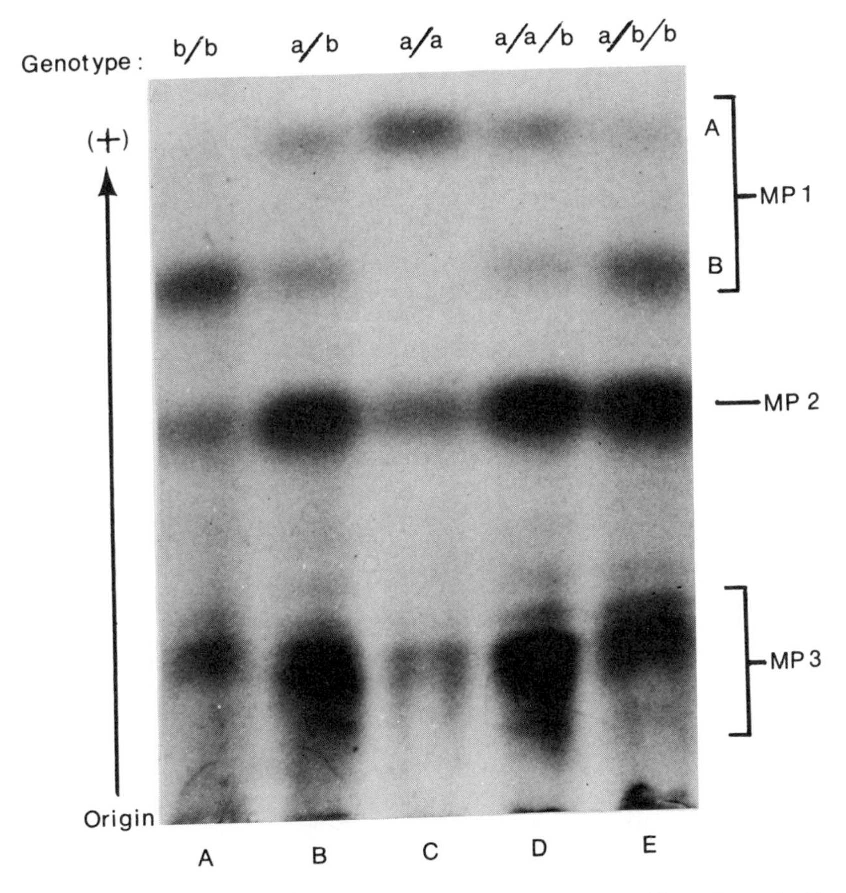

Fig. 2. Muscle protein (MP) patterns in *Poeciliopsis:* A) *P. lucida;* B) *P. monacha-lucida;* C) *P. monacha;* D) *P.* 2 *monacha-lucida;* E) *P. monacha*-2 *lucida.*

---

female "species".

Accurate quantification of isozyme concentrations by scanning densitometry of the starch gels was not available for this study. Nevertheless, visual estimates of staining intensities in the diploid hybrid, *P. monacha-lucida,* heterozygous for MP $1^{a/b}$ indicate the presence of equal concentrations of the monomeric A and B proteins, and therefore suggest that expression of the parental alleles is equal. Both parental ADH alleles also were expressed in *P. monacha-lucida* heterozygous for $ADH^{b/c}$, producing a three-banded pattern for this dimeric molecule. The paternal CC homodimer derived from *P. lucida* stains with slightly greater intensity, however, than is expected if ADH^b and ADH^c allelic expression is equal in the hybrid.

In a previous study, a similar inequality of allelic isozymes was observed in *P. monacha-lucida* heterozygous for the locus encoding the eye-tissue specific isozyme of lactate dehydrogenase (Vrijenhoek, 1972a). Individuals heterozygous for the fast E' and slow E" alleles encoding this tetrameric enzyme, do not have the 1:4:6:4:1 ratio of E'_4, $E'_3 E''_1$, $E'_2E''_2$, $E'_1E''_3$, and E''_4 isozymes that is expected. The tetramers containing the paternal E" subunit derived from *P. lucida* predominate while the purely maternal E'_4 isozyme is barely visible on starch gels. The isozyme patterns for LDH E and ADH suggest partial repression of the maternal alleles in *P. monacha-lucida*.

Unequal expression of allelic isozymes has been reported for diploid interspecific hybrids of other organisms. The delayed synthesis of paternal isozymes during the ontogeny of many hybrids may result from the inability of the paternal alleles to respond to the levels of gene activating molecules in the maternally derived cytoplasmic environment of the egg (Goldberg et al, 1969; Hitzeroth et al, 1968; Klose et al, 1969; Klose and Wolf, 1970; Ohno, 1969). Only a few studies, however, have reported what appears to be partial repression of maternal isozymes in adult tissues of interspecific hybrids (Pipkin and Bremer, 1970; Whitt et al, 1972, 1973). Although the mechanism of maternal repression in these hybrids is not known, the degree of inequality in allelic expression may be correlated with the evolutionary distance between the parental species (Whitt et al, 1973).

The inequalities of allelic expression at enzymatic loci contrast markedly with the apparent equality of allelic expression for the A and B proteins produced by the MP 1 locus in *P. monacha-lucida*. Since this protein is non-enzymatic and a monomer, its expression cannot be altered by differential enzyme activities or non-random polymer assembly. It is not known, however, whether the inequalities of allelic expression at the LDH E and ADH loci in *Poeciliopsis* are due to regulatory mechanisms that affect allelic transcription or to post-transcriptional factors such as non-random polymer assembly, differential enzyme activities, or differential isozyme turnover rates. Ontogenetic studies and in vitro molecular hybridization experiments are needed to distinguish among these alternative possibilities.

Although the mechanisms that favor unequal expression of paternal LDH E_4 and ADH isozymes in *P. monacha-lucida* are not understood, partial repression of maternal alleles might be advantageous if it increases the similarity between *P. monacha-lucida* and females of *P. lucida*, the species upon which this unisexual depends for insemination. Since males of *P. lucida* strongly prefer conspecific mates (McKay, 1971), such increased similarity also would increase the likelihood of unisexuals being inseminated. Sexual selection would favor the accumulation of hypomorphic or even amorphic ("silent") alleles in the *monacha* genome of *P. monacha-lucida,* thus allowing full expression of paternal *lucida* alleles. Such events apparently have occurred since the genes for the dense genital pigmentation pattern typical of females of *P. monacha* are not expressed by the maternal genome of wild caught *P. monacha-lucida* (Vrijenhoek and Schultz, 1974). Perhaps other *monacha* genes that are closely linked with genes controlling the distinctive mate recognition characteristics of this species are also repressed in the maternal genome of *P. monacha-lucida.* Comparative studies of allelic expression among natural *P. monacha-lucida* and laboratory synthesized hybrids (Schultz, 1973) should reveal the degree to which the natural populations have undergone alterations of maternal expression as an adaptation to their sperm dependent relationship with *P. lucida.*

The synthesis of both parental isozymes in the triploid all-female "species" of *Poeciliopsis* indicates that both *monacha* and *lucida* genes are present and functional. The relationship between allelic isozyme expression and genomic dosage is consistent with the hypothesis that *P.* 2 *monacha-lucida* has two doses of *monacha* genes plus one does of *lucida* genes and that *P. monacha*-2 *lucida* has one dose of *monacha* genes plus two doses of *lucida* genes. Muscle protein patterns in the triploids revealed equal contributions of the MP 1^a and MP 1^b alleles in direct proportion to their allelic dosages. Expression of alcohol dehydrogenase isozymes in the triploids showed some slight deviation from the expected patterns, however. Although ADH patterns in both triploids clearly revealed the effects of allelic dosage, a slight inequality in allelic isozyme concentrations suggests that the triploids favored the expression of the ADH^c allele derived from *P. lucida.* This result is consistent with their proposed origin from *P. monacha-lucida* in which expression of the same allele is favored.

The evidence that each of the three parental genes is active in triploid *Poeciliopsis* is supported by previous electrophoretic studies on polyploid organisms. These include the triploid unisexual fish, *Poecilia formosa* (Balsano et al, 1967), triploid unisexual salamanders, *Ambystoma platineum* and *Ambystoma tremblaya* (Uzzell and Goldblatt, 1967), the tetraploid frog, *Odontophyrnus americanus* (Becak et al, 1968), the triploid lizard, *Cnemodophorus tessalatus* (Neaves and Gerald, 1969), and triploids of the fruit fly, *Drosophila melanogaster* (Pipkin and Hewett, 1972). The relationship between gene dosage and the total level of gene products in diploid vs. polyploid organisms is unclear, however. Various workers have obtained a spectrum of results ranging from a distinct increase in gene products of triploid *Drosophila melanogaster* (Pipkin and Hewett, 1972; Lucchesi and Rawls, 1973), to a distinct decrease in the gene products of tetraploid frogs *Odontophyrnus americanus* (Becak and Pueyo, 1970). Others have reported no observable difference between diploids and triploids of *Drosophila melanogaster* (Seecof, 1961) and of chickens (Abdel-Hameed, 1972). Further work is needed to determine the effects of increased gene dosage on gene expression in polyploid organisms.

A close relationship exists among unisexuality, hybridization, and polyploidization (Schultz, 1969). Allopolyploidy has long been overloooked in animals because of the rarity of interspecific crosses and the developmental and reproductive inferiority associated with most interspecific hybrids. The extensive allopolyploidization found in plants is thought to have resulted from their ability to reproduce vegetatively (Stebbins, 1950). Since most animals lack this ability, hybrid populations are considered incapable of lasting long enough for the rare chromosomal doubling that results in allopolyploidization. The recent discovery of various apomictic reproductive modes (parthenogenesis, gynogenesis, and hybridogenesis) among unisexual-hybrid "species" of vertebrates, however, should encourage a reexamination of the proposed autopolyploid origins for certain families of fishes (cf. Ohno et al, 1967; Massaro and Markert, 1968).

The unisexual reproductive modes of *Poeciliopsis* perpetuate the heterozygosity that these fish were endowed by their hybrid origins. Preliminary studies indicate that these unisexual individuals may be heterozygous at greater than 40% of their gene loci (Vrijenhoek, unpublished). This "enforced" heterozygosity results in a multiplicity of proteins shared by each individual in the unisexual population, but does not extract the costs that sexual species must pay in terms of segregational load. Thus, gene duplication in this manner could be an alternative strategy to the allelic variation of sexual species (Fincham, 1966; Sing and Brewer, 1968) and might be of primary importance to the survival of all-female "species".

ACKNOWLEDGMENTS

I wish to thank R. Jack Schultz for making his laboratory strains of *Poeciliopsis* available for this study. This study was supported by N.S.F. institutional grant for science, number GU-3752, to Southern Methodist University.

REFERENCES

Abdel-Hameed, F. 1972. Hemoglobin concentration in normal diploid and intersex triploid chickens: Genetic inactivation or canalization? *Science* 178: 864-865.

Balsano, J. S., R. M. Darnell, and P. Abramoff 1972. Electrophoretic evidence of triploidy associated with populations of the gynogenetic teleost *Poecilia formosa*. *Copeia* 1972: 292-297.

Becak, W. and M. T. Pueyo 1970. Gene regulation in the polyploid amphibian *Odontophrynus americanus*. *Exp. Cell Res.* 63: 448-451.

Becak, W., A. R. Schwantes, and M. L. Schwantes 1968. Polymorphism of albumin-like proteins in the South American tetraploid frog *Odontophrynus americanus* (Salientia: Ceratophrydidae). *J. Exp. Zool.* 168: 473-476.

Cimino, M. C. 1972a. Egg-production, polyploidization and evolution in a diploid all-female fish of the genus *Poeciliopsis*. *Evolution* 26: 294-306.

Cimino, M. C. 1972b. Meiosis in triploid all-female fish (*Poeciliopsis*, Poeciliidae). *Science* 175: 1484-1486.

Fincham, J. R. S. 1966. *Genetic Complementation*. W. A. Benjamin, New York.

Goldberg, E., J. P. Cuerrier, and J. C. Ward 1969. Lactate dehydrogenase ontogeny, paternal gene activation, and tetramer assembly in embryos of brook trout, lake trout, and their hybrids. *Biochem. Genet.* 2: 335-350.

Hitzeroth, H., J. Klose, S. Ohno, and U. Wolf 1968. Asynchronous activation of parental alleles at the tissue specific gene loci observed in hybrid trout during early development. *Biochem. Genet.* 1: 287-300.

Klose, J., H. Hitzeroth, H. Ritter, E. Schmidt, and U. Wolf 1969. Persistence of maternal isozyme patterns of lactate dehydrogenase and phosphoglucomutase system during early development of hybrid trout. *Biochem. Genet.* 3: 91-97.

Klose, J. and Wolf 1970. Transitional hemizygosity of the maternally derived allele at the 6PGD locus during early development of the cyprinid fish *Rutilus rutilis*. *Biochem. Genet.* 4: 87-92.

Lucchesi, J. C. and J. M. Rawls 1973. Regulation of gene function: A comparison of enzyme activity levels in relation to gene dosage in diploids and triploids of *Drosophila melanogaster*. *Biochem. Genet.* 9: 41-51.

Massaro, E. J. and C. L. Markert 1968. Isozyme patterns of

salmonid fishes: evidence for multiple cistrons for lactate dehydrogenase polypeptides. *J. Exp. Zool.* 168: 223-238.

McKay, F. E. 1971. Behavioral aspects of population dynamics in unisexual-bisexual *Poeciliopsis* (Pisces: Poeciliidae). *Ecology* 52: 778-790.

Neaves, W. B. and P. S. Gerald 1968. Lactate dehydrogenase isozymes in parthenogenetic teiid lizards *(Cnemidophorus)*. *Science* 160: 1004-1005.

Ohno, S. 1969. The preferential activation of maternally derived alleles in development of interspecific hybrids. Pp. 137-150. In (Defendi, V., ed.) *Heterospecific genome activation*, the Wistar Institute Press, Philadelphia.

Ohno, S., J. Muramoto, J. Klein, and N. B. Atkin 1967. Diploid-tetraploid relationship in cluploid and salmonoid fish. In (C. D. Darlington and K. R. Lewis, eds.) *Chromosomes Today*, Vol. II, Oliver and Boyd, Edinburgh.

Pipkin, S. B. and T. A. Bremer 1970. Aberrant octanol dehydrogenase isozyme patterns in interspecific *Drosophila hybrids*. *J. Exp. Zool.* 175: 283-296.

Pipkin, S. B. and N. E. Hewitt 1972. Effect of gene dosage on level of alcohol dehydrogenase in *Drosophila*. *J. Heredity* 63: 331-336.

Schultz, R. J. 1967. Gynogenesis and triploidy in the viviparous fish *Poeciliopsis*. *Science* 157: 1564-1567.

Schultz, R. J. 1969. Hybridization, unisexuality, and polyploidy in the teleost *Poeciliopsis* (Poecilliidae) and other vertebrates. *Am. Nat.* 103: 605-619.

Schultz, R. J. 1973. Unisexual fish: laboratory synthesis of a "species." *Science* 179: 180-181.

Seecof, R. 1961. Gene dosage and enzyme activities in *D. melanogaster*. *Genetics* 46: 605-614.

Shaw, C. R. and R. Prasad 1970. Starch gel electrophoresis--A compilation of recipes. *Biochem. Genet.* 4: 297-320.

Sing, C. F. and G. V. Brewer 1969. Isozymes of a polyploid series of wheat. *Genetics* 61: 391-398.

Smith, M., D. A. Hopkinson, and H. Harris 1972. Studies on the subunit structure and molecular size of the human alcohol dehydrogenase isozymes determined by different loci, ADH_1, ADH_2, ADH_3. *Ann. Hum. Genet.* 36: 401-414.

Stebbins, G. L. 1950. *Variation and evolution in plants*. Columbia Univ. Press, New York.

Uzzell, T. and M. Goldblatt 1967. Serum proteins of salamanders of the *Ambystoma jeffersonianum* complex, and the origin of the triploid species of this group. *Evolution* 21: 345-354.

Vrijenhoek, R. C. 1972a. Genetic relationships of unisexual-hybrid fishes to their progenitors using lactate dehydrogenase isozymes as gene markers (*Poeciliopsis*, Poeciliidae). *Am. Nat.* 106: 754-766.

Vrijenhoek, R. C. 1972b. Ph.D. dissertation: Hybrid relationships of diploid and triploid forms of *Poeciliopsis* (Pisces: *Poeciliidae*). Univ. of Connecticut, Storrs.

Vrijenhoek, R. C. 1974. Effects of parasitism on the esterase isozyme patterns of fish eyes. *Comp. Biochem. Physiol.* (B): In Press.

Vrijenhoek, R. C. and R. J. Schultz 1974. Evolution of a trihybrid unisexual fish (*Poeciliopsis;* Poeciliidae). *Evolution* 28: 306-319.

Whitt, G. S., P. L. Cho, and W. F. Childers 1972. Preferential inhibition of allelic isozyme synthesis in an interspecific sunfish hybrid. *J. Exp. Zool.* 179: 271-282.

Whitt, G. S., W. F. Childers, and P. L. Cho 1973. Allelic expression at enzyme loci in an intertribal hybrid sunfish. *J. Heredity* 64: 54-61.

ENZYMIC AND IMMUNOCHEMICAL HOMOLOGIES OF HEXOSE-6-PHOSPHATE DEHYDROGENASE IN FISH

JOHN J. STEGEMAN and TOSHIO YAMAUCHI
Department of Biology
Woods Hole Oceanographic Institution
Woods Hole, Massachusetts 02543 and
Department of Biology
M. D. Anderson Hospital and Tumor Institute
Houston, Texas 77025

ABSTRACT. Isozymes of glucose-6-phosphate dehydrogenase in fish include a soluble and a microsomal form. The soluble enzyme is specific for G-6-P and NADP while the microsomal form, hexose-6-phosphate dehydrogenase (H6PD), catalyzes oxidation of glucose-6-phosphate, galactose-6-phosphate, 2-deoxyglucose-6-phosphate and unsubstituted glucose using either NAD or NADP. H6PD is an autosomally linked dimeric protein in trout, and it responds to the stress of starvation with a slight increase in specific activity in the liver. In a number of species of salmonid fishes H6PD exhibits the same tissue distribution (liver) and virtually identical kinetic properties, with the various substrates. Similarly, the response to pH changes and anions, and characteristics of thermal inactivation are the same in H6PD from different species. Hepatic H6PDs from trout and swordtail exhibit a high degree of antigenic similarity, confirming the homology of the H6PD gene in diverse fishes. In addition, micro-complement fixation data indicate that H6PD and G6PD from trout possess some antigenic determinants in common. This supports the theory that H6PD and G6PD arose from a common ancestral gene.

INTRODUCTION

The heterogeneity of mammalian glucose-6-phosphate dehydrogenase (G6PD), the initial enzyme of the pentose phosphate shunt, was first recognized by Tsao (1960). It has since become obvious that the occurrence of multiple forms of this enzyme is a widespread phenomenon (Thornber et al., 1968; Nobréga et al., 1970; Kamada and Hori, 1970). Numerous investigations have shown G6PD to be an excellent example of an enzyme which variously exhibits several types of heterogeneity: 1) isozymes, resulting from the presence of distinct genes (Ruddle et al., 1968; Shaw and Koen, 1968); 2) allelic isozymes (allozymes) produced by allelic variation at a single gene locus (Yoshida, 1967; Beutler, 1970); 3) multiple molecular forms which result from in vivo or in vitro post-transcriptional alteration of G6PD gene products (Schmukler, 1970; Taketa and Watanabe, 1971; Hori and Noda, 1971).

One of the more intriguing examples of multiple forms of G6PD involves the genetically distinct isozymes localized in the soluble and microsomal fractions of the cell. The soluble form of G6PD is generally an efficient catalyst only of an NADP-linked oxidation of G-6-P. X-linked in mammals, this enzyme is fairly ubiquitous in its phylogenetic appearance and tissue distribution.

The microsomal form of G6PD, now commonly known as hexose-6-phosphate dehydrogenase (H6PD), was first identified in mammals as glucose dehydrogenase, as a result of its property of catalyzing an NAD-linked oxidation of glucose (e.g., Brink, 1953). It was later determined that H6PD occurred in numerous species (Metzgar et al., 1965; Shaw, 1966; Ohno et al., 1966) and was also an active catalyst for both NAD and NADP-linked oxidations of G-6-P, Gal-6-P and 2-deoxyG-6-P, in addition to glucose (Beutler and Morrison, 1967). H6PD is found primarily in the mammalian liver, though it can be detected in kidney, testis, ovary, spleen, and lung (Shaw and Koen, 1968; Srivastava et al., 1969) and its genetic linkage is autosomal (Shaw and Barto, 1965; Ruddle et al., 1968). In addition to subcellular distribution, genetic linkage, and apparent substrate specificity, H6PD is further distinguished from G6PD by differences in its responses to starvation and to the presence of various in vitro activity modulators including steroid hormones (Criss and McKerns, 1968; Mandula et al., 1970).

Despite these differences H6PD and G6PD exhibit a definite functional relationship in that they both catalyze identical oxidations of G-6-P (Srivastava and Beutler, 1969) and utilize the same nucleotide coenzyme. They can thus be expected to share certain structural features, particularly at the active site. The degree of relationship, however, both in terms of evolution and of physiological function, is unclear. Examination of the phylogenetic distribution of the gene for H6PD and characterization of this enzyme in groups other than mammals can provide information relating to both questions.

RESULTS AND DISCUSSION

H6PD IN FISH

The occurrence of H6PD in fish was first suggested by the NAD-linked oxidation of glucose in fish liver extracts (Metzgar et al., 1965) and NADP-linked oxidation of G-6-P and

Gal-6-P by a single electrophoretic band from rainbow trout liver (Ohno et al., 1966). Since then the appearance of one or more of the catalytic activities of H6PD has been observed in diverse species of fishes (Kamada and Hori, 1970; Stegeman and Goldberg, 1971; Shatton et al., 1971). Characteristics of H6PD from fish have been investigated primarily in salmonids, and studies have been conducted on the enzyme in 11 species of the genera *Salmo*, *Oncorhynchus* and *Salvelinus* (Shatton et al., 1971; Stegeman, 1972).

The electrophoretic phenotype of lake trout (*Salvelinus namaycush*) G6PD isozymes is depicted in Figure 1. The presence of H6PD is indicated by the activity of the slower migrating band with both G-6-P and Gal-6-P. The more anodally migrating set of five bands are the lake trout isozymes of soluble G-6-P specific G6PD (Yamauchi and Goldberg, 1973). Unlike H6PD, these isozymes do not appear when substrates other than G-6-P and NADP are employed in the stain.

H6PD is predominantly an hepatic enzyme in trout, although it can be found in low amounts in some non-hepatic tissues such as blood, intestine, kidney, spleen, ovary, testis, and retina (Stegeman, 1972). The enzyme has not been observed in heart, brain, or skeletal muscle (Stegeman and Goldberg, 1971; Shatton et al., 1971). The specific activity of H6PD is greatest in the microsomal fraction of liver cells. However, as pointed out in Stegeman and Goldberg (1971) and Shatton et al., (1971), in trout liver H6PD is also found in nuclear, mitochondrial and soluble fractions, suggesting a rather loose, generalized association with membrane.

Genetic analysis of a population of brook trout *(Salvelinus fontinalis*) indicates that the H6PD gene in at least some salmonid fishes is autosomal (Stegeman and Goldberg, 1971). The gene for G-6-P specific G6PD is also autosomal in trout (Yamauchi and Goldberg, 1973), but the two genes do not appear it be linked. The gene for H6PD is rather variable in some species and there have been as least seven allelic variants described in brook trout from Canadian waters (Stegeman and Goldberg, 1971 and 1972a). From the appearance of three equally spaced electrophoretic bands in fish heterozygous at the H6PD locus it can be assumed that the enzyme in trout is a dimeric protein. H6PD in mammals is also a dimer (Shaw and Koen, 1968).

H6PD in trout liver responds to prolonged starvation by exhibiting a slight increase in specific activity (Yamauchi et al., 1975). Brook trout G6PD on the other hand shows a significant decrease in activity upon starvation and a slow increase upon refeeding (Yamauchi et al., 1975). The H6PD response in trout is similar to the H6PD response observed in starved mammals (Mandula et al., 1970; Kimura and Yamashita,

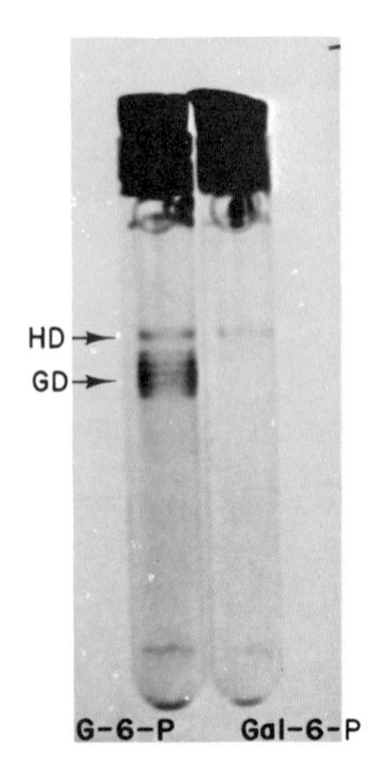

Fig. 1. Polyacrylamide gel electrophoretic patterns of H6PD and G6PD in lake trout liver. The left hand gel was stained at pH 8.0 and 37° with 1.0 x 10^{-3}M G-6-P as substrate and 7.0 x 10^{-4}M NADP as coenzyme. The right hand gel was stained under identical conditions using 1.0 x 10^{-3}M Gal-6-P and 7.0 x 10^{-4}M NADP. The origin (cathode) is at the top. HD, H6PD; GD, G6PD. (After Stegeman and Goldberg, 1971).

1972).

PHYSICO-CHEMICAL PROPERTIES OF FISH H6PD

H6PD has been partially purified from rainbow trout liver *(Salmo gairdneri)* (Shatton et al., 1971) and homodimers have been purified to apparent homogeneity from brook trout and lake trout liver (Stegeman and Goldberg, 1972b). Substrate specificities exhibited by these purified enzymes confirmed what was observed with less pure preparations. H6PD from fish is a single protein catalytically efficient with G-6-P, Gal-6-P, 2-deoxyG-6-P and glucose with either NAD or NADP as H^+ acceptor. The product of glucose oxidation by trout H6PD was found to be gluconic acid with a gluconolactone intermediate (Shatton et al., 1971). These authors also observed activity with galactose and xylose at extremely high substrate concentrations. These activities were not evident at lower concentrations (Stegeman and Goldberg, 1971).

Activity of H6PD in fish toward a given substrate varies

greatly with pH, and with choice of coenzyme. Both of these features are graphically depicted in Figure 2. A pH supporting optimal activity with one substrate is apparently unrelated to the pH optima for the other substrates. It is further quite evident that under most pH conditions the oxidation of G-6-P and Gal-6-P is more efficient with NADP as a coenzyme while that of glucose and 2-deoxyG-6-P is more efficient with NAD. Figure 2 also emphasizes that a comparison between the rate of activity with one substrate or coenzyme and the rate with another must be interpreted with caution.

In general, H6PD's from different species of trout exhibit similar kinetic properties. Substrate K_m values are lowest for G-6-P, around 5×10^{-6} to 2×10^{-5}M. For Gal-6-P, 2-deoxy-G-6-P, and glucose, the values are around 5×10^{-5}M, 7×10^{-4}M, and 2 to 4×10^{-2}M respectively (Stegeman and Goldberg, 1971 and 1972b; Shatton et al., 1971). Higher K_m values for glucose, 1.5×10^{-1}M, have been reported for some species (Metzgar et al., 1965). K_m values for the various substrates have not been found to be greatly affected by the nucleotide used as a hydrogen acceptor, while pH does seem to alter the kinetic properties appreciably (Shatton et al., 1971; Stegeman and Goldberg, 1972b). K_m's for the nucleotide coenzymes, on the other hand, vary somewhat more with choice of sugar substrate. The values are similar for both NAD and NADP, and range from ca. 3×10^{-6}M with G-6-P, up to 2×10^{-5}M or higher with glucose.

Trout H6PD activity is markedly affected by the presence of various ions, and the effects vary with substrate and coenzyme. Mg^{++}, an important cofactor for G6PD, slightly inhibits G-6-P oxidation by H6PD. ATP on the other hand does not affect this oxidation activity by trout H6PD (Stegeman and Goldberg, 1972b), although it does inhibit G-6-P oxidation by G6PD (Avigad, 1966). NAD or NADP-linked oxidations of either G-6-P or Gal-6-P are only marginally affected by the presence of anions. In contrast, anions markedly stimulate glucose oxidation with either nucleotide (Shatton et al., 1971; Stegeman and Goldberg, 1972b) but inhibit the oxidation of 2-deoxyG-6-P. These appear to be general anionic effect as all anions thus far tested produce similar responses.

The isoelectric point for trout H6PD was found to be somewhat high, around pH 6.6 - 7.1 (Stegeman and Goldberg, 1972b). This was higher than most other proteins involved in the analysis, including G6PD, and suggests that trout H6PD contains a relatively smaller percentage of acidic amino acid residues. The slower electrophoretic mobility of H6PD relative to G6PD in a wide variety of vertebrate organisms (Kamada and Hori, 1970) indicates that H6PD is generally a more basic protein

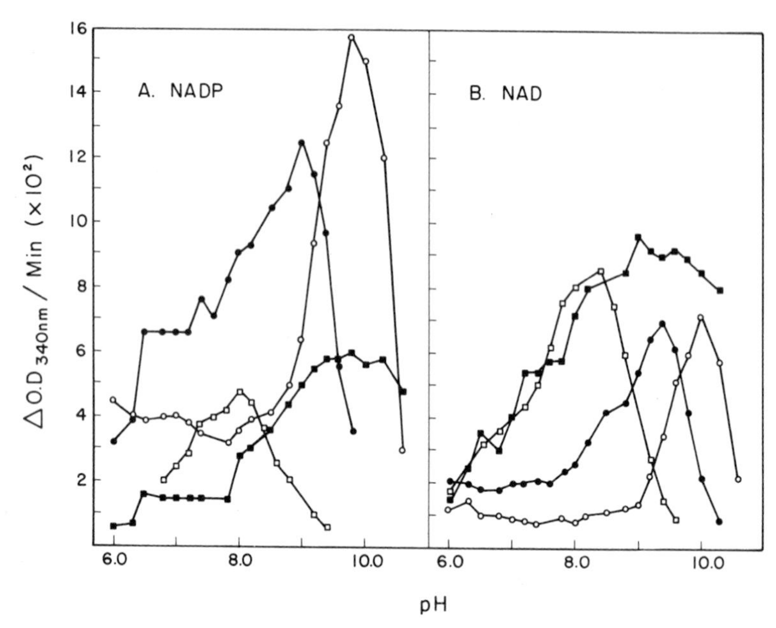

Fig. 2. Effect of pH on the oxidations catalyzed by pure lake trout H6PD. A) Activity with NADP. B) Activity with NAD. Each point represents the average of 3 determinations. G-6-P oxidation ○—○ ; Gal-6-P oxidation ●—● ; 2-dG-6-P oxidation □—□ ; glucose oxidation ■—■

than G6PD.

The molecular weight of H6PD has been determined for several fish species using either gel filtration (Metzgar et al., 1965; Stegeman and Goldberg, 1971) or sedimentation in sucrose gradients (Stegeman, 1972). There are however discrepancies in results achieved with the two methods. Gel filtration yields a molecular weight in the neighborhood of 230,000 while density gradient centrifugation indicates the molecular weight is approximately 140,000. The cause of this discrepancy is unknown, though it possibly results from an association-dissociation phenomenon similar to that frequently observed with G6PD. A further, though perhaps less attractive, explanation is an anomalous association between H6PD and catalase (m.w. 232,000) which was a marker in both gel filtration studies.

H6PD from trout is a far more stable enzyme than G6PD and does not require the presence of added cofactors to maintain a stable, active conformation (Yamauchi, 1972). As one would expect, the enzyme exhibits virtually identical rates of thermal inactivation for all reactions catalyzed (Shatton et al., 1971). Trout H6PD has been reported to retain 50% activity after 15 minutes at 40° (Stegeman, 1972) or after five minutes at 45° (Shatton et al., 1971) and it was completely inactivated after five minutes at 55°. Mochizuki and Hori (1973) on the other hand reported that trout H6PD possessed over 80% of original activity after five minutes at 50° and over 60% activity by the enzyme from two other species. Thermodynamic characteristics of heat inactivation do indicate some species differences in stability of trout H6PD but for the most part results are consistent with data for other proteins from trout and other species (Stegeman and Goldberg, 1972b).

H6PD GENE HOMOLOGY

Antisera prepared against purified brook trout H6PD and purified brook trout G6PD (Yamauchi and Goldberg, 1973) were used to determine whether H6PD from various species of fishes are immunochemically related. Immunological reactions can be detected by enzymic inhibition, which assumes that enzyme and antibody combine so as to impair catalytic efficiency. Electrophoresis followed by histochemical staining can be used to detect such inhibition. (Markert and Holmes, 1969).

This method was used to demonstrate that antiserum to brook trout H6PD selectively inhibited the H6PD activity in brook trout liver and also in swordtail (*Xiphophorus* sp.) liver homogenates (Figure 3A), clearly indicating an antigenetic relationship between the two. By the same token, G6PD antiserum selectively removed the two zones of G6PD activity in extracts of both brook trout and swordtail. (For a discussion of the multiple electrophoretic zones of G6PD in trout, see Yamauchi and Goldberg, 1973).

Though there is moderate interspecific variation in some parameters, the properties of H6PD in various trout species and the above immunological data strongly suggest that the gene for microsomal H6PD is homologous in various species of fishes. Summarizing the characteristics of H6PD (Table 1), we can readily see that the enzyme from fish is also quite comparable to mammalian H6PD in virtually all parameters thus far described, with the possible exception of thermal stability. It therefore appears that the fish and mammalian H6PDs are also homologous and possibly have the same evolutionary history.

Similar NADP-linked oxidations of G-6-P obviously suggest that soluble G6PD and microsomal H6PD in both fish and mammals share certain structural features. This is further suggested

TABLE 1

PROPERTIES OF H6PD IN FISH AND MAMMALS

Parameter		Trout H6PD	Mammalian H6PD
Coenzymes		NAD, NADP (7,12)	NAD, NADP (1)
Substrates		glucose, G-6-P Gal-6-P, 2-dG-6-P (7,12)	glucose, G-6-P Gal-6-P, 2-dG-6-P (1,3)
K_m (molar)	G-6-P	10^{-6} (7,12)	10^{-6} (1)
	Gal-6-P	10^{-5}	10^{-5}
	2-dG-6-P	10^{-4}	10^{-5}
	glucose	10^{-2}	10^{-1}
	NAD	10^{-6}	10^{-6}
	NADP	10^{-5}	10^{-5}
Anion effect (glucose)		stimulation (7,12)	stimulation (1,2)
Reaction product		lactone (7)	lactone (10)
Protein structure		dimer (11)	dimer (9)
Subcellular local		microsomal (7,11)	microsomal (1,9)
Tissue distribution		primarily hepatic (11)	primarily hepatic (5,9)
Genetic linkage		autosomal (11)	autosomal (6,8)
Starvation response		minimal (slight increase) (13)	minimal (4,5)

1) Beutler and Morrison, 1967
2) Horne and Nordlie, 1971
3) Kamada and Hori, 1970
4) Kimura and Yamashita, 1972
5) Mandula et al., 1970
6) Ruddle et al., 1968
7) Shatton et al., 1971
8) Shaw and Barto, 1965
9) Shaw and Koen, 1968
10) Srivastava and Beutler, 1969
11) Stegeman and Goldberg, 1971
12) Stegeman and Goldberg, 1972b
13) Yamauchi et al., 1975

by the appearance of G6PD variants (ostensibly point mutations, e.g., Yoshida, 1967) possessing altered catalytic properties closely resembling those of H6PD (Kirkman et al., 1968). Altered substrate specificity of G6PD molecules modified in vitro (Yoshida, 1973) argues even more persuasively for a structural relationship between G6PD and H6PD. It is clearly plausible to suggest that the two arose from a common ancestral gene (Stegeman and Goldberg, 1971; Srivastava et al., 1972).

Immunochemical studies, using the antisera to trout G6PD described above, have provided further evidence supporting a relationship between H6PD and G6PD. In addition to

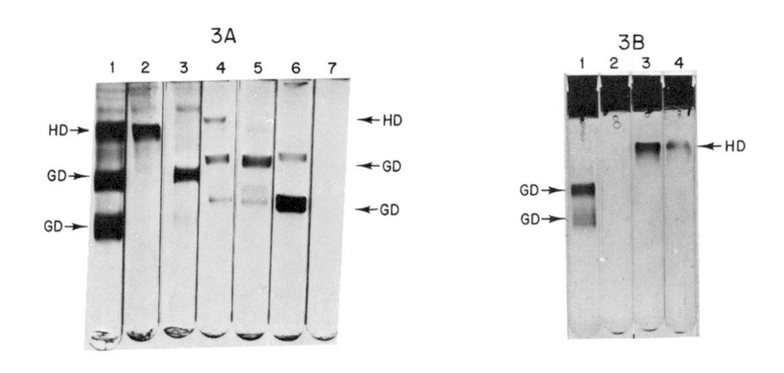

Fig. 3. Immunochemical specificity of G6PD and H6PD isozymes. A. Brook trout or swordtail (*Xiphophorus* sp.) antigens were mixed 1:1 or 1:2 respectively with antisera to brook trout G6PD (G), brook trout H6PD (H), or control serum (C). HD, H6PD; GD, G6PD. After incubation the antigen-antibody mixture was centrifuged at 40,000 xg for 90 minutes and the supernatant applied to a gel. Antigens tested were brook trout liver (1-3), swordtail liver (4,5) and swordtail eye (6,7) extracts. B. Various brook trout antigens were mixed 1:1 with antiserum to brook trout G6PD (G), or with normal rabbit serum (C) and treated as in Fig. 3A. HD, H6PD; GD, G6PD. The antigens tested were purified G6PD (1,2) and purified H6PD (3,4). (Fig. 3B after Yamauchi and Goldberg, 1973).

inhibiting purified brook trout G6PD, the antiserum to G6PD also exhibited some cross-reactivity with purified brook trout H6PD (Figure 3B). This indication of immunochemical relatedness between G6PD and H6PD is substantiated using the more sensitive methodology of microcomplement fixation (Figure 4). Immunological distances (Sarich and Wilson, 1966) were obtained from the MC'F data for brook trout H6PD and lake trout G6PD using the brook G6PD antiserum. These values were 1.8 for pure lake G6PD and 3.5 - 4.6 for brook H6PD.

Srivastava et al. (1972) did not observe any cross-reactivity in double diffusion experiments using human H6PD antigens and antisera to human G6PD. Similarly Kimura and Yamashita (1972) failed to observe cross-reactivity between rat liver H6PD and G6PD. The data from trout however do indicate antigenic relatedness between H6PD and G6PD, lending strong support to the hypothesis that these two enzymes are derived from a common ancestor.

A number of studies indicate that although G6PD is found

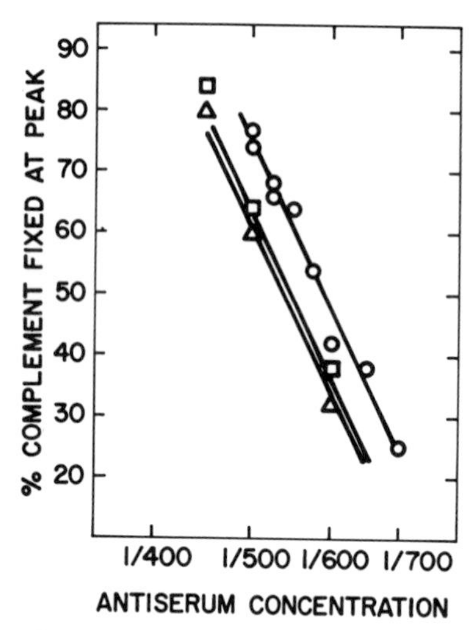

Fig. 4. Microcomplement fixation as a function of antiserum concentration. The antiserum against brook trout G6PD was tested against brook trout G6PD (O), H6PD (Δ), and lake trout G6PD (□). Each point represents the height of a complement fixation curve for a particular antiserum dilution. (After Yamauchi and Goldberg, 1973).

in invertebrate species, H6PD is not (Kamada and Hori, 1970; Mochizuki and Hori, 1973), while both enzymes are clearly present in most classes of vertebrates. Assuming the genes for vertebrate H6PD and G6PD arose from a common ancestral type, this suggests divergence occurred about the time of vertebrate emergence (Stegeman and Goldberg, 1971). More recently, we have failed to find H6PD in representatives of the Agnatha (*Myxine glutinosa, Petromyzon marinus*) or the Chondrichthyes (*Raja erinacea*) while Mochizuki and Hori (1973) have reported NADP but not NAD-linked oxidations of G-6-P and Gal-6-P in the lamprey, *Entosphenus japonicus*. Mochizuki and Hori (1973) have also suggested that certain echinoderms posess an enzyme intermediate in properties to G6PD and H6PD. This suggestion is indeed intriguing and a more thorough analysis of this group and the primitive vertebrates should greatly expand our understanding of the evolution of the H6PD gene.

ACKNOWLEDGEMENTS

This research has been supported by grants from the

United States Public Health Service (NIH) and the National Science Foundation (NSF). Contribution No. 3480 from the Woods Hole Oceanographic Institution.

REFERENCES

Avigad, G. 1966. Inhibition of glucose-6-phosphate dehydrogenase by adenosine 5'-triphosphate. *Proc. Nat. Acad. Sci. USA.* 56: 1453-1547.

Brink, G. 1953. Beef liver glucose dehydrogenase. Purification and properties. *Acta Chem. Scand.* 7: 1081-1089.

Beutler, E. 1970. Genetic variability of glucose-6-phosphate dehydrogenase. *Humangenetik* 9: 250-252.

Beutler, E. and M. Morrison 1967. Localization and characteristics of hexose-6-phosphate dehydrogenase (glucose dehydrogenase). *J. Biol. Chem.* 242: 5289-5293.

Criss, W. E. and K. W. McKerns 1968. Purification and partial characterization of glucose-6-phosphate dehydrogenase from cow adrenal cortex. *Biochem.* 7: 125-134.

Hori, H. and S. Noda 1971. Degradation of glucose-6-phosphate dehydrogenase into enzymatically active forms by trypsin. *J. Histochem. Cytochem.* 19: 299-303.

Horne, N. and R. C. Nordlie 1971. Activation by bicarbonate, orthophosphate, and sulfate of rat liver microsomal glucose dehydrogenase. *Biochim. Biophys. Acta* 242: 1-13.

Kamada, T. and S. H. Hori 1970. A phylogenetic study of animal glucose-6-phosphate dehydrogenases. *Jap. J. Genetics.* 45: 319-339.

Kimura, H. and M. Yamashita 1972. Studies on microsomal glucose-6-phosphate dehydrogenase of rat liver. *J. Biochem.* 71: 1009-1014.

Kirkman, H. N., C. Kidson, and M. Kennedy 1968. Variants of human glucose-6-phosphate dehydrogenase. Studies of samples from New Guinea. p. 126. In: Beutler, E. (ed.), *Hereditary Disorders of Erythrocyte Metabolism,* Grune and Stratton, New York.

Mandula, B., S.K. Srivastava, and E. Beutler 1970. Hexose-6-phosphate dehydrogenase: distribution in rat tissues and effect of diet, age and steroids. *Arch. Biochem. Biophys.* 141: 155-161.

Markert, C. L. and R. S. Holmes 1969. Lactate dehydrogenase isozymes of the flatfish, *Pleuronectiformes*: Kinetic, molecular and immunochemical analysis. *J. Exp. Zool.* 171: 85-104.

Metzger, R. P., S. S. Wilcox and A. N. Wick 1965. Subcellular distribution and properties of hepatic glucose dehydrogenases of selected vertebrates. *J. Biol. Chem.* 240: 2767-

2771.

Mochizuki, Y. and S. Hori 1973. Further studies on animal glucose-6-phosphate dehydrogenases. *Jour. Fac. Sci.* Hokkaido Univ. Ser. VI, Zool. 19: 58-72.

Nobréga, F. G., J. C. C. Maia, W. Colli, and P. H. Saldanha 1970. Heterogeneity of erythrocyte glucose-6-phosphate dehydrogenase (G-6-PD, E.C. 1.1.1.49) activity and electrophoretic patterns among representatives of different classes of vertebrates. *Comp. Biochem. Physiol.* 33: 191-199.

Ohno, S., H. W. Payne, M. Morrison, and E. Beutler 1966. Hexose-6-phosphate dehydrogenase found in human liver. *Science* 153: 1015-1016.

Ruddle, F. H., T. B. Shows, and T. H. Roderick 1968. Autosomal control of an electrophoretic variant of glucose-6-phosphate dehydrogenase in the mouse, *Mus musculus*. *Genetics* 58: 599-606.

Sarich, V. M. and A. C. Wilson 1966. Quantitative immunochemistry and the evolution of primate albumins: Microcomplement fixation. *Science* 154: 1563-1565.

Schmukler, M. 1970. The heterogeneity and molecular transformations of glucose-6-phosphate dehydrogenase in the rat. *Biochim. Biophys. Acta* 214: 309-317.

Shatton, J. B., J. E. Halver, and S. Weinhouse 1971. Glucose (hexose-6-phosphate) dehydrogenase in liver of rainbow trout. *J. Biol. Chem.* 246: 4878-4885.

Shaw, C. R. 1966. Glucose-6-phosphate dehydrogenase: Homologous molecules in deermouse and man. *Science* 153: 1013-1015.

Shaw, C. R. and E. Barto 1965. Autosomally determined polymorphism of glucose-6-phosphate dehydrogenase in *Peromyscus*. *Science* 148: 1099-1100.

Shaw, C. R. and A. L. Koen 1968. Glucose-6-phosphate dehydrogenase and hexose-6-phosphate dehydrogenase of mammalian tissues. *Ann. N. Y. Acad. Sci.* 151: 149-156.

Srivastava, S. K. and E. Beutler 1969. Auxiliary pathways of galactose metabolism: Identification of reaction products of hexose-6-phosphate dehydrogenase and of "galactose dehydrogenase." *J. Biol. Chem.* 244 :6377-6382.

Srivastava, S. K., K. G. Blume, E. Beutler, and A. Yoshida 1972. Immunological difference between glucose-6-P dehydrogenase and hexose-6-P dehydrogenase from human liver. *Nature New Biol.* 238: 240-241.

Stegeman, J. J. 1972. The occurrence, characterization and biological significance of hexose-6-phosphate dehydrogenase in trout. Ph.D. dissertation, Northwestern University, Evanston, Ill.

Stegeman, J. J. and E. Goldberg 1971. Distribution and characterization of hexose-6-phosphate dehydrogenase in trout. *Biochem. Genet.* 5: 579-587.

Stegeman, J. J. and E. Goldberg 1972a. Inheritance of hexose-6-phosphate dehydrogenase polymorphism in brook trout. *Biochem. Genet.* 7: 279-288.

Stegeman, J. J. and E. Goldberg 1972b. Properties of hepatic hexose-6-phosphate dehydrogenase purified from brook trout and lake trout. *Comp. Biochem. Physiol.* 43B: 241-256.

Taketa, K. and A. Watanabe 1971. Interconvertible microheterogeneity of glucose-6-phosphate dehydrogenase in rat liver. *Biochim. Biophys. Acta* 235: 19-26.

Thornber, E. J., I. T. Oliver, and P. B. Scutt 1968. Comparative electrophoretic patterns of dehydrogenases in different species. *Comp. Biochem. Physiol.* 25: 973-987.

Tsao, M. U. 1960. Heterogeneity of tissue dehydrogenases. *Arch. Biochem. Biophys.* 90: 234-238.

Yamauchi, T. 1972. Studies of glucose-6-phosphate dehydrogenase in brook, lake, and splake trout. Ph.D. dissertation, Northwestern University, Evanston, Ill.

Yamauchi, T. and E. Goldberg 1973. Glucose-6-phosphate dehydrogenase from brook, lake, and splake trout: an isozymic and immunological study. *Biochem. Genet.* 10: 121-134.

Yamauchi, T., J. J. Stegeman, and E. Goldberg 1975. The effects of starvation and temperature acclimation on pentose phosphate pathway dehydrogenases in brook trout liver. *Arch. Biochem. Biophys.* (in press).

Yoshida, A. 1967. A single amino acid substitution (asparagine to aspartic acid) between normal (B+) and common Negro variant (A+) of human glucose-6-phosphate dehydrogenase. *Proc. Nat. Acad. Sci. USA* 57: 835-840.

Yoshida, A. 1973. Change of activity and substrate specificity of human glucose 6-phosphate dehydrogenase by oxidation. *Arch. Biochem. Biophys.* 159: 82-88.

ENZYME TYPES OF ATLANTIC COD STOCKS ON THE NORTH AMERICAN BANKS

ALAN JAMIESON
Ministry of Agriculture, Fisheries and Food,
Fisheries Laboratory,
Lowestoft,
Suffolk,
ENGLAND

ABSTRACT. Population samples of Atlantic cod, *Gadus morhua* L., collected off the North American coast were examined for genetic variations at four genetic loci. This census of cod genes is part of a larger current research project using genetic tags to delimit breeding units in cod fisheries. The loci gave the following information.

Cod HbI locus. Hemoglobin showed only slight variation in allele frequency along the north-east coasts of North America. This contrasts with much hemoglobin variation in European cod.

Cod Est_B *locus*. α-butyrate serum esterase allele frequencies showed heterogeneity in the American parts of the species range only, whereas European cod tend to be monomorphic.

Cod Ldh_B *locus*. Cod lactate dehydrogenase-B isoalleles exemplify dynamic polymorphism. North American cod showed the genetic balance seen throughout the species range.

Cod Tf locus. The transferrin frequencies revealed genetic sub-populations of cod at North America, and a striking difference between North American cod and cod elsewhere. The sub-population heterogeneity at North America was so pronounced that it appeared in spite of appreciable stock mixing. As genotyped cod showed no year-class heterogeneity, the continual physical mixing of the sub-populations does not reduce the genetic identities of the sub-populations.

INTRODUCTION

The great bulk of the cod, *Gadus morhua* L., along the North American seaboard occurs between Long Island and Hamilton Inlet, Labrador, but the southern range limit is about Cape Hatteras and the northern limit is where the Hudson Strait meets the Atlantic Ocean. Cod occur inshore along the Gulf of Maine, in the Bay of Fundy, and in the Gulf of St. Lawrence. They also extend seaward over a series of submerged banks stretching north-east from Georges Bank off Cape Cod along the Nova Scotian banks and the Newfoundland banks which culminate in the well-known Grand Bank.

Meristic variants suggest fish races (Heincke, 1898, Sick, 1930) but are not reliable indicators because they are influenced by temperatures at critical developmental stages. Major genes give more elegant descriptions of the genetic architecture of fish populations considered as interbreeding units (Marr, 1957, Cushing, 1964, de Ligny, 1969, Jamieson, 1974). In similar problems of identification, genes are often used to describe ethnic groups, solve forensic problems, and check pedigrees (Harris, 1966, Mourant et al., 1974, Culliford, 1971, Jamieson, 1966a). The zymogram innovation of Hunter and Markert (1957) is now widely used to obtain direct genotypic information.

The present paper is concerned with using certain protein variables in cod tissues to describe the local sub-populations which appear as genetic varieties on the North American banks. This work aims to add new precision to the existing classification of cod stocks (Templeman, 1962). The genetic information about the protein variants supplements the morphometric, meristic, and life-cycle data for cod, giving a better appreciation of their tribal infra-structure.

One example of a simple genetic locus in cod has a pair of co-dominant alleles governing hemoglobins (Sick, 1961). Variation in the proportions of the hemoglobin alleles proved useful for separating some cod stocks in European and Icelandic waters (Sick, 1965a, b, Frydenberg et al., 1965, Møller, 1966a, Jamieson and Jónsson, 1971). A series of co-dominant alleles at another genetic locus governs transferrins in cod sera. This locus has proved useful for characterizing cod stocks over most of the species range (Møller, 1966b, Jamieson, 1967, 1970, Jamieson and Jones, 1967, Jamieson and Jónsson, 1971). A pair of co-dominant alleles at another genetic locus governs lactate dehydrogenase-B subunits. They showed a remarkable similarity in gene frequency between Canadian and Scottish cod (Odense et al., 1966, 1969, Lush, 1970). Subsequent observations show that this same genetic balance extends throughout the range of the species (author's unpublished results). Three alleles at yet another genetic locus for α-butyrate esterases in cod sera showed North American cod stock differences (Jamieson and Thompson, 1972b).

The genotypic data considered in the present paper collate and extend the earlier observations on cod genotypes along the North American states. Hemoglobin, lactate dehydrogenase, esterase, and transferrin data are presented and analyzed.

MATERIAL

The several sources of genotyped cod are shown on a chart outlining North American fishing banks (Fig. 1). Cod were sampled between longitudes 75° and 44° west of Greenwich.

The chronological order of sampling was as follows.

The 310 cod collected in 1962 and 1963 at Ocean City, Woods Hole, and St. Johns were tested for hemoglobin by Sick.

The 508 cod collected in 1964 and 1965 at St. Andrews, Halifax, Grande Rivière, and St. Johns were tested for hemoglobin and lactate dehydrogenase-B by Odense et al. Their Halifax and Grande Rivière samples were tested for transferrin by the author.

The 409 cod collected in the Grand Bank area in 1967 were tested for transferrin. Their otoliths were read for age.

The 474 cod collected in 1971 at Georges Bank, Banquereau, Green Bank, Grand Nord, and Rittu Bank were tested for hemoglobin, lactate dehydrogenase-B, transferrin, and α-butyrate esterase. Otoliths were read.

The above material is thought to have produced all of the presently available genotypic data which permit comparisons between populations of cod sampled along the North American coast. Unfortunately the data obtained for cod at Labrador by Ullrich (1968) could not be directly equated to the genotypes described here, and for that reason could not be included.

METHODS

The method used to type cod hemoglobin was introduced by Sick (1961) and the method used to type cod serum transferrin was introduced by Møller (1966b). The procedure was outlined by Jamieson and Otterlind (1971) and the notations by Jamieson and Jónsson (1971). The method used by Syner and Goodman (1966) gave clear results for LDH-B in cod white muscle extracts. The α-butyrate esterases in cod sera were resolved using the methods of Utter (1969) and described using the notation of Jamieson and Thompson (1972b). Cod otoliths were read for year-classes by the method of Rollefsen (1933).

RESULTS

The genetic data on cod from North American waters are outlined in statements of allele frequencies in Tables 1, 2, 3, and 4. The samples in order from west to east are tabulated in descending columns. The names of the positions where the samples were caught and the times of sampling are shown on the left of the frequencies and can be seen also in Figure 1. Four genetic loci with thirteen alleles are represented in Tables 1 to 4. As some of the material was tested before all the methods for testing different tissues were devised, only the most recently collected specimens were examined for all four genetic loci. A few extremely rare alleles were not tabulated.

A total of 2,202 cod caught off North America have been variously examined by electrophoretic methods revealing their genotypes at up to four genetic loci. A total of 7,244 genes have been scored. The twenty-one cod population samples were taken on twelve banks or localitites. The available repeat samples were at Halifax, St. Johns, and on the Grand Bank. Those repeats were limited to few genetic loci. The results for the different localities were compared, although it was obvious that any comparison showing a difference might be attributed to factors other than position, such as season, year, depth, population age structure or any other uncontrolled variable. All of the available data were included in the statistical tests which follow the presentation of results.

Contingency analyses were applied to the allele frequency data for each of the four genetic loci; the results are shown to the right of the data.

Cod hemoglobin locus, HbI. Twelve populations of cod at eleven localities were tested for hemoglobin. A contingency test, which lumped thirteen scattered examples of rare types along with the allele HbI^2, showed significant heterogeneity, $P < 0.05$, in the hemoglobin frequencies among the populations. The allele frequencies in cod over the whole American region did not vary much, nor did they vary over the decade represented by these data. The lowest frequencies of the allele HbI^1 were seen in the more northerly latitudes, as on both occasions at St. Johns, Newfoundland, on the Newfoundland Grand Bank, and at Grande Rivière in the Gulf of St. Lawrence. This observation gives mild support to the cline suspected by Sick (1965b).

Although this polymorphic locus, cod HbI, has proved an invaluable stock marker in several other Atlantic Ocean areas, so far it makes only a slight contribution to the possible recognition of sub-populations of cod in the American region.

Cod B subunit lactate dehydrogenase locus, Ldh_B. No heterogeneity was detected among the eight population samples of North American cod examined for lactate dehydrogenase-B isoalleles (Table 2). The three cod populations captured in November 1967 and examined by Odense et al. (1969) did not differ from the five populations trawled in February 1971.

TABLE 1

Cod hemoglobin shows genetic polymorphism attributed to the locus HbI. This locus has two familiar co-dominant alleles. The frequencies of the faster allele are tabulated for 12 populations of cod at North America. Thirteen examples of rare alleles found scattered about the region were pooled with the lesser familiar allele in a contingency test. Heterogeneity was due to some of the more northerly samples having the lowest HbI^1 frequencies.

Position	Date	No. of cod	HbI^1		χ^2	Probability	Reference
Ocean City, Maryland	Jan '63	155	0.07		0.51	0.50	a
Woods Hole, Massachusetts	Mar '62	75	0.08		1.06	0.30	a
Georges Bank, Massachusetts	Feb '71	42	0.07		0.21	0.60	d
St Andrews, New Brunswick	Jun '64	70	0.06		1.90	0.20	b
Grand Riviere, Quebec	Sept '64	100	0.03		4.34	<0.05	b
Halifax, Nova Scotia	Jun '64	100	0.06		0.00	1.00	b
Banquereau, Nova Scotia	Feb '71	109	0.08		2.77	0.10	d
Green Bank, Nova Scotia	Feb '71	108	0.08		2.05	0.20	d
St Johns, Newfoundland	Jun '62	80	0.04		0.43	0.50	a
St Johns, Newfoundland	Jun '64	198	0.03		4.45	<0.05	b
Rittu Bank, Newfoundland	Feb '71	108	0.06		0.08	0.80	d
Grand Bank (N.), Newfoundland	Feb '71	108	0.02		2.95	0.10	d
				$\Sigma\chi^2$	20.83	<0.05	

References to sources of data:
a Sick 1965;
b Odense et al., 1966 and 1969;
c Jamieson 1970;
d new data.

TABLE 2

The lactate dehydrogenase B subunit shows genetic polymorphism attributed to the locus Ldh_B. This locus shows two familiar co-dominant alleles of which the faster migrating electrophoretic form is tabulated. No heterogeneity was found in a contingency analysis of this allele frequency data in cod caught at eight positions.

Position	Date	No. of cod	Ldh_B^F		χ^2	Probability	Reference as in Table 1
Georges Bank, Massachusetts	Feb '71	42	0.70		2.02	0.20	d
Grand Rivière, Quebec	Sept '67	100	0.65		0.45	0.50	b
Halifax, Nova Scotia	Jun '67	100	0.63		0.02	0.90	b
Banquereau, Nova Scotia	Feb '71	109	0.61		0.18	0.70	d
Green Bank, Nova Scotia	Feb '71	108	0.61		0.24	0.60	d
St Johns, Newfoundland	Sept '67	98	0.61		0.18	0.70	b
Rittu Band, Newfoundland	Feb '71	108	0.63		0.01	0.90	d
Grand Bank (N.), Newfoundland	Feb '71	108	0.61		0.23	0.90	d
				$\Sigma\chi^2_{(7)}$	3.33	0.80	

TABLE 3

α-butyrate esterase in cod blood sera shows genetic polymorphism attributed to the locus Est_B. This locus contains a series of three co-dominant alleles. The most prevalent allele Est_B^M is recognized by its medium migration rate in electrophoretic separation. A relatively fast allele is called Est_B^F, and a relatively slow allele called Est_B^S are tabulated for the five populations examined at North America. Significant heterogeneity was due to the high value of the fast allele in cod sampled on Green Bank.

Position	Date	No. of cod	Est_B^F	Est_B^S		χ^2	Probability	Reference as in Table 1
Georges Bank, Massachusetts	Feb '71	42	0.12	0.02		0.47	0.80	d
Banquereau, Nova Scotia	Feb '71	109	0.12	0.01		0.78	0.70	d
Green Bank, Nova Scotia	Feb '71	108	0.18	0.01		11.03	<0.01	d
Rittu Bank, Newfoundland	Feb '71	108	0.08	0.03		5.60	0.10	d
Grand Bank (N.), Newfoundland	Feb '71	108	0.06	0.01		5.71	0.10	d
					$\Sigma\chi^2$	23.58	<0.01	

TABLE 4

Transferrin in cod blood sera shows genetic polymorphism attributed to a series of eight co-dominant alleles at the genetic locus Tf. Six Tf alleles occurred in samples of 11 populations of cod examined at North America. A contingency test showed highly significant heterogeneity of the transferrin allele frequencies. Thus transferrin in cod at North America reveals a most striking example of genetic differences occurring between adjacent natural populations within a single species.

Position	Date		No. of cod	Tf^A	Tf^B	Tf^{C1}	Tf^C	Tf^D	Tf^E	χ^2	Probability	Reference as in Tbl. 1
Georges Bank, Massachusetts	Feb	'71	42	0.01	0.18	0.41	0.15	0.24	0.01	16.01	<0.01	d
Grand Rivière, Quebec	Sept	'67	100	0.01	0.43	0.27	0.14	0.14	0.00	6.30	0.30	c
Halifax, Nova Scotia	Jun	'67	100	0.01	0.42	0.18	0.18	0.18	0.04	24.71	<0.001	c
Banquereau, Nova Scotia	Feb	'71	109	0.00	0.31	0.45	0.08	0.15	0.01	21.04	<0.001	d
Green Bank, Newfoundland	Feb	'71	108	0.02	0.22	0.57	0.05	0.11	0.03	64.22	<0.001	d
Rittu Bank, Newfoundland	Feb	'71	108	0.00	0.48	0.33	0.07	0.11	0.01	20.15	<0.01	d
Grand Bank (N.), Newfoundland	Feb	'71	108	0.04	0.42	0.25	0.12	0.15	0.02	15.79	<0.01	d
Grand Bank (N.), Newfoundland	Nov	'67	86	0.00	0.49	0.31	0.03	0.13	0.03	23.85	<0.001	c
Grand Bank (E.), Newfoundland	Nov	'67	174	0.00	0.38	0.33	0.14	0.13	0.03	5.18	0.40	c
Grand Bank (SE.), Newfoundland	Nov	'67	29	0.00	0.55	0.33	0.07	0.05	0.00	28.11	<0.001	c
Flemish Cap, Newfoundland	Nov	'67	120	0.06	0.19	0.28	0.30	0.15	0.01	104.53	<0.001	c
									$\Sigma\chi^2$	329.18	<0.001	

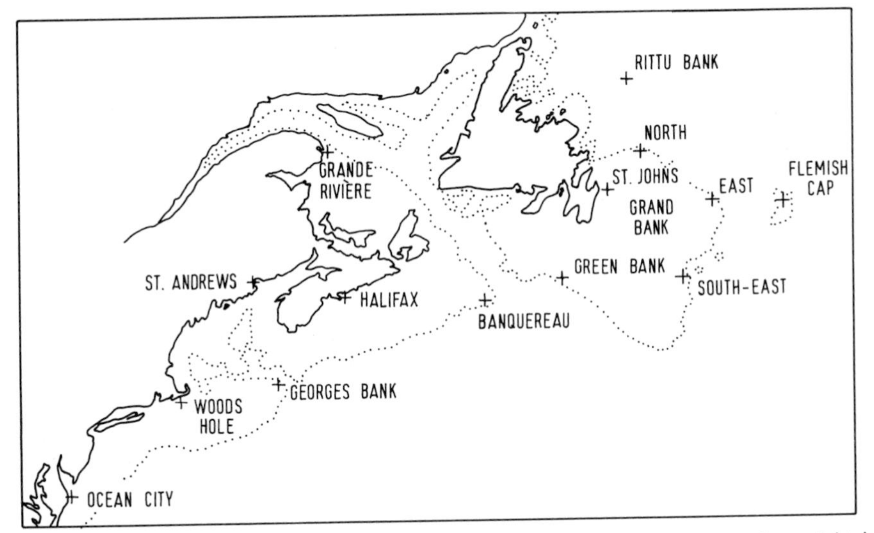

Fig. 1. An outline chart shows those North American localities where population samples of *Gadus morhua* were collected for genetic analyses.

Other data for lactate dehydrogenase-B in cod caught at Aberdeen, Scotland again showed similar allele frequencies (Lush, 1970). The same is true for many cod at Greenland and at Iceland (Jamieson, unpublished data). Among the more recent lactate dehydrogenase-B genotypes of North American cod specimens, the Hardy-Weinberg expectations differed significantly from the observed frequencies for the genotyped cod caught on Rittu Bank. The significant imbalance suggests that the heterozygotes enjoy some selective advantage over the homozygotes.

This locus, Ldh_B, provides geneticists with a satisfying example of a selectively maintained polymorphism. As this balance maintains the two alleles at the same frequency throughout the species range, the locus has no immediate application as a stock marker in practical fisheries research work, but further study of this selection effect could yet be rewarding.

α-butyrate esterase in cod sera, Est_B. The allele frequency estimates in Table 3 show that only five populations of American cod were examined for α-butyrate esterase variants in their blood sera. Fortunately those five samples were taken over a considerable distance and they were almost contemporaneous. That part of Table 3 giving the results of a contingency test on the frequencies of the three esterase isozymes shows significant heterogeneity ($P < 0.01$), largely due to the high value of the allele Est_B^F, 0.18, in cod caught on Green Bank. As may be seen in Table 4, this same sample of cod taken on Green Bank also showed the highest frequency (0.57) of the allele Tf^{Cl} at another locus. As both those alleles are characteristic of North American cod, the Green Bank sample provided the most distinctive American type cod population examined to date.

For the remainder of this species, away from North America, the cod α-butyrate esterases show less variation and tend towards being monomorphic for Est_B^M in European waters (Jamieson and Thompson, 1972b). Nevertheless, in the American cod samples the esterase locus showed more variation than did the hemoglobin locus. Hence it is more valuable as a tool for recognizing sub-populations in the region.

Cod serum transferrin locus, Tf. Six population samples of cod at North America caught in 1967 and five caught in 1971 were examined for transferrin variants. The cod transferrin data in the eleven American cod populations showed considerable heterogeneity. The high significance of this statement is expressed in Table 4 ($P < 0.001$). The observed and expected transferrin genotypes are stated in Table 5. The probability estimates on the right of Table 5 show that the genetic imbalance at the cod Tf locus was widespread and always significant. The cod transferrins demonstrated two interesting features. Firstly, the different American banks showed widely different allele frequencies. Secondly, each bank showed an excess of homozygotes. The simplest interpretation would be to postulate a distinct breeding population of cod on each major fishing bank examined. The appreciable physical mixing of genetically different bank stocks then appears to occur without any effective interbreeding between the stocks. Unfortunately, some of the populations of cod examined on the different banks were captured at different seasons in different years. It was necessary to test the permanence of the regional differences attributed to sub-populations. A large proportion of the data was collected near the Grand Bank of Newfoundland, Fortunately, the age composition of this Grand Bank material was obtained from otoliths; 27 consecutive year-classes of cod were represented and these are tabulated against transferrin genotypes in Table 6. A contingency analysis of gene frequencies failed to detect year-class heterogeneity, and a statistical analysis of the genotypes in Table 6 showed that the considerable excess of homozygote classes persisted for more than a decade. Thus it appears that the form of the transferrin data presented here is not a chance occurrence due to some uncontrolled or temporary effect, but is indicative of an old-established sub-population structure built into the survival strategy of cod stocks of the North American banks.

A visual impression of the variation in Atlantic cod transferrin alleles is given in Figure 2, which illustrates the transferrin allele frequencies presented here for North America and compares transferrin allele frequencies in other parts of the range of this species. The actual frequencies and their variation in the different regions are described in various publications (Jamieson, 1967, 1970, 1974, Jamieson and Jones, 1967, Jamieson and Jónsson, 1971, Jamieson and Otterland, 1971, Jamieson and Thompson, 1972, Møller, 1966a, b, 1968).

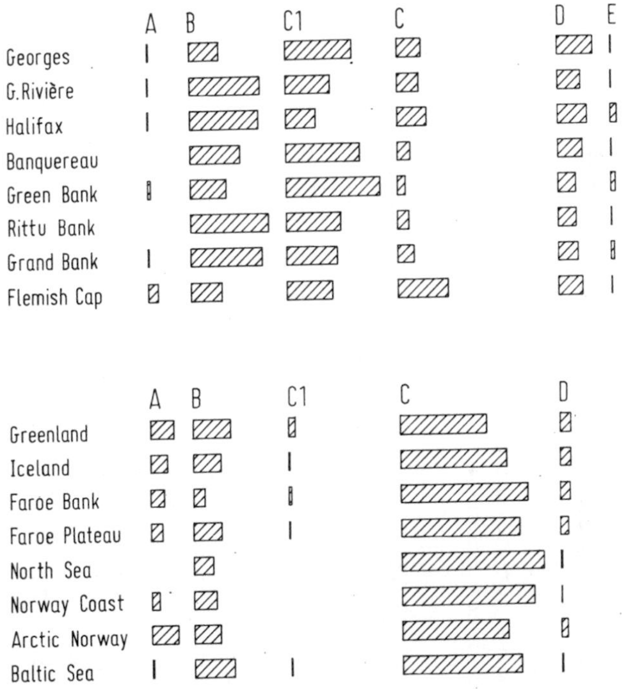

Fig. 2. A visual guide to the regional differences in the proportions of cod transferrin alleles. The sum of the shaded blocks is unity for each named bank or region.

DISCUSSION OF RESULTS

Cod are particularly gregarious at spawning, and are so prolific that the chance spread of a cod mutant by genetic drift alone is improbable. Apart from the rarest examples, the different marker genes in cod are known to be widespread within the species range.

TABLE 5

The observed and expected numbers of Atlantic cod grouped according to their transferrin genotypes. The samples are from eleven positions along the coast of North America. The expected numbers are those calculated assuming genetic equilibrium within each position sample. All showed excessive numbers of homozygotes and corresponding deficient numbers of heterozygotes. This was statistically significant in each sample. Stock mixing is the most likely explanation of this apparent imbalance. The seven most common genotypes are tabulated in detail; thirteen less frequent genotypes were not tabulated but were included in the statistical tests. One rare genotype, Tf^E/Tf^E, did not occur.

Transferrin data

Position	Tf^B/Tf^B		Tf^{Cl}/Tf^{Cl}		Tf^C/Tf^C		Tf^B/Tf^{Cl}		Tf^B/Tf^C		Tf^B/Tf^D		Tf^{Cl}/Tf^D	
	obs	exp	obs	exp	obs	exp	obs	exp	obs	exp	obs	exp	obs	exp
Georges Bank	5	1	7	7	2	1	3	6	0	2	2	4	9	8
Grande Rivière	20	18	12	7	5	2	25	23	12	12	7	12	6	8
Halifax	26	18	9	3	4	3	9	15	14	15	7	15	6	7
Banquereau	24	11	35	22	5	1	7	31	2	5	10	10	16	14
Green Bank	15	5	47	36	1	0	7	27	2	2	7	5	11	13
Rittu Bank	35	25	24	12	3	1	13	34	4	7	14	12	5	8
Grand Bank N.	33	19	17	7	5	2	5	23	7	11	11	14	11	9
Grand Bank N.	26	21	15	8	1	0	14	27	3	2	12	11	9	7
Grand Bank E.	33	25	32	19	11	3	28	43	13	18	20	17	14	15
Grand Bank SE.	11	9	5	3	2	0	8	10	0	2	2	2	1	1
Flemish Cap	10	4	22	10	21	11	5	13	10	14	7	7	7	10

TABLE 5 (continued)

Statistical analysis

Position	Homozygotes		χ^2	Probability
	obs	exp		
Georges Bank	17	11	4.11	0.05-0.02
Grande Rivière	41	29	4.92	0.05-0.02
Halifax	44	27	12.15	0.001 > P
Banquereau	66	36	36.67	0.001 > P
Green Bank	65	42	20.04	0.001 > P
Rittu Bank	64	39	22.81	0.001 > P
Grand Bank N.	57	31	32.74	0.001 > P
Grand Bank N.	43	31	7.25	0.01-0.001
Grand Bank E.	76	50	17.62	0.001 > P
Grand Bank SE.	18	12	5.00	0.05-0.02
Flemish Cap	58	28	36.71	0.001 > P

TABLE 6

The Atlantic cod typed for transferrins in the Newfoundland Grand Bank region were placed in twenty-seven year-classes. The table shows observed numbers of cod genotypes and the expected numbers assuming genetic balance throughout. A very considerable excess of homozygotes present in the data could not be attributed to any year-brood effect such as a temporary influx of migrants. The imbalance was spread proportionately over all year-classes tested. At least a decade was well represented and nine year-classes showed statistically significant imbalance. The seven most common genotypes are tabulated in detail. All genotypes were included in the statistical tests.

Transferrin data

Year-class	Tf^B/Tf^B		Tf^{C1}/Tf^{C1}		Tf^C/Tf^C		Tf^B/Tf^{C1}		Tf^B/Tf^C		Tf^B/Tf^D		Tf^{C1}/Tf^D	
	obs	exp	obs	exp	obs	exp	obs	exp	obs	exp	obs	exp	obs	exp
1942-54	7	9	9	3	3	0	9	14	2	4	4	6	3	2
1955	4	3	5	1	-	-	0	4	0	1	3	1	0	1
1956	3	2	3	1	-	-	0	3	1	1	0	1	3	1
1957	1	3	6	1	1	0	2	4	2	1	1	2	1	1
1958	14	9	6	5	1	0	7	13	2	4	8	5	6	3
1959	8	5	4	2	1	0	4	6	2	2	2	2	0	2
1960	8	5	5	3	2	0	6	8	0	2	3	3	0	2
1961	13	6	4	3	2	0	8	9	0	3	4	3	1	2
1962	13	11	6	6	4	1	12	16	6	5	6	6	6	4
1963	8	6	9	3	1	0	5	10	4	3	3	4	2	3
1964	26	15	14	7	3	1	3	21	6	6	9	8	6	5
1965	15	11	10	6	2	1	7	16	1	5	6	6	4	4
1966	16	9	11	5	2	0	2	13	1	4	6	5	2	3
1967-68	1	0	-	-	-	-	1	1	-	-	-	-	-	-
1942-68	137	96	92	48	22	4	66	135	27	41	55	51	34	36

TABLE 6 (continued)

Statistical analysis

Year-class	Homozygotes		χ^2	Probability
	obs	exp		
1942-54	20	12	9.33	0.01-0.001
1955	9	4	8.75	0.01-0.001
1956	6	3	4.13	0.05-0.02
1957	8	5	2.70	0.10
1958	21	15	3.53	0.10-0.05
1959	13	7	7.39	0.01-0.001
1960	15	9	5.89	0.02-0.01
1961	19	10	11.46	0.001 > P
1962	23	18	2.03	0.20-0.10
1963	18	10	9.18	0.01-0.001
1964	43	24	22.26	0.001 > P
1965	29	18	9.82	0.01-0.001
1966	30	15	22.26	0.001 > P
1967-68	1	1		
1942-68	255	155	94.63	0.001 > P

Recurrent mutations followed by selection and isolation must be largely responsible for the nature of the gene frequency data reported for cod. Unfortunately it is difficult to determine selection intensity, isolation time and the relative importance of those factors influencing the different marker systems in different places. The difficulty exists because the genotypic data describe recent material only, and because no natural population census of cod larval genotypes has been accomplished as yet.

Some general opinions can be reached. It is probable that the few genetic loci studied in cod belong to different chromosomes; if so, each locus provides an independent statement about the possible sub-populations. Repeatable and significant gene frequency differences between population samples require explanation. The simultaneous occurrence of differences at two or more loci could be due to a stock difference. Alternatively the allele frequencies could be due to continuity in the selective forces controlling a balanced polymorphism in several stocks, and this could cause a number of genetic loci to express similarities between material of diverse origin. Herring may provide such an example (Odense et al., 1973). At the opposite end of the range of possible interpretations, temporal selection could show genetic differences even sub-dividing a recognized unit stock, as in the regional gene frequency differences in the American eel (Williams et al., 1973). As the observations on cod show neither the biochemical consistency seen in the herring nor the single spawning aggregate of the eel, a wide choice of interpretations is possible.

The selection pressure which controls the genetic balance of the lactate dehydrogenase-B allozymes (allelic isozymes) in cod tissue appears to act similarly on cod spawned at many different sites in the North Atlantic. Similarly the α-butyrate esterase mutants in cod are held in genetic balance throughout much of the species range, except on the North American banks. The cod hemoglobin variants appear to arrive at different states of genetic balance in different regions. The North American cod population samples were not as heterogeneous at their hemoglobin locus as they were at their esterase locus. Dramatic heterogeneity was seen at their transferrin locus.

The American cod populations showed more variation in tranferrin frequency than for any other locus tested. As the cod populations on the American banks are known to reproduce at different places and at different seasons, contemporary isolation of stocks (Jamieson, 1970) may be a more effective factor in maintaining transferrin variation than directional temporal selection on adjacent banks. Perhaps the large transferrin frequency differences in cod persist over generations, rather like the ancestral transferrins which persist in breeds of domestic livestock distributed about the world (Jamieson, 1966a). This interpretation does not deny selection, but gives more relevance to the contemporary isolation of breeding units in maintaining the observed differences.

The variation in transferrin frequencies among the cod sampled at North America may be described as showing conspicuous peaks in the distribution of the various alleles. The Newfoundland Grand Bank samples, and a sample from Rittu Bank just to the north, showed the highest reported estimates of the allele Tf^{B}. The allele Tf^{C1}, seldom seen east of the Davis Strait (Jamieson, 1967), was well represented at North America and reached a peak occurrence on Green Bank. Flemish Cap, to the east of the Grand Bank, has the most easterly American cod population, and showed the highest frequency of Tf^{C} in the region. Georges Bank yielded the most southerly population of cod examined for transferrins and showed the highest frequency of the allele Tf^{D}. As far as is known, the peaks and troughs in the distribution of cod transferrin allele frequencies are attributed neither to season nor to year-class. They do characterize the cod stocks on different banks. The cod transferrins are valuable genetic markers recommended to fisheries biologists as an effective practical tool for describing the American cod stock units.

To advance the practical application of biochemical-genetic tags for cod stocks it is necessary to find additional marker systems. The literature contains at least three interesting possibilities. Ullrich (1967, 1968) reported an albumin variant in cod collected at Labrador and at Greenland. He also saw variation in serum globulins. The albumin variants and the globulin variants showed differences in the frequency occurrence of electrophoretic patterns for cod on either side of the Davis Strait. Ullrich's globulin groupings are not equated to transferrin types, but could have some common protein species. Tills et al. (1971) looked for erythrocyte enzyme variants in cod blood specimens from the North Sea and from Iceland. Much variation appeared in zymograms showing phosphoglucomutase, but this variation has not been reduced to a simple genetic model suitable for the routine marking of cod. Dando (1974) has described glucosephosphate isomerase isozymes in cod, noting three alleles at one locus and five at another locus, but only English Channel cod have been described so far. As the possible number of isozyme systems is vast, it is probable that additional useful marker systems will turn up.

GENERAL DISCUSSION

The importance of the Atlantic cod fishery in the social history of North America was acknowledged when the House of Representatives in Boston, Massachusetts voted the following motion in 1784: "That leave might be given to hang the representation of a cod-fish in the room where the House sits, as a memorial to the importance of the cod-fishery to the welfare of this Commonwealth, as had been usual formerly." This Atlantic cod, *Gadus morhua* L., has sustained an international fishery along the North American seaboard ever since John Cabot realized its abundance in 1497 and thereby encouraged fishermen to go there early in the sixteenth century. The continued exploitation of this fishery often proved politically divisive during the subsequent commercial development of the maritime states (Innis, 1940).

The current cod fishery along the North American coasts and banks continues to yield 1 million metric tons of cod per annum (Anon., 1972). Presumably the right to share this prize will be discussed at the forthcoming Law of the Sea Conference in Caracas, Venezuela. It is in the best interest of world food-conservation that a clear biological description of cod should be communicated to those who influence the future use of this huge resource.

The geological form of the continental shelf areas and submarine banks, together with hydrographic changes, influence distribution and fluctuations of Atlantic cod in populations about the coasts of Europe, Iceland, Greenland and North America (Tåning, 1953). The cod isolated in the regions show distinctive age structures, growth rates, color phases, migration patterns, feeding habits, and spawning seasons (Wise, 1961). Regional races have been implied in the absence of gene markers (Schmidt, 1930; Svetovidov, 1948).

The vast numbers of cod tagged and recaptured show that remarkably few cod move between the major fisheries. Knowing the fickle movements of the fishing fleets, unexpected tag returns are difficult to verify. Nevertheless, it is possible that a very small number of tagged cod migrate in either direction between West Greenland and Labrador (Templeman, 1962). It is similarly possible, although improbable, that pelagic cod larvae spawned on the Fiskenaes Banke at West Greenland could be carried in that component of the West Greenland current which turns westward and later southward over the Davis Strait towards part of the Labrador cod fishery (Hansen, 1949, 1968, Lee, 1974). The protein type data published here for North American cod and elsewhere for other cod show a wide difference between the cod samples on the two sides of the Davis Strait (Sick, 1965b, Jamieson, 1970). The magnitude of this protein-type difference is so great that it effectively divides the species in two, and anyone familiar with cod transferrins would expect to distinguish a pair of cod taken at random from either side of the Davis Strait.

The recent history of cod at West Greenland suggests that the isolation of the cod on the American coast from other North Atlantic cod may have been geographically more distant in the past. For instance, cod did not live at West Greenland during the cold series of years between 1876 and 1915; recent records show a return to cold conditions in that area (Herman et al., 1973), and the West Greenland cod fishery was seen to be much reduced in November 1973 (Birkett, personal communication).

The practical interest in the data presented here is its capacity to establish the genetic isolation of cod stocks occupying adjacent coastal regions on offshore banks. The significant degrees of genetic isolation seen between the limited material tested to date give a more dramatic demonstration of genetic isolation than would be expected from the known tag returns over many years. Nevertheless, the general pattern of stocks inferred from the tag returns agrees well with the genetic divisions of the North American cod reported here.

The numbers of cod available in any fishery at a given time are much influenced by the survival rates of the year-broods that make up the fishery. This opens the question of whether the successive year-broods in any one area are necessarily derived from the same parental stock or race. For instance, genetic differences between year-classes are continually found in mackerel (Jamieson et al., 1971, de Ligny, personal communication); however, no year-class difference of this kind has so far been reported for cod.

It is thought that a project involving biochemical tagging has at least some advantages over the conventional marking methods, particularly with respect to seeing the units of stock as genetic entities, and even for detecting the presence of mixtures of stocks that could be difficult or impossible to see in a normal set of tag-recapture information. Stock mixing can be implied where the genotypes in natural populations are not in genetic balance (Wahlund, 1928).

If a sufficient number of genetic loci showed variable gene frequencies over all stages in their life-cycle, the stocks would be seen as breeding units in regions where effective barriers or distances ensured isolation. Given this envisaged description of cod, the conventional tagging methods could be used to greater effect. Recognition and identification are not synonymous. A recaptured tagged fish is simply recognized on recapture, whereas a genotyped fish is identified in the first instance. It would be more logical to test genotypes before using mechanical tags, so that each tagging project could be applied to a positively identified breeding unit. Perhaps the most informative tagging experiment should use genetic tags together with conventional tags. The subsequent recapture of genotyped fish could be tested for the possible effects of temporal selection and for the migration preferences of different genotypes. This would test possible stock mixing and homing at the sub-population level.

Within the North American cod fishery, spawning occurs at different times on different grounds, starting with winter spawning in the south and continuing through spring and summer to autumn spawning in the north. The time and locality differences favor the formation of unit breeding stocks (Templeman, 1962), yet genetic isolation of cod on adjacent banks cannot be conclusively established by tag returns alone, because the tags are attached to the cod after they recruit to a local fishery. Unlike rings attached to nesting birds, the tags are attached to post-larval cod and do not trace the origins of recruits. In general the patterns of tag returns support an orderly desire to attribute unit stocks to unit spawning grounds. However, there is no absolute reason why any stock of cod should not spawn in different areas. Some existing genotype data indicate that a unit stock of cod in the North Sea spawns in several areas (Jamieson and Thompson, 1972a). It is also possible that different breeding stocks share a spawning area, and yet remain discrete by assortative breeding at Lofoten (Møller, 1966a) and at south-west Iceland (Jamieson and Jónsson, 1971). Because random mating would eliminate gene frequency differences, the observed differences suggest breeding barriers between cod stocks. Behavior pattern barriers and genetic incompatibilities are well established in diverse phyla.

The range of available protein systems with potential applications is immense, but it is not possible to predict which potential variant ascribed to which genetic locus is going to discriminate two given stocks of fish. Practical experience in genotyping fish and other animals indicates that only a small minority of polymorphisms show population differences and these suggest breeding stock isolates.

The applications of genotypes are qualified by theoretical limitations (Fujino, 1971). A fisheries biologist looking to genes as practical tools would prefer those unusual examples of genetic polymorphism which have arrived at different balanced situations against the genetic backgrounds and environmental niches of particular stocks.

The differences among the results so far obtained from the cod populations on the American banks suggest that deep submarine channels are effective in the genetic isolation of cod along this coast, although pelagic cod eggs and larvae would not necessarily be so limited. A similar stock difference is known where a deep channel separates a stock of cod on Faroe Bank from another on Faroe Plateau (Jamieson and Jones, 1967, Love et al., 1974).

The pelagic phase of cod eggs and larvae lasts about three months, so until drifting larvae can be tagged or typed some sources of recruitment will remain questionable. Obviously, mechanical tags are too cumbersome for larvae, and a better prospect of tracing the youngest cod stages throughout their entire lives is to use their genotypes as ready-made tags. This approach may not assist in the recognition of individual specimens in the way that mechanical tags work in practice, but biochemistry has greater potential for testing the identity of stocks as breeding units. Shifting the emphasis from individual marked fish to general population statistics such as gene frequencies could be advantageous.

For instance, most people see a fish stock as an available resource, whereas the geneticist may recognize a breeding unit (Jamieson, 1974). In attempting to resolve the fish-stock relationships, genotypic information can be critical and is more quickly available than tag returns.

For practical convenience, local stock names continue in common use, and the number of stocks one chooses to recognize is arbitrary. The best practical balance is somewhere between the objectivity in ascertaining real differences and the subjectivity in applying useful names. In the widest sense the unit of stock proves to be an enigma, because isolation is seldom absolute and all stocks are transient. Nevertheless, the applications of genetic tags favor the description of the flow of material through generations in a way that was not possible using non-genetic criteria. The well-tried methods have known capacities and limitations. The recent proliferation of biochemical and genetical information suggests no obvious limit to the technical capacity for describing the genetic make-up of individuals. The humble cod-fish is such an individual, even a tribalist.

CONCLUSIONS

The most incisive gene markers used to detect sub-populations of Atlantic cod at North America are a series of co-dominant allelomorphic genes at the transferrin locus.

Cod hemoglobins have more limited applications as sub-population markers at North America than in European and

Icelandic waters.

Cod α-butyrate serum esterases are more useful as sub-population markers at North America than anywhere else in the species range.

Lactate dehydrogenase B_4 isozymes in cod show dynamic polymorphism, the heterozygotes having a selective advantage.

The protein types in the Atlantic cod suggest an ancient dichotomy of this species now evident along the Davis Strait.

ACKNOWLEDGEMENTS

The author is indebted to Dr. Paul H. Odense et al. for Canadian cod sera and for permission to cite their mimeographs to Committees of the International Council for the Exploration of the Sea, and also to Lowestoft colleagues Mr. Michael J. Holden for Grand Bank cod bloods, Mr. Bernard C. Bedford for reading otoliths and Mr. Robert J. Turner for technical assistance.

REFERENCES

Anon., 1972. *Statist. Bull. Int. Commn NW. Atlant. Fish.*, 239 pp.

Culliford, B. 1971. The examimation and typing of bloodstains in the crime laboratory. Wash., D. C., U.S. Govt. Printing Off. 270 pp. (Nat. Inst. Law Enforc. and Crim. Just. publ. PR 71-77).

Cushing, J. E. 1964. The blood groups of marine animals. *Adv. Mar. Biol.* 2: 85-131.

Dando, P. R. 1974. Distribution of multiple glucose-phosphate isomerases in teleostean fishes. *Comp. Biochem. Physiol.* 47B: 663-679.

Fisher, R. A. and F. Yates 1963. *Statistical Tables for Biological, Agricultural and Medical Research.* Edinburgh, Oliver and Boyd, 6th edition, 146 pp.

Frydenberg, O., D. Møller, G. Naevdal, and K. Sick 1965. Haemoglobin polymorphism in Norwegian cod populations. *Hereditas*, 53: 257-271.

Fujino, K. 1971. Genetic markers in skipjack tuna from the Pacific and Atlantic Oceans. Rapp. P.-v. Réun. *Cons. Int. Explor. Mer*, 161: 15-18.

Hansen, P. M. 1949. Studies on the biology of cod in Greenland waters. Rapp. P.-v. Réun. *Cons. Int. Explor. Mer*, 123, 77 pp.

Hansen, P. M. 1968. Report on cod eggs and larvae. *Spec. Publs Int. Commn NW. Atlant. Fish.*, (7) Pt. 1, 127-137.

Harris, H. 1966. Enzyme polymorphisms in man. *Proc. R. Soc.*

B. 164: 298-310.

Heincke, F. 1898. Naturgeschichte des Herings, *Teil 1, Abh. dt. SeefischVer.*, Band 2, CXXXVI, 128 pp text, 223 pp. tables.

Hermann, F., W. Lenz, and R. W. Blacker 1973. Hydrographic conditions off West Greenland in 1972. *Redbook, Int. Commn NW. Atlant. Fish.*, Part III, 27-32.

Hunter, R. L. and C. L. Markert 1957. Histochemical demonstration of enzymes separated by zone electrophoresis in starch gels. *Science* 125: 1294-1295.

Innis, H. A. 1940. *The Cod Fisheries. The History of an International Economy*. New Haven, Yale University Press, for the Carnegie Endowment for International Peace. 520 pp.

Jamieson, A., 1966. The distribution of transferrin genes in cattle. *Heredity*, Lond., 21 (2), pp 191-218.

Jamieson, A. 1967a. The application of a blood group formula. pp 461-464. In: *10th Eur. Conf. Anim. Blood Grps and Biochem. Polymorph.*, Paris, 1966. Paris, Institut National de la Recherche Agronomique, 544 pp.

Jamieson, A. 1967b. New genotypes in cod at Greenland. *Nature* 215: 661-662.

Jamieson, A. 1970. Cod transferrins and genetic isolates. pp 533-538 In: *11th Eur. Conf. Anim. Blood Grps and Biochem. Polymorph.*, Warsaw, 1968. The Hague, Dr. W. Junk, 607 pp.

Jamieson, A. 1974. Genetic tags for marine fish stocks. pp 91-99. In: Harden Jones, F. R. ed. *Sea Fisheries Research*. London Elek Science, 510 pp.

Jamieson, A. and B. W. Jones 1967. Two races of cod at Faroe. *Heredity*, Lond. 22: 610-612.

Jamieson, A and J. Jónsson 1971. The Greenland component of spawning cod at Iceland. (ICES Special Meeting on the Biochemical and Serological Identification of Fish Stocks, Dublin, 1969). Rapp. P.-v. Réun. *Cons. Int. Explor. Mer*, 161: 65-72.

Jamieson, A., W. de Ligny, and G. Naevdal 1971. Serum esterases in mackerel *Scomber scombrus* L. (ICES Special Meeting on the Biochemical and Serological Identification of Fish Stocks, Dublin, 1969). Rapp. P.-v Réun. *Cons. Int. Explor. Mer*, 161: 109-117.

Jamieson, A. and G. Otterlind 1971. The use of cod blood protein polymorphisms in the Belt Sea, the Sound and the Baltic Sea. (ICES Special Meeting on the Biochemical and Serological Identification of Fish Stocks, Dublin, 1969). Rapp. P.-v Réun. *Cons. Int. Explor. Mer*, 161: 55-59.

Jamieson, A. and D. Thompson 1972a. Blood proteins in North Sea cod (*Gadus morhua* L.) pp 585-591. In: Kovacs, G. and

Papp, M. eds. *12th Eur. Conf. Anim. Blood Grps and Biochem. Polymorph.*, Budapest, 1970. The Hague, Dr. W. Junk, 686 pp.

Jamieson, A. and D. Thompson 1972b. Butyric esterase differences in Atlantic cod, *Gadus morhua* L. *ICES C.M.* 1972/F:32, (mimeo).

Lee, A. J. 1974. Oceanic circulation of the North Atlantic region. pp 1-30, In: Harden Jones, F. R. ed. *Sea Fisheries Research*. London, Elek Science, 510 pp.

de Ligny, W. 1969. Serological and biochemical studies on fish populations. *Oceanogr. and Mar. Biol.*, 7: 411-513.

Love, R. M., I. Robertson, J. Lavety, and G. L. Smith 1974. Some biochemical characteristics of cod, *Gadus morhua* L., from the Faroe Bank compared with those from other fishing grounds. *Comp. Biochem. Physiol.*, 47B: 149-161.

Lush, I. E. 1970. Lactate dehydrogenase isoenzymes and their genetic variation in coalfish, *Gadus virens*, and cod, *Gadus morhua*. *Comp. Biochem. Physiol.* 32: 23-32.

Marr, J. C. 1957. The problem of defining and recognizing sub-populations of fishes. (Contributions to the study of sub-populations of fishes). *Spec. Scient. Rep. U.S. Fish Wildl. Serv.* (208), 1-6.

Møller, D. 1966a. Genetic differences between cod groups in the Lofoten area. *Nature* 212: 824.

Møller, D. 1966b. Polymorphism of serum transferrin in cod. *FiskDir. Skr.*, (Ser. Havunders.), 14: 51-60.

Møller, D. 1968. Genetic diversity in spawning cod along the Norwegian coast. *Hereditas* 60: 1-32.

Morrison, W. J. and J. E. Wright 1966. Genetic analysis of three lactate dehydrogenase isozyme systems in trout: evidence for linkage of genes coding subunits A and B. *J. Exp. Zool.* 163: 259-270.

Mourant, A. E., A. Kopec, and K. Domaniewska-Sobezak (In press). *The Distribution of the Human Blood Groups and Other Biochemical Polymorphisms*. Oxford University Press.

Odense, P. H., T. C. Leung, and T. M. Allen 1966. An electrophoretic study of tissue proteins and enzymes of four Canadian cod populations. *ICES, C.M.* 1966/G:14, 1 p (mimeo).

Odense, P. H., T. C. Leung, T. M. Allen, and E. Parker 1969. Multiple forms of lactate dehydrogenase in the cod, *Gadus morhua* L. *Biochem. Genet.*, 3: 317-334.

Odense, P. H., T. C. Leung, and C. Annand 1973. Isoenzyme systems of some Atlantic herring populations. *ICES, C.M.* 1973/H:21, 5 pp (mimeo).

Rollefsen, G. 1933. The otoliths of the cod. Preliminary report. *FiskDir. Skr.*, (Ser. Havunders.), 4 (3), 14 pp.

Schmidt, J. 1930. Racial investigations. X. The Atlantic cod

(*Gadus callarias* L.) and local races of the same. *C. R. Trav. Lab. Carlsberg,* 18 (6) 71 pp and plates.

Sick, K. 1961. Haemoglobin polymorphism in fishes. *Nature* 192: 894-896.

Sick, K. 1965a. Haemoglobin polymorphism of cod in the Baltic and the Danish Belt Sea. *Hereditas* 54: 19-48.

Sick, K. 1965b. Haemoglobin polymorphism of cod in the North Sea and the North Atlantic Ocean. *Hereditas* 54: 49-69.

Svetovidov, A. N. 1948. Gadiformes. Fauna SSSR, (Ryby) 9 (4), 221 pp and plates. (Translation available: Jerusalem, Israel Progr. for Scient. Transls., for U.S. Nat. Sci. Found., Wash., OTS 63-11071).

Syner, F. N. and M. Goodman 1966. Polymorphism of lactate dehydrogenase in gelada baboons. *Science* 151: 206-208.

Tåning, A. V. 1953. Longterm changes in hydrography and fluctuations in fish stocks. *A. Proc. Int. Commn NW. Atlant. Fish.,* 3: 69-77.

Tills, D., A. E. Mourant, and A. Jamieson 1971. Red-cell enzyme variants of Icelandic and North Sea cod (*Gadus morhua*). (ICES Special Meeting on the Biochemical and Serological Identification of Fish Stocks, Dublin 1969). Rapp. P.-v. Réun. *Cons. Int. Explor. Mer,* 161: 73-74.

Templeman, W. 1962. Divisions of cod stocks in the northwest Atlantic. *Redbook Int. Commn NW. Atlant. Fish.,* 1962, Pt. 3, 79-123.

Ullrich, H. 1967. Erste Mitteilung zur Serumelektrophorese. *Fisch.-Forsch.,* 5 (2) 57-59.

Ullrich, H. 1968. Zweite Mitteilung zur Blutserumelektrophorese von Kabeljau des Nordwestatlantiks. *Fisch.-Forsch.,* 6 (1), 37-39.

Utter, F. M. 1969. Biochemical polymorphisms in the Pacific hake, *Merluccius productus*. Ph.D. Thesis, University of California, Davis, 60 pp.

Wahlund, S. 1928. Zusammensetzung von Populationen und Korrelationserscheinungen von Standpunkt der Vererbungslehre aus betrachtet. *Hereditas,* 11: 65-106.

Williams, G. C., R. K., Koehn, and J. B. Mitton 1973. Genetic differentiation without isolation in the American eel, *Anguilla rostrata. Evolution,* 27: 192-204.

Wise, J. P. 1961. Synopsis of biological data on cod *Gadus morhua* Linnaeus 1758. *FAO Fish Biol. Synops.* (21) 46 pp.

SELECTIVE PRESSURES ON ISOZYMES IN *DROSOPHILA*

CHRISTOPHER WILLS
Dept. of Biology, University of California, San Diego
La Jolla, California 92037

ABSTRACT. The results of a five-year experiment involving inbreeding *Drosophila pseudoobscura* in an attempt to isolate single-gene selective effects at two polymorphic loci are reviewed, along with the rationale for the experiment. Despite 38 generations of brother-sister mating, the effects of background genotype were still apparent in that many of the effects observed were sex-limited. Generally, however, inbred flies segregating for two isozymes of octanol dehydrogenase were affected by octanol or ethanol in the medium, both of which are substrates for this enzyme. Inbred flies segregating for two isozymes of esterase-5 were affected by tributyrin in the medium, a substrate for this enzyme. There was some cross-effect, either because of background or because the pathways in which these two enzymes play a role share a common step. No effect was seen in outbred flies subject to the same conditions. While heterosis was apparent in earlier generations at the ODH locus, it had virtually disappeared by the 38th generation. This does not rule out conditional heterosis as a mechanism for the maintenance of these polymorphisms, however.

The bearing of these results on our understanding of the maintenance of polymorphisms in general is discussed.

INTRODUCTION

The efforts of many experimental population geneticists have been directed in the last few years to gathering evidence to refute what may, from their point of view, be called the non-Darwinian heresy. This was proposed in its present form by Kimura (1968) and elaborated on by King and Jukes (1969), Arnheim and Taylor (1969), and others including many theoreticians associated in one way or another with Kimura. It is, however, much older. Non-Darwinian evolution was proposed to explain the distribution of blood group polymorphisms in human populations and inversion polymorphisms in *Drosophila pseudoobscura* (e.g. Wright 1940), despite evidence to the contrary known at the time (Ford 1940).

The current non-Darwinian proposal is that most electrophoretically-detectable variability found in natural populations and most amino acid substitutions which have occurred during the course of molecular evolution are in fact neutral. This viewpoint is critically reviewed elsewhere in this volume by Powell. Further statistical evidence for neutrality is presented here by Yamazaki and Maruyama, though their approach has recently been strongly criticized by Ayala and Gilpin (1973).

Undoubtedly the truth lies somewhere between the extremes of pan-selectionism and pan-neutrality. Even if neutrality turns out not to be predominant this point of view has been enormously valuable in stimulating a great deal of experimental work, both in the laboratory and in nature. Attempts at refutation have taught us a great deal of new information about the genetic structure of natural populations (e.g., the work of Allard and his group discussed in the present volume).

The original reason for proposing neutrality as an explanation for most genetic variation is now of historical interest only. The difficulty, carefully enunciated by Lewontin and Hubby (1966), lay in the fact that if fitnesses can be considered additive or multiplicative, then homozyogotes for a large number of heterotically-maintained polymorphic loci should have extremely low fitnesses. Proposals for circumventing this difficulty were made in somewhat different ways by Sved, Reed, and Bodmer (1967), King (1967), Milkman (1967), and Wallace (1968); all to one degree or another abandoned the idea of rigid selection coefficients that could be added or subtracted in a regular way. We embodied the approach of Milkman in a computer simulation study (Wills, Crenshaw, and Vitale 1970) which demonstrated that small amounts of a type of truncation selection could maintain large amounts of polymorphism if the following conditions were observed:

a) It is the total number of heterozygous loci for selected functioning polymorphisms that is important in determining the relative success in competition of an organism. Undoubtedly some polymorphisms are much more important than others, and some loci are more likely to be polymorphic than others (for a rationale see Johnson 1974), but when one is dealing with thousands of polymorphisms most of these differences become relatively unimportant compared with the total numbers.

b) The key term in this new understanding of the balanced genetic load (as distinct from the mutational load, which like the poor we have always with us) is relative. If the organism's fitness is not determined by one or a few rare deleterious genes but rather by the number of selected functioning polymorphisms (Wills 1973) it possesses, then that fitness is also determined by:

1) The number of these polymorphisms that are actually selected in the particular environment.

2) The number of polymorphisms possessed by other organisms in the same population. Just as in the country of the blind the one-eyed man is king, in a monomorphic population an organism with one or two selectively useful alleles not possessed by the rest will be at a great advantage. This advantage will be reversed in a population in which the other organisms are more highly heterozygous.

An upper limit to the amount of selected polymorphism that can be maintained will eventually be reached even in a very large population, since the relative selective advantage accruing with each additional polymorphism will become progressively less. Polymorphisms then become lost through fixation. Alternatively, the selective advantage of the polymorphism may be strong enough to overcome the disadvantage of gene dosage problems. If the requisite cytological event occurs, the polymorphism may become permanently enshrined in the form of a gene duplication. The upper limit to the number of polymorphisms that may be maintained is not, be it noted, governed by considerations of additive genetic load but rather by the progressive diminution of the selective forces maintaining each polymorphism as additional ones are added to the population.

This view of the balanced genetic load seemed to us to be testable, and in this paper I would like briefly to review a continuing experiment that has provided evidence concerning some of the hypotheses listed above. The experiment was conducted over a five-year period, and as our knowledge of the organism has increased we have altered our techniques somewhat. While the experimental results do support some of the tenets of what may be called the genotype-environment interaction model of the balanced genetic load, they have not provided strong evidence for certain of the hypotheses listed above, particularly that the polymorphisms we have examined are maintained by conditional heterosis. We suspect that this may be due to the limitations of our experimental design, but it may be telling us something fundamental about the way in which selected functional polymorphisms are maintained in natural populations. Mechanisms other than conditional heterosis are not ruled out by the assumptions of the model given above.

HISTORY OF THE EXPERIMENT

In 1969, it became apparent to us that one way to test the genotype-environment interaction model would be artificially to enhance the selection coefficients associated with particular polymorphisms. This could be done by inbreeding *Drosophila* by brother-sister mating, but forcing the flies to remain heterozygous at a particular locus. Selection coefficients associated with the alleles at that locus should presumably become progressively enhanced as inbreeding continued and that locus became more and more important to the relative survival of the flies. It was essential that naturally-occurring selected functioning polymorphisms be picked for this experiment, since the alleles at these loci will have been subjected to long periods of natural selection. Such alleles would, in homozygotes or heterozygotes, be perfectly capable of performing the biochemical function of that locus, but certain combinations of alleles at that locus might be expected to be at an advantage. This advantage should become more pronounced as inbreeding proceeded.

Two loci of *Drosophila pseudoobscura* were picked which we presumed at the time would proved to be biochemically unrelated. One, coding for the enzyme octanol dehydrogenase, is autosomal, and the other, coding for the most prominent esterase, esterase-5, is sex-linked. The two commonest alleles found in the Strawberry Canyon (Calif.) population of *D. pseudoobscura* (lines of which were kindly provided to us by Dr. S. Prakash) were established in pure line in otherwise outbred stocks, and maintained throughout the experiment in mass culture. The two alleles at each locus will be referred to as F (fast) and S (slow), referring to their electrophoretic mobilities.

Though it was not known to us at the time, Yamazaki (1971) was engaged in a massive study of the same F and S alleles of the Esterase-5 locus. He demonstrated with great thoroughness that on randomized genetic backgrounds and on normal medium no selective differences could be detected between these alleles--or, more accurately, between the blocks of genes in which these alleles were embedded. This work has since been widely quoted as a demonstration that these alleles are in fact neutral, but it only demonstrated that selection could not be detected on a heterozygous genetic background when the flies were grown on a medium which would not be expected *a priori* to interact with the E-5 locus.

Our initial inbreeding results, briefly reported in an abstract (Wills 1970), concerned three independent ODH and three E-5 lines that had been inbred for five generations. Flies within these lines were mated *inter se*, and progeny which were expected to yield regular Mendelian ratios were subjected as adults to three strong environmental stresses: ether, cold, and CO_2 shock. In all the inbred lines the heterozygotes showed the greatest survival and the homozygotes the least. No such effect could be detected when the same experiments were performed on the outbred flies which had been maintained as a control. At the time we considered the possibility that this enhancement of heterosis was due to the ODH and E-5 alleles which we were using as markers, but stated that it was more likely we were picking up enhanced heterosis associated with the blocks of genes in which these markers happened to be embedded. Brother-sister mating is a strong generator of disequilibrium; such blocks would be expected to be of far greater importance in determining the relative survival of the flies than the markers which we happened to be able to detect electrophoretically.

The great difficulty with which we were faced was separating the effects of the single loci we could assay from the blocks of genes that inbreeding inevitably generated. The effects manifested themselves in unexpected ways. Although outbred flies and flies at every level of inbreeding segregated regularly for the F and S ODH and E-5 alleles when they were raised on normal medium, there was a pronounced difference in fertility of the different genotypes. This disappeared as inbreeding proceeded and could therefore properly be attributed, not to the marker alleles, but to loci linked to them. In the course of inbreeding, large numbers of single pair mating were set up. Some of these, particularly in the early stages of inbreeding, proved sterile and were discarded. The genotypes of the flies that were successful in producing offspring were determined by electrophoresis after larvae appeared in the vials. Not all these crosses, in turn, could be used to go on to the next generation; only crosses in which both flies were heterozygous (ODH) or in which the female was heterozygous (E-5) were used. All the other crosses were discarded. For most generations except some of the very earliest, however, enough data was collected to enable comparisons to be made to the expected 1:2:1 or 1:1 ratios in all the flies that produced fertile offspring. The question could be asked: did the flies that produced fertile offspring form a biased sample of the expected genotypes?

Figure 1 shows that, for the ODH flies, the first few

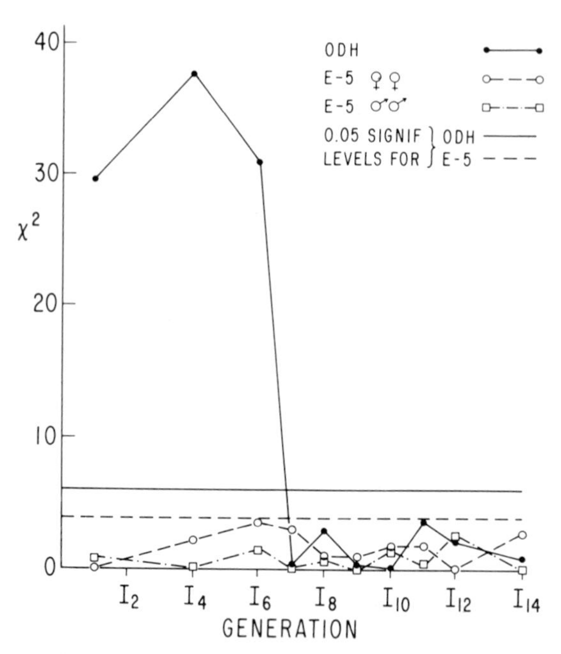

Figure 1. Deviation from genotypic expectation of parents from fertile single-pair matings used to derive flies for the next generation of inbreeding. The deviations are expressed as χ^2 values. The large deviations observed during the first six generations of inbreeding in the ODH lines was due to a deficiency of fertile F/F homozygotes and an excess of fertile S/S fertile S/S homozygotes. The high but non-significant χ^2 values observed for the E-5 females at generations I_4, I_6, and I_7 were due to an excess of fertile heterozygotes. Not all the early data were collected in amounts sufficient for analysis. No significant deviations were observed for generations subsequent to generation I_{14}.

---

generations showed a very large bias away from the expected ratio. The bias was caused by infertility among the F/F homozygotes and an increased fertility among the S/S homozygotes. At the 7th generation this bias suddenly disappeared, and for the remainder of the inbreeding all three genotypes were equally fertile. The E-5 data, in contrast, only border on significance during the early stages of inbreeding. The high Chi-square values seen in generations I_4 through I_7 are caused by an excess of heterozygous females. Though only skirting significance, this excess of heterozygotes was consistent during this period. Males, which were segregating 1:1 for F and S, did not show a noticeable deviation from expectation. This effect, if it is indeed real, disappeared by the 8th generation; after that homozygous females were as likely to be in excess as were heterozygotes.

It is worth noting that the great difference in fertility of ODH genotypes from outbred flies could easily have been interpreted as differences due to the ODH locus itself. Inbreeding, however, enabled us to remove the ODH locus from whatever loci were actually causing the effect -- or else to make them homozygous so that they affected all three ODH genotypes equally.

By generation 12, we hoped that the block effect would have diminished sufficiently to allow single-gene effects to be demonstrated. At that point we were unclear about the substrate affinities of the E-5 enzyme, beyond knowing that it was not a cholinesterase and was probably an aliesterase (see Narise and Hubby 1966, and Berger and Canter 1973, for a detailed examination of this enzyme). Consequently, we grew outbred and 12th generation inbred flies, in crosses arranged to give simple 1:2:1 or 1:1 ratios, on a medium containing 3 mM 1-octanol and a duplicate set (using eggs derived from the same parents) on a medium containing 0.25 M KCl, to which *D. pseudoobscura* is very sensitive. It was our hope to detect effects at the ODH locus in the first case and at neither locus in the second, using the E-5 locus as a control. If no effect on segregation at the E-5 locus was observed, it could be argued that effects seen in flies segregating at the ODH locus could be traceable to that locus; otherwise one would have to argue that genes affected by octanol in the medium just happened to be linked to the ODH locus but similar genes were not linked to the E-5 locus.

Our results (Wills and Nichols 1971) bore out this expectation -- at least in the ODH males. A significant deficiency of S/S homozygotes and excess of S/F heterozygotes was observed in the ODH males grown on octanol (Table 2). No effect was seen on the E-5 flies on either medium.

We presumed that the effect on flies segregating for ODH isoalleles was detected first in the males because the males were already hemizygous for the X chromosome. Polymorphisms which tended to mask the octanol effect would be fewer in males to begin with. Our reasoning was as follows: consider that out of the hundreds or thousands of structural gene loci polymorphic in an outbred population there are a small but appreciable number directly in the pathway of long chain alcohol catabolism and still others in contiguous biochemical pathways. The only one we can currently detect is the ODH locus. An outbred fly heterozygous at the ODH locus might by chance be homozygous at more than the average number of these other biochemically related loci, so that the advantage conferred by heterozygosity at the ODH locus would tend on the average to be cancelled out by homozygosity elsewhere. As the number of these polymorphs is reduced by inbreeding, the ODH locus becomes more and more important in determing relative fitness.

By generation I_{17} (Wills and Nichols 1972), our prediction appeared to be at least in part borne out. A strong deficiency of S/S homozgotes and excess of S/F heterozygotes was observed in the females of one of the inbred ODH lines. Though not reported in that paper for reasons of space, there was still no effect of octanol medium on flies segregating for E-5. At the same time, however, males of one of the inbred lines which had shown a strong effect at generation I_{12} no longer showed an effect at I_{17}. It did, however, show deviations in the same direction as the other lines.

Because our laboratory was moved at generation I_{30}, no full-scale reinvestigation of the effects of further inbreeding was embarked on until generation I_{38}. By this time we had a better understanding of the substrate affinities of the E-5 enzyme. It could be demonstrated by hydrolysis *in situ* in acrylamide gels that E-5 was capable of hydrolyzing short-chain triglycerides such as tributyrin, but not longer-chain triglycerides such as triolein and tripalmitin. Genotype-related differences in hydrolytic activity could also be demonstrated spectrophotometrically (E. Sheldon, in preparation). The details of these assays will be published elsewhere, but they were sufficiently clear-cut to lead us to the conclusion that a suitable stress medium for distinguishing selective differences between E-5 isoalleles might be tributyrin and its shorter-chain relative triacetin. It seemed unlikely to us that such hydrolysis was the primary function of this enzyme in the fly's metabolism. It is a biochemical truism that it is much easier to find the enzyme for a reaction than to find the reaction for an enzyme. But we hoped that by applying selective pressure specifically directed against a known function of E-5 we could demonstrate that the different genotypes were not selectively equivalent in the fly itself. We hoped further that these selective differences would be similar to activity differences exhibited by the different isozymes in vitro, but this hope was not realized.

At the same time it was found that ODH, an enzyme which happens to oxidize octanol rapidly so that this primary alcohol is habitually used to demonstrate its activity, can also slowly oxidize ethanol at high substrate concentrations to give faint bands on an activity-stained gel.

CURRENT EXPERIMENTS

A large-scale experiment was performed at generation I_{38}, using four stress media starting with basic cornmeal-molasses-agar food to which was added 3 mM octanol (dissolved in an equal amount of ethanol to ensure smooth mixing), 6% ethanol, 5% tributyrin or 5% triacetin. All inbred and outbred lines were grown on all media. Because of the severity of the effects of some of the media on the parental flies, it proved necessary to collect eggs by filtration and place the eggs directly on the stress media. All the stress media interfered greatly with the survival of the flies (Table 1). The expectation for the experiment was straightforward: effects on Mendelian ratios should be seen in ODH flies on octanol and ethanol medium but not on the other two; effects on ratios in E-5 flies should be seen on tributyrin and triacetin medium but not the others. Further, it should be expected that if conditional heterosis was the factor maintaining these polymorphisms the heterozygotes, with two isozymic forms of the enzymes, should be at an advantage. These effects should be seen in inbred flies, in which the single-gene effect should have enhanced effect on survival, but not in outbred flies with normally heterozygous genetic background in which the single-gene effects would tend to be masked by other segregating polymorphisms. Further, it would be expected that different stress media would have different effects on the relative fitnesses of the ODH and E-5 genotypes.

As Table 2 indicates, some of these predictions were borne out and some were not. The table summarizes only those crosses and media in which significant deviations from Mendelian expectation were observed, comparing them with the significant deviations from generations I_{12} and I_{17}. The full data from generation I_{38}, including all crosses that produced non-significant deviations, will be published elsewhere (Wills, Phelps, and Ferguson, in press).

RESULTS AND CONCLUSIONS

1) Effects of stress medium were seen only on inbred

TABLE 1
Numbers (and s.e.) of Flies Emerging per Bottle

Medium	Flies per bottle
Normal medium (9% corn meal, 18% molasses, 1% agar)	98.13 ± 8.67
Medium with 6% (v/v) 95% ethanol	16.47 ± 0.38
Medium with 3mM 1-octanol dissolved in equal amount of 95% ethanol	39.78 ± 0.91
Medium with 5% (v/v) tributyrin, polydispersed by 10-sec. Waring Blender treatment	16.08 ± 0.37
Medium with 5% (v/v) triacetin	23.59 ± 0.52

flies. This conforms with our expectation that polymorphisms isolated by inbreeding will become more important to the flies' relative survival than in outbred background.

2) On the whole, the largest and most consistent effects were seen in flies raised on media with additions related to the substrate affinity of the enzymes in question. This division of effects was not complete, however; some effect of tributyrin was observed on ODH flies and some effect of octanol on E-5 flies. We suspect that this is due either to effects of genetic background not yet made homozygous or to the possibility that the catabolic pathways of which ODH and E-5 are steps are not completely independent. Normal oxidation pathways of butyrate and octanol will both pass through butyryl-CoA. This fact was not realized when these two enzymes were chosen at the beginning of the experiment, since the substrate affinities of E-5 were not understood.

3) Triacetin had no effect on flies segregating for either enzyme system. This may be because the products of hydrolysis, acetic acid and glycerol, are both common metabolites produced from other sources.

4) As inbreeding continued, the pronounced heterosis seen in affected flies segregating for the ODH isozymes disappeared. A strong criticism of our earlier results by Yamazaki (1972) was based on the presumption that the heterosis observed at the ODH locus was due to generalized heterotic selection for the block of genes surrounding the ODH locus even after inbreeding. In this view, octanol was simply a

TABLE 2

Summary of significant deviations from Medelian expectation from inbreeding experiment up to generation I_{38}. * = significance at .05 level; ** = significance at .01 level.

	Conforming to expectation
Sex-limited effects	1) I_{12} ODH inbred ♂♂ on octanol, all 3 lines, S/S<S/F>F/F ** (pooled data)
	2) I_{38} ODH inbred ♂♂ on octanol, all 3 lines, S/S<S/F>F/F ** (pooled data)
	3) I_{38} ODH inbred ♀♀ on ethanol, all 3 lines, S/S<S/F<F/F ** (pooled data)
Similar effects in both sexes	1) I_{12} ODH inbred ♀♀ on octanol, one line, S/S<S/F>F/F *
	2) I_{17} ODH inbred ♂♂ and ♀♀ on octanol, 3 out of 6 lines, S/S<S/F>F/F ** (pooled data)
	3) I_{38} E-5 inbred ♂♂ and ♀♀, on tributyrin, all four lines S>F (♂♂) S/S>S/F = F/F (♀♀) ** (pooled data)
	Anomalous
Sex-limited effects	1) ODH inbred ♀♀ on tributyrin, 2 out of 3 lines, S/S<S/F>F/F * (pooled data)

TABLE 2, continued:

	Anomalous
Sex-limited effects	2) E-5 inbred ♂♂ on octanol, one line, S>F **

Two other within-line comparisons showed significant deviations at the .05 level, one parental grown on ethanol and one inbred line grown on triacetin. In view of the total number of within-line comparisons (98) these were probably due to chance.

---

severe stress. The reason the effect was not apparent in outbred flies was that the relative difference in heterozygosity between homozygotes and heterozygotes was not so pronounced in these flies. It seems likely that Yamazaki's criticism of our 12th and 17th generation results has validity at least so far as the heterosis is concerned. Such criticism cannot be applied to the I_{38} results, however, since heterosis is not apparent (except very weakly in inbred ODH females grown on tributyrin) and since the effects seen are not correlated with the severity of the stress as measured by the number of flies surviving on the four stress media.

5) The sex-limited nature of many of the results, and the fact that the order of fitnesses of ODH genotypes on octanol medium has undergone striking changes from I_{17} to I_{38}, are strong indications that background genotype is still playing a role. For the average polymorphism, then, selection coefficients in outbred flies must be very small and extremely dependent on the background genotype.

6) A spectrophotometric assay, involving the clearing of a polydispersed suspension of tributyrin in pH 6.5 phosphate buffer, has been developed. S/F heterozygotes have the most activity per unit protein in crude extracts, followed by F/F homozygotes. This does not correspond to the fitnesses obtained by growth on tributyrin medium, and illustrates the danger of generalizing from kinetic data to fitnesses when the function of the enzyme in metabolism is in doubt.

7) If we have indeed gone some way towards isolating single gene effects, one striking fact which emerges from these data is that genotype-environment interactions are extremely strong. The fitnesses of ODH inbred flies grown on octanol and ethanol were reversed, the former fitness

differences showing up in males and the latter in the females. This is a point which cannot be overemphasized, though it tends to be forgotten by more mathematically-oriented population geneticists who assign a particular fitness to a particular genotype. The relative fitnesses of single loci are just as environment-dependent as those of inversions followed in population cages by Dobzhansky and his co-workers (e.g., Levene, Pavlovsky, and Dobzhansky 1954, 1958). This phenomenon is not confined to highly complex organisms; for example, yeast shows similarly strong genotype-environment interactions (Blatherwick and Wills 1971).

8) Returning to the assumptions we made as a rationale for designing the experiment, it will be noted that there is little evidence from the latest data for conditional heterosis. We have, however, used crude means of selection dictated by our ignorance of the role of these enzymes; it is quite within the realm of possibility that other media could be devised that would strongly select for heterozygotes. Further, the maintenance of polymorphism is not likely to be entirely due to conditional heterosis; we picked it as a likely mechanism to concentrate on because our earlier results indicated that the ODH polymorphism might be maintained in this way. Despite the apparent absence of heterosis at the I_{38} level, the possibility of its prevalence in nature is by no means ruled out.

ACKNOWLEDGEMENTS

I would like to thank Lois Nichols, Elizabeth Kidder, Julia Phelps, Patricia Rorabaugh and Patricia Coughran for valuable assistance in the performance of these experiments, Edward Sheldon for permission to quote unpublished results, and Dr. Peter Carlson for discussions and advice. Supported by Public Health Service Grants GM 14700 and GM 19967.

REFERENCES

Arnheim, N. and C.E. Taylor 1969. Non-Darwinism evolution: consequences for neutral allelic variation. *Nature* 233: 900-902.

Ayala, F.J. and M.E. Gilpin 1973. Lack of evidence for the neutral hypothesis of protein polymorphism. *Jour. Hered.* 64: 297-298.

Berger, E. and R. Canter 1973. The esterases of Drosophila. I. The anodal esterases and their possible role in eclosion. *Dev. Biol.* 33: 48-55.

Blatherwick, C. and C. Wills 1971. The accumulation of genetic variability in yeast. *Genetics* 68: 547-557.

Ford, E.B. 1940. Polymorphism and taxonomy. In: *The New Systematics* (ed. Julian Huxley), pp. 493-513.

Johnson, G.B. 1974. Enzyme polymorphism and metabolism. *Science* 184: 28-37.

Kimura, M. 1968. Genetic variability maintained in a finite population due to mutational production of neutral and nearly neutral isoalleles. *Genet. Res.* 11: 247-269.

King, J.L. 1967. Continuously distributed factors affecting fitness. *Genetics* 55: 483-492.

King, J.L. and T.H. Jukes 1969. Non-Darwinian evolution. *Science* 164: 788-798.

Levene, H., O. Pavlovsky, and Th. Dobzhansky 1954. Interaction of the adaptive values in polymorphic experimental populations of *Drosophila pseudoobscura*. *Evolution* 8: 325-349.

Levene, H., O. Pavlovsky, and Th. Dobzhansky 1958. Dependence of the adaptive values of certain genotypes in *Drosophila pseudoobscura* on the composition of the gene pool. *Evolution* 12: 18-23.

Lewontin, R.C. and J.L. Hubby 1966. A molecular approach to the study of genic heterogeneity in natural populations. II. Amount of variation and degree of heterozygosity in natural populations of *Drosophila pseudoobscura*. *Genetics* 54: 595-609.

Milkman, R.D. 1967. Heterosis as a major cause of heterozygosity in nature. *Genetics* 55: 493-495.

Narise, S. and J.L. Hubby 1966. Purification of esterase-5 from *Drosophila pseudoobscura*. *Biochem. et Biosphy. Acta* 122: 281-288.

Sved, J.A., T.E. Reed, and W.F. Bodmer 1967. The number of balanced polymorphisms that can be maintained in a natural population. *Genetics* 55: 469-481.

Wills, C. 1970. Drosophila isoalleles: enhancement of heterosis by inbreeding. *Genetics* 64: s 66.

Wills, C. 1973. In defense of naive pan-selectionism. *Amer. Natur.* 107: 23-34.

Wills, C., J. Crenshaw, and J. Vitale 1970. A computer model allowing maintenance of large amounts of genetic variability in Mendelian populations.

Wills, C. and L. Nichols 1971. Single gene heterosis in *Drosophila* revealed by inbreeding. *Nature* 233: 123-125.

Wills, C. and L. Nichols 1972. How genetic background masks single gene heterosis in *Drosophila*. *Proc. Nat. Acad. Sci. U.S.* 69: 323-325.

Wright, S. 1940. The statistical consequences of Mendlian heredity in relation to speciation. In: *The New Systematics* (ed. Julian Huxley), pp. 161-183.

Yamazaki, T. 1972. Detection of single gene effect by inbreeding. *Nature New Biol.* 240: 53-54.

Yamazaki, T. 1971. Measurement of fitness at esterase-5 locus in *Drosophila pseudoobscura*. *Genetics* 67: 579-603.

EVOLUTION OF ELECTROPHORETIC MOBILITY IN THE *Drosophila mulleri* COMPLEX

R. H. RICHARDSON, M. E. RICHARDSON,
and P. E. SMOUSE[1]
Department of Zoology
University of Texas at Austin
Austin, Texas 78712

ABSTRACT. Several species of the *Drosophila mulleri* complex were compared for mobility of seven electrophoretically assayed enzyme systems. An analysis of variance established that the variation among species was much greater than that within species. The among-species component was partitioned into an among-clusters component and a within-clusters component, where clusters consisted of sets of sibling species. Most of the among-species variation was attributable to variation among clusters, indicating that mobility per se is a useful taxonomic measure on the macro-level. Detailed analysis, however, resulted in different findings than indicated by karyoptypic analysis. Single-locus analysis established that different functional classes of enzymes did <u>not</u> show appreciably different patterns of allozyme (allelic isozyme) variation among species. Most mobility changes within species and within phyletic clusters appeared to be of the sort caused by simple charge changes. Occasional shifts in larval substrate were accompanied by what appeared to be conformational shifts in certain enzymes. These shifts almost invariably characterize whole taxonomic groups, i.e., species clusters.

INTRODUCTION

The electrophoretic analysis of protein variation was initiated in *Drosophila* by Hubby (1963), Wright (1963), and Beckman and Johnson (1964). The first inter-specific comparisons were made by Hubby and Throckmorton (1965), who used 6 gm mass homogenates of adult flies. The first interspecific comparisons, on the basis of singly-fly assay, were made by Johnson et al (1966a), who were thus able to compare allelic isozyme (allozyme) variation among closely related species (Johnson et al, 1966b). Numerous similar studies have since

---

[1]Present address: Dept. of Human Genetics, University of Michigan, Ann Arbor, Michigan 48104.

been reported. For the most part, these have been concerned with only two aspects of allozyme variation: (1) the number of mobility classes shared by two species, and (2) the population frequencies of those classes shared in common.

The early preoccupation with mobility classes, as qualitative genetic markers, has been superseded by increasing interest in the quantitative properties of the allelomorphs. The most immediately discernible property, of course, is mobility itself. King (1973) has shown that mobility is rather insensitive to random amino acid substitution, since only a small proportion of changes results in a net charge change under normal electrophoretic conditions. From this, one predicts that mobility should tend to be a relatively conservative taxonomic feature of the enzyme. The relationship between electrophoretic separation and phyletic distance should be monotonic, though non-linear, if the observed mobility differences represent the results of simple charge changes.

Electrophoretic mobility is not simply a function of net charge; it also involves molecular weight, conformation, and polymeric complexity. Charge changes will thus, in practice, be confounded with a variety of other alterations. Charge changes should result in regular increments in electrophoretic mobility, but other sorts of alterations lead to less predictable results. Overall, one should expect an increase in the number of classes as the number of taxa considered increases. We are interested here in two separate questions. (1) Is electrophoretic mobility *per se* a reliable criterion of taxonomic affinity? (2) Is it possible to separate mobility difference due to charge changes from those due to conformational or other major shifts?

These questions may be easily addressed by comparing the electrophoretic patterns of species of varying taxonomic affinity. An ideal basis for such comparisons is provided by an assemblage of seventeen (17) closely related species of the *Drosophila mulleri*, and an additional group of eleven (11) related species of the *mulleri*. The first set has been the subject of longstanding studies (Wasserman, 1962; Wasserman et al., 1974; Heed, in preparation), and the phylogeny of these taxa has been reasonably well worked out on cytological, morphological, and reproductive grounds. The second set is karyotypically more diverse and has not yet been as thoroughly studied.

We have attempted to standardize all assay procedures. Identification of mobility classes involves side-by-side comparisons on a single starch gel. Several allozyme standards of known mobility are routinely included on all gels. Relative mobilities of the different classes remain constant

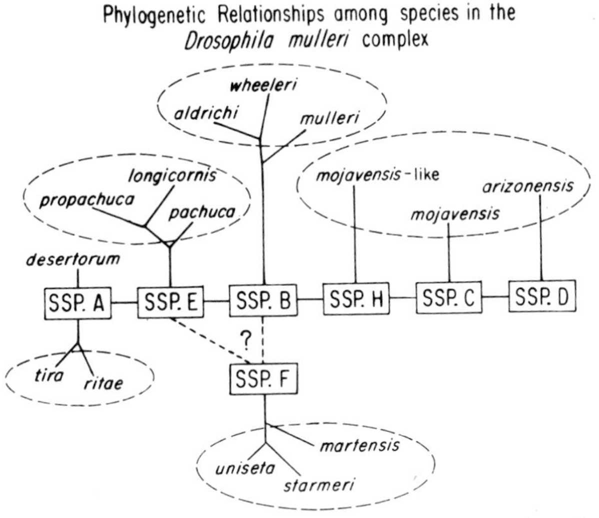

Fig. 1. Phylogenetic relationships among species in the *Drosophila mulleri* complex based on polytene chromosome analysis (modified from Wasserman 1962, and personal communication).

over a range of buffer pH and concentration and over a range of voltage gradients and separation times. Extensive cross-checking allows small differences to be established with confidence. A series of interlocking inter-specific crosses has allowed verification of homology for the loci and species in question.

MOBILITY AS A TAXONOMIC MEASURE

For this analysis, we have restricted attention to twelve (12) intensively sampled species of the *mulleri* complex. The appropriate portion of the phylogeny, based on karyotype (Wasserman, 1962 and personal communication), is presented in Fig. 1. Clusters of sibling species are enclosed within the dashed ovals. *Drosophila desertorum* is often placed with the *pachuca-propachuca-longicornis* cluster. Although there is a karyotypic difference, all four taxa are morphologically similar, and we have included *desortorum* in this cluster for analytical purposes. *Drosophila aldrichi* and *D. wheeleri* are very similar, both morphologically and karyotypically, but can be partially separated on reproductive grounds. Our samples of *D. martensis*, *D. uniseta*, and *D. starmeri* were too small to be included in the analysis.

We have defined the average mobility of a particular locus,

TABLE I
ANALYSIS OF VARIANCE OF ELECTROPHORETIC MOBILITY*

Source	d.f.	SS	MS
Among Species†	11	10,193	927
Among Sibling Clusters	3	8,787	2,929
Within Sibling Clusters	8	1,406	176
Among Localities Within Species	45	189	4

*MDH, PG, AO, ADH, GOT, Est-C, ODH

†*Drosophila tira, D. ritae, D. desertorum, D. hexastigma, D. longicornis, D. pachuca, D. propachuca, D. wheeleri, D. aldrichi, D. mulleri, D. mojavensis, D. arizonensis.*

for a given population, as the weighted average mobility of all alleles found in the population, with the weights being the frequencies of the various mobility classes. A corresponding variance was defined. A weighted analysis of variance was conducted for each locus, and the variation was partitioned into separate "within-species" and "among-species" components. The first component measures the geographic variation in mobility within species. The second component was further partitioned into "within-cluster" and "among-clusters" components. The analysis was pooled over loci (we have included seven multi-allelic loci for this purpose) and is presented in Table I. The average differences among taxa are so well established as to obviate the necessity of formal statistical testing, and we employ the ANOVA format simply to indicate orders of magnitude. The "among-species" mean squares (MS) are two orders of magnitude larger than the "within-species" mean squares. The "within-cluster" mean squares are much smaller than the "among-clusters" mean squares, and average electrophoretic mobility is a very useful taxonomic metric.

On a finer level, the electrophoretic phenogram is shown in Fig. 2. In broad outline, it conforms with Wasserman's phylogeny (Fig. 1) as would be predicted from Table I, but it differs in detail. *Drosophila desertorum* is closer to *D. pachuca* and *D. propachuca* than is either of these latter two to *D. longicornis*, a reversal of the karyotypic pattern. *Drosophila mulleri* is somewhat distant from *D. aldrichi* and *D. wheeleri*, with which it is karyotypically identical. These latter two are all but indistinguishable. The major difference in the two phylogenies, however, is that the *mojavensis-arizonensis* cluster is very different electrophoretically from the *mulleri-aldrichi-wheeleri* cluster, in

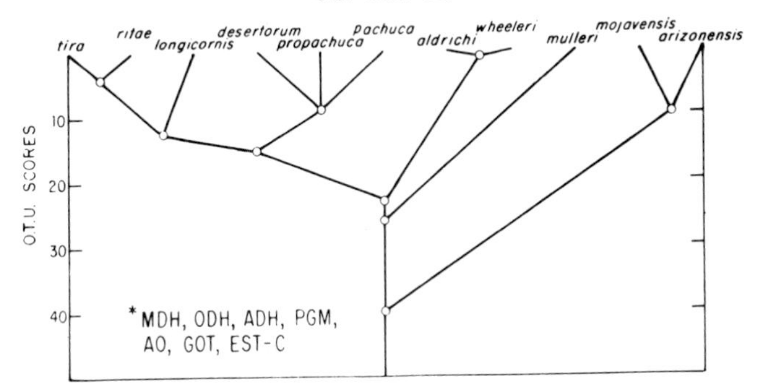

Fig. 2. Phenogram of species based upon average relative mobilities of seven enzyme loci.

distinction to the similarity of the two clusters in Wasserman's phylogeny.

MOBILITY PATTERNS FOR FUNCTIONAL CLASSES OF ALLOZYMES

Kojima, Gillespie, and Tobari (1970) divided electrophoretic enzymes into glycolytic and non-glycolytic classes, and showed that the latter tended to be polymorphic, while the former tended to be strongly monomorphic. Johnson (1974) has reclassified these enzymes according to their substrate specificities. Those with specific substrates were further divided into regulatory and non-regulatory classes. The variation patterns were related to these classes of function. Since closely related species share many physiological properties, interspecific comparisons should be as informative as intra-specific comparisons. Where food sources are the same, digestive enzymes should be similar. Where metabolic processes differ, metabolic enzymes should reflect this differentiation. Enough is known about the ecology of the species of the *mulleri* complex to permit a preliminary examination of patterns for different loci.

Larval substrates are more similar within phyletic clusters than among clusters. Fellows and Heed (1972) and Heed (personal communication) found both *D. mojavensis* and *D. arizonensis* on species of "giant cactus," and have postulated a similar larval substrate for the *martensis-uniseta-starmeri* cluster. We have found the other species of Fig. 1 on "prickly pear cactus" (*Opuntia* spp.). Even within the Opuntia group,

TABLE II
DISTRIBUTION OF ELECTROPHORETIC MOBILITY
DIFFERENCES AMONG TAXA, "VARIABLE SUBSTRATE ENZYMES"

Taxonomic Level	% Sums of Squares	
	Est-C	ODH
Among Species*	99	93
Among Sibling Clusters	97	57
Within Sibling Clusters	2	36
Among Localities Within Species	1	7

**Drosophila tira, D. ritae, D. desertorum, D. hexastigma, D. longicornis, D. pachuca, D. propachuca, D. wheeleri, D. aldrichi, D. mulleri, D. mojavensis, D. arizonensis.*

there is considerable differentiation. Sympatry among species is fairly common, although clusters overlap less than species within a cluster (Patterson and Stone, 1952; Wasserman, 1962). With the exception of *D. aldrichi* and *D. wheeleri* (Patterson and Alexander, 1952; Richardson, unpublished), however, all species have well developed post-zygotic isolating mechanisms, indicating considerable genetic diversification within and among clusters.

Table II lists the percentages of the total sums of squares at different taxonomic levels for two loci characterized as having variable substrates, Est-C and ODH. The major electrophoretic differences are among sibling clusters, although this is more pronounced for Est-C than for ODH. Similar analyses for the other five loci, characterized as being substrate specific, are shown in Table III. There is some question as to whether ADH and GOT are regulatory or not. Electrophoretic variation within clusters is much more pronounced for PGM, AO, ADH, and GOT than for MDH, which follows the pattern of Est-C. There appear to be no consistent patterns of variation within a functional class and no consistent differences among classes.

CHARGE CHANGES AND CONFORMATIONAL SHIFTS

Nei and Chakraborty (1973) have computed the expected frequency of new mutants which would result in a net charge change of 16%, while a two-charge change in one direction has an expected frequency of 0.5%. (These values disregard the possibility of changes in pK, and hence charge, resulting from a neutral substitution adjacent to a charged amino acid.) Charge changes greater than two require either an associated

TABLE III
DISTRIBUTION OF ELECTROPHORETIC MOBILITY
DIFFERENCES AMONG TAXA, "SPECIFIC SUBSTRATE ENZYMES"

Taxonomic Level	% Sums of Squares				
	MDH*	PGM	AO	ADH ?	GOT ?
Among Species	99	99	98	94	100
Among Sibling Clusters,	99	83	54	79	63
Within Sibling Clusters	< 1	16	44	15	37
Among Localities Within Species	1	1	2	6	< 1

*Non-regulatory, others are regulatory.
Drosophila tira, D. ritae, D. desertorum, D. hexastigma, D. longicornis, D. pachuca, D. propachuca, D. wheeleri, D. aldrichi, D. mulleri, D. mojavensis, D. arizonensis.

conformational rearrangement, exposing a charged residue not previously expressed, or multiple amino acid substitutions.

If we examine all mobility classes for a given allozyme which have been found in the *mulleri* complex, one encounters a series of incremental steps in mobility, many of which are uniformly spaced, but some which are irregular. If all of these changes were the result of simple charge changes, spacing should be regular (Henning and Yanofsky, 1963). Conformational shifts would be expected to shift the mobility out of phase. This would either lead to a major displacement of mobility or a small shift, which, when coupled with simple charge changes, would lead to inter-digitating series of mobility classes. These various possibilities are schematized in Fig. 3.

A conformational change is more drastic than a change in surface charge, and would be expected to initiate a "genetic revolution." One might expect such changes to be generally non-adaptive, and that those which survive an evolutionary sieving would be both adaptive and rare. It is therefore natural to anticipate such differences only at the inter-specific level, and possibly only at the inter-cluster level. To examine the actual pattern of changes in mobility within and among species, we have plotted relative mobility against class rank for each enzyme. We shall illustrate the different patterns with three specific examples.

The pattern for MDH-NAD is shown in Fig. 4. The mobility classes are almost equally spaced, and the overall relation for the whole complex is almost a straight line, which might be interpreted as evidence for a series of simple charge changes. When we subdivide the species into phyletic clusters,

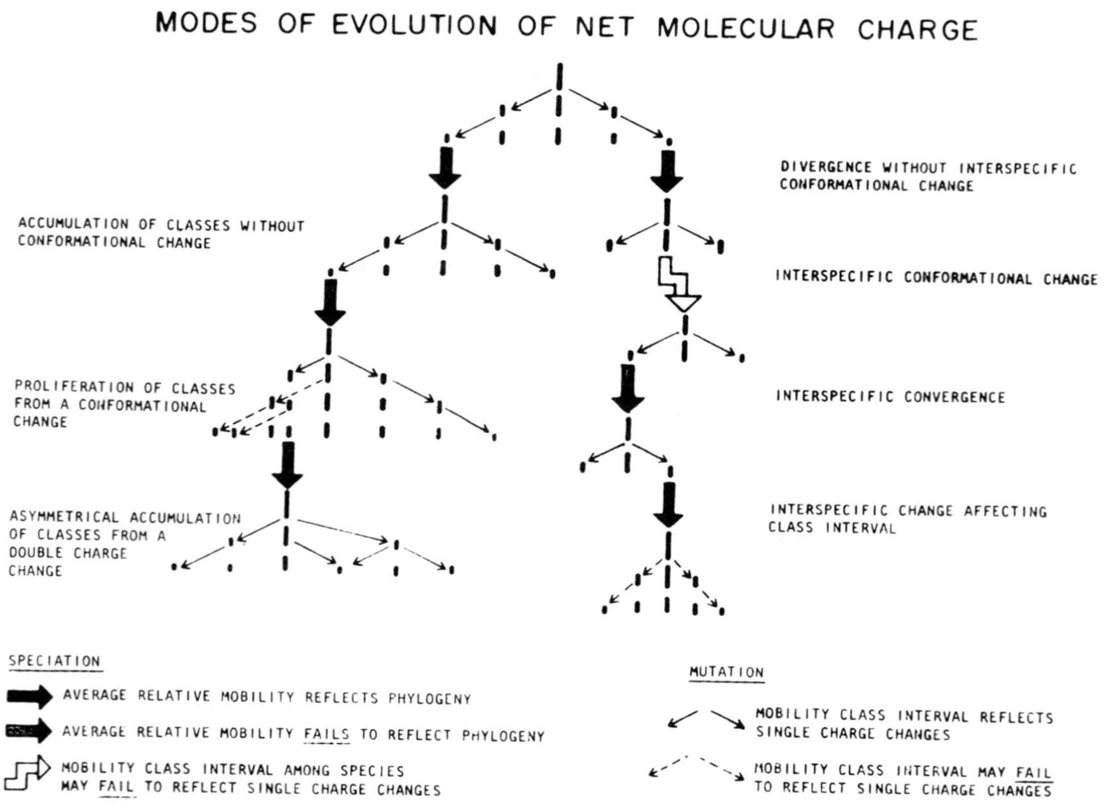

Fig. 3. Modes of evolution of net molecular charge. Mobility intervals are defined as the separation between adjacent classes taken over all species.

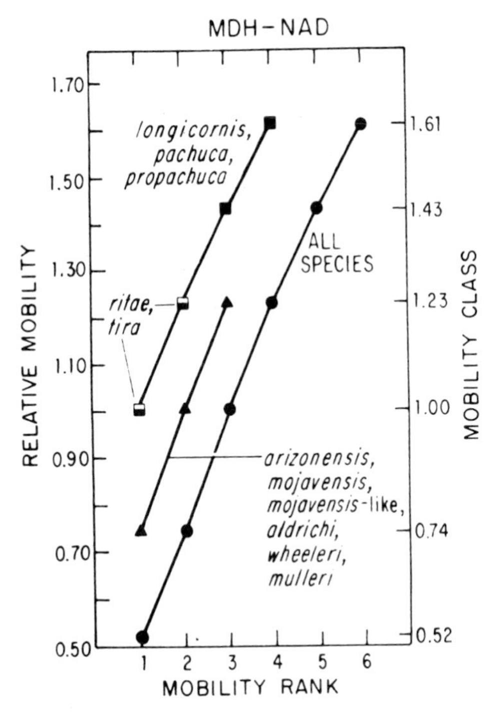

Fig. 4. Relationships of changes in relative mobility for MDH-NAD.

we see that a given species contains alleles in adjacent mobility classes and that adjacent species in the phylogeny contain over-lapping sets of alleles, even though the frequencies of the different classes vary among species. All species of the *arizonensis-mojavensis-mojavensis* like cluster exhibit three classes (.74, 1.00, 1.23) as do the species of the adjacent *aldrichi-wheeleri-mulleri* cluster. Three of the species from the next cluster, *D. pachuca, D. propachuca,* and *D. longicornis,* exhibit classes (1.00, 1.23, 1.43 and 1.61), while *D. desertorum* contains only 1.23. There is no suggestion of conformational alterations, and mobility changes appear to cascade along the phylogeny.

A rather different pattern is evident for Est-C, shown in Fig. 5. The spacing is quite regular below .55 and again above .55, but the step size is abruptly altered at this critical point. The reason for this change in slope is not entirely clear, but one is led to suggest a conformational shift at this point. The shift occurs between the *mojavensis-*

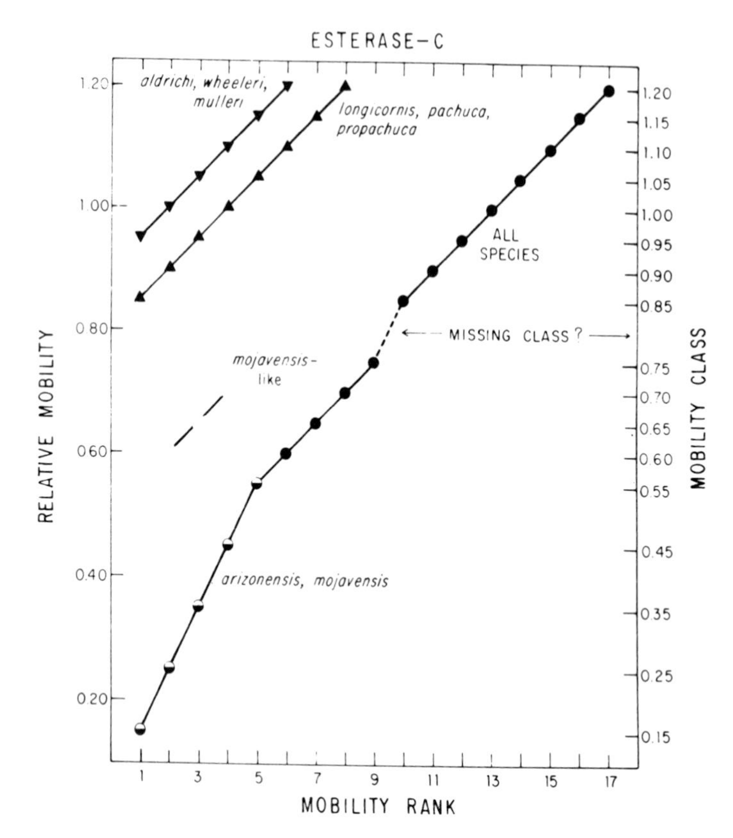

Fig. 5. Relationships of changes in relative mobility for Est-C.

---

arizonensis cluster and the rest of the *mulleri* complex. These two taxa are "connected" to the others by *mojavensis*-like, which straddles the critical class (.55) with no shift in step size. As mentioned earlier, these two taxa are found on "giant-cacti" rather than *Opuntia*. The correspondence is, to say the least, intriguing. The only other complication encountered with Est-C is an apparent "missing class." However, since the adjacent classes on either side were all found within a single species, and since sample size was limited, it is possible that this class was simply missed in sampling.

Still another pattern of mobility changes is exhibited by ODH, shown in Fig. 6. Within a series, step size is somewhat less regular than that for Est-C, but two basic series would

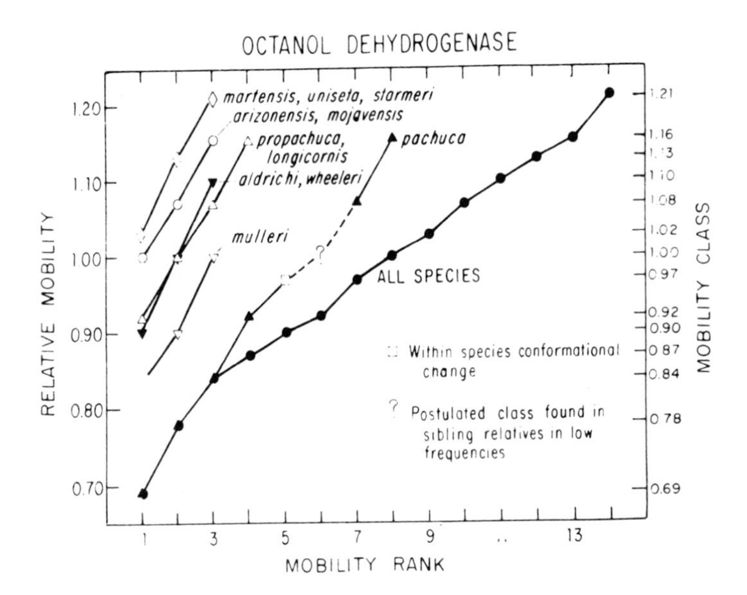

Fig. 6. Relationships of changes in relative mobility for ODH.

seem to be present from the composite curve. When species are separated into smaller groups, however, the step size within each group appears to be more regular. The groups seem to be separated by small, irregular shifts in average position, and it is the resulting inter-digitation which yields the appearance of overall irregularity. Two irregular changes, indicated by open squares in Fig. 6, appear within single species (*D. mulleri* and *D. pachuca*, respectively). We are led to speculate that these irregular shifts may well be due to a series of conformational changes.

Rates of divergence differ markedly among enzymes. Among closely related species, simple charge changes seem to predominate. Major shifts in step size and overall mobility appear to accompany a major change in larval substrate (Est-C), while smaller out-of-phase shifts in average mobility (ODH) appear to have occurred without a major substrate shift. We are inclined to postulate a causal relationship between the substrate shift and the putative conformational change in Est-C. Biochemical procedures are available to determine whether this change really is due to a conformational shift; and it should be possible to determine whether the "change" has any physiological bearing on substrate utilization.

To the extent that mobility differences represent simple charge changes, mobility constitutes a reliable measure of phyletic separation. A major displacement, such as that seen in Est-C, tends to exaggerate the separation. The type of inter-digitation seen with ODH tends to underestimate the separation. Given enough loci, one can expect that these tendencies will become counterbalanced. Certainly, the taxonomy we have constructed is useful in a gross way. Details, however, are more sensitive to failure of the assumption of charge change differences. In the present instance, a reliable *a priori* phylogeny was available, and this has allowed us to detect the patterns in question. In a more general taxonomic vein, uncritical use of mobility can lead to difficulty.

ADDENDUM

We have identified recent collections having little or no post-zygotic isolation between the sibling taxa reported as *D. aldrichi* and *D. wheeleri*, thereby suggesting that our taxa are subspecies. Additionally, W. B. Heed (personal communication) has recently obtained collections which are more similar to the original description of *D. wheeleri*. Thus, it is possible that our populations labeled *D. wheeleri* may be a new subspecies of *D. aldrichi*, and we may not have examined *D. wheeleri*. Since the thrust of our treatment in the foregoing was by sibling clusters rather than by individual species, our interpretation is not changed.

ACKNOWLEDGEMENT

This work was supported by NSF Grant GB 22770, NIH Grant GM 19616-02, and NIH Research Career Development Award GM 47350.

We appreciate the assistance of M. R. Wheeler, W. B. Heed, M. Wasserman, and others in the conduct of this work.

REFERENCES

Beckman, L. and F. M. Johnson 1964. Esterase variations in *Drosophila melanogaster*. *Hereditas* 51: 212.

Fellows, D. P. and W. B. Heed 1972. Factors affecting host plant selection in desert-adapted cactiphilic *Drosophila*. *Ecology* 53: 850-858.

Henning, Y. and C. Yanofsky 1963. An electrophoretic study of mutationally altered A proteins of the tryptophan synthetase of *Escherichia coli*. *J. Mol. Biol.* 6: 16-21.

Hubby, J. L. 1963. Protein differences in Drosophila. I. *Drosophila melanogaster*. *Genetics* 48: 871-879.

Hubby, J. L. and L. Throckmorton 1965. Protein differences in *Drosophila*. II. Comparative species genetics and evolutionary problems. *Genetics* 52: 203-215.

Johnson, F. M., C. G. Kanapi, R. H. Richardson, M. R. Wheeler, and W. S. Stone 1966a. An operational classification of *Drosophila* esterases for species comparisons. In *Studies in Genetics III*, M. R. Wheeler, ed. Univ. Texas Publ. No. 6615: 517-532.

Johnson, F. M., C. G. Kanapi, R. H. Richardson, M. R. Wheeler, and W. S. Stone 1966b. An analysis of polymorphism among isozyme loci in dark and light *Drosophila ananassae* strains from America and Western Samoa. *Proc. Natl. Acad. Sci. USA* 56: 119-125.

Johnson, G. B. 1974. Enzyme polymorphism and metabolism. *Science* 184: 28-37.

King, J. L. 1973. The probability of electrophoretic identity of proteins as a function of amino acid divergence. *J. Mol. Biol.* 2: 317-322.

Kojima, K., J. Gillespie, and Y. N. Tobari 1970. A profile of *Drosophila* species' enzymes assayed by electrophoresis. I. Number of alleles, heterozygosities, and linkage disequilibrium in glucose-metabolizing systems and some other enzymes. *Biochem. Genet.* 4: 627-637.

Nei, M. and R. Chakraborty 1973. Genetic distance and electrophoretic identity of proteins between taxa. *J. Mol. Evol.* 2: 323-328.

Patterson, J. T. and M. L. Alexander 1952. *Drosophila wheeleri*, a new member of the *mulleri* subgroup. In *Studies in the Genetics of Drosophila*. J. T. Patterson, dir. Univ. Texas Publ. No. 5204: 129-136.

Patterson, J. T. and W. Stone 1952. *Evolution in the Genus Drosophila*. The Macmillan Company: New York. 610 pp.

Wasserman, M. 1962. Cytological studies of the repleta group of the genus *Drosophila:* V. The *mulleri* subgroup. In *Studies in Genetics II*, M. R. Wheeler, ed. Univ. Texas Publ. No. 6205: 85-118.

Wasserman, M., H. R. Koepfer and B. L. Ward 1974. Two new Drosophilidae) from Venezuela. *Ann. Ent. Soc. Am.* 66: 1239-1242.

Wright, T. R. F. 1963. The genetics of an esterase in *Drosophila melanogaster*. *Genetics* 48: 787.

FURTHER STUDIES ON ALCOHOL DEHYDROGENASE POLYMORPHISM IN MEXICAN STRAINS OF DROSOPHILA MELANOGASTER

SARAH BEDICHEK PIPKIN, JANE H. POTTER,
SETH LUBEGA, and EUGENIA SPRINGER
Howard University, Washington, D. C. 20001
and
The University of Maryland, College Park, Maryland

ABSTRACT. Further studies have shown that the abrupt change from polymorphism of alcohol dehydrogenase variants in Mexican populations of *D. melanogaster* at Cardel and westward to near monomorphism at Coatzocoalcos and eastward must depend on previous isolation of these strains during their evolutionary history. Synthetic lethality results when an eastern strain chromosome 2, carrying the *Adh* locus, is made homozygous in a fly in which the X and 3rd chromosomes have been replaced from laboratory stocks. That a limited gene exchange between eastern and western strains is taking place in nature is indicated by (1) an eagerness on the part of eastern males to mate with western females in multiple choice tests, although eastern females are reluctant to mate with western males; (2) disorganization of a polygenic character, the mating song of eastern males; (3) the presence of a short cline of the esterase 6^S allele from monomorphism at Coatzocoalcos to 0.07 at Tuxpan, 300 miles northwest; (4) the finding of an excess of homozygotes for three different aldehyde oxidase alleles and a deficiency of two kinds of heterozygotes at Villahermosa, east of Coatzocoalcos. Since ADH can use substrates involved in the metabolism of juvenile hormone (Ursprung and Madhaven, 1971), it is suggested that Adh^I, specifying the enzymatically more active protein, bestows a selective advantage on diapausing larvae in cold regions.

INTRODUCTION

An unusual distribution of allelic frequencies at the alcohol dehydrogenase (ADH) locus in certain Mexican populations of *Drosophila melanogaster* was observed by Pipkin et al., (1973). Whereas a north-south cline in frequency of the Adh^{II} allele occurred in populations of the central highlands, the gulf coastal populations showed an abrupt change from a frequency of 0.51 at Cardel (near Vera Cruz) to 0.99 at Coatzocoalcos, 135 miles southeast in the Isthmus of Tehauntepec. Three other populations east of Coatzocoalcos displayed virtual homozygosity for Adh^{II}, the allele specifying the enzymatically weaker protein. The present study examines reasons for the abrupt change in Adh^{II} frequency east of Cardel. Evidence will be presented that strains of *D. melanogaster* taken from Cardel and north and west of Cardel show signs of isolation in their evolutionary history from the four strains collected east of Cardel. It will be argued that eastern and western strains have evolved into semispecies which are now undergoing limited gene exchange. Implications of this regarding the selective effect of temperature on ADH-polymorphism will be discussed.

MATERIALS AND METHODS

1. *Drosophila stocks*. Founder numbers for stocks of the 12 Mexican and 2 Texan strains of *D. melanogaster* collected June 15 to July 15, 1972, were given by Pipkin et al., (1973). Locations of the collecting stations appear on the map in Fig. 1. Abbreviations are as follows: Mi, Mico, Texas (near San Antonio); CC, Corpus Christi, Texas; Ji, Jiminez (near Ciudad Victoria); Cz, Coatzocoalcos; VH, Villahermosa; Pe, Palenque; Me, Merida; SLP, San Luis Potosi; Ct, Cuatla; At, Acatlan; Ox, Oaxaca. Flies were kept in mass culture on standard corn meal medium enriched with Kellogg K breakfast food except for those used in the mating song recordings. The latter were cultured on Carolina Instant Drosophila Medium and kept at 25°C on a 12-hour light cycle.

2. *Replacement of X and 3rd chromosomes with laboratory strain chromosomes*. The X chromosome of a wildtype Mexican strain was replaced by crossing S/Cy; D/In(3)C3X females with males of a Mexican strain and choosing F_1 males of the genotype Cy/+; D/+. These males were crossed with Cy L/Pm; Sb/In(3)Ubx females. Progeny of the genotype Cy L/+; D/In(3)Ubx were selected and intercrossed. In cases where the X and 3rd chromosomes of the Mexican strain could be replaced, progeny of the genotype +/+; D/Ubx were selected to start a balanced stock homozygous for the chromosome 2 of the Mexican strain but possessing X and 3rd chromosomes derived from laboratory stocks.

3. *Crosses between Mexican strains*. Reciprocal crosses were made between Ca, Pa, Cz, and VH, in all combinations, using 3 virgin flies of each sex to the vial. F_1 males were similarly crossed with F_1 females. Reciprocal backcrosses were made.

4. *Body length*. Body lengths of 25 individuals of each sex of the Pa, Ca, Cz, and VH strains were measured in millimeters with the flies placed on their sides, under a dissecting scope.

5. *Mating success under multiple choice conditions*. Six virgin males and six virgin females, aged 4 to 6 days, taken from each of 2 Mexican strains were placed under a hemispherical viewing chamber similar to that described by Ehrman (1961). Males and females of one of the two strains had previously been fed red

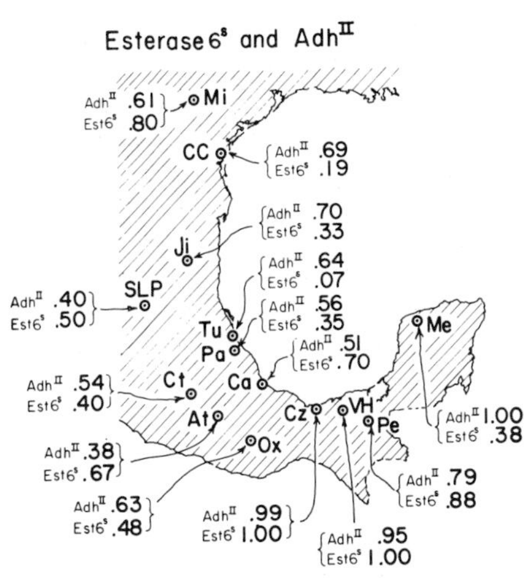

Fig. 1. The locations of collecting stations and the frequencies of Adh^{II} and Est 6^S in *D. melanogaster* of Texan and Mexican strains.

food coloring in sugar water. Combinations studied included Ca, Cz; Ca, Pa; Cz, Pa; Cz, VH. Copulating pairs were removed from the chamber up to a maximum of 30 minutes after the start of the experiment. After each experiment the chamber was cleaned to remove possible odors.

6. *Courtship song*. Adults were separated by sex within 8 hours of eclosion and aged for 3 to 5 days in 25 x 95 mm vials, 10 to 20 flies per vial. Songs were recorded with a ribbon microphone on which rested an observation cage holding a pair of flies. The apparatus was described by Bennet-Clark (1972). To bring a pair together, the female was first aspirated from its vial and transferred to the observation cage. The male was transferred similarly within 45 seconds. The pair were observed under a magnification of 7X and the tape recorder started when the male approached the female. Songs due to the wing movements of the male (Bennet-Clark and Ewing, 1968) were recorded with a Tandberg 9000X or 64X tape recorder, and oscillograms of the tape recording obtained from an oscilloscope equipped with a Kymograph camera. The interpulse interval, measured to the nearest 0.1 mm and converted to milliseconds, represents the elapsed time from the beginning of one sound pulse to the beginning of the next.

7. *Electrophoresis and histochemical staining*. Single females were ground manually in small homogenizers in a drop of

deionized water and placed in slits made in an agar gel. Electrophoresis was carried out on agar gels according to the method of Ursprung and Leone (1965), proceeding for half an hour for alcohol dehydrogenase and esterase 6 and for 50 minutes for aldehyde oxidase. Staining of ADH was performed according to Ursprung and Leone (1965) except that 2-butanol was used as substrate. Staining of esterase 6 was done according to the method of Johnson (1966); of aldehyde oxidase, according to Courtright (1967). All enzyme variants migrate cathodally on agar gel at the pH's used, but the direction of migrations would be reversed on uncharged medium.

RESULTS

1. *Replacement of a part of the genetic background of the Adh locus*. The first clue that the 4 strains of *D. melanogaster* east of Cardel had been isolated in their evolutionary history from other Mexican strains studied came from an attempt to replace the X and 3rd chromosomes of certain strains with chromosomes of laboratory stock origin. This replacemnnt was accomplished in the case of the two central highland stocks attempted; i.e., Ox and SLP, but was unsuccessful for the 4 strains east of Cardel; i.e., Cz, VH, Pe, and Me. Such a replacement was accomplished with 3 *D. melanogaster* stocks recently collected outside of Mexico (Hewitt et al., 1974). Apparently homozygosity of the 2nd chromosome, carrying the $\underline{Adh}^{II}$ allele, from the Cz, VH, Pe, or Me strains, together with X and 3rd chromosomes derived from laboratory stocks, results in a "synthetic lethal," shown graphically in Fig. 2, similar to those described by Dobzhansky and Spassky (1960) in *D. pseudoobscura*. A similar incompatibility between chromosome 2 of the Cz, VH, Pe, and Me strains and chromosomes 3 and/or X of populations of the Cardel region would explain the near homozygosity of $\underline{Adh}^{II}$ in the four eastern Mexican populations.

2. *Crosses between D. melanogaster strains from Ca, Pa, Cz, and VH*. Reciprocal crosses, with no choice of mates, involving all combinations of flies from Pa, Ca, Cz, and VH, taken two at a time, produced F_1 flies, each type of which was fertile *inter se*, respectively, producing viable F_2 adults. Reciprocal backcrosses for each original cross resulted in viable backcross progeny. These results show that there is no incompatibility between a foreign X chromosome placed in a hybrid genotype such as often causes male sterility in hybrids between closely related species (Haldane, 1922).

3. *Sexual isolation*. The results of mating success with

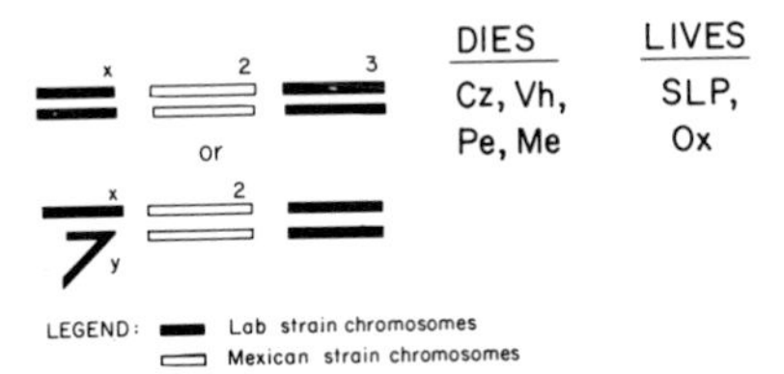

Fig. 2. The effect on viability of replacing the X and 3rd chromosomes of Mexican strains with those of laboratory stocks.

multiple choice between males and females in the Pa-Ca-Cz-VH region, taken two at a time, are presented in Table 1. There

TABLE 1

Test to determine if mating success under multiple choice conditions is at random. The Pa, Ca, Cz, and VH populations were compared in pairwise combinations. Female parent is listed first.

				χ^2	P
Ca x Ca	Pa x Ca	Ca x Pa	Pa x Pa		
25	10	16	18		
3.4818	3.0471	0.0906	0.3261	6.9457	0.1>P>0.05
Ca x Ca	Cz X Ca	Ca x Cz	Cz x Cz		
32	12	52	39		
0.0907	14.0167	9.8685	0.8167	24.7926	P<0.01
Pa x Pa	Cz x Pa	Pa x Cz	Cz x Cz		
19	15	13	13		
1.0667	0	0.2667	0.2667	1.6001	0.7>P>0.5
VH x VH	VH x Cz	Cz x VH	Cz x Cz		
14	15	4	23		
0	0.0714	7.1429	5.7857	13.0000	P<0.01

is a significant preference of the Cz females for the homogametic mating. Cz males, however, show a decided success with Ca females. Thus, on the one hand, there is a barrier to potential gene exchange between these two populations; and, on the other, an incentive, owing to the success of Cz males with Ca females. Cz females are likewise reluctant to mate with VH males. Mating between the more distantly situated Pa and Cz populations takes place at random under our conditions.

4. *Body length*. In conducting experiments on mating success, it was thought that a slight difference in body size could be detected visually between the western strains Pa and Ca on the one hand, and the eastern strains, Cz and VH, on the other. Table 2 shows that small but significant differences in body length occur not only between western and eastern strains; i.e., Pa or Ca compared to Cz, or Pa compared to VH; but even between the two western strains Pa and Ca. No significant difference in body length was found between Ca and VH or between VH and Cz.

TABLE 2
Differences in body length (measured in mm) among eastern (Ca, Pa) and western (Cz, VH) strains of *D. melanogaster*.

Pa - Cz	Pa - VH	Ca - Cz	Ca - VH	Pa - Ca	VH - Cz
0.31* ±0.05	0.23* ±0.05	0.11* ±0.04	0.03 ±0.04	0.20* ±0.05	0.08 ±0.05
0.23* ±0.08	0.18* ±0.08	0.13* ±0.06	0.08 ±0.06	0.10 ±0.07	0.05 ±0.07

* Significant at the 0.05 level.

5. *Courtship song*. Interpulse intervals, measured in milliseconds, of courtship songs from 5 Mexican strains of *D. melanogaster* are presented in Table 3. Fig. 3. shows oscillograms of the songs from males of Cz, Ca, and Me. Measurements of the oscillograms indicate that 4 populations, Tu, Ca, Ox, and Me have similar interpulse intervals with some variation between males in a population. Cz males are more variable, and some individuals show considerable variations in the interpulse intervals within a single song. Means and their variances of the interpulse intervals are presented in Table 3. An F_{max}-test confirmed that the variances of the interpulse interval means were heterogeneous for the Cz males. The critical value, F_{max} .05 (11.15) = 5.77, whereas the maximum variance ratio of the Cz males was 17.17. The other 4 populations have homogeneous variances, $P<0.05$.

6. *Esterase 6 and ADH variants*. In view of the evidence that isolation exists between the eastern and western Mexican strains, it was thought worthwhile to assay the unselected Mexican strains for variants of esterase 6 even though the flies had been in mass culture for a year and a half. Fig. 1 shows frequencies of $Est\ 6^S$ and of Adh^{II}. Assays of ADH were made on freshly collected material (Pipkin et al., 1973). Both Cz and VH were homozygous for the $Est\ 6^S$ allele. Fig. 1

TABLE 3

INTERPULSE INTERVALS (IN MILLISECONDS) OF COURTSHIP SONGS FROM FIVE MEXICAN STRAINS OF *D. MELANOGASTER*

	Tuxpan			Cardel			Oaxaca			Merida			Coatzocoalcos		
Male	n	$\bar{x}$	s^2	n	$\bar{x}$	s^2	n	$\bar{x}$	s^2	n	$\bar{x}$	s^2	n	$\bar{x}$	s^2
1	10	33.75	7.47	26	33.94	11.18	18	32.84	3.17	23	35.22	4.78	22	36.44	17.01
2	23	30.83	3.18	22	33.62	10.57	8	33.22	3.29	19	33.56	5.84	14	28.72	4.28
3	26	34.17	8.00	12	28.97	10.98	8	33.47	3.16	11	35.78	10.22	15	32.85	7.88
4	20	30.76	2.66	14	32.60	8.93	20	33.75	5.57	15	32.85	9.62	18	32.29	4.12
5	14	33.57	5.43	12	34.90	7.13	28	34.81	3.60	16	30.82	8.12	27	32.56	1.35
6	16	34.27	12.23	24	31.65	8.05	19	33.20	4.05	12	32.36	10.62	23	34.78	7.22
7				15	29.55	6.34	13	35.17	1.39	14	33.87	11.34	20	34.36	13.76
8										9	33.29	14.44	12	28.64	17.31
9										8	32.20	4.32	17	34.53	23.18
10													17	30.09	1.55
11													27	33.23	3.13

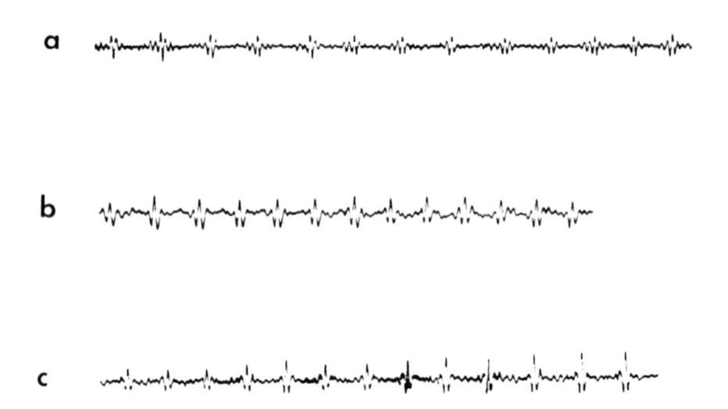

Fig. 3. Oscillograms of the courtship songs of *D. melanogaster* males of the (a) Coatzocoalcos; (b) Cardel; (c) Merida strains. (d) represents a 15 millisecond time marker; each mark equals 15 milliseconds; each space equals 5 milliseconds.

shows a short cline with sharply decreasing frequencies of $\underline{\text{Est } 6^S}$, from 0.70 at Ca to 0.07 at Tu, 135 miles northwest on the gulf coast. However, frequencies of $\underline{\text{Est } 6^S}$ are not distributed clinally in central highland populations (Fig. 1).

No significant differences in frequencies of genotypes carrying Est 6 alleles from those expected on the basis of Hardy-Weinberg equilibrium values are found for 12 of the 14 Mexican and Texan populations (Table 4). An excess of heterozygotes occurs in the Ca and Me populations.

7. Aldehyde oxidase variants. The frequencies of 5 Ald alleles in 10 Mexican strains are given in Fig. 4. $\underline{\text{Ald}^4}$ is a major allele in every population studied except Ox which lacks it. Two alleles, $\underline{\text{Ald}^5}$ and $\underline{\text{Ald}^6}$, of the central highlands, are not found in the coastal populations with the exception of Tu which possesses $\underline{\text{Ald}^5}$. $\underline{\text{Ald}^5}$ is a major allele in every population where it is found. The 5 coastal populations from Pa to Pe are rather similar in their composition and frequency of Ald alleles. $\underline{\text{Ald}^4}$ and $\underline{\text{Ald}^3}$ are the major alleles and $\underline{\text{Ald}^2}$, a minor allele which has a similar frequency in all 5 populations (Fig. 4).

The distribution of Ald genotypes in the Pa, Ca, Cz, and VH populations is far from Hardy-Weinberg equilibrium values (Table 5). At Ca there is a significant excess of two kinds of heterozygotes: $\underline{\text{Ald}^4}/\underline{\text{Ald}^3}$ and $\underline{\text{Ald}^3}/\underline{\text{Ald}^2}$. The latter also

TABLE 4

COMPARISON OF OBSERVED FREQUENCIES OF *ESTERASE 6* GENOTYPES WITH THOSE EXPECTED ON THE BASIS OF HARDY-WEINBERG EQUILIBRIUM VALUES.

Strain	$Est\ 6^S/Est\ 6^S$	$Est\ 6^S/Est\ 6^F$	$Est\ 6^F/Est\ 6^F$	χ^2	P, df = 1
Mi	13	9	0	1.454	$0.3>P>0.2$
CC	0	13	22	1.810	$0.2>P>0.1$
Ji	2	14	11	0.750	$0.5>P>0.3$
Tu	0	4	23	0.175	$0.7>P>0.5$
Pa	3	12	11	0.010	$0.95>P>0.90$
Ca	18	23	1	4.031	$0.05>P>0.02$
Cz	33				
VH	24				
Pe	22	5	1	0.936	$0.5>P>0.3$
Me	0	27	9	12.963	$P<0.01$
SLP	6	13	6	0.040	$0.9>P>0.8$
Ct	4	13	9	0.038	$0.9>P>0.8$
At	11	13	2	0.489	$0.5>P>0.3$
Ox	6	12	7	0.037	$0.9>P>0.8$

TABLE 5

COMPARISON OF ALDEHYDE OXIDASE GENOTYPES WITH THOSE EXPECTED ON THE BASIS OF HARDY-WEINBERG EQUILIBRIUM VALUES. Ald^4, Ald^3, ETC. ARE ABBREVIATED AS 4, 3, ETC.

Genotype	4/4	3/3	2/2	4/3	4/2	3/2	5/5	5/4	5/3	5/2	6/6	6/5	6/4	6/3	χ^2	P
Strain Pa	7	19	0	3	1	7	0	0	0	0	0	0	0	0	19.28	P<0.01 df = 5
Ca	11	0	0	16	2	7	0	0	0	0	0	0	0	0	12.74	0.05>P>0.02 df = 5
Cz	12	6	0	14	0	14	0	0	0	0	0	0	0	0	20.55	P<0.01 df = 5
VH	15	10	3	6	0	5	0	0	0	0	0	0	0	0	24.10	P<0.01 df = 5
SLP	4	10	0	5	0	0	5	0	6	0	0	0	0	0	13.64	0.05>P>0.02 df = 5
Ct	13	1	0	8	0	0	7	5	1	0	2	2	0	0	29.41	P<0.01 df = 7
Tu	0	0	0	2	3	0	5	11	0	0	0	0	0	0	12.04	0.3>P>0.2 df = 9
Ox	0	0	2	0	0	0	25	0	0	22	0	0	0	0	1.13	0.3>P>0.2 df = 1

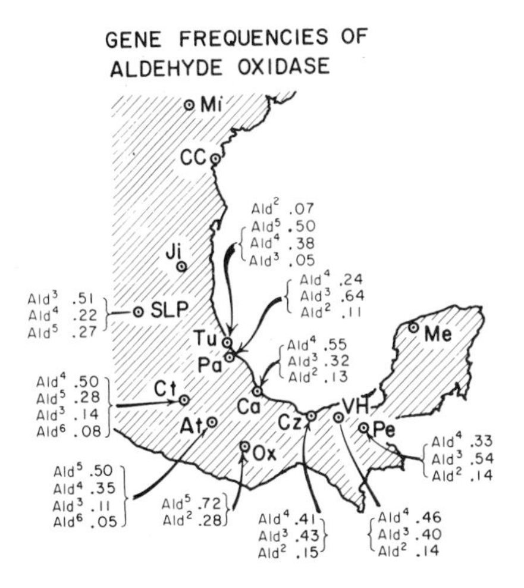

Fig. 4. The distribution of frequencies of aldehyde oxidase alleles in Mexican strains of *D. melanogaster*.

occurs with greater than expected frequency at neighboring Cz. An excess of $\underline{Ald^4}/\underline{Ald^4}$ homozygotes also occurs at Cz. At VH, east of Cz, there is an excess of each of 3 kinds of homozygotes and a deficiency of each of 2 kinds of heterozygotes.

DISCUSSION

Evidence has been presented regarding the most probable explanation for the abrupt change in frequency of $\underline{Adh^{II}}$ from 0.51 at Cardel to 0.95 at Coatzocoalcos located 135 miles southeast in the Isthmus of Tehauntepec. We believe that part of the Cardel 2nd chromosome, carrying the Adh locus is being eliminated from the Coatzocoalcos and Villahermosa populations, which have introgressed to varying degrees with the western strains. Eastern and western strains are considered semispecies of *D. melanogaster*. This conclusion rests upon a variety of evidences. First we have demonstrated an incompatibility between chromosome 2 of Coatzocoalcos and the 3rd and perhaps X chromosomes of laboratory stocks, resulting in synthetic lethality when chromosome 2 of any eastern strain is homozygous in a replaced genotype. Second, the highly variable interpulse interval of the courtship song of the Coatzocoalcos males implies a breakdown in regulation of this polygenic character not seen in the Oaxaca and Tuxpan populations in the west or in the Merida population in the east. Since the interpulse interval is important in stimulating the female to mate (Bennet-Clark and Ewing, 1969), the greater variance of this fitness associated trait in the Coatzocoalcos population suggests gene flow from the western semispecies. Bennet-Clark and Ewing (1969) have shown the interpulse interval of the *D. melanogaster* courtship song to be constant and highly species specific. Potter (unpublished) has found increased variance in the interpulse interval of songs of F_1 males derived from laboratory crosses of widely distant geographic populations of *D. melanogaster*. Third, the demonstration of small but significant differences in body length between the eastern Mexican strains and Paplanta in the west indicate differences in a second polygenic character. Further, the Cardel population, which is located geographically between Paplanta and the eastern strains, possesses an intermediate body length, suggesting gene exchange with its neighbors. Finally, the observation of an excess of homozygotes for 3 aldehyde oxidase alleles and a deficiency of 2 kinds of heterozygotes in the Villahermosa population is both evidence of introgression of parts of chromosome 3, containing the Ald locus from the western into the genome of the eastern strains, and also of the action of natural selection in holding together the coadapted gene complexes of the eastern strain. On the other hand, the short cline of $Est\ 6^S$ from Cardel to Tuxpan suggest introgression into the western strains of part of the third chromosome carrying the $Est\ 6^S$ allele from the eastern strains homozygous for this allele. Our laboratory crosses showed no X-autosome incompatibility in male hybrids between strains of eastern and western semispecies.

The degree of isolation now observed between the Mexican semispecies of *D. melanogaster* suggests a previous more drastic isolation during their evolutionary history. During the upper Pliocene, some 4 to 8 million years ago, Coatzocoalcos and Villahermosa were located on an island off the coast of Mexico according to Fig. 5 (adapted from Schuchert, 1955). Cardel and Paplanta were also on islands nearer the mainland. If the isolated eastern Mexican populations evolved as insular species, it would appear that *D. melanogaster* may have colonized America earlier than Throckmorton (in press) supposed, since the tropical corridor from Asia was closed some 30 million years ago, well after Sophophoran radiation in Asia had occurred (Throckmorton, in press).

Our explanation of the abrupt change in relative frequency of Adh alleles between Cardel and Coatzocoalcos does not negate the observation made by Pipkin et al., (1973) that Adh allelic frequencies are related to minimum temperature. It was shown that a significant regression of the frequency of Adh^{II} on extreme minimum temperature is found when 12 Mexican and 2 Texan populations are considered. Since the 4 eastern

UPPER CENOZOIC PLIOCENE

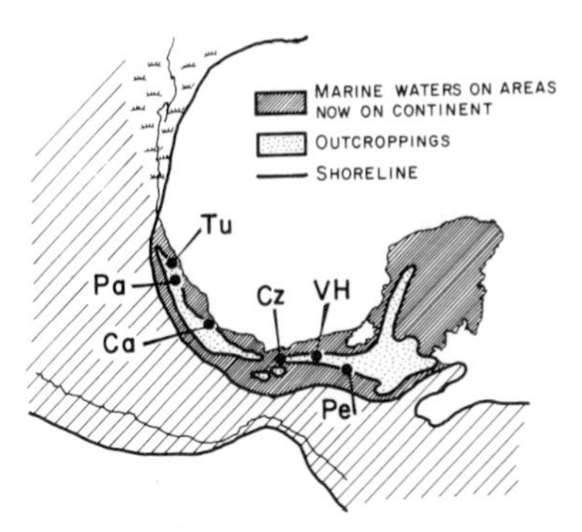

Fig. 5. The gulf coastal area of Mexico during the upper Pliocene (adapted from Schuchert, 1955).

populations show signs of isolation in their evolutionary history, it seems appropriate to exclude them from the calculation of the regression coefficient. When this is done, the regression of $\underline{Adh^{II}}$ frequency on extreme minimum temperature is not significant ($b = 21.7755 \pm 13.06$; $t_s = 1.66734$; df = 9; $0.2>P>0.1$). However, it is important to note that the Adh^{I} allele, specifying a protein from 2 to 4 times as active as that specified by $\underline{Adh^{II}}$ (Hewitt et al., 1974) is found in populations of all areas where extreme minimum temperature falls below 10°C. On the other hand, $\underline{Adh^{II}}$ is almost homozygous in the 3 areas of Mexico where the temperature never falls below 10°C. Similarly, Vigue and Johnson (1973) found the frequency of $\underline{Adh^{II}}$ high in Florida, and decreasing northward. Even within the eastern Mexican semispecies the frequency of $\underline{Adh^{II}}$ fell to 0.79 during June, 1972, at Palenque in the Chiapas Mountains where extreme minimum temperature (mean of the minimum monthly temperature of the coldest month over a 10-year period) is between 5 and 10°C. At 10°C and below, drosophilids starve because they do not forage for food (Pipkin, 1952). Johnson (1974) has pointed out that horse liver ADH in using glyceraldehyde as an internal substrate, contributes to the control of lipid and fatty acid synthesis. That *Drosophila* ADH is involved in lipid and fatty acid metabolism is suggested by the high level of ADH in the adult fat body (Ursprung et al., 1970). Even more important, both horse liver ADH (Waller, 1965) and *Drosophila* ADH (Ursprung and Madhaven, 1971) can use farnesol and farnesal as substrates. These compounds are involved in the metabolism of juvenile hormone. Thus a selective advantage is implied for *Drosophila* carrying an $\underline{Adh^{I}}$ allele. In cool climates, especially where the flies must undergo diapause as larvae or pupae, the enzymatically stronger protein specified by $\underline{Adh}^{I}$ would benefit the relatively few individuals which participate in the revival of adult populations following a prolonged period of cold weather.

ACKNOWLEDGEMENTS

This work was supported by NSF Grant GY-11064, NIH Training Grant GM 02196, NIH Grant 08016 to Howard University, and by the Zoology Department, The University of Maryland.

REFERENCES

Bennet-Clark, H. C. 1972. Microphone and pre-amplifier for recording the courtship song of *Drosophila*. *D. I. S.* 49: 127-128.

Bennet-Clark, H. C. and A. W. Ewing 1968. The wing mechanism involved in the courtship of *Drosophila*. *J. Exp. Biol.* 79: 117-128.

Bennet-Clark, H. C. and A. W. Ewing 1969. Pulse interval as a critical parameter in the courtship song of *Drosophila melanogaster*. *Anim. Behav.* 17: 755-759.

Courtright, J. B. 1967. Polygenic control of aldehyde oxidase in in *Drosophila*. *Genetics*. 57: 25-39.

Dobzhansky, Th. and B. Spassky 1960. Release of genetic variability through recombination V. Breakup of synthetic lethals by crossing over in *D. pseudoobscura*. *Zool. Jahrb.* 88: 57-66.

Ehrman, L. and M. Strickberger 1960. A pictorial study of mating behavior in *Drosophila paulistorum*. *Nat. Hist.* Nov.: 28-33.

Haldane, J. B. S. 1922. Sex ratio and unisexual sterility in animals. *J. of Genetics*. 12: 101-109.

Hewitt, N., S. B. Pipkin, N. Williams, and P. K. Chakrabartty 1974. Variation in ADH activity in strains of *Drosophila* and their hybrids. *J. Hered.* 65: 141-148.

Johnson, F. M. 1966. An isozyme analysis of *Drosophila* speciation and development. Ph.D. Thesis. The University of Texas, Austin, Texas.

Johnson, G. B. 1974. Enzyme polymorphism and metabolism. *Sci.* 184: 28-37.

Pipkin, S. B. 1952. Seasonal fluctuations in *Drosophila* populations at different altitudes in the Lebanon Mountains. *Zeits. f. Indukt. Abstamm. u. Vererb.* 84: 270-305.

Pipkin, S.B., C. Rhodes, and N. Williams 1973. Influence of temperature on *Drosophila* alcohol dehydrogenase polymorphism. *J. Hered.* 64: 181-185.

Schuchert, C. 1955. Atlas of *Paleontological Maps of North America*. John Wiley and Sons, New York. 177 pp.

Throckmorton, L. H. The phylogeny, ecology, and geography of *Drosophila*. In *Handbook of Genetics*. Ed. by R. C. King. New York: Van Nostrand and Reinhold Co. In publication.

Ursprung, H. and J. Leone 1965. Alcohol dehydrogenase: a polymorphism in *D. melanogaster*. *J. Exp. Zool.* 160: 147-154.

Ursprung, H., W. H. Sofer, and N. Burroughs 1970. Ontogeny and tissue distribution of alcohol dehydrogenase in *D. melanogaster*. *Wilhelm Roux Archiv. für Entwicklungsmechanik*. 164: 201-208.

Ursprung, H. and K. Madhaven 1971. Alcohol dehydrogenase and aldehyde oxidase of *Drosophila melanogaster:* farnesol and farnesal serve as substrates. *2nd European Drosophila Conference*, Zürich. Abstract.

Vigue, C. L. and F. M. Johnson 1973. Isozyme variability in species of the genus *Drosophila* VI. Frequency-property-environment relationships of allelic alcohol dehydrogenases in *D. melanogaster*. *Biochem. Gen.* 9: 213-227.

Waller, G., H. Theorell, and J. Sjövall 1965. Liver alcohol dehydrogenase as a 3-β hydroxy-5 β cholanic acid dehydrogenase. *Archiv. Biochem. Biophys.* 111: 671-684.

GENETIC VARIATION IN HAWAIIAN DROSOPHILA II. ALLOZYMIC DIFFERENTIATION IN THE *D. PLANITIBIA* SUBGROUP

W. E. Johnson[1,2], H. L. Carson[1], K. Y. Kaneshiro [1,3],
W. W. M. Steiner [1], and M. M. Cooper[2]
Department of Genetics[1] and Entomology[3]
University of Hawaii
Honolulu, Hawaii
96822

ABSTRACT: The allozymic (allelic isozymic) variation encoded by fifteen loci was examined in sixteen species comprising the *D. planitibia* subgroup of the picture-winged Hawaiian *Drosophila*. The level of heterozygosity within populations is highly variable but in general appears to be slightly lower than that found in continental species of *Drosophila*. Determinations of genetic similarities among species reveals a range of values from 0.11 to 0.99. A phenetic cluster of the matrix of similarity values reveals several distinct clusters. Within these subclusters most dissimilar species are those most likely to have been founded by small numbers of individuals. We interpret this to mean that the kind of founding event will directly affect the rate of allozymic differentiation. Thus, values of genetic similarity between species have limited phylogenetic significance unless supplemental evidence can define the kinds of founding and speciation events that have occurred during the evolution of a group of species.

Present Address: Department of Biology[2]
Western Michigan University
Kalamazoo, Michigan
49001

The application of electrophoretic techniques to the study of natural populations has revealed extensive variation at loci encoding soluble polypeptides (Ayala et al., 1972; Richmond, 1972; Selander and Johnson, 1973; Selander and Kaufman, 1973). The heated debate between the Neo-Darwinian and the Non-Darwinian schools on the adaptive and evolutionary significance of the detectable allozymic (allelic isozyme) variation within and among populations has resulted in a proliferation of studies on variation among conspecific populations (Ayala, 1972; Stebbins and Lewontin, 1972; Bryant, 1974; Kimura, 1968; King and Jukes, 1969; Yamazaki and Maruyama, 1972). Applications of the same techniques for comparing genetic reorganization and differentiation among populations of closely related species for systematic analysis has received considerably less but growing concern (Hubby and Throckmorton, 1968; Kanapi and Wheeler, 1970; Nair et al., 1971; Johnson and Selander, 1971; Lakovaara et al., 1972; Zouros, 1973; Ayala et al., 1974).

The Hawaiian Islands, with a total land area of only 6,435 square miles, are inhabited by an extreme diversity of *Drosophilidae*. The 650 to 700 endemic species representing the present day fauna are believed to be derived from a single successful founder, or possibly from two separate introductions (Throckmorton, 1966). Considering the geological recency of the major Hawaiian Islands, five to six million years old at most (Macdonald and Abbott, 1970), this explosive proliferation of species is truly remarkable. The evolutionary youth of many of these species renders the Hawaiian *Drosophila* particularly suitable for studies of the chromosomal and allozymic differentiation associated with the speciation process.

The purposes of this study were (1) to determine the extent of allozymic variation within species of one subgroup of Hawaiian "picture-winged" *Drosophila;* (2) to determine the patterns and levels of genetic differentiation among these species; (3) to compare the levels and patterns of allozymic versus chromosomal differentiation apparent among species of this group; and (4) to determine the systematic implications of the detectable allozymic variation.

INFORMATION ABOUT THE SPECIES USED IN THIS STUDY

Despite the great morphological and behavioral diversity, the Hawaiian *Drosophiladae* constitute a closely knit evolutionary group with many common characteristics (Throckmorton, 1966).

In general, they are single island endemics, with species numbers that are low relative to widespread continental forms of *Drosophila*. The factors which are believed to have contributed to the proliferation of such large numbers of species and to the small species' numbers have been reviewed by Carson et al.,(1970). The "picture-winged" *Drosophila* include about 110 species and include most of the flies with conspicuously marked wings. The polytene chromosomal relationships determined by Carson and his collaborators (Carson et al., 1970 and references therein) suggest a phylogeny which is in general agreement with the results of investigations of morphology (Throckmorton, 1966; Kaneshiro, 1969), hybridization (Yang and Wheeler, 1969), ovarian transplants (Kambysellis, 1970) and ecology (Heed, 1968; Montgomery, 1972).

The *D. planitibia* subgroup comprises sixteen of the most spectacular and largest species of *Drosophila* ever described. The polytene chromosome relationships and geographic distributions among species of this subgroup (Clayton et al., 1972) are shown in Figure 1. When the Maui complex is considered as a single island (see Carson et al., 1970, for the rationale), each of these species is a single island endemic with only one species found on Kauai, four on Oahu, eight on the Maui complex and three on Hawaii (Clayton et al., 1972). Most of the species occupying the Maui complex appear, in fact, to be restricted to single volcanoes. The only apparent exceptions to this rule are *D. neopicta* and *D. planitibia*, each of which occurs on all three major volcanoes, East Molokai, East Maui and West Maui. However, since the hybridization data now indicates that the "*planitibia*-like" form on Molokai is reproductively isolated from the Maui *D. planitibia* (Craddock, 1974 and unpublished data), it will be treated as a separate taxon.

Fourteen of the sixteen species of the *D. planitibia* subgroup are characterized by the presence of an extra wing-vein in cell R-5. Only *D. picticornis* of Kauai and *D. setosifrons* of Hawaii do not have this distinctive extra wing-vein. The presence of this character only in flies endemic to the geologically recent islands of Maui and Hawaii suggests these flies are in a terminal position in the general phylogeny of the picture-winged group and that they are probably recently evolved (Carson et al., 1970). Carson and Stalker (1968) and Carson (1973) discussed several additional lines of evidence which support these phyletic relationships and concluded that *D. picticornis* of Kauai and *D. obscuripes* of Maui are closest to the presumed ancestral stock on Kauai.

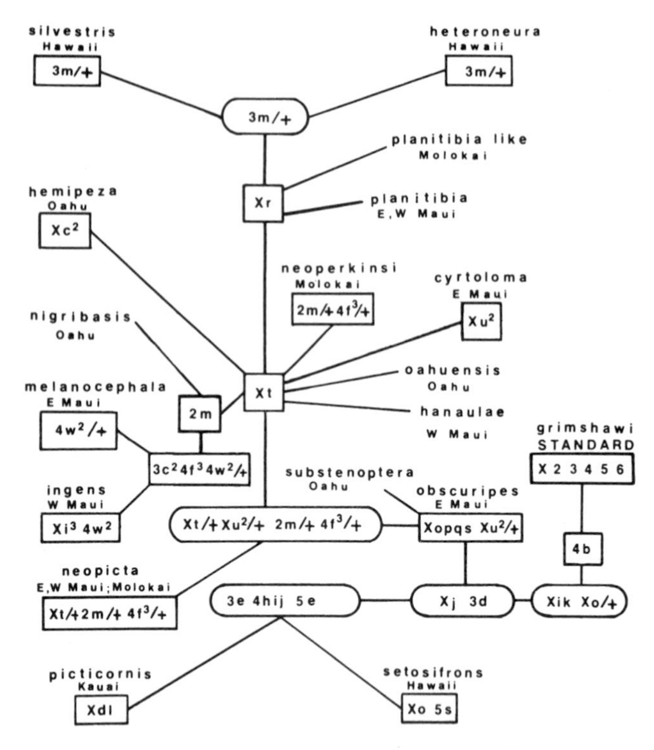

FIG. 1 Chromosome relationships among the 17 species comprising the *D. planitibia* subgroup of the Hawaiian *Drosophila*. The geographic distribution of each species is indicated. All species differ from the *D. grimshawi* Standard (X 2 3 4 5, right, center) by the 4b inversion and two inversions on the X chromosome (i and k). Read the inversion formula for each species of the subgroup additively by following the lines leading from the Xik Xo/+ hypothetical ancestor (lower right-hand side). Letters appearing singly represent fixed inversions; polymorphic inversions are shown e.g. "3m/+".

Presumably the primitive *D. obscuripes* -like stock on the Maui complex proliferated into a number of species, some of which in turn migrated and founded the species on Oahu and Hawaii. With this directionality *D. silvestris* and *D. heteroneura* on the youngest island, Hawaii, would be two of the more recently derived species.

MATERIALS AND METHODS

Specimens representing 15 species were collected between September, 1971. and August, 1973. Nine of these species were represented by at least 30 wild caught individuals. Information for the remaining species is derived from small samples (≤ 30 individuals). Since in most species each population sample is small, we have pooled the samples to obtain the estimate of gene frequency, heterozygosity, and genetic similarity. The time and place of each collection for each species is too extensive to be listed here but this information is available from the authors. The newly captured males and teneral females were used for electrophoresis as soon as possible. Gravid females were isolated and allowed to produce an F_1 for use in salivary gland chromosome preparations, and then used for electrophoresis.

The wild-caught flies were individually homogenized in 0.05 ml of 0.074 M Tris-0.008 M citrate buffer and the supernatant of each fly absorbed by two wicks of Whatman No. 3 filter paper. These wicks were then separated and applied to two horizontal starch gels combining different buffer systems. Following electrophoresis the gels were each sliced horizontally four times, and these slices stained separately for one or more of eleven enzymes. The different enzymes and their locus designations are presented beyond. The two buffer combinations used in the study are as follows: Buffer System A-Gel: pH 8.9, 0.0076 M Tris-0.005 M. citrate; Electrodes: pH 8, 7, 0.269 M borate - 0.1 M sodium hydroxide, and Buffer System C-Gel: pH 8.7 , 0.074 M Tris-0.008 M citrate; Electrodes: pH 8.1, cathode=0.343 M Tris-0.079 M citrate, anode= 0.458 M Tris, 0.0104 M citrate. The staining methods used are similar to those described by Ayala et al., (1972) and Selander et al., (1971). Detailed information regarding the buffer of choice for each enzyme and the modified staining methods are available elsewhere (Steiner and Johnson, 1973). Alleles at a particular locus were numbered according to the relative mobilities of their allozymes on the specified electrophoretic buffer systems, without regard to species. The allele of *D. silvestris* producing the most frequent allozyme was designated 100 (i.e. $\underline{Idh\text{-}1^{100}}$) and alleles corresponding to faster or more slowly migrating bands were arbitrarily assigned values indicative of their respective mobilities. Determinations of identity or non-identity for alleles of different species were made on the basis of side by side comparisons or near juxtaposition on the same gel. These kinds of comparisons often were not available for alleles found at very low frequency, thus some minor alleles were necessarily omitted from the calculations of genetic similarity among species.

For the purpose of assessing genetic relationships, all population samples from each species were pooled and each of these pooled species' samples was considered as an operational taxonomic unit (OTU). Coefficients of genetic similarity based on allozymic frequency data were obtained for all paired combinations of OTU's by the method of Rogers (1972). From the matrices of coefficients for paired combinations of OTU's, cluster analysis was performed by the unweighted pair-group methods, using arithmetic means (UPGMA) (Sneath Sokal, 1973).

RESULTS

ALLOZYMIC VARIATION

The allozymic variation encoded for by fifteen loci were examined in sixteen species. At least eleven species were sampled sufficiently to allow a reasonable approximation of the frequency of alleles and levels of heterozygosity. Four species, *D. oahuensis*, *D. nigribasis*, *D. obscuripes* and *D. substenoptera*, are represented by too few specimens to permit reliable determinations of gene frequency or heterozygosity. However, the common allele at each locus in each species can be inferred by even these small samples. The information on *D. hemipeza* is derived from a single isofemale line maintained in the laboratory. The determination of identity between allozymes of different species is based solely on their mobility on starch gels with the specified buffer systems. Thus, we acknowledge that some of those allozymes which we now identify as being the same may later prove otherwise. This problem must be faced in any comparative study and will probably always remain a major source of error, particularly in view of the proposal of Bernstein et al., (1973) concerning the prevalence of heat-stability differences among allozymes sharing the same electrophoretic mobilities. Among closely related species this source of error should be at a minimum; nevertheless, it is still a factor.

The allele frequencies at seven enzyme loci in samples from natural populations of eleven species of Hawaiian *Drosophila* are given in Table 1. In the following account we will supplement Table 1 by including the major allelic variation seen in the less adequately sampled species.

At Pgm-1 there are four major alleles among species. The common allele in *D. substenoptera* is Pgm-1^{90}. Allele Pgm-1^{95} is the major allele in seven species including *D. obscuripes*, which has a second allele, Pgm-1^{105}, at a frequency of 0.25. Four of the remaining species, *D. silvestris*, *D. heteroneura*, *D. oahuensis*, and *D. nigribasis* have Pgm-1^{100} as their major allele. *D. nigribasis* has a second allele at a frequency of 0.06. In four species, including *D. hemipeza*, Pgm-1^{105} is the most common allele.

Five alleles at Me-1 are commonly found among these species. Allele Me-1^{100} is the major allele in ten species including *D. hemipeza*. The alleles of Me-1 found in the remaining six species are: Me-1^{90} in *D. obscuripes*, Me-1^{98} in *D. substenoptera*, Me-1^{103} in *D. nigribasis*, *D. neoperkinsi* and *D. picticornis*, and Me-1^{108} in *D. oahuensis*.

Two GOT systems are apparent on these buffer systems. We have not determined which is the cytoplasmic form. Three alleles are found commonly among these species; Got-1^{100} is the major allele in five species including *D. hemipeza*, Got-1^{95} is the most frequent allele in eight species including *D. oahuensis*, *D. nigribasis*, and *D. obscuripes*, and Got-1^{91} is the common allele in *D. neopicta* and *D. substenoptera*. *D. picticornis* was not scored for Got-1. At the Got-2 locus the same allele, Got-2^{100}, predominates in all but two of the species sampled. *D. nigribasis* has a unique allele, Got-2^{103} and *D. picticornis* has Got-2^{105} as the most frequent allele. Allele Got-2^{105} also appears to be a low frequency variant in *D. silvestris* (0.023) and *D. heteroneura* (0.022). Another low frequency (0.005) variant is found in *D. cyrtoloma*.

Multiple isozymes of hexokinase are distinguished by electrophoresis. We could consistently score for all species only the most anodal of these forms, HK-1, however, HK-2, HK-3 and HK-4 could also be reliably scored for a few species. These additional hexokinase isozymes were not used for calculations of similarity but will be summarized in the discussion on heterozygosity. Almost no intraspecific variation is found at the HK-1 locus although there are some interspecific differences. Allele HK-1^{90} is the only allele found in samples of *D. hanaulae* and *D. obscuripes*. Allele HK-1^{100} is the predominant allele in *D. silvestris* and *D. heteroneura*. The only allele found in any of the remaining species is HK-1^{95}.

Table 1

Genetic variation at seven loci in natural populations of eleven species of Hawaiian *Drosophila*. The following species are included: sil, *D. silvestris*; het, *D. heteroneura*; plb, *D. planitibia*; pll, *D. planitibia*-like; han, *D. hanaulae*; npk, *D. neoperkinsi*; cyr, *D. cyrtoloma*; mel, *D. melanocephala*; ign, *D. ingens*; npt, *D. neopicta*; pic, *D. picticornis*.

Gene	Genomes Sampled and Alleles	Species										
		sil	het	plb	pll	han	npk	cyr	mel	ign	npt	pic
Pgm-1	Genomes	1246	102	80	10	74	26	216	42	68	152	122
	95					.91	1.00	.90	.90	.88	1.00	.06
	100	.83	.79	.04	.40	.08	-	.03	.10	-	-	.87
	105	.10	.22	.92	.60	-	-	-	-	-	-	.05
	108	-	-	.05	-	-	-	-	-	-	-	.02
	Other	.07(4)	-	-	-	.01	-	.08	-	.12	-	-
Me-1	Genomes	1268	92	340	24	74	26	340	42	68	156	126
	90	-	-	.07	-	-	-	-	-	-	-	-
	100	.79	.98	.93	.96	.96	-	1.00	.95	.97	.97	-
	103	-	-	-	-	-	1.00	-	-	-	-	.95
	105	.20	-	-	.04	-	-	-	-	-	-	-
	Other	-	.02	-	-	.04	-	-	.05	.03	.03	.05
Got-1	Genomes	780	90	350	24	26	26	272	40	68	18	-
	91	-	-	-	-	-	.08	-	.18	-	.95	
	95	.02	.02	-	.08	1.00	.92	.98	.70	1.00	.05	-
	98	-	-	-	-	-	-	.01	.12	-	-	-
	100	.66	.64	.97	.92	-	-	-	-	-	-	-
	105	.32	.33	.02	-	-	-	-	-	-	-	-
	Other	-	-	.01	-	-	-	.01	-	-	-	-

TABLE 1 (Continued)

Idh-1	Genomes	1280	100	348	24	74	26	332	42	68	141	76
	94	-	-	-	-	-	-	-	-	-	.99	-
	96					1.00	1.00	.997	.98	1.00	-	-
	100	.99	.98	.998	.96	-	-	-	-	-	-	-
	103	-	-	-	-	-	-	-	-	-	-	.72
	106	-	-	-	-	-	-	-	-	-	-	.28
	Other	.01	.02	.002	.04	-	-	.003	.02	-	.01	-
Adh-1	Genomes	1238	98	352	24	74	26	284	42	68	156	126
	95	.23	.24	.50	-	-	-	-	-	-	-	-
	97	-	-	-	-	-	-	-	-	-	-	.99
	100	.77	.74	.50	-	-	-	-	-	-	-	-
	102	-	-	-	-	-	-	-	-	-	.99	-
	105	.005	-	-	1.00	1.00	1.00	1.00	1.00	1.00	-	-
	Other	-	.02	-	-	-	-	-	-	-	.01	.01
Mdh-1	Genomes	1290	90	348	24	74	26	332	42	68	156	126
	91	-	-	-	-	1.00	1.00	.997	1.00	.91	-	-
	93	-	-	-	-	-	-	.003	-	-	.99	.95
	95	.003	.05	-	-	-	-	-	-	-	-	-
	100	.996	.95	.996	1.00	-	-	-	-	-	-	-
	105	.001	-	.004	-	-	-	-	-	-	-	-
	Other	-	-	-	-	-	-	-	-	.09	.01	.05
Mdh-2	Genomes	1260	90	68	10	74	8	136	32	12	152	108
	93	-	-	-	-	1.00	1.00	1.00	1.00	1.00	-	-
	100	.995	1.00	1.00	1.00	-	-	-	-	-	1.00	.95
	105	.005	-	-	-	-	-	-	-	-	-	-
	110	-	-	-	-	-	-	-	-	-	-	.04
	Other	-	-	-	-	-	-	-	-	-	-	.01

Most of the better sampled species display low intraspecific variability at the isocitrate dehydrogenase locus (Idh-1) but only *D. picticornis* appears to be highly polymorphic. Six different alleles are commonly found. Allele Idh-1^{96} predominates in six species including *D. nigribasis*, while Idh-1^{100} is the major allele in five species including *D. hemipeza*. Allele Idh-1^{94} is shared by *D. neopicta* and *D. substenoptera* while unique alleles are found in *D. obscuripes* (Idh-1^{90}) and *D. oahuensis* (Idh-1^{98}).

Although the alcohol dehydrogenase locus (Adh-1) shows considerable intrapopulation variation in both *D. silvestris* and *D. heteroneura*, Adh-1^{100} predominates in both species. Additionally, at the Adh-1 locus in *D. silvestris* there is significant interpopulation variation (Craddock and Johnson, in preparation). The populations of *D. planitibia* from East and West Maui show "fixed" differences between them (Adh-1^{95} and Adh-1^{100} respectively) but little or no intrapopulation variation. Although considerable interspecific differentiation has occured, Adh-1 is essentially invariant within all of the remaining species. Allele Adh-1^{105} is the predominant form in seven species including *D. oahuensis*. For the remaining species the major alleles are: Adh-1^{103} in *D. nigribasis* and *D. substenoptera*, Adh-1^{104} in *D. obscuripes*, and Adh-1^{93} in *D. hemipeza*.

Of the two MDH systems present, we have not determined which is cytoplasmic and which is mitochondrial. A pattern of low intrapopulation variation combined with considerable interspecific differentiation is found at the Mdh-1 locus. Allele Mdh-1^{100} (=MDH-D^{3}-Rockwood et al., 1971) is the most frequent allele in four species, Mdh-1^{93} (=MDH-D^{4}) predominates in five species including *D. oahuensis*, *D. nigribasis* and *D. hemipeza*, while Mdh-1^{91} (=MDH-D^{5}) is the major allele in another five species. Mdh-1^{92} and Mdh-1^{94} are the common alleles in *D. substenoptera* and *D. obscuripes* respectively. Variation at Mdh-2 is found in only two species, *D. silvestris* and *D. picticornis*. Two alleles predominate among species; Mdh-2^{100} is found in ten species including *D. nigribasis*, *D. hemipeza*, *D. obscuripes* and *D. substenoptera*, and Mdh-2^{93} in six species including *D. oahuensis*.

At the α glycerophosphate dehydrogenase locus, α Gpd-1^{100} predominates in all species. This allele is presumed to be the same as α GPDH^{2} described by Rockwood et al., (1971) and is found commonly among even diverse and distantly related Hawaiian *Drosophila*. Low frequency variants are found in five species, αGpd-1^{90} at .014 in *D. hanaulae*,

Gpd-1^{95} at .001 in *D. silvestris*, α Gpd-1^{98} at .006 in D. neopicta, α Gpd-1^{105} at .008 in *D. picticornis* and an unnamed variant at .015 in *D. ingens*.

Two additional loci (Lap-1 and Odh-1) were surveyed to obtain estimates of the levels of variability but they were not used for determinations of similarities coefficients. These loci were omitted because the variable quality of resolution made interspecific comparisons unreliable.

ANALYSIS AND DISCUSSION

LEVEL OF VARIABILITY

Table 2 summarizes the allozymic variation found at twelve loci in each of ten species of Hawaiian *Drosophila*. The frequencies of heterozygous individuals at each locus (h) were obtained by direct count or by calculation on the assumption of a Hardy-Weinberg equilibrium. Within species the distribution of variation is similar to that found in most other forms studied (Ayala, 1972; Selander and Johnson, 1973). We have found variation at every locus where the number of genomes sampled was sufficiently large; however, the amount of polymorphism varies from locus to locus and from species to species. For instance, in *D. silvestris* the percent of heterozygous individuals at the various loci ranges from essentially zero at α Gpd-1 to about 45% at Got-1. The amplitude of the range of variation does not equal that found in many species (summarized in Johnson 1974), however, the class of enzymes (esterases) for which there are generally such high h values have not been included in this study. A study of the level of variation at loci among intraspecific populations of *D. silvestris*, reported elsewhere (Craddock and Johnson, in preparation), indicates that in Hawaiian *Drosophila* the familiar pattern of relatively small changes in h among populations is again repeated. Some loci (Pgm-1, Me-1, Got-1) are significantly polymorphic within many species, while others may be highly polymorphic within only one or two species and essentially without variation in the majority. For instance, more than forty percent of the individuals of *D. heteroneura* are heterozygous at the Adh-1 locus whereas the majority of the species are essentially invariant at the same locus. Similarly, Idh-1 is marginally polymorphic in most species but highly polymorphic (h=.29) in *D. picticornis*.

TABLE 2

Proportion of individuals heterozygous (h) at each of 12 loci in ten species of Hawaiian *Drosophila*. The following species are included: sil, *D. silvestris*; het, *D. heteroneura*; plb, *D. planitibia*; pll, *D. planitibia*-like; han, *D. hanaulae*; cyr, *D. cyrtoloma*; mel, *D. melanocephala*; ign, *D. ingens*; npt, *D. neopicta*; pic, *D. picticornis*.

Locus	Species and (h)										
	sil	het	plb	pll*	han	cyr	mel	ign	npt	pic	$\overline{HET}$
(Adh-1) [1]	.050	.428	NV	NV	NV	NV	NV	NV*	.013	.016	.063
(Got-1) [2]	.447	.377	.091	.167	NV*	.036	.500	NV	.011*	-	.241
(Got-2) [3]	.059	.044	NV	NV	NV*	.011	NV	NV	NV*	-	.018
(αGpd-1) [4]	.000	NV	NV	NV	.027	NV	NV	.029	.013	.016	.010
(Hk-1) [5]	.003	NV	NV	NV	NV	NV	NV	NV	NV	NV	.000
(Idh-1) [6]	.012	.048	.004	.083	NV	.006	.048	NV	.014	.289	.047
(Lap-1) [7]	.027	-	NV	NV	.042	.194	NV	-	NV	.189	.057
(Mdh-1) [8]	.009	.108	.078	NV	NV	.006	NV	.176	.013	.095	.054
(Mdh-2) [9]	.010	NV	NV	NV	NV	NV	NV*	NV	NV	.093	.012
(Me-1) [10]	.318	.043	.113	.081	.081	NV	.095	.059	.064	.063	.093
(Pgm-1) [11]	.312	.313	.250	.400	.135	.174	.190	.235	NV	.246	.206
(Odh-1) [12]	.041	.048	NV	NV	-	.037	NV	.200	.029	.029	.048
($\bar{H}$) [13]	.117	.128	.045	.061	.032	.046	.076	.070	.015	.104	

* Sample size was less than twenty individuals. These loci were not used to calculate $\bar{H}$ or $\overline{HET}$. HET = unweighted mean across species.

1. Alcohol dehydrogenase 2. Glutamate oxaloacetate transaminase 3. Glutamate oxalocetate transaminase 4. αGlycerophosphate dehydrogenase 5. Hexokinase 6.Isocitrate dehydrogenase 7. Leucine amino peptidase 8. Malate dehydrogenase 9. Malate dehydrogenase 10. Malic enzyme 11. Phosphoglucomutase 12. Octanol dehydrogenase 13. Mean Individual Heterozygosity.

Because of the great heterogeneity of h values among loci, heterozygosity over all loci ($\bar{H}$ values) based on small numbers of such loci must be interpreted cautiously. The $\bar{H}$ values in the ten species listed in Table 2 range from 0.015 in *D. neopicta* to 0.128 in *D. heteroneura*. These values are considerably lower than those found for continental species (Ayala, 1972; Richmond, 1972); however, these estimates are not comparable since those for the "picture-winged" flies do not include contributions from the loci encoding for the esterases, acid phosphatase, and some of the other more variable enzymes.

No obvious relationship exists between the levels of allozymic chromosomal polymorphism. For instance both *D. picticornis* and *D. silvestris* are highly polymorphic allozymically (0.104 and 0.117 respectively) and yet chromosomally *D. silvestris* is highly polymorphic while *D. picticornis* is not. However, this similarity of genic heterozygosity may be fortuitous and simply reflect the choice of enzymes studies. One method of minimizing the sampling error due to the small number of loci involved is to obtain an average of h values at each locus across species. Table 3 lists and contrasts the mean heterozygosity ($\overline{HET}$) across species at each of 16+ loci of Hawaiian *Drosophila* with those of thirteen non-Hawaiian *Drosophila* species (summarized in Johnson, 1974). The loci have been grouped by their regulatory functions according to the rationale of Johnson (1974). The variable substrate loci in Hawaiian *Drosophila* have a mean $\overline{HET}$ of .18 which is lower than the comparable value of .27 for the non-Hawaiian flies but considerably higher than the mean $\overline{HET}$ for the regulatory and non-regulatory enzymes. The $\overline{HET}$ values for regulatory enzymes of Hawaiian versus non-Hawaiian species are about the same. (.11 versus 0.13 respectively). The $\overline{HET}$ value for non-regulatory enzymes are likewise similar. Thus, by this method of calculation, the mean levels of heterozygosities ($\overline{HET}$) among Hawaiian *Drosophila* are comparable to those of non-Hawaiian species although the $\bar{H}$ values for these same Hawaiian species are generally lower than those of continental forms. The overall values suggest agreement with Johnson's hypothesis that variation at the more important regulatory enzymes is more responsive to some form of balancing selection. However, in the Hawaiian *Drosophila* data, it is only the esterases which elevate the variable substrate overall HET; and the $\overline{HET}$ values of only three of the seven regulatory enzyme loci (Pgm-1, Hk-2 and Hk-3) values really support Johnson's contention.

TABLE 3

The average proportion of individuals estimated to be heterozygous at each 16$^+$ loci in species of Hawaiian *Drosophila* and 13 Non-Hawaiian *Drosophila* Species[1]

	Hawaiian *Drosophilia*			Non-Hawaiian *Drosophila*
Locus	Genomes Sampled	Number of Species	$\overline{HET}$	HET[1]
		Variable Substrate		
Est.	50	2	.43[2]	.43
Lap-1	1022	7	.06	.17
Odh-1	1478	8	.05	.22
Overall			.18	.27
		Regulatory Enzymes		
Adh-1	2438	9	.06	.28
Hk-1	536	9	.00	.08
Hk-2	174	2	.26	.08
Hk-3	284	4	.12	.08
Hk-4	154	2	.00	.08
Me-1	2506	10	.09	.08
Pgm-1	2104	9	.21	.20
Overall			.11	.13

TABLE 3 (Continued)

Locus	Genomes Sampled	Hawaiian *Drosophila* Number of Species	Hawaiian *Drosophila* HET	Non-Hawaiian *Drosophila* HET[1]
Non-Regulatory Enzymes				
αGpd-1	2530	9	.01	.04
Idh-1	2462	9	.05	.14
Mdh-1	2526	9	.05	.02
Mdh-2	1888	7	.01	.02
Got-1	1560	6	.24	.05[3]
Got-2	1272	6	.02	.00[3]
Overall			.07	.05

[1]From Johnson (1974)

[2]From Rockwood et al., (1971)

[3]These values are the averages reported in vertebrate forms (Selander and Johnson, 1973)

Of the non-regulatory enzymes, the high variability of Got-1 is obviously contradictory to Johnson's hypothesis. Although we would agree with Johnson (1974) that the functional and structural properties of individual enzymes are important in the determination of the amount of variability maintained, we feel that his classification is somewhat arbitrary and is, at present, unjustified.

GENETIC DIFFERENTIATION AMONG SPECIES

Coefficients of genetic similarity (Table 4) were derived by the method of Rogers' (1972) from the variation at ten of the twelve enzyme loci listed in Table 2 (Odh-1 and Lap-1 excluded). These coefficients may assume values from 0 to 1, with 1 indicating genetic identity. Another method of calculating genetic similarity proposed by Nei (1972) has been used extensively in similar studies (Zouros, 1973; Ayala et al., 1974). Rogers' formula and that of Nei have been shown to rank populations in the same order and to be highly correlated, although Nei's method usually gives similarity values that are slightly greater (Nevo et al., 1974). The mean similarity among species (S=.444) approximates the estimates of average genetic similarity between closely related non-Hawaiian *Drosophila* (Hubby and Throckmorton, 1968; Nair et al., 1971; Lakovaara et al., 1972) although the amount of genetic differentiation between species is highly variable (Table 4). Little differentiation has occurred between some species (i.e. between *D. silvestris* and *D. heteroneura* and among *D. melanocephala*, *D. ingens*, and *D. cyrtoloma* values of S are all greater than .95) whereas extensive differentiation has occurred between others (i.e. between *D. hanaulae* and *D. picticornis*, S=.11).

Clustering of the similarity coefficients (Figure 2) indicates two major groupings of species, one involving *D. silvestris*, *D. heteroneura*, *D. planitibia*, *D. planitibia*-like, and *D. hemipeza* and the other *D. hanaulae*, *D. cyrtoloma*, *D. ingens*, *D. melanocephala*, *D. neoperkinsi* and *D. oahuensis*. The phyletic relationships within the first group will be discussed in depth elsewhere (Craddock and Johnson, in preparation). The high allozymic similarity of *D. hemipeza* to other members of that cluster is somewhat surprising since *D. hemipeza* lacks the Xr inversion common to the other species. One could argue that because *D. hemipeza* is poorly represented this closeness is spurious; however, it is interesting that of all the species of the *D. planitibia* subgroup only *D. silvestris*, *D. heteroneura*, *D. planitibia*, *D. planitibia*-like

TABLE 4

Coefficients of genetic similarity (Rogers' S) between 16 species of Hawaiian *Drosophila*.

Species	sil	het	plb	pll	han	oah	nig	hem	npk	cyr	mel	ign	hpt	obs	sub	pic
silvestris	1.00	.96	.74	.71	.32	.31	.30	.56	.23	.32	.33	.30	.39	.33	.30	.22
heteroneura		1.00	.76	.72	.33	.29	.29	.58	.22	.33	.34	.31	.41	.33	.30	.23
planitibia			1.00	.85	.30	.30	.30	.78	.30	.40	.40	.39	.49	.34	.40	.39
planitibia-like				1.00	.42	.45	.35	.74	.41	.52	.54	.50	.50	.34	.40	.36
hanaulae					1.00	.51	.31	.30	.78	.89	.87	.89	.39	.48	.20	.11
oahuensis						1.00	.50	.40	.59	.60	.58	.60	.40	.30	.30	.30
nigribasis							1.00	.40	.49	.41	.38	.40	.40	.30	.40	.50
hemipeza								1.00	.30	.40	.40	.40	.40	.32	.40	.48
neoperkinsi									1.00	.88	.87	.88	.41	.37	.31	.30
cyrtoloma										1.00	.96	.99	.50	.38	.30	.21
melanocephala											1.00	.96	.51	.35	.32	.21
ingens												1.00	.50	.38	.30	.20
neopicta													1.00	.38	.59	.39
obscuripes														1.00	.30	.23
substenoptera															1.00	.29
picticornis																1.00

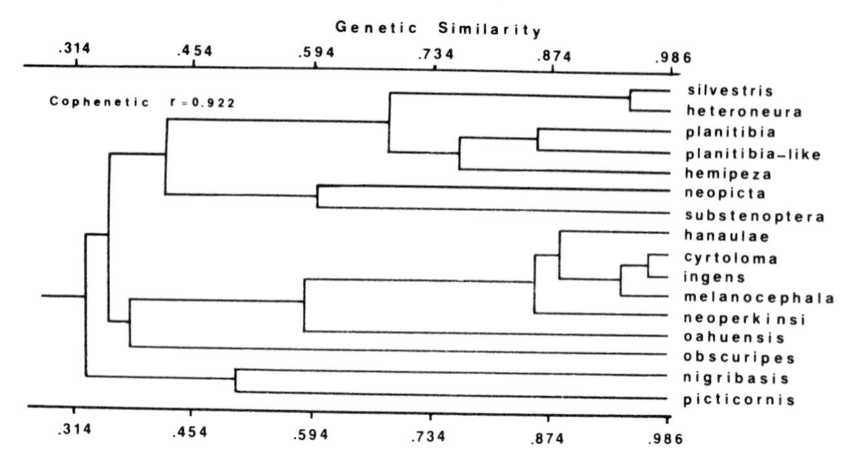

FIG. 2. Similarity dendrogram among species of the *D. planitibia* subgroup derived from a matrix of Rogers' coefficients by the unweighted pair-group method, using arithmetic means (Sneath and Sokal, 1973).

and *D. hemipeza* have been found to utilize plant species of the family *Lobeliaceae* as ovipositing sites (Montgomery, 1972). Most of the other species of the *D. planitibia* subgroup that have been reared from rotting plants, taken from the field, appear to utilize species of Aralioids as hosts. In view of the high degree of specificity for specialized ovipositing sites shown by Hawaiian *Drosophila*, this host-plant similarity strongly supports the contention of a close genetic relationship of *D. hemipeza* to the other members of this subcluster. Two alternatives could explain the absence of the Xr inversion in *D. hemipeza*. *D. hemipeza* may have been founded from the ancestral population before the Xr inversion arose or, alternatively, was founded from a population that was polymorphic for Xr and the founder individuals of *D. hemipeza* either failed to carry the inversion or lost it subsequently. One other possibility, although seemingly unlikely, cannot be dismissed. Could the close allozymic similarity be a result of convergence because of the host plant commonality and substrate affinities? Any further consideration of these possibilities at this time would be highly speculative.

The phenogram also indicates a close allozymic relationship among *D. cyrtoloma*, *D. oahuensis*, and four other species. These relationships are fairly consistent with those derived from the inversion data although the minimal allozymic divergence among species occupying the Maui complex was unexpected in view of the numerous inversion differences and differences in the metaphase chromosomes found among these species (Clayton et al., 1972). This high allozymic similarity probably resulted as a consequence of allopatric speciation involving a wide distribution of populations and subsequent partitioning of these populations by a strong geographic isolating barrier (in this case, the ocean channels or a void of suitable host plants). The islands of Maui and Molokai are presently separated by a shallow ocean channel but at one or more times during the Pleistocene they were connected by a land bridge (Stearns and Macdonald, 1942). Additionally, the cool and wet climate associated with the glacial periods (Stearns, 1966) would have promoted a wider distribution of host plants, thus periodically diminishing or completely eliminating the isolating barriers between Maui and Molokai and between East and West Maui as well. Only a few phases of alternating periods of isolation and dispersal would be required for the formation of the present species distribution on the Maui complex. Thus speciation among these forms may not involve small founder populations nor would the species necessarily have undergone the accompanying "genetic revolution" associated with founding events. Under these conditions the allelic divergence would be minimal unless these isolates were subjected to strong differential and directional selection at enzyme loci or at chromosomal arrangements with associated enzyme alleles. *D. oahuensis* on Oahu was probably founded by a single or a few individuals and, as a consequence, has undergone the "flush" (Carson, 1968) and "genetic revolution" typical of most of the Hawaiian *Drosophila*. The phenogram also illustrates that within the other subclusters the most dissimilar species are those most likely to have been founded by small numbers of individuals. These results provide some suggestive evidence that the rate of genetic divergence varies tremendously as a direct result of the kind of speciation event involved. Thus, values of genetic similarity between species have little phylogenetic significance unless supplemental data can define the kinds of founding and speciation events that have occurred during the formation of a species group.

ACKNOWLEDGEMENTS

This work was supported by Grants G.B. 27586 and G.B. 29288 from the National Science Foundation to the University of Hawaii. We also acknowledge the cooperation of the Island Ecosystems IRP/IBP Hawaii (NSF G.B. 23230).

REFERENCES

Ayala, F.J. 1972. Darwinian versus non-Darwinian evolution in natural populations of *Drosophila*. *Proc. Sixth Berkeley Symp. Math. Stat. Prob.* 5: 211-236.

Ayala, F. J., J. R. Powell, M. L. Tracey, C. A. Mourao and S. Perez-Salas 1972. Enzyme variability in the *Drosophila willistoni* group. IV. Genetic variation in natural populations of *Drosophila willistoni*. *Genetics* 70:113-130.

Ayala, F. J., M. L. Tracey, L. G. Barr, and J. G. Ehrenfeld 1974. Genetic and reproductive differentiation of the subspecies, *Drosophila equinoxialis caribbensis*. *Evolution* 28: 24-41.

Bernstein, S. C., L. H. Throckmorton, and J. L. Hubby 1973. Still more genetic variability in natural populations. *Proc. Nat. Acad. Sci. USA* 70: 3928-3921.

Bryant, E. H. 1974. On the adaptive significance of enzyme polymorphisms in relation to environmental variability. *Amer. Natur.* 108: 1-19.

Carson, H. L. 1968. The population flush and its genetic consequences. In *Population Biology and Evolution*, R. C. Lewontin, ed., Syracuse Univ. Press, New York, N. Y.

Carson, H. L. 1973. Ancient chromosomal polymorphism in Hawaiian *Drosophila*. *Nature* 241: 200-202.

Carson, H. L., D. E. Hardy, H. T. Spieth and W. S. Stone 1970. The evolutionary biology of the Hawaiian *Drosophilidae*. In: *Essays in Evolution and Genetics in Honor of Theodosius Dobzhansky* (M. K. Hecht and W. C. Steere, eds.), Appleton-Century Crofts, New York.

Carson, H. L. and H. D. Stalker 1968. Polytene chromosome relationships in Hawaiian species of *Drosophila*. II. The *D. planitibia* subgroup. Studies in *Genetics* IV. Univ. Texas Publ. 6818: 355-365.

Clayton, F. E., H. L. Carson, and J. E. Sato 1972. Polytene chromosome relationships in Hawaiian species of *Drosophila*. VI. Supplemental data on metaphases and gene sequences. *Studies in Genetics* VII. Univ. Texas Publ. 7213: 163-177.

Craddock, E. M. 1974. Reproductive relationships between homosequential species of Hawaiian *Drosophila*. *Evolution:* in press.

Craddock, E. M. and W. E. Johnson. Genetic variation in Hawaiian *Drosophila*. III. Chromosomal and allozymic diversity in *Drosophila silvestris* and its homosequential species. In preparation.

Hardy, D. E. 1969. Notes on Hawaiian "idiomyia" (*Drosophila*). *Studies in Genetics* V. Univ. Texas Publ. 6918: 71-77.

Heed, W. B. 1968. Ecology of the Hawaiian *Drosophilidae*. *Studies in Genetics* IV. Univ. Texas Publ. 6818: 387-419.

Hubby, J. L. and T. H. Throckmorton 1968. Protein differences in *Drosophila* IV. A study of sibling species. *Amer. Natur.* 102: 193-205.

Johnson, G. B. 1974. Enzyme polymorphism and metabolism. Science 184: 28-37.

Johnson, W. E. and R. K. Selander 1971. Protein variation and systematics in kangaroo rats (genus *Dipodomys*). *Syst. Zool.* 20: 377-405.

Kambysellis, M.P. 1970. Compatibility in insect tissue transplantations. I. Ovarian transplantations and hybrid formation between *Drosophila* species endemic to Hawaii. *J. Exp. Zool.* 175: 169-180.

Kanapi, C. G. and M. R. Wheeler 1970. Comparative isozyme patterns in three species of the *Drosophila nasuta* complex. *Tex. Rep. Biol. Med.* 28: 261-278.

Kaneshiro, K. Y. 1969. A study of the relationships of Hawaiian *Drosophila* species based on external male genitalia. *Studies in Genetics* V. Univ. Texas Publ. 6918: 55-70.

Kimura, M. 1968. Genetic variability maintained in a finite population due to mutational production of neutral and nearly neutral isoalleles. *Genet. Res.* 2: 247-269.

King, J. L. and T. H. Jukes 1969. Non-Darwinian evolution. *Science* 164: 788-798.

Lakovaara, S., A. Saura, and C. T. Falk 1972. Genetic distance and evolutionary relationships in the *Drosophila obscura* group. *Evolution* 26: 177-184.

Macdonald, G. A. and A. T. Abbott 1970. *Volcanoes in the Sea: the Geology of Hawaii.* University of Hawaii Press, Honolulu.

Montgomery, S. L. 1972. Comparative breeding site ecology and the adaptive radiation of picture-winged *Drosophilidae* (*Diptera: Drosophilidae*) in Hawaii. M. S. Thesis, Univ. of Hawaii.

Nair, P. S., D. Brncic, and K. Kojima 1971. Isozyme variations and evolutionary relationships in the *mesophragmatica* group of *Drosophila*. *Studies in Genetics* VI. Univ. Texas Publ. 7103: 17-28.

Nei, M. 1972. Genetic distance between populations. *Amer. Natur.* 106: 283-292.

Nevo, E., Y. J. Kim, C. R. Shaw, and C. S. Thaeler, Jr. 1974. Genetic variation, selection and speciation in *Thomomys talpoides* pocket gophers. *Evolution* 28: 1-23.

Richmond, R. C. 1972. Enzyme variability in the *Drosophila willistoni* group. III. Amounts of variability in the super-

species, *D. paulistorum*. *Genetics* 70: 87-112.
Rockwood, E. S., C. G. Kanapi, M. R. Wheeler, and W. S. Stone 1971. Allozyme changes during the evolution of the Hawaiian *Drosophila*. *Studies in Genetics* VI. Univ. Texas Publ. 7103: 193-212.
Rogers J. S. 1972. Measures of genetic similarity and genetic distance. *Studies in Genetics* VII. Univ. Texas Publ. 7213: 145-153.
Selander, R. K. and W. E. Johnson 1973. Genetic variation among vertebrate species. *Ann. Rev. Ecol. Syst.* 4: 75-91.
Selander, R. K., and D. W. Kaufman 1973. Genetic variability and strategies of adaptation in animals. *Proc. Nat. Acad. Sci. USA* 70: 1875-1877.
Selander, R. K., M. H. Smith, S. Y. Yang, W. E. Johnson, and J. B. Gentry 1971. Biochemical polymorphism and systematics in the genus *Peromyscus*. I. Variation in the old-field mouse (*Peromyscus polionotus*). *Studies in Genetics* VI. Univ. Texas Publ. 7103: 49-90.
Sneath, P. H. A. and R. R. Sokal 1973. *Numerical Taxonomy*. W. H. Freeman and Co., San Francisco. 573 pp.
Stearns, H. T. 1966. *Geology of the State of Hawaii*. Pacific Books, Palo Alto, Calif.
Stearns, H. T. and G. A. MacDonald 1942. Geology and ground water resources of the island of Maui, Hawaii. Hawaii Division of Hydrography, Bulletin 7, 344 pp.
Stebbins, G. L. and R. C. Lewontin 1972. Comparative evolution at the levels of molecules, organisms, and populations. *Proc. Sixth Berkeley Symp. Math. Stat. Prob.* 5: 23-42.
Steiner, W. W. M. and W. E. Johnson 1973. Techniques for electrophoresis of Hawaiian *Drosophila*. Tech. Report No. 30. US/IBP Island Ecosystems IRP. 21 pp.
Throckmorton, T. H. 1966. The relationships of the endemic Hawaiian *Drosophilidae*. *Studies in Genetics* III. Univ. Texas Publ. 6615: 335-396.
Yang, H. Y. and M. R. Wheeler 1969. Studies on interspecific hybridization within the picture-winged group of endemic Hawaiian *Drosophila*. *Studies in Genetics* V. Univ. Texas Publ. 6918: 133-170.
Yamazaki, T. and T. Maruyama 1972. Evidence for the neutral hypothesis of protein polymorphism. *Science* 178: 56-58.
Zouros, E. 1973. Genic differentiation associated with the early stages of speciation in the *mulleri* subgroup of *Drosophila*. *Evolution* 27: 601-621.

PURIFIED DROSOPHILA α-AMYLASE ISOZYMES: GENETICAL, BIOCHEMICAL, AND MOLECULAR CHARACTERIZATION

WINIFRED W. DOANE, IRENE ABRAHAM, M. M. KOLAR, RUSSELL E. MARTENSON and GLADYS E. DEIBLER
Department of Biology, Kline Biology Tower
Yale University, New Haven, Connecticut 06520
and
National Institutes of Health, Bethesda, Maryland

ABSTRACT. The genetic control of α-amylase in *D. melanogaster* and *D. hydei* is compared. Electrophoretic variants from *Zurich* (Amy^7) and *Chile* (Amy^8) strains of *hydei* were used in recombination experiments. The linkage group for *Amy* in *hydei* is V, the homologue of the *Amy*-containing 2R element in *melanogaster*; *Amy* lies 6 map units right of *cn* in *hydei*. Evidence for a tightly linked duplication of *Amy*, as found in *melanogaster*, is wanting in *hydei*. However, a newly described male-specific enzyme, probably an α-amylase but under independent genetic control and suggestive of a duplication, has been identified as limited to mature accessory glands in adult males. α-Amylase-7 was purified from adults of the *Zurich* strain of *hydei*. The MW of its polypeptide chain, determined by SDS gel electrophoresis, is 54,500. A purified sample of α-amylase-8 from the *Chile* strain gave similar results, as did a partially purified α-amylase-2 extract from larvae of *melanogaster*. An amino acid analysis was performed on amylase-7 from *hydei*, and rabbit antibodies were prepared against it. The antisera cross-reacted readily with the amylase-7 and -8 antigens from *hydei* but gave little or no reaction with amylase-2 from *melanogaster*. No precipitation could be detected between antisera to the *Drosophila* amylase and amylase in human saliva; neither could a cross-reaction be gotten between antibodies against human salivary amylase isozymes and our *Drosophila* antigens. The ease with which amylase may be purified provides an excellent opportunity to do amino acid sequencing of its genetic variants in *Drosophila*. It also adds to the many advantages of using the amylase system to analyze genetic regulatory mechanisms in this eukaryote.

INTRODUCTION

Among the advantages of analyzing the genetic control of α-amylase (EC 3.2.1.1, α-1,4-glucan-4-glucanohydrolase) isozymes in *Drosophila* is the relative ease with which the enzyme may be purified from crude fly homogenates (Doane et al., 1973a, b, 1974). Stability, even at room temperature, and simplicity of routine assays for specific activity, whether in the test tube or in polyacrylamide gels following electrophoresis, are other convenient attributes of the enzyme (Doane, 1967a, 1969). More important, the potential for genetic analysis offered by *Drosophila* is unrivaled by comparable vertebrate systems utilizing amylase isozymes such as those described at this conference, i.e. chick pancreatic amylase (Lehrner and Malacinski, 1974) and human salivary amylase (Karn et al., 1974). This is primarily due to the sophisticated tools of genetic engineering developed over the years for *Drosophila*. In addition, the uncomplicated banding patterns produced by amylase from homozygous strains of "fruit" flies" have not necessitated the involved genetic and epigenetic interpretations ascribed to vertebrate systems.

With our present ability to purify amylases from *Drosophila* in milligram quantities, we anticipate the resolution of many of the biochemical difficulties encountered in the use of so small a eukaryote to study genetic regulatory mechanisms. The bulk of the work described here relates to the system being developed for *Drosophila hydei*. First the genetics of α-amylase in *Drosophila melanogaster* will be reviewed, since we shall return to that species toward the end of the paper.

GENETICS OF AMYLASES IN *DROSOPHILA MELANOGASTER*

In view of the simplicity of the genetic interpretation of amylase zymograms in *Drosophila*, it is ironic that one of the first species to be analyzed electrophoretically, *D. melanogaster*, should possess a duplication of the structural gene for the enzyme. This is revealed by a complexity of banding patterns not found in any other *Drosophila* species examined. Kikkawa's (1960) original recombination studies on strain with "strong" and "weak" amylase activities placed the *Amy* locus in the middle of chromosome 2R in *D. melanogaster*. His (Kikkawa, 1964) electrophoretic analysis suggested the structural gene for the enzyme might be duplicated in some strains, with the duplicated loci so tightly linked that they would be inherited as a single unit in classical recombination experiments.

A number of lines of evidence indicate this is indeed the situation (Doane, 1970, and below). First, all amylase isozymes investigated have similar biochemical characteristics, e.g. pH optimum, substrate specificities, activators, inhibitors, molecular weights, etc. Second, comparison of amylase banding patterns produced by flies from laboratory and natural populations of 42 species of *Drosophila*, and the closely related *Zaprionus vittiger*, reveals a great deal of polymorphism; yet, in no case other than *D. melanogaster* is evidence found for duplication of the controlling structural gene. (A possible exception in *D. hydei* will be described here.) In addition, the most commonly observed banding pattern in laboratory and natural populations of *D. melanogaster* is a single band in position No. 1 (see also McCune, 1969). The duplication, on the other hand, is phenotypically expressed in certain strains by two major α-amylase isozymes with different electrophoretic mobilities; here, the *Amy* alleles at the two tightly linked loci are distinct. Of particular significance is Doane's (1970) evidence that the duplicated *Amy* genes or their products are, in part, independently regulated during development and in response to dietary changes.

Finally, genetic recombination between alleles at the two linked *Amy* loci provides evidence for the presence of the duplication in *D. melanogaster* (Bahn, 1967; Doane, 1967b, 1969). Bahn's data indicate a linkage intensity on the order of one recombinant in 13,000. This places the duplicated loci only 0.008 map units apart! Pooled data (Kikkawa, 1964; Bahn, 1967; Doane, 1969) place the *Amy* "region" at 2-77.7 on the genetic map; its location on the salivary chromosome map is between sections 54B and 55 (Bahn, 1971a, b).

A summary of the banding patterns of the genetically determined amylase isozymes from homozygous strains of *D. melanogaster* is presented in Figure 1. In addition to strains previously described (see Doane, 1970), two new types are shown, Amy^{5} and $Amy^{5,6}$ (Puijk and de Jong, 1972). There is a total of 13 patterns composed of 6 different isozymes based on mobilities, 8 multiple molecular forms based on mobilities plus heat resistance, and at least 11 forms if one considers specific activities as well. Eventually a proximal and distal *Amy* locus must be designated by separate symbols for strains with the duplication. This cannot be done yet for all strains so the former symbols which indicate the *Amy* "region", rather than a specific locus(i), continue to be used with banding patterns depicted by superscripts.

STRUCTURAL GENE FOR α-AMYLASE IN *DROSOPHILA HYDEI*

The major site of amylase synthesis in *Drosophila* is the midgut; in larvae of *D. melanogaster* and *D. hydei*, this site is restricted to certain cells in the posterior midgut region (Doane, 1970). These secretory cells contain polytene chromosomes which, in *D. melanogaster*, are too small to be readily analyzed cytologically. In *D. hydei*, by contrast, the level

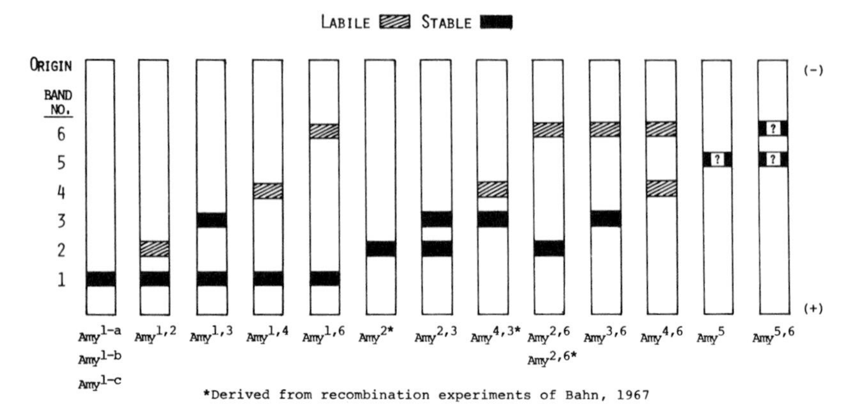

Figure 1. Banding patterns of amylase isozymes in *Amy* strains of *D. melanogaster*. Relative activities are not shown, but Amy^1 strains with three levels of specific activity are indicated by a, b, and c, and two $Amy^{2,6}$ strains are figured whose isozymes differ in relative activities.

---

of polyteny is sufficient for analysis with the light microscope, thus providing an opportunity to correlate enzyme synthesis and/or activation with visible signs of transcription, i.e. chromosomal "puffs" (Doane, 1971a). The genetic control of amylase is simpler in *D. hydei* than in *D. melanogaster* because it is not complicated by the type of duplication described above for the latter species. This again makes *D. hydei* more appropriate for "puffing" studies than *D. melanogaster*, so the genetic location of *Amy* was undertaken in *D. hydei* (Doane, 1970, 1974).

INHERITANCE OF AMYLASE VARIANTS IN D. HYDEI

Amylases from 8 strains were analyzed electrophoretically in polyacrylamide by the starch-iodine method of Doane (1967a, c). Included were four "wild" strains from the Yale stock collection: *Chile*, *New Haven*, *Vera Cruz*, and *Zürich*. Mutant stains included: *Lobe*; *bb*, *p*, *vg*; *st*, *jv*, *sca*; and *p*, *cn*. All strains produced a single major amylase band in position No. 7, except *Chile*, which produced a No. 8 isozyme (Figure 2). *Zurich* and *Chile* strains were reciprocally crossed, F_1 females backcrossed to both parental types, and testcross

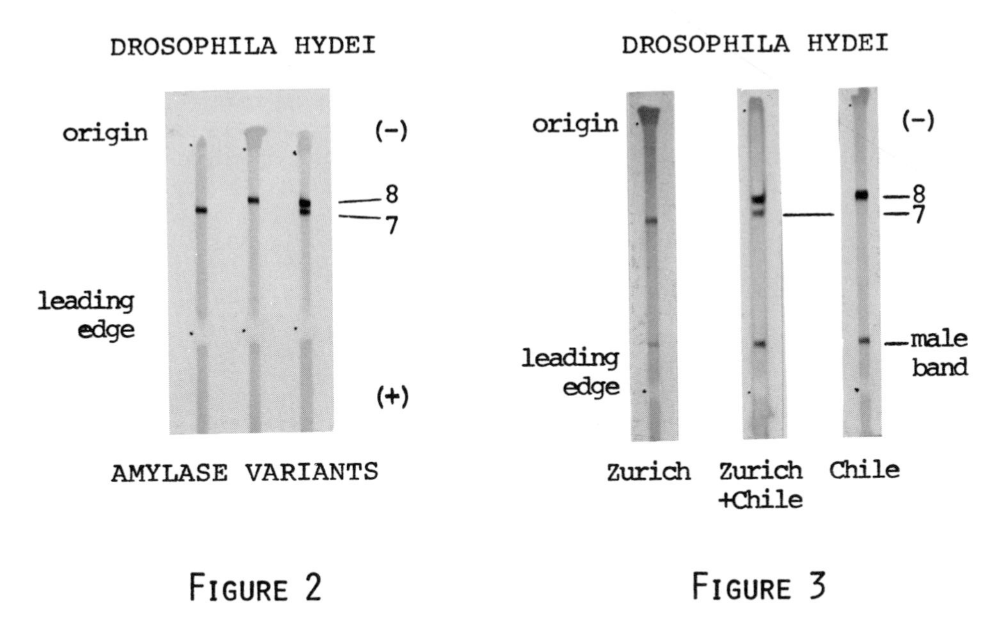

FIGURE 2

FIGURE 3

Figure 2. Contact print of a starch-iodine film impression of amylase isozymes from week-old adults of *Zurich* (Amy^7) and *Chile* (Amy^8) strains of *D. hydei* and their hydrid. Disc gels were incubated ½ hr against a 1.5% starch-in-acrylamide film.
Figure 3. Amylase patterns produced by 4 pairs of pooled accessory glands from males, 11 days, of the *Zurich* and *Chile* strains and by a mixture of 4 pairs from each strain, incubated 1½ hrs against a 0.5% starch film.

---

progeny analyzed electrophoretically. Results indicate the amylases are controlled by codominant alleles at a single autosomal locus in a manner consistent with Mendelian inheritance. The alleles were designated Amy^7 and Amy^8. All other strains were homozygous for alleles which are electrophoretically indistinguishable from Amy^7.

As in *D. melanogaster*, heterozygotes for the *Amy* alleles in *D. hydei* produced an additive banding pattern; both parental types appeared with no "hybrid" band. Derivative amylase bands (minor or "pseudoisozymes") occasionally showed at more anodal positions than the genetically determined isozymes generating them. These insignificant forms never exceeded about two percent of the total amylase activity.

IDENTIFICATION OF THE AMY LINKAGE GROUP

Three marker strains were used to identify the linkage group to which *Amy*, the presumed structural gene for amylase in *D. hydei* belongs: *bb, p, vg; st, jv, sca;* and *p, cn*. All are homozygous for Amy^7 and contain recessive mutants which span four of the five linkage groups in this species. Thus, *bb* is in the X linkage group, *p* and *st* are in II, *jv* in IV, and *cn* (or $red\ C\text{-}1_{Sp}$), *sca* and *vg* in V (cf. Spencer, 1949; Berendes, 1963; Gregg and Smucker, 1965; Horikx-Jacob, 1968; and Gloor, 1971).

Females from each marker strain were outcrossed to Amy^8 males and their F_1 female progeny backcrossed to males of the respective multiple recessive strain. Testcross progeny were aged a week under standardized conditions and classified. Each recombinant class was homogenized en masse (20 flies or less) in an equivalency of 1 fly/5 μl water, centrifuged, and analyzed by disc electrophoresis, using 10 μl samples of supernatants. When both amylase isozymes characterized the extract, quantitation (Doane 1967a) of the relative activity of each isozyme provided an estimate of the frequency of the two *Amy* alleles among the flies extracted.

The results of crosses between *bb, p, vg* (Amy^7) and Amy^8 are in Table 1A. It is clear from the last column that *Amy* is associated with linkage group V. Crosses with the other marker strains verified this conclusion. In *D. hydei*, linkage group V belongs to chromosome 5, which is homologous to 2R in *D. melanogaster* (Berendes, 1963).

GENETIC RECOMBINATION EXPERIMENTS

Two strains, *sca cn* Amy^7 and *sca cn* Amy^7*vg*, were synthesized from the marker strains of the preceeding section. These were outcrossed to Amy^8 in three- and four-point crossover analyses, respectively. Testcross progeny were aged one week and tested electrophoretically as follows: 1) recombinants between recessive markers were tested individually; 2) parental types were tested en masse as described before.

Table 1B summarizes the data from recombination experiments. Corrections are for the lowered viability of parental types with multiple recessive markers. The data indicate that *Amy* lies between *cn* and *vg* at about 6 maps units to the right of *cn*. Crossover distances between *sca*, *cn*, and *vg* agree fairly well with the data of Dr. H. R. Kobel (unpublished; Gloor, 1971), who found 20% crossing-over between *sca* and *cn*, and 10% between *cn* and *vg*. Our data do not give a reliable estimate of the distance between *Amy* and *vg* because of the

TABLE I
GENETIC CONTROL OF *AMYLASE* IN *DROSOPHILA HYDEI*[1]
A. Linkage Group of *Amy* Locus

Crosses: *bb, p, vg* (Amy^7)♀ X *Chile* (Amy^8)♂
F_1♀ X *bb, p, vg* (Amy^7)♂

Testcross Progeny			Isozyme No. 7 Activity/
Phenotype	Females	Males[2]	Total Amylase Activity
+, +, +	69	125	0.47
bb, +, +	72	-	0.52
+, p, +	55	125	0.50
bb, p, +	61	-	0.53
+, +, vg	63	118	1.00
bb, +, vg	57	-	0.95
+, p, vg	41	106	0.97
bb, p, vg	62	-	1.00

B. Summary of Recombination Experiments

Testcrosses: I - *sca cn* Amy^7/Amy^8♀ X *sca cn* Amy^7♂
II - *sca cn* Amy^7 *vg*/Amy^8♀ X *sca cn* Amy^7 *vg*♂
III - Amy^8/*sca cn* Amy^7 *vg*♀ X *sca cn* Amt^7 *vg*♂

Expt. No.	Total Progeny	Recombinants Between Markers *sca-cn*	*sca-Amy*	*sca-vg*	*cn-Amy*	*cn-vg*	*Amy-vg*
I	927	186	210		42		
II	715	122	166	182	52	112	76
III	877	150	190	222	48	100	72
% Crossing over:							
Over-all	-	18.1	22.5	25.4	5.8	13.6	9.4
Corrected	-	17.2	21.4	23.9	5.5	12.8	8.9

[1]Doane, 1970, 1974.
[2]In males of *D. hydei*, *bb* is not visibly expressed.

variable expression of *vg* despite attempts to control it; *vg* is in undoubtedly closer to *Amy* than Table 1, B suggests, judging from Dr. Kobel's data and our own subsequent cytogenetic studies (Doane, 1971b, unpublished).

PURIFICATION AND CHARACTERIZATION OF α-AMYLASES

To date, no synthesis of a specific enzyme has been conclusively correlated with puffing at a given structural gene in *Drosophila*. Even in the most thoroughly analyzed diptern, *Chironomus tentans*, where the formation of Balbiani Ring 2 (a giant puff) coincides with the synthesis of a discrete messenger RNA molecule (Daneholt, 1973; Daneholt and Hosick,

1973), there remains some ambiguity about the polypeptide(s) for which the message codes (Grossbach, 1969; reviews, Beermann, 1972; Berendes, 1973).

As a secretory protein of polytene cells, amylase offers an excellent prospect for chromosomal puffing studies. Amylase activity can be specifically enhanced in larvae and adults by placing them on a starch diet (Doane, 1970, 1971a; Hosbach et al., 1972). Preliminary studies indicate that this increase in activity correlates with the formation of a puff in posterior midgut cells at what appears to be the structural gene for amylase in *D. hydei* (Doane, 1971a, b, unpublished). It remains to be seen if this puffing also correlates with *de novo* synthesis of amylase. Hence, it was necessary to purify the enzyme (Doane et al., 1973a, b, 1974; Doane and Kolar, 1974) in order to prepare antibodies to it for use in labeling studies designed to quantitate synthesis of the amylase polypeptide chain and to resolve certain questions about its molecular characteristics.

α-*AMYLASE-7 FROM* D. HYDEI

Purification. Amylase was extracted from mature adults of the *Zurich* strain, reared on starch-yeast. Partial purification was achieved (Table 2) by the method of Loyter and Schramm (1962), through which α-amylase may be specifically precipitated from crude extracts as a glycogen-enzyme complex insoluble in 40% ethanol. The protocol employed is outlined in Figure 4.

"Native glycogen" was hydrolyzed by autodigestion in crude, whole fly extracts (S_1) to prevent loss of amylase in the first ethanol precipitation due to its binding to that glycogen. Proteins precipitable by 40% ethanol were removed prior to the addition of an optimal amount of repurified shellfish glycogen. Since the subsequent ethanol precipitation of the complex also brought down some contaminating material, it was necessary to separate the amylase from the contaminants by slab gel electrophoresis. The apparatus of Roberts and Jones (1972) was used in a preparative manner with 5% acrylamide, 0.1 M Tris-borate buffer, pH 9.4, and a constant voltage of 450 V for 90 min at 7°C. The amylase band, identified by its amylolytic activity and lability to α-amylase inhibitor, was cut out, eluted, dialyzed, and lyophilized to dryness.

Purified amylase-7 samples were tested for impurities by disc electrophoresis; gels were stained with Coomassie Brilliant Blue. No protein contaminants were detected (Fig. 5). Estimates indicate that the amylase makes up about 0.1% of the total soluble protein in the original crude extract of

TABLE 2

PARTIAL PURIFICATION OF AMYLASE FROM *D. HYDEI*[1]

71.3 g of flies

Extract	Total Protein[2]	Reduction Assay[3]			Starch-Iodine Assay[4]			Protein Yield
		Specific Activity	Amylase Yield	Purification Factor	Specific Activity	Amylase Yield	Purification Factor	
	mg	*MU/µg*	%		*SU/µg*	%		%
S_1	4,118.00	4.28	100.0		2.16	100.0		100.0
S_7	12.18	975.52	67.5	227.9	453.45	62.2	209.9	0.3

[1]Doane et al,, 1973, 1974.

[2]Lowry et al., 1951.

[3]3,5-Dinitrosalicylic acid reduction assay: 1 *MU* is 10^{-4} µMoles maltose equivalents/min at 25°C (see Doane, 1969).

[4]Starch-iodine assay: 1 *SU* is 10^{-4} mg starch hydrolyzed/min at 25°C (see Doane, 1969).

FIGURE 4

PROTOCOL FOR AMYLASE PURIFICATION

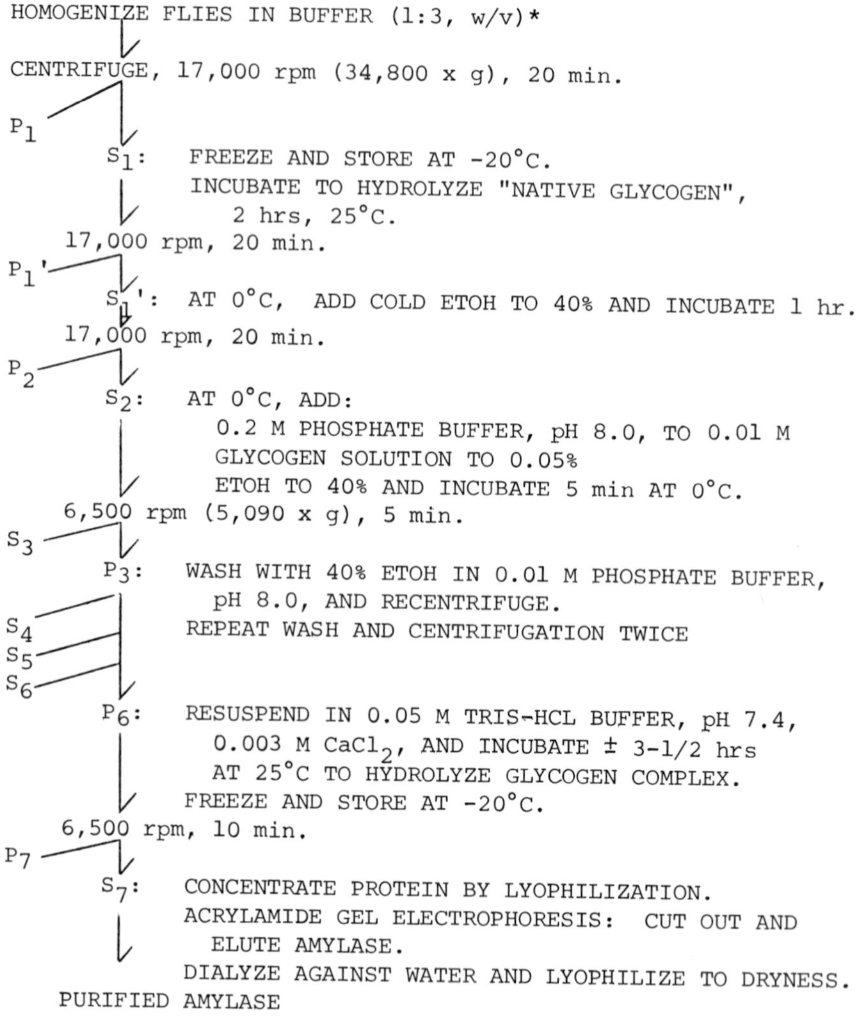

HOMOGENIZE FLIES IN BUFFER (1:3, w/v)*

CENTRIFUGE, 17,000 rpm (34,800 x g), 20 min.

P_1

S_1: FREEZE AND STORE AT -20°C.
INCUBATE TO HYDROLYZE "NATIVE GLYCOGEN", 2 hrs, 25°C.

17,000 rpm, 20 min.

P_1'

S_1': AT 0°C, ADD COLD ETOH TO 40% AND INCUBATE 1 hr.

17,000 rpm, 20 min.

P_2

S_2: AT 0°C, ADD:
0.2 M PHOSPHATE BUFFER, pH 8.0, TO 0.01 M
GLYCOGEN SOLUTION TO 0.05%
ETOH TO 40% AND INCUBATE 5 min AT 0°C.

6,500 rpm (5,090 x g), 5 min.

S_3

P_3: WASH WITH 40% ETOH IN 0.01 M PHOSPHATE BUFFER, pH 8.0, AND RECENTRIFUGE.
REPEAT WASH AND CENTRIFUGATION TWICE

S_4
S_5
S_6

P_6: RESUSPEND IN 0.05 M TRIS-HCL BUFFER, pH 7.4, 0.003 M $CaCl_2$, AND INCUBATE ± 3-1/2 hrs AT 25°C TO HYDROLYZE GLYCOGEN COMPLEX.
FREEZE AND STORE AT -20°C.

6,500 rpm, 10 min.

P_7

S_7: CONCENTRATE PROTEIN BY LYOPHILIZATION.
ACRYLAMIDE GEL ELECTROPHORESIS: CUT OUT AND ELUTE AMYLASE.
DIALYZE AGAINST WATER AND LYOPHILIZE TO DRYNESS.

PURIFIED AMYLASE

* Buffer: 0.004 M Phosphate, 0.004 M NaCl, 0.001 M $CaCl_2$, PTU to saturation, pH 6.9.

Fig. 4. Outline of protocol used in the purification of α-amylase-7 from week-old adults of the *Zurich* strain of *D. hydei*. The homogenization buffer is shown at the bottom; phenylthiourea (PTU) prevented melanin formation which is pronounced in adult extracts of this species.

flies.

Molecular weight determination. The MW of amylase-7 was determined in 0.1% sodium dodecyl sulfate (SDS) gels according to the method of Weber and Osborn (1969), using the buffer system of Dunker and Rueckert (1969). The polypeptide chain of cytochrome C served as reference for calculating the relative mobilities of other MW markers and amylase-7 (Figure 8). Before introduction into gel tubes, sample proteins were reduced in 1% solutions of SDS containing 1% 2-mercaptoethanol, with or without 4 M urea; these were boiled 5 min in a water bath. In one series, the amylase was carboxymethylated by substituting iodoacetamide for 2-me; in another, the potent reducing agent, dithiothreitol (0.0015 M), was used in the pretreatment and the polyacrylamide gel subjected to a 2-3 hr prerun to remove any oxidizing effects of residual ammonium persulfate.

The MW of the amylase-7 polypeptide chain was calculated to be 54,500 daltons. It migrated as a single band in all cases, with no sign of it giving rise to lower MW material (cf. Robyt et al., 1971). Earlier MW estimates of the active amylase molecule from *D. melanogaster* and *D. hydei* (Doane, 1970, unpublished), based on alterations of pore size in polyacryamide disc gels, had given values of roughly 50,000 daltons. The active molecule thus appears to consist of a single polypeptide chain. This is consistent with the results of the genetic analysis. The overwhelming evidence from studies of other α-amylases points to the enzyme being a monomer. (cf. Malacinski and Rutter, 1969; Cozzone et al., 1970; Keller et al., 1971; Lehrner and Malacinski, 1974; Karn et al., 1974).

Amino acid content. A standard amino acid analysis was performed on amylase-7. The average moles percent from duplicate runs is listed for each residue in the second column of Table 3. For comparison, the moles percent of amino acids in α-amylases from various vertebrate sources was calculated from the literature (columns 3-5). The value for tryptophan is relatively high for amylase-7 from *D. hydei*, but not so in comparison with other amylases. Oddly, tyrosine, which is also quite high among amylases, did not appear in the *Drosophila* material. Possibly, this is due to residual acrylamide in our purified samples (Lees and Paxman, 1973). Further comparison reveals real differences in lysine, arginine, serine, and alanine between our material and the other amylases listed. The content of aspartic acid, glycine, and phenylalanine appears somewhat different as well. Methionine and histidine are characteristically low in all amylases listed.

A small amount of galactosamine, judging from its chromatographic behavior, was detected in the amylase-7 samples.

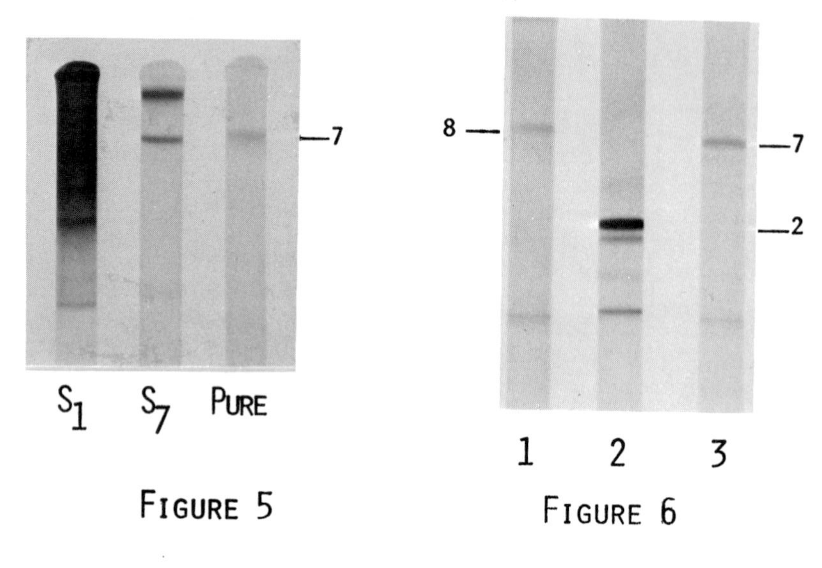

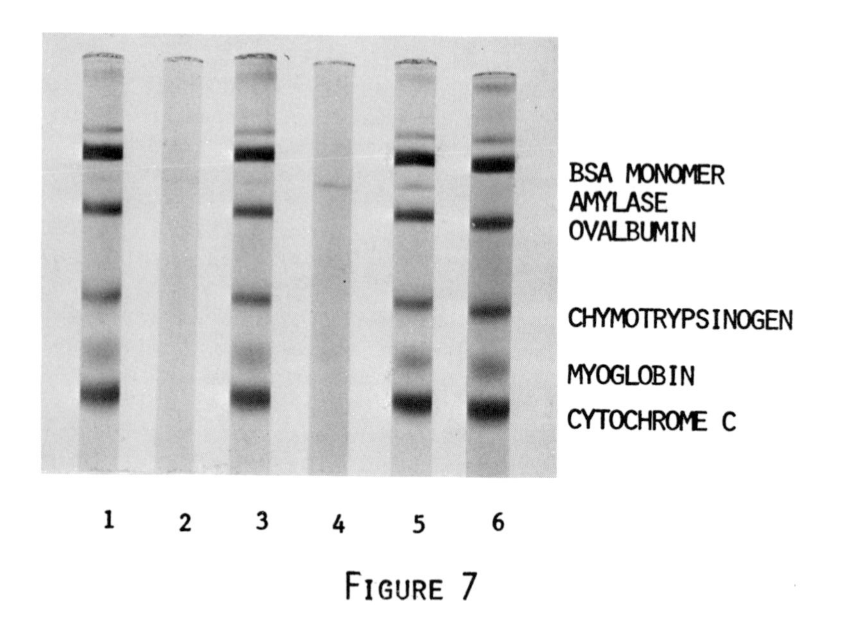

Figure 5. Disc gels containing extracts from various stages in the amylase purification (Figure 4), stained with Coomassie Blue. Figure 6. Disc gels containing purified samples of amylase-7 and -8 from adults of *D. hydei* (3 and 1, respectively), and of a partially purified (S_7) fraction of amylase-2 from larvae of *D. melanogaster* (2), stained with Coomassie Blue.

Figure 5,6, and 7 legends (cont.) Figure 7. Drosophila amylases and MW markers in 0.1% SDS gels stained with Coomassie Blue. Proteins were pretreated by boiling 5 min in a solution of 1% SDS, 1% 2-mercaptoethanol and 4 M urea. Samples, left to right, were: 1) amylase-7 plus MW markers, 2) amylase-8, 3) amylase-8 plus markers, 4) amylase-2, 5) amylase-2 plus markers, and 6) markers alone.

This was not quantitated, but carbohydrates have been reported to be associated with certain amylase molecules, e.g. human parotid (Kauffman et al., 1970; Keller et al., 1971; Watanabe and Keller, 1974), though not with all (cf. Lehrner and Malacinski, 1974). Amylase-7 can just barely be seen in polyacrylamide gels stained by the periodic acid-Schiff method (Fairbanks et al., 1971), but we do not know the degree to which this represents carbohydrate that is an intergral part of the amylase molecule or residual dextrins bound to it as

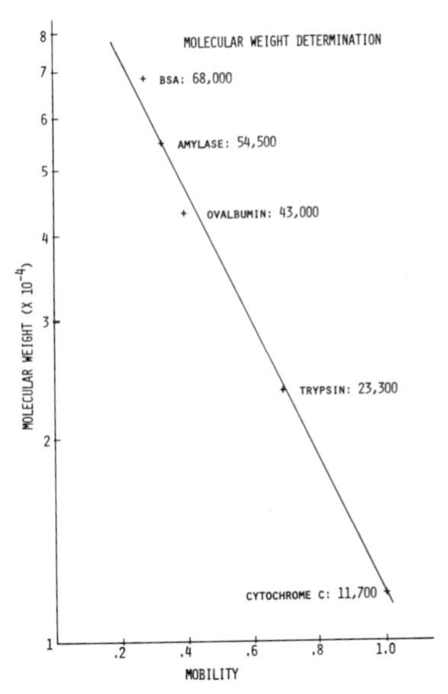

Figure 8. MW determination for amylase-7 from *D. hydei* in SDS gels. Relative mobility of polypeptide chains, with cytochrome C as reference, is plotted against the log of the MW for each. Four independent estimates of the MW of the amylase -7 chain were averaged; deviation in any given estimate was less than 5%. Within each estimate, mobilities were based on 2 to 6 replications for each protein.

a result of using glycogen precipitation during purification. The extent to which the amino sugar, galactosamine, and perhaps other carbohydrates may alter the migration rate of Drosophila amylase in SDS gels remains to be clarified.

OTHER DROSOPHILA AMYLASES

α-Amylase-8 from D. hydei. Amylase-8 was purified from the *Chile* strain of flies, with adjustments made for its specific activity in crude extracts being almost double that of *Zurich* extracts. Purified amylase-7 and -8 are shown in Figure 6. In the amylase-8 gel shown, two weakly stained derivative bands appeared in the No. 7 and 6 positions, but these do not photograph well. Only a single derivative stained in the amylase-7 gel in the No. 6 position. One may speculate that these derivative bands result from deamidation of the amylase-8 and -7, respectively. Certain vertebrate amylases generate "pseudoisozymes" toward the anode in a similar fashion, apparently by deamidation of asparagine and/or glutamine residues (Keller et al., 1971; Karn et al., 1974; Lehrner and Malacinski, 1974). However, it is clear that Drosophila amylases do not generate anodal bands to nearly the same degree under routine handling. Figure 9 shows zymograms produced by prolonged incubation of disc gels containing amylase-8 against a starch film subsequently stained with iodine. Although no inference should be drawn about the relative activities of the amylase bands figured, it is clear that amylase-8 may generate as many as 4-5 anodal bands, as well as one toward the cathode. Amylase-7, on the other hand, is not so prone; only 2-3 anodal derivatives have been detected so far by starch-I staining. Coomassie blue failed to reveal all of the anodal bands, indicating that they do not account for a large percentage of the total protein.

The rate of deamidation of asparaginyl residues depends strongly on the nature of neighboring amino acids (Robinson et al., 1970). If deamidation is responsible for the anodal bands generated by amylase-7 and -8, then the difference between their ability to produce these "pseudoisozymes" may be interpreted as further evidence of differences in their amino acid sequences. Why Drosophila amylases should differ so greatly from vertebrate amylases in their overall tendency to generate anodal "families" is not clear. The moles percent of aspartic acid is only slightly lower for amylase-7 than for the others in Table 3, and glutamic acid appears similar to the others.

The MW of the polypeptide chain of amylase-8 was determined in SDS gels, using a slightly different set of MW markers

TABLE 3
AMINO ACID CONTENT OF α-AMYLASES FROM *D. HYDEI* AND VARIOUS VERTEBRATES[1]

Amino Acid	D. hydei Amylase-7	Chick (Pa2) Pancreatic[2]	Rabbit (P_2) Pancreatic[3]	Human (B) Parotid[4]
Tryptophan	3.3	3.2	1.6	3.2
Lysine	7.5	4.4	4.1	4.5
Histidine	2.4	2.6	2.0	2.3
Arginine	2.8	6.5	5.5	5.9
Aspartic Acid	12.1	14.5	14.7	15.3
Threonine	5.4	4.9	4.7	4.1
Serine	10.4	6.1	6.7	6.3
Glutamic Acid	7.4	7.3	7.2	6.7
Proline	3.8	4.3	4.1	5.2
Glycine	13.0	10.2	11.6	9.9
Alanine	8.0	6.4	5.4	4.7
½-Cystine or Cysteic Acid	2.7	1.1	2.1	3.4
Valine	6.4	7.3	8.1	7.2
Methionine	1.0	2.5	1.3	2.0
Isoleucine	4.0	4.6	5.6	5.4
Luecine	5.9	4.4	6.0	5.0
Tyosine	0	4.4	4.3	4.1
Phenylalanine	3.9	5.3	4.8	5.2
Total:	100	100	100	100
MW:	54500	55000	54000	56000[5]

[1]Amino acid residues are expressed as moles percent. A standard amino acid analysis was performed on amylase-7. Samples containing 0.5-1 mg protein were hydrolyzed for 24 hr at 110°C in constantly boiling HCl in an evacuated desiccator, with or without 4% thioglycolic acid to preserve tryptophan. Hydrolyzates were analyzed on a Beckman 121 Amino Acid Analyzer operated in the high sensitivity mode employing a Beckman System AA Computing Integrator. Analyses for basic amino acids, tryptophan, and amino sugars were carried out on an 8 cm PA-35 resin column; neutral and acidic amino acids were analyzed on the standard AA-15 resin column. Cystine + cysteine and methionine were determined as cysteic acid and methionine sulfone, respectively, after performic acid oxidation.
[2]Lehrner and Malacinski, 1974.
[3]Malacinski and Rutter, 1969.
[4]Kauffman et al., 1970.
[5]Keller et al., 1971.

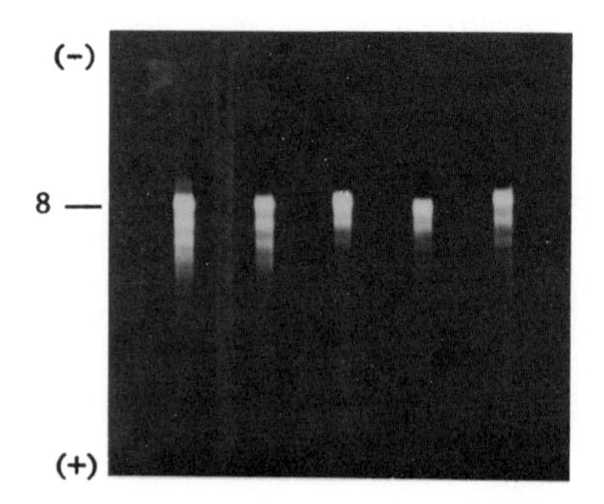

FIGURE 9

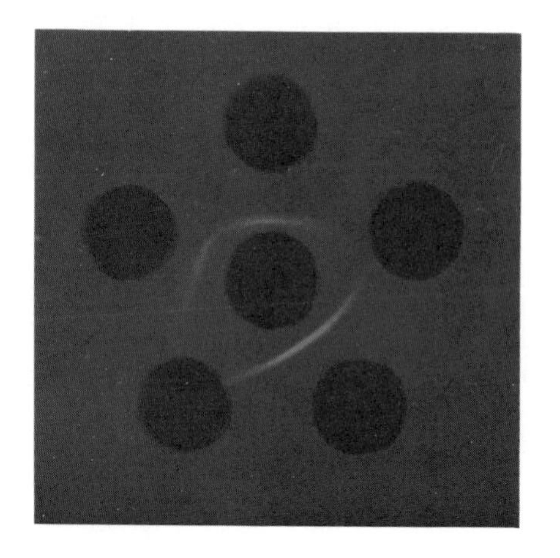

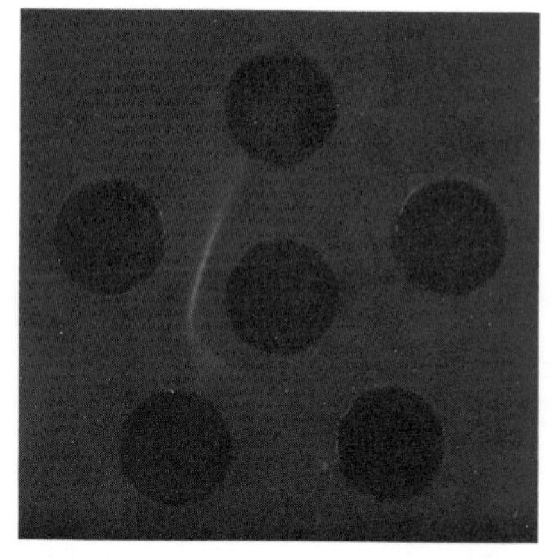

FIGURE 10

FIGURE 11

Figure 9. Photograph of 0.5% starch-in-acrylamide film, stained with iodine-KI reagent. Purified amylase-8 samples, separated by disc electrophoresis, were incubated 1½ hrs against the film to bring out "pseudoisozymes" in addition to isozyme No. 8. Figure 10. Ouchterlony plate showing cross-reactions between rabbit antiserum to amylase-7 (central well) and various concentrations of amylase-8 (top and lower right wells) and of amylase-7 (remaining outer wells). Figure 11. Ouchterlony plate showing cross-reactions between rabbit antiserum to amylase-7 (central well) and various concentrations of amylase-7 (outer left wells). The amylase-2 antigen from

Figure 9,10, and 11 legends (cont.) *D. melanogaster* (remaining outer wells) failed to form precipitin lines here over a range of concentrations which produced cross-reactions with the *D. hydei* antigens.

---

than those used for amylase-7. Figure 7 shows the position to which amylase-8 migrated. Its MW and molecular structure appear to be similar to those of amylase-7.

α-Amylase-2 from D. melanogaster. A strain of *D. melanogaster* characterized by a single major amylase band, *Amy*^{2} was chosen for purification. In this species, amylase activity is greatest in the late third instar larva, which produces the same banding pattern, qualitatively, as the adult. Crude larval extracts were pruified to the S_7 step (Figure 4). This partially purified amylase, when tested electrophoretically, was found to be almost pure for amylase-2 plus an anodally generated band in the No. 1 position (Figure 6). It did not contain major protein contaminants as was observed in S_7 fractions from adults of *D. hydei*. Glycogen precipitation is apparently far more specific for amylase in larval extracts than in adults; this was verified for *D. hydei* as well.

The amylase-2 fraction was sufficiently pure to use for MW determinations in SDS gels. As can be seen in Figure 7, the migration rate of both it and its derivative band was the same, so that a single amylase band appeared between the polypeptide chains of bovine serum albumin and ovalbumin. The MW of amylase-2, then, is indistinguishable from that of the amylases in *D. hydei*. Again, the evidence favors a monomeric structure. This being the case, Bahn's (1967) calculations on the size of the *Amy* gene in *D. melanogaster* and the number of nucleotide pairs separating it from the duplication in certain strains must be re-examined.

CROSS-REACTIVITY OF ANTISERA TO α-AMYLASE-7 WITH OTHER AMYLASES

Antibodies were prepared by injecting rabbits with purified samples of amylase-7 from *D. hydei*. Cross-reactions on Ouchterlony plates were readily obtained between antisera (or immunoglobulin fraction) to amylase-7 and both amylase isozymes from *D. hydei* (Figures 10, 11). Attempts to demonstrate a precipitin reaction between these antisera (from two rabbits) and the amylase-2 antigen from *D. melanogaster*, however, proved ambiguous. Despite a wide range of antigen/antibody concentrations tested, in only one experiment were extremely pale precipitin lines observed. Three attempts to repeat these results failed, so, if an interspecific cross-reaction does occur, it is very weak. This suggests a greater difference between the *hydei* and *melanogaster* amylases than anticipated on the basis of the many other features they have in common.

We also tested our amylase antigens from *Drosophila* against immunoglobulins from rabbit serum prepared against the "odds" and "evens" series of human salivary amylase isozymes (kindly provided by Dr. R. C. Karn). Conversely, the antisera to amylase-7 was tested against human saliva. Thus far, no cross-reactivity has been observed in either case.

MALE-SPECIFIC ENZYME WITH AMYLOLYTIC ACITVITY

Until recently, it was assumed that there was no duplication of the *Amy* locus in *D. hydei*, at least among the strains examined. Upon finding (Doane and Abraham, 1974) a sex-limited enzyme with extremely weak amylolytic activity in this species, the picture is no longer so simple. This male-specific enzyme is restricted to the mature accessory glands (paragonia) of adults (Figure 3) and can be demonstrated in extracts of axenically reared males. Curiously, it is also present in minute amounts in mature larval extracts, but the tissue source here is unknown.

The male-specific enzyme is apparently another α-amylase, judging from the features it shares in common with amylase-7 and -8 (Table 4). Its mobility in polyacrylamide gel electrophoresis, however, is much faster than other Drosophila amylases. Its controlling locus is also different, as can be deduced from the middle zymogram in Figure 3. Electrophoretic variants have not yet been found, but we have developed inbred lines which show quantitative differences for it. Because of its limited activity in single flies, it has proven difficult to analyze. Hope lies in the fact that it can be brought down by glycogen precipitation in much the same way as amylase-7 and -8. The restriction of this new starch-degrading enzyme to the mature male accessory glands is of interest from a functional point of view. Its relationship to the other amylases, also present in these glands, is provocative. So, too, is the prospect that it may represent a duplication of the *Amy* gene in *D. hydei*.

SUMMARY

A promising system to analyze genetic regulatory mechanisms is emerging from the analysis of amylase isozymes in *Drosophila*. With purification of the enzyme and production of antibodies to it, we are now in a position to study genetic

TABLE 4

STARCH DEGRADING ENZYMES IN *D. HYDEI*

Characteristic	α-Amylase-7 and -8		Male Enzymes	
	Chile	Zurich	Chile	Zurich
Electrophoresis - R_f	0.300 (*Amy*^{8})	0.343 (*Amy*^{7})	0.813	0.813
Tissue Specificities	Midgut, hemolymph, fat body, ovary (not testis), several other tissues, paragonia		Mature paragonia	Mature paragonia (very weak)
Substrates	Starch, amylose, amylopectin, glycogen, β-limit dextrins		Starch, glycogen (no others tested yet)	
Inhibitors	EDTA, EGTA, glutathione, α-amylase inhibitor from wheat grain		EGTA (presumably EDTA), α-amylase inhibitor from wheat grain	

control of the enzyme at the translational level in *D. hydei*. Dietary induced puffing at the *Amy* locus in larval midgut cells is expected to provide a handle for the analysis of transcriptional regulation. Procedures are being developed to selectively screen for mutations of "regulatory genes" which may act at either of these broadly defined levels in the expression of *Amy*. Eventually the system will be adapted to *D. melanogaster*.

Molecular characterization of Drosophila amylases has proven of interest for comparison with other amylases. It remains to be seen if the genetically determined isozymes reflect differences in their amino acid sequences, but a start has been made. There is an extensive literature in population genetics and evolutionary biology dependent upon the assumption that isozymes reflect such differences, which in turn result from base pair alterations in the structural genes for them. Considering this and the relative ease with which α-amylase may be purified, its isozymes offer an excellent opportunity to test this assumption in one of the prime targets of theoretical predictions, *Drosophila*.

ACKNOWLEDGEMENTS

This work was supported by NSF grants GB 8607 and GB 29276 and NIH GM 18729 and GM 397-14. We wish to thank Mrs. Ida Foster for her technical assistance, and Dr. R. C. Karn for supplying us with antibodies to human salivary amylase isozymes.

REFERENCES

Bahn, E. 1967. Crossing over in the chromosomal region determining amylase isozymes in *Drosophila melanogaster*. *Hereditas* 58: 1-12.

Bahn, E. 1971a. Cytogenetical localization of the *Amylase* region in *Drosophila melanogaster* by means of translocations. *Hereditas* 67: 75-78.

Bahn, E. 1971b. Position-effect variegation for an isoamylase in *Drosophila melanogaster*. *Hereditas* 67: 79-82.

Beerman, W. 1972,ed. *Developmental Studies on Giant Chromosomes*. Springer-Verlag, New York. 227 pp.

Berendes, H. J. 1963. The salivary gland chromosomes of Drosophila hydei Sturtevant. *Chromosoma* (Berl.) 14: 195-206.

Berendes, H. J. 1973. Synthetic activity of polytene chromosome. *Int. Rev. Cytol.* 35: 61-116.

Cozzone, P. and G. Marchis-Mouren 1972. Use of pulse labeling technique in protein structure determination: Ordering of the cyanogen bromide peptides from porcine pancreatic α-amylase. *Biochim. Biophys. Acta* 257: 222-229.

Daneholt, B. 1973. The giant RNA transcript in a Balbiani Ring. In: *Molecular Cytogenetics*, B. Hamkalo and J. Papaconstantinou, eds. Plenum Press, New York. pp. 155-165.

Daneholt, B. and H. Hosick 1973. Evidence for transport of 75S RNA from a discrete chromosome region via nuclear sap to cytoplasm in *Chironomus tentans*. *Proc. Nat. Acad. Sci.*, Wash. 70: 442-446.

Doane, W. W. 1967a. Quantitation of amylases in *Drosophila* separated by acrylamide gel electrophoresis. *J. Exp. Zool.* 164: 363-378.

Doane, W. W. 1967b. Cytogenetic and biochemical studies of amylases in *Drosophila melanogaster*. *Amer. Zool.* 7: 780.

Doane, W. W. 1967c. A novel approach to the analysis of enzymes separated by disc electrophoresis. *SABCO Jour.* (Japan) 3: 63-68.

Doane, W. W. 1969. *Amylase* variants in *Drosophila melanogaster:* Linkage studies and characterization of enzyme extracts. *J. Exp. Zool.* 171: 321-342.

Doane, W. W. 1970. *Drosophila* amylases and problems in cellular differentiation. In: *RNA in Development*, E. W. Hanly, ed. *Int. Symp. Prob. Biol.*, I, 1969. Univ. Utah Press, Salt Lake City. pp. 73-109.

Doane, W. W. 1971a. Isoamylases in Drosophila hydei: a system for the analysis of gene-specific puffing activity. *Drosophila Inform Serv.* 47: 100.

Doane, W. W. 1971b. X-ray induced deficiences of the *Amylase* locus in *Drosophila hydei*. *Isozyme Bull.* 4: 46-48.

Doane, W. W. 1974. Structural gene for α-amylase in *Drosophila hydei:* Genetics and Enzyme characteristics. (in preparation)

Doane, W. W. and I. Abraham 1974. Starch-degrading enzyme in Drosophila hydei restricted to mature male accessory gland. *Isozyme Bull.* 7: 38.

Doane, W. W. and M. M. Kolar 1974. Molecular weight of the polypeptide chain of α-amylase from an *Amy*^{7} strain of *Drosophila hydei*. *Isozyme Bull.* 7: 32-33.

Doane, W. W., M. M. Kolar, and P. M. Smith 1973a. Purification of α-amylase from *Drosophila*. *Genetics* 74: s64.

Doane, W. W., M. M. Kolar, and P. M. Smith 1973b. Purified α-amylase from D. hydei. *Drosophila Inform. Serv.* 50:50.

Doane, W. W., M. M. Kolar, and P. M. Smith 1974. Purification and molecular weight determination of α-amylase from *Drosophila hydei*. (in preparation)

Dunker, A. K. and R. R. Rueckert 1969. Observations of molecular weight determinations on polyacrylamide gel. *J. Biol. Chem.* 244: 5074-5080.

Fairbanks, G., T. L. Steck, and D. F. H. Wallach 1971. Electrophoretic analysis of the major polypeptides of the human erythrocyte membrane. *Biochemistry* 10: 2606-2617.

Gloor, H. 1971. Report: New mutants, hydei. *Drosophila Inform. Serv.* 47: 47-52.

Gregg, T. G. and L. A. Smucker 1965. Pteridines and gene homologies in the eye color mutants of *Drosophila hydei* and *Drosophila melanogaster*. *Genetics* 52: 1023-1034.

Grossbach, U. 1969. Chromosomen-Aktivität und biochemische Zelldifferenzierung in den Speicheldrüsen von Camptochironomus. *Chromosoma* 28: 136-187.

Horikx-Jacobs, A. 1968. Report on localization experiments with *Drosophila hydei*. *Genetics Lab., Univ. Leiden,* The Netherlands. (unpublished)

Hosbach, H. A., A. H. Egg, and E. Kubli 1972. Einfluss der Futterzusammensetzung auf Verdauungsenzym-Aktivitäten bei *Drosophila melanogaster*-Larven. *Rev. Suisse Zool.* 79: 1049-1060.

Karn, R. C., B. B. Rosenblum, and A. D. Merritt 1975. Genetic and post-transcriptional mechanisms determining human amylase isozyme heterogeneity. *Isozymes IV. Genetics and Evolution,* C. L. Markert, editor, Academic Press, New York. pp. 745-761.

Kauffman, D. K., N. I. Zagar, E. Cohen, and P. J. Keller 1970. The isoenzymes of human parotid amylase. *Arch. Biochem. Biophys.* 137: 325-339.

Keller, P. J., D. L. Kauffamn, B. J. Allan, and B. L. Williams 1971. Further studies on the structural differences between the isoenzymes of human parotid α-amylase. *Biochemistry* 10: 4867-4874.

Kikkawa, H. 1960. Further studies on the genetic control of amylase in *Drosophila melanogaster*. *Jap. J. Genet.* 35: 382-387.

Kikkawa, H. 1964. An electrophoretic study on amylase in *Drosophila melanogaster*. *Jap. J. Genet.* 39: 401- 411.

Lees, M. B. and S. A. Paxman 1973. Amino acid composition of myelin proteins recovered from polyacrylamide gels. *J. Neurochem.* 21: 1031-1034.

Lehrner, L. M. and G. M. Malacinski 1975. Genetic and structural studies of chicken α-amylase isozymes and their modified forms, and structural studies of hog amylase. *Isozymes IV. Genetics and Evolution.* C. L. Markert, editor, Academic Press, New York. pp. 727-743.

Lowry, O. H., N. J. Rosebrough, A. L. Farr, and R. J. Randall 1951. Protein measurement with the folin phenol reagent. *J. Biol. Chem.* 193: 265-275.

Loyter, A. and M. Schramm 1962. The glycogen-amylase complex as a means of obtaining highly purified α-amylases.

Biochim. Biophys. Acta 65: 200-206.

Malacinski, G. M. and W. J. Rutter 1969. Multiple molecular forms of α-amylase from the rabbit. *Biochemistry* 8: 4382-4390.

McCune, T. 1969. Amylase isozymes in natural populations of Drosophila melanogaster. *Drosophila Inform. Serv.* 44: 77-78.

Puijk, K. and G. de Jong 1972. α-Amylases in a population of D. melanogaster from Dahomey. *Drosophila Inform. Serv.* 49: 61.

Roberts, R. M. and J. S. Jones 1972. Improved apparatus for vertical gel electrophoresis. *Analyt. Biochem.* 49: 592-597.

Robinson, A. B., J. H. McKerrow, and P. Cary 1970. Controlled deamidation of peptides and proteins: An experimental hazard and a possible biological timer. *Proc. Nat. Acad. Sci.*, Wash. 66: 753-757.

Robyt, J. F., C. G. Chittenden, and C. T. Lee 1971. Structure and function of amylases. *Arch. Biochem. Biophys.* 144 : 160-167.

Watanabe, S. and P. J. Keller 1974. Isolation and partial characterization of glycopeptides from human parotid amylase A. *Biochim. Biophys. Acta* 336: 62-69.

Weber, K. and M. Osborn 1969. The reliability of molecular weight determinations by dodecyl sulfate-polyacrylamide gel electrophoresis. *J. Biol. Chem.* 244: 4406-4412.

ELECTROPHORETIC VARIANTS AS A TOOL IN THE ANALYSIS OF GENE ORGANIZATION IN HIGHER ORGANISMS

ARTHUR CHOVNICK, MARGARET McCARRON, WILLIAM GELBART AND JANARDAN PANDEY

Genetics and Cell Biology Section, Biological Sciences Group, The University of Connecticut
Storrs, Connecticut 06268

ABSTRACT. The rosy locus of chromosome 3 in *Drosophila melanogaster* controls the biosynthesis of the enzyme xanthine dehydrogenase. This report describes our progress in an investigation of the structural and functional organization of this locus as a model system for the study of higher organism gene organization. A number of wild-type isoalleles of the rosy locus have been isolated which are associated with the production of electrophoretically distinguishable XDH molecules. Large scale recombination experiments were carried out involving null enzyme mutants induced on electrophoretically distinct wild-type isoalleles. The genetic sites responsible for the mobility differences are followed as unselected markers in these crosses. Since electrophoretic variants represent alterations in the portion of the locus that encodes structural information, the resultant map of electrophoretic sites defines the minimal limits of that portion of the rosy locus. The present report additionally presents evidence that the structural element of the rosy locus is a single, uninterrupted sequence coding for a polypeptide molecule that is present in XDH as a dimer.

The significance of the present results for the general problem of higher organism gene organization is considered.

INTRODUCTION

During the past 20 years, investigations with microbial systems have provided a model for the functional organization of the genetic material which, in many respects, is clearly applicable to higher organisms. Nevertheless, recent evidence of major organizational features of the genetic material unique to higher eukaryotes has developed from several independent research directions, and the topic has become a major focus of current research. The reader is directed to several excellent recent reviews of this topic (Beermann, 1973; Davidson and Britten, 1973; Georgiev, 1972; Laird, 1973). Emerging from these studies is the proposition that the genetic unit in higher organisms is a much larger entity than its counterpart in prokaryotes, and that the difference involves order of magnitude quantities of DNA. Since eukaryote mRNA is monocistronic, and the size range of polypeptide gene products of higher organisms is similar to that of prokaryotes, considerable attention has been directed to the major portion of the higher organism gene, that which does not code for proteins. Several intriguing models of higher organism gene organization have been proposed (Britten and Davidson, 1973; Crick, 1971; Georgiev, 1972; Paul, 1972). A common feature of these models is that they postulate a large regulatory segment, an order of magnitude larger than the contiguous structural element under its control.

For the purpose of examining the validity of this generalized model of gene organization, the rosy locus of *Drosophila melanogaster* is an appealing system. Rosy is a genetic unit which controls the enzyme xanthine dehydrogenase (XDH), and which has been restricted to salivary chromosome region 87D8-12 (Lefevre, 1971a). Three observations place the structural information for XDH in or near rosy. (1) The rosy eye color mutants exhibit no detectable XDH activity (Glassman and Mitchell, 1959). (2) Heterozygotes possessing one dose of $\underline{ry}^{+}$ exhibit approximately 50% of normal enzyme activity while flies carrying three doses of $\underline{ry}^{+}$ have 150% activity (Grell, 1962). (3) Isoalleles differing in their XDH electrophoretic mobility map to rosy or its immediate vicinity (Yen and Glassman, 1965).

Large scale recombination studies involving tests of rosy mutant heteroalleles are facilitated by the addition of purine to the culture medium which permits only rare $\underline{ry}^{+}$ progeny to complete development. From such experiments, a fine structure map (Figure 1) of null enzyme rosy mutants was elaborated (Chovnick, Ballantyne and Holm, 1971), which has served as the basis for experiments bearing upon the mechanism of recombination in higher organisms. Since none of these rosy mutant alleles (Figure 1) has a detectable altered XDH product, they provide no information as to their structural (amino acid coding) or control (transcription and translation regulating) roles. Our strategy then, is to identify variants of standard $\underline{ry}^{+}$ alleles for which structural or control categorization is possible. These variants will be used to partition the locus into its structural and control components. Several classes of such variants are presently under investigation in our laboratory. They include wild-type isoalleles which exhibit variation in electrophoretic mobility, heat stability, and level of XDH activity, as well as mutants which exhibit interallelic complementation or altered sensitivity to purine, an inhibitor of XDH. In addition to distinguishing between models of gene organization, these classes of mutants serve as raw material for investigations of the genetic control of XDH during development.

The present report describes the identification of sites within the rosy locus responsible for variation in the electrophoretic mobility of XDH. Such sites surely are responsible for differences in charged amino acids in the XDH peptide. By superimposing these sites on the known genetic map of rosy, we will have a minimum estimate of the structural portion of the locus. At face value, the general class of models discussed above would predict that the electrophoretic sites will cluster in a small portion of the rosy map. Such prediction is based upon the assumption that the genetic map of X-ray

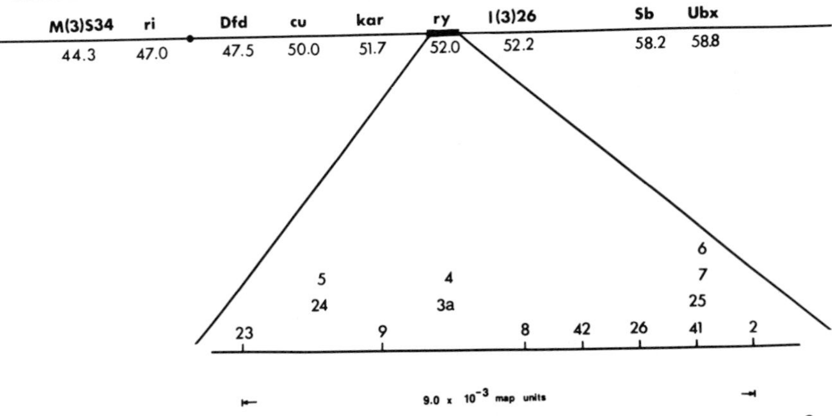

Fig. 1. A genetic map of the rosy region of chromosome 3. The map positions of various mutants used in this study are indicated, and the genetic fine structure of the rosy locus is summarized.

induced XDH-null mutants (Figure 1) includes the entire rosy functional unit. However, the results support the idea that most, if not all, of this map encodes structural information.

RESULTS

ELECTROPHORETIC VARIANTS AND WILD-TYPE ISOALLELES OF ROSY

As previously noted (Yen and Glassman, 1965; Charlesworth and Charlesworth, 1973), electrophoretic variants of XDH, which map to the rosy locus, are readily isolated from laboratory stocks and natural populations of *Drosophila melanogaster*. We have established a number of stable lines which exhibit single bands of XDH of uniform character upon electrophoresis

(Figure 2). Five discretely different mobilities of XDH have

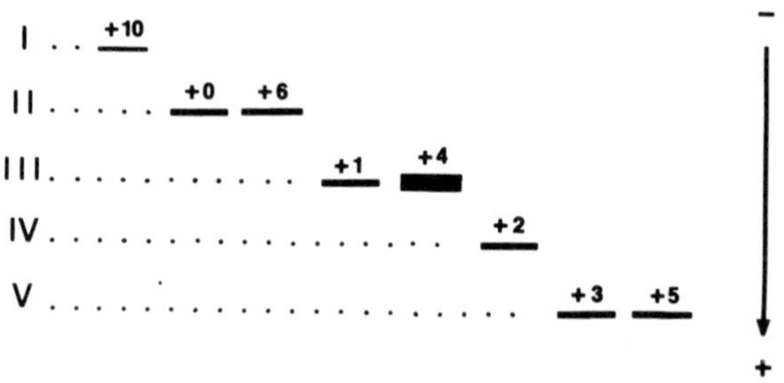

I . . +10
II . . . . . +0 +6
III . . . . +1 +4
IV . . . . +2
V . . . . +3 +5
– +

Fig. 2. The ry^+ isoalleles used in this study, diagrammatically arranged according to increasing electrophoretic mobility.

been identified. They are designated by Roman numeral superscripts in order of increasing mobility, XDH^{I} through XDH^{V}. Several mobility classes are each represented by two different wild type alleles. These alleles are derived from different sources, and might well possess different coding sequences leading to the same net charge on the XDH molecule.

FINE STRUCTURE MAPPING OF ELECTROPHORETIC SITES

In the absence of a selective procedure, it is impractical to directly map the genetic sites responsible for differences in electrophoretic mobility. Our experimental approach has been to induce null enzyme mutants upon each of the ry^+ isoalleles, and then to utilize the purine selective system to recover wild type recombinants in mutant heteroallele mapping experiments. Wild type recombinants recovered from experiments involving mutants of identical ry^+ ancestry invariably exhibit the parental electrophoretic class of XDH. Recombination experiments involving mutants induced on different ry^+ alleles permit us to follow the electrophoretic sites as unselected markers. Electrophoretic classification of the wild type recombinant survivors permits localization of these electrophoretic sites as well as the null enzyme mutant sites (McCarron, Gelbart, and Chovnick, 1974). In this experimental system, the ry^{+0} isoallele is our standard, and null enzyme mutants of ry^{+0} have been mapped relative to each other (Figure 1). Our experimental logic is illustrated in Figure 3. Utilizing a previously localized null enzyme rosy mutant of the ry^{+0} isoallele as a fixed reference point (ry^{8}), fine structure recombination tests are carried out against a series of null enzyme mutants induced on a different isoallele (ry^{+1}) . Some of these mutants are located to the left of our reference point (ry^{10X}), and others to the right (ry^{10Y}). In mutant hetero-

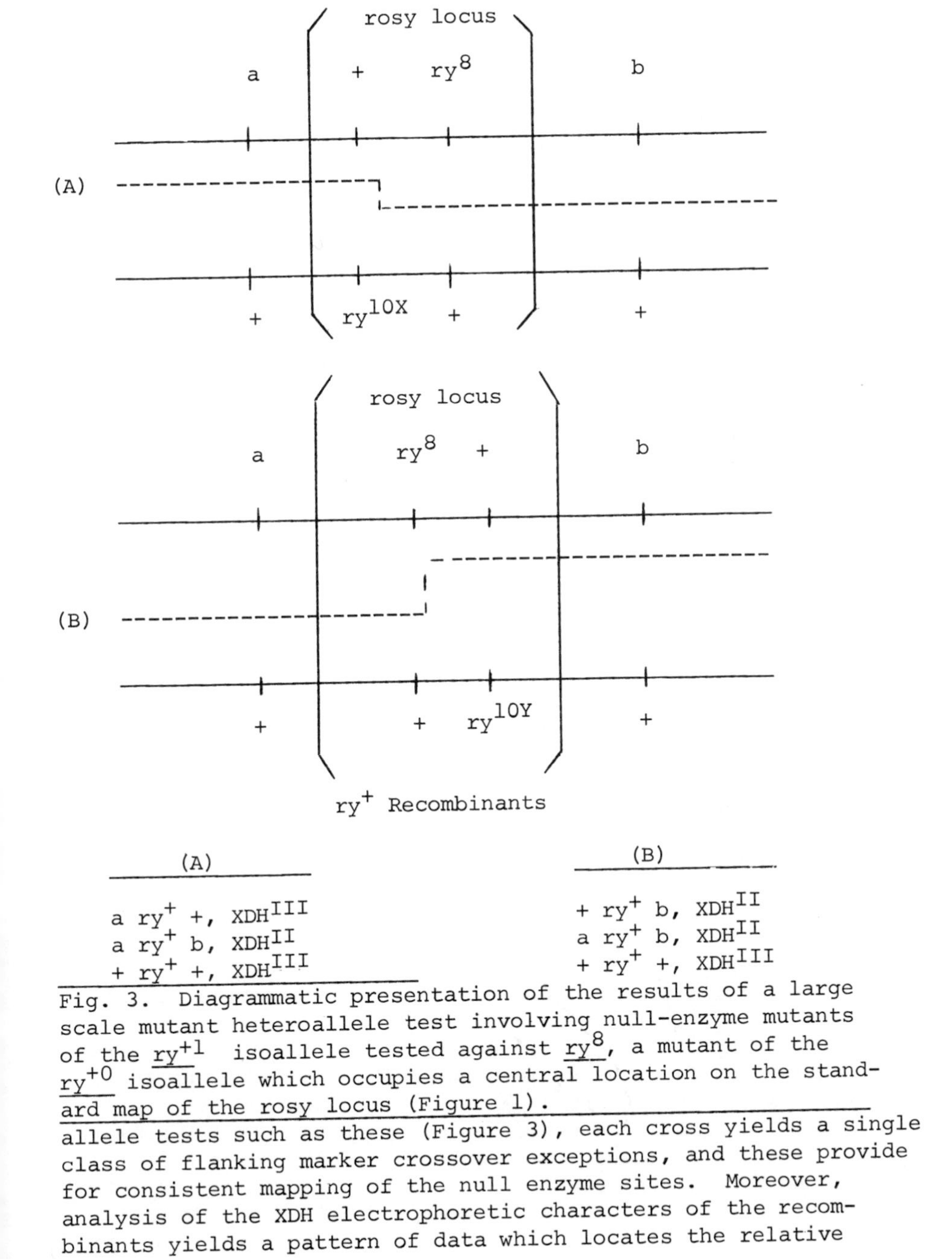

Fig. 3. Diagrammatic presentation of the results of a large scale mutant heteroallele test involving null-enzyme mutants of the $\underline{ry^{+1}}$ isoallele tested against $\underline{ry^{8}}$, a mutant of the $\underline{ry^{+0}}$ isoallele which occupies a central location on the standard map of the rosy locus (Figure 1).

allele tests such as these (Figure 3), each cross yields a single class of flanking marker crossover exceptions, and these provide for consistent mapping of the null enzyme sites. Moreover, analysis of the XDH electrophoretic characters of the recombinants yields a pattern of data which locates the relative position of the electrophoretic site(s) as well. Thus, for the crosses described in Figure 3 involving ry^{8}, a mutant of the ry^{+0} isoallele (XDH^{II}), and a series of null enzyme mutants of the ry^{+1} (XDH^{III}) isoallele, the pattern of results indicates that the genetic basis for the electrophoretic difference maps to the right side of ry^{8}.

Figure 4 illustrates the results obtained from the effort to further localize the electrophoretic site inferred from the results illustrated in Figure 3. In these experiments, four different ry^{100} series mutants were tested against ry^{41}, the rightmost mutant of the ry^{+0} isoallele (Figure 1). All of the ry^{100} series mutants were located to the left of ry^{41}, and the results of these crosses are summarized in Figure 4. The major features of these data are: (1) The crossovers are all XDH^{III} indicating that the electrophoretic site lies to the right of all crossover points, (2) the frequent occurrence (20/63) of coincident conversion (co-conversion) of the electrophoretic site with conversion of ry^{41} is taken as a measure of proximity of the electrophoretic site to ry^{41}. (3) Conversion of ry^{41} does occur without conversion of the electrophoretic site, indicating that the two sites, in fact, are distinct. Hence we define the existence of a site, ry^{e111}, responsible for the electrophoretic difference between the ry^{+0} and ry^{+1} isoalleles. In addition, we are able to describe ry^{+0} as possessing the e111s (slow) alternative at this site, in contrast to ry^{+1} which we describe as e111f (fast).

The observation of coincident conversion described in the present analysis (Figure 4) is an important, and potentially most useful feature of recombination involving exceedingly short genetic intervals. The significance of the phenomenon of co-conversion was recognized quite early from fungal studies as demonstrating that the conversion event involves a segment of DNA rather than a site. Fogel, Hurst, and Mortimer (1971) have studied the frequency of co-conversion as a function of distance between the sites, and have shown a linear relationship inversely proportional to distance, and thus a direct measure of proximity.

Following the experimental approach illustrated in Figures 3 and 4, we have carried out additional experiments designed to identify genetic sites responsible for the electrophoretic mobility differences between the ry^{+0} standard, and the isoalleles ry^{+2}, ry^{+3}, ry^{+4}, and ry^{+5}. A detailed description of this work will be presented elsewhere. Figure 5 summarizes the results of these experiments in the form of a genetic map of the rosy locus in which ry^{5} and ry^{41} represent the left and right ends of the pre-existing map of null enzyme mutants (Figure 1). The relative positions of seven identified electrophoretic sites are indicated. It is of some interest that two

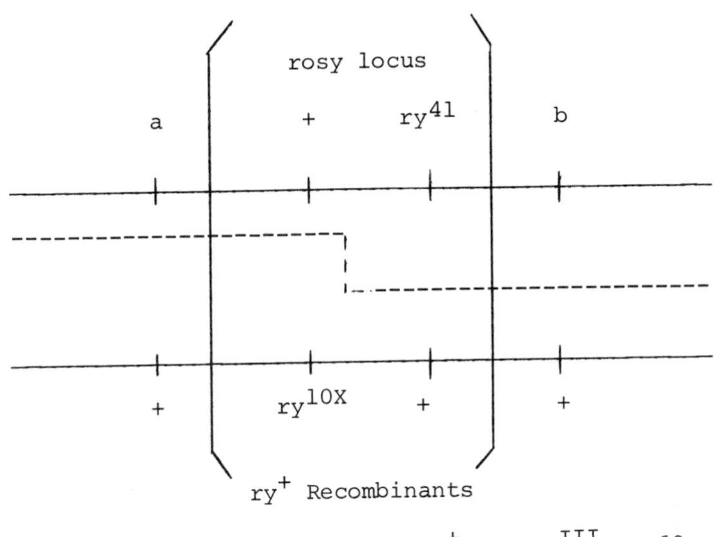

	ry^{+} Recombinants	N
Crossover	a ry^{+} +, XDH^{III}	60
Convertant, ry^{10X}	+ ry^{+} +, XDH^{III}	47
Convertant, ry^{41}	a ry^{+} b, XDH^{II}	43
Co-conversion, ry^{41} + e	a ry^{+} b, XDH^{III}	20
Total Exceptions		170 $7.03{\times}10^{6}$

Fig. 4. Diagrammatic presentation of the results of a large scale mutant heteroallele test involving null-enzyme mutants of the ry^{+1} isoallele tested against ry^{41}, a mutant of the ry^{+0} isoallele which occupies a marginal position on the standard map of the rosy locus (Figure 1).

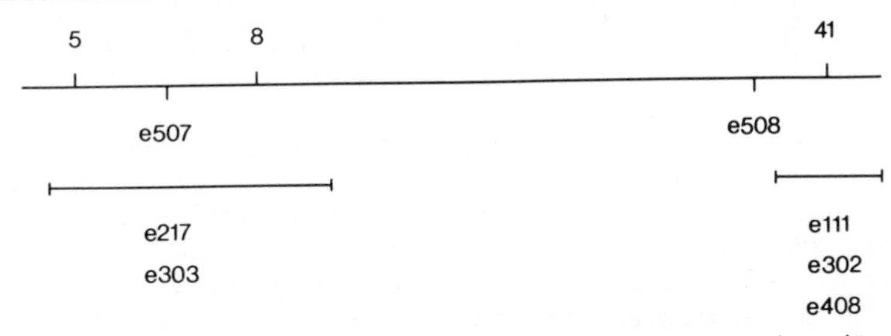

Fig. 5. The positions of the known electrophoretic sites relative to the standard null-XDH mutants ry^{5}, ry^{8}, and ry^{41} (see Figure 1).

site differences (ry^{e507} and ry^{e508}), distinguishing ry^{+0} and ry^{+5} have been identified. Similarly, two site differences were identified from the ry^{+0} - ry^{+3} tests (ry^{e303} and ry^{e302}).

ONE OR TWO STRUCTURAL ELEMENTS?

Consider next the distribution of the electrophoretic sites elaborated in the present report (Figure 5). Clearly, structural sites within the rosy locus are not confined to a small portion of the map. Indeed, at first glance, one might conclude that most of the standard rosy locus map (Figure 1) represents a single structural element coding for the XDH peptide. However, a comparison of the distribution of the seven identified electrophoretic sites (Figure 5) with the distribution of known null activity mutant alleles (Figure 1) suggests that the genetic basis for electrophoretic variation in XDH may be restricted to two quite separable sectors of the standard map. The possibility exists that the structural information for XDH resides in two elements, perhaps separated by a control region of undefined length. On this model, each structural element codes for a different peptide, which we designate as the α and β subunits of XDH. An active XDH molecule would be a heterodimer consisting of one molecule of each subunit. Let the leftmost group of electrophoretic sites (Figure 5) reside in the coding element for the α subunit, and the rightmost group reside within the β coding element. Let α^S and α^F represent subunits of slower and faster mobilities, respectively, and let β subunits be similarly designated. For consideration of this model, the XDH moiety produced by $\underline{ry^{+0}}$ would then be $\alpha^S\beta^S$, while that produced by $\underline{ry^{+5}}$ would be $\alpha^F\beta^F$. The XDH electrophoretic pattern of $\underline{ry^{+0}/ry^{+5}}$ heterozygotes consists of a heavy band of intermediate mobility and light bands corresponding in mobility to $\underline{ry^{+0}}$(XDH^{II}) and $\underline{ry^{+5}}$ (XDH^{V}) homozygotes. Observations such as these had led previously to the suggestion that *Drosophila* XDH was a dimer consisting of two units of a single peptide, the $\underline{ry^{+}}$ product (Yen and Glassman, 1965). We refer to this model as the single subunit model. On the one subunit model, in order of increasing mobility, these bands correspond to: XDH^{II} - XDH^{II} homodimers, XDH^{II} -XDH^{V} heterodimers and XDH^{V} - XDH^{V} homodimers. On the two subunit model, the slowest migrating form is $\alpha^S\beta^F$ and $\alpha^F\beta^S$ molecules and the fastest form is $\alpha^F\beta^F$. Thus, $\underline{ry^{+0}}$/$\underline{ry^{+5}}$ heterozygotes exhibit identical electrophoretic patterns on either model. However, a distinction between these two models is possible.

The two subunit model predicts that $\alpha^S\beta^S/\alpha^F\beta^F$ and $\alpha^S\beta^F/\alpha^F\beta^S$ individuals will generate identical three-banded hybrid electrophoretic patterns. On the one subunit model, these same heterozygotes would be designated XDH^{II}/XDH^{V} and XDH^{int}/XDH^{int}, respectively, the XDH^{II}/XDH^{V} heterozygote would be expected to produce the three-banded hybrid electrophoretic pattern while the XDH^{int}/XDH^{int} individual would produce a single band of XDH^{int} mobility. Essentially, the one subunit model predicts a cis-trans difference whereas the two subunit model predicts no such difference (See Figure 6).

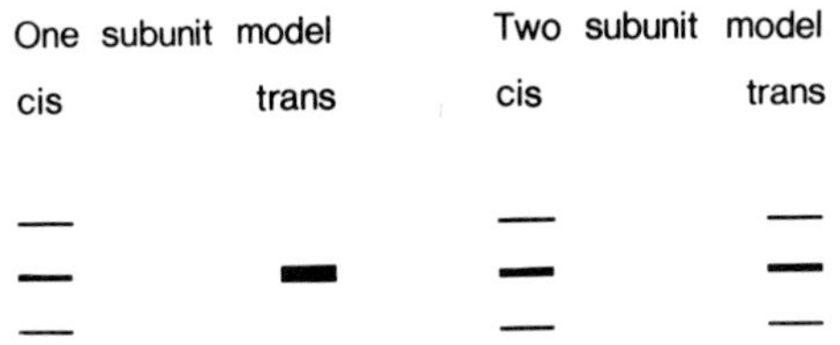

One subunit model		Two subunit model	
cis	trans	cis	trans

Fig. 6. The predictions of the one and two subunit models. Diagrams of the electrophoretic patterns expected on either model are included. The cis configuration is $ry^{e507S}\ ry^{e508S}/ry^{e507F}ry^{e508F}$ trans is $ry^{e507S}\ ry^{e508F}/ry^{e507F}\ ry^{e508S}$.

For the sake of this discussion, the ry^{+0}/ry^{+5} heterozygous genotype may be rewritten as $ry^{e507S}\ ry^{e508S}/ry^{e507F}\ ry^{e508F}$. This represents the cis configuration of electrophoretic sites. From chromosomes recovered as co-conversions in recombination tests, we were able to generate ry^+ allele combinations $ry^{e507S}\ ry^{e508F}$ and $ry^{e507F}\ ry^{e508S}$, and thus to carry out the cis-trans test. Examination of XDH electrophoretic patterns revealed a clear-cut cis-trans difference entirely consistent with the single subunit model (Fig. 7). Hence, we conclude that most, if not all of the standard genetic map of the rosy locus (Fig. 1) represents a single, uninterrupted DNA sequence which is the XDH structural element. Moreover, we submit that this structural element consists of a unique DNA sequence of some 3 X 10^3 nucleotide pairs in length. That the structural element is a unique DNA sequence is documented by our total experience with fine structure recombination experiments involving some 2 X 10^3 ry^+ recombinants analyzed. We have seen no evidence of the "unequal" crossing over that would occur if repeat sequences were present. The physical length of the structural element derives from the molecular weight of *Drosophila* XDH which is estimated at 250,000 daltons (Glassman et al., 1966). Assuming the biologically active XDH molecule to consist of two identical subunits (Yen and Glassman, 1965) we are led to a subunit size of some 1,000 amino acids, which requires a coding sequence of some 3 X 10^3 nucleotide pairs. It is of interest to note that this estimate of the physical length of the XDH structural element is strikingly similar to an estimate based upon recombination data. Lefevre (1971b) correlates 0.01 map units with a length of 3.7-3.8 X 10^3 nucleotide pairs in *Drosophila melanogaster*. The standard map of

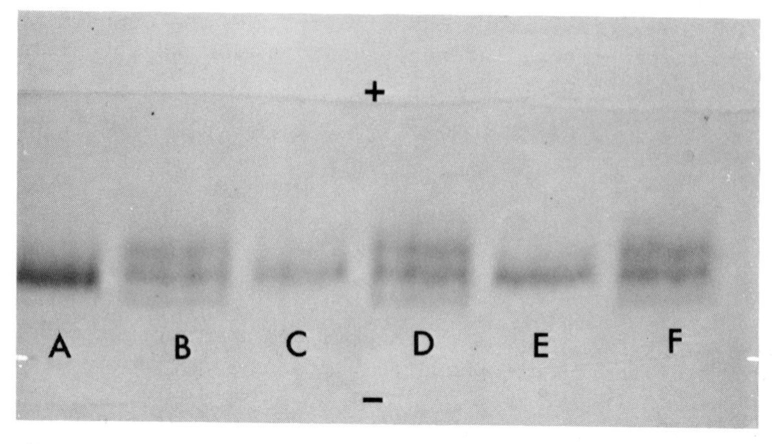

Fig. 7. The observed XDH electropherogram of alternating trans (A,C,E) and cis (B,D,F) heterozygotes.

the rosy locus (Figure 1) extends over 0.009 map units and thus 3.3-3.4 x 10^3 nucleotide pairs.

DISCUSSION

Let us return to the problem of gene organization and the general model under test. Clearly, the present data do not support the model which anticipated a cluster of electrophoretic sites localized to a short sector of the total map. In formulating this model, we tacitly assumed that the distribution of mutants would directly reflect the relative sizes (both physical and recombinational) of the postulated control and structural elements. With the failure of the present data to support the model, we should either reject the general model or question the validity of our underlying assumptions.

Consider our assumptions. (1) The distribution of mutants within the control and structural elements directly reflects the relative lengths of their nucleotide sequences. Consider the nature of rosy eye color mutants which comprise the standard map of the rosy locus. Since even low levels of XDH activity permit expression of the ry^+ phenotype, the visual selective procedure requires that these mutants lack all activity. Clearly, single base-pair alterations in the structural element have the potential to eliminate all XDH activity. In the control element, as Crick (1971), Paul (1972) and others perceive it, such alterations generally should have a less drastic effect on the enzyme activity. These workers intimate that control specificity resides in extensive nucleotide sequences, in contrast to the triplet codon specificity of structural regions. Thus, while most single base-pair alterations in control elements might cause slight quantitative changes in XDH activity, they would not be expected to produce the complete disruption of activity necessary for expression of the mutant eye color. In contrast to prevailing scientific folklore, we believe single site alterations to comprise the bulk of intragenic events produced by X-irradiation, the agent that generated these rosy mutants. Our assertion is best documented by the work of Malling and De Serres (1973), involving a study of X-ray induced ad-3B mutants in *Neurospora crassa*. Less extensive, but supportive data is available in Drosophila as well. Thus, three of the five X-ray induced maroon-like mutants in *Drosophila melanogaster* are inferred to be base substitutions since they participate in interallelic complementation (Chovnick et al., 1969). Extrapolating from these observations, we presume most of the rosy mutants also to be single base-pair alterations. Inadvertently, the conspiracy of X-ray mutagenesis and the visual selection of null-XDH eye color mutants restricted the lesions recovered to the rosy structural element. Thereby, we are able to account for the concordance of the XDH-null and electrophoretic maps. If this explanation is correct, the use of alternative mutagens and more sensitive selective procedures may yet permit us to examine variation within the control element.

(2) The recombination rates within control and structural elements are identical per unit of nucleotide length. In fact, the possibility exists that most exchange may take place within the structural element, thus producing the biased distribution of mutant map sites which we have observed. Why would a mechanism limiting exchange in this manner evolve? Control elements have been implicated as regions containing middle repetitive DNA. If so, recombination in these regions will be characterized by unequal exchanges. Assuming an optimal size and sequence for a particular control element, we envisage a selective advantage favoring the evolution of mechanisms which inhibit exchange in this element, thereby stabilizing it. Analogous disadvantages accorded unequal exchange may have participated in limiting the positions of highly redundant loci (18S and 28S rRNA [bobbed] genes -- Ritossa and Spiegelman, 1965; histone genes -- Pardue et al., 1972) to centric regions, where the frequencies of exchange are reduced. Recall also that most rapidly annealing DNA localizes to centric regions (Gall, Cohen, and Polan, 1971). Moreover, transposition of the bobbed locus to a more distal position via inversion of the X-chromosome results in increased recombination and the unequal exchange associated with such redundancy (Atwood, 1969; Schalet, 1969). We note, as an exception to

the above pattern, that the redundant 5S rRNA genes are located in the distal polytene region 56E-F on the right arm of chromosome 2 (Wimber and Steffenson, 1970). Despite this exception, we believe that mechanisms exist for regional control of recombination frequency. Similarly, on the intralocus level, exchange within a control element may be inhibited, thereby creating the illusion that most of the locus is structural. The pertinence of this argument for the present study of the rosy locus is obvious, and assessment of this possibility is under investigation.

Let us next consider the possibility that, in fact, the assumptions underlying the present experiment are valid, and that the general model is incorrect. We are then left with the fact that higher organisms possess an order of magnitude more DNA than is needed to code for proteins. We must search elsewhere for the *raison d'etre* of this excess DNA.

Other functions may involve control of DNA replication and chromosome behavior. Both of these processes involve the interactions of large numbers of chromosomal sites (Plaut, Nash, and Fanning, 1966; Baker and Carpenter, 1972). Certainly, a variety of initiator, binding, and receptor sites must be required for these processes to operate normally; the excess DNA may serve as such sites.

Our long range interest is in the organization of a genetic unit in higher organisms, and in particular, the control mechanisms underlying differential gene function. Establishment of the limits of the rosy structural element is a necessary prerequisite to the identification of mutants in the control process. The work reported herein represents our progress in delimiting said structural element. Further localization of electrophoretic and other classes of structural variants will aid in clearly defining these limits.

ACKNOWLEDGEMENT

This investigation was supported by a research grant, GM-09886, from the Public Health Service.

REFERENCES

Atwood, K.C. 1969. Some aspects of the bobbed problem in *Drosophila*. *Genetics* (Suppl.) 61: 319-327.

Baker, B.S. and A.T.C. Carpenter 1972. Genetic analysis of sex chromosomal meiotic mutants in *Drosophila melanogaster*. *Genetics* 71: 255-286.

Beerman, W. 1973. Chromomeres and genes. *Cell Differentiation* 4: 1-33.

Charlesworth, B. and D. Charlesworth 1973. A study of linkage disequilibrium in populations of *Drosophila melanogaster*. *Genetics* 73: 351-359.

Chovnick, A., G.H. Ballantyne, and D.G. Holm 1971. Studies on gene conversion and its relationship to linked exchange in *Drosophila melanogaster*. *Genetics* 69: 179-209.

Chovnick, A., V. Finnerty, A. Schalet, and P. Duck 1969. Studies on genetic organization in higher organisms. I. Analysis of a complex locus in *Drosophila melanogaster*. *Genetics* 62: 145-160.

Crick, F.H.C. 1971. General model for the chromosomes of higher organisms. *Nature* 234: 25-27.

Davidson, E.H. and R.J. Britten 1973. Organization, transcription, and regulation in the animal genome. *Quart. Rev. Biol.* 48: 565-612.

Fogel, S., D.D. Hurst, and R.K. Mortimer 1971. Gene conversion in unselected tetrads from multipoint crosses. In: *Stadler Genetics Symposia*. Vol. 1 and 2. G. Kimber and G.P. Redei, editors. Agricultural Experimental Station, University of Missouri, Columbia, Missouri. pp. 89-110.

Gall, J.G., E.H. Cohen, and M. Polan 1971. Repetitive DNA sequences in *Drosophila*. *Chromosoma* 33: 319-344.

Georgiev, G.P. 1972. The structure of transcriptional units in eukaryotic cells. *Current Topics in Develop. Biol.* 7: 1-60.

Glassman, E. and H.K. Mitchell 1959. Mutants of *D. melanogaster* deficient in xanthine dehydrogenase. *Genetics* 44: 153-162.

Glassman, E., T. Shinoda, H.M. Moon, and J.D. Karam 1966. In vitro complementation between non-allelic Drosophila mutants deficient in Xanthine Dehydrogenase. IV. Molecular weights. *J. Molec. Biol.* 20: 419-422.

Grell, E.H. 1962. The dose effect of $\underline{ma\text{-}l^+}$ and $\underline{ry^+}$ on xanthine dehydrogenase activity in *Drosophila melanogaster*. *Z. Vererb.* 93: 371-377.

Laird, C. 1973. DNA of Drosophila chromosomes. *Ann. Rev. Genet.* 7: 177-204.

Lefevre, G., Jr. 1971a. Cytological information regarding mutants listed in Lindsley and Grell 1968. *D.I.S.* 46: 40.

Lefevre, G., Jr. 1971b. Salivary chromosome bands and the frequency of crossing over in *Drosophila melanogaster*. *Genetics* 67: 497-513.

Malling, H.V. and F.J. De Serres 1973. Genetic alterations at the molecular level in X-ray induced ad-3B mutants of *Neurospora crassa*. *Radiat. Res.* 53: 77-87.

McCarron, M., W. Gelbart, and A. Chovnick 1974. Intracistronic mapping of electrophoretic sites in *Drosophila melanogaster*:

Fidelity of information transfer by gene conversion. *Genetics* 76: 289-299.

Pardue, M.L., E. Weinberg, L.H. Kedes, and M.L. Birnstiel 1972. Localization of sequences coding for histone messenger RNA in the chromosomes of *Drosophila melanogaster*. *J. Cell Biol.* 55: 199a (Abstr.).

Paul, J. 1972. General theory of chromosome structure and gene activation in eukaryotes. *Nature* 238: 444-446.

Plaut, W., D. Nash, and T. Fanning 1966. Ordered replication of DNA in polytene chromosomes of *Drosophila melanogaster*. *J. Molec. Biol.* 16: 85-93.

Ritossa, F.M. and S. Spiegelman 1965. Localization of DNA complementary to ribosomal RNA in the nucleolus organizer region of *Drosophila melanogaster*. *Proc. Natl. Acad. Sci. U.S.A.* 53: 737-745.

Schalet, A. 1969. Exchanges at the bobbed locus of *Drosophila melanogaster*. *Genetics* 63: 133-153.

Wimber, D.E. and D.M. Steffensen 1970. Localization of 5S RNA genes on *Drosophila* chromosomes by RNA-DNA hybridization. *Science* 170: 639-642.

Yen, T.T.T. and E. Glassman 1965. Electrophoretic variants of xanthine dehydrogenase in *Drosophila melanogaster*. *Genetics* 52: 977-981.

THE EVOLUTION OF REGULATORY MECHANISMS STUDIES ON THE MULTIPLE GENES FOR LYSOZYME

NORMAN ARNHEIM
Department of Biochemistry
State University of New York
Stony Brook, N. Y., 11790

ABSTRACT. Two forms of lysozyme are known in birds. They differ markedly in molecular weight, amino-terminal amino acid sequence, amino acid composition and immunological cross reactivity; and they appear to be coded for by two distinct genetic loci.

A number of birds contain the genes for both of these lysozymes but they differ with respect to whether one or both lysozymes are present in egg white. The species-specific controls over egg white lysozyme synthesis in the oviduct must have arisen during the evolution of these birds. Studies are currently being carried out on the mechanisms behind these control processes, since information on changes in gene expression patterns during phylogeny will lead to new insights into the molecular basis of evolution.

Lysozymes are enzymes which break down the peptidoglycan of the bacterial cell wall. Enzymes with this activity have been found in microorganisms, plants, invertebrates and vertebrates (Fleming, 1922; Takeda et al., 1966; Jollès, 1969; Ghuysen, 1968; Glazer et al., 1969; Tsugita, 1971; Imoto et al.,1972; Powning and Davidson, 1973).

Until recently only one form of lysozyme was known in vertebrates. This form is typified by the much studied chicken egg white lysozyme whose complete amino acid sequence and three-dimensional structure is known (Canfield, 1963; Jollès et al., 1963; Blake et al., 1965). Amino acid sequence studies on the lysozymes found in Man (Canfield, 1968; Canfield et al., 1971; Jollès and Jollès, 1971), other primates (Herman et al., 1973), rodents (Riblet, 1974; Mross et al., unpublished studies) and the egg white of a number of birds besides the chicken (see Dayhoff, 1972) have been carried out. The results have shown these enzymes to be homologous with chicken egg white lysozyme, and I will refer to them as "chick type" lysozymes.

Several years ago it was unexpectedly found that one bird, the Embden goose, apparently lacked a "chick type" lysozyme in its egg white and had in its place a radically distinct form of lysozyme (Dianoux and Jollès 1967; Canfield and McMurry, 1967). I will refer to this type of enzyme as the "goose type".

STRUCTURAL COMPARISONS OF "GOOSE TYPE" AND "CHICK TYPE" LYSOZYMES

Since the initial discovery of the "goose type" lysozyme in the Embden goose, this type of enzyme has also been found in several other bird species (Arnheim and Steller, 1970; Arnheim, 1974; Prager et al., 1974). Extensive structural studies have been carried out on highly purified "goose type" lysozymes from the Embden goose and the black swan. The "goose type" lysozymes have a higher molecular weight than the "chick type" enzymes. Amino- terminal analysis and SDS acrylamide gel electrophoresis indicate that the "goose type" enzymes have a molecular weight of 21,500 (Arnheim et al., 1973b) in contrast to a molecular weight of around 14,500 for "chick type" enzymes (Imoto et al., 1972).

Immunological studies on the two types of egg white lysozyme also indicate that they differ radically in overall structure. Antibodies directed against chicken egg white lysozyme will not react with any "goose type" enzyme, nor will antibodies directed against Embden goose egg white lysozyme react with any of the "chick type" enzymes (Arnheim and Steller, 1970; Hindenburg et al., 1974; Prager et al., 1974).

Amino acid composition data on a typical "chick type" and "goose type" lysozyme are shown in Table I. The data indicate two radically different structures. Amino-terminal amino acid sequence analysis has been carried out on two "goose type" lysozymes (Canfield et al., 1971; Arnheim et al., 1973b). A comparison of the first 20 amino acid residues of the black swan "goose type" lysozyme with chicken egg white lysozyme is shown in Table II. Little, if any, amino acid sequence similarity can be recognized. In summary, the striking differences in chemical and immunological properties, in addition to some limited genetic data (Arnheim, 1974), suggest that these two lysozymes are the products of two distinct genetic loci. Whether or not they arose by gene duplication and divergence or by convergence must await more data on the amino acid sequence and three dimensional structure of a "goose type" enzyme and its comparison with the already known structure of chicken egg white lysozyme.

SPECIES VARIATION IN EGG WHITE LYSOZYME CONTENTS AND ITS ADAPTIVE SIGNIFICANCE

During our studies on egg white lysozymes from different bird species, we discovered a new phenotype with respect to the content of the two forms of lysozyme. Prior to this finding only one of the two lysozyme types had been found in the egg white of a single species. Some species, like the chicken and Peking duck, contain only "chick type" lysozymes; while the

TABLE I

AMINO ACID COMPOSITION OF EGG WHITE LYSOZYMES[1]

	Black Swan "Goose Type"	Chicken
Trp	3	6
Lys	18	6
His	5	1
Arg	11	11
Asp	20	21
Thr	13	7
Ser	9	10
Glu	15	5
Pro	5	2
Gly	20	12
Ala	15	12
Half-Cys	4	8
Val	10	6
Met	3	2
Ile	11	6
Leu	7	8
Tyr	9	3
Phe	3	3
Total	181	129

1) The composition data comes from Arnheim et al (1973b)("Goose type") and Canfield (1963) (Chicken enzyme).

Embden goose apparently contains only a "goose type" enzyme. Our studies revealed that the black swan contains both a "chick type" and "goose type" lysozyme in its egg white (Arnheim and Steller, 1970).

The adaptive significance of this species variation in egg white lysozyme content is unknown. The "goose type" and "chick type" enzymes differ appreciably in their reaction with inhibitors and their catalytic activity on small molecular weight substrates (Charlemagne and Jollès, 1967; Dianoux and Jollès, 1967, 1969; Jollès et al., 1968; McKelvy et al., 1970). Both lysozymes are, however, muramidases and catalyze the hydrolysis of the glycosidic linkage between N - acetylmuramic acid and N - acetyglucosamine in the bacterial peptidoglycan (Arnheim et al., 1973a). When whole bacterial cells *(Micrococcus lysodeikticus)* are used as substrate, the "goose type" enzyme is found to be 5 to 10 times more active on a molar basis (Canfield and McMurry, 1967; Dianoux and Jollès, 1967; Arnheim et al., 1973b).

TABLE II

AMINO-TERMINAL AMINO ACID SEQUENCE OF BLACK SWAN "GOOSE TYPE" LYSOZYME AND CHICKEN EGG WHITE LYSOZYME[1]

G	NH_2 -	Arg -	Thr -	Asp -	Cys -	Tyr -	Gly -
C	NH_2 -	Lys -	Val -	Phe -	Gly -	Arg -	Cys -
		1				5	
G	Asn -	Val -	Asn -	Arg -	Ile -	Asp -	Thr -
C	Glu -	Leu -	Ala -	Ala -	Ala -	Met -	Lys -
				10			
G	Thr -	Gly -	Ala -	Ser -	Cys -	Lys -	Thr -
C	Arg -	His -	Gly -	Leu -	Asp -	Asn -	Tyr -
		15					20

1) The sequence data for the "Goose Type" enzyme (G) was taken from Arnheim et al (1973b) and the data for the chicken egg white enzyme (C) from Canfield (1963).

Evidence also exists which suggests that the "goose type" and "chick type" lysozymes may differ in their recognition of certain structural features of the bacterial peptidoglycan. More specifically it has been argued that the "goose type" enzymes may have a preference for N - acetylmuramic acid residues which are substituted with a peptide moiety (Arnheim et al., 1973a). Such differences in specificity among lysozymes could be adaptively significant, since bacterial species vary with respect to the amino acid composition of the peptide moeity and the frequency with which the N - acetylmuramic acid residues are substituted with a peptide (Osborn, 1969). Further enzymological studies on both types of lysozymes using purified bacterial peptidoglycans are needed to support these hypotheses.

GENETIC BASIS FOR THE VARIATION IN EGG WHITE LYSOZYME CONTENT AMONG SPECIES

The presence or absence of either of the forms of lysozyme in the egg white of a species could be due to the presence or absence of the gene for the enzyme in the species. Alternatively the gene for both enzymes might be present in each species but the presence or absence of either lysozyme in egg white could be due to different patterns of expression of the two lysozyme genes in different species.

In order to distinguish between these two hypotheses we sought to find evidence for the presence of a second form of lysozyme in a species which has only a single lysozyme in its egg white. Our first study culminated in the finding of the "goose type" enzyme in chicken white blood cells which have an antibacterial function: the polymorphonuclear leukocytes (PMN).

Large cytoplasmic granules were isolated from chicken PMN by Dr. John Spitznagel (Brune and Spitznagel, 1973). An extract of the granules was found to contain a "chick type" and a "goose type" lysozyme. The two forms of lysozyme can be separated by acrylamide gel electrophoresis (Hindenburg et al., 1974) or ion exchange chromatography (Figure 1). The identification of the separated lysozymes as "chick type" or "goose type" was made by making use of the fact, mentioned above, that antibodies against one of these forms of lysozyme

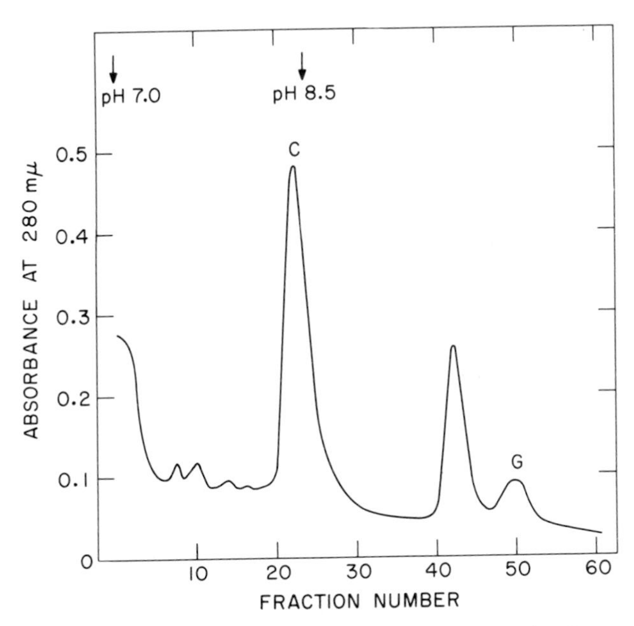

Figure 1. Bio Rex 70 chromatography of partially purified lysozymes from chicken bone marrow. The lysozymes from an extract of chicken bone marrow were partially purified by absorption to a carboxymethylcellulose column at pH 9.4 (ammonium acetate, 0.1M) and elution with 0.4M ammonium carbonate at the same pH. The lysozymes were dialyzed against 0.2M sodium

Figure 1. Cont'd. phosphate buffer pH 5.8 and applied to a Bio Rex 70 column equilibrated with the same buffer. A stepwise elution procedure was carried out by washing the column with 0.2M sodium phosphate buffer pH 7.0 followed by the same buffer adjusted to pH 8.5. The arrows in the figure indicate the buffer changes. The peaks labeled C and G had lysozyme activity and were found to contain "chick type" and "goose type" lysozymes, respectively, by immunological methods.

are specific for that form and will not react with the other type of lysozyme.

Studies on the Peking duck which has only "chick type" lysozymes in its egg white (Jollès et al, 1965; Imanishi et al., 1966; Prager and Wilson, 1971, 1972) have shown that this species also has a gene coding for a "goose type" lysozyme. Gel filtration studies on an extract of pus from a bird with an arthritic lesion showed a broad elution profile of lysozyme activity which could be resolved immunologically into a lower molecular weight "chick type" lysozyme region and a slightly higher molecular weight "goose type" lysozyme region (Fig. 2).

Finally, studies on the Embden goose have revealed that this species, which apparently contains only a "goose type"

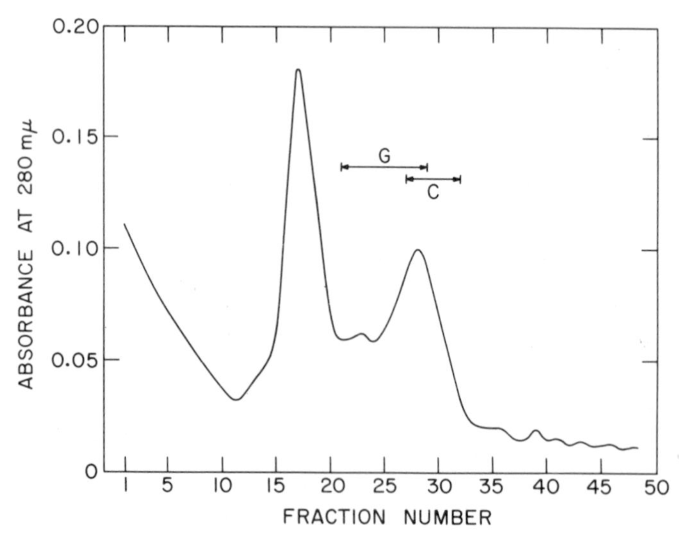

Figure 2. Sephadex G-75 chromatography of partially purified lysozymes from duck pus. The lysozymes from an extract of duck pus were partially purified by absorption and elution from a carboxymethylcellulose column at pH 9.4. The material eluted by 0.4M ammonium carbonate was dialyzed against 0.2M sodium phosphate buffer pH 7.18, placed on a Bio Rex 70 column

Figure 2. Cont'd. and washed with the same buffer. One fraction that was eluted contained both a "chick type" and "goose type" lysozyme. This fraction was then placed on a Sephadex G-75 column equilibrated with 0.1M ammonium acetate buffer pH 9.4. A broad peak of lysozyme activity was eluted and is indicated by the partially overlapping bars labeled G and C. The indentification of the higher molecular weight enzyme as a "goose type" (G) and the lower molecular weight enzyme as a "chick type" (C) was made immunologically.

---

lysozyme in its egg white, has a "chick type" enzyme in its bone marrow (Hindenburg et al., 1974).

EVOLUTION OF THE REGULATION OF EGG WHITE LYSOZYME SYNTHESIS

Among the species we have studied the "goose type" and "chick type" lysozyme loci are both present. Different species, however, have different gene expression patterns in the cells which carry out egg white lysozyme synthesis. These differences in gene expression patterns must have arisen during the evolution of these species from their common ancestors. Further studies on the mechanisms by which these changes occurred are of special interest since the ways in which gene expression patterns can change during evolution may be of central importance to our understanding of evolutionary processes.

We have recently initiated studies designed to learn more about what molecular factors control the appearance of the two lysozyme types in egg white. We happen to be in a fortunate position for carrying out such studies since not only is a good deal known about the induction of egg white protein synthesis in the chicken, but large amounts of material are available for biochemical analysis. Studies have shown that estrogen stimulates the cytodifferentiation of the immature oviduct and that one of the newly differentiated cell types, the tubular gland cell, is responsible for the synthesis of the major egg white proteins including lysozyme (Kohler et al., 1969; Oka and Schimke, 1969). It has also been shown that the induction of the synthesis of these proteins is a consequence of the synthesis of their respective messenger RNAs (Chan et al., 1973; Palmiter, 1973). One of these messenger RNAs. that coding for ovalbumin, has been highly purified and DNA copies of this message have been synthesized (Harris et al., 1973; Sullivan et al., 1973). Our laboratory is actively seeking to isolate the messenger RNA which codes for chicken egg white lysozyme. Once it is purified, a DNA copy of this message can also be made using reverse transcriptase. This DNA copy of the "chick type" lysozyme message can be used to determine whether the block in the appearance of the "chick type" enzyme in Embden goose egg is due to a block at the level of transcription or translation in the cells of the oviduct.

Within the next few years molecular biologists and biochemists will provide a variety of techniques and tools for studying the control and regulation of protein synthesis in eukaryotic species. Application of advances in our understanding of these processes to the question of how differences in gene expression patterns developed during phylogeny will provide new insights into the molecular basis of evolution.

REFERENCES

Arnheim, N. 1974. *Multiple Genes for Lysozyme in"Lysozyme"* (Osserman, E. F., R. E. Canfield and S. Beychok eds.) pp. 153-161, Academic Press, New York.

Arnheim, N., A. Hindenburg, G. Begg, and F. J. Morgan. 1973b. Multiple genes for lysozyme in birds. *J. Biol. Chem.* 248: 8036-8042.

Arnheim, N., M.Inouye, L. Law, and A. Laudin 1973a. Chemical studies on the enzymatic specificity of goose egg white lysozyme. *J. Biol. Chem.* 248: 233-236.

Arnheim, N. and R. Steller. 1970. Multiple genes for lysozyme in birds. *Arch. Biochem. Biophys.* 141: 656-661.

Blake, C. C. F., D. F. Koenig. G. A. Mair, A. C. T. North, D. C. Phillips, and V. R. Sarma. 1965. Structure of hen egg white lysozyme. A three dimensional fourier synthesis at 2 Å resolution, *Nature*. 206: 757-763.

Brune, K. and J. K. Spitznagel. 1973. Peroxidaseless chicken leukocytes: Isolation and characterization of antibacterial granules. *J. Infect. Diseases.* 127: 84-93.

Canfield, R. E. 1963. The amino acid sequence of egg white lysozyme. *J. Biol. Chem.* 238: 2698-2707.

Canfield, R. E. 1968. Cited in L. N. Johnson, D. C. Phillips, and J. A. Rupley. The activity of lysozyme: An interim review of crystallographic and chemical evidence. *Brookhaven Symp. Biol.* 21: 136-137.

Canfield, R. E.,S. Kammerman, J. H. Sobel, and F. J. Morgan. 1971. Primary structure of lysozymes from man and goose. *Nature New Biol.* 232: 16-17.

Canfield, R. E. and S. McMurry, 1967. Purification and characterization of lysozyme from goose egg white. *Biochem. Biophys. Res. Commun.* 26: 38-42.

Chan, L., A. Means, and B. W. O'Malley. 1973. Rates of induction of specific translatable messenger RNAs for ovalbumin and avidin by steroid hormones. *Proc. Nat. Acad. Sci. U.S.A.* 70: 1870-1874.

Charlemagne, D. and P. Jollès. La Specificité De Divers Lysozymes vis-a-vis De Substrats De Faible Poids Moléculaire Provenant De La Chitine. *Bull. Soc. Chim. Biol.* 49:1103-1113.

Dayhoff, M.O. 1973. *Atlas of Protein Sequence and Structure.* Volume 5. Nat. Biomed. Res. Found., Silver Spring, Maryland.

Dianoux, A. C. and P. Jollès. 1967. Etude d'un Lysozyme Pauvre en Cystine et en Tryptophane: le Lysozyme de Blanc d'Oeuf d'Oie. *Biochim. Biophys. Acta.* 133: 472-479.

Dianoux, A. C., P. Jollès. 1969. Differences dans le Comportement des Lysozymes de Blanc d'Oeuf de Poule et d'Oie vis-a-vis de *Micrococcus lysodeikticus. Bull. Soc. Chem. Biol.* 51: 1559-1564.

Fleming, A. 1922. On a remarkable bacteriolytic element found in tissues and secretions. *Proc. Roy. Soc. Ser. B.* 93: 306-317.

Ghuysen, J. M. 1968. Use of bateriolytic enzymes in determination of wall structure and their role in cell metabolism. *Bact. Rev.* 32: 425-464.

Glazer, A. N., A. O. Barel, J. B. Howard, and D. M. Brown. 1969. Isolation and characterization of fig lysozyme. *J. Biol. Chem.* 244: 3583-3589.

Harris, S. E., A. R. Means, W. M. Mitchell and B. W. O'Malley. 1973. Synthesis of [3_H] DNA complementary to ovalbumin Messenger RNA: Evidence for limited copies of the ovalbumin gene in chick oviduct. *Proc. Nat. Acad. Sci. U.S.A.* 70: 3776-3780.

Hermann, J., J. Jolles, D.H. Buss, and P. Jollès 1973. Amino acid sequence of lysozyme from baboon milk. *J. Molec. Biol.* 79: 587-595.

Hindenburg, A., J. Spitznagel, and N. Arnheim 1974. Isozymes of lysozyme in leukocytes and egg white: Evidence for the species-specific control of egg white lysozyme synthesis. *Proc. Natl. Acad. Sci. U.S.A.* 71: 1653-1657.

Imanishi, M., S. Shinka, N. Miyagawa, T. Amano, and A. Tsugita 1966. Amino acid compositions of duck and turkey egg white lysozymes. *Biken J.* 9: 107-114.

Imoto, T., L.N. Johnson, A.C.T. North, D.C. Phillips, and J.A. Rupley 1972. Vertebrate lysozymes. In: *The Enzymes,* third edition (P.D. Boyer, ed.), Academic Press, N.Y. . pp. 665-868.

Jollès, P. 1969. Lysozymes: A chapter of molecular biology. *Angewandte Chemie,* International Edition 8: 227-294.

Jollès, P., J.S. Blancard, D. Charlemagne, A.C. Dianoux, J. Jollès, and J. LeBaron 1968. Comparative behavior of six different lysozymes in the presence of an inhibitor. *Biochem. Biophys. Acta.* 151: 532-534.

Jollès, J., J. Jauregui-Adell, I. Bernier, and P. Jollès 1963. La Structure Chimique du Lysozyme de Blanc d'Oeuf de Poule: Etude Detaillee. *Biochem. Biophys. Acta.* 78: 668-689.

Jollès, J., G. Spotorno, and P. Jollès 1965. Lysozymes characterized in duck egg white: Isolation of a histidine-less lysozyme. *Nature* 208: 1204-1205.

Kohler, P. O., P. M. Grimley, and B. W. O'Malley. 1969. Estrogen induced cytodifferentiation of the ovalbumin-secreting glands of the chick oviduct. *J. Cell Biol.* 40: 8-27.

McKelvy, J. F., Y. Eshdat, and N. Sharon. 1970. Action of specific and irreversible inhibitors of hen egg white lysozyme on lysozymes from four other sources. *Fed. Proc.* 29, p. 532 (Abs.)

Mross, G. A., T. H. White, and A. C. Wilson (Unpublished data).

Oka, T., and R. T. Schimke. 1969. Interaction of estrogen and progesterone in chick oviduct development. *J. Cell Biol.* 41: 816-831.

Osborn, M. J. 1969. Structure and biosynthesis of the bacterial cell wall. *Ann. Rev. Biochem.* 38: 501-538.

Palmiter, R. D. 1973. Rate of ovalbumin messenger ribonucleic acid synthesis in the oviduct of estrogen-primed chicks. *J. Biol. Chem.* 248: 8260-8270.

Powning, R. F. and W. J. Davidson. 1973. Studies on insect bacteriolytic enzymes- I. Lysozymes in hemolymph of *Galleria Mellonella* and *Bombyx Mori*. *Comp. Biochem. Physiol.* 45B: 669-686.

Prager, E.M., and A. C. Wilson. 1971. Multiple lysozymes of duck egg white. *J. Biol. Chem.* 246: 523-530.

Prager, E., A.C. Wilson, and N. Arnheim 1974. Widespread distribution of lysozyme g in egg white of birds. *J. Biol. Chem.* (In press).

Riblet, R. 1974. Sequence studies of mouse lysozyme in *Lysozyme* (Osserman, E.F., R.E. Caufield, and S. Beychok, eds.) Academic Press, New York. pp. 89-93.

Sullivan, D., R. Palacios, J. Stavenzer, J.M. Taylor, A.J. Faras, M.L. Kiely, N.M. Summers, J.M. Bishop,and R.T. Schimke 1973. Synthesis of a deoxyribonucleic acid sequence complementary to ovalbumin messenger ribonculeic acid and quantification of ovalbumin genes. *J. Biol. Chem.* 248: 7530-7539.

Takeda, H., G.A. Strasdine, D.R. Whitaker, and C. Roy 1966. Lytic enzymes in the digestive juices of *Helix Pomatia*. *Canad. J. Biochem.* 44: 509-518.

Tsugita, A. 1971. Phage lysozyme and other lytic enzymes in *The Enzymes*, P.D. Boyer, ed.) 3rd ed., Vol. 5, pp. 343-411, Academic Press, New York.

SIMULTANEOUS GENETIC CONTROL OF THE STRUCTURE AND RATE OF SYNTHESIS OF MURINE GLUCURONIDASE

ROGER E. GANSCHOW
The Children's Hospital Research Foundation
Elland and Bethesda Aves.
Cincinnati, Ohio 45229

ABSTRACT. Inbred strains of mice can be divided into two classes based upon liver and kidney levels of β-glucuronidase. Most strains exhibit liver activity levels between 40 and 100 activity units per gram in contrast to a few strains with values between 3 and 6. Glucuronidases from high and low activity strains also exhibit a difference in structure as determined by heat sensitivity. Previous genetic studies, as well as examination of available inbred strains of mice, suggest that the genetic control of the difference in activity level as well as the difference in structure resides in the glucuronidase structural gene (<u>Gus</u>). Immunochemical studies clearly show that the difference in tissue activity levels is directly related to the number of glucuronidase active sites and not to a change in the quality of the active sites. Using combined immunochemical and isotope incorporation techniques it is demonstrated that the difference in the number of active sites between high and low activity strains is a function of the rate of synthesis of glucuronidase. Several models are considered to explain structural gene control of the rate of glucuronidase synthesis, including the concept of autogenous regulation of gene expression.

INTRODUCTION

The regulation of the rate of production as well as the quality of the protein by structural genes has been clearly demonstrated in bacteria (see review by Goldberger, 1974). This control, termed "autogenous regulation", is the result of direct interaction between the protein product of a structural gene and the operon within which the structural gene resides.

A similar mechanism for the control of gene expression in animal cells is suggested by the work of Stevens and Williamson (1972; 1973) in which myeloma protein was found to interact specifically with the m-RNA for immunoglobulin heavy chains. This interaction may play an important role in regulating levels of myeloma protein by repressing the translation of heavy chain m-RNA as the intracellular level of myeloma protein increases.

Structural gene alterations have been reported in animal cells which appear to control enzyme levels by influencing the rate of enzyme synthesis. Among these are a human variant of glucose-6-phosphate dehydrogenase (G-6-PD Hektoen) in which a single amino acid change in the enzyme is associated with a substantial increase in G-6-PD levels (Yoshida, 1970) and several human hemoglobin variants in which altered hemoglobin structures are associated with increased hemoglobin levels (Shibata et al., 1964; Pootrakul et al., 1967; Miyaji et al., 1968). None of these structural changes appears to affect the rate of degradation of the protein in question. In each case, however, further analysis of the relationship between structural alteration and increased level of protein is somewhat restricted because of limitations on human experimentation.

Analysis of the control of gene expression in animal cells has been enhanced in recent years by the availability of mutants in strains of inbred mice which exhibit altered enzyme phenotypes. We have chosen to study a strain of mice which exhibits a difference in both the structure and tissue activity levels of the enzyme β-glucuronidase (EC 3.2.1.31) when compared to most other strains of mice.

Livers of most sublines of the C3H strain exhibit about 10 per cent of the glucuronidase activity levels observed in other inbred strains (Morrow et al., 1950; Paigen and Noell, 1961; Ganschow and Paigen, 1968). Low liver activity is accompanied by lowered activity in spleen and kidney. However, the effect on spleen activity is substantially less than on liver or kidney activity. Genetic analyses reveal that these differences are under the control of a single Mendelian autosomal factor (Law et al., 1952; Paigen, 1961b; Paigen and Ganschow, 1965).

Glucuronidase activity from strains with high levels of the enzyme is stable at 71°, while that of low activity strains is rapidly inactivated at this temperature (Paigen, 1961a; Ganschow and Paigen, 1968). Presumably this difference represents a change in enzyme structure since mixtures of the stable and labile forms exhibit heat-inactivation kinetics identical to the kinetics calculated for a mixture behaving as the sum of the individual components. This conclusion is strengthened by the fact that a substantial difference exists between high and low activity strains in the reactivity of glucuronidase with goat anti-beef liver glucuronidase antibody (Ganschow, 1973). The structural gene controlling this difference has not been successfully separated from the genetic element controlling the tissue activity difference (Ganschow and Paigen, 1967; Paigen, personal communication;

Ganschow and Kepferle, unpublished data), suggesting either identity or very close linkage.

Our initial attempts to understand the interrelationship of these altered parameters of glucuronidase expression reveal that the strain differences in tissue activity levels reflect a disparity in the quantity of enzyme which can be accounted for by a difference in the rate of glucuronidase synthesis.

MATERIALS AND METHODS

Animals. All mice were obtained from the production colony of the Jackson Laboratory (Bar Harbor, Maine) and fed Purina Mouse Chow (St. Louis) *ad libitum*. Animals were between 60 and 90 days of age when used experimentally.

Assay of enzyme activity. β-Glucuronidase was assayed according to the procedure of Meisler and Paigen (1972) with 1.0 mM p-nitrophenylglucuronide (Cyclo Chem. Co., Los Angeles) as substrate in a volume of 1.0 ml. One unit of enzyme activity is defined as that amount of enzyme which releases 1 μmole of p-nitrophenol per hour at 56°.

Preparation of tissue extracts. In addition to its localization in lysosomes, approximately 40% of murine liver and kidney glucuronidase is bound within the membranes of the endoplasmic reticulum (Paigen, 1961a; Swank and Paigen, 1973). Since the immunoprecipitation of labeled glucuronidase requires that the enzyme first be solubilized and be in a milieu from which only glucuronidase is precipitated, the following tissue extraction precedure was developed.

Animals were killed by cervical dislocation; tissues were rapidly removed, weighed, and then chilled. Tissue homogenates were prepared to a final concentration of 10% (w/v) in 0.1 M TRIS-0.15N NaCl, pH 7.5, with a Polytron homogenizer (Kinematica GMBH, Lucerne). Aliquots were then removed for glucuronidase assay and for assay of radioactivity in total tissue protein. To the remaining homogenate was added 10% sodium deoxycholate with stirring to a final concentration of 0.5%. After incubation for 30 min on ice, 1.0M acetic acid was added until the pH was lowered to 4.6 and the remaining precipitate was removed by centrifugation for 45 min at 39,000 g in a Sorval RG2-B centrifuge (Norwalk, Conn.) with the SS-34 rotor. The supernatant was adjusted to pH 7.5 with 2.0 M TRIS, heated at 56° for 30 min and the resulting precipitate removed by sedimenting as above. A second heat step was performed as above after lowering the pH of the supernatant to 4.6 and again the precipitate was removed by centrifugation. To the supernatant from the second heat step was added solid $(NH_4)_2SO_4$ to give a 50% saturated solution. After standing for two hours on ice, the precipitate was collected by centrifugation for 45 min at 12,000 g in the Sorvall RC2-B centrifuge and dissolved in water up to a final volume equivalent to 5% of the original homogenate volume. For every 9 volumes of dissolved precipitate, 1 volume of 1.0 M TRIS-1.5 N NaCl, pH 7.5, was added and allowed to stand at room temperature for 1 hour. This mixture was then filtered through a Gelman Acrodisc AN-1200 (Ann Arbor) and from the resulting filtrate all immunochemical isolations of glucuronidase were accomplished. Glucuronidase activity recoveries ranged between 85 and 95% of the original homogenate activity.

Isotope incorporation and electrophoresis. Rates of synthesis of liver glucuronidase were measured by pulse labeling each of 10 C3H mice with 200 μCi of ^{3}H-leucine (Amersham-Searle; 46 Ci/mmole and each of 10 B/6 mice with 10 μCi ^{14}C-leucine (Amersham-Searle; 348 mCi/mmole). The isotope was administered intraperitoneally and 2 hours later the animals were killed. Pooled liver homogenates from each strain were prepared as above, assayed for glucuronidase activity, and 0.1 ml set aside for measurement of radioactivity in total protein. The two pooled homogenates were combined and a tissue extract containing both C3H and B/6 glucuronidase was prepared as above. A sufficient quantity of goat antimouse glucuronidase antiserum was added to the extract to achieve complete precipitation of glucuronidase activity after incubation at room temperature for 30 min and then at 4° overnight. The resulting immunoprecipitate was washed twice with 0.15 N NaCl, dissolved by boiling for 1 min in a solution of 1% sodium dodecyl sulfate-1% β-mercaptoethanol (SDS-ME), and the polypeptides separated electrophoretically in polyacrylamide gels by the method of Weber and Osborn (1969). Gels were stained for protein in Coomassie Brilliant Blue (Sigma) and unbound stain removed electrophoretically as suggested by Weber and Osborn (1969). The stained gels were then sliced into 2 mm slices. Each slice was placed into an 8 ml glass scintillation vial and covered with 0.5 ml of a solution of 5% water-95% Soluene 100 (Packard Inst. Co., Downers Grove, Ill.) and heated at 56° for 3 hours to solubilize the protein from the gel. Toluene-based scintillator liquid was added to each vial and the radioactivity in the resulting mixture was determined with an Isocap 300 liquid scintillation spectrometer (Searle Analytic; Des Plaines, Ill.). At least 2000 total counts were obtained in each channel for each slice.

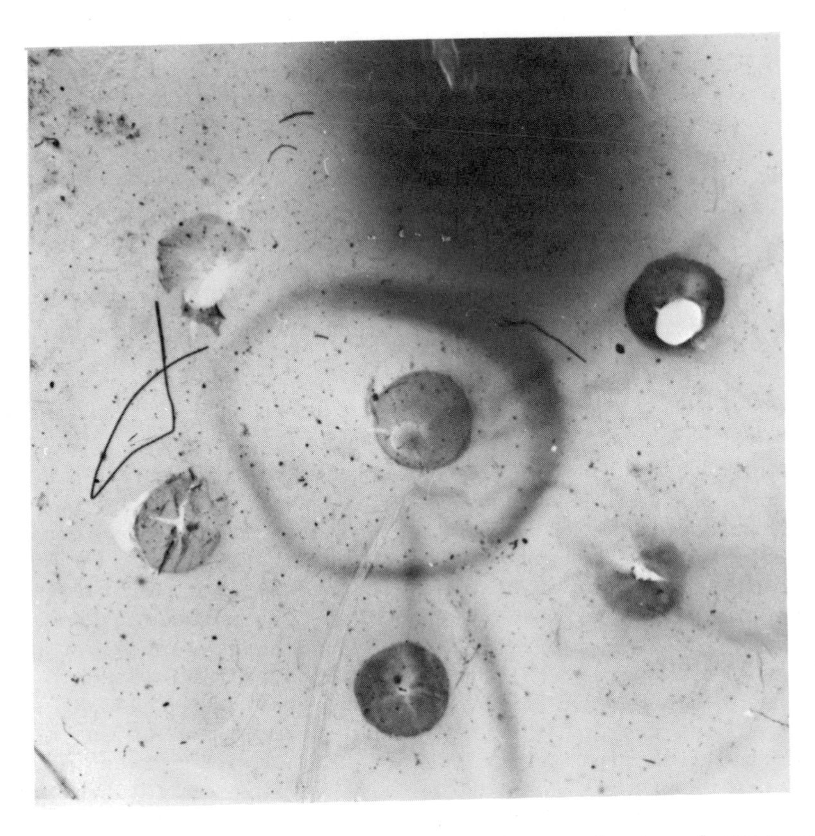

Figure 1. Ouchterlony double-diffusion analysis of goat anti-mouse liver glucuronidase antiserum and kidney glucuronidase. To the center well was added B/6 kidney extract containing 3 activity units of glucuronidase. To the well at the top of the figure was added 40 μl of undiluted antiserum; and 40 μl of increasing serial dilutions were added to the other outside wells in a clockwise direction. The plate was incubated at 37° for 48 hours, and unprecipitated proteins were leached with several changes of 0.1 M TRIS-0.15 N NaCl, pH 7.5. Precipitated proteins were stained with Coomassie Brillant Blue and unbound stain removed with several changes of a solution containing 450 ml methanol and 100 ml glacial acetic acid per liter.

A measure of the radioactivity incorporated into liver protein was obtained by collecting the precipitate formed after the addition of 5 ml of 5% trichloroacetic acid (TCA) to 0.1 ml of the original homogenate. The precipitate was washed twice with cold 5% TCA, dissolved in 1.0 ml of Soluene, and counted in a toluene-based scintillator. The rate of glucuronidase synthesis was estimated as the distintegrations per minute (dpm) in immunoprecipitated enzyme relative to the dpm in TCA precipitable protein.

Characterization of antiserum. Goat antimouse glucuronidase antiserum was prepared as outlined by Ganschow (1973). This antiserum was monospecific by several criteria including Ouchterlony double-diffusion, as shown in Figure 1, and completely precipitates all glucuronidase activity in extracts of mouse liver and kidney.

RESULTS

Tissue activity levels. Inbred strains of mice can be divided into two classes based upon liver and kidney activity levels of β-glucuronidase (Ganschow and Paigen, 1968). The vast majority of strains exhibit mean values ranging from 38 to 102 activity units per gram wet weight of liver. This group is clearly distinguished from another, smaller group whose activity levels range between 3 and 6 activity units per gram. In the present study the C57BL/6J(B/6) strain was used to represent the high activity group and the C3HeB/FeJ(C3H) strain to represent the low activity group. Comparison of the mean liver and kidney glucuronidase activity levels between the two strains reveals that C3H mice accumulate only 10% of the liver and 14% of the kidney activity found in mice of the B/6 strain (Table 1).

Heat inactivation of glucuronidase. In order to substantiate the reported structural difference in glucuronidase between strains with high and strains with low tissue levels of the enzyme (Paigen, 1961a; Ganschow and Paigen, 1968), crude B/6 and C3H kidney homogenates were heated at 71° and the rate of inactivation of glucuronidase measured in each (Figure 2). As expected, enzyme activity from the low activity C3H strain was much more labile than that from the high activity B/6 strain. These data, together with data from previous genetic analyses of this structural difference (Ganschow and Paigen, 1967; Ganschow and Kepferle, unpublished data; Paigen, personal communication), permit us to conclude that mice of the C3HeB/FeJ strain carry the Gus^h allele and C57BL/6J the Gus^b allele of the glucuronidase structural gene located on chromosome 5. The nomenclature used is a revision of a previous

TABLE 1

LIVER AND KIDNEY ACTIVITY LEVELS OF B/6 AND C3H MICE

STRAIN	GLUCURONIDASE ACTIVITY LEVEL (MEAN ± S. E.)	
	LIVER	KIDNEY
	μmoles pNP/hr/gm	
C57BL/6J(B/6)	53.0±3.5	18.2±1.8
C3HeB/FeJ(C3H)	5.6±0.4	2.5±0.2

Table 1. Liver and kidney glucuronidase activity levels of B/6 and C3H mice. Glucuronidase activity levels in liver and kidney homogenates were determined for 5 animals of each strain as described in MATERIAL AND METHODS. Values are expressed as the mean ± the standard error of the mean (S.E.).

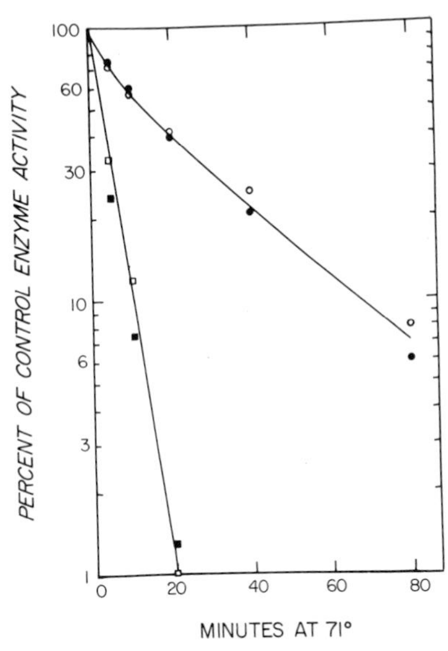

Figure 2. Heat-inactivation kinetics of kidney glucuronidase from B/6 and C3H mice. Kidney homogenates (10% w/v) were prepared in water, brought to pH 4.6 with acetate buffer and heated at 71° for times up to 80 min. Percentage of surviving glucuronidase activity is plotted as an exponential function of time of heating. Open and closed symbols refer to different experiments. O, ●, B/6; □, ■, C3H.

system which designates the allele carried by C3H mice as g and the allele in C57BL/6 mice as G. This change has been approved by the Committee on Standardized Genetic Nomenclature for Mice.

Immunochemical analysis. The disparity in tissue levels of glucuronidase activity between mice of the C3H and B/6 strains could result from a qualitative change in the active site of glucuronidase or from an altered concentration of glucuronidase molecules. To distinguish between these alternatives, increasing amounts of kidney glucuronidase activity from B/6 and from C3H were completely precipitated with goat anti-mouse glucuronidase antiserum. The washed precipitates were solubilized by boiling in an SDS-ME solution, the component polypeptides separated electrophoretically, and the gels stained for protein. The quantity of protein in each electrophoretic component was determined by densitometry of the stained gels. Each gel contained three stained components, the heavy and light polypeptide chains of immunoglobulin G and the 70,000 dalton subunit of glucuronidase (for details of the size characteristics of murine glucuronidase, see Ganschow, 1973). No differences in molecular weight were measurable between the glucuronidase subunits of the two strains. The quantity of glucuronidase protein in each gel was estimated by densitometry of the stained gels and integration of the area under the peak corresponding to the glucuronidase subunit.

The results, plotted on Figure 3, show that within the range examined, staining intensity is proportional to the amount of immunoprecipitated enzyme activity. Moreover, the fact that the color yield per enzyme activity unit is identical for glucuronidase of each strain shows that the difference in tissue activity levels of glucuronidase between C3H and B/6 mice is directly related to the number of glucuronidase molecules and not to a change in catalytic properties.

Rates of glucuronidase synthesis. The reduced concentration of glucuronidase in tissues of C3H mice could result from a genetic change in either the rate of synthesis or the rate of degradation of the enzyme. To distinguish between these possibilities experimentally, combined immunochemical and isotope incorporation studies were performed. The relative rate of liver glucuronidase synthesis was determined for each strain by measuring the short-term incorporation of radioactive leucine into immunoprecipitates of liver glucuronidase and into liver protein of B/6 and C3H mice.

Since the radioactivity incorporated into glucuronidase was measured using immunoprecipitated enzyme, it was necessary

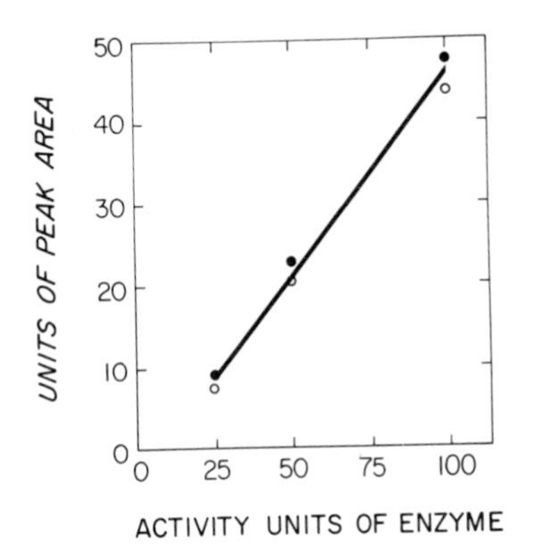

Figure 3. Staining intensity of immunoprecipitated B/6 and C3H glucuronidase protein in SDS-ME polyacrylamide gels as a function of the amount of glucuronidase activity precipitated. Electrophoresis of polypeptides from dissolved immunoprecipitates of glucuronidase was performed in SDS-ME polyacrylamide gels as described in MATERIALS AND METHODS. Densitometry of the stained gels was accomplished at 600 nm using a Gilford 2400 recording spectrophotometer (Gilford Inst. Co., Oberlin, Ohio) with a linear transport device attached. The peak area corresponding to the 70,000 dalton glucuronidase polypeptide was integrated for 3 different profiles of each gel. The mean value for each peak is plotted as a function of the amount of glucuronidase activity in each immunoprecipitate. ●, B/6; O, C3H.

to establish that radioactivity in the immunoprecipitates was derived only from glucuronidase. The profile of counts in an SDS-ME gel after electrophoresis of immunoprecipitates of labeled liver glucuronidase shows that the only radioactive peak corresponds exactly to the position of the glucuronidase polypeptide (Figure 4). In order to further minimize the introduction of spurious label, only the counts in the region of the gel containing the glucuronidase polypeptides were used in assessing the relative rates of glucuronidase synthesis.

Values for the relative rates of enzyme synthesis are given in Table 2. The radioactivity in the enzyme is normalized to the total counts incorporated into liver protein and the data expressed as the fraction of total dpm in glucuronidase. It is apparent from the data of Table 2 that B/6 liver glucuronidase is synthesized at a rate approximately 7 times that of C3H. This difference is sufficient to completely account for the 7-fold difference in liver activity levels of the enzyme. Similar results have been obtained in other ex-

TABLE 2

RELATIVE RATE OF SYNTHESIS AND ACTIVITY LEVELS OF LIVER GLUCURONIDASE IN B/6 AND C3H MICE

STRAIN	GLUCURONIDASE RADIOACTIVITY	TOTAL PROTEIN RADIOACTIVITY	FRACTION OF TOTAL dpm IN GLUCURONIDASE	LIVER GLUCURONIDASE ACTIVITY LEVEL
	dpm/gm	liver $X10^{-6}$	$X10^{6}$	μmoles pNP/hr/gm
B/6 (^{14}C)	20	1.8	11.1	53
C3H (^{3}H)	42	27	1.6	7.6
	B/6/C3H		6.9	7.0

Table 2. The relative rate of synthesis and activity levels of liver glucuronidase in B/6 and C3H mice. Combined immunochemical and isotope incorporation techniques were used to determine the relative rates of synthesis of liver glucuronidase in B/6 and C3H mice as described in MATERIALS AND METHODS. Activity levels were determined on the pooled homogenates of each strain. An interstrain comparison of the rates of synthesis and the activity levels are presented as ratios of B/6 values to C3H values.

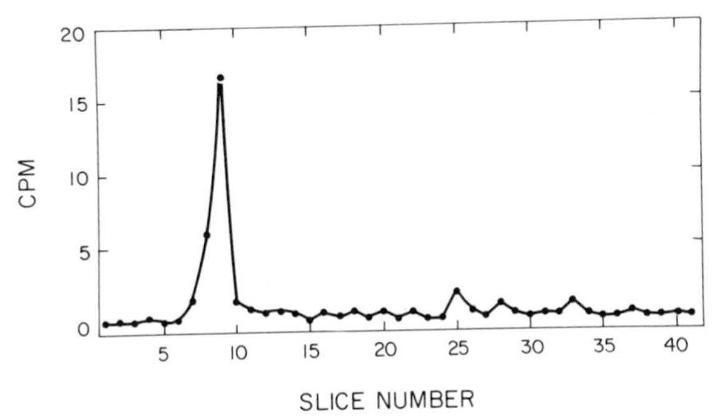

Figure 4. Radioactivity profile of an SDS-ME gel after electrophoresis of immunoprecipitates of labeled glucuronidase. An extract was prepared from livers of B/6 mice previously pulsed for 2 hours with 25 μCi of ^{3}H-leucine (for details see MATERIALS AND METHODS). Glucuronidase was completely precipitated from this extract with goat antimouse glucuronidase antiserum. A washed immunoprecipitate containing 100 activity units of glucuronidase was dissolved in SDS-ME and the polypeptides separated electrophoretically on an SDS-ME polyacrylamide gel. The gel was stained, sliced and counted as described in MATERIALS AND METHODS.

periments in which the isotopes administered to each strain were reversed or in which 200 μCi of ^{3}H-leucine were given to animals of each strain and the immunoprecipitates of each strain prepared separately.

DISCUSSION

The results of our studies show that the strain differences in tissue activity levels of glucuronidase are a function of a difference in the rate of glucuronidase synthesis. It is possible that the genetic control of glucuronidase synthetic rates resides in the glucuronidase structural gene on chromosome 5, since no recombinants between tissue levels and heat susceptibility of glucuronidase activity have been found (Ganschow and Paigen, 1967; Paigen, personal communication; Ganschow and Kepferle, unpublished data). We are presently conducting experiments which extend the genetic analysis to large numbers of animals in order to determine whether these parameters of glucuronidase expression are under separate control. It should be emphasized that the failure to demonstrate recombination in animal systems between two genetic elements does not permit the conclusion that these elements are identical. However, among all inbred mouse strains and sublines with low tissue levels of glucuronidase, this enzyme exhibits thermolability (Ganschow and Paigen, 1968).

Within the C3H strain, the subline of origin, C3H/St, has high activity levels and thermostable enzyme, while all sublines derived either directly (and at different times) or indirectly from C3H/St exhibit low enzyme levels and thermolability. If our assumptions regarding the origins of the C3H sublines are correct, then either residual heterozygosity existed in the ancestry of currently maintained C3H sublines for glucuronidase alleles which were subsequently fixed in a non-random manner among the numerous sublines, or a back mutation(s) to the high activity-thermostable phenotype occurred and were fixed in the C3H/St mice after the last subline was derived. If the latter explanation is correct, two back-mutations would have been required if activity levels and thermolability of glucuronidase are under separate genetic control. This is highly improbable. Another strain, AKR, has low activity, thermolabile enzyme and no known ancestral relationship to the C3H strain (Ganschow and Paigen, 1968). These facts argue strongly for identical genetic control of the differences in structure and rates of synthesis of glucuronidase.

Another effect of presumably the $\underline{Gus}^{h}$ allele is the marked difference which exists between the post-natal time course of glucuronidase accumulation in livers of high and low activity strains (Paigen, 1961b). High activity mice accumulate liver glucuronidase throughout growth whereas C3H mice, born with a 3-fold lower level of enzyme, accumulate the enzyme only until the twelfth post-natal day, after which a progressive decline in activity level occurs until the substantially lowered adult value is reached. Analogous patterns are observed for the enzyme in other tissues but the time at which the alteration in accumulation occurs is different for each tissue.

Genetic studies show the timing of glucuronidase accumulation to be under the control of a single gene (Paigen, 1961b). In view of the results presented here, it is likely that this gene is the glucuronidase structural gene (Gus) since lowered adult levels of glucuronidase activity are a partial function of the altered post-natal accumulation pattern of glucuronidase in $\underline{Gus}^{h}$ homozygotes.

How does the Gus gene control the rate of glucuronidase synthesis? Two possibilities have already been suggested in the introduction to this paper. Both are based upon the concept of autogenous regulation of gene expression first proposed by Maas and McFall (1964) and later formalized and experimentally verified by Goldberger (1974). In autogenously regulated systems, a protein controls the expression of its own structural gene by direct interaction with the structural gene or an associated genetic element. This mechanism is operative in the repression of his operon in bacteria by the first enzyme in the biosynthetic pathway of histidine (Blasi et al., 1973). Another type of autogenous regulation, that of repression of specific m-RNA translation by its protein product, is suggested by studies on the interaction of myeloma protein with its heavy polypeptide chain m-RNA (Stevens and Williamson, 1972; 1973). However, in both examples of autogenous regulation, the control is negative, in that the protein represses gene expression. To speculate that glucuronidase interacts with the Gus gene or its message in order to repress glucuronidase synthesis would require that Gus^h glucuronidase be a substantially better repressor than the Gus^b product, since Gus^h homozygotes synthesize glucuronidase at approximately 10% of the rate and accumulate only 10% of the activity found in Gus^b homozygotes.

Other possible sites at which the control of glucuronidase synthesis by the Gus locus could be exerted include processing or transfer of the glucuronidase message between the nucleus and cytoplasm, stability of the glucuronidase message, and translation of the glucuronidase message.

Whatever models are invoked to explain our results, they must also explain our as yet unpublished results which show that the Gus^h allele, while responsible for a reduction in the basal rate of kidney glucuronidase synthesis, does not exert this control over the androgen-induced rate of synthesis in kidney (Ganschow et al., manuscript in preparation).

ACKNOWLEDGEMENTS

This work was supported by U.S. Public Health Service grants AM-14470 and HD-05221. The technical assistance of Susan M. Snyder is gratefully acknowledged.

REFERENCES

Blasi, F., C. B. Bruni, A. Avitabile, R. G. Deeley, R. F. Goldberger, and M. M. Meyers 1973. Inhibition of transcription of the histidine operon in vitro by the first enzyme of the histidine pathway. *Proc. Nat. Acad. Sci. U.S.A.* 70: 2692-2696.

Ganschow, R. E., 1973. The genetic control of acid hydrolases. In: *Metabolic Conjugation and Metabolic Hydrolysis*, Vol. 3. W. J. Fishman, editor, Academic Press, New York, pp.

189-207.

Ganschow, R. E., and K. Paigen 1967. Separate genes determining the structure and intracellular location of hepatic glucuronidase. *Proc. Nat. Acad. Sci. U.S.A.* 58: 938-945.

Ganschow, R. E., and K. Paigen 1968. Glucuronidase phenotypes of inbred mouse strains. *Genetics* 59: 335-349.

Goldberger, R. R. 1974. Autogenous regulation of gene expression. *Science* 183: 810-816.

Law, L. W., A. G. Morrow, and E. M. Greenspan 1952. Inheritance of low liver glucuronidase activity in the mouse. *J. Nat. Cancer Inst.* 12: 909-916.

Maas, W. K., and E. McFall 1964. Genetic aspects of metabolic control. *Ann. Rev. Microbiol.* 18: 95-110.

Meisler, M., and K. Paigen 1972. Coordinated development of β-glucuronidase and β-galactosidase in mouse organs. *Science* 177: 894-896.

Miyaji, T., Y. Oba, K. Yamamoto, S. Shibata, I. Iuchi, and H. B. Hamilton 1968. Hemoglobin Hijiyama: A new fast-moving hemoglobin in a Japanese family. *Science* 159: 204-206.

Morrow, A. G., E. M. Greenspan, and D. M. Carroll 1970. Comparative studies of liver glucuronidase activity in inbred mice. *J. Nat. Cancer Inst.* 10: 1199-1203.

Paigen, K. 1961a. The effect of mutation on the intracellular location of β-glucuronidase. *Exp. Cell Res.* 25: 286-301.

Paigen, K. 1961b. The genetic control of enzyme activity during differentiation. *Proc. Nat. Acad. Sci. U.S.A.* 47: 1641-1649.

Paigen, K. and W. K. Noell 1961. Two linked genes showing a similar timing of expression in mice. *Nature* 190: 148-150.

Paigen, K. and R. Ganschow 1965. Genetic factors in enzyme realization. *Brookhaven Symp. Biol.* 18: 99-115.

Pootrakul, S., P. Wasi, and S. Na-Nakorn 1967. Haemoglobin J-Bangkok: A clinical haematological and genetical study. *Brit. J. Haemat.* 13: 303-309.

Shibata, S., T. Miyaji, I. Iuchi, S. Veda, and I. Takeda 1964. Hemoglobin Hikari: A fast-moving hemoglobin found in two unrelated Japanese families. *Clin. Chim. Acta* 10: 101-105.

Stevens, R. H. and A. R. Williamson 1972. Specific Ig^G mRNA molecules from myeloma cells in heterogeneous nuclear and cytoplasmic RNA containing Poly-A. *Nature* 239: 143-146.

Stevens, R. H. and A. R. Williamson 1973. Isolation of messenger RNA coding for mouse heavy-chain immunoglobulin. *Proc. Nat. Acad. Sci. U.S.A.* 70: 1127-1131.

Swank, R. T. and K. Paigen 1973. Biochemical and genetic evidence for a macromolecular β-glucuronidase complex in microsomal membranes. *J. Mol. Biol.* 77: 371-389.

Weber, K. and M. Osborn 1969. The reliability of molecular weight determinations by dodecyl sulfate-polyacrylamide gel electrophoresis. *J. Biol. Chem.* 244: 4406-4412.

Yoshida, A. 1970. Amino acid substitution (histidine to tyrosine) in a glucose-6-phosphate dehydrogenase variant (G6PD) Hektoen) associated with over-production. *J. Mol. Biol.* 52: 483-490.

EXPRESSION OF ENZYME PHENOTYPES IN HYBRID EMBRYOS

DAVID A. WRIGHT
Biology Department
The University of Texas System Cancer Center
M. D. Anderson Hospital and Tumor Institute
Texas Medical Center
Houston, Texas 77025

ABSTRACT. Analysis of the offspring of natural heterozygotes and of hybrid frogs produced in the laboratory have established the genetic basis of fifteen electrophoretic enzyme variants in the *Rana pipiens* complex. A loose linkage (recombination frequency = 0.375) was found between the loci for 6-phosphogluconate (6PGD) and glucose-6 phosphate dehydrogenases (G6PD). The times of expression of twelve of these loci in F_1 hybrid embryos have been determined. Early embryos through the neurula stages have only maternal enzyme patterns, reflecting maternal control of enzyme synthesis during oogenesis. The earliest gene-controlled hybrid enzyme pattern to be expressed is 6PGD at the neural tube stage. Other enzyme loci are expressed subsequently, the supernatant isocitrate dehydrogenase (IDH-s) at tail bud, lactate dehydrogenase B (LDHB) at muscular response, and so forth.

Actinomycin-D will prevent the appearance of hybrid 6PGD patterns in hybrid embryos only if treatment precedes the time of expression (neural tube) by more than twenty hours (neural plate). Puromycin treatment 8 hours before the neural tube stage will block appearance of a hybrid 6PGD pattern. In a lethal hybrid *Rana pipiens* x *R. macroglossa* 97% reach the neural plate stage and then exogastrulate. These embryos do not express hybrid 6PGD or LDH patterns. While drug treatment probably affects the molecular events in gene expression, defective enzyme patterns in lethal hybrids are probably caused by disruption of processes influencing cellular viability.

INTRODUCTION

Electrophoretic analysis of enzymes from hybrid embryos resulting from crosses between parents of different electrophoretic phenotypes allows us to evaluate the expression of parental alleles during development. Eggs and early embryos might be expected to have enzyme patterns determined by the maternal genome. After a particular gene locus is activated a hybrid enzyme would be expected in the heterozygous embryos.

The nature of an F_1 hybrid enzyme pattern depends on subunit structure; tetramers produce 5 band patterns, dimers 3 band patterns, etc., (Shaw 1964). Expression of parental enzyme phenotypes have been examined using electrophoretic techniques in a number of animal systems including *Drosophila* (Courtright 1967, Wright and Shaw 1969, 1970) sea urchins (Ozaki and Whitely 1970, Whitely and Whitely 1972), fish (Hitzeroth et al., 1968, Goldberg et al., 1968, Klose et al., 1969, 1970, Whitt et al., 1972, Yamauchi and Goldberg 1974), frogs (Wright and Moyer 1966, Johnson and Chapman 1971), and mammals (Chapman et al., 1971, Epstein et al., 1972). Because of the ease of manipulation of nuclear and cytoplasmic components in amphibian embryos we have found them very useful in evaluating the nature of maternal cytoplasmic effects on enzyme patterns in early frog development (Wright and Subtelny, 1971, 1973).

Some difficulties, especially with claims of asynchrony of allelic expression, have been pointed out elsewhere (Siciliano et al., 1974). Several procedures are needed to evaluate the results of electrophoretic analysis of enzyme phenotypes in hybrid embryos in terms of the temporal sequence of genetic activation and expression. FIrst the genetic nature of electrophoretic differences between parental types must be determined. The time of expression of a particular enzyme locus might then be defined as the time of detection of a hybrid enzyme pattern upon electrophoresis. An experimental or genetic system should be devised to distinguish between maternal effects which are cytoplasmic and maternal effects due to preferential allelic expression. The time course of gene expression for a particular enzyme might be better understood by treatment of embryos with drugs that inhibit processes of RNA or protein synthesis. Likewise analysis of the expression of enzyme genes in various abnormal combinations of nuclear and cytoplasmic elements that result in lethal embryos might help to understand the mechanism of control of the particular enzyme locus.

This paper describes genetic analysis of several enzyme loci in the *Rana pipiens* complex, appearance of hybrid enzyme patterns in F_1 hybrids, the effects of inhibitors of macromolecular synthesis on the appearance of such patterns and the expression of enzyme genes in a lethal hybrid combination.

GENETICS OF ENZYME PATTERNS

Methods for vertical starch gel electrophoresis and specific enzyme staining are essentially those reported by Shaw and Prasad 1970. In the course of electrophoretic analysis of wild caught frogs, animals that are heterozygous at one or more loci are obtained. Crosses are made using artificial ovulation and fertilization (Rugh 1934). The crosses can be analyzed as tadpoles for inheritance of enzyme patterns. Examples of several patterns are shown in Fig. 1. Data accumulated from a number of crosses demonstrate Mendelian patterns of inheritance and independent assortment for a number of loci including phosphoglucomutase (PGM), lactate dehydrogenase (LDH), mannosephosphate isomerase (MPI), α-glycerophosphate dehydrogenase (αGPD) and glutamate oxaloacetate transaminase (GOT).

Geographically separated populations of frogs of the *Rana pipiens* complex are a rich source of electrophoretic variants useful in developmental and genetic studies. We were successful in raising an F_1 hybrid resulting from a *Rana pipiens* female from Wisconsin and a variety of *Rana pipiens* from Sinaloa Mexico (Bagnara and Stackhouse 1973). The parental types and adult F_1 hybrid are shown in Figs. 2 and 3.

The F_1 hybrid male in Fig. 3 was backcrossed to a Wisconsin female and the offspring analyzed as tadpoles for enzyme phenotypes. A few enzyme patterns are shown in Fig. 4. The data are summarized in Table 1. Each locus yields the expected 1:1 ratio of hybrid and pipiens type patterns with the exception of the LDH-B locus in which an excess of hybrid patterns is obtained. This might be due to an unknown reduced viability gene linked to the pipiens LDH-B locus in the hybrids. This analysis also indicated a loose linkage of 6-phosphogluconate dehydrogenase (6PGD) to glucose-6 phosphate dehydrogenase (G6PD), 36 recombinant types and 60 parental types (χ^2=5.34, P=.03) for a recombination frequency of 0.375. Loci for Esterases 1 and 3 are either very tightly linked or are controlled by a single locus since no recombinants were seen in the 65 individuals analyzed for both. All other loci appear to assort randomly.

TIME OF EXPRESSION OF PATERNAL ALLELES

Analysis of enzyme patterns was performed on hybrid embryos. Embryos were staged according to Shumway's (1940) series for *Rana pipiens*. Ten embryos of a stage were removed from the jelly membranes, placed on a piece of parafilm and most of the adhering water removed with filter paper. The embryos were frozen in homogenizing buffer (5λ/embryo) thawed and homogenized with the tip of a Pasteur pipette in a 1 ml conical centrifuge tube. The homogenate was centrifuged at 14,000 x g for 20 min. and the supernatant fluid was applied to the starch gel.

The time of expression of paternal alleles for several enzymes in viable F_1 hybrid embryos is summarized in Table 2.

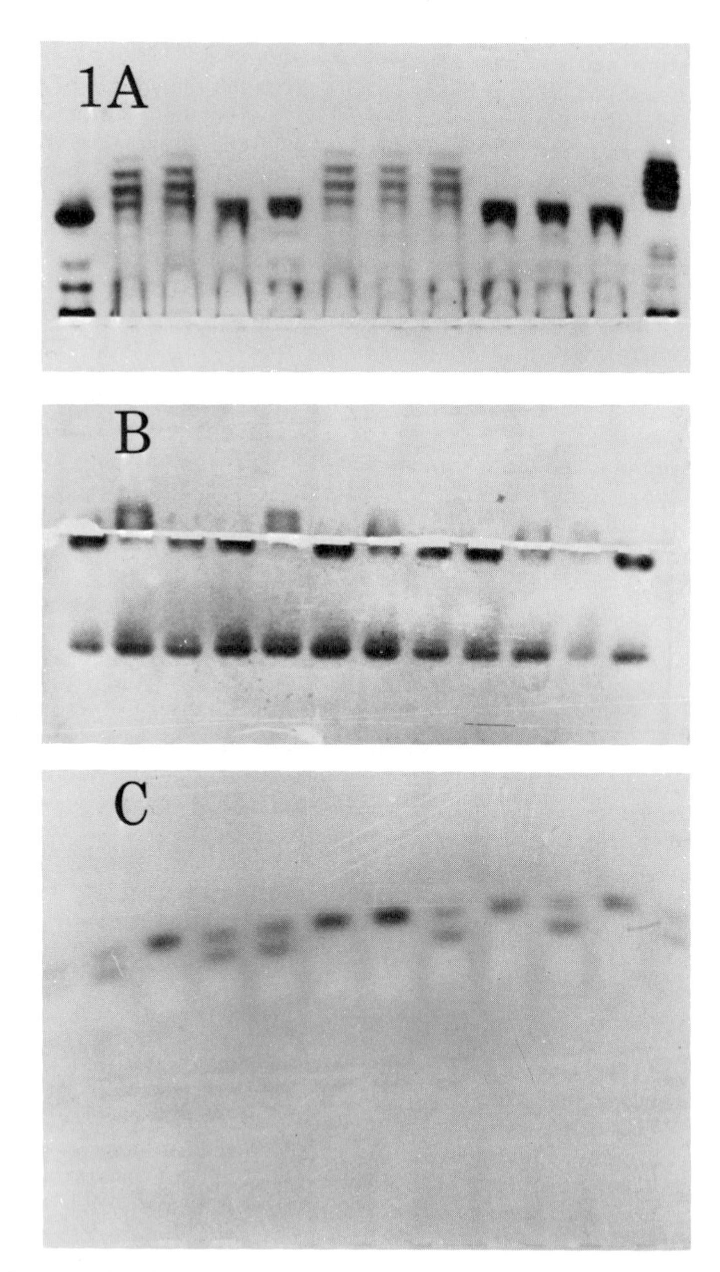

Fig. 1. Analysis of enzyme patterns among offspring of naturally occurring heterozygotes. A. Photograph of starch gel stained for LDH. Patterns on the far left and far right are from heart extracts of a female *Rana pipiens pipiens* (Wisconsin)

Fig. 1 legend continued. and a male *R. p. sphenocephala* (Tennessee). The latter was heterozygous at the LDH-B locus while the female was homozygous for a third allele. The 10 patterns in between are from their offspring showing two types of heterozygous patterns. B. Segregation of parental type patterns for glutamate oxaloacetate transaminase (GOT) in 12 offspring of a heterozygous female and homozygous male at the GOT-1 locus. Note that GOT-2, the cathodal isozyme, is not affected by the GOT-1 genotype. C. Analysis of mannosephosphate isomerase (MPI) in the same samples and order as in 1B above. Note the 2 band patterns in the heterozygotes suggesting a monomer structure for this enzyme and the apparent independent assortment of the GOT-1 and MPI loci.

TABLE 1

ANALYSIS OF BACKCROSS HYBRID ENZYME PATTERNS

Enzyme Locus	Observed hyb pip	Expected hyb pip	χ^2	P
6PGD	71 : 88	79.5:79.5	1.82	>.15
LDH-B	96 : 64	80 : 80	6.40	<.02
MDH-s	64 : 70	67 : 67	0.27	>.60
IDH-s	64 : 62	63 : 63	0.03	>.80
MDH-m	51 : 69	60 : 60	2.70	≃.10
G6PD	59 : 65	62 : 62	0.29	≃.60
PGM	61 : 67	64 : 64	0.28	≃.60
Ald.	34 : 28	31 : 31	0.58	>.40
Est 1	37 : 28	32.5:32.5	1.25	>.25
Est 3	61 : 63	62.0:62.0	0.03	>.80
Est 4	31 : 42	36.5:36.5	1.66	≃.20
Est 5	44 : 52	48.0:48.0	.67	>.40

This is a summary of results using reciprocal crosses between geographically separated frogs within the *Rana pipiens* complex. Several enzymes listed here have been reported previously including 6PGD, supernatant isocitrate dehydrogenase (IDH-s), mitochondrial malate dehydrogenase (MDH-m) and LDH (Wright and Subtelny 1971, Wright and Moyer 1966, Siciliano et al., 1974). The time of appearance of 6PGD, IDHs and PGM is earlier than reported previously. This is due to the present use of a 0.01 M Tris HCl buffer containing mM EDTA and mM β-mercaptoethanol for homogenization of embryos, instead of a 1/10 dilution of buffered Ringer's solution used previously. Time of expression of LDH-B and MDH-m loci are the same regardless of the type of homogenizing buffer.

Examples of this type of analysis are shown in Fig. 5. A test for localization of expression of the IDH and 6PGD loci is also demonstrated in Fig. 5. No differences are seen in the

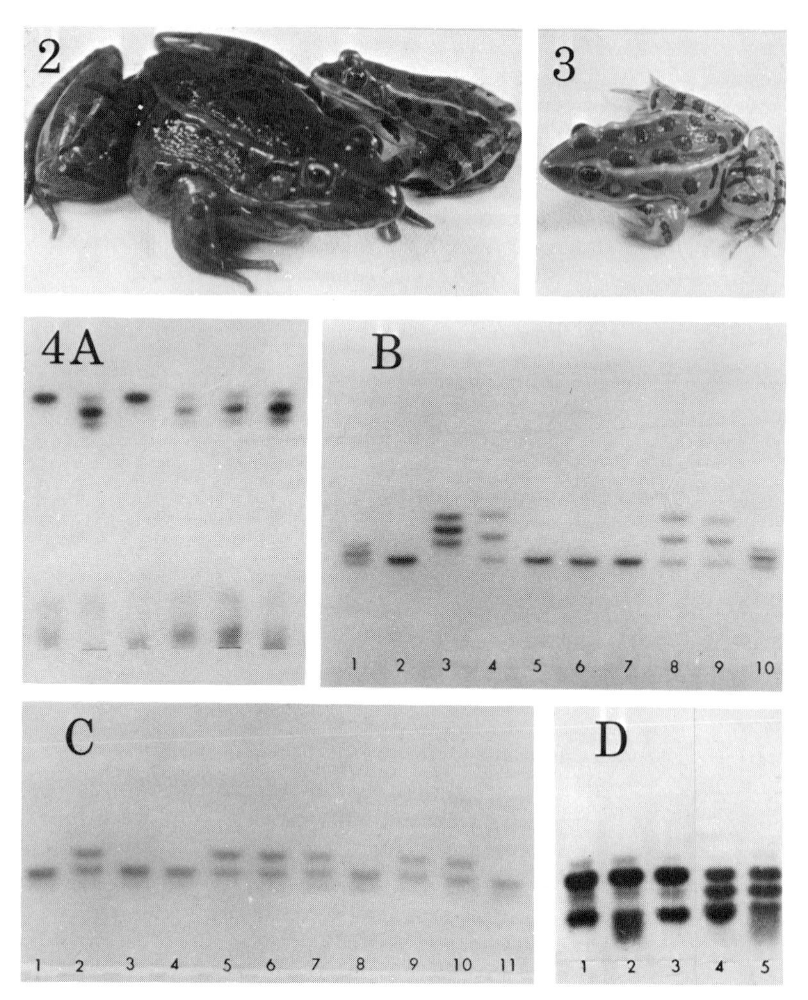

Fig. 2. Frogs of the *Rana pipiens* complex from Mexico, on the left, and *R. p. pipiens* from Wisconsin on the right.

Fig. 3. F_1 hybrid male reared from fertilized egg in the laboratory.

Fig. 4. Analysis of enzyme patterns in backcross hybrid progeny of a *R. p. pipiens* female and the F_1 hybrid male shown in Fig. 3. A. Segregation of pipiens, (single band) and hybrid (three band) patterns of IDH-s. B. Analysis of patterns of 6PGD in the backcross hybrids was complicated by the fact that the pipiens female had a heterozygous pattern like the first sample while the F_1 hybrid had a pattern like the fourth sample. "Hybrid" patterns (3rd, 4th, 8th, 9th) were distinguished from "pipiens" patterns (1st, 2nd, 5-7th, 10th) by the presence

Fig. 4 legend cont. of the most anodal enzyme band characteristic of the Mexican frogs. C. Segregation of pipiens type PGM as in the 1st, 3rd, 4th, 8th and 11th samples from hybrid patterns as in the 2nd, 5-7th, 9-10th samples. D. Independence of loci for supernatant malate dehydrogenase (MDH-s) and mitochondrial MDH (MDH-m). All samples are backcross hybrid offspring. The first sample is the same as the pipiens pattern while the 5th is the same as the F_1 hybrid. Samples 1 and 3 are homozygous for the supernatant locus, most anodal isozyme, and also homozygous for the mitochondrial isozyme; sample 2 homozygous MDH-s, heterozygous MDH-m; sample 4 heterozygous MDH-s, homozygous MDH-m; sample 5 heterozygous at both loci.

TABLE 2

TIME OF EXPRESSION OF PATERNAL GENE PRODUCTS

Enzyme Locus	Stage	Hrs. at 18°C
6PGD	16 - neural tube	72
IDH-s	17 - tail bud	84
LDH-B	18 - muscular response	96
MDH-m	19 - heart beat	118
PGM-2	20 - gill circulation	140
G3PD	20 - gill circulation	140
GPI	21 - mouth open	162
T. O.	21 - mouth open	162
ME-1 (NADP-MDH)	21 - mouth open	162
Aldolase-C	22 - tail fin circulation	192
MPI	23 - opercular fold	216
ADH	24 - operculum closed on right	240

Abbreviations of enzymes not used in the text, G_3PD = glyceraldehyde-3-phosphate dehydrogenase, GPI = glucosphosphate isomerase, T. O. = tetrazolium oxidase, ME = malic enzyme, ADH = alcohol dehydrogenase.

time of expression of 6PGD or IDH-s in dorsal vs. ventral portions of the embryo. The IDH-m isozyme seems to be more intensely stained in the dorsal region of the embryo.

DISSECTION OF THE STEPS IN GENE EXPRESSION

Before the expression of a paternal and hybrid enzyme forms can be seen in a hybrid there are a series of events which must take place. These include the following:

1. Activation of the gene locus - i.e. transcription of the m-RNA specific for the enzyme in question.
2. Transport of the message to the cytoplasm, availability of

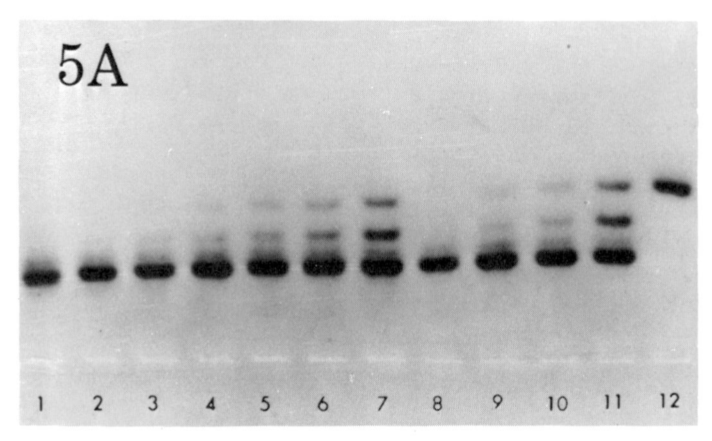

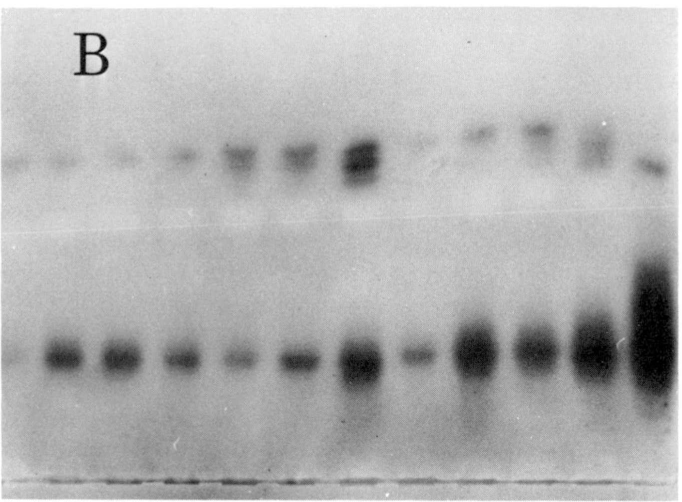

Fig. 5. Expression of paternal and hybrid enzyme patterns in heterozygous embryos, resulting from a cross of Wisconsin female x Mexican male. Samples 1-3 whole embryos; 1. stage 13, 2. stage 14, 3. stage 15. Samples 4-7 ventral region of embryos; 4. stage 16, 5. stage 17, 6. stage 18, 7. stage 20. Samples 8-11 dorsal region; 8. stage 15, 9. stage 16, 10. stage 17, 11. stage 20. Sample 12 heart extract of male in the cross. A. Starch gel slice stained for 6PGD. B. Gel slice stained for IDH. Parental differences are seen only in the more anodal IDH-s isozyme.

---

ribosomes, t-RNAs and enzymes necessary for the next step.
3. Synthesis of the polypeptide specified by the m-RNA.

4. Assembly of the polypeptide subunits into active enzyme molecules that can be detected after gel electrophoresis.

In order to understand the time course of these events experiments using inhibitors of RNA and protein synthesis, actinomycin D and puromycin respectively, were employed. Attempts to inhibit RNA and protein synthesis in frog embryos by emersion of whole embryos in solutions of the drugs actinomycin D and puromycin were largely unsuccessful, apparently due to the fact that the frog embryo is largely a closed system and the drugs do not penetrate the ectodermal layer. This problem was solved by injecting the embryos using a micropipette or by cutting the embryo in an isotonic medium and incubating in the presence of drugs, radioactive precursors, etc.

Cuts were made that extended through the embryo 1/3 to 1/2 from the anterior end. Embryos so cut were incubated at 16°C, 5 to a dish, in 5 ml isotonic medium, containing 0, 50 and 100 μg/ml actinomycin-D, and 0, 50 and 100 μg/ml puromycin. The isotonic medium contained 3.5 g/l NaCl, 0.13 g/l $CaCl_2 \cdot 2H_2O$, 0.12 g/l $MgCl_2 \cdot 6H_2O$, 0.07 g/l $KHCO_3$.

Incorporation of [^{14}C] leucine (0.5 μCi/ml, 312 mc/mmole) or [^{3}H] uridine (5 μCi/ml, 8 Ci/mmole) into 5% trichloroactic acid insoluble - 0.6N NCR (Amersham Searle) soluble material was measured by liquid scintillation spectrometry.

In control and actinomycin treated embryos the wound would heal partially by 24 hours. Cutting the embryos does not adversely affect development since operated and unoperated controls develop normally (Fig. 6). Actinomycin treatment of neurula stage 14 and 15 embryos allowed them to develop in 24 hours to a neural tube stage which survives several days in an arrested neural tube-tail bud stage (Fig. 7). Puromycin treatment completely prevents wound healing, only partial morphogenesis takes place and embryos treated at high doses do not live much longer than 24 hours. Incorporation of [^{14}C] leucine into the embryos was inhibited 55% and 73% by the 50 μg/ml and 100 μg/ml respectively of puromycin. Incorporation of [^{3}H] uridine was inhibited 59% and 92% by the 50 μg/ml and 100 μg/ml doses of actinomycin respectively.

The effects of these drugs on the expression of three enzyme loci were studied. 6PGD is normally expressed at stage 16 (neural tube), IDH-s is normally expressed at stage 17 (tail bud) and MDH-m is normally expressed at stage 19 (heart beat) in hybrids between Wisconsin and Mexican *R. pipiens*. We first considered the effects on 6PGD. If the hybrid embryos were treated at stage 15 (neural folds closed) and allowed to develop until the controls were past stage 16, the controls expressed a hybrid 6PGD pattern and actinomycin and puromycin treated embryos also expressed a hybrid pattern. Treatments beginning

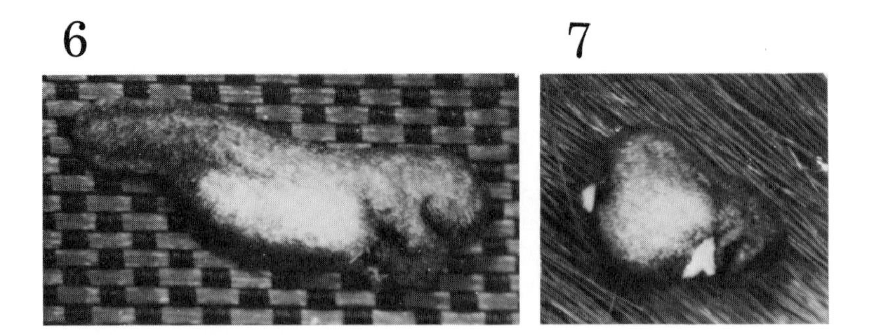

Fig. 6. A hybrid embryo resulting from a cross of a Wisconsin *R. p. pipiens* with a Mexican *R. pipiens*. The embryo had been cut at stage 14 as described in the text and grown in an isotonic medium to stage 19. The ventrally directed bend in the tail is characteristic of this cross.
Fig. 7. A hybrid embryo from the same cross as the embryo in Figure 6 cut in the same manner but incubated in 100 μg/ml actinomycin for the same period of time. Note arrested morphological development.

at stage 14 (neural fold some 8 hours earlier) and allowing the control embryos to develop past stage 16 showed that puromycin treatment prevented expression of a hybrid 6PGD pattern but that actinomycin treatment merely decreased the relative intensity of the hybrid bands. If treatment was begun earlier at stage 13 (neural plate), and the embryos allowed to develop until controls were at stage 16 (28 hours), hybrid patterns were seen in controls, but no hybrid patterns were seen in the actinomycin or puromycin treated embryos.

These results suggest the following events occur in the expression of the genetic locus for 6PGD. The gene is activated between stages 13 and 14 to produce an m-RNA. This message is transported to the cytoplasm where it directs the synthesis of the 6PGD subunits beginning between stages 14 and 15. Assembly of the first active enzyme from these subunits occurs between stages 15 and 16.

For the locus which controls IDH-s each of these events is apparently delayed by 7 to 10 hours, corresponding to the a later expression of IDH-s at the tail bud stage (stage 17) rather than neural tube (stage 16). Stage 15 embryos cut and allowed to develop to stage 17 demonstrated a hybrid pattern, those treated with puromycin did not express a hybrid IDH-s pattern while actinomycin-D treatment allowed a hybrid pattern to be expressed. Actinomycin treatment of hybrid embryos at stage 14 or earlier did prevent the appearance of hybrid patterns. The events in expression of MDH-m patterns (normally expressed by stage 19 in hybrid embryos) were even further delayed. A summary of these experiments is seen in Table 3.

TABLE 3
EFFECTS OF INHIBITORS OF MACROMOLECULAR SYNTHESIS ON EXPRESSION OF HYBRID ENZYME PHENOTYPES

Stages of Treatment		Drug	6PGD	IDH-s	MDH-m
Exp. 1 St.	13-16	A 100	-	-	-
	14-17	A 100	+	-	-
	14-17	P 100	-	-	-
	15-17½	A 100	++	+	-
	15-17½	P 100	+	-	-
	13-17½	0	+++	++	-
Exp. 2	14-19	A 50	+	+	-
	14-19	A 100	+	-	-
	15-19	A 50	++	++	+
	15-19	A 100	++	+	-
	15-19	P 50	+	-	-
	16-19	A 50	+++	+	++
	16-19	A 100	+++	++	++
	16-19	P 50	+++	++	+
	16-19	P 100	++	++	-
	14-19	0	++++	+++	+++

A 50 = Actinomycin D 50 μg/ml
A 100 = Actinomycin D 100 μg/ml
P 50 = Puromycin 50 μg/ml
P 100 = Puromycin 100 μg/ml

GENE EXPRESSION IN A LETHAL HYBRID

Analysis of lethal combinations of nuclear and cytoplasmic components in *Rana* indicate that if an arrested hybrid remains alive, even though it does not progress morphologically, genes for specific enzymes may be expressed on time with respect to control embryos. Johnson (1971) has reported this in arrested *R. pipiens* x *R. sylvatica* hybrids for LDH. In androgenetic haploid hybrids which arrest earlier than non-hybrid haploids the time expression of 6PGD, IDH-s and LDHB loci is the same as diploid controls for a number of nuclear-cytoplasmic combinations within the *Rana pipiens* complex (Wright and Subtelny 1971, 1973).

An interesting comparison can be made with results of gene expression in drug-treated hybrids with the androgenetic hap-

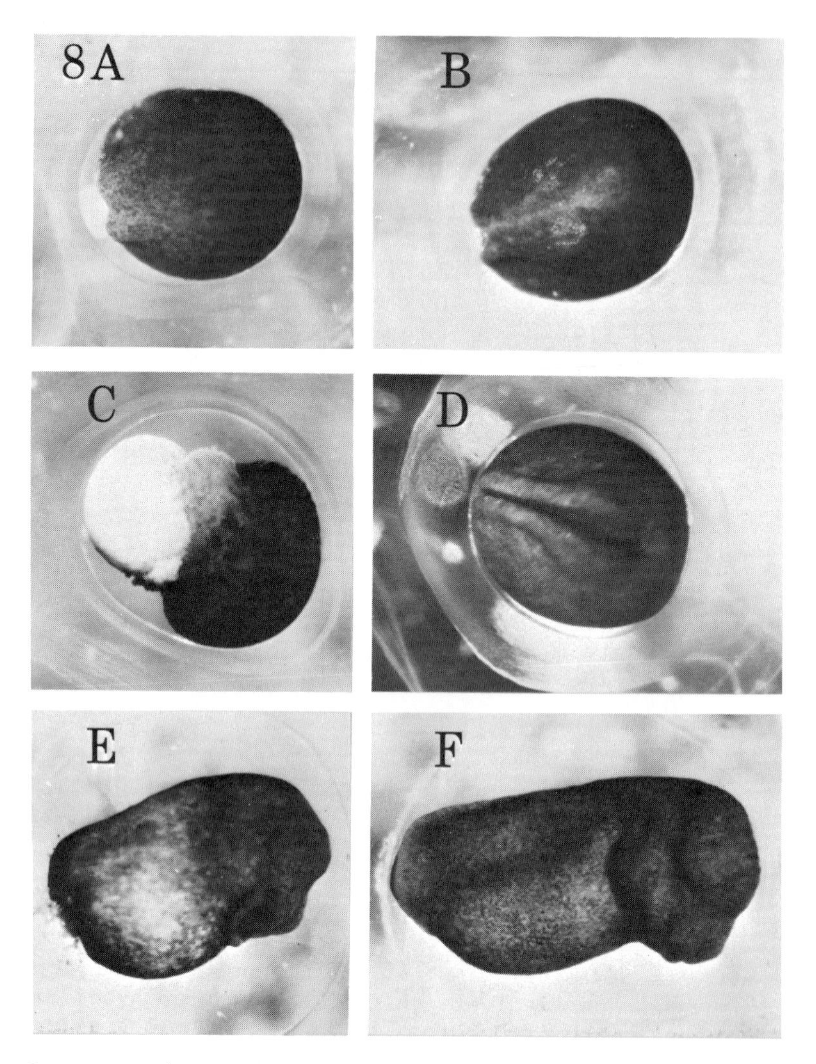

Fig. 8. Comparison of development of a lethal hybrid combination, *Rana pipiens* female x *R. macroglossa* male (A,C+E) with normal *R. pipiens* x *R. pipiens* control embryos (B,D+F). A and B compare hybrid and controls as controls are forming neural folds. C and D show exogastrulation in the hybrid as controls show neural fold closure. E and F compare the morphologies of the rare (3%) of hybrid embryos that survive past neurula to a tail bud control embryo of the same age.

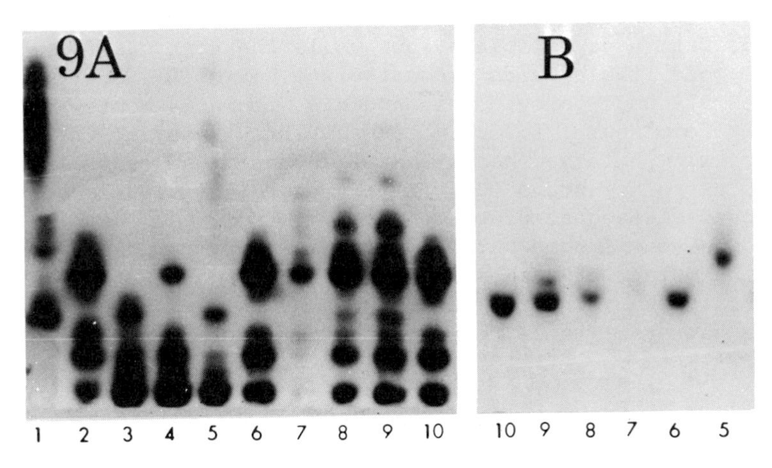

Fig. 9. Expression of hybrid enzyme phenotypes in hybrid *R. pipiens* x *R. macroglossa* embryos that survive neurulation. Samples are 1. *R. macroglossa* heart; 2. *R. pipiens* heart; 3. *R. macroglossa* liver; 4. *R. pipiens* liver; 5. *R. macroglossa* kidney; 6. *R. pipiens* stage 22 control embryo; 7-9. individual *R. pipiens* x *R. macroglossa* embryos of increasing normality of the same age as control embryos (stage 22); 10. repeat control *R. pipiens* embryo stage 22. A. Gel slice stained for LDH. B. Gel slice stained for 6PGD.

loid hybrids reported by Wright and Subtelny (1971). In one cross the androgenetic haploids survived only until the diploids were at stage 19 (heart beat). In these embryos the nuclear gene for 6PGD was expressed while the form of MDH-m controlled by the foreign haploid nucleus was not expressed. In another cross of the same type but involving different parents the haploid hybrids showed the same kind of abnormalities (abnormal gastrulation, formation of abnormal neural folds) but the embryos lived until the controls were beyond stage 20 (gill circulation). In this cross both 6PGD and MDH-m loci were expressed on time with the diploid controls. Other androgeneic haploid hybrids such as those involving a *R. palustris* haploid nucleus in *R. pipiens* cytoplasm died before the expression of genes for several enzymes were due to be expressed.

In this regard the results of a cross between *Rana pipiens* (Wisconsin) and a species of frogs collected in Guatamala tentatively identified as *Rana macroglossa* (Stuart 1948) are of interest. Eggs of *Rana pipiens* were fertilized with *R. macroglossa* sperm, controls consisted of the same pipiens eggs fertilized with pipiens sperm. The hybrids developed until the midgastrula stage without any difference from the controls. Gastrulation of the hybrid seems to be impeded in some way, with a pronounced yolk plug at the end of gastrulation. The hybrids form a neural plate, but while the controls form the neural fold, the hybrids exogastrulate, (Fig. 8). Of these embryos 97% undergo cytolysis and die within 24 hours. Three percent form neural folds and subsequently abnormal tail bud embryos with enlarged heads that live until the controls have opercular folds (stage 23). Electrophoretic analysis of the embryos that exogastrulate compared to those that continue to develop, even if somewhat abnormally, are what would be expected based on molecular events indicated from experiments with inhibitors above. Those that do not neurulate express no hybrid enzyme patterns; those that survive demonstrate hybrid patterns (Fig. 9).

ACKNOWLEDGEMENTS

This work has been supported by N. I. H. grant HD 07021.

REFERENCES

Bagnara, J. T. and H. L. Stackhouse 1973. Observations on Mexican *Rana pipiens*. *Amer. Zool.* 13: 139-143.

Chapman, V. M., W. K. Whitten, and F. H. Ruddle 1971. Expression of paternal glucose phosphate isomerase-1 (Gpi-1) in preimplantation stages of mouse embryos. *Devel. Biol.* 26: 153-158.

Courtright, J. B. 1967. Polygenic control of aldehyde oxidase in *Drosophila*. *Genetics* 57: 25-39.

Epstein, C. J., J. A. Weston, W. K. Whitten, and E. S. Russell 1972. The expression of the isocitrate dehydrogenase locus (Id-1) during mouse embryogenesis. *Devel. Biol.* 27: 430-433.

Goldberg, E., J. P. Cuerrier, and J. C. Ward 1968. Lactate dehydrogenase ontogeny, paternal gene activation, and tetramer assembly in embryos of brook trout, lake trout and their hybrids. *Biochem. Genet.* 2: 335-350.

Hitzeroth, H., J. Klose, S. Ohno, and U. Wolf 1968. Asynochronous activation of parental alleles at the tissue specific gene loci observed on hybrid trout during early development. *Biochem. Genet.* 1: 287-300.

Johnson, Kurt E. 1971. A biochemical and cytological investigation of differentiation in the interspecific hybrid amphibian embryo *Rana pipiens* ♀ x *Rana sylvatica* ♂. *J. Experimental Zool.* 177: 191-206.

Johnson, K. E. and V. M. Chapman 1971. Expression of paternal genes during embryogenesis in the viable interspecific hybrid amphibian embryo *Rana pipiens* ♀ x *Rana palustris* ♂

- Electrophoretic analysis of five enzyme systems. *J. Exp. Zool.* 178: 313-318.

Klose, J., H. Hitzeroth, H. Ritter, E. Schmidt, and U. Wolf 1969. Persistence of maternal isoenzyme patterns of the lactate dehydrogenase and phosphoglucomutase system during early development of hybrid trout. *Biochim. Genet.* 3: 91-97.

Klose, J. and U. Wolf 1970. Transitional hemizygosity of the maternally derived allele at the 6PGD locus during early development of the cyprinid fish *Rutilus rutilus*. *Biochem. Genet.* 4: 87-92.

Ozaki, H. and A. H. Whitely 1970. L-Malate dehydrogenase in the development of the sea urchin *Strongylocentrotus purpuratus*. *Devel. Biol.* 21: 196-215.

Rugh, R. 1934. Induced ovulation and artificial fertilization in the frog. *Biol. Bull.* 66: 22-29.

Shaw, C. R. 1964. The use of genetic variants in the analysis of isozyme structure. *Brookhaven Symp. Biol.* 17: 117-129.

Shaw, C. R. and R. Prasad 1970. Starch gel electrophoresis of enzymes -- a compilation of recipes. *Biochem. Genet.* 4: 297-320.

Shumway, W. 1940. Stages in the normal development of *Rana pipiens*. *Anat. Rec.*, 78: 139-147.

Siciliano, M. J., D. A. Wright, and C. R. Shaw 1974. Biochemical polymorphisms in animals as models for man. *Progress in Medical Genetics* Vol. X ch. 2, pp. 17-53.

Stuart, L. C. 1948. The amphibians and reptiles of Alta Verapoz, Guatamala. Misc. Pub. Museum of Zool., U. Mich. 69: 1-109.

Whitely, A. H. and H. R. Whiteley 1972. Replication and Expression of maternal and paternal genomes in a blocked echinoid hybrid. *Develop. Biol.* 29: 183-198.

Whitt, G. S., P. L. Cho, and W. F. Childers 1972. Preferential inhibition of allelic isozyme synthesis in an Interspecific Sunfish hybrid. *J. Exptl. Zool.* 179: 271-282.

Wright, D. A. and F. H. Moyer 1966. Parental influences on lactate dehydrogenase in the early development of hybrid frogs in the genus *Rana*. *J. Exp. Zool.* 163: 215-230.

Wright, D. A. and C. R. Shaw 1969. Genetics and ontogeny of α-glycerophosphate dehydrogenase isozymes in *Drosophila melanogaster*. *Biochem. Genet.* 3: 343-353.

Wright, D. A. and C. R. Shaw 1970. Time of expression of genes controlling specific enzymes in *Drosophila* embryos. *Biochem. Genet.* 4: 385-394.

Wright, D. A. and S. Subtelny 1971. Nuclear and cytoplasmic contributions to dehydrogenase phenotypes in hybrid frog embryos. *Develop. Biol.* 24: 119-140.

Wright, D. A. and S. Subtelny 1973. Effects of haploidy and hybridization on the activities of four dehydrogenases in frog embryos. *Develop. Biol.* 32: 297-308.

Yamauchi, T. and E. Goldberg 1974. Asynchronous expression of glucose-6-phosphate dehydrogenase in splake trout embryos. *Develop. Biol.* 39: 63-68.

THE STATUS OF THE EVOLUTIONARY THEORY OF LACTATE DEHYDROGENASE ISOZYME FUNCTION IN RELATION TO FROG LDHs

STANLEY N. SALTHE
Department of Biology
Brooklyn College of the City University of New York
Brooklyn, New York 11210

ABSTRACT. Given the conditions of temperature, pH, and substrate concentration that appear to be present in nature, LDH kinetics and variability in frogs are consistent with an evolutionary theory of LDH function and tissue distribution. This theory is paraphrased and slightly elaborated. Attention is paid as well to the adaptive theory of K_m alteration with temperature and to the natural selective theory of allelic isozyme variability.

One of the aims of the evolutionary biologist in dealing with primary gene products is to consider to what degree and in what manner natural selection is involved with their functional (presumably adaptive) properties. The functional properties of enzymes are, primarily at least, their catalytic and kinetic capabilities. The kinetic properties of LDHs have been the subject of controversy in the last decade, particularly with respect to the functional meaning, if any, of substrate inhibition (Wuntch et al., 1970a; Everse et al., 1970; Wuntch et al., 1970b). In addition, some suggestions have been made concerning the functional meaning and evolutionary significance of changes in K_m with temperature in poikilotherms (Somero, 1969; Hochachka and Somero, 1973).

For some years we have been involved in studying the LDHs of frog populations, with the primary intention being to discover the extent and kinds of inter- and intrapopulation variability. During this period we have accumulated kinetic data that can be brought to bear on these general questions of the adaptive meaning of properties such as substrate related changes in reaction velocity and temperature related changes in K_m.

It should be noted that the dilute kinetics studies of B_4LDH referred to herein derive from studies of crude homogenates, while the A_4LDH kinetics were observed using purified enzymes, as were the high enzyme concentration studies on B_4LDH. It is important to realize that the tissues of these frogs do not typically contain the usual spectrum of five isozymes, because the B subunits are to one degree or another unable to form heteromultimers (Salthe, 1969).

Table 1 shows the relevant variables forming the initial and boundary conditions in the enzymatic environment together

TABLE 1

IMPORTANT ENVIRONMENTAL VARIABLES AFFECTING LDH KINETICS. Velocity refers to the initial rate of the reaction.

VARIABLE	HAS AN EFFECT UPON:
Substrate concentration	velocity
Enzyme concentration	degree of substrate inhibition; velocity
Coenzyme concentration	degree of substrate inhibition
pH	velocity; K_m; V_{max}; substrate concentration
Temperature	velocity; K_m; V_{max} substrate concentration; degree of substrate inhibition

with the kinetic properties that are modified when these critical variables are altered. The relationships between the kinetic properties of these LDHs are shown in the substrate inhibition curve in figure 1 for B_4LDH of *Rana pipiens*. In this and subsequent figures the substrate axis is plotted logarithmically so as to better visualize events at low substrate concentrations. In the absence of gross changes in the shape of this curve under different conditions, a change in K_m will move the entire curve along the substrate axis, while a change in V_{max} will move the curve vertically. A change in degree of substrate inhibition will register as a change in the slope of the inhibited portion of the curve.

Figure 2 demonstrates the effect of increasing the concentration of enzyme. Of necessity the kinetics of the upper curve were monitored using a stopped-flow apparatus. The lower curve is the same (standard) curve shown in figure 1. There is an eight-fold difference in V_{max} between these curves. Theoretically, there should be no change in K_m when enzyme concentration is increased, hence I have shown none here. Our data (Eby et al., 1973), however, suggest that some changes may occur. If this should be confirmed then there is some reason to suspect that dilute kinetics K_m values are artifactual, possibly resulting from altered subunit relationships. Increasing enzyme concentration minimizes substrate inhibition, but, unlike the situation in mammals (Wuntch et al., 1970) and birds (Everse et al., 1970) it is not eliminated in the B_4 isozyme at this concentration of enzyme, which is considered to be physiological for mammals (Wuntch et al., 1970). Obviously, there exists in principle some concentration of enzyme great enough so that the

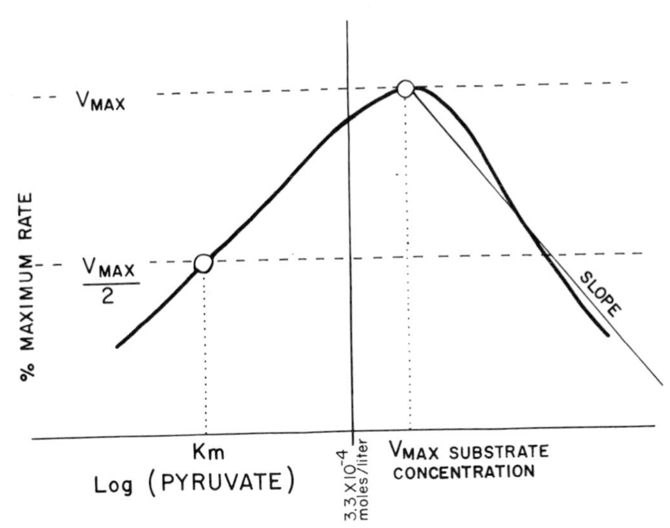

Fig. 1. The geometry of an LDH substrate inhibition curve based on that of B_4LDH-1 of *R. pipiens* at standard conditions (25°, pH 7.6). The substrate concentration 3.3 X 10^{-4} moles/liter will be a reference point. The rates are initial rates of reaction.

initial reaction rate will be so fast that no significant amount of the inhibiting abortive ternary complex (Arnold and Kaplan, 1974) will be formed during the period of monitoring the reaction. In mammals and birds that concentration appears near to that found in cells. Unless enzyme concentrations in frogs are very much greater, that concentration is not reached in frog cells.

We have also found (Eby et al., 1973) that, while increasing the concentration of coenzymes (NADH for the pyruvate reductase reaction) decreases the degree of substrate inhibition, it requires amounts of NADH well beyond those considered physiological to achieve this effect. At physiological levels of 0.1 mM NADH (Kaplan, 1960), substrate inhibition is not supressed even at high enzyme concentrations. We used 0.1 mM NADH in all the dilute enzyme assays to be reported here. Thus, the *Rana pipiens* B_4LDH under reasonably natural concentrations of enzyme and coenzyme, shows substrate inhibition even without preincubation with NAD (Everse et al., 1970). The latter procedure is capable of restoring substrate inhibition of homeotherm LDHs at high enzyme concentration, and so would presumably increase that inhibition in the frog system. There is some reason to believe that such preincubation mimics events in the cells

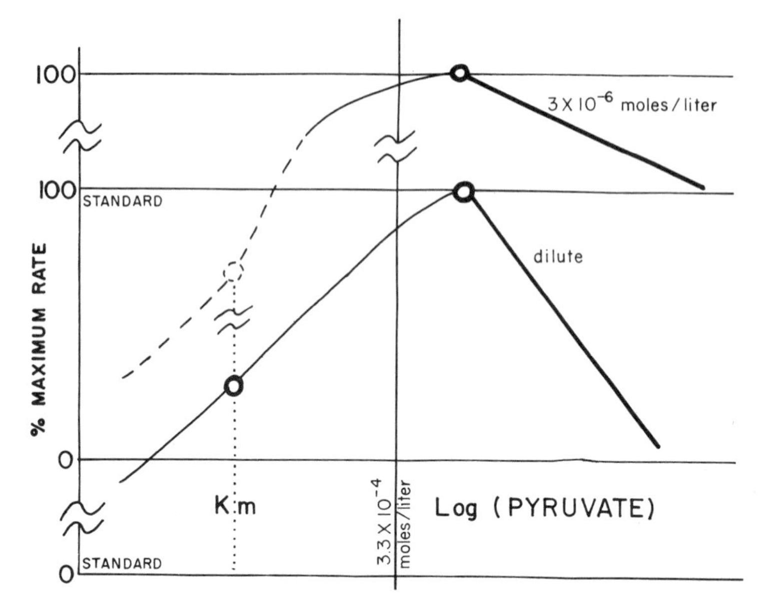

Fig. 2. The effect of enzyme concentration on the substrate inhibition curve of figure 1 (data from Eby et al., 1973). The V_{max}s of the curves are related by percent difference, but the curves themselves are drawn to scale independently, each having its own abscissal base line.

because of the relatively high concentrations of NAD found in cells. The remainder of this paper will concern itself with dilute enzyme kinetics. In all these studies the enzyme concentration was adjusted to give an initial rate of 100 absorption units/ 60 seconds under standard conditions (pH 7.6, 25°) at 0.33 mM pyruvate.

Figures 3 and 5 show the effects on the substrate inhibition curve of modifying the pH. A slight change in the degree of inhibition occurs. V_{max} is altered upward as pH is increased. But because of rather large changes in the K_m the initial velocity at any given concentration of substrate may increase or decrease with pH, giving rise to pH optima in the region between 0.1 and 1 mM pyruvate. Thus, at 0.33 mM the optimum is pH 7 (Levy and Salthe, 1971). The effect of pH on reaction velocity is in part explained by the fact that ionic interactions are important at the active site. Thus, a histidine is the source of a proton during the conversion of pyruvate to lactate (Adams et al., 1973). The transfer of that proton may reason-

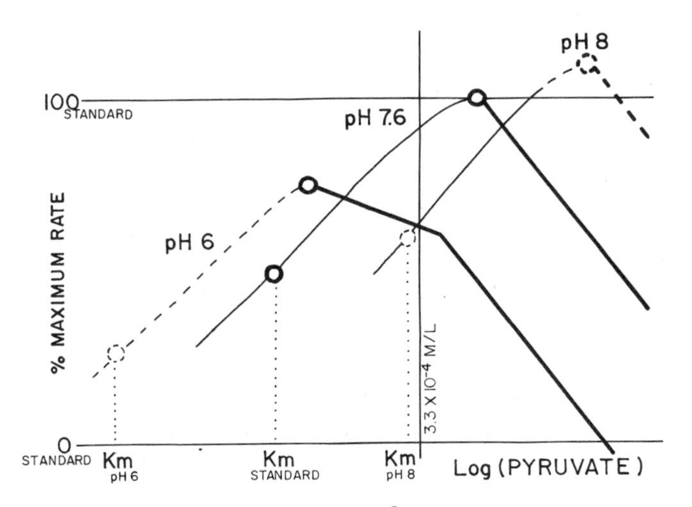

Fig. 3. The effect of pH (at 25°) on the substrate inhibition curve of figure 1 (data from Salthe, 1965; *Comp. Biochem. Physiol.* 16: 393 and Levy and Salthe, 1971). The V_{max}s of the curves are related by percent difference.

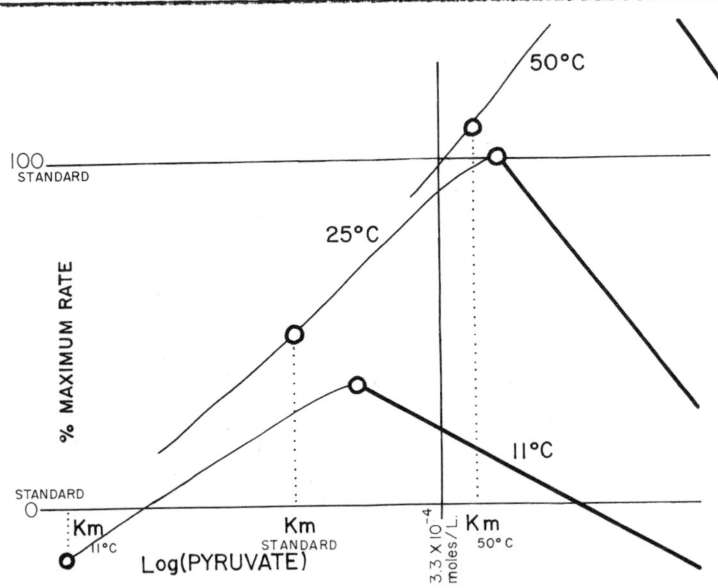

Fig. 4. The effect of temperature (at pH 7.6) on the substrate inhibition curve of Figure 1 (data as in Figure 3). The V_{max}s of the curves are related by percent difference.

ably be imagined to be inhibited at low pH. Consistent with these findings, Hochachka and Lewis (1971) found decreased activation energies (ΔH) under alkaline conditions in the LDHs of a fish.

The rather marked effect of pH on K_m, however, is not so easily explained if we assume that K_m is primarily a measure of enzyme-substrate affinity, a possibility not yet ascertained for these enzymes. Substrate and coenzyme are adequately ionized for binding above pH 6. Arginine, of which several are involved in substrate binding (Adams et al., 1973), is essentially protonated below pH 11. Aspartic acid, also involved, is adequately ionized above about pH 4. We have some very preliminary data suggestive of an enzymatic pK somewhere around pH 7.5 for *R. pipiens* B_4LDH. Similar changes in K_m with pH have also been recorded in mammalian LDHs by Fritz (1967) and Amarasingham and Uong (1968), and for the LDHs of a snake by Gerez De Burgos et al., (1973).

Figure 4 shows the effects on the substrate inhibition curve of modifying temperature. V_{max} changes drastically with temperature. The rather large changes in K_m, however, result in

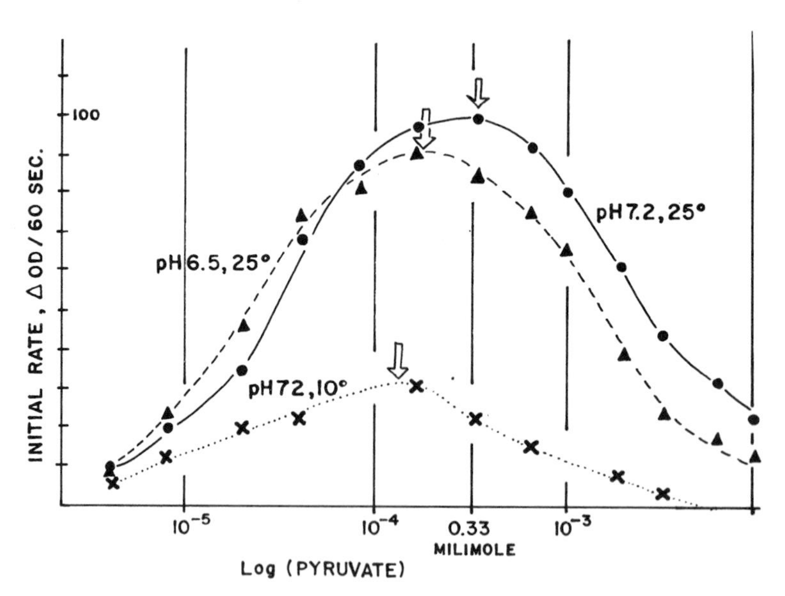

Fig. 5. The effects of temperature and pH on the substrate inhibition curve of the B_4LDH of the frog *R. palustris*. The arrows point to the approximate position of V_{max}.

significant differences in Q_{10} at different substrate concentrations. Figure 5 shows similar data for the B_4LDH of the closely related frog *R. palustris* plotted as initial velocity rather than as percent maximum rate, showing that even at quite low substrate concentrations Q_{10} does not become less than unity. Hochachka and Somero (1973) have suggested that these relatively smaller changes in initial rate at low substrate concentrations, caused by temperature-induced changes in K_m, can be considered to represent an adaptation of poikilotherm enzymes such that reaction velocities at low temperatures are not as low as they would be without this change in K_m, thus compensating somewhat for the decrease in velocity necessitated by lower temperaturss. To the extent that the decreases in K_m are due to a general increased interaction of chemical bonds at low temperatures, it is not clear how this phenomenon can be considered to be an adaptation. If, on the other hand, this effect on K_m is a result of altered relative rates of breakdown of the ES complex into enzyme and substrate and into enzyme and product, the possibility for specific adaptations would seem to be better.

So far, then, we may summarize that in these frog LDHs, increases in both temperature and pH result in increases in both V_{max} and K_m within the limits of these critical environmental variables investigated. It is of some interest,then, to note that studies monitoring changes in blood pH with changes in temperature in poikilotherms (including ranid frogs) (Reeves, 1969; Howell, 1970) show an inverse relationship between the two. Thus, at 25°, blood pH is around 7.6, while at 10° it is near 8. Gerez De Burgos et al., (1973) suggested that because of this correlation the decrease in K_m initiated by decreasing temperature might be opposed by the simultaneous increase in pH, which would favor an increased K_m. I will return to this point later.

I would like here to paraphrase and slightly elaborate the evolutionary theory of the origin of kinetic differences between the B_4 and A_4LDH isozymes that was presented by Cahn et al. in 1962, and that was more fully explored by Kaplan et al. in 1968. and Everse et al. in 1970. Working with the products of duplicated genes, natural selection has favored the evolution of two basically different responses to increased substrate (looking at the pyruvate reductase activity favored by the equilibrium constant of the reaction linking pyruvate and lactate). In anaerobic tissues subject to bursts of activity, such as skeletal muscle, it was favorable to obtain increased lactate when the concentration of pyruvate increased, as during stressful activity. This allows a quick source of energy from the glycolytic pathway. At the same time an oxygen debt is accumulated in the form of lactate, which is later paid by converting this back to pyruvate, usually in the liver. The reaction in the direction of lactate dehydrogenation is accomplished here despite the unfavorable equilibrium by very rapid utilization of pyruvate in other pathways. In aerobic tissues, on the other hand, it was not desirable that an increase in pyruvate concentration lead to increased lactate production; the pyruvate is more efficiently utilized in the Krebs citric acid cycle. The isozyme specialized to operate in this situation was selected so that as pyruvate increases in concentration the enzyme is increasingly inhibited. Some pyruvate reductase activity is desirable in such tissues, however, as a source of NAD for the glycolytic pathway, which must continue to produce some of the pyruvate. Since inhibition of the enzyme is brought about by a fortuitous resemblance of the enol form of pyruvate to lactate (Arnold and Kaplan, 1974) it was probably present in the earliest enzymes catalyzing this pathway, given that a certain amount of enolization is inevitable at biological pHs. Minimally, two properties of the LDHs need to be modified in order to accomplish these ends. First, a change in K_m in at least one isozyme is required so that at physiological substrate concentrations the aerobic enzyme will be functioning on the inhibited portion of its activity curve while the anaerobic one is functioning on the ascending limb of its curve (figure 6).

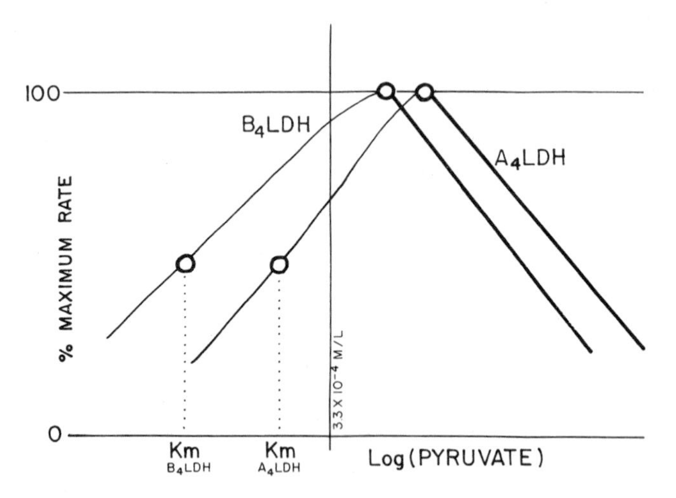

Fig. 6. Substrate inhibition curves of B_4- and A_4LDH from *R. pipiens* at standard conditions (25°, pH 7.6). Data from Levy and Salthe, 1971; Levy and Salthe, 1974.

Second, some change is required in the rates of synthesis and/or degradation of these enzymes in different tissues such that the isozyme appropriate to aerobic tissues comes to predominate in those tissues during ontogeny while the one suited to anaerobic metabolism has the probability of its presence increased in anaerobic tissues. Note that there is an assumption here that on the overall average the concentration of pyruvate will not be different in different tissues at any given moment. Recent studies with lizards suggest that this may be an oversimplification (Bennett and Licht, 1972). However, even extreme compartmentalization of tissues would not alone be fatal to this theory.

Looking again at our frog system, questions must be asked at this point about substrate concentrations. First, is there a range of substrate concentrations over which the B_4 and A_4 isozymes are functioning as required by the theory --that is, with the former becoming increasingly inhibited as the concentration of pyruvate increases while the latter increases its activity? As can be seen in figure 6, there is. The range is between 0.6 mM, where the B_4 enzyme has its V_{max}, and 1.0 mM, where the A_4 isozyme has its V_{max} under conditions of pH (7.6) and temperature (25°) that can be taken as good approximations of natural summer conditions within the frog. Next, it is important to ask whether pyruvate in fact exists at these concentration in frogs at this temperature. From the studies of Sacks et al. (1954) it appears that in active frog muscle at 20° pyruvate can reach 0.7 mM, while there is about a two-fold increase in pyruvate concentration in resting muscle as the temperature increases from 2° to 20°. Thus, we may infer that at 25° pyruvate may reach higher concentrations than 0.7mM in muscle, while at 10° we could reasonably extrapolate to a concentration of 0.33 mM in these muscles. Therefore, at reasonable summer conditions of temperature, pH, and pyruvate concentration (and, we may extrapolate, enzyme and coenzyme concentrations as well) in frogs we find the LDH isozymes operating in ways consistent with the requirements of the evolutionary theory. It should again be pointed out that in these frogs (*Rana pipiens* and *R. palustris*) there are no significant amounts of heteropolymeric isozymes formed by hybridization of the two major gene products (Salthe, 1969); thus the kinetic curves shown in this paper would be characteristic of the LDHs found in the tissues of the frog (see Wright and Moyer, 1973, for confirmation of the identity of LDH subunits in *R. pipiens*).

What about winter conditions? At 10° substrate can reach 0.33 mM; pH would be about 8. We do not yet have kinetic measurements at these conditions but I can offer some predictions based on work already done. The effect of temperature will be greater on V_{max} than will be the effect of pH; V_{max} for both enzymes will decrease significantly. Since Gerez De

Burgos, et al. (1973) found a smaller decrease in K_m for the A_4 isozyme of a snake than for the B_4 isozyme at pH 8, we can predict that the effects of pH on the K_m of the A_4 isozyme will more significantly oppose the effects of temperature, so that the distance between the two activity curves will increase at lower temperatures, the A_4 curve remaining about where it is in warmer conditions with respect to substrate concentration while the B_4 curve moves into the lower substrate range. Figure 7 shows some relevant data for these isozymes from *R. palustris* at warmer, more acid conditions and at cooler, more alkaline conditions (neither of which are physiological, however). The first prediction appears to be sound. The second, however, does not. Both A_4 and B_4 curves moved less into the lower substrate range than they would have on the basis of temperature alone, showing that the effect of increasing pH on the K_m opposes the effect on K_m of decreasing temperature to about the same degree for both isozymes. Note that the movement of the kinetic curves with temperature in figure 7 in relation to changes in substrate concentration is such that the relationships between these curves is not much altered by changing temperature.

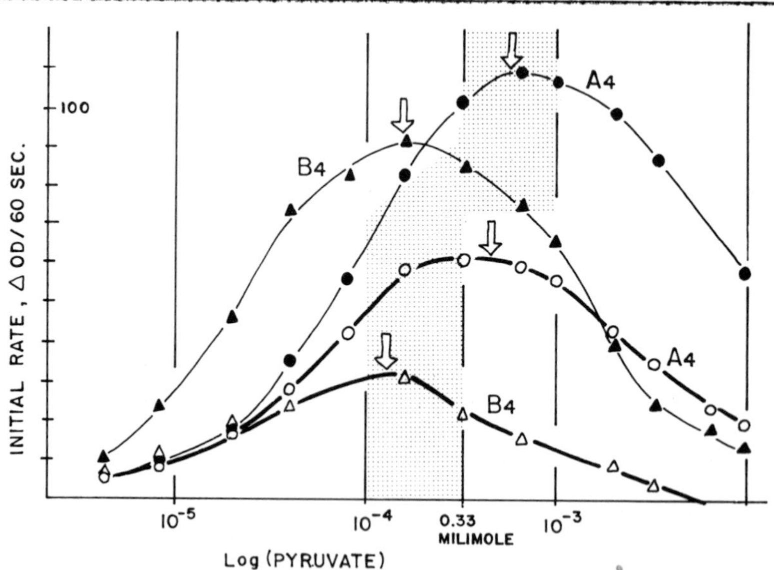

Fig. 7. Substrate inhibition curves for *R. palustris* B_4- and A_4 LDHs under different conditions of temperature and pH. The thin curves with filled symbols represent 25°, ph 6.5 while the thick curves with open symbols represent 10°, pH 7.2. The arrows point to the approximate positions of V_{max}. The shaded area indicates the expected ranges of substrate concentration under the two sets of environmental conditions.

Thus, we may predict that in winter conditions, also, the kinetics of the isozymes of these frogs will probably be found to conform to the requirements of the evolutionary theory of LDH isozyme distribution.

Finally, I would like to assess the implications of intrapopulational variability in these enzymes for our understanding of their physiological function. The A_4 isozyme of *R. pipiens* shows no intrapopulational variability and no geographic variability in either electrophoretic or immunological properties (Levy and Salthe, 1974). It appears identical in these respects to the A_4 enzyme of *R. palustris*. In addition, we have been unable to detect significant kinetic variability, so that for all intents and purposes we may conclude that there is a single wild type allele at the LDH-A locus in these frogs, except for rare mutants. On the other hand the B_4 isozyme shows considerable electrophoretic and immunological variability (Salthe, 1969); but no significant kinetic variability accompanies these other types of variability (Levy and Salthe, 1971). The B_4 isozyme of *R. palustris* is electrophoretically and immunologically distinct from any of the *R. pipiens* variants, but cannot be distinguished kinetically from any of these variants. It is clear from these data that natural selection is operating differentially on these isozymes, a prediction that one would have made using the evolutionary theory of LDH isozyme distribution, but which would of course be consistent with <u>any</u> theory involving differential function.

Using figure 8, we can test the *R. pipiens*, and also the *Acris crepitans* (Salthe and Nevo, 1969), intrapopulational variability against a theory (Johnson, 1971) relating the intensity of balancing selection to the equilibrium constants of the reactions which the enzymes in question mediate. According to this theory, selection should detect, and, therefore affect, enzymes catalyzing rate-limiting steps in biochemical pathways. The equilibrium constant of a reaction measures the rate-limiting potentiality of an enzyme, while the number of allozymes present in a population may be taken as an indication of the intensity of the (balancing) selection operating (Kimura and Crow, 1964). It is clear from the figure that the B_4 isozyme is about as variable as this theory would predict, and so we may conclude that balancing selection of some sort may be operating for the LDH-B locus. The A_4 isozyme, however, is very much less variable than we would expect on this theory. It might perhaps be subject instead to intense normalizing selection.

In conclusion, it seems that the evolutionary theory of LDH isozyme tissue distribution, based on the substrate inhibition properties of these enzymes, can be useful in organizing our thinking about their physiological functions and about

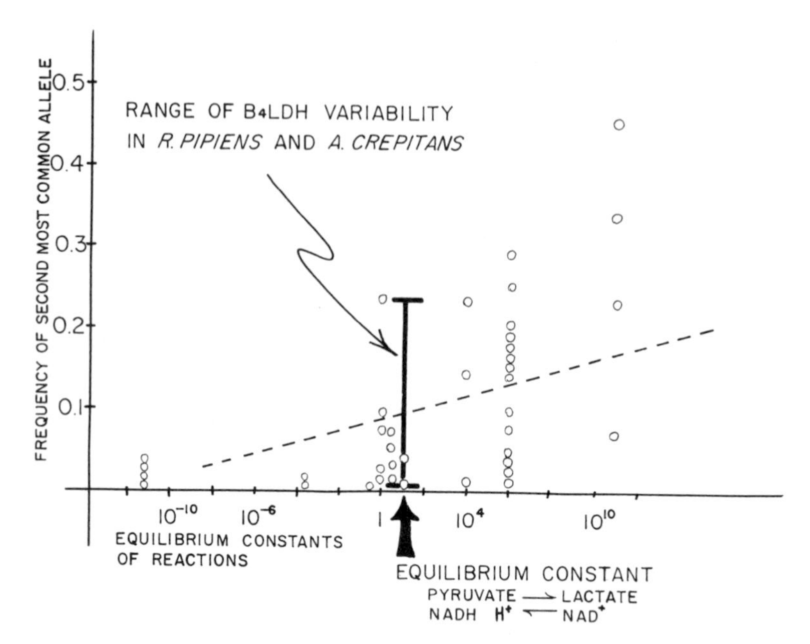

Fig. 8. Correlation between equilibrium constants of reactions and the variability of *Drosophila* enzymes mediating the reactions (from Johnson, 1971), with frog LDH data superimposed. Each circle represents data from one population.

respiratory temperature adaptations in amphibians, and perhaps in poikilotherms in general.

REFERENCES

Adams, M. J., M. Buehner, K. Chandrasekhar, G. C. Ford, M. L. Hackert, A. Liljas, M. G. Rossman, I. E. Smiley, W. S. Allison, J. Everse, N. O. Kaplan, and S. S. Taylor 1973. Structure-function relationships in lactate dehydrogenase. *Proc. Nat. Acad. Sci.* 70: 1968-1972.

Amarasingham, C. R. and A. Uong 1968. An examination of reaction kinetics in evaluating the differential distribution of lactate dehydrogenase isozymes. *Ann. N.Y. Acad. Sci.* 151: 424-428.

Arnold, L. J. and N. O. Kaplan 1974. The structure of the abortive diphosphopyridine nucleotide-pyruvate-lactate dehydrogenase ternary complex as determined by proton magnetic resonance analysis. *J. Biol. Chem.* 249: 652-655.

Bennett, A. F. and P. Licht 1972. Anaerobic metabolism during activity in lizards. *J. Comp. Physiol.* 81: 277-288.

Cahn, R. D., N. O. Kaplan, L. Levine, and E. Zwilling 1962. Nature and development of lactic dehydrogenases. *Science* 136: 962-969.

Eby, D., S. N. Salthe, and A. Lukton 1973. Frog lactate dehydrogenase kinetics at physiological enzyme levels. *Biochim. Biophys. Acta* 327: 227-232.

Everse, J., R. L. Burger, and N. O. Kaplan 1970. Physiological concentrations of lactate dehydrogenases and substrate inhibition. *Science* 168: 1236-1238.

Fritz, P. J. 1967. Rabbit lactate dehydrogenase isozymes: effect of pH on activity. *Science* 156: 82-83.

Gerez De Burgos, N. M., C. Burgos, M. Gutierrez, and A. Blanco 1973. Effect of temperature upon catalytic properties of lactate dehydrogenase isoenzymes from a poikilotherm. *Biochim. Biophys. Acta* 315: 250-258.

Hochachka, P. W. and J. K. Lewis 1971. Interacting effects of pH and temperature on the K_m values for fish tissue lactate dehydrogenases. *Comp. Biochem. Physiol.* 39B: 925-933.

Hochachka, P. W. and G. N. Somero 1973. *Strategies of Biochemical Adaptation*. W. B. Saunders Co. Philadelphia, 358 pp.

Howell, B. J. 1970. Acid-base balance in transition from water breathing to air breathing. *Feder. Proc.* 29: 1130-1134.

Johnson, G. B. 1971. Metabolic implications of polymorphism as an adaptive strategy. *Nature* 232: 347-349.

Kaplan, N. O. 1960. The pyridine coenzymes. In Boyer, P., H. Lardy, and K. Myrbäck (Eds.) *The Enzymes*, 2nd Ed., Volume 3: 105-169.

Kaplan, N. O., J. Everse, and J. Admiraal 1968. Significance of substrate inhibition of dehydrogenases. *Ann. N. Y. Acad. Sci.* 151: 400-412.

Kimura, M. and J. F. Crow 1964. The number of alleles that can be maintained in a finite population. *Genetics* 49: 725-738.

Levy, P. L. and S. N. Salthe 1971. Kinetic studies on variant heart-type lactate dehydrogenases in the frog, *Rana pipiens*. *Comp. Biochem. Physiol.* 39B: 343-355.

Levy, P. L. and S. N. Salthe 1974. Studies on the variability of muscle-type lactate dehydrogenase in the frog, *Rana pipiens*. *Comp. Biochem. Physiol.* 47B: 355-378

Reeves, R. B. 1969. Role of body temperature in determining the acid-base state in vertebrates. *Feder. Proc.* 28: 1204-1208.

Sacks, J., R. V. Ganslow, and J. T. Diffees 1954. Lactic and pyruvic acid relations in frog muscle. *Amer. J. Physiology* 177: 113-114.

Salthe, S. N. 1969. Geographic variation of the lactate dehydrogenases of *Rana pipiens* and *Rana palustris*. *Biochem.*

Genet. 2: 271-303.

Salthe, S. N. and E. Nevo 1969. Geographic variation of lactate dehydrogenase in the cricket frog, *Acris crepitans*. *Biochem. Genet.* 3: 335-341.

Somero, G. N. 1969. Enzyme mechanisms of temperature compensation: immediate and evolutionary effects of temperature on enzymes of aquatic poikilotherms. *Amer. Natur.* 103: 517-530.

Wright, D. A. and F. H. Moyer 1973. Immunochemistry of frog lactate dehydrogenase (LDH) and the subunit homologies of amphibian LDH isozymes. *Comp. Biochem. Physiol.* 44B: 1011-1016.

Wuntch, T., R. F. Chen, and E. S. Vesell 1970a. Lactate dehydrogenase isozymes: kinetic properties at high enzyme concentrations. *Science* 167: 63-65.

Wuntch, T., R. F. Chen, and E. S. Vesell 1970b. Lactate dehydrogenase isozymes: further kinetic studies at high enzyme concentration. *Science* 169: 480-481.

GENETIC VARIATION IN THE GENUS *BUFO* II. ISOZYMES IN NORTHERN ALLOPATRIC POPULATIONS OF THE AMERICAN TOAD, *Bufo americanus*.

SHELDON I. GUTTMAN
Department of Zoology, Miami University
Oxford, Ohio 45056

ABSTRACT. Electrophoretically demonstrable variation was analyzed in enzymes encoded by fourteen structural gene loci in 620 individuals of northern allopatric *Bufo americanus* representing 25 populations. Average individual heterozygosity was 11.6 percent, mean percentage of loci polymorphic was 25.7 percent, and average number of alleles detected per locus was 1.45. Extensive clinal variation in allelic frequencies and genic variability was observed. Marginal populations of this species appear to be as variable as central populations. The American toad possesses more genic variability than any other vertebrate species reported to date. This may be a reflection of an adaptive strategy unique to toads.

INTRODUCTION

The recent development of electrophoretic techniques for demonstrating genic variability at enzyme loci in natural populations (see reviews in Gottlieb, 1971; Selander and Johnson, 1973) has permitted researchers to attempt to clarify many basic questions of evolutionary biology. Among these are genic variation within and between populations, degree of heterozygosity within a population and genetic variation in peripheral versus central populations. Most investigations of either some or all of the above problem areas have been with *Drosophila* (Lewontin and Hubby, 1966; Ayala, et al., 1972a, 1972b; Richmond, 1972) lizards (McKinney, et al., 1972; Webster, et al., 1972) or mammals (Nevo and Shaw, 1972; Selander, et al., 1969a, 1969b, 1971; Smith, et al., 1973). However, genic variability has not been studied as well in lower vertebrates. Avise and Selander (1972) investigated three cave and six surface populations of a fish (*Astyanax mexicanus*), Webster (1973) studied one population of a salamander (*Plethodon cinereus*), Dessauer and Nevo (1969) examined 32 populations of two species of cricket frogs (*Acris gryllus* and *A. crepitans*), and Rogers (1973) reported the variation in eleven populations of two species of toads (*Bufo cognatus* and *B. speciosus*). In total, genic variability estimates are available for 25 species of vertebrates.

SHELDON I. GUTTMAN

For the past five years the major thrust of my research has dealt with two closely related species of toads, the American toad (*Bufo americanus*) and Fowler's toad (*B. fowleri*). The former species is a widespread form, occurring from the Maritime Provinces to southeastern Manitoba, south to Mississippi and northeastern Kansas (Wright and Wright, 1949). Three subspecies are presently recognized: *B. a. copei* in the extreme northern portion of the range, *B. a. charlesmithi* in the southwestern portion of the range and *B. a. americanus* throughout the remainder of the range. The second species, *B. fowleri*, is also widespread and is found from central New England to the Gulf Coast and west to Michigan, northeastern Oklahoma, and eastern Texas (Conant, 1958). Often where the two species occur in sympatry they hybridize and introgression may occur (Blair, 1941; Volpe, 1952 and 1955). Approximately 2,500 toads from eighty populations of the two species have been collected in an attempt to determine the genic variation within and between populations, degree of heterozygosity within populations, genetic variation in peripheral versus central populations, degree of interspecific versus intraspecific variation, and the amount of introgression that has occurred between the two species. The present paper deals with isozymic variation in 25 northern allopatric populations of the American toad.

MATERIALS AND METHODS

A total of 620 individuals (499 males, 121 females) of *Bufo americanus*, representing 25 local breeding aggregations, were collected during the spring of 1971. The localities sampled are shown in Figure 1 and listed in Table 1.

Upon receipt in the laboratory samples of liver and kidney tissues were removed from each specimen, and homogenized together in a volume of distilled water equal to twice the volume of tissue. The homogenates were centrifuged for thirty minutes at 5130 g and 4°C. After centrifugation the supernatant was stored at -20°C until subjected to electrophoresis.

The fourteen loci coding for enzymes studied are as follows: Glutamic oxaloacetic transaminases, two loci (GOT-1, GOT-2); phosphoglucomutases, two loci (PGM-2, PGM-3); malate dehydrogenases(NAD-dependent), two loci(MDH-1, MDH-2); isocitrate dehydrogenases (NADP-dependent), two loci (IDH-1, IDH-2); alcohol dehydrogenase (ADH-1); xanthine dehydrogenase (XDH-1); alpha-glycerophosphate dehydrogenase (alpha GPD-1); 6-phosphogluconate dehydrogenase (6-PGD-1); lactate dehydrogenase (LDH-2); and peptidase (PEP-1).

Techniques of horizontal starch-gel electrophoresis and

TABLE I
ALLELE FREQUENCIES AT THE GOT-1 LOCUS

POPULATION	SAMPLE SIZE	ALLELES a	b	c	e	f	PROPORTION of HETERO-ZYGOTES
1. HANTSPORT, N.S.	23				1.00		0.000
2. EVANDALE, N.B.	10			0.20	0.80		0.400
3. OAK POINT, N.B.	12			0.25	0.75		0.333
4. PRICEVILLE, N.B.	8			0.06	0.94		0.125
5. L'AUX GRANDES POINTES, QUE.	39			0.04	0.96		0.077
6. CHUTE DES PASSES QUE.	32			0.09	0.91		0.188
7. CHIBOUGAMU, QUE.	10			0.05	0.95		0.100
8. BISHOP MILLS, ONT.	7	0.14	0.07	0.07	0.71		0.571
9. WASWANIPI, QUE.	8			0.25	0.75		0.500
10. MATAGAMI, QUE.	65			0.33	0.65	0.02	0.277
11. COCHRANE, ONT.	17			0.53	0.47		0.588
12. DRIFTWOOD, ONT.	12			0.46	0.54		0.583
13. HEARST, ONT.	12			0.63	0.37		0.583
14. PESHTIGO, WISC.	30		0.05	0.37	0.58		0.400
15. UNDERHILL, WISC.	28		0.12	0.14	0.74		0.393
16. PENSAUKEE, WISC.	32		0.03	0.36	0.61		0.438
17. PELLA, WISC.	31			0.27	0.73		0.226
18. LITTLE SUAMICO, WISC.	30		0.04	0.28	0.66	0.02	0.567
19. SHIOCTON,WISC.	27		0.01	0.32	0.67		0.519
20. WRIGHTSTOWN, WISC.	31	0.02	0.03	0.29	0.66		0.452
21. FREMONT, WISC.	29		0.05	0.31	0.64		0.379
22. RUSH LAKE, WISC.	40		0.02	0.34	0.64		0.475
23. MT. CALVARY, WISC.	28			0.34	0.66		0.321
24. KEWASKUM, WISC.	30			0.17	0.83		0.267
25. RACINE, WISC.	29			0.15	0.85		0.172

protein staining were similar to those described by Selander, et al. (1971) with the following modifications: MDH and PGM were demonstrated with a continuous Tris-citrate (pH 8.0) buffer (buffer system 5 of Selander, et al., 1971); ADH, alpha-GPD, IDH, LDH, and XDH with Tris-maleate (buffer system 9). The Tris-maleate buffer system was also used for PEP which was detected using a modification of the staining procedure of Lewis and Harris (1967).

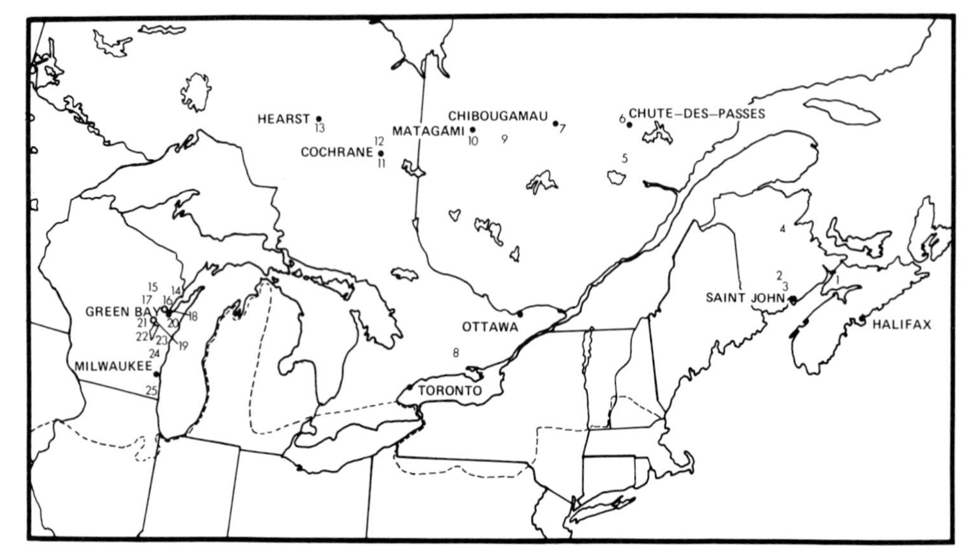

Fig. 1. Map of northeastern United States and Canada showing localities of toads used in analyzing electrophoretic variation. Locality names are listed in Table 1. Dashed line indicates northern extent of range of *Bufo fowleri*.

RESULTS

The abbreviations used to designate each enzyme are indicated in the Materials and Methods section. The genes coding for the enzyme are represented by italicizing the abbreviations. If several forms of the same enzyme are present and each is controlled by a separate gene locus, a hyphenated numeral is added to the symbol of the enzyme. The enzyme with the greatest anodal migration is designated one, the next two, and so on. When allelic variation occurs, the allele with the greatest anodal migration is called a, the next b, and so on. A letter(s) may be skipped; this indicates that those alleles are not present in the populations presently under consideration. Alleles are written as superscripts over the symbol representing the locus.

Fourteen enzyme loci have been studied in northern allopatric *B. americanus*. The frequency distributions for the five polymorphic loci are presented in Tables 1-5 and the electrophoretic variation in these proteins is described below.

GOT-1. Five alleles are represented at this locus (Fig. 2, Table 1). In all populations except two (Cochrane and Hearst, Ontario), the $GOT\text{-}1^{e}$ allele is present in highest

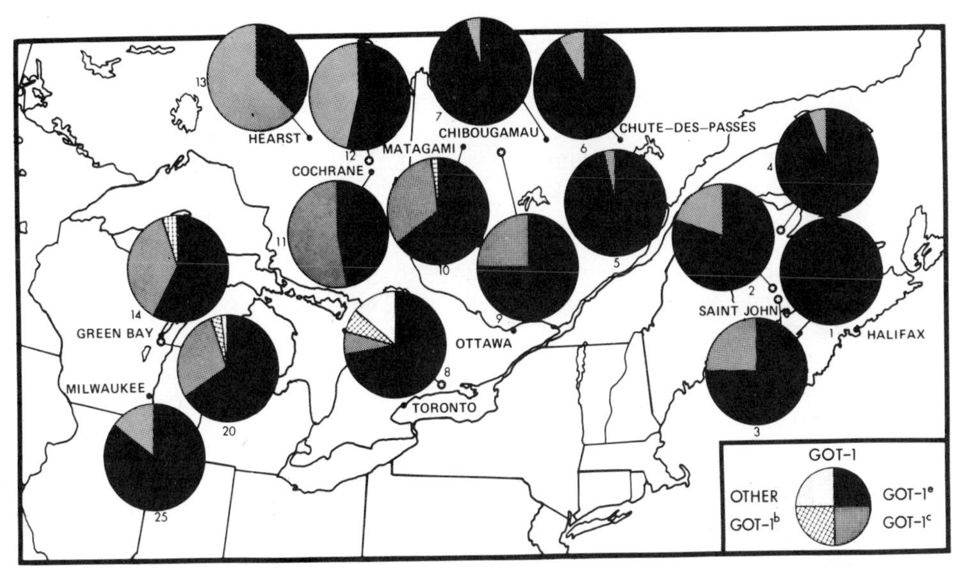

Fig. 2. Geographic distribution of GOT-1 isozymes among populations of *Bufo americanus*. Pie graphs are centered over collecting sites and numbers refer to populations in Fig. 1 and Table 1. Lines from certain pie graphs lead to localities. Because of space only three Wisconsin populations are included.

frequency. The other major allele is $GOT\text{-}1^c$, which is present at a frequency greater than 50% in the above two populations. $GOT\text{-}1^a$ and $GOT\text{-}1^f$ are rare, each being found in only two populations. The frequency of the fifth allele, $GOT\text{-}1^b$, ranges from 0.01 to 0.12; it is found in nine of the 25 populations sampled. $GOT\text{-}1^c$ increases with increasing longitude ($p < .01$) while $GOT\text{-}1^e$ decreases as the longitude increases ($p < .01$).

PGM-2. All samples are polymorphic for two alleles, $PGM\text{-}2^b$ and $PGM\text{-}2^c$ (Table 2). A third allele, $PGM\text{-}2^d$ is present in frequencies ranging from 0.03 to 0.32 in all populations except those from Nova Scotia, New Brunswick, and Bishop Mills, Ontario. The frequency of $PGM\text{-}2^d$ increases from east to west ($p < .01$).

PGM-3. The Canadian samples from east of the northern Quebec-Ontario border have a high frequency of $PGM\text{-}3^c$; ranging from 0.80 to 1.00 (Fig. 3, Table 3). In the Cochrane, Driftwood, and Hearst, Ontario populations, $PGM\text{-}3^c$ approaches equilibrium with $PGM\text{-}3^b$. Eleven of the twelve Wisconsin samples possess at least one of two alleles not found in the Canadian samples ($PGM\text{-}3^a$ and $PGM\text{-}3^d$). The frequencies of $PGM\text{-}3^a$, $PGM\text{-}3^b$, and $PGM\text{-}3^d$ increase with increasing longitude ($p < .01$); $PGM\text{-}3^c$ decreases from east to west ($p < .01$).

TABLE 2

ALLELE FREQUENCIES AT THE PGM-2 LOCUS

POPULATION	SAMPLE SIZE	ALLELES			PROPORTION of HETEROZYGOTES
		b	c	d	
1. HANTSPORT, N.S.	22	0.46	0.54		0.727
2. EVANDALE, N.B.	10	0.56	0.44		0.400
3. OAK POINT, N.B.	11	0.59	0.41		0.818
4. PRICEVILLE, N.B.	8	0.19	0.81		0.375
5. L'AUX GRANDES POINTES, QUE.	39	0.71	0.21	0.08	0.410
6. CHUTE DES PASSES, QUE.	32	0.48	0.42	0.10	0.500
7. CHIBOUGAMAU, QUE.	9	0.44	0.50	0.06	0.889
8. BISHOP MILLS, ONT.	7	0.64	0.36		0.714
9. WASWANIPI, QUE.	8	0.63	0.25	0.12	0.625
10. MATAGAMI, QUE.	65	0.75	0.22	0.03	0.400
11. COCHRANE, ONT.	17	0.47	0.35	0.18	0.647
12. DRIFTWOOD, ONT.	12	0.58	0.25	0.17	0.333
13. HEARST, ONT.	12	0.54	0.21	0.25	0.667
14. PESHTIGO, WISC.	30	0.40	0.42	0.18	0.633
15. UNDERHILL, WISC.	29	0.41	0.41	0.18	0.931
16. PENSAUKEE, WISC.	32	0.30	0.52	0.18	0.719
17. PELLA, WISC.	31	0.47	0.34	0.19	0.484
18. LITTLE SUAMICO, WISC.	30	0.40	0.43	0.17	0.767
19. SHIOCTON, WISC.	27	0.33	0.48	0.19	0.704
20. WRIGHTSTOWN, WISC.	31	0.36	0.48	0.16	0.645
21. FREMONT, WISC.	29	0.36	0.36	0.28	0.724
22. RUSH LAKE, WISC.	40	0.46	0.39	0.15	0.650
23. MT. CALVARY, WISC.	28	0.36	0.48	0.16	0.643
24. KEWASKUM, WISC.	30	0.42	0.47	0.11	0.633
25. RACINE, WISC.	29	0.28	0.40	0.32	0.828

MDH-1. The supernatant form of NAD-dependent malate dehydrogenase is polymorphic with a maximum of three alleles being present in any one population (Table 4). All Canadian samples and seven Wisconsin samples are essentially monomorphic for $MDH\text{-}1^b$. Two Canadian populations have $MDH\text{-}1^a$ present in low frequency. $MDH\text{-}1^c$ is unique to the Wisconsin populations; its frequency ranges from 0.00 to 0.17.

IDH-2. The mitochondrial form of NADP-dependent isocitrate dehydrogenase is represented by three alleles (Fig. 4, Table 5).

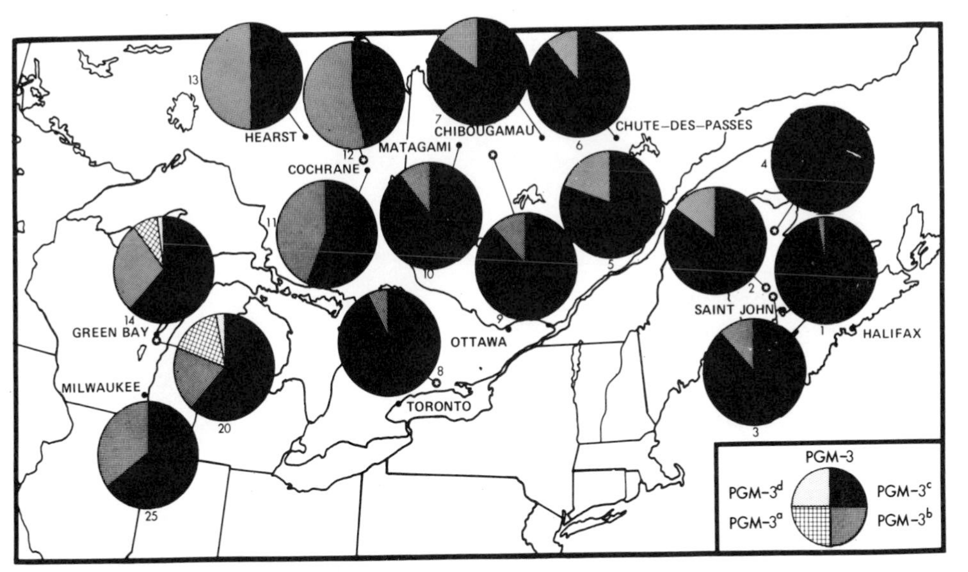

Fig. 3. Geographic distribution of PGM-3 isozymes among populations of *Bufo americanus*. (see legend of Fig. 2)

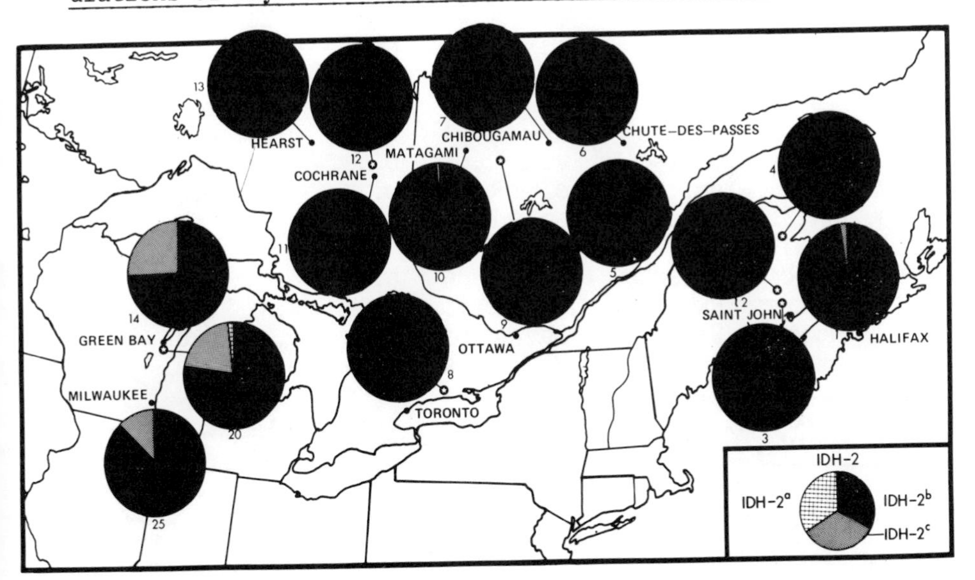

Fig. 4. Geographic distribution of IDH-2 isozymes among populations of *Bufo americanus*. (see legend of Fig. 2).

TABLE 3

ALLELE FREQUENCIES AT THE PGM-3 LOCUS

POPULATION	SAMPLE SIZE	ALLELES				PROPORTION of HETERO-ZYGOTES
		a	b	c	d	
1. HANTSPORT, N.S.	23		0.02	0.98		0.044
2. EVANDALE, N.B.	10		0.15	0.85		0.300
3. OAK POINT, N.B.	12		0.12	0.88		0.250
4. PRICEVILLE, N.B.	8			1.00		0.000
5. L'AUX GRANDES POINTES, QUE.	39		0.20	0.80		0.308
6. CHUTES DES PASSES, QUE.	32		0.11	0.89		0.219
7. CHIBOUGAMAU, QUE.	10		0.15	0.85		0.300
8. BISHOP MILLS, ONT.	7		0.07	0.93		0.143
9. WASWANIPI, QUE.	8		0.12	0.88		0.250
10. MATAGAMI, QUE.	65		0.11	0.89		0.200
11. COCHRANE, ONT.	16		0.44	0.56		0.625
12. DRIFTWOOD, ONT.	12		0.54	0.46		0.583
13. HEARST, ONT.	12		0.50	0.50		0.500
14. PESHTIGO, WISC.	30	0.08	0.28	0.62	0.02	0.533
15. UNDERHILL, WISC.	29	0.04	0.29	0.64	0.03	0.448
16. PENSAUKEE, WISC.	32	0.05	0.38	0.53	0.04	0.500
17. PELLA, WISC.	31	0.07	0.37	0.53	0.02	0.742
18. LITTLE SUAMICO, WISC.	30	0.07	0.33	0.60		0.367
19. SHIOCTON, WISC.	27	0.06	0.37	0.50	0.07	0.778
20. WRIGHTSTOWN, WISC.	31	0.11	0.23	0.61	0.03	0.548
21. FREMONT, WISC.	29	0.03	0.14	0.79	0.03	0.310
22. RUSH LAKE, WISC.	40	0.08	0.23	0.68	0.01	0.375
23. MT. CALVARY, WISC.	28	0.14	0.15	0.71		0.571
24. KEWASKUM, WISC.	30	0.07	0.22	0.68	0.03	0.433
25. RACINE, WISC.	29		0.35	0.65		0.517

The Canadian samples are monomorphic or essentially so for $IDH\text{-}2^b$. However, in Wisconsin, $IDH\text{-}2^c$ is present in all populations and ranges in frequency from 0.09 to 0.31. The Wrightstown, Wisconsin sample is the only one possessing $IDH\text{-}2^a$. The frequency of $IDH\text{-}2^b$ increases with increasing latitude and decreases with increasing longitude ($p < .01$) while $IDH\text{-}2^c$ decreases with increasing latitude and increases as longitude increases ($p < .01$).

TABLE 4

ALLELE FREQUENCIES AT THE MDH-1 LOCUS

POPULATION	SAMPLE SIZE	ALLELES			PROPORTION of HETEROZYGOTES
		a	b	c	
1. HANTSPORT, N.S.	23		1.00		0.000
2. EVANDALE, N.B.	10		1.00		0.000
3. OAK POINT, N.B.	12		1.00		0.000
4. PRICEVILLE, N.B.	8		1.00		0.000
5. L'AUX GRANDES POINTES, QUE.	39	0.06	0.94		0.128
6. CHUTE DES PASSES, QUE.	32		1.00		0.000
7. CHIBOUGAMAU, QUE.	10		1.00		0.000
8. BISHOP MILLS, ONT.	7		1.00		0.000
9. WASHWANIPI, QUE.	8		1.00		0.000
10. MATAGAMI, QUE.	65		1.00		0.000
11. COCHRANE, ONT.	17		1.00		0.000
12. DRIFTWOOD, ONT.	12	0.04	0.96		0.083
13. HEARST, ONT.	12		1.00		0.000
14. PESHTIGO, WISC.	30		1.00		0.000
15. UNDERHILL, WISC.	29		0.83	0.17	0.345
16. PENSAUKEE, WISC.	32		0.84	0.16	0.313
17. PELLA, WISC.	31	0.03	0.95	0.02	0.097
18. LITTLE SUAMICO, WISC.	30		0.83	0.17	0.333
19. SHIOCTON, WISC.	27		0.85	0.15	0.296
20. WRIGHTSTOWN, WISC.	31	0.03	0.92	0.05	0.161
21. FREMONT, WISC.	29		0.97	0.03	0.069
22. RUSH LAKE, WISC.	40		0.98	0.02	0.050
23. MT. CALVARY, WISC.	28		1.00		0.000
24. KEWASKUM, WISC.	30		0.97	0.03	0.067
25. RACINE, WISC.	29		0.98	0.02	0.035

Monomorphic enzymes. The following enzymes are monomorphic in the populations sampled: ADH, XDH, alpha-GPD, GOT-2, 6-PGD, LDH-2, MDH-2, IDH-1, and peptidase.

TABLE 5

ALLELE FREQUENCIES AT THE IDH-2 LOCUS

POPULATION	SAMPLE SIZE	ALLELES a	ALLELES b	ALLELES c	PROPORTION of HETEROZYGOTES
1. HANTSPORT, N.S.	23		0.98	0.02	0.044
2. EVANDALE, N.B.	10		1.00		0.000
3. OAK POINT, N.B.	12		1.00		0.000
4. PRICEVILLE, N.B.	8		1.00		0.000
5. L'AUX GRANDES POINTES, QUE.	39		1.00		0.000
6. CHUTE DES PASSES, QUE.	32		1.00		0.000
7. CHIBOUGAMAU, QUE.	10		1.00		0.000
8. BISHOP MILLS, ONT.	7		1.00		0.000
9. WASWANIPI, QUE.	8		1.00		0.000
10. MATAGAMI, QUE.	66		0.99	0.01	0.015
11. COCHRANE, ONT.	17		1.00		0.000
12. DRIFTWOOD, ONT.	12		1.00		0.000
13. HEARST, ONT.	12		1.00		0.000
14. PESHTIGO, WISC.	29		0.74	0.26	0.379
15. UNDERHILL, WISC.	30		0.83	0.17	0.267
16. PENSAUKEE, WISC.	32		0.81	0.19	0.375
17. PELLA, WISC.	30		0.87	0.13	0.200
18. LITTLE SUAMICO, WISC.	30		0.78	0.22	0.300
19. SHIOCTON, WISC.	26		0.71	0.29	0.423
20. WRIGHTSTOWN, WISC.	31	0.02	0.77	0.21	0.387
21. FREMONT, WISC.	29		0.69	0.31	0.414
22. RUSH LAKE, WISC.	40		0.84	0.16	0.225
23. MT. CALVARY, WISC.	28		0.91	0.09	0.179
24. KEWASKUM, WISC.	30		0.83	0.17	0.333
25. RACINE, WISC.	29		0.88	0.12	0.172

DISCUSSION

The results of the survey of allozyme variation in northern allopatric populations of the American toad, *B. americanus*, are summarized in Table 6. For each population, the mean percentage of loci heterozygous per individual and polymorphic per population, as well as the mean number of alleles per locus per population, is listed. The fourteen loci were examined in all populations.

TABLE 6

ESTIMATES OF GENIC VARIABILITY AT 14 LOCI IN *BUFO AMERICANUS*

POPULATION	SAMPLE SIZE	MEAN PERCENTAGE OF LOCI: HETERO-ZYGOUS PER INDIVIDUAL	MEAN PERCENTAGE OF LOCI: POLYMORPHIC PER POPULATION[1]	MEAN NUMBER OF ALLELES PER LOCUS PER POPULATION
1. HANTSPORT, N.S.	23	5.8	7.1	1.21
2. EVANDALE, N.B.	10	7.9	21.4	1.21
3. OAK POINT, N.B.	12	10.0	21.4	1.21
4. PRICEVILLE, N.B.	8	3.6	14.3	1.14
5. L'AUX GRANDES POINTES, QUE.	39	6.6	21.4	1.36
6. CHUTES DES PASSES, QUE.	32	6.5	21.4	1.29
7. CHIBOUGAMAU, QUE.	10	9.2	21.4	1.29
8. BISHOP MILLS, ONT.	7	10.2	21.4	1.36
9. WASWANIPI, QUE.	8	9.8	21.4	1.29
10. MATAGAMI, QUE.	66	6.4	21.4	1.43
11. COCHRANE, ONT.	17	13.3	21.4	1.29
12. DRIFTWOOD, ONT.	12	11.3	21.4	1.36
13. HEARST, ONT.	12	12.5	21.4	1.29
14. PESHTIGO, WISC.	29	13.9	28.6	1.57
15. UNDERHILL, WISC.	30	17.0	35.7	1.64
16. PENSAUKEE, WISC.	32	16.8	35.7	1.64
17. PELLA, WISC.	30	12.5	35.7	1.64
18. LITTLE SUAMICO, WISC.	30	16.7	35.7	1.64
19. SHIOCTON, WISC.	26	19.4	35.7	1.64
20. WRIGHTSTOWN, WISC.	31	15.7	35.7	1.86
21. FREMONT, WISC.	29	13.5	28.6	1.64
22. RUSH LAKE, WISC.	40	12.7	28.6	1.64
23. MT. CALVARY, WISC.	28	12.8	28.6	1.57
24. KEWASKUM, WISC.	30	12.4	28.6	1.57
25. RACINE, WISC.	29	12.3	28.6	1.43

[1] 95% criterion for polymorphism

Figure 5 and Table 6 show that the percentage of the loci heterozygous in an average individual varies extensively from locality to locality. Of the populations studied, Priceville, New Brunswick has the lowest value, 3.6%, and Shiocton, Wisconsin has the highest, 19.4%. Mean heterozygosity for all populations sampled is 11.6%. The mean percentage of loci polymorphic per population ranges from 7.1% in the Nova Scotia toads to 35.7% in six Wisconsin populations. The mean percentage of loci polymorphic for all populations is 25.7%. The mean number of alleles per locus per population ranges from 1.14 in the Priceville, New Brunswick population to 1.86 in the Wrightstown, Wisconsin sample. The mean number of alleles for all populations is 1.45.

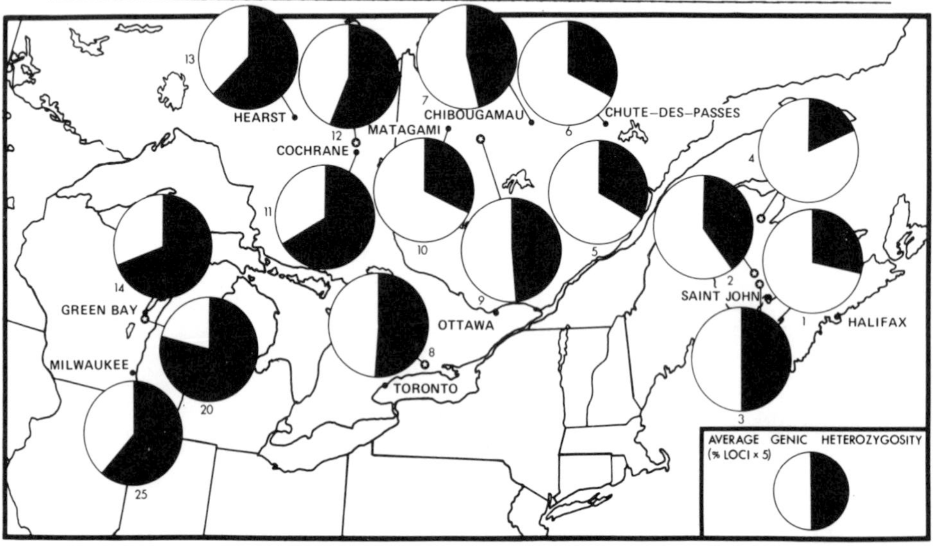

Fig. 5. Geographic variation in average heterozygosity in *Bufo americanus*. Dark areas of circles representing samples are proportional to percentages of loci heterozygous (Table 6) X 5. (see legend of Fig. 2)

Perhaps more meaningful data than the above can be realized by subdividing the genetic variability estimates into regional and subspecific categories. The Nova Scotia and New Brunswick samples are from the extreme northeastern portion of the range of *B. a. americanus* (Conant, 1958; Cook, pers. comm.). For these populations, genic heterozygosity averages 6.8%, loci polymorphic per population averages 16.1%, and mean number of alleles per locus averages 1.19. These figures are the lowest for any possible natural subdivision of the populations examined. The Priceville, New Brunswick population is unusually interesting because for two of the three measures of genic variability (mean percentage of loci heterozygous per individual and mean number of alleles per population) the lowest values were obtained in this sample (Table 6). The Priceville area had been subjected to extensive spraying of DDT for control of the spruce budworm. Gorham (pers. comm.) collected toads in this area in 1967 and they "showed a fairly high DDT concentration". Strong selection for DDT resistant types could be leading towards homozygosity.

The only populations that can be unanimously designated as representing *B. a. copei* are those from L'Aux Grandes Pointes, Chute des Passes, Chibougamau, and Matagami, Quebec (Conant, 1958; Ashton, et al., 1973; Cook, pers. comm.). Conant's (1958) range for the subspecies is designated as the coast of Labrador to James Bay, Ontario and includes only the four above populations. Ashton, et al. (1973) include all the Quebec and Ontario populations, with the exception of the Bishop Mills sample, as falling within the range of *B. a. copei*. Cook's (pers. comm.) distribution map is a hybrid of the above two: *B. a. copei* is found alone in the area delimited by Conant (1958), both subspecies are found together south to the range suggested by Ashton, et al. (1973), and only *B. a. americanus* is found south of the line. If the conservative approach is taken and only the four northernmost Quebec populations are considered to be *B. a. copei*, then mean heterozygosity per individual is 7.2%, mean loci polymorphic per population is 21.4%, and mean number of alleles per locus per population is 1.34(Table 7). Using the southernmost range proposed for *B. a. copei*, the mean heterozygosity per individual is 9.5%, mean loci polymorphic per population is 21.4%, and mean number of alleles per locus per population is 1.33 (Table 7). The above data in conjunction with those presented in Table 6 indicate that genic variability estimates do not significantly differ between the populations within the various distributions proposed for *B. a. copei*. In addition, although estimates of genic variability for the two subspecies appear to be significantly different (Table 7), the fact that the lowest and highest values for each category occurred in *B. a. americanus* (Table 6) would render the differences meaningless.

In the Canadian samples there is a trend toward increasing mean heterozygosity per individual west of the northern Quebec-Ontario border (Table 7). The highest percent heterozygosity per individual east of this border is approximately 10%. In western Ontario, the range is from 11.3% to 13.3%. This trend is accentuated when the Wisconsin populations are considered. The range of individual heterozygosity per population is 12.3%

to 19.4% with a mean of 14.6%. This is substantially higher than the mean of 7.6% for toad populations east of the northern Quebec-Ontario border. The mean percentage of loci polymorphic per population and mean number of alleles per locus per population also follow this trend (Table 7).

TABLE 7

SUPSPECIFIC AND REGIONAL ESTIMATES OF MEAN GENIC VARIABILITY

GROUPING	MEAN PERCENTAGE OF LOCI: HETEROZYGOUS PER INDIVIDUAL	MEAN PERCENTAGE OF LOCI: POLYMORPHIC PER POPULATION[1]	MEAN NUMBER OF ALLELES PER LOCUS PER POPULATION
B. a. copei (smaller range)	7.2	21.4	1.34
B. a. copei (larger range)	9.5	21.4	1.33
B. a. americanus (exclusive of populations designated as *B.a. copei* by Ashton et al., 1973)	12.5	27.7	1.51
East of Northern Quebec-Ontario Border (Populations 1-10)	7.6	19.3	1.28
Western Ontario (Populations 11-13)	12.4	21.4	1.31
Wisconsin (Populations 14-25)	14.6	32.2	1.62
TOTAL MEAN FOR ALL POPULATIONS	11.6	25.7	1.45

[1] 95% criterion for polymorphism

Although central populations of *B. americanus* are not under consideration in this paper, an investigation of the same fourteen loci in these populations indicates that the parameters of genic variability are within the range of the northern populations (Guttman, in prep.). In fact, the means for the Wisconsin samples exceed those for the central populations. These data indicate that geographic marginality in populations of *B. americanus* does not necessarily result in increased genic homozygosity. Similar lack of correlation between marginality and genic homozygosity has been recently documented (Prakash, et al., 1969; Ayala, et al., 1971, 1972a; Richmond, 1972).

Evidence has been accumulating supporting the idea that populations of vertebrates are on the average less genetically variable than populations of invertebrates (McKinney, et al., 1972; Selander and Kaufman, 1973; Selander and Johnson, 1973). The most recent of these papers summarize published and some unpublished data on genic heterozygosity in animal populations. Selander and Kaufman (1973) note an interesting pattern: invertebrate genic heterozygosities range from 9.7% to 25.0% with a mean of 15.1% while vertebrate genic heterozygosities range from 1.0% to 11.2% with a mean of 5.8%. They increased published and unpublished values of genic heterozygosity of the horseshoe crab, sparrow, and two rodent genera by 70% to compensate for the fact that esterases, which are highly polymorphic loci, were not included in the loci assessed. They then attempt to correlate the differences between invertebrates and vertebrates with Levins' (1968) theory of adaptive strategies in relation to environmental uncertainty or "grain". For vertebrate populations which experience the environment as fine grained, the optimum strategy would often be a single phenotype specialized to the most frequently encountered set of conditions. However, for the invertebrates the coarse grained environment more often necessitates a strategy in which specialized types occur in proportions dependent upon the frequency of the various environmental patches. Levins' theory, therefore, predicts that invertebrates would have greater heterozygosity, more polymorphic loci, and a larger number of alleles per locus than vertebrates.

Genic variability for the American toad does not lend support to the above findings. Mean genic heterozygosity per population for the 25 populations in the present study is 11.6% (Table 7). Since esterases are not included in this calculation, if the mean genic heterozygosity were increased by 70%, the new mean becomes 19.7%, a figure which is above the mean of the invertebrate range. In addition, if we consider that transferrin and albumin loci are highly variable in this species (Guttman and Wilson, 1973), genic heterozygosity values would probably surpass 25.0% and the American toad would rank as the species with the greatest mean percentage of loci heterozygous per individual reported to date. The mean percentage of loci polymorphic per population is also high in comparison with most vertebrate populations, which range from 10.0% to 20.0% (Selander and Kaufman, 1973), and would be higher with the addition of the above loci. However, it does not approach many invertebrate means (25.0% to 50.0%).

Rogers (1973) used ten loci, including esterases, to determine genic variability in *B. cognatus* and *B. speciosus*. He found that the mean proportion of heterozygous loci per individual per population per species was 12.1% and 10.7% respectively; the heterozygosities are probably underestimates due to the presence of a null allele at one of the esterase loci. These values, while less than those reported here for *B. americanus*, are higher than any vertebrate considered by Selander and Kaufman (1973). In addition, for *B. cognatus* and *B. speciosus* Rogers (1973) found that the average percentages of the ten loci which are polymorphic in each species is 52.0% and 42.0% respectively. Only a few species of *Drosophila* have been shown to be more polymorphic. Whether this unusual genic variability is linked to a peculiar adaptive strategy in some species of toads, is common to all toads, is common to anurans with similar strategies, or some other explanation is involved, is presently unknown.

Coefficients of genic similarity (Rogers, 1972) were calculated for the 25 *Bufo* populations. Similarity values ranged between 0.89 and 0.99; conspecific populations of other vertebrates are generally within this range (Selander and Johnson, 1973). Comparisons of similarity values between the two subspecies and/or regional groupings of populations do not reveal any major differences in these values.

ACKNOWLEDGEMENTS

It is with a great deal of pleasure that I gratefully acknowledge the aid of the following people in collecting toads: S. Gorham, F. Cook, R. Ashton, P. Buckley, M. Ewert, E.G. Hoffman and Sons. Assistance in the laboratory was provided by C. Garland, G. Graf, R. MacBride, D. Anson, P. Ward, and J. Patton. S. Rogers provided the computer program for the genetic similarity analyses, K. Wilson and M. Farrell helped with computer programming, and Miami University donated computer time. C. McKinney and D. Merkle aided in modifying some of the techniques. This study was funded by NSF Grant GB 23601.

REFERENCES

Ashton, R.E., S.I. Guttman, and P. Buckley 1973. Notes on the distribution, coloration, and breeding of the Hudson Bay toad, *Bufo americanus copei* (Yarrow and Henshaw). *J. Herp.* 7:17-20.

Avise, J.C. and R.K. Selander 1972. Evolutionary genetics of cave-dwelling fishes of the genus *Astyanax*. *Evolution* 26:1-19.

Ayala, F.J., J.R. Powell, and Th. Dobzhansky 1971. Enzyme variability in the *Drosophila willistoni* group, II. Polymorphisms in continental and island populations of *Drosophila willistoni*. *Proc. Natl. Acad. Sci. U.S.A.* 68:2480-2483.

Ayala, F.J., J.R. Powell, and M.L. Tracey 1972a. Enzyme variability in the *Drosophila willistoni* group, V. Genic variation in natural populations of *Drosophila equinoxialis* . *Genet. Res.* 20:19-42.

Ayala, F.J., J.R. Powell, M.L. Tracey, C.A. Mourao, and S. Perez-Salas 1972b. Enzyme variability in the *Drosophila willistoni* group, IV. Genic variation in natural populations of *Drosophila willistoni*. *Genetics* 70:113-139.

Blair, A.P. 1941. Variation, isolation mechanisms, and hybridization in certain toads. *Genetics* 26:398-417.

Conant, R. 1958. *A Field Guide to Reptiles and Amphibians*. Houghton Mifflin Co., Boston.

Dessauer, H.C. and E. Nevo 1969. Geographic variation of blood and liver proteins in cricket frogs. *Biochem. Genet.* 3:171-188.

Gottlieb, L.D. 1971. Gel electrophoresis: New approach to the study of evolution. *Bioscience* 21:939-944.

Guttman, S.I. and K.G. Wilson 1973. Genetic variation in the Genus *Bufo*, I. An extreme degree of transferrin and albumin polymorphism in a population of the American toad (*Bufo americanus*). *Biochem. Genet.* 8:329-340.

Levins, R. 1968. *Evolution in Changing Environments*. Princeton Univ. Press, Princeton, N.J.

Lewis, W.H.P. and H. Harris 1967. Human red cell peptidases. *Nature* 215:351-355.

Lewontin, R.C. and J.L. Hubby 1966. A molecular approach to the study of genic heterozygosity in natural populations, II. Amount of variation and degree of heterozygosity in natural populations of *Drosophila pseudoobscura* . *Genetics* 54:595-609.

McKinney, C.O., R.K. Selander, W.E. Johnson, and S.Y. Yang 1972. Genetic variation in the side-blotched lizard (*Uta stansburiana).* *Univ. Texas Publ.* 7213:307-318.

Nevo, E. and C.R. Shaw 1972. Genetic variation in a subterranean mammal, *Spalax ehrenbergi.* *Biochem Genetics* 7:235-241.

Prakash, S., R.C. Lewontin, and J.L. Hubby 1969. A molecular approach to the study of genic heterozygosity in natural populations, IV. Paterns of genic variation in central, marginal, and isolated populations of *Drosophila pseudoobscura.* *Genetics* 61:841-858.

Richmond, R.C. 1972. Enzyme variability in the *Drosophila willistoni* group, III. Amounts of variability in the superspecies, *D. paulistorum.* *Genetics* 70:87-112.

Rogers, J.S. 1972. Measures of genetic similarity and genetic distance. *Univ. Texas Publ.* 7213:145-153.

Rogers, J.S. 1973. Protein polymorphism, genic heterozygosity, and divergence in the toads *Bufo cognatus* and *B. speciosus.* *Copeia* 1973:322-330.

Selander, R.K., W.G. Hunt, and S.Y. Yang 1969a. Protein polymorphism and genic heterozygosity in two European subspecies of the house mouse. *Evolution* 23:379-390.

Selander, R.K. and W.E. Johnson 1973. Genetic variation among vertebrate species. *Ann. Rev. Ecol. Systematics* 4:75-91.

Selander, R.K. and D.W. Kaufman 1973. Genic variability and strategies of adaptation in animals. *Proc. Natl. Acad. Sci. U.S.A.* 70:1875-1877.

Selander, R.K., M.H. Smith, S.Y. Yang, W.E. Johnson, and W.E. Gentry 1971. Biochemical polymorphism and systematics in the Genus *Peromyscus*, I. Variation in the old-field mouse *(Peromyscus polionotus).* *Univ. Texas Publ.* 7103:49-90.

Selander, R.K., S.Y. Yang, and W.G. Hunt 1969b. Polymorphism in esterases and hemoglobin in wild populations of the house mouse (*Mus musculus).* *Univ. Texas Publ.* 6918:271-338.

Smith,M.H., R.K. Selander, and W.E. Johnson 1973. Biochemical polymorphism and systematics in the Genus *Peromyscus*, III. Variation in the Florida deer mouse (*Peromyscus floridanus)*, a Pleistocene relict. *J. Mammal.* 54:1-13.

Volpe, E.P. 1952. Physiological evidence for natural hybridization of *Bufo americanus* and *Bufo fowleri.* *Evolution* 6:393-406.

Volpe, E.P. 1955. Intensity of reproductive isolation between sympatric and allopatric populations of *Bufo americanus* and *Bufo fowleri.* *Amer. Nat.* 89:303-317.

Webster, T.P. 1973. Adaptive linkage disequilibrium between two esterase loci of a salamander. *Proc. Natl. Acad. Sci. U.S.A.* 70:1156-1160.

Webster, T.P., R.K. Selander, and S.Y. Yang 1972. Genetic variability and similarity in the *Anolis* lizards of Bimini. *Evolution* 26:523-535.

Wright, A.H. and A.A. Wright 1949. *Handbook of Frogs and Toads of the United States and Canada*. Comstock Publ. Co., Ithaca, N.Y.

THE ATROPINESTERASE-COCAINESTERASE SYSTEM OF ISOZYMES IN RABBITS: DISTRIBUTION OF PHENOTYPES AND THEIR GENETIC ANALYSIS

CLYDE STORMONT AND YOSHIKO SUZUKI
Serology Laboratory, Department of Reproduction,
School of Veterinary Medicine, University of California
Davis, California 95616

ABSTRACT. In an earlier note in *Science* the authors described a multizoned, six-phenotype system of rabbit serum esterases and showed that the atropinesterase and cocainesterase activity of rabbit serum are properties of that system. Phenotypes A and AF exhibited both activities. Phenotypes F, P, and S exhibited cocainesterase activity alone, whereas the serum of rabbits of phenotype M was incapable of hydrolyzing either substrate. Isozyme zone A, found only in phenotypes A and AF, is synonymous with atropinesterase. Isozyme zone S, common to phenotypes A, AF, F, P, and S, but lacking in phenotype M, is synonymous with cocainesterase.

In this report we examine the inheritance of the six phenotypes in 91 litters with a total of 425 offspring, and also present data on the distribution of the six phenotypes in various populations, including inbreds. The data are readily interpretable on the basis of 5 autosomal alleles designated A, F, M, P, and S. Alleles A and F act as codominants and are dominant to the remaining three alleles. Allele P can be interpreted as a duplication which incorporates alleles M and S. Accordingly, it acts as a dominant to alleles M and S. It follows that genotype MS is one of four which produces phenotype P. An alternative model entertains three alleles, M, P, and S, at one locus, and three alleles A, F, and a at a neighboring locus, with allele a acting as a silent allele.

INTRODUCTION

As shown in a previous study (Stormont and Suzuki, 1970), the serum of rabbits (*Oryctolagus cuniculus*) can be grouped into one or another of six, multizoned, esterase phenotypes. Under the conditions described in that report, the six phenotypes, designated A, AF, F, M, P, and S, appeared in a region of the gels just anodal to serum albumin. Of particular interest was the observation that the long-known atropinesterase activity of rabbit serum and the more recently described cocainesterase activity of rabbit serum are properties of that system.

Atropinesterase activity was restricted to phenotypes A and AF where it appeared to be localized in a rapidly and strongly staining zone designated A. All phenotypes except M exhibited cocainesterase activity which appeared to be restricted to a strongly staining zone labelled S, the most cathodal of all the zones in the system. Thus, with respect to atropinesterase and cocainesterase activity, only three kinds of rabbits were observed: 1) those with both enzymes (zymogram phenotypes A and AF), 2) those with cocainesterase alone (zymogram phenotypes F, P and S) and 3) those which lacked both enzymes (zymogram phenotype M). The six phenotypes and the positions of zones A (atropinesterase) and S (cocainesterase) relative to the other zones are shown in Fig. 1.

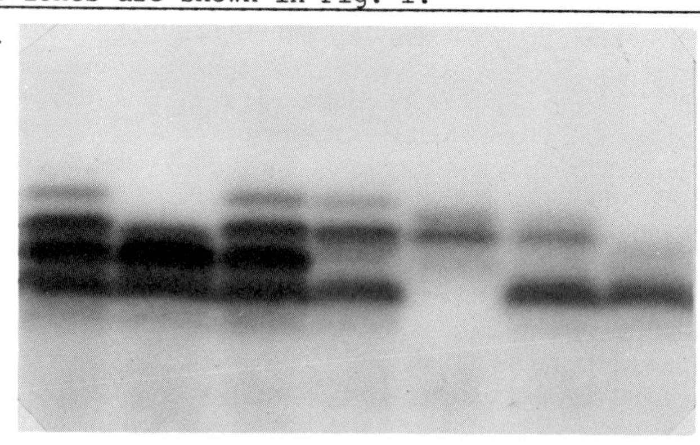

Fig. 1. This photo of a gel-slice, stained with α-naphtyl acetate as the substrate, shows the six phenotypes A, AF, F, M, P, and S in the atropinesterase-cocainesterase system of rabbit serum esterases. Phenotype AF is repeated on the left margin in order to show the positions of the major isozyme zones F (fast), M (medium), A (atropinesterase) and S (cocainesterase or slow). Zone A is the most rapidly and intensely staining of all the zones when using α-naphthyl acetate as substrate. When using α-naphthyl butyrate as substrate, zones A, S and M stain at about the same speed and degree of intensity. This gel exaggerates the difference between phenotypes P and S inasmuch as the minor zones of phenotype S are barely visible.

Marked differences in the staining capacities of the various zones were also noted when using a variety of esters as substrates. There were also marked differences in the thermal stabilities of the various zones. For example, zone A and atropinesterase activity were inactivated by heating the serum for 1 hour at 56°C but such treatment had no effect on zone S and cocainesterase activity. Since the time of the foregoing report, van Zutphen (1972) has confirmed the differences in the thermal stabilities of atropinesterase and cocainesterase and, like us, has yet to observe rabbits whose serum possesses atropinesterase in the absence of cocainesterase. Thus, it would appear that the synthesis of atropinesterase, or zone A, is somehow contingent upon the synthesis of cocainesterase, or zone S.

The inheritance of atropinesterase was thoroughly analyzed by Sawin and Glick (1943). They showed that the presence of that enzyme vs. its absence could be accounted for on the basis of a single pair of autosomal alleles, designated here as _A_ and _a_, such that rabbits of genotypes _AA_ and _Aa_ possess the enzyme whereas rabbits of genotype _aa_ lack it. Moreover, they showed that the level of atropinesterase activity in rabbits of genotype _AA_ is, on the average, more than twice as high as that in the serum of _Aa_ rabbits, thereby indicating that allele (or alleles) _a_ competes effectively with allele _A_. Insofar as we are aware, there has been no published data on the inheritance of cocainesterase and its genetic relationship to atropinesterase. However, we mentioned in the foregoing report that the six zymogram phenotypes, and thereby atropinesterase and cocainesterase, could be accounted for on the basis of a model that involves five autosomal alleles designated _A_, _F_, _M_, _P_, and _S_. Since then we have bolstered considerable our data on the inheritance of the six zymogram phenotypes and their distributions in various populations of rabbits. The present report is concerned with the presentation of those data.

MATERIALS AND METHODS

Blood samples were obtained from 916 rabbits among which were 91 litters with a total of 425 offspring. The sources of the rabbits are shown in Table I. With the exception of 82 out of the 129 rabbits in the partially inbred JAX populations, from which serum samples alone were obtained, two blood samples were collected from each rabbit, one in anticoagulant (2% Na citrate and 0.5% NaCl) and the other without.

Serum obtained from each sample without anticoagulant was examined for atropinesterase and cocainesterase activity and was classified with respect to the six zymogram phenotypes in accordance with previous procedure (Stormont and Suzuki, 1970). All serum samples were kept frozen when not in use.

Sonicates of washed blood cells were electrophoresed and classified with respect to zymogram phenotypes in two systems of red cell esterases (_Es-1_ and _Es-3_) and a system of platelet esterases (_Es-2_) in accordance with the methods of Schiff and Stormont (1970). All sonicates were kept frozen when not in use.

TABLE I

DISTRIBUTION OF SIX ZYMOGRAM PHENOTYPES IN VARIOUS POPULATIONS OF RABBITS

Breed or Line *	Source *	Phenotypes and Counts						Totals
		A	AF	F	M	P	S	
NZW	ARS	46	7	9	18	34	10	124
NZW	LSS	61	10	21	23	43	14	172
NZW	SSI	113	0	0	9	18	10	150
Panmictic	COH	113	24	56	62	60	26	341
Inbred WH	JAX	5	0	0	0	0	0	5
Inbred AC	JAX	2	0	0	0	0	3	5
Inbred AX	JAX	0	0	0	0	0	16	16
Inbred AXbubu	JAX	0	0	0	0	0	32	32
Inbred IIIc	JAX	0	0	0	7	7	5	19
Inbred III	JAX	0	0	0	0	34	0	34
Inbred IIIvo	JAX	0	0	0	0	18	0	18
Totals		340	41	86	119	214	116	916

*NZW (New Zealand White), Panmictic (see text), Inbred (see text), ARS (Animal Resources Service, School of Veterinary Medicine, this University), LSS (Lee's Small Animal Supply, Santa Rosa, California), SSI (Small Stock Industries, Pea Ridge, Arkansas), COH (City of Hope National Medical Center, Duarte, California) and JAX (Jackson Laboratory, Bar Harbor, Maine).

RESULTS

The serum of each rabbit fell into one or another of the six zymogram phenotypes (Fig. 1) as summarized in Table I. In total, there were 340 type A rabbits, 41 type AF, 86 type F, 119 type M, 214 type P, and 116 type S. Or, with respect to atropinesterase and cocainesterase alone, there were 381 rabbits (340 type A + 41 type AF) with both enzymes, 416 with cocainesterase alone (86 type F + 214 type P + 116 type S) and 119 (all type M) which lacked both enzymes. Significantly, there were no rabbits among the 916 which possessed atropinesterase in the absence of cocainesterase.

THE INHERITANCE OF THE SIX ZYMOGRAM PHENOTYPES

Data on the inheritance of the six zymogram phenotypes are summarized in Table II. All except three of the 21 possible mating types were encountered.

TABLE II

INHERITANCE OF THE SIX ZYMOGRAM PHENOTYPES IN 91 LITTERS WITH A TOTAL OF 425 OFFSPRING

No. of Litters	Mating Types	Offspring of Types A	AF	F	M	P	S
9	A X A	45	0	0	0	0	0
1	A X A	3	0	0	1	0	0
2	A X A	11	0	0	0	5	0
1	A X A	2	0	0	0	0	4
2	A X AF	8	3	0	0	0	0
1	A X AF	1	0	1	0	0	0
1	A X F	2	1	1	2	0	0
1	A X F	2	1	0	1	0	0
2	A X F	0	6	4	0	0	0
4	A X M	7	0	0	12	0	0
1	A X M	2	0	0	0	2	0
1	A X M	0	0	0	2	0	0
3	A X P	12	0	0	0	0	0
4	A X P	14	0	0	6	7	0
3	A X P	9	0	0	0	7	0
2	A X P	6	0	0	0	5	4
1	A X P	0	0	0	1	1	0
1	A X P	0	0	0	0	7	0
2	A X S	15	0	0	0	0	0
2	A X S	6	0	0	0	6	0
2	A X S	5	0	0	0	0	4
1	AF X F	0	2	1	0	0	0
1	AF X M	1	0	3	0	0	0
3	AF X P	8	0	6	0	0	0
1	F X F	0	0	1	1	0	0
1	F X M	0	0	1	0	4	0
1	F X P	0	0	6	0	1	0
1	F X S	0	0	2	0	0	3
3	M X M	0	0	0	8	0	0
4	M X P	0	0	0	13	13	0
1	M X P	0	0	0	0	2	0

TABLE II (continued)

No. of Litters	Mating Types	A	AF	F	M	P	S
		Offspring of Types					
12	P X P	0	0	0	0	42	0
1	P X P	0	0	0	1	2	0
2	P X P	0	0	0	0	6	8
1	P X S	0	0	0	0	1	7
12	S X S	0	0	0	0	0	38
Totals 91		149	13	26	54	115	68

In order to reduce the amount of tabular material in the presentation of the family data, matings of a kind which produced offspring of the same phenotypes were pooled. For example, there were 9 matings among 13 A X A which produced nothing but type A offspring.

The family data (Table II) are wholly consistent with a model based on five autosomal alleles (A, F, M, P and S), with genotype-phenotype relationships as shown in Table III. This model was derived in part by visual inspection of the phenotypes and in part by performing runs on 50:50 mixtures of serum from rabbits of the various phenotypes. According to the model,

TABLE III

GENOTYPE-PHENOTYPE RELATIONSHIPS BASED ON THE ONE-LOCUS, FIVE-ALLELE MODEL

Phenotypes	Genotypes
A	AA, AM, AP, and AS
AF	AF
F	FF, FM, FP, and FS
M	MM
P	MP, MS, PP, and PS
S	SS

alleles A and F act as codominants, and each, in effect, is dominant to the remaining three alleles. This effect can be simulated by performing zymogram runs on 50:50 mixtures of serum. For example, runs performed on mixtures of the serum of rabbits of types A and F yield a zymogram pattern indistinguishable from type AF. Thus, on that basis, alone it can be inferred that phenotype AF signifies genotype AF. Similarly, performing runs on mixtures of A and M, A and P and A and S yields patterns that are indistinguishable from type A. This signifies that rabbits of type A could be of genotypes AA, AM, AP, and AS. Likewise, performing runs on mixtures of F and M, F and P, and F and S yields patterns indistinguishable from type F. This signifies that rabbits of phenotype F could be of genotypes FF, FM, FP, and FS.

When zymogram runs are performed on mixtures of P and M and P and S, the resulting patterns are indistinguishable from phenotype P. This signifies that allele P is dominant to alleles M and S and that rabbits of phenotype P could be of genotypes PP, MP and PS.

When runs are performed on mixtures of M and S, the resulting pattern is indistinguishable from phenotype P and, therefore, we must include genotype MS among those responsible for phenotype P. Thus, neither of the alleles M and S is dominant to the other. The fact that their combined effect is equivalent to the action of allele P alone suggests that allele P could be a duplication which incorporates a part or all of alleles M and S.

The reader will, to satisfy his curiosity, superimpose the zones of phenotype A upon those of phenotype F, or the converse, to generate phenotype AF. He will also superimpose the zones of phenotype A upon those of phenotypes M, P, and S, and so on, to satisfy himself that the resulting patterns are those expected when performing runs on 50:50 mixtures of serum of the various phenotypes.

If the genotype-phenotype relationships are correctly depicted (Table III), then the following results are to be expected in examining the zymogram phenotypes of offspring from the 21 possible kinds of matings:

1. A X A matings should yield no more than 2 types of offspring per litter and none, of course, of types AF and F.

There were 9 matings of A X A which yielded offspring only of type A, a total of 45. There was 1 mating which yielded 3 type A and 1 type M offspring, 2 which yielded 11 type A and 5 type P offspring, and 1 which yielded 2 type A and 4 type S offspring. Thus, in the 4 segregant matings there were 16 type A and 10 non-A offspring.

Although the ratio of type A to non-A where both parents are known to be heterozygous should approximate 3 type A to 1 non-A offspring, it should be kept in mind that some of the matings which produced only type A offspring were very likely matings of heterozygous parents. Therefore, when considering only the segregant matings, bias is introduced in favor of an excess of non-A segregants. The observation of 16 type A to 10 non-A offspring suggests such bias. The reader should keep this point in mind when examining not only the A X A matings but also F X F and P X P matings.

2. A X AF matings should yield offspring only of types A, AF

and F. If the A parent is homozygous, a ratio of 1 type A to 1 type AF offspring is expected. If the A parent is heterozygous (genotype AM, AP or AS) a ratio of 2 type A to 1 type AF to 1 type F is expected.

There were two matings which yielded only A and AF offspring, and in a ratio of 8:3. There was one mating which yielded only 2 offspring, one of type A and the other of type F.

3. A X F matings are the only matings of the 21 possible kinds of matings which could yield up to 4, but no more than 4, types of offspring per litter. A X F matings are the only matings which can produce, in composite, offspring of all 6 phenotypes.

There was one mating which yielded offspring of 4 types, namely, 2 type A, 1 type AF, 1 type F and 2 type M offspring. According to the 5-allele model, the A parent was of genotype AM and the F parent was of genotype FM. The expected ratio would be 1:1:1:1 of those four types. There was one mating which produced offspring of 3 types (2 type A, 1 type AF and 1 type M). Here, again, the A parent must have been of genotype AM and the F parent of genotype FM in order to produce a type M offspring. There was one mating which produced 6 type AF and 4 type F offspring, which approached the 1:1 ratio expected when the A parent is of genotype AA and the F parent is of genotype FM or FP or FS.

4. A X M matings should yield no more than 2 types of offspring per litter and none of types AF, F and S.

The results in all 6 litters were consistent with expectation.

5. A X P matings should yield offspring of no more than 3 types per litter and none of type AF and F.

Although there were no exceptions to these expectations in any of the 14 litters, there was one unusual litter in that all 7 offspring were of type P. This could happen if the A parent were of genotype AM or AP or AS and the likely prospect that the P parent was of genotype PP. The probability of a litter of 7 rabbits, all of type P, from such a mating is $(1/2)^7$.

6. A X S matings should yield offspring of no more than 2 types per litter and none of types AF, F and M.

The results were consistent with expectation.

7. AF X AF matings should yield offspring of types A, AF, and F in a 1:2:1 ratio.

There were no matings of that kind.

8. AF X F matings should yield only AF and F offspring in a 1:1 ratio.

There was one such litter with 2 AF and 1 F offspring.

9. AF X M, AF X P and AF X S matings should yield only A and F offspring in a 1:1 ratio.

There were no AF X S matings. There was 1 AF X M and 3 AF X P matings with a total of 18 offspring, 9 of type A and 9 of type F.

10. F X F matings should yield no more than 2 types of offspring per litter and none of types A and AF.

There was only 1 F X F mating with 2 offspring of type F and one of type M. Accordingly, both parents must have been of genotype FM.

11. F X M matings should yield no more than 2 types of offspring per litter and none of types A, AF and S.

There was 1 F X M mating with 1 offspring of type F and 4 of type P. The parent of type F could have been of genotype FP or FS, and the expected ratio is 1 type F to 1 type P.

12. F X P matings should yield no more than 3 types of offspring per litter and none of types A and AF.

There was one F X P mating with 6 offspring of type F and 1 of type P.

13. F X S matings should yield offspring of no more than 2 types per litter and none of types A, AF, and M.

There was one F X S mating with 2 type F and 3 type S offspring. The expected ratio is 1:1.

14. M X M matings should yield nothing but type M offspring.

There were 3 M X M matings with a total of 8 offspring, all of type M.

15. M X P matings should yield offspring all of type P when the parent of type P is of genotype PP or PS, and should yield offspring of types M and P in a 1:1 ratio when the parent of type P is of genotype MS or PM.

There was one M X P mating which yielded 2 offspring, both of type P. There were four M X P matings which yielded 13 type M and 13 type P offspring. Here, again, bias is a factor when comparing observed with expected ratios.

16. M X S matings should yield only offspring of type P.

There were no M X S matings.

17. P X P matings should yield offspring of no more than three phenotypes per litter and none of types A, AF, and F.

There were only 3 segregant matings out of 15, the remaining matings producing nothing but type P offspring, a total of 42. One of the segregant matings produced 1 type M and 2 type P offspring. There were two segregant matings which produced offspring of types P and S, a total of 6 type P and 8 type S. Offspring of types P and S are expected only when both parents are of genotype PS or one parent is of genotype MS and the other PS, but in all such matings the expected ratio is 3 type P to 1 type S offspring. The observed ratio differs significantly from expectation.

18. P X S matings would be expected to produce only type P offspring when the parent of type P is of genotype MP or PP. When the parent of type P is of genotype MS or PS offspring of types P and S are expected in a 1:1 ratio.

There was only one P X S mating which produced 1 type P

and 7 type S offspring. Here, again, the ratio differed considerably from expectation.

19. S X S matings should yield nothing but type S offspring. There were 12 S X S matings with a total of 38 offspring, all of type S.

There were, of course, a number of does which produced more than one litter and a number of bucks which sired more than one litter. Accordingly, it was possible to establish the precise genotypes of some of the parents. Unfortunately, space does not permit the presentation of the results of the progeny tests.

THE DISTRIBUTION OF THE SIX ZYMOGRAM PHENOTYPES

The distribution of the six zymogram phenotypes in various breeds and partially inbred strains is given in Table I.

All six phenotypes were represented in the two California populations (ARS and LSS) of rabbits of the New Zealand White (NZW) breed and in proportionately similar numbers. The Arkansas population (SSI) of NZW rabbits had no representatives of phenotypes AF and F, signifying the absence of allele F in that population.

The populations designated Panmictic was truly a panmix. That breeding population was assembled by Dr. Charles H. Todd for the main purpose of propagating all of the known gamma globulin allotypes in rabbits. Accordingly, there were no restrictions on breeds, color types and color patterns when assembling that colony.

In our initial sampling of the Panmictic colony we limited the collection to 100 samples, which were drawn only from adult males and females being used as breeders. Not only were all 6 esterase alleles represented among those 100 rabbits, but also a variety of color and pattern genes.

The results of a gene-frequency analysis on the esterase types of those 100 rabbits are presented in Table IV. In that analysis we used the approach suggested by Cotterman (1954) in deriving the frequencies of the three major alleles which control the ABO blood groups of man. In other words, we treated phenotypes M, P and S as blood group O and let phenotypes A, AF and F simulate blood groups A, AB and B. Thus, we arrived at a frequency of 0.1930 (or 0.19) for allele A, 0.1224 (or 0.12) for allele F and 0.06846 (or 0.69) for the composite of alleles M, P and S. Using those estimates to obtain expected numbers of rabbits in each of the phenotypes A, AF, F and "O", where "O" represents the sum of all rabbits of phenotypes M, P and S, we found that there was good agreement between observed and expected numbers, as shown in Table IV.

TABLE IV

A GENE-FREQUENCY ANALYSIS BASED ON THE ZYMOGRAM TYPES OF 100 RANDOMLY SELECTED BREEDERS IN THE PANMICTIC COLONY, TABLE I*

Phenotypes	Observed		Expected	
A	29		30.2	
AF	6		4.7	
F	17		18.3	
M	21	48		46.9
P	20			
S	7			

*Frequencies p, q and r of alleles A, F and "O", where "O" represents alleles M, P and S, are estimated to be 0.1930, 0.1224 and 0.6846, respectively using the method suggested by Cotterman (1954). Chi-square for the deviation between observed and expected is 0.6 which corresponds to a probability, P,>0.30.

With respect to the seven inbred lines, all from the Jackson Laboratories (JAX), Bar Harbor, Maine, perhaps the main thing to be noted is the marked reduction of phenotypic variability in contrast with the non-inbred populations. Only five serum samples were obtained from the WH line, representing two parents and their 3 offspring. All were of type A, thereby suggesting that allele A could be fixed in that line. Although only five samples were obtained from rabbits in the AC line, it is clear that at least two alleles, A and S, were still segregating in that line. All 16 samples of line AX were of phenotype S, suggesting that allele S may be fixed in that line. Similarly, the 32 samples from line AXbubu were of type S, indicating that allele S may be fixed in that line. In line IIIc it is evident from the distribution of types (7M, 7P and 5S) that at least 2 alleles (M and S) were still segregating in that line at the time of sampling. The 34 samples from line III and the 18 samples from line IIIvo were all of type P indicating that allele P was at or near fixation in those lines. The inbreeding coefficients for the seven JAX strains at the time they were sampled in 1969 were: WH (0.70), AC (0.64), AX (0.54), AXbubu (0.59), IIIc (0.85), III (0.93), and IIIvo (0.92).

DISCUSSION

Although all of our data agree well with the model based on a series of five alleles, there is the difficulty of explaining how one allele, as A, can code for two enzymes, namely, atropinesterase and cocainesterase, with different substrate specificities and markedly different thermostabilities. Perhaps this will be no problem when all the steps in their synthesis, including post-ribosomal events, are fully understood. And then, again, such knowledge could force us to reject the 5-allele model. In the meantime, of course, it would be well to consider other possible models, and this we have done.

One of several two-locus models which seemed to be the most plausible is based on alleles P, M and S at one locus and alleles A, F and a at a neighboring locus. The dominance or "masking" relationships would be essentially the same as those based on the 5-allele model, the main difference being the postulation of a sixth gene, namely, allele a of a second locus which would act essentially as an amorph or silent allele. Thus, the 36 possible genotypes could generate the observed phenotypes.

However, with any degree of crossing over between the two loci, there are 9 kinds of matings which would be expected to produce offspring of more phenotypic classes per litter than those expected under the model based on one locus with five alleles. For example, in A X A matings where both parents are of genotype AaMS offspring of types, A, M, P and S could be expected whereas under the one-locus model the maximum number of expected types is two per litter. Table V summarizes these expectations for the 9 kinds of matings.

The failure to find recombinants, or phenotypes not expected under the one-locus model, does not mean that we should reject all two-locus models at this time. We do, however, suggest that the simplest model, namely, that based on a single series of five autosomal alleles, be used as a working model, or until there is convincing genetic and/or biochemical evidence to reject it.

The literature on atropinesterase of rabbits traces to a paper published over 100 years ago. In spite of the many papers on the pharmacology, biochemistry and genetics of that enzyme there is still much to be learned regarding its biosynthesis and genetic control, especially in view of its relationship to cocainesterase, as set forth here and in our previous report (1970), and also the studies of van Zutphen (1972). We can all wonder why rabbits evolved the capacity to produce atropinesterase and cocainesterase. Was it to permit some of them to survive on leaves of plants of the families Solanaceae and

TABLE V

MAXIMUM NUMBERS OF EXPECTED PHENOTYPES IN THE OFFSPRING OF INDIVIDUAL LITTERS WHEN CONSIDERING THE ONE-LOCUS AND TWO-LOCUS MODELS (See Discussion).

Mating types	One locus	Two loci
A X A	2	4
A X F	4	6
A X M	2	3
A X P	3	4
A X S	2	3
F X F	2	4
F X M	2	3
F X P	3	4
F X S	2	3

Erythroxylaceae and, if so, why has the polymorphism persisted so long after that selection pressure became relaxed? Obviously, the picture promises to become even more intriguing when unique substrates like atropine and cocaine are found for the other isozymes (e.g., zones F and M in Fig. 1) of this system.

In a report to be published elsewhere, we shall show that the phenotypes in this system are strongly correlated with those of the ES 2 or platelet system described by Schiff and Stormont (1970).

ACKNOWLEDGEMENTS

We are indebted to Mr. V. L. Isley, Mr. Robert W. Dubbell, Dr. Charles W. Todd, and Dr. Richard R. Fox (RRF supported by RR-00251) in connection with the assistance they provided in obtaining blood samples from the respective rabbit populations of sources LSS, SSI, COH, and JAX (Table I), and to Mr. B. G. Morris and our former student, Dr. Robert Schiff, for their assistance in collecting blood samples from the ARS, LSS, and COH populations.

REFERENCES

Cotterman, C.W. 1954. Chapter 35, pp. 449-465 in *Statistics and Mathematics in Biology*. Edited by Kempthorne, O., T.A. Bancroft, J.W. Gowen and J.L. Lush. The Iowa State College Press, Ames, Iowa.

Sawin, P. and D. Glick 1943. Atropinesterase, a genetically determined enzyme in the rabbit. *Proc. Nat. Acad. Sci. USA*,

29:55-59.
Schiff, R. and C. Stormont 1970. The biochemical genetics of rabbit blood esterases: two new esterase loci. *Biochem. Genet.* 4:11-24.
Stormont, C. and Y. Suzuki 1970. Atropinesterase and cocainesterase of rabbit serum: localization of the enzyme activity in isozymes. *Science* 167:200-202.
van Zutphen, L.F.M. 1972. Qualitative and quantitative detection of atropinesterase and cocainesterase in two breeds of rabbits. *Enzymologia* 42:201-218.

BIOCHEMICAL CHARACTERIZATION OF ESTERASE ISOZYMES OF THE MARINE SNAIL, *LITTORINA LITTOREA*

MARTHA R. MATTEO
Biology Department
University of Massachusetts/Boston
Harbor Campus
Boston, Massachusetts 02125

ABSTRACT. *Littorina littorea* is a hardy intertidal snail which can withstand a broad range of environmental conditions, including pollution of various kinds. Its enzymes are being studied as part of an exploration of this adaptability. The esterases of *L. littorea* have been characterized according to substrate and inhibitor specificities and are evaluated as potential markers for population studies involving this organism. Eight regions of esterase activity are described, five of which are biochemically unique and at least seven of which are polymorphic. The phenotypes of these regions are presented and possible interpretations discussed.

ADAPTABILITY OF LITTORINA LITTOREA

Littorina littorea (Gastropoda; edible periwinkle) is a common intertidal organism on both coasts of the Atlantic Ocean. It is found from Labrador to Maryland (Wells, 1965) and has recently been studied off the coast of Virginia (Kraueter, 1973). This distribution is historically recent; *L. littorea* did not appear in collection records until the mid-1800's and then only in Canada and Maine. From there, it apparently spread steadily southward, at a rate of about 10 miles per year (Wells, 1965), probably by means of its planktonic larvae which float in the currents for about a week (Tattersall, 1920; Fretter and Graham, 1962). In this respect, *L. littorea* differs from other North American Littorines which bear live young and whose populations are probably more localized geographically (see Berger, 1973, for discussion).

Our interest in *L. littorea* stems from its ability to adapt to many environments. It can tolerate a broad range of temperatures (Fraenkel, 1960), salinities (Fischer, 1948) and hydration, extremes of which might be expected in its intertidal habitat. More important to us, however, is its apparent ability to tolerate various forms of pollution (Fretter and Graham, 1962). We have found *L. littorea* populations in Boston Harbor dwelling a few hundred yards downstream of industrial pipes and within a few feet of sewer discharges. Ireland

(1973) examined zinc levels in tissues of several marine organisms downstream of a fluvial source of zinc pollution and found 0.028 to 0.274 mgZn/gm dry weight of *L. littorea* tissue, the particular amount being proportional to the distance of the sampled snails from the Zn source. These results suggest that the enzymes of these *L. littorea* were exposed to high levels of zinc and that these levels were not toxic to the point of lethality.

Enzymes, in general, tend to be active and stable over more limited ranges of environmental variation than those experienced by *L. littorea*. We have therefore focused on enzymes for an explanation of *L. littorea*'s adaptability. One possible source of broad range tolerances would be the existence of enzymes of such primary sequence that they are stable over the entire necessary range of environmental conditions. Such a solution has evolved for acetylcholinesterases of some migratory fish exposed during their lifetime to large changes in water temperature (Somero, these proceedings; see Hochachka and Somero, 1973, for discussion). The other main sources of enzymatic flexibility are isozymes differing in kinetic and stability properties.

ROLE OF ISOZYMES IN ADAPTION

Isozymes may arise from a variety of genetic and epigenetic sources (see Markert and Whitt 1968, and Kaplan, 1968 for discussion). The two major genetically-based types, non-allelic (coded for by different genetic loci) and allelic (coded for by different alleles at the same locus) isozymes, would have different adaptive consequences for the survival of *L. littorea* populations. Non-allelic isozymes would be present in every individual of every *L. littorea* population. These isozymes could together provide every individual with the necessary metabolic flexibility to withstand all the observed types of environmental stress. Somero (these proceedings) has found non-allelic cholinesterase isozymes with different and overlapping thermal properties in those migratory fish which do not have a single broad range cholinesterase. However, allelic isozymes ("allozymes", Prakash et al., 1969), with particular properties may have adaptive value; only those individuals able to code for those allelic isozymes would benefit. If more than two such alleles are involved for any given locus, no single individual *L. littorea* would be able to adapt to the entire range of conditions in which the species as a whole can survive. *L. littorea* populations subjected to different selective forces might therefore be expected to have different allozyme profiles.

Allozymes (allelic isozymes) have been implicated in the adaptation of several organisms. Tsakas and Krimbas (1970) correlated particular esterase allozymes with pesticide resistance in the olive fruit fly. Tripathi and O'Brien (this conference) have identified cholinesterase allozymes with altered pesticide-binding capacities which confer pesticide resistance on houseflies. Koehn and Mitton (1972) demonstrated parallel changes in allozyme frequencies in homologous loci of two mussel genera (*Modiolus demissus* and *Mytilus edulis*) which inhabited different niches within the same habitat. Schopf and Gooch (1971) reported a cline in a bryozoan leucine aminopeptidase allele associated with a 6° change in water temperature. Finally, Powell (1971) found significantly higher levels of heterozygosity with regard to 22 different loci in Drosophila populations exposed to constantly changing environments compared to those raised in stable environments. Despite these examples, the adaptive value of heterozygosity is the subject of energetic debate (see Lewontin, 1973 for review).

L.LITTOREA ESTERASES

We have begun an investigation of enzyme polymorphisms in *L.littorea* as a way of approaching the question of its resistance to and possible benefit from pollution of various kinds. Enzymes were selected from three categories, esterases, carbohydrases, and enzymes of intermediary metabolism, to optimize our chances of identifying significant variation and to balance our view of overall heterozygosity. Esterases offer the opportunity for studying many loci at once, are highly polymorphic in most organisms studied, and are often responsive to low levels of molecules which might be present in industrial and municipal wastes, such as heavy metals and pesticides. Carbohydrases must be important in food processing of this herbivorous snail and should therefore interact directly with the environment. Enzymes involved in intermediary metabolism have been postulated by Kojima et al. (1970) to be among the least variable in a population and were chosen for comparison of heterozygosity with the other enzyme categories.

Vuilleumier and Matteo (1972) began with the esterases, examining zymograms of soft body parts of 300 snails from 7 U. S. and French populations by acrylamide gel electrophoresis. They found 5 multi-band regions of esterase activity, three of which could be sorted into genetically interpretable phenotypes. An analysis of phenotype frequencies at these regions revealed esterase variation within and between populations and suggested a north-south cline for one phenotype along the northeastern United States coastline.

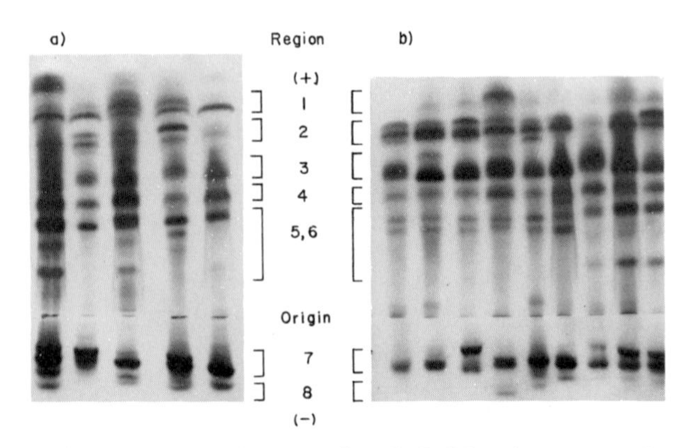

Fig. 1. Typical patterns obtained following starch gel electrophoresis of *L. littorea* extracts and staining for non-specific esterases by standard procedures (Schiff and Stormont, 1970) with (a) α naphthyl acetate (αNaAc) and (b) α naphthyl propionate (α naphthyl propionate (α NaPro) as substrates. EDTA (3 x 10^{-3} M), required for resolution of region 2, was present during a 30 min preincubation as well as during staining. Regions were defined by a combination of electrophoretic mobility and substrate and inhibitor specificities (see text). (Reproduced by permission of Pergamon Press).

It is desirable to have genetic data to support interpretations of zymogram patterns involving possible allozymes. This is particularly important for esterases because the substrates generally used to reveal bands of esterase activity following electrophoresis are rather non-specific. Many totally unrelated enzymes such as lipases, carbonic anhydrases, and cholinesterases may be uncovered by these techniques. Since breeding studies are not presently feasible for *L. littorea* (annual breeding cycle, planktonic larvae; Fretter and Graham, 1962), it was decided to establish allozyme relationships among the various esterase bands by biochemical means. The approach, described by Allen and Weremiuk (1970), is based on the assumption that major functional differences between esterases, such as in substrate and inhibitor specificities, reflect coding by different genetic loci. Esterase bands having similar electrophoretic and kinetic properties are assigned to the same region and considered to be components of a single genetic system. It is then reasonable to attempt to interpret the banding patterns of one genetic system in terms of allelic isozymes.

Starch gel electrophoresis (Schiff and Stormont, 1970) was

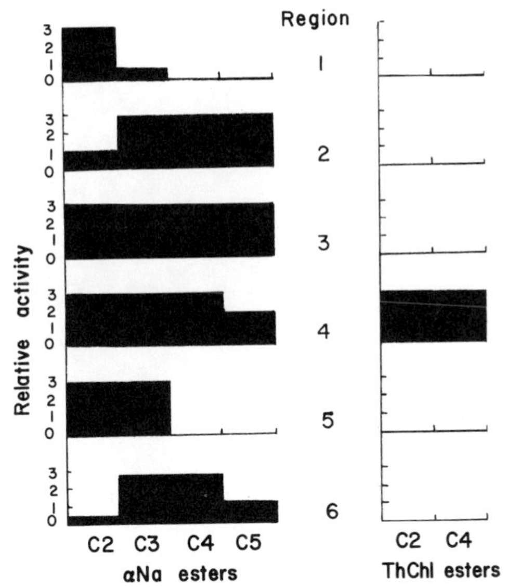

Fig. 2. Relative substrate specificities of *L. littorea* esterases with (a) α naphthyl (αNa) esters and (b) thiocholine iodide (ThChI) esters of varying carbon chain length. Specificities were based on staining intensities graded from 0 (no reaction) to +3 (strongest reaction) among the isozymes with a given substrate, as described by Schiff (1970). (Reproduced by permission of Pergamon Press).

found to give better esterase resolution (more and sharper bands) than the acrylamide system of Vuilleumier and Matteo (1972) and was used in a characterization study by Matteo et al. (1975). Figure 1 shows two typical gels, one stained in the presence of α naphthyl acetate (αNaAc; 1a) and the other with α naphthyl propionate (αNaPro; 1b). Bands which reacted with αNaAc but not αNaPro were assigned to region 1. On the basis of such comparisons, band specificities for several substrates were determined and are summarized for the anodal esterases in Figure 2. As indicated in this Figure, region 4 was unique in its ability to react with thiocholine iodide (ThChI) substrates (see Fig. 3). The cathodal esterases (regions 7 and 8) differed from each other only in the intensity of their staining and were identical in substrate specificity to region 5. These data identify region 4 as a probable cholinesterase and reveal the existence of two overlapping regions (5 and 6) of distinctive substrate specificity (Fig. 4). The esterases were further classified by the use of inhibitors: eserine, diisopropylfluorophosphate (DFP), acetazolamide, iodoacetamide, $ZnCl_2$, p-hydroxymercuribenzoate (pHMB), and the chelating agent ethylenediaminetetraacetate (EDTA).

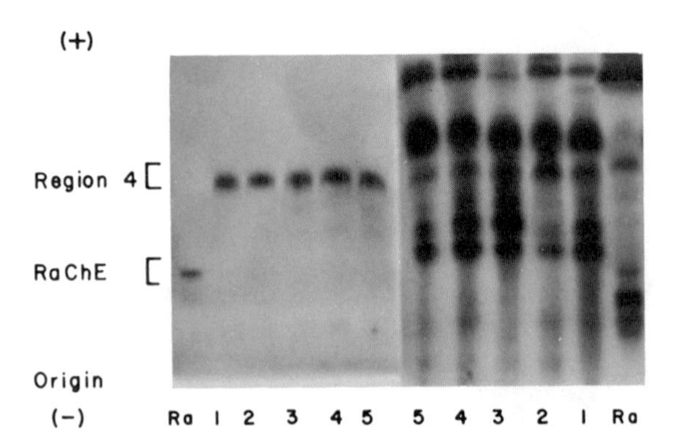

Fig. 3. Cholinesterase activity of *L. littorea* esterases. The gels shown are mirror image slices stained in the presence of AcThChI (left) as described by Brewer (1970) except for the omission of tetraisopropylphosphoramide or αNaAc (right) as described in the legend of Figure 1. Rabbit serum (Ra) was included on the gel for comparative purposes. (Reproduced by permission of Pergamon Press).

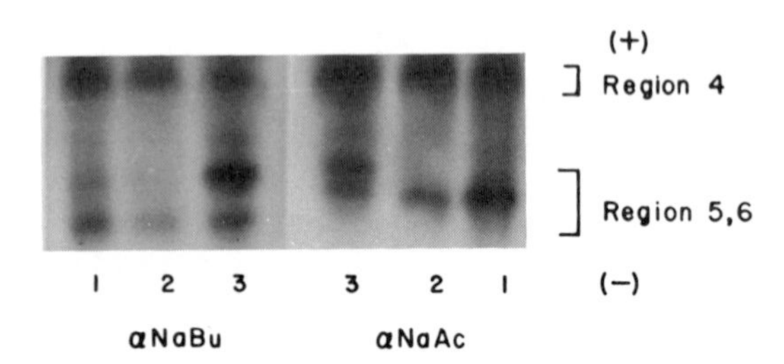

Fig. 4. Specificity of *L. littorea* esterase regions 4-6. These mirror image gel slices were stained in the presence of αNaBu (left) and αNaAc (right) as described in the legend of Figure 1. (Reproduced by permission of Pergamon Press).

In these tests, identical gel slices were pre-incubated for 30 minutes in ten-fold dilutions of the test solution followed by the addition of substrate (αNaAc or AcThChI) and dye couple Inhibitor specificities are summarized qualitatively in Table I.

The most striking features of these results are 1) the usefulness of substrate and inhibitor specificities to distinguish among the various esterase regions, 2) the tentative identification of region 4 as a cholinesterase, based on its ability to hydolyze thiocholine substrates and its inhibition

TABLE I

INHIBITOR SENSITIVITIES OF *L. LITTOREA* ESTERASES

Region	Effective Inhibitors[a,b]
1	pHMB (10^{-2}M)
2	eserine (10^{-5}M), DFP[c]
3	DFP ($<10^{-7}$M), pHMB (10^{-3}M), $ZnCl_2$ ($\leq 10^{-3}$M)
4	eserine (10^{-6} to 10^{-5}M), DFP (10^{-5}M), $ZnCl_2$ (10^{-3}M)
5	insensitive to compounds tested.
6	not surveyed.
7, 8	insensitive to compounds tested.

[a]Concentration indicated in parentheses is that which yielded 70% inhibition, compared to the uninhibited control, after 30 minutes pre-incubation of the gel slice with inhibitor alone followed by one hour in the presence of both inhibitor and substrate (αNaAc).

[b]All regions were insensitive to up to 10^{-2}M $MgCl_2$ and up to 10^{-3}M acetazolamide and iodoacetamide.

[c]Exact DFP concentration for 70% inhibition could not be determined since all concentrations tested obliterated the control region 2. It is assumed, therefore, that volatilized DFP was sufficient to inhibit this region.

by low concentrations of eserine, and 3) the sensitivity of several esterases to an organophosphate compound (DFP; regions 2, 3, and 4) and heavy metals (pHMB, regions 1 and 3; $ZnCl_2$, regions 3 and 4). These properties point firmly to esterases as valuable markers in adaptation studies involving polluted waters.

The esterase regions were characterized phenotypically using αNaAc as substrate. The observed phenotypes for regions 1, 2, 7, and 8 are summarized in Figure 5. Examples of these phenotypes are shown in Figure 1a (region 1), Figure 6 (region 2) and Figure 7 (regions 7 and 8). The frequencies of these phenotypes in one population (Nahant, Massachusetts; relatively unpolluted site) are presented in Table 2. Region 4 appeared monomorphic (1 band phenotype) in this study but has been found to be polymorphic (two-band phenotype) in a subsequent examination involving a Boston Harbor population and an unusually long electrophoretic run. Regions 5 and 6 were not included in the Nahant study but representative phenotypes from a Scituate, Massachusetts population are diagrammed in Figure 8. Regions 5, 6, 7, and 4, if the latter is indeed

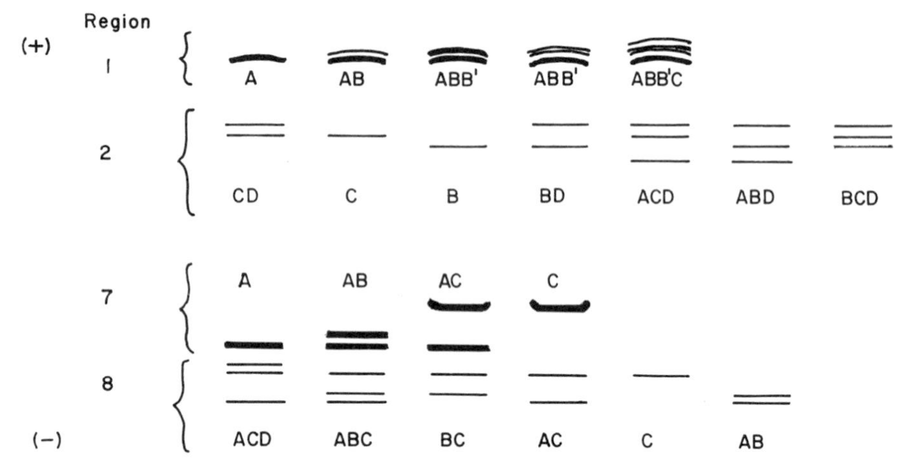

Fig. 5. Phenotypes of *L. littorea* esterase regions 1, 2, 7, and 8 observed among 300 individuals from one Massachusetts population (Nahant), using αNaAc as substrate as described in the legend of Fig. 1. Region 1 is the most anodal, migrating just behind the buffer front. Region 2 is the next most anodal region (see Fig. 1). Regions 7 and 8 are cathodal esterases.

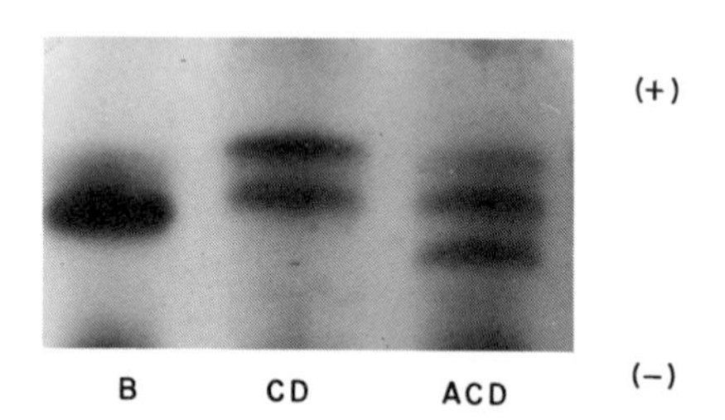

Fig. 6. Representative phenotypes of *L. littorea* region 2 esterase obtained with αNaPro as substrate as described in the legend of Fig. 1. This gel confirmed the existence of 4 separate bands (A, B, C, and D) which comprise region 2 phenotypes.

polymorphic, lend themselves readily to interpretation as dimeric systems with multiple alleles: two alleles in regions 4 and 5, four alleles in region 6 and three in region 7. Region 3, although biochemically distinct and apparently polymorphic, is not resolved sufficiently on our gels to be examined phenotypically.

Efforts are now underway in our laboratory to compare *L. littorea* esterase phenotype frequencies of Boston Harbor

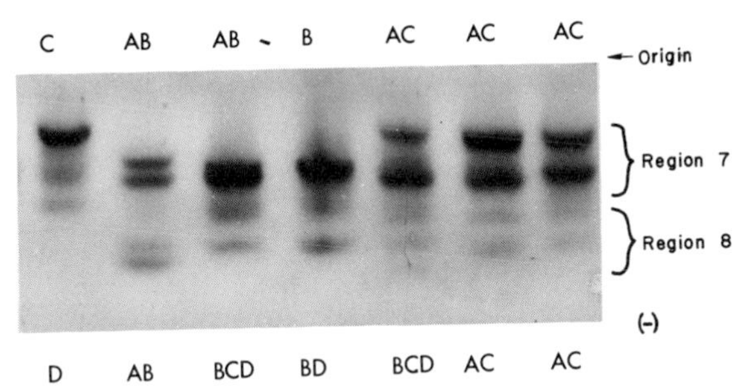

Fig. 7. Representative phenotype of *L. littorea* esterases of regions 7 and 8 with αNaAc as substrate, obtained as described in the legend of Fig. 1.

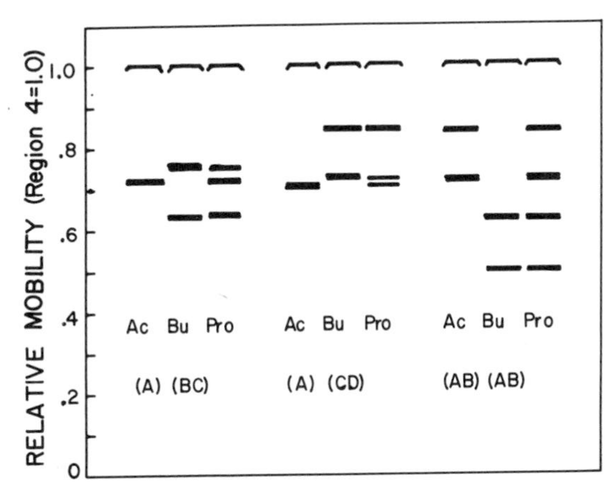

Fig. 8. Representative phenotype of *L. littorea* esterases of regions 5 and 6 with αNaAc, αNaPro and αNaBu as substrates, obtained as described in the legend of Fig. 1. "Relative mobility" is the ratio of migration from the origin (in cm) of the region 5 or 6 band divided by the migration of the region 4 cholinesterase. Region 5 and 6 phenotypes are revealed with α naphthyl acetate (Ac) and α naphthyl butyrate (Bu) respectively. Both regions react with α naphthyl propionate (Pro). Since band B of region 5 and band D of region 6 have identical mobilities, both αNaAc and αNaBu must be used to analyze these regions.

sites of varying water quality. The observed differences will presumably reflect differences in environmental selection pressure.

TABLE 2

PHENOTYPE FREQUENCIES OF *L. LITTOREA* ESTERASES IN A NAHANT, MASSACHUSETTS POPULATION

Region	Phenotype	Number of Individuals (N)	Frequency (N/t)
1	A	141	.642
	AB[b]	63	.286
	ABB'[c,d]	12	.055
	ABB'C[e]	4	
		t=220[a]	
2[f]	B	4	.0233
	C	20	.117
	C'[c]	2	.0117
	BD	8	.046
	CD[g]	98	.576
	ABD	2	.0117
	ACD	23	.135
	A'CD[c]	1	.0059
	BCD[h]	12	.0709
		t=120[a]	
7	A	80	.414
	B	6	.031
	C	2	.013
	AB	68	.351
	AC	35	.181
	BC	2	.013
		t=193[a]	
8[i]	A	6	.048
	B	6	.048
	C	47	.376
	AB	1	.008
	AC	20	.160
	BC	22	.176
	BD	1	.008
	CD	1	.008
	ABC	3	.024
	ACD	3	.024
	BCD	1	.008
	none	13	.104
	BB'CC'[c]	1	.008
		t=125[a]	

[a]A total of 300 snails were examined. For each region, t is the number of snails for which phenotype could be determined.

(Table 2 footnotes cont.)

[b]One dark band (A) plus one adjacent more anodal fine band (B).

[c]The prime (') notation was used to indicate a band which migrated very close to the unprimed band but which could be reproducibly distinguished from it.

[d]Includes two dark bands of equal intensity and one dark band (A) plus two fine bands (BB').

[e]Includes several phenotypes, all with one dark band (A), a fine or dark band (B or BB') anodal to A and a fine band (C) anodal to the B bands.

[f]Phenotypes not observed were: A, D, AB, AC, AD, BC, ABC, ACC'D.

[g]Included phenotypes with two bands of equal intensity as well as those having one strong band (either C or D) and one weak band.

[h]Included all phenotypes in which BCD bands were represented. All had intense C bands, but some had weak B or D bands.

[i]Phenotypes not observed were: D, AD, ABD.

REFERENCES

Allen, S. L. and Weremiuk 1971. Intersyngenic variations in the esterases and acid phosphatases of *Tetrahymena pyriformis*. *Biochem. Genet.* 5: 119-133.

Berger, E. 1973. Gene-enzyme variation in three sympatric species of *Littorina*. *Biol. Bull.* 145: 83-90.

Brewer, G. J. 1970. *Introduction to Isozyme Techniques*. 186 pp. Academic Press, N. Y. (pp. 122-123).

Fischer, P. H. 1948. Données sur la resistance at le vitalite des mollusques. *J. Conchyliol.* 88: 100-140.

Fraenkel, G. 1960. Lethal high temperatures for three marine invertebrates: *Limulus polyphemus*, *Littorina littorea* and *Paeurus longicarpus*., *Oikos* 11/2: 171-182.

Fretter, V. and A. Graham 1962. *British Prosobranch Molluscs*. 755 pp. Ray Society, London (pp. 387,539).

Hochachka, P. and G. Somero 1973. *Strategies of Biochemical Adaptation*. 358 pp. W. B. Saunders Co., Phil., Pa.

Ireland, M. P. 1973. Result of zinc pollution on the zinc content of littoral and sub-littoral organisms in Cardigan Bay, Wales. *Environ. Pollution* 4: 27-35.

Kaplan, N. O. 1968. Nature of multiple molecular forms of enzymes. In: *Multiple Molecular Forms of Enzymes*. Ann. N. Y. Acad. Sci., Vol. 151 (Art. 1; 689 pp.), pp. 382-389.

Koehn, R. K. and J. G. Mitton 1972. Population genetics of marine Pelecypods. I. Ecological heterogeneity and evolutionary strategy at an enzyme locus. *Amer. Natural.* 106: 47-56.

Kojima, K., J. Gillespie, and Y. N. Tobari 1970. A profile of *Drosophila* species' enzymes assayed by electrophoresis. I. Number of alleles, heterozygosities, and linkage disequilibrium in glucose-metabolizing systems and some other enzymes. *Biochem. Genet.* 4: 627-637.

Kraueter, J. 1973. Offshore currents, larval transport, establishment of southern populations of *L. littorea linne* along the U. S. Atlantic coast. (in press). (Paper delivered at Thalassia Jugoslavica conference on Marine Invertebrate Larvae, Sept. 1973, Rovinj, Yugoslavia).

Lewontin, R. C. 1973. Population genetics. *Ann. Rev. Genet.* Vol. 7, pp. 1-17. Academic Press, N. Y.

Markert, C. L. and G. S. Whitt 1968. Molecular varieties of isozymes. *Experientia* 24: 977-991.

Matteo, M. R., R. Schiff, and L. Garfield 1975. The non-specific esterases of the marine snail: *Littorina littorea.* Histochemical characterization. *Comp. Biochem. Physiol.* 50A: 141-147.

Powell, J. 1971. Genetic polymorphism in varied environments. *Sci.* 174: 1035-1036.

Prakash, S., R. C. Lewontin, and J. L. Hubby 1969. A molecular approach to the study of genic heterozygosity in natural populations. IV. Patterns of genic variation in central, marginal and isolated populations of *Drosophila pseudo-obscura.* *Genetics* 61: 841-848.

Schiff, R. 1970. The biochemical genetics of rabbit erythrocyte esterases-histochemical classifcation. *J. Histochem. Cytochem.* 18: 709-721.

Schiff, R. and C. Stormont 1970. The biochemical genetics of rabbit erythrocyte esterases - two new esterase loci. *Biochem. Genet.* 4: 11-24.

Schopf, T. and J. Gooch 1971. Gene frequencies in a marine ectoproct: a cline in natural populations related to sea temperature. *Evol.* 25: 286-289.

Somero, G.N. 1975. The roles of isozymes and allozymes in adaptation to varying temperatures. In: *Isozymes II: Physiological Function.* C. L. Markert, ed., Academic Press, N.Y. pp. 221-234.

Tattersall, W. M. 1920. Notes on the breeding habits and life history of the periwinkle. *Sci. Invest. Fish. Br.* Ire. 1: 1-11.

Tsakas, S. and C. B. Krimbas 1970. The genetics of *Dacus oleae.* IV. Relation between adult esterase genotypes and survival to organophosphate insecticides. *Evol.* 24: 807-815.

Vuilleumier, F. and M. R. Matteo 1972. Esterase polymorphisms in European and American populations of the periwinkle, *Littorina littorea* (Gastropoda). *Experentia* 28: 1241-1242.

Wells, H. W. 1965. Maryland records of the Gastropod, *Littorina littorea*, with a discussion of factors controlling its distribution. *Chesapeake Sci*. 6: 38-42.

GENETIC AND STRUCTURAL STUDIES OF CHICKEN α-AMYLASE ISOZYMES AND THEIR MODIFIED FORMS, AND STRUCTURAL STUDIES OF HOG AMYLASE

L.M. LEHRNER AND G.M. MALACINSKI
Department of Zoology, Indiana University
Bloomington, Indiana 47401

ABSTRACT. In the chicken population at large three electrophoretically distinct pancreatic α-amylase (EC 3.2.1.1) isozymes were described and designated Pa 1, Pa 2, and Pa 3. The local population of chickens, however, possessed only isozymes Pa 2 and Pa 3, present as three phenotypes: Amy_2 B, consisting of isozyme Pa 2; Amy_2 BC, consisting of isozymes Pa 2 + Pa 3; and Amy_2 C, consisting of isozyme Pa 3.

Analysis of pancreatic biopsies permitted the establishment of breeding flocks with defined amylase phenotypes. Matings of these flocks established that these amylases are inherited as codominant alleles at a single genetic locus. Further, there was no evidence of ontogenetic modification of the amylase isozymes.

It was observed both in vitro and in vivo that amylase isozymes Pa 2 and Pa 3 each generated a family of at least three faster migrating amylolytic proteins. These post-translationally modified amylases were designated Pa Xa, Pa Xb, and Pa Xc where X represents the number of the progenitor amylase. The in vitro and in vivo process(es) of amylase modification appear to differ markedly.

Structural analyses of purified amylases demonstrated that all amylase isozymes are nonglycosidated, monomeric molecules of molecular weight 55,000. The data are consistent with the hypothesis that the modified amylases are produced by deamidation of asparagine and/or glutamine residues.

In addition, structural analyses of hog pancreatic amylases demonstrated that they also are monomeric molecules.

INTRODUCTION

Knowledge of the relationship between the structure, physiological function, and regulation of the enzyme alpha-amylase (alpha-1,4-glucan-4-glucanohydrolyase, EC 3.2.1.1) should have an important impact in several areas of contemporary biology. This enzyme plays a central role in digestive metabolism of virtually all vertebrates by hydrolyzing the C_1-O glycoside bond of starch. It is also produced in a very limited number of highly differentiated cells within an organism. As a secretory product of the acinar cells of the pancreas and salivary glands, its synthesis and regulation probably involve a number of highly specialized mechanisms.

In order to establish the structural, functional, and regulatory properties of amylase, an analysis of several aspects of amylase biochemistry, including several features of the genetic, protein chemistry, and native properties of the enzyme was undertaken.

A survey of several potential experimental systems was carried out before embarking on these studies. A system was sought that possessed the following attributes: 1) easily maintained laboratory animal, 2) quickly recognizable and unambiguous amylase phenotypes, 3) high concentration of amylase so that the enzyme could be purified for structural analysis, 4) potential for genetic and embryological analyses, and 5) potential for in vivo confirmation and evaluation of the biological significance of in vitro observations.

For various reasons several laboratory mammals and insects were eliminated from consideration. Some mammals, including the rabbit, which we had previously employed, have low concentrations of pancreatic amylase (Malacinski and Rutter, 1969). Others, including the rat, presented ambiguous phenotypes. The hog is not easily used for genetic analysis, and the same is true for man.

The chicken, on the other hand, appeared to fulfill most of the above requirements. The only apparent deficiency was the absence of amylase synthesis in its salivary glands, which was not a major limitation for these studies.

MATERIALS AND METHODS

Assays. Amylase was assayed by the method of Bernfeld (1955). Protein was determined by the method of Lowry et al. (1951).
Purification. Amylase was purified by the Loyter and Schramm (1962) glycogen precipitation technique as modified by Malacinski and Rutter (1969). Final purification was achieved by preparative disc gel electrophoresis. A 5% polyacrylamide gel in a Canalco (Rockville, Md.) "Prep-Disc-PD2/70" unit was employed. All procedures were as described in the Canalco Operation Manual.
Analytical Disc Gel Electrophoresis. Electrophoresis was performed at pH 8.3 in Tris-glycine buffer in the 7.5% polyacrylamide gel system devised by Davis (1964).
SDS/Urea Disc Gel Electrophoresis. The urea gel system of Raff et al. (1971) was made 0.2% in sodium dodecylsulfate (SDS). Samples were dissolved in Renaud's buffer (Renaud et al., 1968). Electrophoresis was performed at pH 8.3 in Tris-glycine buffer. Gels were stained and destained by the method of Weber and

Osborn (1969).

Amino Acid Composition. Samples were hydrolyzed in redistilled 6N hydrochloric acid in evacuated, sealed tubes maintained at 110°C. Amino acid analysis was performed according to the procedure of Spackman et al. (1958). Tryptophan content was determined by the method of Liu (1972).

Peptide Maps. The procedures for reduction, carboxymethylation, and proteolysis by trypsin have been previously described (Malacinski and Rutter, 1969). Chromatography and electrophoresis were performed by the method of Low (1970). The peptides were located by the staining method of Dreyer and Bynum (1967).

RESULTS AND DISCUSSION

The three amylase isozymes are distributed into four different phenotypes, as seen in commercially available chickens (Fig. 1). The isozymes, following the nomenclature developed

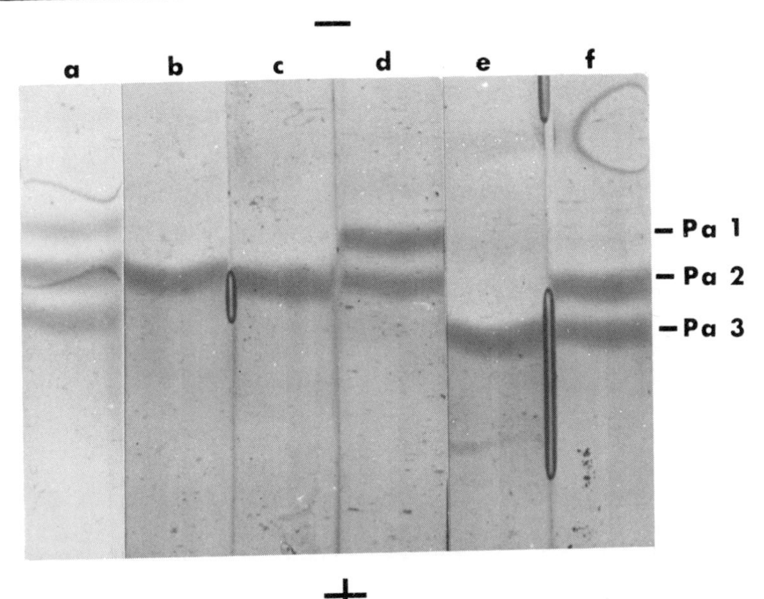

Fig. 1. Analytical polyacrylamide disc gel electrophoresis of micro-glycogen purified (Schramm and Loyter, 1966) samples of amylase. Proteins were stained with Coomassie blue. a) BL (Pa 2) + BS (Pa 2) + OEDG (Pa 1 + Pa 2) + RIR (Pa 3) + WLH (Pa 2 + Pa 3), b) BL (Pa 2), c) BS (Pa 2), d) OEDG (Pa 1 + Pa 2), e) RIR (Pa 3), f) WLH (Pa 2 + Pa 3). Chicken breed abbreviations: BL=Black Langsham, BS=Black Silkie, OEDG=Old English Duck Game, RIR=Rhode Island Red, WLH=White Leghorn.

for human amylase isozymes (Merritt et al., 1973) are designated Pa 1, Pa 2, and Pa 3 in order of increasing electrophoretic mobility toward the anode. The phenotypes are designated: Amy_2 AB consisting of isozymes Pa 1 + Pa 2, Amy_2 B consisting of isozyme Pa 2, Amy_2 BC consisting of isozymes Pa 2 + Pa 3, and Amy_2 C consisting of isozyme Pa 3. Chickens available from local suppliers displayed only isozymes Pa 2 and Pa 3 distributed in three phenotypes: Amy_2 B, Amy_2 BC, and Amy_2 C.

From chickens seven days post-hatching onward it was possible to obtain a pancreatic sample by biopsy, following which the animals could be raised for a normal life span. Using this technique it was determined that an animal's phenotype remained constant for its entire life span.

In addition, this knowledge of the animal's phenotype made possible breeding experiments and analysis of the inheritance of amylase. The genetic data presented in Table I are consistent with the interpretation that co-dominant alleles at a single genetic locus represent the basis of chicken amylase inheritance.

TABLE I

SEGREGATION ANALYSIS OF PANCREATIC AMYLASE PHENOTYPE

Mating Amy_2 Phenotypes		No. Matings	Total Progeny	Progeny Amy_2 Phenotypes				P*
Male	Female			B	BC	C	x^2	
BC	BC	4	37	10	17	10	.2432	.9>P>.5
B	BC	2	56	27	29	0	.0714	.9>P>.5
B	B	2	19	19	0	0	-	-
C	B	4	39	0	39	0	-	-
C	C	3	30	0	0	30	-	-
BC	C	1	5	0	1	4	-	.156**

* P calculated by chi square method assuming condominant alleles.

** Calculated by binomial expansion

Besides permitting matings for genetic analysis, knowledge of the animal's phenotype proved to be indispensable for the purification of the isozymes. Matings were arranged to produce offspring with a single amylase isozyme, and it was thus possible to purify each isozyme individually.

Following purification, the structural properties of the isozymes were studied. Carbohydrate analysis was performed by the method of Rheinhold (1972) as modified by W. Hirs (personal communication). No bound carbohydrate was detected in either isozyme Pa 2 or Pa 3.

In dissociating SDS gels both isozymes behaved identically and a molecular weight of approximately 55,000 was calculated for each.

The amino acid compositions revealed a marked similarity. Approximately 50 lysine plus arginine residues are present in each isozyme. Peptide maps were prepared from tryptic digests of reduced and carboxymethylated isozymes, which were completely soluble. Peptide maps produced from individual samples of isozymes Pa 2 or Pa 3 revealed approximately 50 peptides. However, a map produced from a mixture of isozymes Pa 2 + Pa 3 revealed approximately 57 peptides (Fig. 2). This observation suggests that the primary sequences of Pa 2 and Pa 3 differ by several amino acids.

In addition to the amylase isozymes extractable from the chicken pancreas, a protein which migrated immediately ahead of the amylase isozyme was observed in each preparation. It is readily apparent (Fig. 3) that the leading edge of the amylase peak obtained from preparative electrophoresis is composed of two proteins, while the trailing edge of the peak is a single homogeneous protein. Direct amylase activity staining (Ward, et al., 1971)of the analytical gels revealed that both proteins in the leading edge of the peak possessed amylolytic activity.

Other laboratories studying human salivary amylases have observed that upon incubation at room temperature, the more anodic amylase forms increase in intensity while the cathodic amylase forms decrease in intensity (Keller et al., 1971; Karn et al., 1973). Chicken amylase purified by preparative electrophoresis was, therefore, incubated in 0.3M Tris buffer, pH 9, at 37°C for 10 days. Not only was there a gradual increase in the intensity of the faster migrating amylase, but in addition two still faster migrating protein species were generated.

Each of these faster migrating proteins retained amylolytic activity (Fig. 4). As will be detailed later, the generation of these faster migrating amylases is a non-proteolytic, spontaneous post-translational modification of the gene product. These faster migrating amylolytic proteins do not meet the suggestions of the IUPAC-IUB Commission on Biochemical Nomenclature (1971) for designation as "isozymes". They are referred to therefore, in this text as "modified amylases".

The modified amylases are designated Pa Xa, Pa Xb, and Pa Xc, in order of increasing electrophoretic mobility toward the anode. X represents the number of the amylase isozyme which generated the modified amylases.

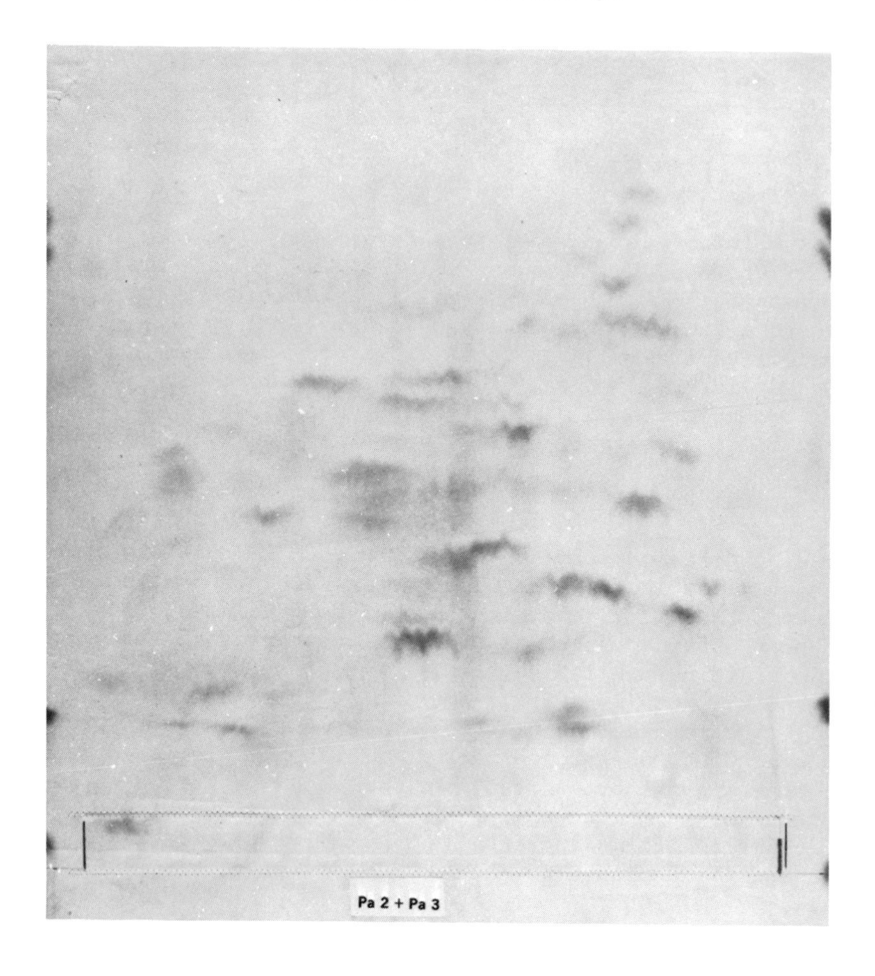

Fig. 2. Peptide map obtained from chromatography right to left, (butanol-acetic acid-water) and electrophoresis, bottom to top, (pyridine-acetic acid-water pH 3.7) of trypsin digestion products of amide-carboxymethylated derivatives of amylase isozymes Pa 2 + Pa 3.

Since modified amylase Pa 2a has an electrophoretic mobility almost identical to that of isozyme Pa 3, if a mixture of amylase isozymes had been employed as starting material for purification it would have been virtually impossible to purify isozyme Pa 3 from modified amylase Pa 2a. However, as mentioned previously by arranging proper matings of the genetically defined chickens it was possible to obtain material for purification which contained only a single isozyme.

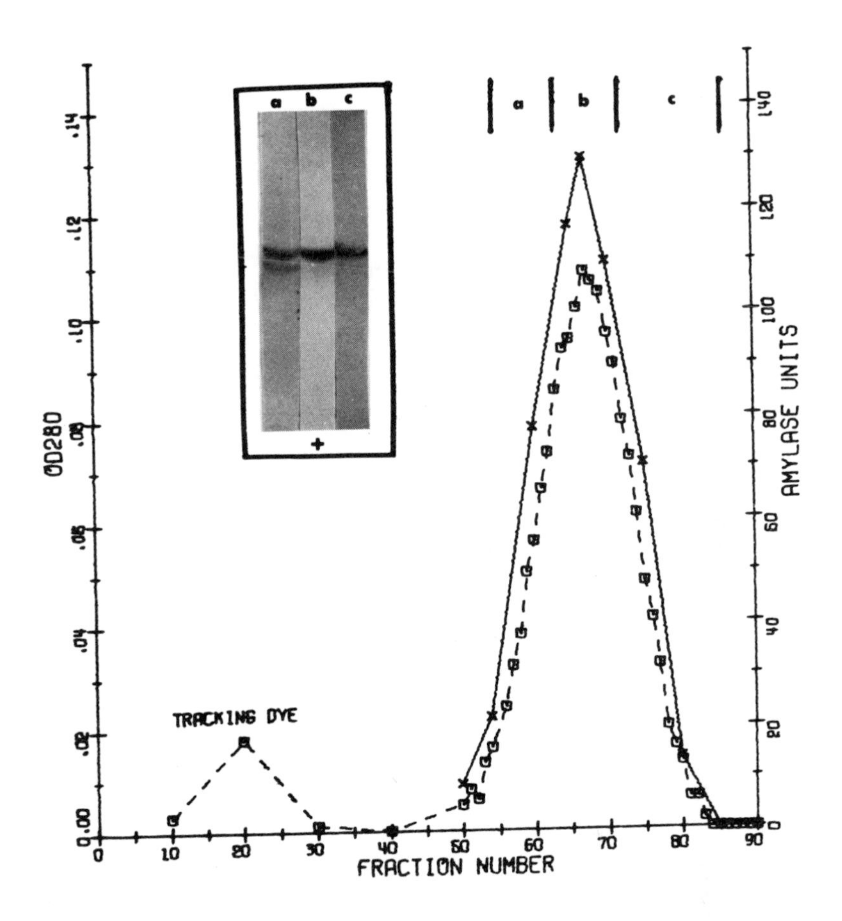

Fig. 3. Elution profile of a representative preparative electrophoresis run of Pa 3 type amylase. Vertical lines represent the fractions that were pooled. OD280 - □ - □ - □ ; Amylase units -X-X-X- (a) Group I (b) Group II (c) Group III.

A comparison of the structural properties of the modified amylases and amylase isozymes was undertaken. Carbohydrate analyses revealed that all forms of chicken pancreatic amylase lack bound carbohydrate. Some amylases from other sources, however, have been reported to contain bound carbohydrate (Stiefel and Keller, 1973; Beaupoil-Abudie et al., 1973; Keller et al., 1971).

SDS disc gel electrophoresis was employed to determine if the modified amylases possessed a molecular weight different from the parent isozymes. Samples Pa 2, Pa 3, Pa 2 family, Pa 3 family and reduced and carboxymethylated samples of Pa 2 and Pa 3 families were subjected to SDS gel electrophoresis.

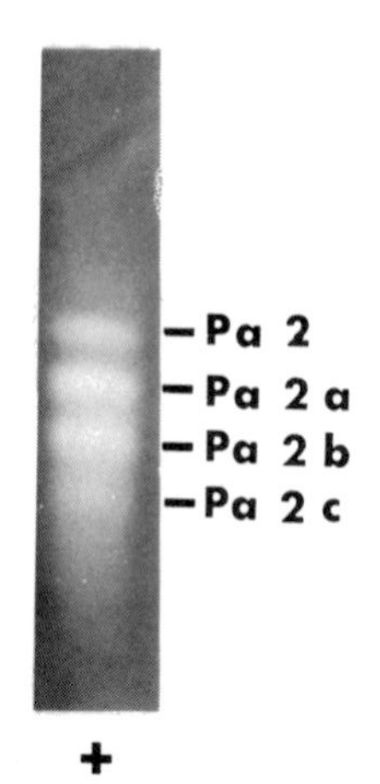

Fig. 4. Disc gel electrophoresis of sample Pa 2 after 10 days incubation in 0.3M Tris (pH 9) at 37°C. Enzyme activity stained by KI - I_2 method.

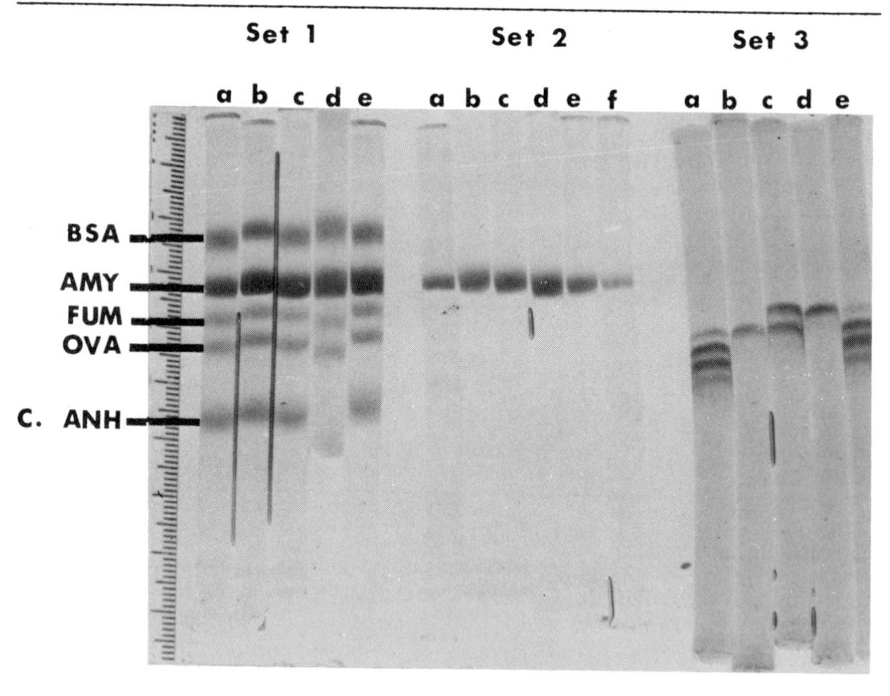

Fig. 5. Disc gel electrophoresis of purified amylase.
Set 1 - SDS/Urea gels with molecular weight markers (Bovine serum albumin, Fumerase, Ovalbumin, and Carbonic anhydrase)
a) Pa 2 + Pa 2 family, b) Pa 2 + Pa 3 + Pa 2 family + Pa 3 family,

(Fig. 5 legend, continued)
c) Pa 3 + Pa 3 family, d) Pa 2 family + RCM Pa 2 family + Pa 3 family + RCM Pa 3 family, e) Pa 2 + Pa 3.
Set 2 - SDS/Urea gels a) RCM Pa 3 family, b) Pa 3 family, c) Pa 3, d) Pa 2, e) Pa 2 family, f) RCM Pa 2 family.
Set 3 - pH 8.3 non-SDS, non-urea gels a) Pa 3 family, b) Pa 3 c) Pa 3 + Pa 2, d) Pa 2, e) Pa 2 family.

The data in Fig. 5 indicate that all amylase species, including reduced and carboxymethylated amylases, migrate as a single homogeneous band. By co-electrophoresis of various amylase samples and molecular weight standards a molecular weight of 55,000 ± 600 was calculated for each amylase species.

In addition amylase isozymes and their families were subjected to digestion with trypsin, at a ratio of 50:1 of trypsin to amylase. While a loss of activity was observed, analyses of aliquots on non-dissociating analytical gels before and after trypsin exposure revealed that trypsin was not able to convert amylase isozymes to modified forms nor convert one modified amylase to another modified amylase. From the results of these two experimental procedures it is concluded that modified amylases can not be produced by proteolysis.

The amino acid compositions of the families were determined. A marked similarity to the pure isozymes was observed. As was the case in the analyses of the pure isozymes, approximately 50 lysine plus arginine residues were detected in each amylase family. In the process of generating the amylase families for peptide mapping, the progenitor amylase isozyme was reduced to an insignificant concentration. Therefore, in order to have all family members represented in a family map, a mixture of the progenitor isozyme and its modified amylases was mapped together (eg. isozyme Pa 2 + Pa 2 family and isozyme Pa 3 + Pa 3 family). Each of these maps revealed approximately 55 peptides.

It has been suggested by Karn et al. (1973) that modified amylases are produced by successive deamidation of glutamine and/or asparagine residue(s) to glutamic and/or aspartic acid residue(s). Thus the increase of approximately 3-5 peptides in the family map versus the single isozyme map agrees with the prediction that deamidation is the basis of the amylase modification.

From the strucutral analyses of the isozymes and modified amylases it can be stated that the number of tryptic peptides observed on maps equals the number of lysine plus arginine residues present in the chicken pancreatic amylase molecule. Therefore, each isozyme is either a monomeric molecule or consists of non-duplicate, i.e. unique peptides. However, there is no evidence that any amylase sample contains material below

55,000 molecular weight on dissociating SDS gels. This fact strongly implies that chicken pancreatic amylase is a monomeric molecule.

In order to determine if modified amylases might possess a physiological function the lumen contents of chicken intestines were examined. In vitro it required approximately 10 days to generate a significant amount of modified amylases. Therefore, since the reported maximum passage time of material through the chicken intestinal tract is only 12 hours (Hill, 1971) the in vitro experiments implied that little, if any, modified amylase would be present in the intestinal lumen.

The intestinal tract of a freshly sacrificed animal was divided into eight regions: descending duodenum, ascending duodenum, jejunum, proximal ileum, distal ileum, cecae, large intestine, and cloaca. The lumen contents from each region were extruded, homogenized, and analyzed for the presence of amylase. In addition, a sample of pancreatic homogenate was analyzed. It is obvious (Fig. 6A and B) in both activity and protein stained gels that modified amylase Pa 2b reaches a significant concentration in the jejunum and maintains its relative concentration throughout the remaining length of the intestine. However, Pa 2a never is present in more than a trace amount. In contrast, in vitro Pa 2b only reached significant levels after Pa 2a had been present at a high concentration.

It was further observed (Fig. 6C) that a distribution pattern of family members, identical to that observed in an Amy_2 B animal, is also present in the intestinal tract of Amy_2 BC and Amy_2 C animals.

Thus, it is evident that the slow process of modified amylase formation in vitro is not an accurate indication of the extremely rapid process of modified amylase formation in vivo. If this nearly complete non-correlation of in vitro results with the actual in vivo situation represents a general phenomenon of amylase form and function, then extreme caution must be employed in attempting to assess the in vivo significance of data obtained from in vitro amylase experimentation.

One of the major questions remaining to be resolved in the field of amylase research relates to the actual structure of the molecule. The genetic data and peptide map data obtained from chicken pancreatic amylase easily fit a model in which amylase is a monomeric molecule. Several other laboratories also report the monomeric structure of other vertebrate amylases (Malacinski and Rutter, 1969; Keller et al., 1971; Robinovitch and Sreebny, 1970; and Cozzone et al., 1970a).

However, there is a report that hog pancreatic amylase purchased from a commercial supplier (Worthing Biochemical Corp.) is a dimer composed of two identical subunits (Robyt, et al., 1971).

It has been reported by Cozzone et al. (1970b) that hog pancreatic amylase contains eight methionine residues. Therefore, cleavage by cyanogen bromide would be expected to yield nine fragments if hog amylase is monomeric and five fragments if hog amylase consists of identical subunits. Cozzone's laboratory (Cozzone et al., 1971) reported the isolation of nine CNBr fragments from hog pancreatic amylase supporting the concept that amylase is a monomeric molecule.

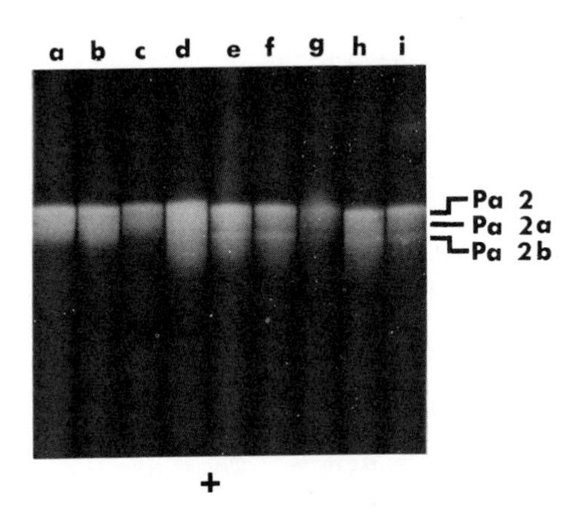

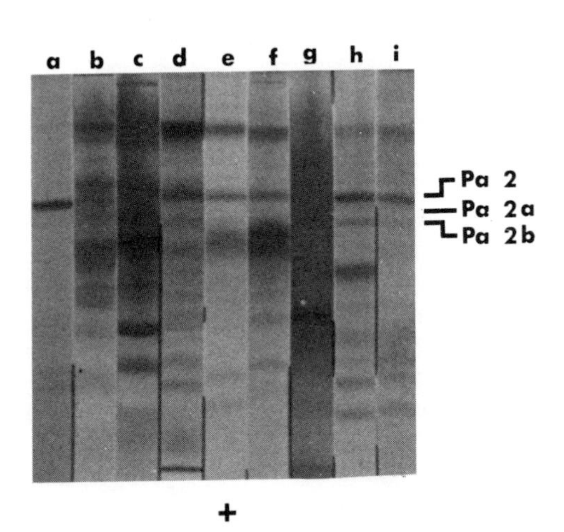

Fig. 6A and 6B. Disc gel electrophoresis of crude lumenal homogenates. Amylolytic activity was resolved by the KI - I_2 method. (A) Proteins were stained with Coomassie blue (B). a) Pancreas, b) Descending duodenum, c) Ascending duodenum, d) Jejunum, e) Proximal ileum, f) Distal ileum, g) Cecae, h) Large intestine, i) Cloaca.

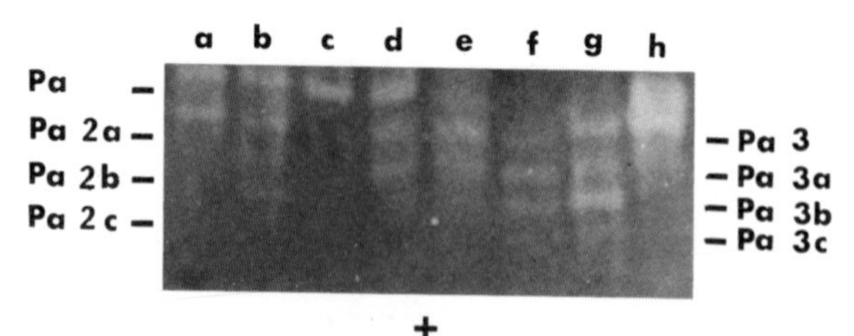

Fig. 6C. Slab gel electrophoresis of crude lumenal homogenates, crude pancreatic homogenates and families of modified amylases generated in vitro. Amylolytic activity was revealed by the KI - I_2 method. (a) Pancreas Amy_2 BC animal, (b) Proximal ileum Amy_2 BC animal, (c) Pancreas Amy_2 BC animal, (d) Proximal ileum Amy_2 B animal, (e) Pa 2 family, (f) Pa 3 family, (g) Proximal ileum Amy_2 C animal, (h) Pancreas Amy_2 C animal.

However, a critical piece of experimental data, a comparison of the tryptic peptide maps, for both the monomeric and purported subunit forms of amylase has never been reported by Robyt (Robyt et al., 1971). Since the purported subunits had been observed in a commercially available enzyme, identical hog pancreatic amylase purchased from Worthington Biochemical Corp. was employed in this study. This sample contained not only two forms of hog pancreatic amylase as had previously been reported (Marchis-Mouren and Pasero, 1967; Rowe et al., 1968) but four protein species. Direct activity stain revealed that each of the four proteins possesses amylolytic activity.

In preparation for peptide mapping the hog pancreatic amylase mixture was reduced and carboxymethylated. Contrary to the results obtained with chicken pancreatic amylase, upon dialysis to remove urea, 80-90% of the hog sample precipitated.

The precipitate was collected by centrifugation at 10,000 grams and aliquots of the supernatant and pellet subjected to SDS gel electrophoresis. The supernatant was composed of a somewhat heterogenous mixture of low molecular weight peptides but contained a distinct peptide of approximate molecular weight 20,000. The pellet contained three distinct components: a major component of approximately 53,000 and a minor component of approximately 39,000 molecular weight. In addition the pellet also contained the identical low molecular weight components present in the supernatant.

Based upon the model of hog pancreatic amylase proposed by Cozzone and Marchis-Mouren (1972) it can be seen that if a proteolytic nick had been present in the molecule at arrow 1 Fig. 7 then reduction of the inter-peptidic disulfide bridge would permit the molecule to separate into two components: a fragment of approximately 20,000 molecular weight with

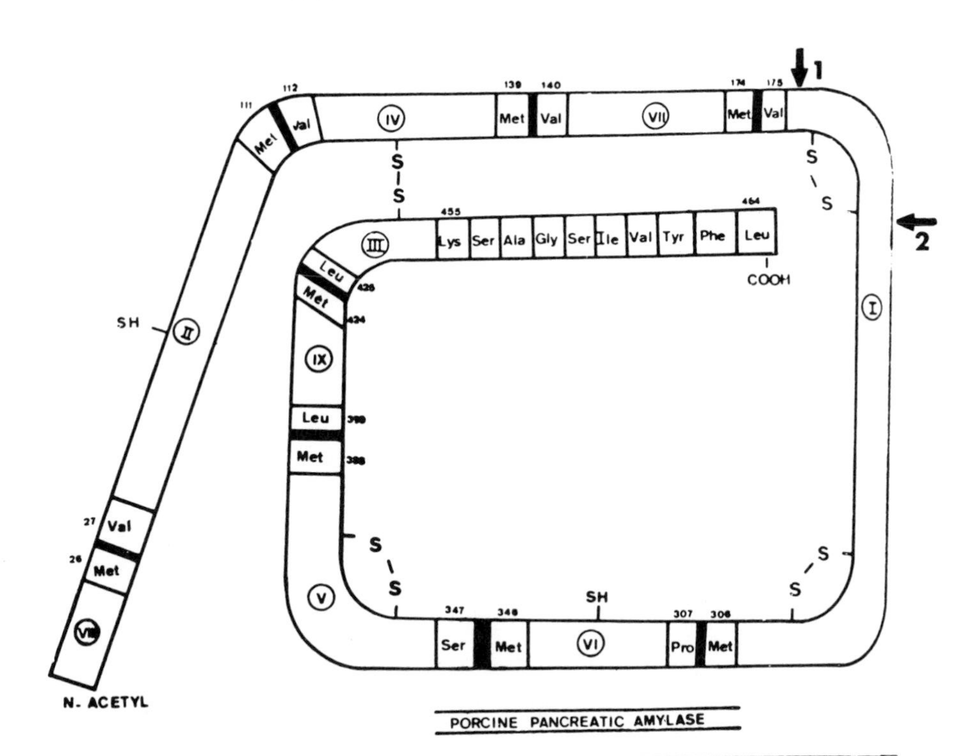

CNBr Fragment	Molecular Weight	Lysine	Arginine
I	15,103	5	8
II	9,443	3	5
III	4,433	2	2
IV	3,201	1	1
V	5,682	3	3
VI	4,441	2	2
VII	3,881	2	1
VIII	3,189	1	2
IX	3,297	1	1

Model taken from Cozzone and Marchis-Mouren (1972)
CNBr fragment data taken from Cozzone et. al. (1971)

Fig. 7. Model of hog pancreatic amylase.

approximately 16 lysine plus arginine residues and a fragment of approximately 32,000 molecular weight.

As can readily be seen in Fig. 8a the tryptic peptide map of the supernatant, which contains a peptide of approximate

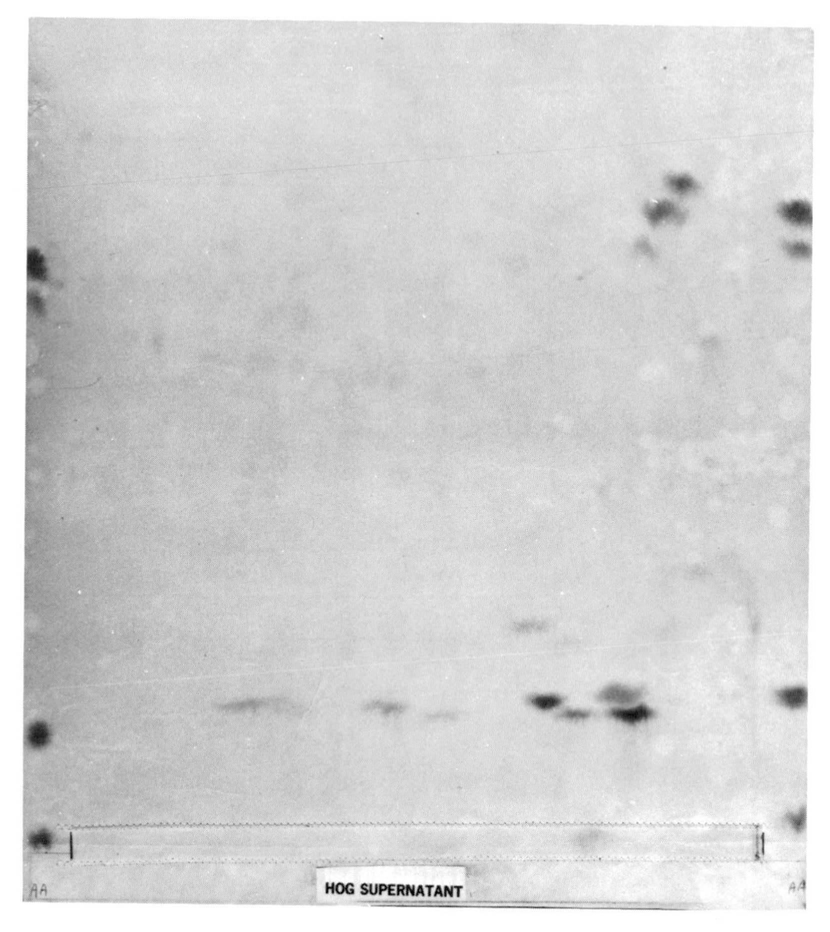

Fig. 8a. Peptide maps obtained from chromatography, right to left, (butonol-acetic acid-water) and electrophoresis, bottom to top, (pyridine-acetic acid-water pH 3.7) of trypsin digestion products of amide-carboxymethylated derivatives of hog pancreatic amylases. (a) Supernatant.

molecular weight 20,000 revealed approximately 19 tryptic peptides which is in agreement with the model in Fig. 7. However, the tryptic peptide map of the pellet (Fig. 8b) revealed approximately 42 peptides which corresponds closely to the number of lysine plus arginine residues present in hog pancreatic amylase (Cozzone,et al., 1971). It is concluded therefore, that all vertebrate amylases are monomeric molecules and limited proteolysis of the commercially available hog pancreatic molecule is the basis for the presence of the reported subunits.

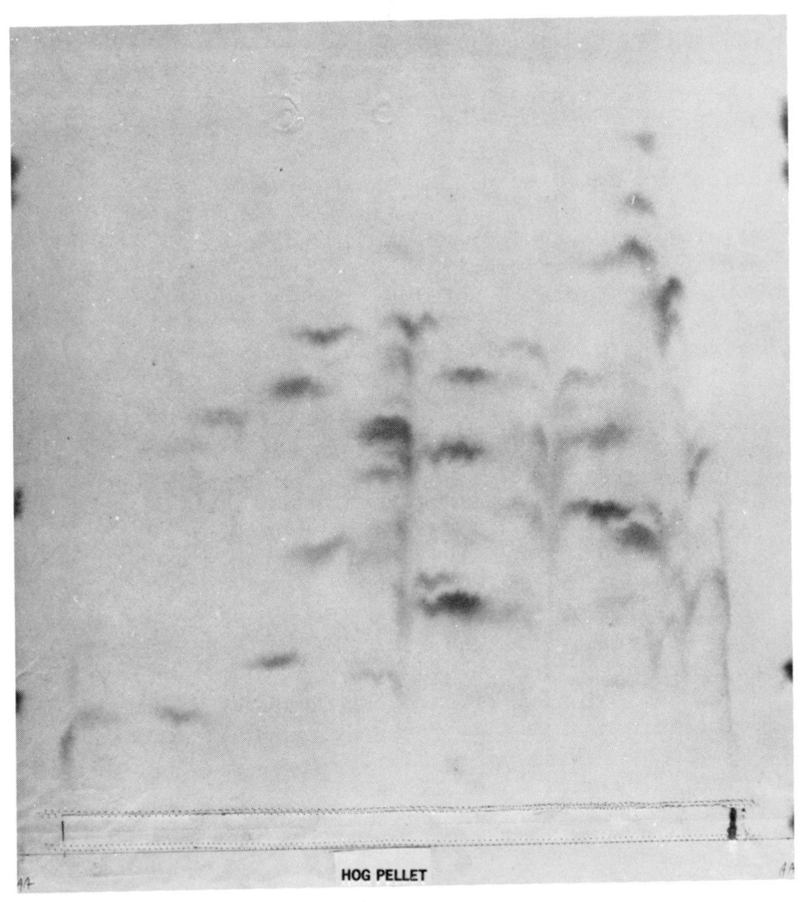

Fig. 8b. Peptide maps obtained from chromatography (see legend for Fig. 8a). 8b is hog pellet.

ACKNOWLEDGEMENTS

This investigation was supported in part by NSF grant no. GB 36973.

REFERENCES

Beaupoil-Abadie, B., M. Raffalli, P. Cozzone, and G. Marchis-Mouren 1973. Determination of the carbohydrate content of porcine pancreatic amylase. *Biochim. Biophys. Acta* 297: 436-440.

Bernfeld, P. 1955. Amylases, α and β. *Methods Enzymol.* 1: 149-158.

Cozzone, P., L. Pasero, B. Beaupoil, and G. Marchis-Mouren 1970a. Characterization of porcine pancreatic isoamylases chemical and physical studies. *Biochim. Biophys. Acta* 207:490-504.

Cozzone, P., L. Pasero, B. Beaupoil, and G. Marchis-Mouren 1970b. Characterization of porcine pancreatic isoamylases: separation and amino acid composition. *Biochim. Biophys. Acta* 200:590-593.

Cozzone, P., L. Pasero, B. Beaupoil, and G. Marchis-Mouren 1971. Isolation and characterization of the cyanogen bromide peptides of two forms of porcine pancreatic amylase. *Biochemie* 53:957-968.

Cozzone, P. and G. Marchis-Mouren 1972. Use of pulse labeling technique in protein structure determination: ordering of the cyanogen bromide peptides from porcine pancreatic α-amylase. *Biochim. Biophys. Acta* 257:222-229.

Davis, B.J. 1964. Disc electrophoresis II. Method and application to human serum proteins. *Ann. N.Y. Acad. Sci.* 121: 404-427.

Dreyer, W.J. and E. Bynum 1967. High-voltage paper electrophoresis. *Methods Enzymol.* 11:32-39.

Hill, K.J. 1971. In *Physiology and Biochemistry of the Domestic Fowl*. ed. Bell, D.J. and B.M. Freeman. Academic Press, vol. 1, chapter 2, pp. 25-47.

IUPAC-IUB Commission on Biochemical Nomenclature 1971. The nomenclature of multiple forms of enzymes recommendations. *Arch. Biochem. Biophys.* 147:1-3.

Karn, R.C., J.D. Shulkin, A.D. Merritt, and R.C. Newell 1973. Evidence for post-transcriptional modification of human salivary amylase (Amy_2) Isozymes. *Biochemical Genetics* 10:341-350.

Keller, P.J., D.L. Kauffman, B.J. Allan, and B.L. Williams 1971. Further studies on the structural differences between the isoenzymes of human parotid α-amylase. *Biochemistry* 10: 4867-4874.

Liu, T.H. 1972. Determination of tryptophan. *Methods Enzymol.* 25:44-55.

Low, T.L. and C.K. Low 1970. The amino acid sequences of porcine and bovine serum albumins. *Ph.D. Dissertation* The University of Texas at Austin. pp. 26-35.

Lowry, O.H., N.J. Rosenbrough, A.L. Farr, and R.J. Randall 1951. Protein measurement with the folin phenol reagent. *J. Biol. Chem.* 193:265-275.

Loyter, A. and M. Schramm 1962. The glycogen-amylase complex as a means of obtaining highly purified α-amylases. *Biochim. Biophys. Acta* 65:200-206.

Malacinski, G.M. and W.J. Rutter 1969. Multiple molecular forms of α-amylase from the rabbit. *Biochemistry* 8:4382-4390.

Marchis-Mouren, G. and L. Pasero 1967. Isolation of two amylases in porcine pancreas. *Biochim. Biophys. Acta* 140: 366-368.

Merritt, A.D., M.L. Rivas, D. Bixler, and R. Newell 1973. Salivary (Amy_1) and pancreatic (Amy_2) amylase: Electrophoretic characterizations and genetic studies. *Am. J. Hum. Genet.* 25: 510-522.

Raff, R.A. , G. Greenhouse, K.W. Gross, and P.R. Gross 1971. Synthesis and storage of micro tubule proteins by sea urchin embryos. *J. Cell Biol.* 50:516-527.

Reinhold, V.N. 1972. Gas-liquid chromatographic analysis of constituent carbohydrate in glycoproteins. *Methods Enzymol.* 25:244-249.

Renaud, F.L., A.J. Rowe, and I.R. Gibbons 1968. Some properties of the protein forming the outer fibers of cilia. *J. Cell Biol.* 36:79-90.

Robinovitch, M.R. and L.M. Sreebny 1972. On the nature of the molecular heterogeneity of rat parotid amylase. *Arch. Oral Biol.* 17:595-600.

Robyt, J.F., C.G. Chittenden, and C.T. Lee 1971. Structure and function of amylases I. The subunit structure of porcine pancreatic α-amylase. *Arch. Biochem. Biophys.* 144:160-167.

Rowe, J.J.M., J. Wakim, and J.A. Thoma 1968. Multiple forms of porcine pancreatic α-amylase. *Anal. Biochem.* 25:206-220.

Schramm, M. and A. Loyter 1966. Purification of α-amylases by precipitation of amylase-glycogen complexes. *Methods Enzymol.* 8:533-537.

Spackman, D.H., W.H. Stein, and S. Moore 1958. Automatic recording apparatus for use in the chromatography of amino acids. *Anal. Chem.* 30:1190-1206.

Stiefel, D.J. and P.J. Keller 1973. Preparation and some properties of human pancreatic amylase including a comparision with human parotid amylase. *Biochim. Biophys. Acta* 320:345-361.

Weber, K. and M. Osborn 1969. The reliability of molecular weight determinations by dodecyl sulfate-polyacrylamide gel electrophoresis. *J. Biol. Chem.* 244:4406-4412.

Ward, J.C., A.D. Merritt, and D. Bixler 1971. Human salivary amylase: genetics of electrophoretic variants. *Amer. J. Hum. Genet.* 23:403-409.

GENETIC AND POST-TRANSLATIONAL MECHANISMS DETERMINING HUMAN AMYLASE ISOZYME HETEROGENEITY

R. C. KARN, BARNETT B. ROSENBLUM,
JEWELL C. WARD and A. DONALD MERRITT
Department of Medical Genetics
Indiana University Medical Center
Indianapolis, Indiana 46202

ABSTRACT. Human α-amylase expressions in tissue and secretion sources yield complex phenotypes when subjected to discontinuous polyacrylamide sheet electrophoresis. Two loci, Amy_1 active in the salivary glands and Amy_2 active in the pancreas, produce different electrophoretic phenotypes. Genetic studies show Mendelian segregation of the multiple alleles at each locus; however, studies involving double heterozygotes show that the two loci are closely linked with the variant alleles being either in coupling or repulsion. Population data showing the distributions of the Amy_1 and Amy_2 phenotypes are also presented. In addition to genetic studies, biochemical separation of amylase isozymes and a modified electrophoretic system have allowed further resolution of amylase phenotypes. Data are presented demonstrating that the complex isozyme patterns are the result of post-translational modifications: glycosidation and deamidation, both occurring in the expression of salivary phenotypes and deamidation in pancreatic phenotypes. Salivary and pancreatic amylases were also compared in immunologic analyses employing double diffusion and found to be immunologically indistinguishable. We have combined this observation with the conclusion that the two loci are closely linked and advance the hypothesis that the loci are the result of a duplication.

INTRODUCTION

Studies of electrophoretic enzyme polymorphisms began with the identification of the first multiple molecular enzyme system (Hunter and Markert, 1957). The isozyme pattern, which may also be referred to as the phenotype, observed in electrophoretic systems can be complex since isozyme heterogeneity may be created at several levels. Human α-amylase (α-1,4-glucan 4-glucanohydrolase; E.C. 3.2.1.1) is proving to be an excellent model with which to study protein polymorphisms in man since variation in phenotype can be shown to originate via the mechanisms shown in Table 1. Two loci, Amy_1 active in the salivary glands and Amy_2 in the pancreas, produce completely different phenotypes in the tissues and secretion

TABLE 1

AMYLASE ISOZYME HETEROGENEITY
(Phenotypic Variation)

A. Locus

1. Amy_1
2. Amy_2

B. Gene

1. Amy_1^A - $(B, C \ldots r)$
2. Amy_2^A - (B, C)

C. Post-translational Modification

1. Glycosidation
2. Deglycosidation
3. Deamidation

sources studied (Kamaryt and Laxova, 1966; Ward et al., 1971; Merritt et al., 1973a). Further variation arises from the existence of multiple alleles at each locus. Then, modifications of the gene product following translation result in multiple isozymes from the single protein coded for by each allele (Karn et al., 1973a; Karn et al., 1973b). It is our purpose to discuss the contributions of these mechanisms to human amylase heterogeneity.

MATERIALS AND METHODS

Sample Collection and Treatment. Saliva, urine, serum, and pancreas specimens were collected as described previously (Ward et al., 1971; Merritt et al., 1973a). Saliva and serum samples were either tested immediately or frozen for storage. Urine samples were refrigerated and tested within a week because frozen samples usually proved unsatisfactory for electrophoresis. Human pancreases obtained at autopsy were either used fresh or stored frozen. Amylase activity was determined by the method of Bernfeld (1955).

Enzyme Purification. High molecular weight glycogen used in amylase purification was prepared using Sigma type II shellfish glycogen (Malacinski and Rutter, 1969). Biogels P10 and P100 (Bio-Rad Laboratories) were used for gel chromatography of purified salivary amylase. Salivary and pancreatic amylase were purified by the glycogen method of Loyter and Schramm (1962), and the resulting suspension was passed through a Biogel P10 column to remove some of the limit dextrins. The heavier and lighter molecular weight families of isozymes of salivary amylase were then separated by the method of Keller et al. (1971). The carbohydrate contents of the isozyme families were estimated using the method of Dubois et al. (1956). Additional details of the purification procedure are described elsewhere (Karn et al., 1974).

Electrophoresis. The vertical sheet polyacrylamide electrophoretic system employing a discontinuous buffer developed in this laboratory (Merritt et al., 1973a) was used in routine screening of saliva, serum, urine, and pancreatic samples. Modifications of this system are reported elsewhere (Rosenblum et al., submitted for publication).

Antibody Production, Purification and Analysis. Antibodies were prepared using adult, white male, New Zealand rabbits and the immunoglobulin G fraction subsequently purified from the rabbit serum by ammonium sulfate fractionation and DEAE cellulose chromatography.

The immunologic relationship of salivary and pancreatic amylase was analyzed by double diffusion experiments. Agarose (Fisher Scientific Co.), 1% in 0.05$\underline{M}$ TRIS-HCl, pH 8.2 buffer, was layered on 8 x 10 cm glass slides and then wells (7 mm diam) were cut 11 mm apart and 10 mm from the center well (10 mm diam) using a Shandon die. After precipitin lines had developed amylase was identified with a starch-iodine staining technique with or without prior deproteinization (Karn et al., 1974).

RESULTS AND DISCUSSION

TISSUE AND BODY FLUID EXPRESSION OF AMYLASE PHENOTYPES

Human amylase isozyme phenotypes as expressed in a number of tissues and secretions including saliva, urine, serum, and pancreas have been studied and are reviewed by Merritt et al. (1973a). In early studies amylase zymograms were obtained with agar electrophoresis (Kamaryt and Laxova, 1966; Ogita, 1966). Later an improved electrophoretic system employing a polyacrylamide sheet gel allowed the resolution of additional isozyme species (Boettcher and de la Lande, 1969; Ward et al., 1971). The phenotypes obtained from the various sources were studied, each of which gives a complex isozyme pattern when stained for amylase activity, are diagrammed in Fig. 1. It is apparent that urine, serum, and pancreas zymograms are similar while saliva contains a more complex pattern having roughly twice the number of isozymes as the other three sources. It should be noted that small amounts of the major Amy_1 isozyme species, most frequently Amy_1 A2, are seen on occasion in serum and urine. Although the liver has been implicated as a source of amylase in other organisms (Arnold and Rutter, 1963), we have ascribed the Amy_2 isozymes seen in serum and urine to the pancreas because neither cystic fibrosis patients with pancreatic insufficiency nor a child who had

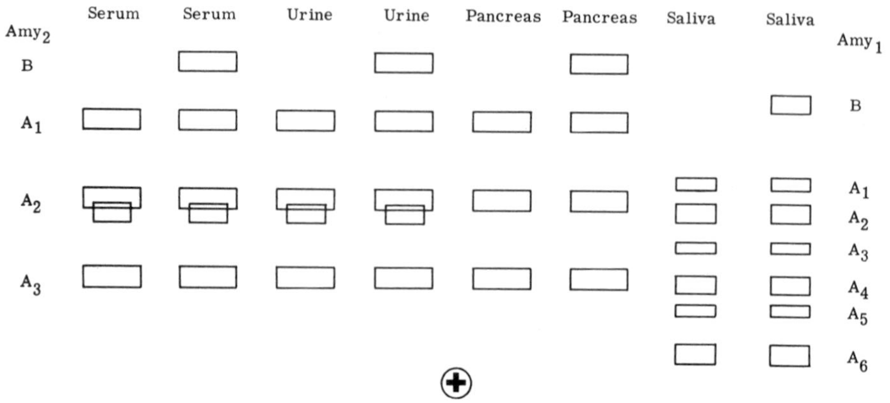

Serum Serum Urine Urine Pancreas Pancreas Saliva Saliva

Amy_2: B, A_1, A_2, A_3

Amy_1: B, A_1, A_2, A_3, A_4, A_5, A_6

Fig. 1. Diagrammatic representation of amylase phenotypes expressed in tissue and secretion sources. Saliva and pancreas samples show Amy_1 and Amy_2 phenotypes respectively, while serum and urine Amy_2 phenotypes frequently have a salivary contribution (Amy_1 A2 shown). Amy_2 A alternates with Amy_2 B in serum, urine, and pancreas phenotypes. Amy_1 A and Amy_1 B are the saliva phenotypes on the right.

---

undergone a complete pancreatectomy have Amy_2 isozymes in their sera and urines. Similar findings have been reported by other investigators (Fridhandler et al., 1972).

The common phenotype at each locus has been designated A and variant phenotypes have been assigned subsequent letters of the alphabet in the order in which they were discovered. The variant phenotypes at both loci were originally described as having only a more-slowly migrating isozyme in addition to an apparently normal isozyme pattern (Merritt et al., 1973a). A new interpretation of variant patterns is discussed in the section Electrophoretic Phenotypes of Amylase. Although isozymes are conventionally numbered beginning with the species migrating farthest from the origin, we have numbered the amylase isozymes beginning with the one nearest the origin, since faster migrating forms appear as specimens age.

GENETIC STUDIES OF HUMAN AMYLASE

The inheritance of the variant phenotypes at both loci has been studied in a number of laboratories (reviewed by Merritt et al., 1973a) and the phenotypes have been shown to be under the control of alleles segregating at each locus in a simple Mendelian manner. Further, when variant alleles occur at both loci in the same individual, they do not segregate

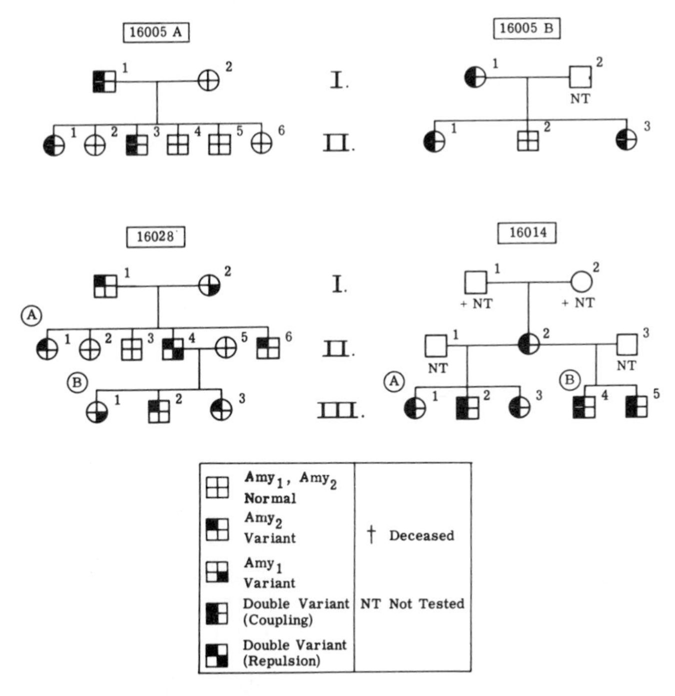

Fig. 2. Pedigrees of families with Amy_1/Amy_2 double heterozygotes.

independently and it has been concluded that the two loci are linked, the variant alleles being either in coupling or repulsion (Fig. 2). Although family studies have been done on a number of individuals variant at both loci, no crossovers have been observed between the two loci, suggesting that they are closely linked or adjacent (Merritt et al., 1973b). Family studies involving these loci and other markers show that the amylase loci are linked to the Duffy blood group and other markers on chromosome 1 (Merritt et al., 1973b; Meyers et al., 1973).

The frequency of the variant phenotypes observed at both loci in various populations has been studied by screening serum, urine, and saliva obtained from individuals in several ethnic groups (Merritt et al., 1973a). Although we have observed 8 different variant Amy_1 phenotypes (See section: Electrophoretic Phenotypes of Amylase) the combined frequency of heterozygotes in the white American population is less than 1 percent. The combined frequency of variant Amy_1 phenotypes in the black population, however, is roughly ten times that. Of the two variant phenotypes observed at the Amy_2 locus,

Amy_2 B has been observed in 10 percent of white Americans and Amy_2 C in roughly 33 percent of black Nigerians. Both alleles are found however, in the black American population, an observation which, from the frequency of the Amy_2^B allele, allows an estimate of 27 percent racial admixture. Only two variants, both Amy_1 F, have been observed among 53 Orientals included in the study.

POST-TRANSLATIONAL CONTRIBUTIONS TO THE AMYLASE PHENOTYPES

Although human amylase phenotypes show simple Mendelian segregation at both loci, it was not clear until recently how the expression of a single locus resulted in the complex multiple isozyme patterns observed. The original biochemical studies of Keller and her colleagues (1971) showed that while purified pancreatic amylase consists of a single peak on gel filtration, purified salivary amylase consists of two peaks of slightly different molecular weights. They further demonstrated that the two salivary peaks were two families of isozymes which appeared to be very similar or identical by amino acid analysis and polypeptide fingerprinting. However, the higher molecular weight family contained carbohydrate while the lower molecular weight family did not. The electrophoretic migration of the higher molecular weight family of isozymes on polyacrylamide is slightly retarded resulting in two series of isozymes. These are designated odds isozymes (Amy_1 A1, A3,...) and evens isozymes (Amy_1 A2, A4,...) (Karn et al., 1973a) since Keller's original designation of families A and B conflicts with the established procedure of designating variant phenotypes with capital letters.

In a later report, Stiefel and Keller (1973) claimed to have detected a small amount of carbohydrate associated with pancreatic amylase. This material, however, does not appear to be differentially present in pancreatic isozymes since they obtained a single peak by gel filtration and a single series of isozymes by electrophoresis. It has also been suggested that the more anodal isozymes within each series, the two salivary families (Keller et al., 1971) as well as the single pancreatic series (Karn et al., 1973b), might be explained by deamidation of slower migrating isozymes. Now deamidation has been proposed for chicken pancreatic amylase (Lehrner and Malacinski, 1975) and for *Drosophila* amylase (Doane et al., 1975).

Although the amino acid composition and polypeptide fingerprints of the two salivary isozyme families appear to be similar or identical in their analyses, Keller and her colleagues (1971) found that no conversion was ever observed between the isozymes of the two families. For this reason they proposed that there might be a polypeptide alteration in one salivary isozyme family allowing glycosidation. If this were the case a minimum of two genes would be required to code for salivary amylase, one for the odds isozymes and one for the evens isozymes. In a recent report, we described an enzyme isolated from the oral bacterial flora which converts the odds isozymes to evens isozymes (Karn et al., 1973a). Carbohydrate analyses performed on the odds isozymes before and after the conversion to evens isozymes showed carbohydrate contents consistent with those observed for odds and evens isozymes isolated from purified parotid saliva. The converted isozymes retained enzymatic activity and it was concluded that the action of the salivary amylase modifier was probably restricted to the removal of carbohydrate (deglycosidation).

We repeated Keller's experiments in which faster isozymes in both salivary families were created, apparently at the expense of slower isozymes, under conditions of elevated pH (9.0) and temperature (37°C). Although faster species appeared under these conditions, it was not clear whether they were actually created sequentially from the slower species, whether some conversion in the reverse direction might be occurring at the same time, or both.

To answer these questions, purified isozymes were subjected to two-dimensional electrophoresis in polyacrylamide sheet gels (Karn et al., 1973b; Rosenblum et al., submitted). Since deamidation is promoted by high temperature, the first dimension gel system was modified to allow short runs at high voltage (27 V/cm) and low temperature (4°C) (cold gel). The second dimension gel was either a standard (warm) gel (16 hr run at 30°C, 11 V/cm) or a cold gel as shown in Fig. 3. It is apparent that in a warm second dimension gel, faster isozymes can be generated from slower isozymes in the purified salivary isozyme families but the reverse conversion, faster to slower, does not occur. We have observed this unidirectional conversion in both salivary and pancreatic amylase isozyme series. It is also evident that this conversion can be retarded or halted completely if the second dimension run is performed in a cold gel.

In order to obtain more direct evidence for deamidation preliminary studies have been performed in which ammonia released from purified amylase in the outer well of a Conway dish was trapped in the inner well and subsequently detected by the Nessler's reaction (Rosenblum et al., submitted for publ.). More ammonia is released at higher temperature (37°C) and/or pH (11.5) than at moderate temperature (25°C) and pH (5.0), consistent with Flatmark's (1966) observations on the deamidation of horse cytochrome c. These results strongly suggest that the faster isozymes in human salivary and pancreatic amylase phenotypes are the result of deamidation of slower

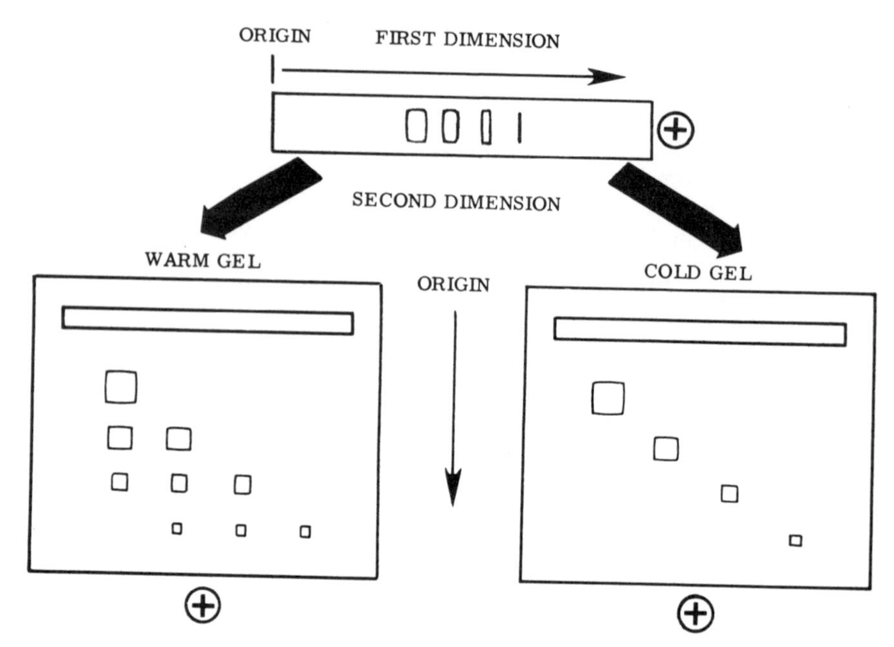

Fig. 3. Diagram of a two-dimensional gel electrophoresis showing conversion of slower-migrating amylase isozyme species to faster-migrating species. The first dimension was run in a cold gel, the second dimension was run in either a warm or cold gel.

isozymes.

We previously presented a model which demonstrated that the complex salivary isozyme phenotype could result from the action of a single gene if the primary gene product was modified by post-translational processes, glycosidation and deamidation (Karn et al., 1973a). At least one of these, deamidation, apparently is occurring in the generation of pancreatic phenotypes. In order to determine whether pancreatic isozymes might also be deglycosidated, purified preparations of pancreatic amylase were treated with the salivary amylase modifier previously described. No new isozyme species were created in these experiments and we conclude that the carbohydrate reported by Stiefel and Keller (1973) is not amenable to removal by this enzyme. Thus the only post-translational process responsible for the complex pancreatic phenotype appears to be deamidation.

Fig. 4. Double diffusion analysis of pancreatic amylase and odds and evens salivary amylase isozyme families. Wells 1 and 4: evens isozymes antibody; well 2: odds isozymes; wells 3 and 5: pancreatic amylase; well 6: odds salivary isozymes; center well: odds isozymes antibody.

IMMUNOLOGIC INTERACTIONS OF Amy_1 *AND* Amy_2 *ISOZYMES*

Rabbit antibodies were prepared against the separated odds and evens isozyme families of normal salivary amylase (Amy_1 A) and against normal pancreatic amylase (Amy_2 A). These were used in double diffusion experiments to test the cross-reactivities among human amylase isozymes (Karn et al., 1974).

Fig. 4 shows an experiment in which the salivary amylase isozyme families and pancreatic amylase have been tested with antibodies prepared against the odds isozyme family of salivary amylase. A continuous precipitin line with no spurs was obtained between pancreatic and both families of salivary amylase, indicating immunologic identity. Identical results were obtained when either evens isozymes antibodies or pancreatic isozymes antibodies were used. In no case was spurring ever observed. Complete cross-reactivity was also obtained when the odds and evens isozyme families of Amy_1 A were tested against each other and against the isozyme families of a variant phenotype, Amy_1 B (Fig. 5). Immunologic identity of human

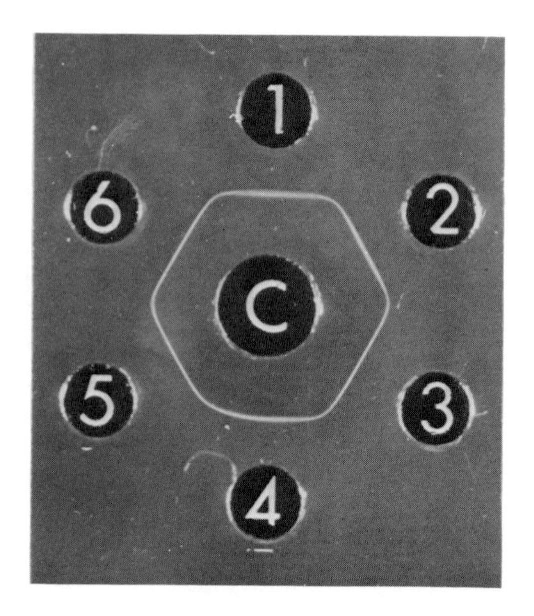

Fig. 5. Double diffusion analysis of normal and variant odds and evens salivary isozymes families. Well 1: normal homozygote (Amy_1^A/Amy_1^A) odds isozymes; wells 2 and 4: heterozygote (Amy_1^A/Amy_1^B) odds isozymes; wells 3 and 6: heterozygote evens isozymes; well 5: normal homozygote evens isozymes. Center well: odds isozymes antibody.

salivary and pancreatic amylases has been observed by others employing different techniques (McGeachin and Reynolds, 1961; Ogita, 1966).

ELECTROPHORETIC PHENOTYPES OF AMYLASE

The amylase phenotypes which have been observed to date, employing the polyacrylamide sheet gel systems used in this laboratory, are shown diagrammatically in Fig. 6. The phenotypes are categorized in two groups for the two loci established in humans, Amy_1 for salivary and Amy_2 for pancreatic amylase. We have now separated the isozyme families of salivary amylase from Amy_1 variants employing gel filtration and have improved the resolution of Amy_2 variant patterns with a new electrophoretic system (Karn et al., 1974; Rosenblum et al., submitted for publication). These modifications in technique allowed the observation of other variant isozymes at both loci. Fig. 6 shows additional variant isozymes observed

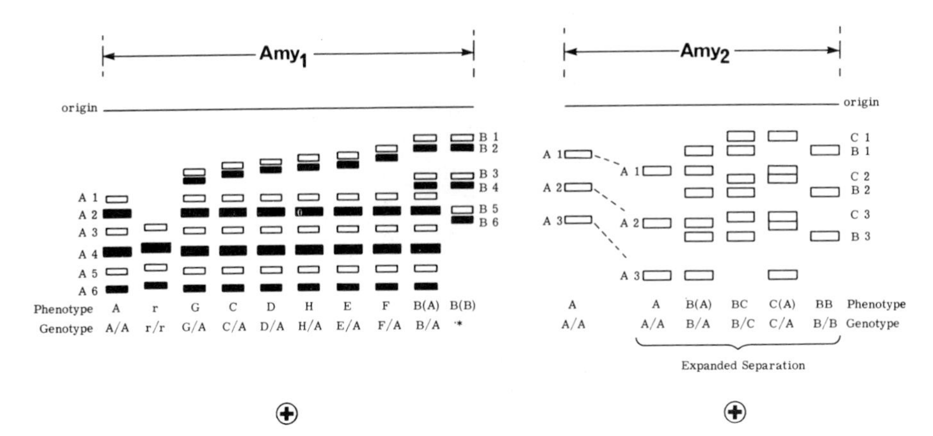

Fig. 6. Diagram of Amy_1 and Amy_2 phenotypes showing additional variant isozymes. Separation of Amy_1 isozymes was by the method of Ward et al. (1971) and Amy_2 isozymes by the method of Rosenblum et al. (submitted for publication). * Postulated appearance of Amy_1^B/Amy_1^B homozygote (see text).

following separation of the odds and evens isozymes from the purified salivary amylase of an Amy_1 B variant. In addition to the Amy_1 A isozymes expected in the heterozygote, two or more variant species appear in each isozyme family. Additional variant isozymes in Amy_2 phenotypes are shown in Fig. 6. Heterozygotes for the normal allele have a series of normal isozymes and a series of variant isozymes. The B/C doubly mutant heterozygote has two variant series but lacks Amy_2 A isozymes.

We conclude from these observations that each of the two complex amylase expressions in humans is controlled by a single locus. If that were not true, then the complete variant patterns observed could only have arisen via multiple mutations. The probability of such an occurrence would be the joint probability of mutation at all loci hypothesized for each expression. Thus it is more likely that the salivary and pancreatic phenotypes are primarily controlled by separate but single gene loci, and that the complex zymograms observed for each allele are the result of post-translational modifications.

A new phenotype, Amy_1 r, has been observed which is apparently recessively inherited. Both parents of Amy_1 r individuals demonstrate the Amy_1 A phenotype which would be expected to obscure the Amy_1 r phenotype by overlapping it. Although the Amy_1 r isozymes are slightly slower on electrophoresis than the corresponding Amy_1 A isozymes, we have not

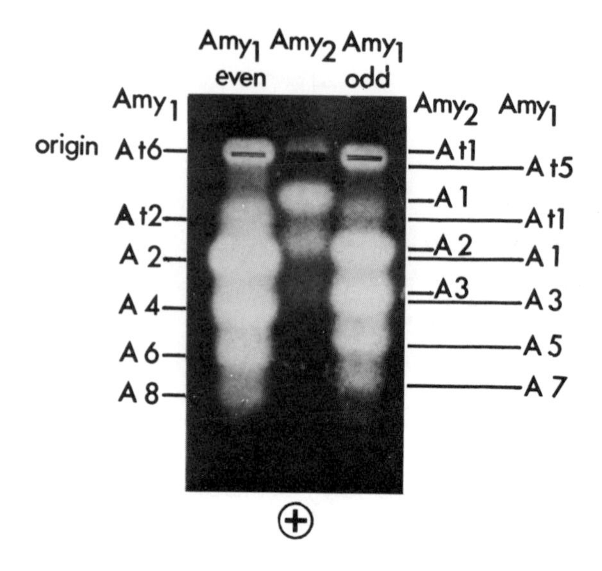

Fig. 7. Transitional isozymes (t) observed in Amy_1 odds and evens isozymes and Amy_2 isozymes.

yet been able to resolve them in presumed obligate heterozygotes, Amy_1^A/Amy_1^r.

TRANSITIONAL AMYLASE ISOZYME SPECIES

The development of new electrophoretic buffer systems (Rosenblum et al., submitted) has resulted in the identification of quasi-stable precursors of human amylase isozymes. These species migrate more slowly than the regular pattern and rapidly convert to the faster-migrating, relatively-stable series under conditions which promote deamidation. Several transitional isozymes (denoted with a "t") have been identified in salivary amylase and one in pancreatic amylase as shown in Fig. 7.

The slowest of the transitional species in the salivary and pancreatic samples migrates cathodally. It is apparent that the cathodally-migrating pancreatic species (A t1) moves farther than either of the cathodally-migrating salivary species (A t5) and (A t6) and that the odds salivary isozyme (A t5) is the slowest of the cathodal species. These observations are consistent with the relationships observed for the anodally-migrating species in the 3 series (Fig. 6).

Fig. 8 shows an updated version of our original model (Karn et al., 1973a) for human amylase isozyme heterogeneity. The new model includes both the Amy_1 and Amy_2 loci and shows

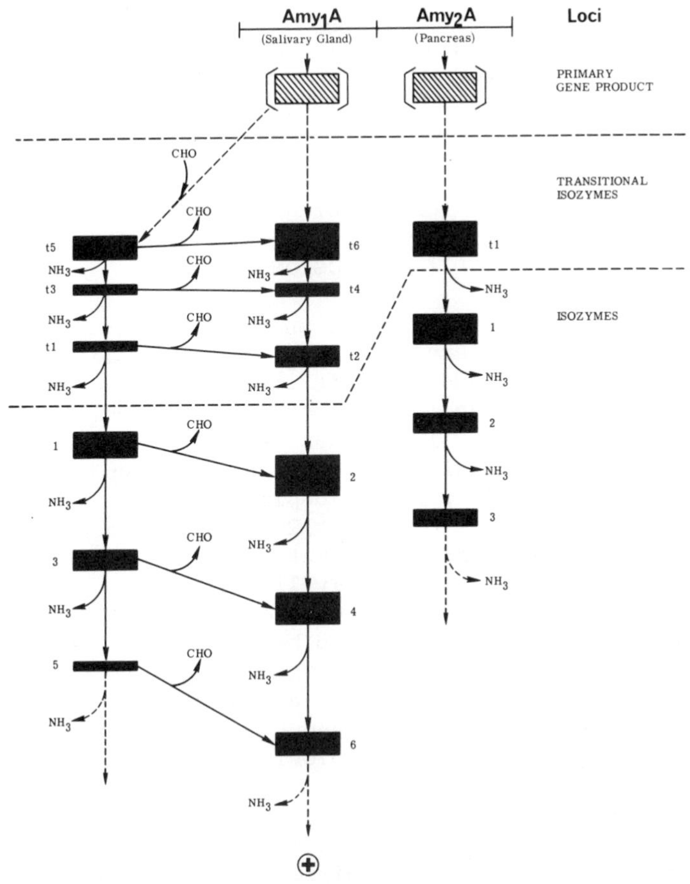

Fig. 8. Model for the multiple isozymes expressed by Amy_1 and Amy_2 phenotypes. The position of t isozymes, not usually seen in standard electrophoretic systems, is enclosed by horizontal dashed lines.

the primary (unmodified) gene products and transitional isozymes set off from each other and from the relatively-stable anodal series by dashed lines. Three post-translational modifications are visualized for the primary salivary amylase gene product: glycosidation of a fraction of the amylase protein population, followed by deamidation of both glycosidated and nonglycosidated species, and eventually deglycosidation of the glycosidated series in whole saliva. The primary pancreatic amylase gene product, on the other hand, undergoes only one post-translational modification: deamidation.

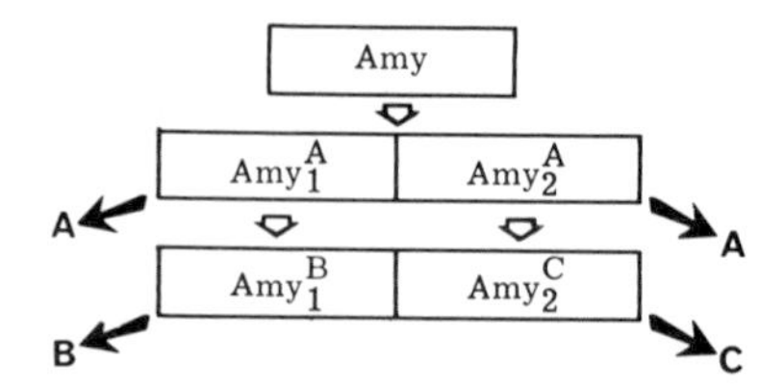

Fig. 9. An hypothesis for the origin of the amylase loci.

CONCLUSIONS

Human amylase appears to be under the control of two closely-linked loci, Amy_1 and Amy_2 on chromosome 1 (Merritt et al., 1973b). In addition, it is apparent from our work (Karn et al., 1974) and that of others (McGeachin and Reynolds, 1961; Ogita, 1966) that salivary and pancreatic amylases are immunologically identical. We hypothesize that the two loci are the result of duplication of a single primordial amylase locus (Fig. 9). The variant phenotypes are envisioned as having arisen by mutations at the two loci following duplication. Two closely-linked amylase loci have also been demonstrated in *Drosophila melanogaster* (Bahn, 1968; Doane, 1967) and are proposed to be the result of a duplication (Doane, 1969). Other organisms which appear to have different amylase expressions in the pancreas and saliva are the rabbit (Malacinski and Rutter, 1969) and the mouse where duplication has been proposed for the two closley-linked amylase loci (Sick and Nielsen, 1964).

Although the genetic evidence strongly suggests separate loci for salivary and pancreatic amylases, no crossovers have been observed to date. In addition, the two enzymes appear very similar in biochemical analyses which have been reported (Stiefel and Keller, 1973). In order to determine whether more than one amylase locus exists in humans, we are pursuing further biochemical analyses as well as extending our efforts to ascertain crossovers between the hypothesized loci.

Keller and her colleagues have reported finding only single polypeptide chains for human salivary and pancreatic amylases (Keller et al., 1971; Stiefel and Keller, 1973). We have repeated those experiments and obtained similar results (Karn, unpublished) and, in addition, our genetic data do not support an association of monomers as a source of isozyme heterogeneity. It is apparent, on the other hand, that each of the two human amylase expressions, salivary and pancreatic, is under the control of a single gene and that the post-translational contributions described above account for the complex phenotypes observed with electrophoresis.

ACKNOWLEDGEMENTS

This investigation was supported in part by PHS research grants GM 19178 and AM 13428. Dr. R. C. Karn and Mr. Barnett B. Rosenblum were supported by PHS Training Grant DE 119. Dr. Jewell C. Ward was supported by PHS Training Grant GM 1056.

NOTE ADDED IN PROOF

We have now identified odds isozymes of pancreatic (Amy_2) amylase which may account for the carbohydrate reported by Stiefel and Keller (1973). These isozymes behave in all respects as those described for salivary (Amy_1) amylase by Karn et al. (1973).

REFERENCES

Arnold, M. and W. J. Rutter 1963. Liver Amylase III. Synthesis by the perfused liver and secretion into the perfusion medium. *J. Biol. Chem.* 238: 2760-2765.

Bahn, E. 1968. Crossing over in the chromosomal region determining amylase isozymes in *Drosophila melanogaster*. *Hereditas* 58: 1-12.

Bernfeld, P. 1955. Amylases, α and β. *Meth. Enzymol.* 1: 149-158.

Boettcher, B. and F. A. de la Lande 1969. Electrophoresis of human saliva and identification of inherited variants of amylase isozymes. *Aust. J. Exp. Biol. Med. Sci.* 47: 97-103.

Doane, W. W. 1967. Cytogenetic and biochemical studies of amylases in *Drosophila melanogaster*. *Amer. Zool.* 7: 780.

Doane, W. W. 1969. *Drosophila* amylases and problems in cellular differentiation. In: *Problems in Biology: RNA in Development*. (Hanley, E. W., ed.), pp. 73-109, University of Utah Press, Salt Lake City.

Doane, W. W., I. Abraham, M. M. Kolar, R. E. Martensen, and G. E. Deibler 1975. Purified *Drosophila* α-amylase isozymes: Genetical, biochemical, and molecular characterization. *Isozymes IV. Genetics and Evolution*. C. L. Markert, editor, Academic Press, New York. pp. 585-607.

Dubois, M., K. A. Gilles, J. K. Hamilton, P. A. Rebers, and F. Smith 1956. Colorimetric method for determination of sugars and related substances. *Anal. Chem.* 28: 350-356.

Flatmark, T. 1966. On the heterogeneity of beef heart cytochrome c III. A kinetic study of the non-enzymatic deamidation of the main subfractions (Cy I-Cy III). *Acta Chem. Scand.* 20: 1487-1496.

Fridhandler, L., J. E. Berk, and M. Ueda 1972. Isolation and measurement of pancreatic amylase in human serum and urine. *Clin. Chem.* 18: 1493-1497.

Hunter, R. L. and C. L. Markert 1957. Histochemical demonstration of enzymes separated by zone electrophoresis in starch gels. *Science* 125: 1294-1295.

Kamaryt, J. and R. Laxova 1966. Amylase heterogeneity variants in man. *Humangenetik* 3: 41-45.

Karn, R. C., J. D. Shulkin, A. D. Merritt, and R. C. Newell 1973a. Evidence for post-transcriptional modification of human salivary amylase (Amy_1) isozymes. *Biochem. Genet.* 10: 341-350.

Karn, R. C., B. B. Rosenblum, and A. D. Merritt 1973b. A biochemical explanation for the complex isozyme patterns of salivary amylase (Amy_1) and pancreatic amylase (Amy_2). *Amer. J. Hum. Genet.* 25: 39A.

Karn, R.C., B.B. Rosenblum, J.C. Ward, A.D. Merritt, and J.D. Shulkin 1974. Immunological relationships and post-translational modifications of human salivary amylase (Amy_1) and pancreatic amylase (Amy_2) isozymes. *Biochem. Genet.* 12: 485-499.

Keller, P. J., D. L. Kauffman, B. J. Allan, and B. L. Williams 1971. Further studies on the structural differences between the isoenzymes of human parotid α-amylase. *Biochemistry* 10: 4867-4874.

Lehrner, L. M. and G. M. Malacinski 1975. Genetic and structural studies of chicken α-amylase isozymes and their modified forms, and structural studies of hog amylase. *Isozymes IV. Genetics and Evolution.* C. L. Markert, editor, Academic Press, New York. pp. 727-743.

Loyter, A. and M. Schramm 1962. The glycogen-amylase complex as a means of obtaining highly purified α-amylases. *Biochim. Biophys. Acta* 65: 200-206.

Malacinski, G. M. and W. J. Rutter 1969. Multiple molecular forms of α-amylase from the rabbit. *Biochemistry* 8: 4382-4390.

McGeachin, R. L. and J. M. Reynolds 1961. Serological differentiation of amylase isozymes. *Ann. N. Y. Acad. Sci.* 94: 996-1001.

Merritt, A. D., M. L. Rivas, D. Bixler, and R. Newell 1973a. Salivary and pancreatic amylase: electrophoretic characterizations and genetic studies. *Amer. J. Hum. Genet.* 25: 510-522.

Merritt, A. D., E. W. Lovrien, M. L. Rivas, and P. M. Conneally 1973b. Human amylase loci: genetic linkage with the Duffy blood group locus and assignment to linkage group I. *Amer. J. Hum. Genet.* 25: 523-538.

Meyers, D. A., M. L. Rivas, A. D. Merritt, E. W. Lovrien, R. S. Huntzinger, and D. Bolling 1973. Four-point linkage analysis of the Amy, Fy, PGM and Rh loci. *Amer. J. Hum. Genet.* 25: 52A.

Ogita, Z. 1966. Genetico-biochemical studies on the salivary and pancreatic amylase isozymes in humans. *Med. J. Osaka Univ.* 16: 271-286.

Rosenblum, B.B., R.C. Karn, and A.D. Merritt. The pleiotropic isozyme expression from deamidation of human salivary amylase. *J. Biol. Chem.* Submitted for publication.

Sick, K. and J. T. Nielsen 1964. Genetics of amylase isozymes in the mouse. *Hereditas* 51: 291-296.

Stiefel, D. J. and P. J. Keller 1973. Preparation and some properties of human pancreatic amylase including a comparison with human parotid amylase. *Biochim. Biophys. Acta* 302: 345-361.

Ward, J. C., A. D. Merritt, and D. Bixler 1971. Human salivary amylase: genetics of electrophoretic variants. *Amer. J. Hum. Genet.* 23: 403-409.

INDUCTION AND ISOLATION OF ELECTROPHORETIC MUTATIONS IN MAMMALIAN CELL LINES

MICHAEL J. SICILIANO, RONALD HUMPHREY, E.J. MURGOLA,
AND MARION C. WATT
Biology and Physics Departments
The University of Texas System Cancer Center
M.D.Anderson Hospital and Tumor Institute
Texas Medical Center Houston, Texas 77025

ABSTRACT. We have attempted to indicate the feasibility of generating significant information on mutagenesis in mammalian cells using a non-selective system. The key points in the procedure involve the use of a cell line with high plating efficiency, early cloning following exposure to mutagen, and screening clonal isolates for electrophoretic variants at a large number of enzyme loci. The Chinese hamster ovary cell line (CHO) was used. We characterized the electrophoretic mobility of the products of 36 enzyme loci in this line and also found that many enzyme loci, active in various Chinese hamster tissues, were not expressed in the cell line. 124 clones, isolated immediately after exposure to various doses of UV and MNNG, were examined at an average of 27.5 enzyme loci for electrophoretic shifts.

Phenotypic variances were seen in 3 different clonal isolates. One of these appears to be a true point mutation at the locus coding for acid phosphatase (AP), the second appears as a likely point mutation at an esterase locus, and the third a modification in which the products of several esterase loci, normally unexpressed in CHO cells, become expressed.

Subcloning of the line containing the AP mutant revealed that a cloning delay following exposure to mutagen, to permit DNA strand and chromatid segregation, would likely allow the detection of additional mutants and the calculation of mutation frequencies.

INTRODUCTION

In bacterial and phage systems, the mutational approach (studying spontaneous and induced inherited variability) has contributed greatly to our understanding of a wide range of biological phenomena (Yanofsky 1971). Now we are seeking ways to study many of these same phenomena in mammalian somatic cell systems. In this respect the mutational approach may be valuable in generating markers for linkage study and in understanding problems related to the control of gene expression, protein structure and function, molecular evolution, gene dosage, and mechanisms of mutation and gene repair.

The methods of approach in detecting mammalian somatic cell mutants have basically been borrowed from bacterial systems wherein, due to the rarity of mutational events, selective techniques are employed in isolating mutants. Most selective techniques are based upon mutations which allow viability in the presence of metabolic inhibitors or in the absence of a normally essential and specific component of the culture medium. Such methods select for recessive mutations for drug resistance, auxotrophy, or temperature sensitivity. For a thorough review of these procedures as applied to mammalian cells the reader is referred to Thompson and Baker (1973).

In bacterial systems, where the medium may be specifically defined and cells are haploid, the selective techniques have proven very adequate in detecting and isolating mutations in a wide range of enzyme loci. In mammalian cell systems the situation is quite different. The serum requirement for growth of mammalian cells and subsequent difficulty in establishing a defined medium places serious problems on developing a range of auxotrophic mutants. Furthermore, the diploid state of most genetic loci often necessitates multi-step selection in order to isolate drug resistant or temperature sensitive mutants which depend upon homo- or hemizygous presence of certain specific recessive alleles. Such methods are either so severe as to leave doubt as to the true genetic nature of the surviving clones or limit one to studying certain sex-linked loci in male derived cultures. Even results from the most hypothetically successful selective system, loss of the activity of sex linked hypoxanthine-guanine phosphoribosyltransferase (HGPRT) from male Chinese hamster lines, has been challenged as not being a true gene mutation since the frequency of azaguanine resistant clones (the selective system for that mutation) did not drop with an increase in ploidy (Harris 1971).

While it seems clear that at least some cells surviving a particular selective condition represent true mutations (Beaudet et al., 1973) the real frequency of such genetic changes may often be clouded by such additional phenomena as metabolic cooperation between mutant and wild cell types (Zeeland et al., 1972), cell concentration (Morrow 1972), karyotype variability in mammalian cell lines (Bootsma et al., 1973), possible mitotic segregation (Roufa et al., 1973), variable viability of mutants under non-permissive conditions, and degree of phenotypic expression at the time of selection (Thompson and Baker 1973).

Therefore, selective systems for studying mutations in mammalian cell lines are quite limited not only in the confidence in the genetic nature of the variation but also in the number and distribution of loci that can be screened for. Another limitation is in the type of mutations produced. All one step mutagenesis experiments in such systems demand recessive genetic changes which result in loss of enzyme activity. Presumably these would be caused by changes in the DNA leading to a nonsense codon with premature chain termination, to a missense codon resulting in an amino acid substitution affecting the active site or 3 dimensional structure of the enzyme, to a frame shift producing an anomalous protein, or to gross structural damages such as deletions, chromosome loss or rearrangement. But what of the whole class of missense mutations leading to amino acid substitutions which do not necessarily adversely effect enzyme activity? A significant proportion of these might be ones in which an acidic residue is replaced by a basic or neutral one, a neutral by an acidic or basic, or a basic by acidic or neutral.

Such a mutation might be detectable electrophoretically (Henning and Yanofsky 1963) but the very activity of such a mutated protein, or the activity of the polypeptide specified at the unmutated homologous locus, defies detection. So the question becomes whether or not we would expect such electrophoretic mutations to take place in sufficiently high frequency to justify looking for them non-selectively. To answer this we ask what is the comparative frequency of single base pair mutations which we would expect to result in an electrophoretic change in a protein as opposed to one which would lead to the loss of enzyme activity. We have carried out such a calculation using the α chain of the bacterial enzyme tryptophan synthetase (TS) as a model.

Yanofsky's group (see 1971 review) not only determined the complete sequence of the 268 amino acids which make up this polypeptide but also identified the specific amino acid changes (missense mutations) which result in the loss of enzyme activity. Considering the possible codons which code for each amino acid and considering every possible single base change in each codon which would result in a substitution of an acidic, basic, or neutral amino acid for one of another kind gives one a measure of the total number of single base changes in the DNA which should result in an electrophoretic shift in the enzyme. Aspartate and glutamate were considered acidic, lysine and arginine basic, and the rest neutral. Due to the degeneracy of the code there are many sites at which more than one codon could specify the same amino acid. In such a case the number of base pair changes for each possible codon which could lead to an electrophoretic shift was determined and the actual number of such changes at that position was considered as an average. By this system 504 single base pair changes in the gene coding for the α chain of TS would theoretically produce a protein with a different electrophoretic mobility than the wild type. The same methodology reveals 78 base changes would lead to a nonsense codon which would result in the loss of enzyme activity. Only 11 base pair changes would result in amino acid substitutions causing the loss of enzyme activity (Yanofsky 1971).

It appears as though the number of single base changes in a gene which would result in an electrophoretic shift in a protein outnumber the changes which result in the complete loss of enzyme activity by at least 5 to 1 (504 : 11 + 78). While these figures represent only theoretical estimates, there are data consistent with the calculation which brought them about. For instance, Henning and Yanofsky (1963) predicted 8 mutant sequenced TS proteins should have been electrophoretically different from wild type. Seven were different. Also Yanofsky (1971) in reviewing 600 mutants leading to no activity found that only a small percentage were of the missense type. Both of these results are consistent with our calculations, i.e., (1) that one can make predictions as to which amino acid substitutions will cause an electrophoretic shift in the protein and,(2) most single base pair substitutions causing loss of activity will be of the nonsense rather than the missense type.

Although a large number of assumptions were made for these calculations, they do indicate that electrophoretic changes in proteins represent a significant class of mutations, the frequency of which if determined could result in a greater generalization of mutagenesis in mammalian cells. Such mutants if isolated could also become a rich source of markers within a single cell line. The calculated high frequency of such mutations also justifies the development of non-selective systems for detecting and isolating them.

Shaw et al., (1973) looked for induced mutations in

E. coli affecting the electrophoretic mobility of enzymes. Colonies were picked after exposure to nitrosoguanidine (MNNG), grown up as isolates and analyzed for electrophoretic shifts in proteins specified by 13 enzyme loci. A total of 1400 isolates were examined. Five electrophoretic mutations were detected in the total of 18,200 genes analyzed (1400 isolates x 13 loci) for a mutational frequency of 2.7×10^{-4}. Examination of this original work suggests that diploid mammalian systems might actually be more efficient in detecting electrophoretic mutations. In the haploid bacterial system, an electrophoretic mutant which also had reduced enzyme activity might not survive whereas if such a mutation took place at a diploid locus the wild type protein would still be present so that the cell and its mutation would not be lost. Our laboratory has uncovered such a situation for a null allele which produces a polypeptide with different electrophoretic mobility at the 2 locus for glyceraldehyde-3-phosphate dehydrogenase (G3PD) in certain platyfish (genus *Xiphophorus*). Heterozygotes produce only a 4 banded pattern in a 1:3:3:1 ratio for this tetramer (Siciliano et al., 1974) instead of the 5 banded 1:4:6:4:1 as seen and expected in heterozygotes for 2 electrophoretic alleles with normal activity (Wright et al., 1972). Homozygotes for the variant allele with reduced activity often do not survive (Wright and Siciliano 1973).

Since mamalian cell systems are diploid for most loci, such mutants would be detected. We therefore have a paradoxical situation in which the increased ploidy of mammalian cell systems actually increases the probability of detection of these electrophoretic mutations. This is true not only because of the complementarity of the homologous locus but also because diploidy increases the number of genes at which an electrophoretic mutation can take place within a single cell. Since a mutant gene can be detected in a heterozygote due to the codominance of electrophoretic mutations, diploidy of mamalian cell systems is no detraction from the advantages described above.

In general the experimental protocol we have used for the detection and isolation of mutations resulting in electrophoretic shifts in specific enzymes is a modification of the method introduced for mammalian cells by Povey et al., (1973). They exposed a human lymphoblastoid cell line to mutagen, ultraviolet light (UV) or MNNG, allowed the cells to recover for a few weeks, plated out "clones" orginating from 3 to 4 cells, grew each "clone" up to sufficient quantity for starch gel electrophoretic analysis of variants at 26 enzyme loci. They examined a total of 115 treated "clones."

The cloning delay after mutagen exposure and multiple cells used to form a "clone" were necessitated by the low plating efficiency of the cell line used. Such methods make the mutation frequency impossible to determine. Also it is likely that many mutations obtained may have been lost by being overgrown by non-mutagenized, surviving cells (putting mutated cells in lower frequency than actually generated). Nevertheless they report a probable point mutation in one of the disomic loci for 6 phosphogluconate dehydrogenase (6PGD). The mutant "clone" had the typical 3 banded pattern expected for a heterozygote at a locus specifying a dimeric enzyme. They also report in 2 other clones the expression of an isozyme form not seen in the parental line. This will be discussed in conjunction with our results.

In our initial experiments reported here, we have tried to overcome some of the cloning difficulties experienced by Povey et al., (1973) by using a well established mammalian cell line with high plating efficiency. Chinese hamster ovary (CHO) cells were used since these cells have the additional advantages of having a stable karyotype with a low chromosome number near diploid and can be frozen for storage. We therefore hope to increase the efficiency of detecting and isolating mutants by cloning immediately after mutagen exposure and increasing to 36 the number of enzyme loci screened by starch gel electrophoresis.

MATERIALS AND METHODS

Mutagen Treatment and Cloning. CHO cells were used in all of this work. The CHO line has been in culture since 1958 (Tijo and Puck 1958) and maintained in our laboratory from 1968. The cells are grown as a monolayer on glass or plastic growth vessels in McCoy's 5a medium supplemented with 20% fetal calf serum (Gibco, Inc.).

In the present experiments the following protocol was followed. 1.5×10^6 stock cells were introduced into petri dishes containing growth medium. The dishes were incubated about 18 hours at 37°C in an atmosphere of 5% CO_2 and 95% air. For exposure to UV the medium was decanted from each dish and replaced with Hank's salt solution without phenol red. Cells at the inside edge of the dish were removed by a sterile cotton swab since they are protected from UV light by the rim of the dish. The Hank's solution was decanted and each dish exposed to a UV source (two 15 watt Sylvania germicidal bulbs). The light source to cell surface distance was adjusted so that exposures of 4.5, 9, and 13.5 seconds yielded survival levels of 90%, 45%, and 9% respectively.

For exposure to MNNG individual dishes received either 0.15 μg/ml for one hour (50% survival) or 0.25 μg/ml for one hour (10% survival).

Immediately after irradiation or chemical treatment the cell monolayer was trypsinized from the surface, diluted into medium, and the cell concentration per ml determined. The single cell suspension was then diluted and 0.25 ml was delivered into each of the individual cloning wells of a plastic cloning dish (Falcon Plastics, Inc.). Two cloning dishes were done for each of 3 doses of UV or 2 doses of MNNG. Suspensions were diluted so that for untreated controls, (with 100% survival) and UV exposed samples with 90% survival, one cell would be expected to be delivered/well. For the experiments with 45% or 50% survival and 9% survival 2 and 10 cells/well were plated respectively.

The cloning dishes were incubated for 8 days and wells containing a single colony were marked. The single colony from each well was trypsinized from the surface and transferred to a growth vessel. As soon as a monolayer of cells was formed a portion of cells was frozen at -70°C for future stock source and the remainder processed for enzyme assay. For processing through the whole battery of enzymes approximately 40×10^6 cells were needed from each clone.

Electrophoresis. Vertical starch-gel electrophoresis (Shaw and Prasad 1970) followed by histochemical staining for 28 enzymes was conducted on homogenates from stock CHO cells, mutagen exposed clones; A3 cells (a different Chinese hamster line obtained from T. Caskey, M.D., Baylor College of Medicine, Houston, Texas); and spleen, testis, liver, muscle, brain, kidney, and heart tissues removed from Chinese hamsters. We characterized this material for the enzymes listed in Table I.

RESULTS AND DISCUSSION

Characterization of products of enzyme loci active in CHO cells. Consistent with the work of others who have worked on Chinese hamster material (Meera Khan 1971; Westerveld, et al., 1971; Someren, et al., 1972; Westerveld and Meera Kahn 1972; Osterman and Fritz 1973: Someren and Henegouwen 1973) we resolve the products of single loci after electropheresis of CHO supernatants for 6PGD, G6PD, TO, GPI, PGK, HA, PK, ADA, LDH, and AK; 2 loci for MDH and IDH; and 3

TABLE 1

Enzymes Stained for After Starch-Gel Electrophoresis[1]

6 Phosphogluconate Dehydrogenase (6PGD)	Peptidase (PEP)
Glucose-6-Phosphate Dehydrogenase (G6PD)	Pyruvate Kinase (PK)[6]
Mannose Phosphate Isomerase (MPI)	Phosphoglyceric Acid Mutase (PGAM)[8]
Isocitrate Dehydrogenase (IDH)	Phosphoglycerate Kinase (PGK)[7]
NADP-Linked Malate Dehydrogenase (ME)	Malate Dehydrogenase (MDH)
Tetrazolium Oxidase (TO)	Lactate Dehydrogenase (LDH)
Nucleoside Phosphorylase (NP)[2]	Adenylate Kinase (AK)
Adenosine Deaminase (ADA)[2]	Fumarase (FUM)
Glucosephosphate Isomerase (GPI)	Acid Phosphatase (AP)
Phosphoglucomutase (PGM)	Esterase (EST)
Triosephosphate Isomerase (TPI)	β-Glucuronidase (GLU)[8]
α Mannosidase (MAN)[3]	Hexoseaminidase (HA)[9]
Glutamate-Oxaloacetate Transaminase (GOT)[4]	Enolase (ENO)[7]
Glyceraldehyde-3-Phosphate Dehydrogenase (G3PD)[5]	Aldolase (ALD)

[1] Shaw and Prasad, 1970.
[2] Spencer et al., 1968.
[3] Peonaru and Dreyfus, 1973.
[4] DeLorenzo and Ruddle, 1970.
[5] Wright et al., 1972.
[6] Susor and Rutter, 1971.
[7] Omenn and Cohen, 1971.
[8] Swank and Paigen, 1973.
[9] Okada et al., 1971.

loci for peptidases. While Someren et al., (1972) report only one GOT we resolve what appears to be both the supernatant and mitochondrial forms, as seen in mouse tissues (DeLorenzo and Ruddle 1970), in all our material (Fig. 1). We also resolve the products of 3 PGM loci (as indicated by isozyme distribution in Chinese hamster tissues) in CHO and A3 cells whereas others have reported only one in A3 cells (Westerveld, et al., 1971). Three loci for PGM have also been recognized in human material (Hopkinson and Harris 1968).

CHO and A3 cells also express the products of single loci for ME, NP, ENO, MAN, G3PD, TPI, MPI, PGAM, FUM, AP, GLU, and ALD. The electrophoretically separable products of 2 EST loci are also resolvable (see below). For all the above loci, the electrophoretic mobilities of the products from the CHO stock cells, A3 cells, and Chinese hamster tissues were identical except for ADA. In that case the enzyme from CHO had a slower mobility (Fig. 3). We therefore can examine mutagen exposed CHO clones for electrophoretic shifts in the characterized products of 36 enzyme loci.

Comparison of CHO and A3 cells with Chinese hamster tissues reveals additional loci for ALD, LDH, AK, PK, and EST expressed in the tissues, the products of which are not seen in the cell lines (e.g. - Fig. 2). Since Deaven and Peterson (1973) indicate CHO cells, though slightly hypodiploid (20 to 22 chromosomes), contain at least one complete haploid set of chromosomes (n=11) we suspect that the genes coding for the unexpressed products exist in our cells in a repressed state. Therefore, in our mutagenesis experiments it appears that we can not only screen for genetic changes resulting in electrophoretic shifts in the products of expressed loci, but also for the "turn on" or reexpression of gene products of repressed loci.

Isozyme analysis of clones exposed to mutagen. Due to the fact that not all 26 histochemical staining methods were available during the entire period of these experiments, less than 36 loci products were screened for in each clone exposed to mutagen. The average number of loci products tested for electropheretic shifts/clone was 27.5. To date, from the UV experiments with 90, 45, and 9 per cent survivals we have tested 26, 59, and 18 clonal isolates respectively. From the MNNG experiments with 50 and 10 per cent survivals we have tested 11 and 10 clonal isolates respectively. Phenotypic variants were detected in 3 clones and are discussed below.

1. Clone 50-66. This clone was isolated after exposure to UV in which 50% of the cells survived. It was typical for all gene products screened with the exception of AP. The typical AP pattern in both CHO and A3 is a single clear cathodal band. Clone 50-66 demonstrated a 3 banded pattern with the slowest band equivalent to wild type (Fig. 4). The spacing of the bands was consistent with that expected in a heterozygote for a dimer molecule. The dimer structure of AP in mammalian cells has been demonstrated from electrophoretic data (Shows and Lalley 1974). However, the intensity of the bands was not the expected 1:2:1 but instead skewed in the parental band direction.

The pattern is most easily interpreted as being due to a point mutation in one of the 2 AP homologous loci producing a heterozygous cell. If the mutation had taken place during the post-DNA synthesis period (G_2) of the cell cycle, the first division after cloning would have produced 2 cells in the well with different genotypes - one heterozygous for the mutant AP and the other homozygous normal. Therefore because of immediate cloning after mutagen exposure our "clone" could be composed of cells of 2 different genotypes for AP accounting for our skewed pattern.

An alternate explanation was that we had a heterozygote clone but that the variant allele had much reduced activity. The frozen stock of this clone was regrown and tested again. The three banded pattern was still apparent but this time less skewed in favor of the parental band. It appeared that clone 50-66 was a mixed clone in which the heterozygotes were outgrowing the homozygote normals. To test this possibility clone 50-66 was subcloned. Of the 8 subclones initially grown up and tested for AP, 7 showed a 3 banded pattern much more typical of a heterozygote for a dimer enzyme although the mutant gene seemed to have less activity. One subclone (50-66-5) showed the wild type single band (Fig. 5). Clone 50-66 was subcloned again and twenty additional subclones all showed the heterozygous pattern. These results are consistent with the hypothesis that the mutation took place when the cell was in G_2, giving rise to a mixed clone in which the mutant cells for AP outgrew the wild type. To check on the possibility that 50-66-5 did not by chance lose the chromosome carrying the variant AP and was in fact most likely a segregant not receiving the mutant gene, it was karyotyped along with subclones showing the heterozygous pattern and shown to have the same number of chromosomes.

Clearly we have demonstrated the feasibility of the technique to detect electrophoretic mutants. However certain complications have arisen especially with regard to the cloning methods. Early cloning (directly after mutagen exposure) does not allow time for DNA strand or chromatid segregation. Thus several cell divisions would be required to segregate a mutated DNA strand from unaffected ones. In the case of clone 50-66 the cell carrying the mutant gene appeared to outgrow normal segregants. It is entirely possible, indeed likely, that many more electrophoretic mutations were induced but did not compete well against their sibs while growing up to 40×10^6 cells.

2. Clone 50-30. This clone was also isolated after exposure to UV in which 50% of the cells survived. It produced a slow band in the more cathodal esterase region (Fig. 6). Of the 124 clones analyzed it was the only one to show this band. Its original homogenate showed it consistently on repeat runs. Upon regrowing the frozen stock of this clone the band was not found. There are many trivial explanations for the appearance of the band in the original supernatant. However there are reasons to consider it as the product of a mutant allele. Many esterase heterozygotes display only a 2 banded pattern due to the monomeric nature of many esterases. The weakness of the variant band compared to its likely homologue is once again reminiscent of the mixed clone situation seen for AP. In this case, however, the mutant cells were perhaps the slower growing cells so that after regrowing up the "clone" the mutants became lost. The repeatability of the result from the original supernatant and the complete absence of this band from all other clones support this view.

3. Clone 0.15-47. This clone was isolated after exposure to 0.15 μg/ml MNNG (50% survival). Its phenotype for esterase was completely different from all other clones. As can be seen from Fig. 7, instead of having only 2 bands of esterase activity it has 6 additional bands. Suspecting that these extra bands did not represent electrophoretic variants of the 2 bands seen in the parental CHO cells, the supernatant from clone o.15-47 was run on the same gel along with extracts of various Chinese hamster tissues (Fig. 8). As can be seen the extra esterase bands in this clone all have counterparts in the Chinese hamster tissue. This fact and the finding that no atypical bands were seen for the other 34 loci rules out the possibility that these bands were produced by some contaminant in the culture. The appearance of Chinese hamster gene products not normally expressed in CHO cells is suggested. It should be noted that this effect in 0.15-47 is specific for esterase. Additional loci for PK, AK, ALD, and LDH remain unexpressed in this clone.

PLATE I. ZYMOGRAMS, ANODE IS ALWAYS TO THE TOP

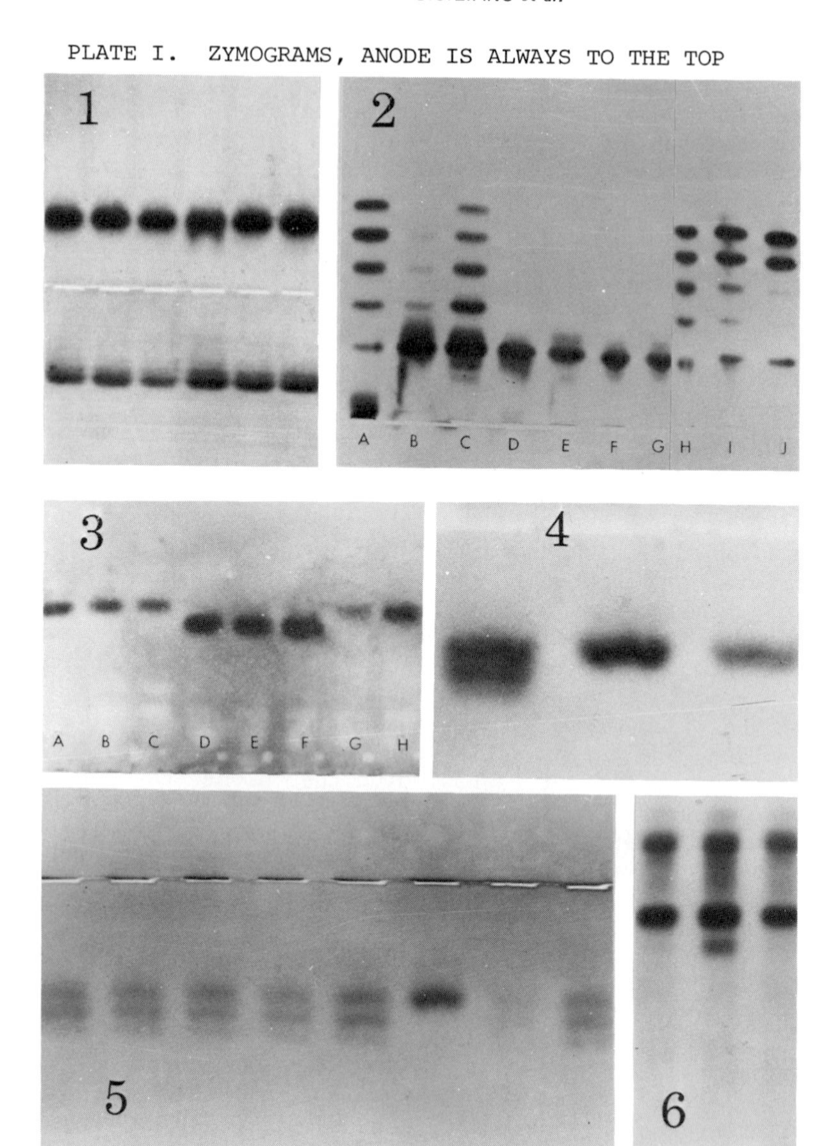

Fig. 1. GOT. Anodal and cathodal forms from CHO clonal isolates.

Fig. 2. LDH. Forms from Chinese hamster testis (A), liver (B), muscle (C), CHO cells (D and E), A3 cells (F and G), brain (H), kidney (I), and heart (J). Cell lines express only the A polypeptide. Neither LDH-X near origin, from testis nor LDH-B, most anodal form seen in several tissues, are expressed in either cell line.

Legends for Plate I., continued:

Fig. 3. ADA. Forms from Chinese hamster testis (A), liver (B), muscle (C), CHO cells (D,E,F) and A3 cells (G and H).

Fig. 4. AP. Three banded pattern of clone 50-66. Single banded patterns of 2 other clones, representative of original CHO cells and all other clonal isolates, are to the right.

Fig. 5. AP. Eight initial subclones of clone 50-66. Seven show typical heterozygote 3 banded pattern. Only one (subclone 50-66-5) retains single band pattern of original CHO cells.

Fig. 6. Esterases. Supernatant from clone 50-30 is in the middle. On either side are extracts from 2 other clones representative of the normal CHO pattern. The slow band in 50-30 was seen only in that clone.

Povey et al., (1973) suggested a reappearance of an enzyme (pep-D) in 2 "clonal" isolates after exposure of human lymphoblastoid cell line F137D to mutagens. However some pep-D activity could be seen in their F137D cells. This could be due to some cells in the line with that activity, 2 of which were cloned out in their experiment. This possibility is particularly likely since the F137D line is actually a subline of F137 where pep-D activity was very strong. Therefore their results could be explained by contamination of F137D by some F137 cells with the pep-D activity and the subsequent chance cloning out of 2 such contaminant cells. The likelihood of our results being due to such a series of events is extremely low due to the complete absence of the extra bands in our original CHO cells.

At this point in our studies we can only speculate about the meaning of the expression of the extra bands. Of particular interest in these speculations are the 2 closely spaced most anodal bands of clone 0.15-47. These correspond perfectly with fast migrating bands from liver and kidney. These bands appear to occupy the same relative position on the gel and to have the same tissue specificity as est-2 in mice described by Klebe et al., (1970). These workers, using somatic cell hybridization, have described a genetic control system for the expression of that locus in certain mouse cell lines. While no genetic data is available indicating that the extra bands in clone 0.15-47 are truly products of different genetic loci, the tissue specificity of the expression of the corresponding bands in the Chinese hamster material appears to indicate so (Fig. 8). The simultaneous re-expression of several genetic loci with shared physiological function in response to a single event

PLATE II. ZYMOGRAMS. ANODE IS ALWAYS TO THE TOP

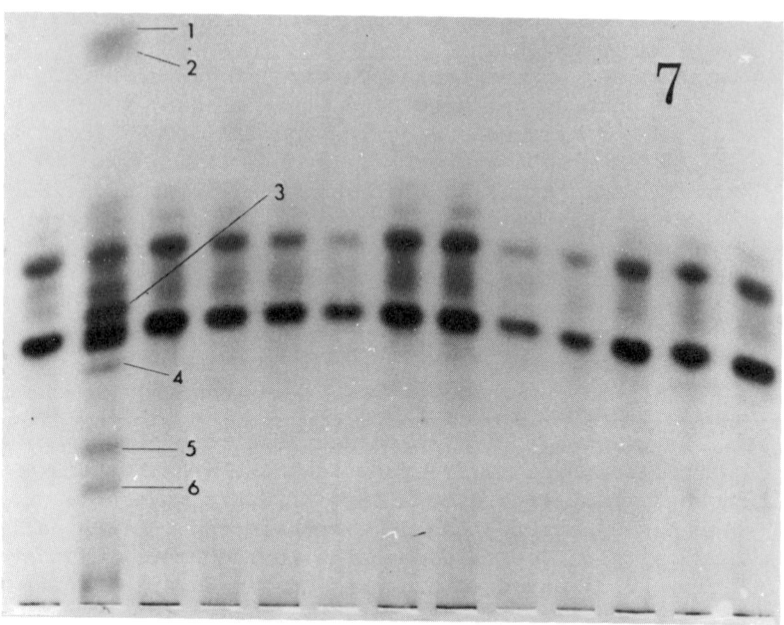

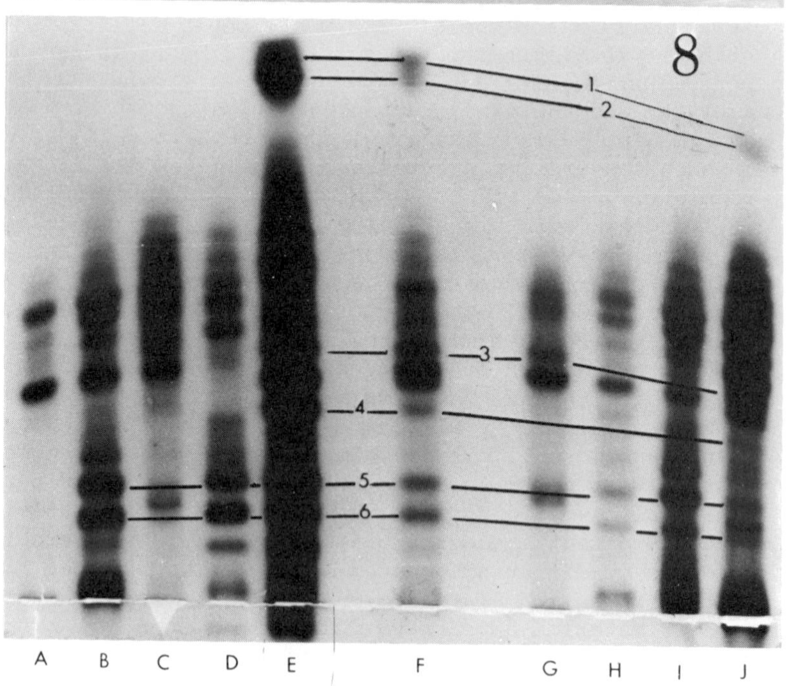

Lengends for Plate II:

Fig. 7. Esterases. Supernatants from 13 clonal isolates of CHO cells exposed to mutagen. Second slot contains extract from clone 0.15-47. Six additional bands, not seen in other clones nor original CHO cells are marked.

Fig. 8. Esterase. Forms from Chinese hamster testis (B), spleen (C), muscle (D), liver (E), brain (G), eye (H), lung (I), and kidney (J), extract from CHO control cells is in A. Bands in the tissues which correspond to the extra bands in clone 0.15-47 (F), are connected by lines.

---

seems to be a reasonable interpretation of our findings. Whether or not the fact that many esterase loci appear to be linked on the same chromosome (Petras and Biddle 1967) in mammals has any bearing on this phenomenon is something that future work in this system may elucidate.

In conclusion, the methods described here allow one to detect variants for altered expression in enzymes coded for by multiple loci as well as electrophoretic shifts in gene products. Since these changes appear to make up a significantly large and important proportion of total mutations that can effect a cell and since the traditional selective systems for detecting and isolating mutants would miss these, the value of the method is indicated. Due to the problems of early strand and chromatid segregation following mutagenesis with the subsequent differential growth capability of the segregant cells, it is not possible to calculate mutation frequencies in this work. Future work will attempt to correct for this by altering the time of cloning after mutagen exposure.

ACKNOWLEDGEMENTS

This work was supported in part by grants CA 04484 and GM 19513 from the National Institutes of Health, as well as grant ACS IN-43-N from the American Cancer Society and an award from the Ramona Reed Memorial Fund.

REFERENCES

Beaudet, A.L.. D.J. Roufa, and C.T. Caskey 1973. Mutations affecting the structure of hypoxanthine: guanine phosphoribosyl-transferase in cultured Chinese hamster cells. *Proc. Nat. Acad. Sci., U.S.A.* 70: 320-324.

Bootsma, D., W. Keijzer, W.J. Kleijer, E.A. de Weerd-Kastelein, J. de Wit, P. Meera Khan, H. Van Someren, and A. Westerveld 1973. Mammalian cell culture studies. *Genetics Supplement* 73: 167-179.

Deaven, L.L. and D.F. Peterson 1973. The chromosomes of CHO, an aneuploid Chinese hamster cell line: G-band, and autoradiographic analysis. *Chromosoma* 41: 129-144.

DeLorenzo, R.J. and F.H. Ruddle 1970. Glutamate oxalate transaminase (GOT) genetics in Mus musculus: Linkage, polymorphism, and phenotypes of the Got-2 and Got-1 loci. *Biochem. Genet.* 4: 259-273.

Harris, M. 1971. Mutation rates in cells at different ploidy levels. *J. Cell. Physiol.* 78: 177-184.

Henning, V. and C. Yanofsky 1963. An electrophoretic study of mutually altered A proteins of the tryptophan synthetase of *Escherichia coli*. *J. Mol. Biol.* 6: 16-21.

Hopkinson, A.H. and H. Harris 1968. A third phosphoglucomutase locus in man. *Ann. Hum. Genet.*, London 31: 359-367.

Klebe, R.J., T. Chen, and F.H. Ruddle 1970. Mapping of a human genetic regulator element by somatic cell genetic analysis. *Proc. Nat. Acad. Sci., U.S.A.* 66: 1220-1227.

Meera Khan, P. 1971. Enzyme electrophoresis on cellulose acetate gel: zymogram patterns in man-mouse and man-Chinese hamster somatic cell hybrids. *Arch. Biochem. Biophys.* 145: 470-483.

Morrow, J. 1972. Population dynamics of purine and pyrimidine analog sensitive and resistant mammalian cells grown in culture. *Genetics* 71: 429-438.

Okada, S., M.L. Veath, J. Leroy, and J.S. O'Brien 1971. Ganglioside GM_2 storage diseases: hexosaminidase deficiencies in cultured fibroblasts. *Amer. J. Hum. Genet.* 23: 55-61.

Omenn, G.S. and P.T. Wade Cohen 1971. Electrophoretic methods for differentiating glycolytic enzymes of mouse and human origin. *In Vitro* 7: 132-139.

Osterman, J. and P.J. Fritz 1973. Pyruvate kinase isozymes: a comparative study in tissues of various mammalian species. *Comp. Biochem. Physiol.* 44B: 1077-1085,

Petras, M.L. and F.G. Biddle 1967. Serum esterases in the mouse, *Mus musculus*. *Can. J. Genet. Cytol.* 9: 704-710.

Poenaru, L. and J.C. Dreyfus 1973. Electrophoretic heterogeneity of human α-mannosidase. *Biochem. Biophys. Acta* 303: 171-174.

Povey, S. S.E.Gardiner, B. Watson, S. Mowbray, H. Harris, E. Arthur, C.M. Steel, C. Blenkinsop, and H.J. Evans 1973. Genetic studies on human lymphoblastiod lines: isozyme analysis on cell lines from forty-one different individuals and on mutants produced following exposure to a chemical mutagen. *Ann. Hum. Genet.*, London 36: 247-266.

Roufa, D.J., and B.N. Sadow, and C.T. Caskey 1973. Derivation of TK^- clones for revertant TK^+ mammalian cells. *Genetics* 75: 515-530.

Shaw, C.R., J.N. Baptist, D.A. Wright, and T.S. Matney 1973. Isolation of induced mutations in *E. coli* affecting the electrophoretic mobility of enzymes. *Mutation Res.* 18: 247-250.

Shaw, C.R., and R. Prasad 1970. Starch gel electrophoresis of enzymes--A compilation of recipes. *Biochem. Genet.* 4: 297-320.

Shows, T.B. and P.A. Lalley 1974. Control of lysosomal acid phosphatase expression in man-mouse cell hybrids. *Biochem. Genet.* 11: 121-139.

Siciliano, M.J., D.A. Wright, and C.R. Shaw 1974. Biochemical polymorphisms in animals as models for man. *Prog. in Med. Genet.* 10: 17-53.

Someren, H. van and H.B. van Henegouwen 1973. Independent loss of human hexosaminidases A and B in man-Chinese hamster somatic cell hybrids. *Humangenetik* 18: 171-174.

Someren, H. van, P. Meera Khan, A. Westerveld, and D. Bootsma 1972. Two new linkage groups in man, both carrying different loci for lactate dehydrogenase and glutamic-pyruvic transaminase. *Nature New Biology* 240: 221-223.

Spencer, N., D.A. Hopkinson, and H. Harris 1968. Adenosine deaminase polymorphism in man. *Ann. Hum. Genet.*, London 32: 9-14.

Susor, W.A., and W.J. Rutter 1971. A method for the detection of pyruvate kinase, aldolase and other pyridine nucleotide-linked enzyme activities after electrophoresis. *Anal. Biochem.* 43: 146-155.

Swank, R.T. and K. Paigen 1973. Biochemical and genetic evidence for a macromolecular β-glucuronidase complex in microsomal membranes. *J. Mol. Biol.* 77: 371-357.

Thompson, L.H. and R.M. Baker 1973. Isolation of mutants of cultured mammalian cells in *Methods in Cell Biology*, D.M. Prescott, Ed., Vol. VI, Academic Press, New York, pp. 209-281.

Tijo, J.H. and T.T. Puck 1958. Genetics of somatic mammalian cells. II. Chromosomal constitution of cells in tissue culture. *J. Exptl. Med.* 108: 259-268.

Westerveld, A. and P. Meera Khan 1972. Evidence for linkage between human loci for 6-phosphogluconate dehydrogenase and phosphoglucomutase$_1$ in man-Chinese hamster somatic cell hybrids. *Nature* 236: 30-32.

Westerveld, A., R.P.L.S. Visser, P. Meera Khan, and D. Bootsma 1971. Loss of human genetic markers in man-Chinese hamster somatic cell hybrids. *Nature New Biology* 234: 20-24.

Wright, D.A. and M.J. Siciliano 1973. Cellular differentiation and genetic regulation as studied through analysis of a lethal enzyme gene in heterozygotes. *Genetics* 74: s300 (Abstract).

Wright, D.A., M.J. Siciliano, and J.N Baptist 1972. Genetic evidence for the tetramer structure of glyceraldehyde-3-phosphate dehydrogenase. *Experientia* 28: 889.

Yanofsky, C. 1971. Tryptophan biosynthesis in *Escherichia coli*. Genetic determination of the proteins involved. *J.A.M.A.* 218: 1026-1035.

Zeeland, A.A. Van, M.C.E. Van Diggelen, and J.W.I.M. Simons. 1972. The role of metabolic cooperation in selection of hypoxanthine-guanine-phosphoribosyl-transferase (HG-PRT)-deficient mutants from diploid mammalian cell strains. *Mutation Res.* 14: 355-363.

GENETICS OF HUMAN SUPEROXIDE DISMUTASE ISOZYMES

G. BECKMAN
L. BECKMAN
Department of Medical Genetics
University of Umeå
901 85 Umeå, Sweden

ABSTRACT. Before the function of the enzyme was recognized by Dr. Fridovich and his group, superoxide dismutase was studied under different names such as erythrocuprein, hepatocuprein, indophenol oxidase, tetrazolium oxidase, and orgotein. In man there are two chemically and immunologically distinct isozymes: an anodally faster moving isozyme A, which is found in the cytosol, and a slower moving isozyme B, bound to the mitochondria. Erythrocytes lack isozyme B and polymorphonuclear leucocytes appear to lack isozyme A. Three different electrophoretically distinct phenotypes SOD 1, SOD 2-1 and SOD 2 have been found in isozyme A, while no variations so far have been observed in isozyme B. Family data are presented which show that the three SOD phenotypes are controlled by two autosomal alleles SOD^1 and SOD^2. The SOD^2 gene appears to be associated with a somewhat decreased enzyme activity. The SOD^2 gene is rare in most human populations and has so far been found to be polymorphic only in northern Sweden and Finland. The function of the superoxide dismutase isozymes is discussed, especially the possible importance of the lack of isozyme A for the bactericidal action of phagocytes.

The enzyme superoxide dismutase was first recognized and described by McCord and Fridovich (1968, 1969). The enzyme was found to catalyze the dismutation of $O_2^{\cdot-}$ free radicals to yield hydrogen peroxide and oxygen according to the following reaction.

$$O_2^{\cdot-} + O_2^{\cdot-} + 2H^+ \longrightarrow O_2 + H_2O_2$$

Whether or not the enzyme also has the ability to quench singlet oxygen, which has been proposed by some investigators (Finazzi Agró et al, 1972) remains to be verified. Evidence against the quenching ability of superoxide dismutase has been presented (Marklund and Marklund, 1974).

The physiological function of superoxide dismutase is apparently to protect organisms metabolizing oxygen against the potentially deleterious effect of superoxide free radicals (McCord et al, 1971). Further studies have verified the protective effect of superoxide dismutase against O_2 radicals (cf. Lavelle et al, 1973). Superoxide free radicals, which are known to be hazardous, have been found to be generated in a number of biological and chemical reactions not previously anticipated (cf. Fridovich, 1972, 1974a). Superoxide dismutase is expected to be generally distributed among oxygen metabolizing organisms, which also seems to be the case. McCord et al (1971) have shown that strictly anaerobic microorganisms contain no superoxide dismutase activity, while aerobic microorganisms do. The aerotolerant anaerobes show superoxide dismutase activity but this activity is generally lower than among the aerobic microorganisms. The chemical and biological properties of superoxide dismutase have recently been reviewed by Fridovich (1974a).

Synonymous names for superoxide dismutase

McCord and Fridovich (1969) reported that the enzyme responsible for the dismutation of $O_2^{\overline{\cdot}}$ free radicals viz. superoxide dismutase was a copper containing protein identical with the previously known copper containing proteins bovine hemocuprein and human erythrocuprein. No enzymatic function had been ascribed to these two proteins earlier. Simultaneously Carrico and Deutsch (1969) reported that human hepatocuprein, cerebrocuprein and erythrocuprein were the same protein. As the three proteins were found to be identical Carrico and Deutsch (1969) suggested that the protein should be given the common name cytocuprein. Bovine hepatocuprein and cerebrocuprein were originally found and described as separate proteins by Mann and Keilin (1939) and Porter and Folch (1957) respectively.

An enzyme with the property to inhibit the reduction of NBT (nitro blue tetrazolium) to blue formazan was described by Brewer(1967). The name indophenol oxidase was suggested for the enzyme until it could be further identified. Results, which have appeared during the last years (Beauchamp and Fridovich, 1971), imply that the enzyme by Brewer (1967) referred to as indophenol oxidase is superoxide dismutase (cf. Beckman, 1973). Tetrazolium oxidase, "oxidase" and "white patch enzyme" are other names used by some investigators instead of indophenol oxidase to describe an enzyme with the same property viz. inhibition of the reduction of NBT to blue formazan (Baur and Schorr, 1969; Selander et al, 1969; Kirjarinta et al, 1969; Johnson et al, 1973; Johnson and Selander, 1971; Lakovaara and Saura, 1971; Utter, 1971; Ayala et al, 1972; Burnet, 1972; Levan and Fredga, 1972; Wahren and Tegelström, 1973; Page and Whitt, 1973; Whitt et al, 1972, 1973).

A metalloprotein with exceptional pharmacodynamic properties, i.e. a pronounced general anti-inflammatory effect, has been the object of extensive pharmacological studies during the 1960's. This protein, called orgotein, has been shown to be identical with superoxide dismutase (cf. Huber, 1972).

Hemocuprein, erythrocuprein, hepatocuprein, cerebrocuprein, cytocuprein, indophenol oxidase, tetrazolium oxidase or "oxidase", "white patch" enzyme and orgotein are all names coined before the enzymatic function of the protein was known. After the discovery of the enzymatic function of the protein, superoxide dismutase seems to be the proper name to use.

Superoxide dismutase isozymes in Man

Two isozymes of superoxide dismutase, referred to as SOD A and SOD B, have been found in extracts of human organs, tissues and cell cultures. Isozyme B is lacking in erythrocytes and isozyme A in polymorphonuclear leucocytes. All other specimens show both isozymes but in variable proportions (cf. Beckman et al., 1973a). The enzyme appears not to be present in any body fluids (cf, Huber, 1972; Beckman et al., 1973a). Differential centrifugation of cell homogenates has revealed that isozyme A is localized in the soluble phase and isozyme B in fractions enriched in mitochondria (Beckman et al., 1973a). A specific mitochondrial isozyme of superoxide dismutase has also been observed in chicken liver (Fridovich, 1973; Weisiger and Fridovich, 1973a) and in bovine liver (Frantz, 1973 and Marklund, 1973). The mitochondrial superoxide dismutase from chicken liver is localized in the matrix space of the mitochondria (Weisiger and Fridovich, 1973b). The same localization of isozyme B, namely to the matrix space, has also been observed in mitochondria from human material (Beckman and Lundgren, 1974), and preliminary data indicate that isozyme B is about twice as large as isozyme A (Beckman, 1974). This seems to be the case also with the superoxide dismutase isozymes from chicken liver (Weisiger and Fridovich, 1973a) and bovine liver (Marklund, 1973).

Differences in the immunological specificity of the two isozymes have been observed (Beckman et al, 1974a). The electrophoretic separation of a supernatant solution after mixing anti-liver antibody with liver extract, followed by staining for superoxide dismutase activity, revealed no superoxide dismutase activity in the positions of the A and B isozymes. Enzyme activity was however found in an anodally slow moving enzyme-antienzyme complex. Hence the antisera against crude human liver extracts contained either one or more antibodies with the ability to react with both isozymes A and B of superoxide dismutase, or specific antibodies directed towards isozymes A and B, respectively. When erythrocytes and a KB cell fraction containing only isozyme A were used to absorb the antiserum only antibody activity against isozyme B was left, viz. the absorbed antiserum removed isozyme B and left isozyme A untouched. When extracts of polymorphonuclear leucocytes and a KB cell fraction containing only isozyme B were used to absorb the antiserum only antibody activity against isozyme A was left. These reciprocal absorption experiments suggest that isozymes A and B differ immunologically.

The fact that the two isozymes of superoxide dismutase differ immunologically must be considered in working out immunological methods for the determination of the level of superoxide in various tissues. This means that the immunological method used by Hartz et al (1973), for the determination of the level of superoxide dismutase in various tissues measures only the level of isozyme A, since the antiserum was made by immunizing with superoxide dismutase from erythrocytes, which contain only isozyme A.

Electrophoretic variations in isozyme A (Fig. 1) are not reflected in isozyme B indicating that the two isozymes are controlled by genes at separate gene loci (Brewer, 1967; Beckman et al, 1973a). Somatic cell hybridization experi-

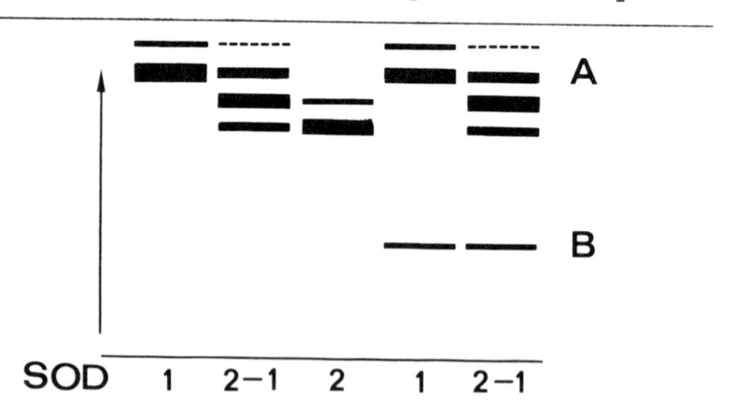

Fig. 1. Schematic drawing showing the two isozymes A and B and the three phenotypes SOD 1, 2-1, and 2 in extracts of (a) red cells, and (b) placental tissue. Buffer system: tris-borate EDTA, pH 8.6.

ments have shown that the gene controlling the fast moving isozyme of "tetrazolium oxidase" is located on the 21st chromosome, while the gene controlling the electrophoretically slow moving isozyme is located on the 6th chromosome (cf. Ruddle, 1973). Since the SOD A gene is located on the 21st chromosome one would expect a gene dosage effect on the SOD level in individuals with Down's syndrome.

Phenotypes of superoxide dismutase A

Three different phenotypes of the soluble form of superoxide dismutase, isozyme A, have been found. These are referred to as SOD 1, SOD 2-1, and SOD 2. The two genes which control the three phenotypes are referred to as SOD^1 and SOD^2 (Beckman, 1973). The electrophoretic patterns of the three superoxide dismutase phenotypes are shown in Fig. 1. The common homozygous phenotype SOD 1 shows one major zone and one minor anodally faster moving zone with superoxide dismutase activity. The rare homozygous phenotype, SOD 2, shows a similar isozyme pattern but with an overall slower electrophoretic mobility. The heterozygote, SOD 2-1, shows the parental zones and in addition a hybrid enzyme with intermediate electrophoretic mobility, which is in agreement with the biochemical findings that the enzyme is a dimer (cf. Beckman et al, 1973a). The soluble form of superoxide dismutase shows also in other species a hybrid enzyme in the heterozygote, for example in the dog (Baur and Schorr, 1969), the Kangaroo rat *Dipodomys* (Johnson and Selander, 1971) and the slug *Arion ater rufus* (Burnet, 1972). Assuming free recombination of monomers and equal activity contributions of both alleles, the staining intensities of the three main zones of the heterozygote, SOD 2-1, are expected to occur in the proportion 1:2:1. Visual inspection of the electrophoretic patterns of a series of SOD 2-1 heterozygotes suggests a somewhat weaker staining of the SOD 2 zone and thus that the SOD^2 gene may be associated with a somewhat lower enzyme activity (Beckman et al, 1973b). Preliminary results from an investigation of the specific activity of superoxide dismutase determined on purified enzyme from SOD 1 and SOD 2 individuals support this observation (Marklund et al, 1974).

Family data

The formal genetics of superoxide dismutase has been studied in family material from northern Sweden. The results verify the hypothesis (Beckman, 1973; Beckman et al, 1973a) that the three SOD phenotypes 1, 2-1, and 2 of isozyme A are controlled by two autosomal alleles (Beckman et al, 1974b). Fig. 2a and b shows two pedigrees illustrating the

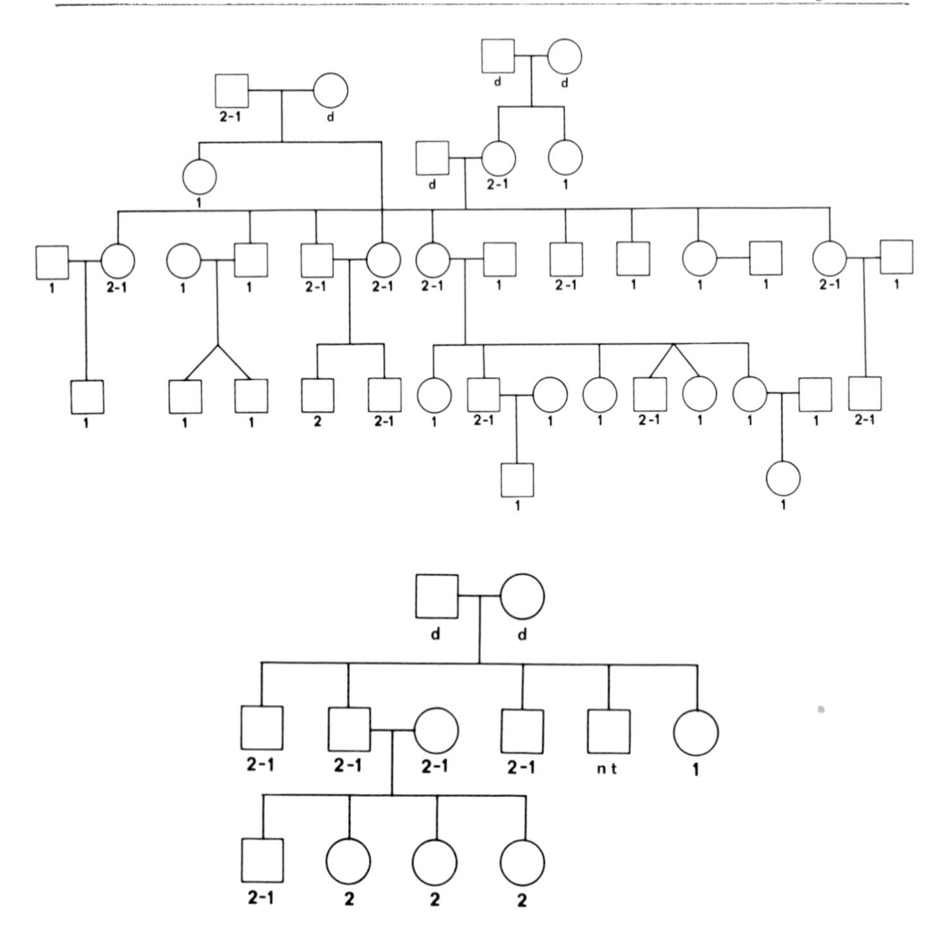

Fig. 2a and b. Pedigrees showing the inheritance of the SOD^2 gene. d = deceased, nt = not tested

segregation of the SOD genes. The inheritance of the SOD types in selected families is shown in Table 1. The reciprocal crosses 1 x 2-1 and 2-1 x 1 showed no difference and have therefore been pooled. The observed numbers show no statistically significant deviations from the expected numbers calculated according to the a priori method, but there is a tendency towards an overrepresentation of the SOD 2-1 and SOD 2 types. Thus in the back-cross the differ-

TABLE 1
INHERITANCE OF SOD PHENOTYPES IN SELECTED FAMILIES

Parental mating combination		SOD types of offspring			No. of offspring
		1	2-1	1	
1 x 1	observed	18	0	0	18
	expected	18	0	0	
1 x 2-1	observed	39	57	0	96
	expected	47.7	48.3	0	
2-1 x 2-1	observed	0	4	4	8
	expected	2	4	2	

ence between observed and expected numbers gives a X^2 of 3.38, $0.1 > P > 0.05$. The results give no support for a negative selection against the SOD 2-1 and SOD 2 types. Preliminary results have indicated that the SOD^2 gene is associated with a lower enzyme activity and therefore a priori one could suspect an association between this gene and diseases particularly those who stem from an increased instability of cells and/or cellular organelles. An examination of hospital records showed no accumulation of any particular disease either manifest or in the past in individuals with the SOD 2-1 and SOD 2 types (Beckman et al, 1974b).

Population data

Beckman (1973) reported a rather high frequency of the SOD 2-1 phenotype in the county of Norrbotten in northern Sweden. The third and anticipated phenotype SOD 2 was later found and described (Beckman and Pakarinen, 1973; Beckman et al, 1973b). In the county of Norrbotten the frequency of the SOD 2-1 type is about 2% and in the area close to the Finnish border (Tornedalen) the frequency is as high as 5%. In the last mentioned area the frequency of the SOD^2 allele is 0.0251, which is significantly higher than 0.01, a frequency commonly used to delineate polymorphic genes.

The population of northern Sweden is known to be quite heterogeneous and composed of ethnic groups of Lappish, Finnish, West-European origins. It is therefore of interest to try to trace the high frequency of the SOD^2 gene to any particular population. Out of 210 Lapps tested, one was found to have the SOD 2-1 phenotype. The mother of the propositus was of Lappish descent, while the father was not. Hence the frequency of the SOD^2 allele among the Swedish

TABLE 2

SUPEROXIDE DISMUTASE PHENOTYPES AND SOD^2 GENE FREQUENCY IN DIFFERENT POPULATION GROUPS

Population	Material	SOD phenotype				SOD^2 gene frequency	References
		1	2-1	2	n		
Swedes (Tornedalen)	RBC	1,628	80	2	1,710	0.0246	Beckman et al (1973)
Swedes (Norrbotten except Tornedalen)	RBC	1,475	12	0	1,487	0.0040	Beckman (1973)
Swedes (Västerbotten)	RBC	2,358	8	0	2,366	0.0017	Beckman (1973)
Swedes (Umeå)	PL	1,029	1	0	1,030	0.0005	Beckman and Pakarinen (1973)
Swedes (Uppsala)	PL	1,004	3	0	1,007	0.0015	Beckman and Pakarinen (1973)
Finland (northern Finland)	RBC	309	5	1	315	0.0111	Beckman and Pakarinen (1973)
Finland	RBC	398	8	0	406	0.0099	Kirjarinta et al (1969)
Finland (Swedish-speaking Finns)	RBC	397	1	0	398	0.0013	Kirjarinta et al (1969)
Lapps (Sweden)	RBC	209	1	0	210	0.0023	Beckman (1973)
Germany	RBC	4,094	6	0	4,100	0.0007	Ritter and Wendt (1971)
Southern Ireland	RBC	260	0	0	260	0.0000	Welch and Mears (1972)
Westray (Orkneys)	RBC	394	12	0	406	0.0148	Welch and Mears (1972)
Ethiopia	RBC	160	0	0	160	0.0000	Welch and Mears (1972)
Negros (Nigeria)	PL	132	0	0	132	0.0000	Beckman and Pakarinen (1973)
Chinese (Hawaii)	PL	67	0	0	67	0.0000	Beckman and Pakarinen (1973)

Japanese (Hawaii)	PL	348	0	0	348	0.0000	Beckman and Pakarinen (1973)
Filipino (Hawaii)	PL	83	0	0	83	0.0000	Beckman and Pakarinen (1973)
Polynesian (Hawaii)	PL	23	0	0	23	0.0000	Beckman and Pakarinen (1973)
Mixed population (Hawaii)	PL	1,028	0	0	1,028	0.0000	Beckman and Pakarinen (1973)
Papua	RBC	301	0	0	301	0.0000	Welch and Mears (1972)
Bengalis	RBC	271	0	0	271	0.0000	Das et al (1970)

RBC = Red blood cells; PL = placental extracts

Lapps is rather low and the Lapps are apparently not the source of the SOD^2 gene. A Finnish origin seems more likely. The frequency of the SOD 2-1 phenotype was found to be significantly higher among individuals with Finnish surnames than that among individuals with non-Finnish surnames ($P < 0.005$). The SOD^2 gene frequency was found to be about 1% in a series of blood donors from Oulu in northern Finland (Beckman and Pakarinen, 1973). The high frequency of the SOD^2 gene in the Tornedalen area has been confirmed in a second study of this population (Beckman et al, 1973b). In this study a good agreement was found between the observed and expected numbers of the three SOD phenotypes.

A number of deviating phenotypes of superoxide dismutase has been observed in different parts of the world. They all appear to be electrophoretically identical to the phenotype described by Brewer (1967) and hence to the SOD 2-1 type except for two types found in Australia (Kirk, personal communication). Brewer found one individual with a deviating type out of several thousand individuals tested from different parts of the world. No information was given as to the origin of the different populations studied. Welch and Mears (1972) and Welch et al (1973) reported a relatively high frequency (about 3%) of what appears to be the SOD 2-1 phenotype on the Westray Island of the Orkneys. The original settlers on this island were of Scandinavian descent. Ritter and Wendt (1971) found the SOD 2-1 phenotype in six individuals out of 4,100 tested in a population from southern Germany. Kirjarinta et al (1969) reported the occurrence of a "white patch" variant (SOD 2-1) in 8 out of 406 non-related Finns, in 1 among 398 Swedish-speaking Finnish subjects and in 11 members of two families found in an examination of 151 Skolt Lapps. The SOD 2-1 phenotype has also been observed in the Australian and Iranian populations (Kirk, personal communication). The population data so far available (Table 2) show that the frequency of the SOD^2 gene is low in most parts of the world with the exception of northern Sweden and Finland.

Functional aspects

The physiological function of superoxide dismutase implies that lack of the enzyme is not compatible with life for an oxygen metabolizing organism (McCord et al, 1971). The reason is the lack of a protective mechanism against the very reactive O_2^- radical produced in biological oxidations involving oxygen (cf. Fridovich, 1972, 1973). The discovery of superoxide dismutase as a scavenger for O_2^- radicals has been followed by a burst of investigations trying to evaluate the role of the O_2^- radicals in various biological processes and the protective effect of superoxide dismutase.

The protective effect of SOD against oxygen toxicity and ionizing radiation was proposed by Fridovich (1972). The mechanism behind oxygen toxicity is not clearly understood even though the production of free radicals is suspected to be part of the tissue damaging effect associated with an increase in oxygen tension (cf. Hauugaard, 1968). Oxygen radicals are, for example, known to be involved in lipid peroxidation, a phenomenon known to be associated with oxygen toxicity. Zimmerman et al (1973) have shown that superoxide dismutase has the ability to protect against lipid peroxidation. Ionizing radiation leads to the production of oxygen radicals, a process which is facilitated by increased oxygen tension. It is therefore of interest to study the radiosensitivity of cells with different SOD levels.

A useful material for radiosensitivity studies should be individuals with the SOD 2 phenotype and with Down's syndrome. The SOD 2 type appears to be associated with lower enzyme activity, and since the SOD gene is localized on the 21st chromosome patients with Down's syndrome are expected to show a 50% increase in their SOD activity.

PHA-stimulated lymphocytes have been shown to express an increased resistance against radiation damage. Preliminary data from our laboratory suggest an increased SOD level in PHA stimulated lymphocytes. The observation that polymorphonuclear leucocytes apparently lack isozyme A of superoxide dismutase (Beckman et al, 1973a) may open some interesting new aspects on the bacteriocidal mechanism related to the PMN's. It was earlier reported by Huber (1972) that PMN's lack superoxide dismutase, a situation which, however, would seem incompatible with life. The observation by Beckman et al (1973a) that PMN's lack isozyme A but not the mitochondrial form of the enzyme, isozyme B, provides an explanation to this apparent paradox. De Chatelet et al (1974) propose that the cyanide-insensitive form of superoxide dismutase viz. isozyme B, is present also in the cytosol in polymorphonuclear leucocytes, a result not in agreement with the results presented by Weisiger and Fridovich (1973b). They found the cyanide-insensitive form of superoxide dismutase localized only to the mitochondria. Further studies will be needed to verify the observation by De Chatelet et al (1974).

The exact mechanism behind the bacteriocidal function of the polymorphonuclear leucocytes is not known with certainty. Theories have been put forward, many adding valuable information to the overall picture of the bacteriocidal mechanism, but no complete picture has emerged (cf. Klebanoff, 1971).

The fact that PMN's appear to contain only the mitochondrial form of superoxide dismutase, isozyme B, opens the door for speculations as to the possible role of the lack of the cytosol isozyme A in PMN's. The bactericidal action of phagocytes and the possible role of the superoxide radical and of superoxide dismutase has been discussed by Fridovich (1974b). Johnston et al (1973) have shown that superoxide dismutase can inhibit the phagocytic killing of bacteria.

REFERENCES

Ayala, F. J., J. R. Powell, M. L. Tracey, C. A. Mourão, and S. Pérez-Salas 1972. Enzyme variability in the *Drosophila Willistoni* group. IV. Genetic variation in natural populations of *Drosophila Willistoni*. *Genet.* 70: 113-139.

Baur, E. W. and R. T. Schorr 1969. Genetic polymorphism of tetrazolium oxidase in dogs. *Science* 166: 1524-1525.

Beauchamp, C. and I. Fridovich 1971. Superoxide dismutase: improved assays and an assay applicable to acrylamide gels. *Analyt. Biochem.* 44: 276-287.

Beckman, G. (unpublished data)

Beckman, G. 1973. Population studies in northern Sweden. VI. Polymorphism of superoxide dismutase. *Hereditas* 73: 305-310.

Beckman, G. and E. Lundgren. (unpublished data).

Beckman, G. and A. Pakarinen 1973. Superoxide dismutase: a population study. *Hum. Hered.* 23: 346-351.

Beckman, G., L. Beckman, and S. Holm 1974a. Immunological differences between human superoxide dismutase isozymes. (in prep.).

Beckman, G., L. Beckman, and L.-O. Nilsson 1973b. A rare homozygous phenotype of superoxide dismutase, SOD 2. *Hereditas* 75: 138-139.

Beckman, G., L. Beckman, and L.-O. Nilsson 1974b. Genetics of human superoxide dismutase. *Hereditas* (subm. for publ.).

Beckman, G., E. Lundgren, and A. Tärnvik 1973a. Superoxide dismutase isozymes in different human tissues, their genetic control and intracellular localization. *Hum. Hered.* 23: 338-345.

Brewer, G. J. 1967. Achromatic regions of tetrazolium stained starch gels: Inherited electrophoretic variation. *Amer. J. Hum. Genet.* 19: 674-680.

Burnet, B. 1972. Enzyme protein polymorphism in the slug *Arion ater*. *Genet. Res.* 20: 161-173.

Carrico, R. J. and H. F. Deutsch 1969. Isolation of human hepatocuprein and cerebrocuprein, their identity with erythrocuprein. *J. Biol. Chem.* 244: 6087-6093.

Das, S. R., B. N. Mukherjee, and S. K. Das 1970. The distribution of some enzyme group systems among Bengalis. *Indian J. Med. Res.* 58: 866-875.

De Chatelet, L. R., C. E. McCall, L. C. McPhail, and R. B. Johnston, Jr. 1974. Superoxide dismutase activity in leucocytes. *J. Clin. Invest.* 53: 1197-1201.

Finazzi Agro, A., C. Giovagnoli, P. De Sole, L. Calabrese, G. Rotilio, and B. Mondovi 1972. Erythrocuprein and singlet oxygen. *FEBS Letters* 21: 183-185.

Frants, R. 1973. Isolation of superoxide dismutase from bovine liver. *Acta Academiae Aboensis, ser. B.* 33: 1-6.

Fridovich, I. 1972. Superoxide radical and superoxide dismutase. *Accounts of Chemical Research,* 5: 321-326.

Fridovich, I. 1973. Superoxide radical and superoxide dismutase. *Biochem. Soc. Transac.* 1: 48-50.

Fridovich, I. 1974a. Superoxide dismutase. In *Molecular mechanisms of oxygen activation.* Ed. O. Hayaishi, Acad. Press

Fridovich, I. 1974b. Superoxide radical and the bactericidal action of phagocytes. *N. Engl. J. Med.* 290:624-625.

Hartz, J. W., S. Funakoshi, and H. F. Deutsch 1973. The levels of superoxide dismutase and catalase in human tissues as determined immunochemically. *Clin. Chim. Acta* 46: 125-132.

Haugaard, N. 1968. Cellular mechanisms of oxygen toxicity. *Physiol. Rev.* 48: 311-373.

Huber, W. 1972. Orgotein: a metalloprotein drug, investigational brochure for the physician. DDI report.

Johnson, W. E., and R. K. Selander 1971. Protein variation and systematics in Kangaroo rats (genus *Dipodomys*). *System. Zool.* 20: 377-405.

Johnson, A. G., F. M. Utter, and H. O. Hodgins 1973. Estimate of genetic polymorphism and heterozygosity three species of the rockfish (genus *Sebastes*). *Comp. Biochem. Physiol.* 44: 397-406.

Johnston, R. B., Jr., B. Keele, L. Webb, D. Kessler, and K. V. Rajagopalan 1973. Inhibition of phagocytic bactericidal activity by superoxide dismutase: a possible role for superoxide anion in the killing of phagocytized bacteria. *J. Clin. Invest.* 52: 44a (abstr.).

Kirjarinta, M., J. Fellman, C. Gustafsson, E. Keisala, and A. W. Eriksson 1969. Two rare electrophoretic variants of erythrocyte enzymes in Finland. *Scand. J. Clin. Lab. Invest.* 23 suppl. 108: 46.

Klebanoff, S. J. 1971. Introleucocytic microbicidal defects. *Ann. Rev. Med.* 22: 39-62.

Lakovaara, S. and A. Saura 1971. Genetic variation in natural

populations of *Drosophila Obscura*. *Genet*. 69: 377-384.
Lavelle, F., A. M. Michelson, and L. Dimitrijevic 1973. Biological protection by superoxide dismutase. *Biochem. Biophys. Res. Commun*. 55: 350-357.
Levan, G. and K. Fredga 1972. Isozyme polymorphism in three species of land snails. *Hereditas* 71: 245-252.
Mann, T. and D. Keilin 1939. Haemocuprein and hepatocuprein, copper-protein compounds of blood and liver in mammals. *Proc. Roy. Soc. B*. 126: 303-315.
Marklund, S. 1973. A novel superoxide dismutase of high molecular weight from bovine liver. *Acta Chem. Scand*. 27: 1458-1460.
Marklund, S. and G. Marklund 1974. Involvement of the superoxide anion radical in the autoxidation of pyrogallol and a convenient assay for superoxide dismutase. *Europ. J. Biochem*. (subm. for publ.).
Marklund, S., G. Beckman, L.-O. Nilsson, and T. Stigbrand. (unpubl. data).
McCord, J. M. and I. Fridovich 1968. The reduction of cytochrome C by milk xanthine oxidase. *J. Biol. Chem*. 243: 5753-5760.
McCord, J. M. and I. Fridovich 1969. Superoxide dismutase. An enzymic function for erythrocuprein (hemocuprein). *J. Biol. Chem*. 244: 6049-6055.
McCord, J. M., B. B. Keele, Jr., and I. Fridovich 1971. An enzyme-based theory of obligate anaerobiosis: the physiological function of superoxide dismutase. *Proc. Nat. Acad. Sci., Wash*. 68: 1024-1027.
Page, L. M. and G. S. Whitt 1973. Lactate dehydrogenase isozymes, malate dehydrogenase isozymes and tetrazolium oxidase mobilities of darters *(Etheostomatini)*. *Comp. Biochem. Physiol*. 44: 611-623.
Porter, H. and J. Folch 1957. Cerebrocuprein. I. A copper containing protein isolated from brain. *J. Neurochem*. 1: 260.
Ritter, H. and G. G. Wendt 1971. Indophenol oxydase variability. *Humangenetik* 14: 72.
Ruddle, F. H. 1973. Linkage analysis in man by somatic cell genetics. *Nature* 242: 165-169.
Selander, R. K., W. G. Hunt, and S. Y. Yang 1969. Protein polymorphism and genetic heterozygosity in two European subspecies of the house mouse. *Evol*. 23: 379-390.
Utter, F. M. 1971. Tetrazolium oxidase phenotypes of rainbow trout (*Salmo gairdneri*) and Pacific salmon (*Oncorhynchus spp*.). *Comp. Biochem. Physiol*. 39B: 891-895.
Wahren, H. and H. Tegelstrom 1973. Polymorphism of esterases and tetrazolium oxidases in the Roman snail, *helix*

pomatia: a study of populations from Sweden and Germany. *Biochem. Genet.* 9: 169-174.

Weisiger, R. A. and I. Fridovich 1973a. Superoxide dismutase. Organelle specificity. *J. Biol. Chem.* 248: 3582-3592.

Weisiger, R. A. and I. Fridovich 1973b. Mitochondrial superoxide dismutase. *J. biol. Chem.* 248: 4793-4796.

Welch, S. G. and G. W. Mears 1972. Genetic variants of human indophenol oxidase in the Westray Island of the Orkneys. *Hum. Hered.* 22: 38-41.

Welch, S. G., J. V. Barry, B. E. Dodd, P. D. Griffiths, R. G. Huntsman, G. C. Jenkins, P. J. Lincoln, M. McCathie, G. W. Mears, and C. W. Parr 1973. A survey of blood group, serum protein and red cell enzyme polymorphisms in the Orkney Islands. *Hum. Hered.* 23: 230-240.

Whitt, G. S., P. I. Cho, and W. F. Childers 1972. Preferential inhibition of allelic isozyme synthesis in an interspecific sunfish hybrid. *J. Exp. Zool.* 179: 271-282.

Whitt, G. S., W. F. Childers, J. Tranquilli, M. Champion 1973. Extensive heterozygosity at three enzyme loci in hybrid sunfish populations. *Biochem. Genet.* 8: 55-72.

Zimmerman, R., L. Flohe, U. Weser, and H.-J. Hartmann 1973. Inhibition of lipid peroxidation in isolated inner membrane of rat liver mitochondria by superoxide dismutase. *FEBS Letters* 29: 117-120.

MOLECULAR SYSTEMATICS AND CLINAL VARIATION IN MACAQUES

LINDA L. DARGA[1], MORRIS GOODMAN[1], MARK L. WEISS[1],
G. WILLIAM MOORE[1], WILLIAM PRYCHODKO[2],
HOWARD DENE[2], RICHARD TASHIAN[3], and ANN KOEN[1]

[1]Department of Anatomy
Wayne State University Medical School
Detroit, Michigan 48202

[2]Department of Biology
Wayne State University
Detroit, Michigan 48202

[3]Department of Human Genetics
University of Michigan
Ann Arbor, Michigan 48104

ABSTRACT. Gene frequency data were gathered on seven genetic loci from 17 populations of ten species of macaques from known geographic areas. The data were analyzed via two computer methods, each generating dendrograms depicting phylogenetic relationships. These relationships are discussed in terms of the traditional taxonomy of macaques. *M. fascicularis*, *M. mulatta*, *M. cyclopis* and *M. fuscata* form a closely related assemblage. *M. speciosa*, the Celebes and pig-tailed macaques are the most divergent groups. The Celebes populations are joined together and form an assemblage with the *M. nemestrina* groups, a geographically reasonable union. Clinal patterns of genetic variation suggest the operation of natural selection.

INTRODUCTION

The systematics of the genus *Macaca* has been a problem for primate biologists. There is no general agreement on the number of species, estimates varying between a minimum of 11 (Ellerman and Morrison-Scott, 1951) and a maximum of 16 (Kellogg, 1945).

Until recently the taxonomic and, by inference, phylogenetic relationships within this genus were based primarily on morphological characteristics such as cranial measurements, tail length, pattern of coloration and length and direction of growth of hair on the crown and face. The use of simply inherited genetic characteristics has become widely accepted as an adjunct to the classic morphological traits. This study utilizes several genetic markers in an attempt to clarify the

TABLE I
SPECIES AND GEOGRAPHIC ORIGIN

Species and origin	Common Name	Abbreviation	Number of Samples
M. fascicularis Thailand	Irus macaque	M. fasc Th	197
M. fascicularis Malaya		M. fasc Ma	263
M. fascicularis Philippines		M. fasc Ph	78
M. fascicularis Java		M. fasc Ja	22
M. mulatta Thailand	Rhesus macaque	M. mul Th	28
M. mulatta Bangladesh		M. mul Ba	55
M. mulatta India		M. mul In	26
M. nemestrina Thailand	Pig-tailed macaque	M. nem Th	92
M. nemestrina Malaya		M. nem Ma	162
M. nemestrina Sumatra		M. nem Su	203
M. speciosa Thailand	Stump-tailed macaque	M. spec Th	201
M. cyclopis Formosa	Formosan macaque	M. cyc Fo	118
M. fuscata Japan	Japanese macaque	M. fusc Ja	236
M. radiata India	Bonnet macaque	M. rad In	59
M. maura Celebes	Moor macaque	M. ma Ce	9
M. niger Celebes	Black ape	M. ni Ce	10
M. sinica Ceylon	Toque monkey	M. sin Cy	71

systematics of the macaques.

To date, gene frequency data on 17 populations of ten species of macaques from known geographic areas have been gathered by starch and urea gel electrophoreses and immuno-electrophoresis; this represents over 1700 individuals sampled. Table I lists the species, their common names and number of samples tested.

MATERIALS AND METHODS

Seven genetic loci were analyzed: hemoglobin alpha and beta chains (Ishimoto et al, 1970), two erythrocyte enzymes--6-PGD (Prychodko et al, 1971) and carbonic anhydrase I (Tashian et al, 1971), a gene which controls the expression of CA I (DeSimone et al, 1973)--and two serum proteins, thyroxine binding prealbumin (Weiss et al, 1971) and transferrin (Prychodko et al, 1971) and the analysis of new samples by the authors, this paper. Tables II through VII list the gene frequency data for each population analyzed.

Usage of the term "gene frequencies" can be problematical since the assumption of allelic variation may be incorrect. Breeding data for some species of macaques support the assumption that allelic variants of the structural locus of CA I (Tashian et al, 1971), transferrin (Goodman and Wolf, 1963) and prealbumin (Bernstein et al, 1970) are inherited in a Mendelian fashion. Hemoglobin presents a more complex picture. We describe our hemoglobin results by assuming a single alpha chain locus with four alleles and a single beta chain locus with three allelic variants.

The analysis of genetic distance was carried out via the use of an index of dissimilarity (ID): one-half the sum of the differences in allele frequencies at one locus between two populations. Algebraically ID is: $\sum_{i-1}^{n} \frac{|X_{ij} - X_{ik}|}{2}$

where X_{ij} is the frequency of the ith allele in the jth population, and X_{ik} is the frequency of that allele in the kth population. Therefore, complete identity of allele frequencies yields an ID of zero while complete dissimilarity yields an ID of one. ID values comparing two populations for several loci can be obtained by averaging.

The resulting dissimilarity matrix was analyzed by two different computer programs, the unweighted pair group method (UWPGM) of Sokal and Michener (1958) based on the divergence hypothesis, and the maximum parsimony method (patterned after the approach used with protein sequence data of Moore et al, 1973), which have proved useful in inferring the course of

TABLE II
ALLELIC FREQUENCIES FOR MACAQUE CARBONIC ANHYDRASE I AND THE CONTROL LOCUS

Species	Sample Size	CA I Alleles							CA I Control Alleles	
		A	A_2	B	C	D	D_2	D_3	X^-	X^+
M. fasc Th	(136)	.978			.007		.015			1.0
M. fasc Ma	(261)	.979		.002	.019					1.0
M. fasc Ph	(74)	1.0								1.0
M. fasc Ja	(7)	1.0								1.0
M. mul Th	(15)	.767					.233			1.0
M. mul Ba	(26)	1.0								1.0
M. mul In	(24)	1.0								1.0
M. nem Th	(63)	.109		.829		.072			.554	.446
M. nem Ma	(165)	.556		.416	.028				.568	.432
M. nem Su	(73)	.390		.610					.700	.300
M. spec Th	(145)	.293			.707					1.0
M. cyc Fo	(92)	.995					.005			1.0
M. fusc Ja	(137)	.842					.158			1.0
M. rad In	(52)				1.0					1.0
M. ma Ce	(7)		.643					.357		1.0
M. ni Ce	(6)		.083					.917		1.0
M. sin Cy	(55)	.490			.510					1.0

evolution from gene frequency data; each generates a dendogram depicting the degrees of genetic similarity.

The divergence hypothesis, which assumes relatively uniform rates of evolution in all lines, states that the more ancient the common ancestor for a pair of OTUs the greater is the genetic distance for that pair of species. This model builds from the smallest to the largest branches joining together two members with the lowest average dissimilarity value between them; a phenogram is generated showing which populations are phenetically closest. If marked nonuniformities in evolutionary rates are present in any lineage, fallacious phylogenies may result. The more closely the actual evolutionary rates fit the assumption of uniform rates, the higher the probability that the results describe the true cladogeny of the group.

A stronger method for depicting phylogenetic relationships, is based on the hypothesis of parsimonious evolution. In constructing a phylogenetic tree, evolutionary rates in different lines need not be uniform. The assumptions of the method are: 1) that evolutionary changes take place in the fewest steps needed to account for them; 2) that the ancestral gene

TABLE III

FREQUENCY OF TRANSFERRIN ALLELES IN MACAQUES

Species	Tf Alleles														Sample Size
	B	C	C'	D	D'	D''	E	F	F'	F''	G	G'	H	H'	
M. fasc Th	.036	.277	.055	.439	.005	.008	.051	.036	.051	.003	.058			.033	(197)
M. fasc Ma	.013	.117		.674		.002	.025	.025	.042		.023	.004		.076	(263)
M. fasc Ph				1.0											(78)
M. fasc Ja		.175	.025	.550	.025		.100				.050		.075		(20)
M. mul Th	.071	.214		.036			.071	.286			.304			.018	(28)
M. mul Ba	.082	.354		.118	.073	.027	.100	.091		.018	.109		.018	.009	(55)
M. mul In	.039	.615		.115	.039			.058		.019	.096			.019	(26)
M. nem Th	.005	.120		.196		.005	.011	.130			.533				(92)
M. nem Ma		.056		.361			.003	.275	.019		.287				(162)
M. nem Su		.067		.350			.002	.230			.342			.010	(203)
M. spec Th		.003							.445					.552	(201)
M. cyc Fo		.920								.081					(118)
M. fusc Ja								.973			.027				(236)
M. rad In		.568						.407			.025				(59)
M. ma Ce								.500	.056		.444				(9)
M. ni Ce								.950			.050				(10)
M. sin Cy	.058	.246		.051			.283	.022		.007	.333				(69)

TABLE IV
ALLELIC FREQUENCIES FOR MACAQUE PREALBUMIN

Species	TBPA Alleles		Sample size
	F	S	
M. fasc Th	.996	.004	(119)
M. fasc Ma	.889	.111	(356)
M. fasc Ph	.940	.060	(58)
M. fasc Ja	.841	.159	(22)
M. mul Th	.714	.286	(14)
M. mul Ba	.885	.115	(39)
M. mul In	.929	.071	(28)
M. nem Th	.983	.017	(57)
M. nem Ma	1.0		(231)
M. nem Su	.984	.016	(95)
M. spec Th	1.0		(292)
M. cyc Fo	.711	.289	(159)
M. fusc Ja	.003	.997	(158)
M. rad In	.705	.295	(56)
M. ma Ce	1.0		(9)
M. ni Ce	1.0		(9)
M. sin Cy	.385	.615	(65)

frequencies, at the forks of the tree, should yield the fewest changes possible in gene frequencies over the entire tree. Thus, we are required to choose that network topology and those ancestral frequencies which minimize the IDs summed throughout the network. In the execution of this program, the initial tree is transformed into a network, which lacks a root, to facilitate length calculations.

Applied to gene frequency data of extant populations, the assumption of evolutionary change occurring in the fewest possible steps could become strained. However, no evidence concerning biological evolution and macaque speciation would necessarily invalidate this premise.

The basis for this approach is that similar allele frequencies tend to reflect common ancestry rather than haphazard variation. Therefore, the network that minimizes the amount of transition due to haphazard variations in allele frequencies is the best estimate of true ancestral history.

Thus, in the analysis of macaque systematics, this paper reflects two new features: additional samples, from populations representing new geographic areas, were included, and a more powerful method for depicting phylogenetic relationships was developed.

TABLE V
ALLELIC FREQUENCIES FOR MACAQUE 6-PGD

Species	6-PGD Alleles					Sample size
	A	B	C	D	S	
M. fasc Th	.967		.026	.007		(136)
M. fasc Ma	.854		.105	.042		(263)
M. fasc Ph	1.0					(78)
M. fasc Ja	1.0					(21)
M. mul Th	1.0					(15)
M. mul Ba	.840	.160				(25)
M. mul In	.729	.271				(24)
M. nem Th	.992				.008	(63)
M. nem Ma	.978		.022			(162)
M. nem Su	.951		.021		.028	(72)
M. spec Th	.426			.574		(168)
M. cyc Fo	.994				.006	(88)
M. fusc Ja	1.0					(148)
M. rad In	1.0					(54)
M. ma Ce	1.0					(7)
M. ni Ce	1.0					(6)
M. sin Cy	1.0					(57)

GENE FREQUENCY ANALYSIS VIA UWPGM AND MAXIMUM PARSIMONY

Fig. 1 shows the dendrogram generated by UWPGM. The two rhesus populations from India and Bangladesh join together early in the runs, soon joined by the *M. cyclopis* population. The pairwise dissimilarity distance between the Thai and Indian populations of *M. mulatta* actually exceeds that between *M. cyclopis* and each of the three rhesus populations. Thai *M. mulatta* then joins these populations.

M. fascicularis populations join together early in the agglomerative cycles, with the Thai and Javan groups uniting first, then joined by the Malayan and the Philippine populations. Interspecifically, the irus and rhesus then join. The close phenetic relationship of these two species groups could be accounted for by recent common ancestry or by interbreeding, some evidence of which has been discovered by Fooden (1964). Indeed, the genetic distance between Thai irus and Bangladesh rhesus is equal to the distance between the conspecific Thai and Philippine irus populations (see Table VIII).

M. radiata and *M. sinica* are united, illustrating their phenetic similarity, and this assemblage is then joined to the

TABLE VI

TABLE VII

ALLELIC FREQUENCIES FOR MACAQUE HEMOGLOBIN ALPHA AND BETA CHAINS

Species	HB Alpha Alleles				HB Beta Alleles			Sample size
	M	M'	H	P	M	H	O	
M. fasc Th	.185		.715	.099		1.0		(197)
M. fasc Ma	.401		.532	.067		1.0		(263)
M. fasc Ph	.500		.500			1.0		(59)
M. fasc Ja	.139		.861			1.0		(18)
M. mul Th			1.0			1.0		(27)
M. mul Ba			1.0			1.0		(47)
M. mul In			1.0			1.0		(26)
M. nem Th	.612		.388			1.0		(85)
M. nem Ma	1.0					1.0		(160)
M. nem Su	.993		.007			1.0		(71)
M. spec Th	.495		.505			1.0		(193)
M. cyc Fo			1.0			1.0		(102)
M. fusc Ja			1.0			1.0		(218)
M. rad In	.275		.725			1.0		(49)
M. ma Ce	.786	.214			.143	.143	.714	(7)
M. ni Ce	.833	.167			.500		.500	(6)
M. sin Cy	.049		.951			1.0		(71)

TABLE VIII
INDEX OF DISSIMILARITY MATRIX FOR MACAQUE OTUs BASED ON 7 LOCI

			2	3	4	5	6	7	8	9	10	11	12	13	14	15	16	17
1	MI	Th	.10	.14	.09	.20	.14	.16	.36	.33	.38	.36	.19	.34	.30	.51	.54	.27
2	MI	Ma		.09	.12	.25	.19	.21	.36	.33	.38	.33	.24	.37	.33	.52	.54	.33
3	MI	Ph			.13	.27	.23	.24	.34	.32	.36	.34	.25	.37	.35	.49	.51	.35
4	MI	Ja				.17	.12	.16	.38	.37	.41	.40	.16	.30	.30	.55	.57	.24
5	MM	Th					.14	.18	.39	.38	.42	.43	.15	.21	.25	.51	.56	.17
6	MM	Ba						.06	.41	.42	.45	.41	.14	.30	.31	.56	.59	.23
7	MM	In							.43	.43	.47	.40	.12	.32	.30	.58	.60	.27
8	MN	Th								.18	.16	.45	.46	.56	.41	.46	.54	.44
9	MN	Ma									.06	.47	.47	.53	.46	.44	.49	.46
10	MN	Su										.50	.51	.58	.49	.47	.52	.49
11	MS	Th											.44	.54	.34	.55	.58	.40
12	MC	Fo												.27	.25	.59	.61	.23
13	MF	Ja													.36	.62	.57	.27
14	MR	In														.49	.51	.25
15	MMa	Ce															.20	.58
16	MNi	Ce																.64
17	MSi	Cy																.00

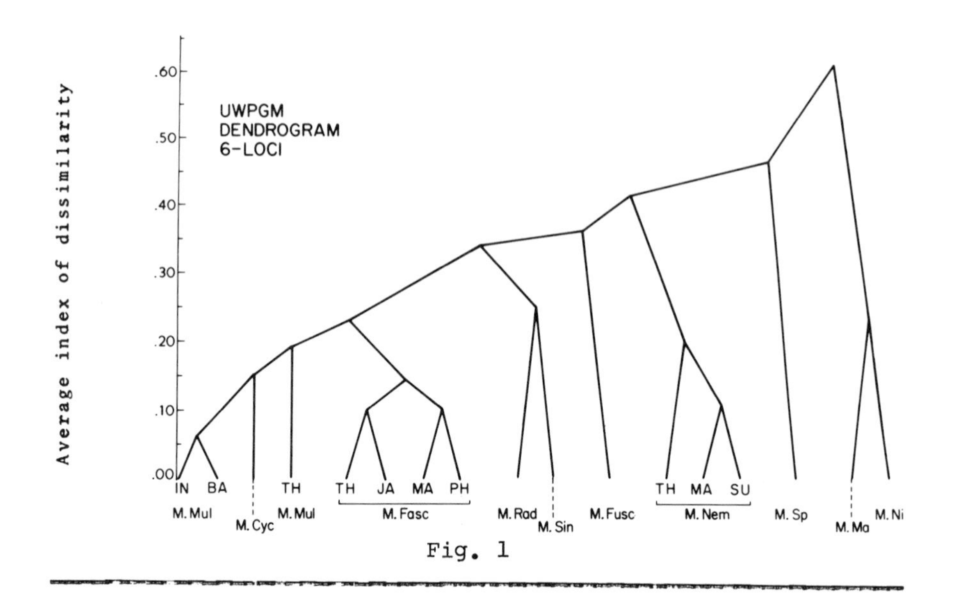

Fig. 1

rhesus-irus assemblage, followed by the union of the Japanese macaque, the ID between *M. fuscata* and *M. mulatta* being sizeable. However when *M. fuscata* is compared just to Thai *M. mulatta*, a much smaller ID value is obtained (see Table VIII). Finally, *M. speciosa* joins this assemblage.

The union of the Malayan and Sumatran populations of *M. nemestrina* is joined by the Thai *M. nemestrina*. The pigtailed macaques show no particularly close relationship to the rhesus or irus groups, even when compared to sympatric populations of the other two species.

The genetic dissimilarity separating *M. maura* from *M. niger* only slightly exceeds that separating the Thai and Sumatran populations of the single species, *M. nemestrina*. Thus, the Celebes varieties are united first and then join the rest of the macaques as the most divergent of the populations analyzed.

Fig. 2 illustrates the dendrogram generated by the maximum parsimony approach. The difference between the two trees is a function of the different assumptions behind the two methods of constructing the dendrograms.

This dendrogram depicts the amounts of change in gene frequency in each lineage by the length of the line representing that lineage. It becomes apparent that the rates of evolution in different lineages have varied considerably.

Again, the Malayan and Philippine irus join together early.

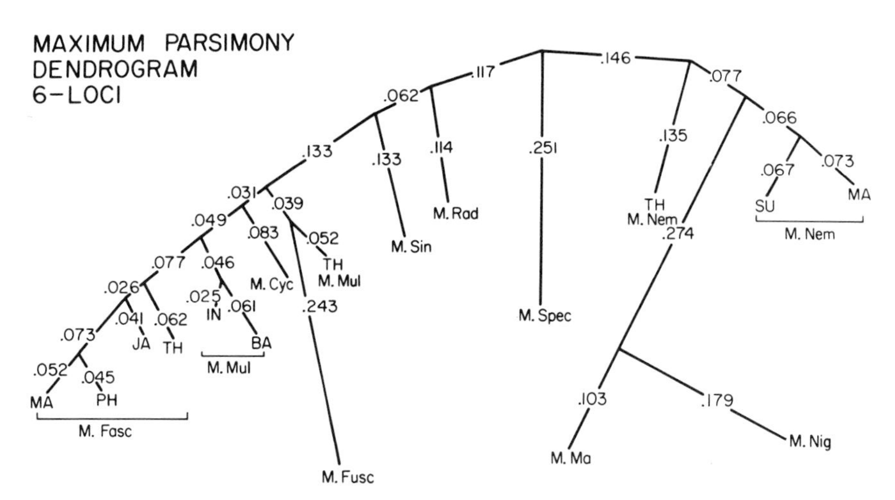

Fig. 2. Link lengths, based on the amount of change in gene frequencies in each line, are indicated.

Next, the Javan irus population and finally, the Thai population join the other irus groups.

As before, Indian and Bangladesh rhesus join each other. They next join the irus populations and then *M. cyclopis* joins this assemblage. Next, *M. fuscata* and Thai *M. mulatta* join each other and are then united with the assemblage of irus, Indian and Bangladesh rhesus, and Formosan macaques. There has been considerable evolutionary change in the Japanese macaque line, as shown in Fig. 2, however the total length of the tree is decreased by its union with the Thai *M. mulatta*. This supports the hypothesis that *M. fuscata* evolved from an isolated rhesus population.

M. sinica next joins the irus-mulatta-like assemblage followed by *M. radiata*. These two species are no longer joined together before uniting with the rhesus and irus groups. *M. sinica* seems to have undergone slightly more evolutionary change than *M. radiata*, accounted for by change at the transferrin and prealbumin loci.

The most ancient split in this genus is depicted as having occurred at the point where *M. speciosa* diverged from pig-tailed and Celebes macaques on the one hand and from the irus-mulatta-like-sinica-radiata assemblage on the other; it is here that the root of the tree was placed.

M. speciosa is one of the more divergent species analyzed, its branch length indicating an increased amount of evolutionary change, relative to *M. radiata* and *M. sinica*, since divergence from these other lineages. This difference is largely due to greater change at the transferrin and 6-PGD loci.

Again *M. niger* and *M. maura* join together with *M. niger* showing the greater evolutionary change since divergence from the common ancestor, mainly due to the transferrin, hemoglobin beta chain and carbonic anhydrase I loci. The Celebese macaques then join the previously united assemblages, followed by the union of the *M. nemestrina* species. Within the pig-tailed macaques, the Malayan and Sumatran populations join together and then unite with the Thai *M. nemestrina* population.

A close relationship of the Celebes and *M. nemestrina* populations is geographically reasonable and is supported by the hypothesis of Fooden (1969) that the Celebes varieties evolved from an isolated population of *M. nemestrina*.

In comparing the results of genetic analysis with the traditional taxonomy, some previous conclusions are challenged. Ellerman and Morrison-Scott (1951) placed *M. fuscata* with *M. speciosa* on the basis of their short tails and other physical characteristics, while *M. cyclopis* has at times (Kellogg, 1945) been placed within the irus group. However, the allelic data for *M. cyclopis*, *M. fuscata*, *M. mulatta* and *M. fascicularis* suggest that they form a closely related assemblage.

Fiedler (1956), among others, places the Celebes macaques in two distinct genera. However, the molecular data support their inclusion within the genus *Macaca*. The close relationship between the two Celebes species may be exemplified by their hemoglobin loci. They show increased variability at this locus, unusual for island species, although only a total of 13 samples was available for study. Initially, each species showed a unique variant by starch gel electrophoresis, more anodal than the hemoglobins found in the other macaques. It was ascertained, by urea starch gel analysis, that an electrophoretically identical beta chain was responsible for this phenotype, in both species, as illustrated in Fig. 3. In addition another beta chain variant, which remained near the origin, was discovered to be responsible for their slow phenotype on starch gels. This slow phenotype was electrophoretically identical to the slow type possessed by the other macaque species, and it was therefore thought to be due to a variant alpha chain. Finally, on one *M. niger* sample and two *M. mauras*, there is a variant alpha chain, slightly less cathodal than the alpha chain found in the other individuals of these two species, which may be discriminated on urea gels due to the finer banding of the protein.

Thus *M. maura* and *M. niger*, often placed in distinct genera, share three hemoglobin alleles which are unique among the macaques. From the phenetic point of view, the ID separating

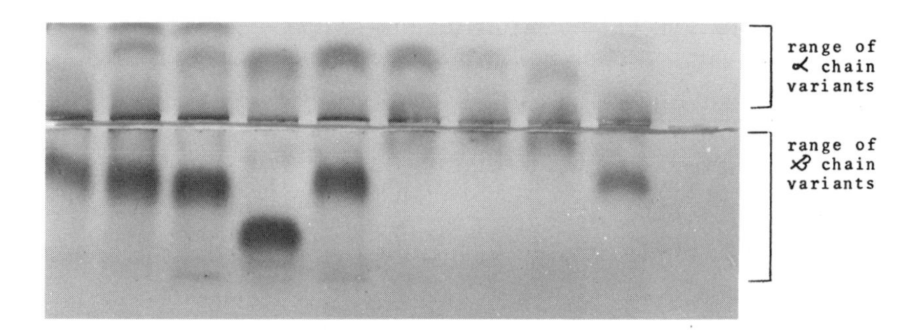

Fig. 3. Separation of macaque hemoglobin by urea starch gel electrophoresis, oriented with cathode toward the top and anode toward the bottom. First and last samples are human. 2 and 3 are *M. fascicularis* heterozygotes with alpha chain variants. 4 through 8 are from *M. maura*, illustrating the two beta chain variants not found in any other macaque samples, except for *M. niger*.

---

these two species is within the range of ID separating conspecific populations of *M. nemestrina* and is less than the ID separating *M. radiata* and *M. sinica*. Within the genus the phylogenetic relationships suggested by maximum parsimony indicate that the Celebes species shared a common ancestor with *M. nemestrina* after the point at which the pig-tailed macaques had diverged from *M. speciosa*.

ISLAND AND CLINAL EFFECTS

A study of the gene frequency data suggests apparent trends in the macaques. Island populations tend to be homozygous or show low levels of variability, for particular alleles. At the transferrin locus, Tf^c predominates in *M. cyclopis*, Tf^f in *M. fuscata* and *M. niger*, and Tf^d in Philippine *M. fascicularis*. In each case the allele exceeds 90%. Other loci exhibit this trend to lesser degrees. Thus at the 6-PGD locus, the Celebes and Japanese macaques, along with *M. radiata* and the Philippine *M. fascicularis* are characterized by fixation or near fixation of one allele. The Thai population of *M. nemestrina* and *M. fuscata* also possess high levels of a single allele. Although the founder effect is a likely explanation of this pattern, one cannot discount the possibility of altered selective factors operating on small populations in island

TABLE IX
GENE FREQUENCY EVIDENCE FOR CLINAL VARIATION

TRANSFERRIN LOCUS

Tf^c			Tf^d		
M. rad	In	.568	M. fasc	Ph	1.0
M. mul	In	.615		Ja	.550
	Ba	.354		Ma	.674
	Th	.214		Th	.439
M. fasc	Th	.277	M. nem	Su	.350
	Ma	.117		Ma	.361
				Th	.196

Tf^f			Tf^g		
M. fusc	Ja	.973			
M. mul	Th	.286	M. rad	In	.025
	Ba	.091	M. mul	In	.096
	In	.058		Ba	.109
				Th	.304

PREALBUMIN LOCUS

TBPA		Fast	Slow
M. mul	In	.983	.017
	Ba	.929	.071
	Th	.714	.286
M. fusc	Ja	.003	.997
M. fasc	Th	.996	.004
	Ma	.889	.111
	Ph	.940	.060
	Ja	.841	.159

environments (Mayr, 1967).

Clinal variation is geographically widespread. Table IX shows several alleles of the transferrin and prealbumin locus, which illustrate this trend. Tf^c attains high levels in *M. radiata* in south India, and *M. mulatta* of India, occurring between 56-61% in the two groups. Allele C then decreases in frequency moving eastward through the range of *M. mulatta*, rises a little upon entering the range of Thai *M. fascicularis* and continues to decrease in *M. fascicularis* in Malaysia. Allele Tf^d is fixed in the Philippine *M. fascicularis*, decreases in Malayan and Javan populations and continues to decrease moving north into Thailand. This cline is repeated in *M. nemestrina*, where the Sumatran and Malayan populations have the highest frequencies of 35% and 36%, while allele D decreases moving north in *M. nemestrina* in Thailand.

Tf^f is the most predominant allele in *M. fuscata*, reaching 97% frequency. The eastern Thai population of *M. mulatta* has the highest frequency of this allele in the rhesus populations, and the frequency drops moving westward through the range of *M. mulatta*.

Finally, Tf^g has a low frequency in *M. radiata* of India, 2.5%, rises slightly to 9.5% in India *M. mulatta* and continues to increase in frequency eastward through the range of rhesus populations, reaching 30% in the Thai population.

The prealbumin locus also shows a cline. TBPA fast reaches 98% in Indian *M. mulatta*, decreases, moving eastward, to 71% in Thailand rhesus; continuing the trend *M. cyclopis* has a frequency of 71%, while the eastern-most population, *M. fuscata* possesses a frequency of only 0.3% of the fast allele.

M. fascicularis also shows decreasing frequencies of the fast allele going from Thailand where the fast allele attains a frequency of 99.6%, through Malaya, the Philippines, and Java, where the frequency is 84%.

Clinal variation suggests the operation of natural selection over wide geographic areas at these loci. However, gene flow could contribute to the results seen here, so that, while the results are suggestive, no conclusion can be drawn.

ACKNOWLEDGMENTS

This research was supported by grant GB-36157 from the National Science Foundation and grant GM-15409 from the U.S. Public Health Service. The excellent help of Mrs. Ya-shiou L. Yu in the carbonic anhydrase typing is gratefully acknowledged. We wish to thank Mr. Jeheskel Shoshani of Wayne State University for his aid in collecting the blood samples. Finally, the cooperation and assistance of the following people is greatly appreciated: In Indonesia, Dr. Sarwonono Prawirohardjo, Director, The Bureau for International Relations (L.I.P.I.), Djakarta, and Mrs. Sjamsjah Achmad, also from L.I.P.I., Mr. H. M. Kamil Oesman and Mr. L. Darsono, Fa. PRIMEX, Djakarta, and Mr. B. Galstaun, Director, Djakarta Zoo; in Ceylon, Prof. P. Seneviratna, Dean, Faculty of Veterinary Science, University of Ceylon, Dr. T. A. Bongso and Dr. M. R. Jainudeen, Peradeniya, Ceylon; and Drs. Carolyn and Wendell Wilson, University of Washington, Seattle, Washington.

BIBLIOGRAPHY

Bernstein, R. S., J. Robbins, and J. E. Rall 1970. Polymorphism of monkey Thyroxine-binding prealbumin (TBPA): mode of inheritance and hybridization. *Endocrinology* 86: 383-390.

DeSimone, J., E. Magid, and R. E. Tashian 1973. Genetic Variation of the Carbonic anhydrase isozymes of macaque monkeys. II. Inheritance of Red Cell carbonic anhydrase levels in different carbonic anhydrase I genotypes of the pig-tailed macaque, *Macaca nemestrina*. *Biochem. Genet.* 8: 165-174.

Ellerman, J. R. and T. C. S. Morrison-Scott 1951. Checklist of paleoarctic and Indian Mammals 1758-1946. London: British Museum of Natural History.

Fiedler, W. von. 1956. Ubersicht uber das system der primates. *Primatologia* 1: 179-182.

Fooden, J. 1964. Rhesus and crab-eating macaques: intergradation in Thailand. *Science* 143: 363-365.

Fooden, J. 1969. Taxonomy and evolution of the monkeys of Celebes. *Bibliotheca Primatologica* 10: 1-148.

Goodman, M. and R. C. Wolf 1963. Inheritance of Serum Transferrins in Rhesus Monkeys. *Nature* 197: 1128.

Ishimoto, G., T. Tanaka, H. Nigi, and W. Prychodko 1970. Hemoglobin variation in macaques. *Primates* 11: 229-241.

Kellogg, R. 1945. Macaques. in *Primate Malaria* (Aberle, ed). Issued by the office of Medical Information.

Mayr, E. 1967. Evolutionary challenges to the mathematical interpretation of evolution. in *Mathematical Challenges to the Neo-Darwinian Interpretation of Evolution*. (Moorhead and Kaplan, eds). Philadelphia, Wistar Inst.

Moore, G. W., J. Barnabas and M. Goodman 1973. A method for constructing maximum parsimony ancestral amino acid sequences on a given network. *J. Theor. Biol.* 38: 459-485.

Prychodko, W., M. Goodman, B. M. Singal, M. L. Weiss, G. Ishimoto, and T. Tanaka 1971. Starch-gel electrophoretic variants of erythrocyte 6-Phosphogluconate dehydrogenase in Asian macaques. *Primates* 12: 175-182.

Sokal, R. R. and C. D. Michener 1958. A statistical method for evaluating systematic relationships. *Kansas Univ. Sci. Bull.* 38: 1409-1438.

Tashian, R. E., M. Goodman, V. E. Headings, J. DeSimone, and R. H. Ward 1971. Genetic variation and evolution in the red cell carbonic anhydrase isozymes of macaque monkeys. *Biochem. Genet.* 5: 183-200.

Weiss, M. L., M. Goodman, W. Prychodko, and T. Tanaka 1971. Species and geographic distribution patterns of the macaque prealbumin polymorphism. *Primates* 12: 75-80.

GENETIC CONTROL AND ISOZYMIC COMPOSITION OF DRUG ACETYLATING ENZYMES

W. W. WEBER, G. DRUMMOND, J. N. MICELI and R. SZABADI
Guttman Laboratory for Human Pharmacology and Pharmacogenetics, Department of Pharmacology
New York University, New York, N. Y. 10016

ABSTRACT. Observations in families of slow and rapid isoniazid acetylator rabbits indicate the existence of a new drug acetylating polymorphism in this species. Crosses between rabbits of both isoniazid acetylator phenotypes show that the variation in p-aminobenzoic acid (PABA) N-acetyltransferase (NAT) activity in peripheral blood is genetically determined and controlled by a pair of allelic autosomal genes. The regulatory systems controlling the enzymes for PABA acetylation in blood and isoniazid acetylation in liver are interdependent since the highest blood PABA NAT activities are associated with the lowest liver isoniazid NAT activities, and vice versa. Comparisons of isozyme patterns and other biochemical properties of partially purified NAT's from blood and liver indicate that NAT from slow isoniazid acetylator liver and NAT's from blood of both acetylator phenotypes may possess certain structural features in common. The isozyme pattern and biochemical characteristics of NAT from rapid isoniazid acetylator liver differ from those of either slow isoniazid acetylator NAT or blood NAT. The possible significance of these findings to the evolutionary origin of drug acetylating enzymes in rabbit blood and liver is discussed.

INTRODUCTION

Genetic studies in man and rabbit support the hypothesis that the inherited variation in the rate of isoniazid metabolism is transmitted as a single gene effect. Individuals termed slow acetylators of this and certain other drugs (sulfamethazine, hydralazine, dapsone) that undergo acetylation prior to excretion are homozygous for a recessive gene, r. Rapid acetylators are either homozygous or heterozygous for a dominant gene, R. Both genes are prevalent in human and rabbit populations, and studies in man show that the proportion of slow and rapid acetylator individuals varies widely from one population to another depending upon their ethnogeographic origin.

Investigators have sought to account for this pharmacogenetic trait in terms of an intrinsic difference in the physiological and biochemical attributes of rapid and slow acetylators. Biochemical studies of tissues obtained from slow and rapid acetylators of both man and rabbit have established that the N-acetyltransferase (NAT) of liver is a major determinant of this trait in both species. Homozygous rr individuals have low levels of liver isoniazid NAT activity while homozygous RR and heterozygous (Rr) individuals have relatively high levels of this enzyme activity. Extensive examination of the enzymes isolated from liver of both acetylator phenotypes of both species by standard enzymological techniques has not disclosed any difference between them. We have suggested from comparisons of electrophoretic patterns of rabbit liver NAT's from slow and rapid acetylators on polyacrylamide-agarose gels that the acetylator phenotypic difference might be associated with an electrophoretic difference. The patterns obtained were quite variable. Recoveries of NAT activity were low, and as such they could not be accepted as conclusive evidence for such a difference without corroboration (Weber, 1973).

The scope of our studies on drug metabolizing NAT's has been extended recently to include NAT's contained in the extrahepatic tissues as well as those in liver. Lines of rapid and slow acetylator rabbits established and maintained in our laboratory during the course of this work have been very useful. Early observations on this rabbit population suggested that the level of NAT activity in peripheral blood was dependent in a rather unexpected way upon the acetylator phenotype. Slow acetylator rabbits, who have low levels of liver NAT for isoniazid and for sulfadiazine, another drug which is used to classify rabbits according to acetylator phenotype (Frymoyer and Jacox, 1963), appeared to have relatively high levels of p-aminobenzoic acid (PABA) NAT activity in the blood compared to rapid acetylator rabbits (Figure 1). The possibility that unknown factors underlying this correlation might advance our understanding of heritable determinants which modify NAT levels in tissues prompted us to investigate the genetic and biochemical relationships between liver and this extrahepatic NAT in greater detail.

Our initial studies on the extrahepatic NAT's were focused mainly on the tissue distribution and some biochemical characterizations of these enzymes obtained from slow and rapid acetylator rabbits (Hearse and Weber, 1973). The extrahepatic tissues were found to contribute at least two-thirds of the total acetylating capacity to the acetylation of a polymorphic substrate such as sulfamethazine in the slow acetylator rabbit. Also, their studies showed that multiple forms of NAT existed in different tissues, and that certain tissues such as the liver contained more than one form. Finally, they showed that the NAT's of liver and certain extrahepatic tissues including the gut mucosa, spleen, kidney, and peripheral blood possessed some biochemical features in common.

Comparisons have been extended now to include characterization of the drug acetylating enzymes by the technique of isoelectric focusing. Our most recent studies of NAT's isolated from peripheral blood and liver of slow and rapid acetylator rabbits are described briefly in this report. They indicate the isozymic composition of the multiple forms of NAT present in these tissues and delineate further the nature of similarities and differences between them.

RESULTS

Genetic studies. The rabbit population referred to in Figure 1 consisted of healthy, adult animals of both sexes, some of which were related familially and some not. In view of the correlation observed between PABA NAT specific activity and acetylator phenotype, it was of interest to determine whether the PABA NAT activity levels in peripheral blood are a characteristic of the individual. Samples of peripheral blood were collected from a series of rabbits with different PABA acetylating capacity on two different occasions separated by several months, and assayed for PABA NAT activity. The high correlation observed between initial and repeat assays on the same rabbit is evident in Figure 2. The constancy of levels between the two assays indicated that the level of PABA NAT activity could be taken as an individual characteristic of healthy adult rabbits.

Specific matings were carried out between rabbits of known acetylator phenotype to obtain information on the mode of transmission of genetic factors controlling PABA NAT activity levels. The levels of PABA NAT activity in a few selected rabbit families are presented in Figure 3. A genotype of RR is proposed from pedigree analysis for the sample of rapid acetylator rabbits included there. The correlation between the level of PABA NAT activity and acetylator phenotype suggested by the data on the population of related and unrelated rabbits in Figure 1 is much more apparent when only familially related rabbits of both phenotypes are considered (Figure 3).

Additional data of a similar sort are presented as a frequency distribution in Figure 4. These data were derived from six slow (rr x rr) and thirteen rapid (RR x RR, and Rr) acetylator families consisting of 77 rabbits. The association of high levels of PABA NAT activity with slow polymorphic acetylation is apparent again. In addition, there is also a

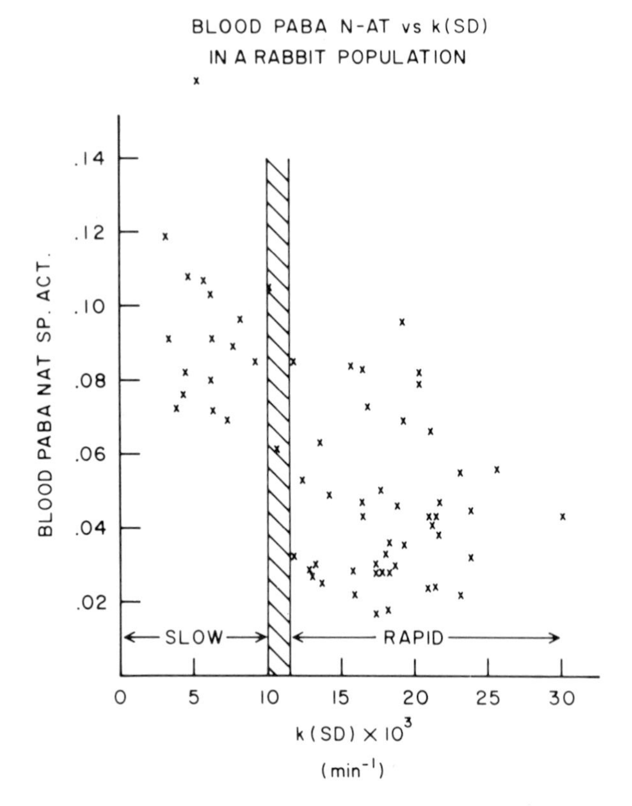

Figure 1. PABA NAT activity in Figures 1 and 2 and 3 and 4 and Tables 1 and 2 was determined by the microprocedure of Hearse and Weber (1973) and expressed as μmole AcPABA formed/min/ml blood. The half-life of sulfadiazine elimination was determined by the method of Hearse and Weber (1973) and expressed as the first order elimination constant, k SD (min^{-1}).

---

region of overlap between the slow and rapid acetylators. The mean values and their standard deviations of the first order sulfadiazine elimination rate constant, (k SD), and the PABA NAT specific activity of blood characteristic of these rabbits, are given in Table 1. The mean value of (k SD) for 22 slow acetylator rabbits was about 0.006 min^{-1} while the mean value for the rapid acetylators was about three times greater. The corresponding mean values of peripheral blood PABA NAT activity differ in the opposite direction by approxi-

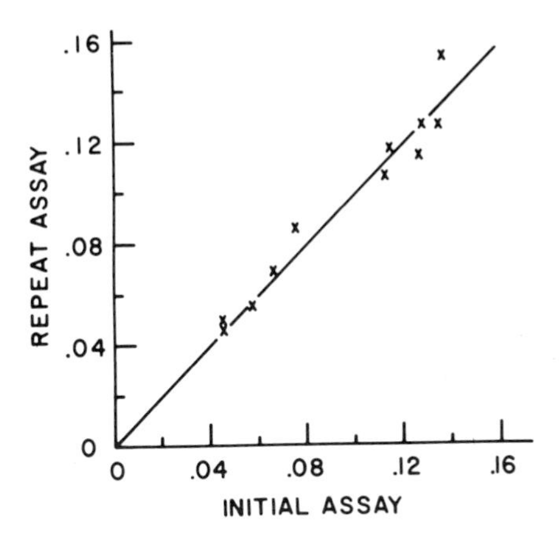

Figure 2. See legend for Figure 1.

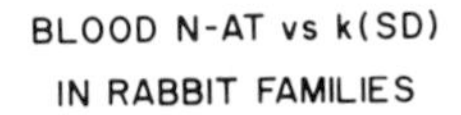

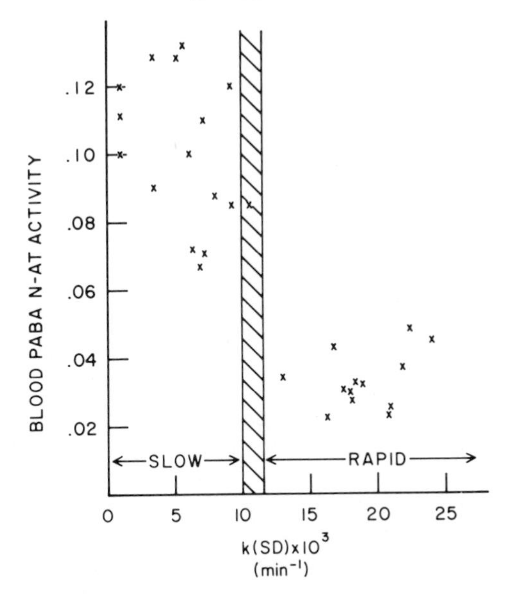

Figure 3. See legend for Figure 1.

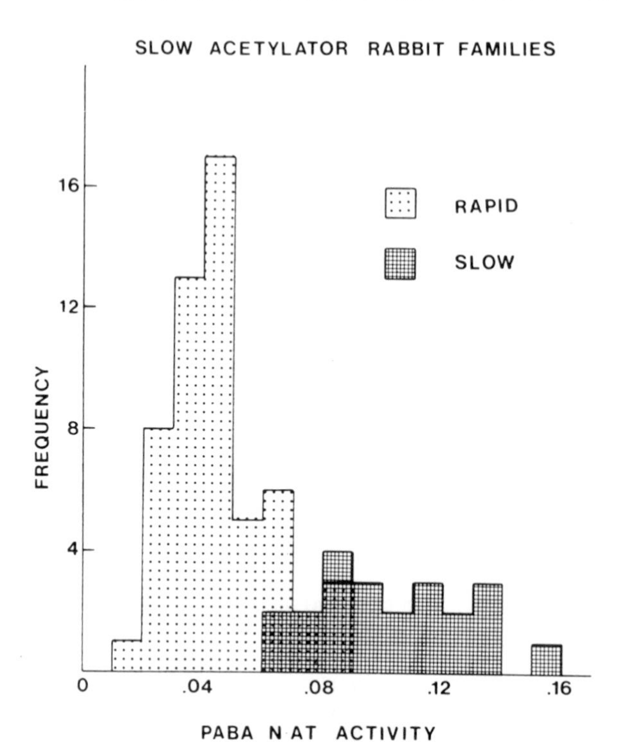

Figure 4. See legend for Figure 1.

---

mately two to three fold. Inspection of Figure 4 shows that slow acetylator rabbits tend to have specific activities centering about a value of 0.12 μmole AcPABA formed/min/ml blood while the corresponding value for rapid acetylators is approximately 0.04. The value for rabbits at the midpoint of the region of overlap is approximately 0.08 μmole AcPABA formed/min/ml blood. This value also equals (½ x 0.04 + ½ x 0.12) and is what might be expected if a gene-dosage effect were operating to determine the level of PABA NAT activity in peripheral blood of rabbits.

Data on parents and F_1 offspring from three rabbit crosses are presented in Table 2 as additional support for the concept of a gene-dosage effect. It can be seen that blood PABA NAT values of offspring from crosses between parents with high

TABLE 1
RATE OF SULFADIAZINE ELIMINATION AND BLOOD NAT ACTIVITY IN 19 RABBIT FAMILIES

Parental Acetylator Phenotype	k SD (min^{-1})	Blood PABA NAT Activity (μmole Ac-PABA formed/min/ml)
Slow	0.0057±0.0031 (22)	0.10±0.024 (23)
Rapid	0.018±0.0040 (57)	0.045±0.017 (52)

Number of rabbits observed in parentheses. k SD determined as in Figure 1.

(Family I) and low (Family III) blood PABA NAT levels tend to cluster about parental values, suggesting homozygosity for genes controlling this enzyme in each set of parents. Offspring from Family II, a cross between high and low blood PABA NAT rabbits, had activities which were intermediate between those of the parents.

Biochemical and isoelectric focusing studies. The association between peripheral blood PABA NAT activity levels and drug acetylator phenotype suggests that peripheral blood PABA NAT levels and liver isoniazid NAT levels vary inversely, i.e., that when the level of PABA NAT in blood is high, the level of isoniazid NAT in liver is low, and vice versa. Thus, it was important to establish this relationship by direct study of rabbit liver in vitro. The levels of NAT activity coexisting in these two tissues examined in a series of slow and rapid (RR) acetylator rabbits are shown in Figure 5. Specific activities of liver isoniazid NAT are expressed in terms of the conversion of the alternate polymorphic substrate, sulfamethazine, to its acetylated derivative. These results show that they can vary over a several hundred fold range and that homozygous rapid acetylator rabbits have at least ten times more liver NAT activity for this polymorphic substrate than slow acetylator rabbits. Also, it is apparent that an inverse relationship does exist between the levels of peripheral blood PABA NAT and liver isoniazid NAT.

Some of the properties of peripheral blood PABA NAT and liver isoniazid NAT are presented for comparison in Table 3. The isoniazid (or sulfamethazine) acetylating activity is distributed primarily in liver and duodenal mucosa. PABA acetylating activity is distributed widely throughout many tissues including liver and gut mucosa. Differences in

TABLE 2
F_1 OFFSPRING FROM RAPID AND SLOW ISONIAZID ACETYLATOR RABBITS

Proposed Acetylator Genotype	k SD (min^{-1})	Blood PABA NAT Activity (μmole Ac-PABA formed/min/ml
Family I.		
Parents		
rr	0.0011	0.13
rr	0.0016	0.13
F_1		
rr	0.0048	0.095
rr	0.0055	0.16
rr	0.0069	0.11
rr	0.0070	0.13
Family II.		
Parents		
RR	0.024	0.027
rr	0.0032	0.13
F_1		
Rr	0.010	0.069
Rr	0.012	0.076
Rr	0.012	0.080
Rr	0.016	0.098
Rr	0.016	0.056
Family III.		
Parents		
RR	0.024	0.037
RR	0.018	0.041
F_1		
RR	0.016	0.022
RR	0.016	0.043
RR	0.017	0.030
RR	0.018	0.028
RR	0.019	0.030
RR	0.021	0.024

Parents were: I, M-136 x F-201; II, M-EE x F-201; III, M-EE x F-F19. k SD and blood PABA NAT activity determined as in Figure 1.

substrate specificity, stability in storage and thermostability, pH optima and inhibitory characteristics are indicated there also. It is evident that the distribution of NAT activity for sulfamethazine and PABA do overlap. In addition, they

TABLE 3

COMPARISON OF DRUG ACETYLATING NAT'S

	Isoniazid NAT	PABA NAT
Tissue	Liver, duodenal mucosa	Widespread (including liver and duodenal mucosa)
Drug Substrates	Isoniazid, sulfamethazine, PABA	PABA
Stability in Storage	Not Required	Dithiothreitol and Mg++ required
Thermostability ($t_{1/2}$ at 50°)	28 minutes	7-8 minutes
pH Optima	5.6, 7.2	6.6, 7.2
Inhibition		
PCMB (5×10^{-6}M)	10 percent	100 percent
Fe++ (1×10^{-4}M)	0	30 percent

Isoniazid NAT activity determined with sulfamethazine and PABA NAT activity determined with PABA according to the procedure in Figure 1.

TABLE 4
ISOELECTRIC FOCUSING OF RABBIT NAT ISOZYMES

Isoniazid Acetylator Phenotype	Isoelectric pH	
	Liver	Blood
Slow	5.30	5.42, 6.28
Rapid	5.31, 6.67	5.17, 6.44

resemble each other very closely in certain respects and differ with respect to certain others. The pH optimum profiles in Figure 6 illustrate this point very well. Profiles for both tissues have two peaks of activity and are completely superimposable above pH 7.0. The profile for PABA NAT activity in liver is significantly broader than the profile in blood. The former has a second peak at pH 5.6, while the latter has a second peak at 6.6.

Thus, it appears that NAT activity of liver and peripheral blood contains more than one component, possibly isozymes. Electrophoretic studies described below confirmed this possibility.

Isoelectric focusing patterns prepared from NAT's partially purified from these tissues obtained from slow and rapid

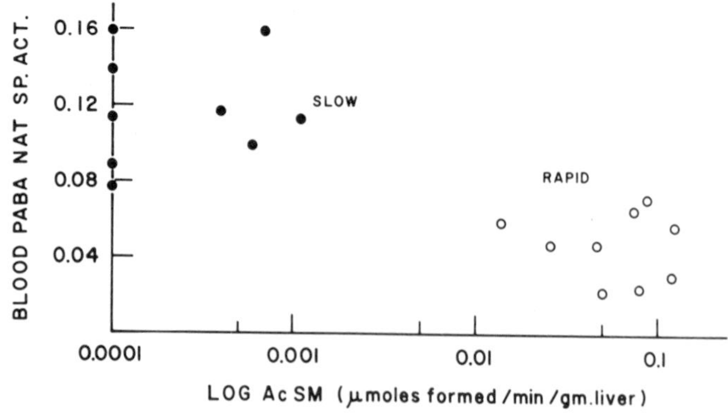

Figure 5. PABA NAT activity determined as in Figure 1. Liver NAT activity determined by the procedure of Hearse and Weber (1973) on 100,000g supernatant fraction of a 25 percent homogenate with sulfamethazine.

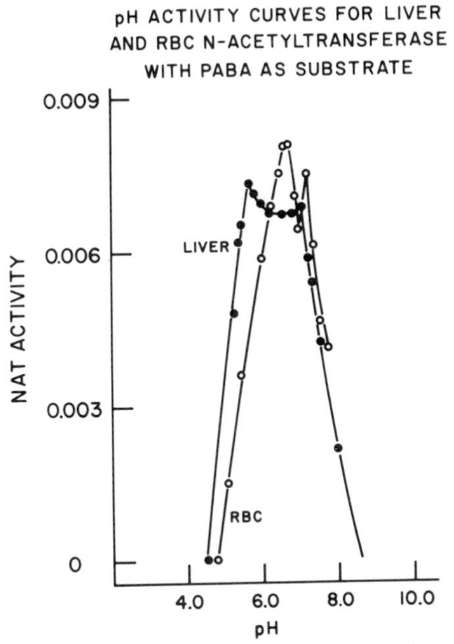

Figure 6. Liver NAT partially purified by the procedure of Weber (1971). Blood NAT partially purified approximately 1000-fold over a 100,000g supernatant fraction by anion and cation exchange column chromatography. All incubations performed at 27°C as in Figure 1.

---

acetylator rabbits using PABA as the substrate are shown in Figure 7. The pattern from slow acetylator liver (Figure 7a) shows that only one NAT isozyme was detectable. The pattern from rapid acetylator liver (Figure 7b) shows an isozyme which focuses at the same pH. In addition, this pattern contains a second isozyme which focuses about one pH unit higher. The patterns of activity from whole blood of slow and rapid acetylator rabbits (Figure 7c and 7d) both have two isozymes. Comparison of the patterns from peripheral blood of several rabbits of each acetylator phenotype did not show any consistent difference in the isozyme patterns. Comparison of the isozyme patterns from peripheral blood and liver indicated that the isozyme of peripheral blood which focuses at the lower pH coincides with the liver isozyme which is common to both acetylator phenotypes. This is the predominant isozyme in all four patterns. In contrast, the second isozyme from

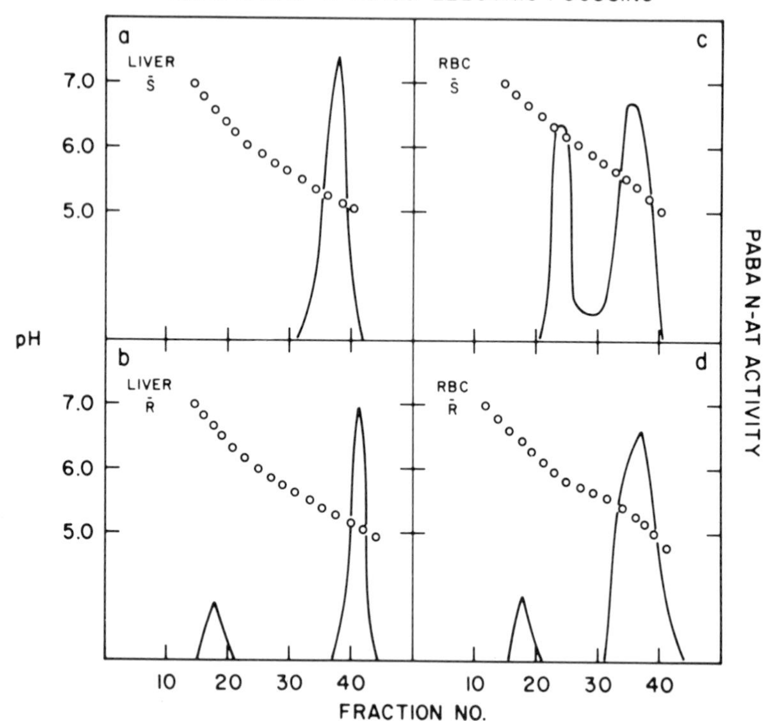

Figure 7. Isoelectric focusing performed according to procedure of Haglund (1971). Narrow range ampholines, pH 5.00-7.00, were required for optimal resolution of partially purified NAT's. Focusing was performed at 4°C for 48 hours. Fractions (2.5 ml) were collected, and the pH of each was determined at 4°C. PABA NAT activity in each fraction was determined as in Figure 1.

---

rapid acetylator liver consistently focused at a somewhat higher pH (6.67) than the corresponding isozyme from whole blood of both acetylator phenotypes (6.28 - 6.44) (Table 4).

Further tests of individual isozymes isolated from both tissues of both acetylator phenotypes by this technique showed other interesting properties (Table 5), particularly in their substrate specificities. The isozyme obtained from slow acetylator rabbit liver and the predominant isozyme from

TABLE 5
COMPARISON OF NAT ISOZYMES FROM RABBIT LIVER AND BLOOD

	Liver			Blood			
	Slow	Rapid		Slow		Rapid	
Isoelectric pH	5.30	5.31,	6.67	5.44,	6.28	5.18,	6.44
Stability	S	S	U	S	U	S	U
PABA/ Sulfamethazine	+/-	+/+	+/+	+/-	+/-	+/-	+/-

The mean PABA /Sulfamethazine activity ratio was 0.64 for the pH 5.30 isozyme for rapid acetylator liver, and 0.61 for the pH 6.67 isozyme. The ratios in the other isozymes tend toward ∞ as the activity with sulfamethazine was very low.

peripheral blood of both acetylator phenotypes exhibited high activity with PABA and low activity with sulfamethazine. The corresponding isozyme from rapid acetylator liver contained high activity for both PABA and sulfamethazine.

DISCUSSION

The explanation of the marked genetic variation in the rate of drug acetylation has been the subject of considerable study. Jenne (1965) was first to explore the possible basis for polymorphic acetylation of isoniazid in man and the lower NAT activity in slow acetylators. He found that partially (30-fold) purified NAT from several human slow isoniazid acetylators had low but detectable levels of NAT activity. NAT preparations from slow and rapid isoniazid acetylators had the same K_m for isoniazid, the same K_i for the competitive inhibitor, hydralazine, and the same temperature stability characteristics. The low level of activity in the slow acetylator could not be attributed to the presence of an enzyme inhibitor. Thus, Jenne suggested that slow and rapid acetylators might differ only in respect to the amount of a single species of NAT molecule.

Further comparisons on more highly purified liver NAT preparations from man (Weber, Cohen, and Steinberg, 1968; Steinberg, 1970) and rabbit (Weber, Cohen, and Steinberg, 1968) confirmed and extended these findings. Total liver NAT for isoniazid was found to be at least an order of magnitude greater in the liver of the rapid acetylator than in the slow acetylator. NAT's purified 300-500 fold from the two acetylator phenotypes were indistinguishable by their pH characteristics, heat stability, kinetic properties, and specificity for a variety of drugs and other substrates. Electrophoretic patterns on polyacrylamide-agarose gels did indicate that different forms of NAT might exist in slow and rapid acetylator liver (Szabadi, 1970; Hearse, Szabadi, and Weber, 1970; Szabadi, Drummond, and Weber, 1972). The interpretation of these results was clouded by instability of the enzyme during electrophoresis, and we felt that additional study with a better procedure for separating the enzyme components was needed before concluding that the two acetylator phenotypes were characterized by different forms of liver NAT.

Isoelectric focusing proved to be suitable for this purpose as it provided high resolution and permitted the use of stabilizing agents during the focusing run. The isozyme patterns in Figure 7a and 7b clearly show that slow and rapid acetylator rabbit livers do contain different forms of NAT. The isozyme pattern from slow acetylator liver NAT consists of a single component while the pattern from rapid acetylator liver NAT contains two components. We believe that this is the first definitive demonstration of an intrinsic biochemical difference associated with the isoniazid acetylation polymorphism.

The isozyme which is present in slow acetylator liver NAT (Figure 7a) focuses at the same pH (5.3) as the predominant isozyme in rapid acetylator liver NAT. It is tempting to suggest that this electrophoretic component corresponds to a gene product which is common to liver NAT's of both acetylator phenotypes. The relatively greater specificity that the predominant isozyme (5.3) from rapid acetylator liver manifests for sulfamethazine (Table 5) makes this possibility unlikely. This suggests that there is still an intrinsic difference between these isozymes from the two acetylator phenotypes which is not resolved by isofocusing under the conditions of the experiment. Further studies are required to determine the nature of this difference.

An additional observation concerning the levels of slow acetylator liver NAT activity can be made from the data in Figure 5. Sulfamethazine acetylating activity of some slow acetylator rabbits was virtually undetectable while it was present at low levels in others. Thus it appears to be distributed into two groups. This could signify the presence of different amounts of a single species of NAT in the livers from rabbits of each subgroup, or it could indicate that there are different species of NAT in the two subgroups. A plausible interpretation of these findings is that the slow acetylator rabbit population is genetically heterogeneous and includes rabbits with different slow acetylator alleles, i.e., r', r", etc.

The family data in Figures 3 and 4 and in Table 2 indicate that genetic factors have an important influence on the level of PABA NAT activity in rabbit blood. These data can be accounted for genetically by the hypothesis that PABA acetylation in this tissue is controlled by two allelic autosomal genes. Assuming that this is a valid hypothesis, rabbits with a high level of blood PABA NAT would be classified as homozygous for one of these alleles, and rabbits with a low level of blood PABA NAT would be classified as homozygous for the other allele. Also, rabbits with intermediate blood PABA NAT values, illustrated by the F_1 offspring from Family II in Table 2, would correspond to the heterozygous class.

It is also evident from the inverse relation between blood PABA NAT activity and liver isoniazid (sulfamethazine) NAT activity illustrated in Figure 5 that the regulatory systems controlling these enzyme activities are related in some way. An explanation for this phenomenon is unknown. Previous studies of NAT activity in these tissues by Hearse and Weber (1973) have directed attention to the possible existence of structural features common to both hepatic and extrahepatic NAT's. The biochemical data in Table 3 and Figure 6 and the isozyme patterns in Figure 7 for NAT's derived from slow acetylator liver and blood provide additional support for this possibility. The NAT isozyme of slow acetylator liver (Figure 7a) focuses at the same pH (5.3) as the predominant isozyme of blood from both acetylator phenotypes (Figure 7c and 7d). The similarity of these isozymes with respect to thermostability and relative substrate specificity for PABA and sulfamethazine in Table 5 are further indications of their close resemblance.

We suggest that the evidence presented here and in an earlier study by Hearse and Weber (1973) support the concept that drug acetylating NAT's in the rabbit may have been derived from a primordial gene product corresponding to the isozyme which occurs in slow acetylator liver. To account for the different forms of NAT in blood and in rapid acetylator liver, this gene product must have undergone modification in some individuals through mutation and environmentally determined effects which are tissue-specific. If this occurred, modification in blood of both acetylator phenotypes has led to the formation of an additional minor isozyme which migrates more rapidly than the predominant isozyme. In liver of rapid acetylators, modification has led to an additional minor isozyme with migratory properties which are slightly different from those of the corresponding isozyme in blood, and to an altered capacity to catalyze the acetylation of drugs such as isoniazid.

ACKNOWLEDGEMENTS

This work was partially supported by PHS grants GM-15064 and GM-17184.

REFERENCES

Frymoyer, J. W. and R. F. Jacox 1963. Investigation of the genetic control of sulfadiazine and isoniazid metabolism in the rabbit. *J. Lab. Clin. Med.* 62: 891-904.

Haglund, H. 1971. Isoelectric focusing in pH gradients - A technique for fractionation and characterization of ampholytes. *Methods of Biochemical Analysis* 19: 1-104.

Hearse, D. J., R. Szabadi, and W. W. Weber 1970. Enzyme multiplicity in the N-acetylation of drugs. *The Pharmacologist* 12: 274.

Hearse, D. J. and W. W. Weber 1973. Multiple N-acetyltransferase and drug metabolism. Tissue distribution, characterization and significance of mammalian N-acetyltransferase. *Biochem. J.* 132: 519-526.

Jenne, J. W. 1965. Partial purification and properties of the isoniazid transacetylase in human liver. Its relationship to the acetylation of p-aminosalicylic acid. *J. Clin. Invest.* 44: 1992-2002.

Steinberg, M. S. 1970. Purification and properties of human liver N-acetyltransferase from rapid and slow isoniazid acetylators. Ph. D. Thesis, New York University.

Szabadi, R. 1970. Electrophoretic properties of rabbit liver N-acetyltransferase from rapid and slow INH acetylators. M. S. Thesis, New York University.

Szabadi, R., G. Drummond, and W. W. Weber 1972. Further evidence of drug acetylating enzyme multiplicity: electrophoresis of liver INH-NAT from rapid and slow acetylators. *Fifth International Congress on Pharmacology* p. 225.

Weber, W. W. 1971. N-acetyltransferase (mammalian liver) in metabolism of amino acids and amines. *Methods in Enzymology*. Eds. H. Tabor and C. W. Tabor. Academic Press, New York. 17B: 805-812.

Weber, W. W. 1973. Acetylation of drugs. *Metabolic Conjugation and Metabolic Hydrolysis*. Ed. W. H. Fishman, Academic Press, New York. pp. 249-296.

Weber, W. W., S. N. Cohen, and M. S. Steinberg 1968. Purification and properties of N-acetyltransferase from mammalian liver. *N.Y. Acad. Sci.* 151: 734-741.

GENETICS OF THE URINARY PEPSINOGEN ISOZYMES

LOWELL R. WEITKAMP
and
PHILIP L. TOWNES
Division of Genetics, Departments of Anatomy
and Pediatrics, University of Rochester School
of Medicine and Dentistry, Rochester, New York 14642

ABSTRACT. Human urinary pepsinogen isozymes, Pg 2, Pg 3, Pg 4, and Pg 5, have been scored for their absence, presence as a weak band, or presence as an intense band in 652 family members. Pg 3 through Pg 5 were independently classified by the two authors with agreement on a specific category for for 87% of the individuals. Considering all four isozymes, there were, among the 94 black and 65 white unrelated people with distinct patterns, 18 different phenotypes. From the distribution of the intensities of the four bands in a given individual, it is apparent that control of the intensity of one isozyme is not unrelated to that of others. Segregation analysis for each isozyme separately, assuming an intense band is dominant to a weak band and the latter is dominant to absence, demonstrates familial clustering for each type of variant for each isozyme. For some variants there are no exceptions to dominant Mendelian inheritance, but for others the fit is less good. Preliminary data suggest a locus controlling Pg 5 intensity may be linked to the HL-A loci, but control of Pg 2 and 4 is apparently not closely linked to the HL-A region.

Seven isozymes of pepsinogen have been distinguished in extracts of gastric mucosa by alkaline agar gel electrophoresis (Samloff and Townes, 1970a). Four of these, designated Pg 2 through Pg 5 in order of decreasing anodal mobility, are regularly found in urine when present in gastric mucosa. Some individuals, however, lack Pg 5 in both mucosa and urine. From family studies the presence of Pg 5 (phenotype A) is controlled by a dominant gene, Pg^a, and by inference its absence by an allele, Pg^b, at a single autosomal locus (Samloff and Townes, 1970b). In Caucasians the frequency of the recessive B phenotype (absence of Pg 5) is 14% (Samloff and Townes, 1970b) and in American Negroes 20% (Samloff et al., 1973; Townes and White, 1974). The B type has not been observed in American Chinese, Japanese or Filipinos (Samloff et al., 1973).

Another phenotype, characterized by an intense Pg 4 band, has been reported in 6% of 424 school girls (Bowen et al.,

1972). It was proposed to result from a third allele, Pg^c, described as co-dominant to Pg^a and Pg^b. The absence of Pg 4 as well as 5 has been noted in about 20% of American Negroes with the B phenotype (Townes and White, 1974). This pattern has been labeled B^1 and hypothesized to result from an allele Pg^{b1}, recessive to Pg^b.

There are in fact a large number of pepsinogen patterns which differ from each other in the relative activity of the four urinary isozymes. We have observed high activity of Pg 3, 4 or 5 either individually or in various combinations and together with the absence of one or more of the four isozymes. (The absence of all four bands, although common enough, can usually be overcome by concentrating the urine up to ten fold.) In the course of several years we have been impressed both by familial occurrence of various patterns and by the difficulty in some instances of precisely categorizing individuals with respect to pattern type. Description of these patterns and their genetic control is the subject of this report.

In terms of our subjective impression of pattern differences, the presence of absence of the intensely active Pg 4 band is usually evident. Reading the presence or absence of Pg 5 appears generally consistent, although not always easy; it is particularly difficult if Pg 4 is intense. The presence or absence of Pg 2 and Pg 4 is often problematic, with recognition of the absence of the latter possible only in the absence of Pg 5. In some families in which all individuals have both Pg 3 and 5, striking differences in the relative activity of these isozymes is apparent. In other families discrimination between such differences is difficult.

Over the past several years we have typed urinary pepsinogen on a number of families studied primarily for other purposes and not included in previous reports on pepsinogen. During this period our perception of phenotypic differences has grown. In an effort to estimate the validity of typing for the various patterns as we now see them and to determine to what extent they may be heritable we have recently reclassified 652 such family members. Using Kodachrome slides, which provide an excellent copy of the original gel, the two authors have independently and without knowledge of family relationships categorized individuals with respect to absent, present or intense Pg 3, 4 and 5 bands. One author (P.L.T.) additionally read the presence or absence of Pg 2. Excluding this latter character, agreement was reached for 574 of the family members, there being nonagreement or ambiguity for the remaining 78. Including all four bands, there were 18 different patterns among 159 unrelated, classifiable individuals in these families (Table 1).

TABLE 1

PEPSINOGEN PHENOTYPES IN 159 UNRELATED INDIVIDUALS IN THE AMERICAN WHITE AND BLACK POPULATIONS

Phenotypes[1]	White		Black	
	Number	%	Number	%
2③	1	1.5	2	2.1
2③4	2	3.1	15	16.0
23④	0		1	1.1
2③④	3	4.6	1	1.1
234⑤	3	4.6	0	
2③45	27	41.5	40	42.6
2③④5	0		1	1.1
2③4⑤	5	7.7	11	11.7
③	0		1	1.1
34	1	1.5	0	
3④	1	1.5	0	
345	5	7.7	2	2.1
3④5	2	3.1	4	4.3
34⑤	10	15.4	7	7.4
③④5	0		1	1.1
③4⑤	3	4.6	8	8.5
4⑤	1	1.5	0	
④⑤	1	1.5	0	
	65	99.8	94	100.2

[1]A circle around the number indicates the designated isozyme is more active relative to other isozymes in the same individual and to a lesser extent in comparison to other isozymes on the same gel.

The phenotypes in Table 1 have been arranged to show a progression from anodally migrating to cathodally migrating bands. The combination of relative intensities of the various isozymes in a given individual does not appear to be random. Thus, an individual with Pg 2 present is more likely to have an intense Pg 3 than a weak Pg 3, the opposite being true when Pg 2 is absent (cf. Table 2). If Pg 2 is present there is a greater percentage of weak or absent Pg 5 bands than if Pg 2 is absent.

The frequencies of the previously described categories, A, B, B^1, AC, and BC (or C), are shown for the American black and white populations in Table 3. Our limited frequency data are consistent with the frequencies reported in the surveys cited above. Additionally, C occurs in blacks as well as whites, in our sample with a frequency of 8-10% for both

TABLE 2

INTERRELATIONSHIPS OF Pg ISOZYMES:
PRESENCE OR ABSENCE OF Pg 2 CONTRASTED TO THE RELATIVE INTENSITIES OF Pg 3 AND Pg 5[1]

Population	2+		2-	
	③	3+	③	3+(3-)
White	38	3	3	21
Black	70	1	10	13

Population	2+			2-		
	5-	5+	⑤	5-	5+	⑤
White	6	27	8	2	7	15
Black	19	41	11	1	7	15

[1] O = intense band; + = weak band; - = absent band

TABLE 3

PEPSINOGEN PHENOTYPE FREQUENCIES IN THE AMERICAN BLACK AND WHITE POPULATIONS

Population	A		B		B^1		AC		BC (C)	
	No.	%	No.	%	No.	%	No.	%	No.	%
White	54	83.1	3	4.6	1	1.5	3	4.6	4	6.1
Black	68	72.3	16	17.0	2	2.1	6	6.4	2	2.1

populations. The B^1 phenotype, previously reported to have a frequency of 4-5% in American Negroes, has been found in one of the 65 whites listed in Table 3 as well as in another unrelated Caucasian in the family material.

As an initial approach to the inheritance of variation in the activity of the pepsinogen isozymes, we have calculated "gene" frequencies for each of the four isozymes separately on the assumption that presence is dominant to absence and an intense band is dominant to a weak band. The dominant "genes" with frequencies low enough to provide several informative matings were Pg^{2+} (presence of Pg 2), Pg③ (intense Pg 3), Pg④ (intense Pg 4), Pg⑤ (intense Pg 5) and Pg $^{5+}$ (weak Pg 5). The frequencies of these "genes" in the white and black populations in our study are listed in Table 4.

Segregation analyses for matings in which both parents were unequivocally typed are given in Tables 5 through 8 for Pg^{2+}, Pg③, Pg④, Pg⑤ and Pg^{5+}. The expected number of offspring, shown in parentheses, is only approximate. First, the calculation of the expected number assumes random mating, but in

TABLE 4

PEPSINOGEN "GENE" FREQUENCIES IN THE AMERICAN BLACK AND WHITE POPULATIONS FOR THE INDICATED CHARACTERS[1]

Population	Gene Frequency				
	Pg^{2+}	Pg③	Pg④	Pg⑤	Pg^{5+}
White	.39	.39	.05	.20	.45
Black	.51	.68	.05	.15	.39

[1]See text for explanation of symbols.

TABLE 5

SEGREGATION FOR PRESENCE OR ABSENCE OF Pg 2[1]

Mating Type	Number	Offspring Phenotype		
		2+	2-	?
White				
2+ X 2+	6	10(12.9)	5(2.1)	2
2+ X 2+/2-	3	5(5.9)	2(1.1)	0
2+ X 2-	15	22(28.0)	23(17.0)	4
2+/2- X 2-	1	0(1.0)	2(1.0)	0
2- X 2-	7	1(0)	16(17)	2
Black				
2+ X 2+	29	80(80.0)	10(10.0)	4
2+ X 2+/2-	4	7(8.3)	3(1.7)	2
2+/2- X 2+/2-	1	1(.75)	0(.25)	0
2+ X 2-	12	21(22.6)	13(11.4)	5
2- X 2-	2	0(0)	4(4)	0

[1]Expected numbers calculated from population gene frequencies assuming 2+ is dominant to 2- are enclosed in parentheses.

fact several of the families are subsets of larger pedigrees. Second, the "gene" frequencies, in addition to being based on the small number of individuals included in this study, pretend that the variability in each of the isozymes is unrelated to that in others; this is clearly unlikely (Table 2). Another source of discrepancy between observed and expected values, especially for Pg^{2+}, is possible bias toward either the positive or negative categories among the offspring assigned an uncertain type. Nevertheless, this approach provides a framework for determining at the least whether control of the relative intensities of the pepsinogen isozymes is familial. All individuals have been typed for 20 or more marker systems and offspring inconsistent with stated parentage eliminated from the study.

TABLE 6

SEGREGATION FOR PRESENCE OR ABSENCE OF Pg ③

Mating type	Number	Offspring Phenotype		
		③	3	?
White				
③ x ③	14	42(38.5)	3(6.5)	1
③/3 x ③	1	1	2	0
③ x 3	5	9(6.2)	1(3.8)	0
③/3 x 3	1	0	2	0
Black				
③ x ③	35	106(108.3)	9(6.7)	2
③/3 x ③	2	3	1	0
③ x 3	6	10(9.8)	3(3.2)	0

TABLE 7

SEGREGATION FOR PRESENCE OR ABSENCE OF Pg ④

Mating type	Number	Offspring Phenotype		
		④+	④-	?
④+ x ④+	2	4(5.3)	3(1.7)	0
④+ x ④-	12	25(21.0)	15(19.0)	0
④+/④- x ④-	9	9(12.0)	15(12.0)	0
④- x ④-	66	0(0)	207(207)	0

TABLE 8

SEGREGATION FOR ⑤, 5+ and 5-

(Excluding individuals who are ④)

Mating	No.	Offspring Phenotype			
		⑤	5+	5-	?
White					
⑤ x ⑤	7	15(12.0)	0(2.4)	0(.6)	0
⑤ x 5+	9	12(14.5)	14(10.0)	0(1.5)	0
5+ x 5+	2	1(0)	8(9.1)	1(0.9)	0
Black					
⑤ x ⑤	3	6(6.3)	2(1.2)	0(0.5)	0
⑤ x 5+	14	24(26.5)	15(18.2)	10(4.3)	1
⑤ x 5-	1	0(3.2)	6(1.3)	0(1.5)	0
5+ x 5+	11	3(0)	10(19.3)	9(2.7)	0
5+ x 5-	10	0(0)	10(18.9)	19(10.1)	0
5- x 5-	1	0(0)	0(0)	5(5)	0

Segregation for the presence or absence of Pg 2, shown in Table 5, demonstrates that the character is familial and approximates the inheritance of a dominant trait. There was one 2+ child among 21 offspring of nine 2- x 2- matings. If the occurrence of the trait were random, then the expected number of 2+ children, based on 146 in a total sample of 204 children, would be 15. The observed difference is highly significant. Similarly, the frequency of 2- children is much higher in 2+ by 2- matings than in 2+ by 2+ matings (Table 5). On the other hand, in adiition to the exceptional child from the 2- by 2- matings, there is for the remaining matings, i.e. those with at least one 2+ parent, an excess of 2- children. The observed number was 146 Pg 2+ and 58 Pg 2- compared to expected values of 159.5 and 44.5, respectively. The difference is significant at $p<0.025$.

There were unfortunately no 3+ by 3+ matings (3+ being recessive to ③), but the number of ③ and 3+ offspring from other matings fits expectation rather well. The total number of ③ and 3+ children was 171 and 21 compared to expected values of 169.7 and 22.3.

The intense Pg ④ band (Pg C phenotype) is quite clearly a dominantly inherited character. There were no ④+ children among 207 offspring of 66 ④- by ④- matings. Assuming a 5% gene frequency in both blacks and whites the number of ④+ by ④- children expected from backcross matings was 33 and 31, respectively, compared to 34 and 30 observed. We have not encountered any B^1 by B^1 matings (B^1 being the absence of Pg 4 as well as Pg 5). Among 73 individuals in 10 pedigrees in which this type did occur, 15 were B^1, 19 B, 6 B or B^1, 30 A and 3 A or B. There was one instance in which the types as scored were inconsistent with the notion that B^1 is the product of an allele recessive to Pg^b; a mating of two A individuals produced one A, one B and one B^1 offspring. Thus, although the B^1 phenotype is difficult to distinguish from the B phenotype, there is nevertheless an indication that it is familial.

Segregation for $Pg^{⑤}$ and Pg^{5+} (Pg^a) and Pg^{5-} (Pg^b) is shown in Table 8. Families segregating for Pg ④ have been omitted because of the difficulty in distinguishing between 5+ and 5- in the presence of the ④ band. In concordance with the findings of Samloff and Townes (1970B) all five offspring of the one 5- by 5- mating are 5-. However, viewed in terms of Pg^a and Pg^b only, there is an excess of Pg^b offspring from the remaining matings. The 39 Pg^b offspring observed compares to an expected of 22.1, a difference significant at $p<0.01$. The excess occurs entirely in black families, our data in white families (one B offspring from 18 A x A matings compared to an expected number of 3) being consistent with the segregation ratios reported by Samloff and Townes (1970a). From inspection of Table 8 the excess of B offspring stems primarily from matings of weak Pg^a (5+) rather than intense Pg^a (⑤) parents. In our sample 89% of the white matings had at least one ⑤ parent compared to 45% of the black matings. This fact may account at least in part for the apparent racial difference in the goodness of fit to the expected segregation ratios obtained in this study compared to that of Samloff and Townes, although, according to the gene frequency estimates in Table 4, the number of matings in which there is one ⑤ parent in whites should only be on average 59%. Excluding the one 5- by 5- mating, there are 23 families in which ⑤ is not present in either parent. There are, however, four ⑤ children among the 61 offspring. The segregation of ⑤, 5+ and 5- in the remaining matings is 58, 45, and 11, respectively, compared to expected values of 62.5, 42.2, and 9.3 ($p>.6$). We conclude that varying intensity of the Pg 5 isozyme is familial and approximates the dominant scheme proposed, although there is difficulty in distinguishing between ⑤ and 5+ on the one hand and 5+ and 5- on the other.

Up to this point we have treated each of the isozymes independently, even though their varying intensities are clearly interrelated. It is possible to construct schemes using a large number of alleles at a single locus to account for the inheritance of the various patterns, but the exercise has little point since, with a hierarchial series of dominant alleles there would be few critical matings. On a smaller scale we have in Table 9 examined the allelism of Pg^a, Pg^b and Pg^c proposed by Bowen et al. (1972). (There are 2 more A x A matings in this Table than in Table 8 because these matings were excluded from the latter Table due to our inability to distinguish between ⑤ and 5+ in a parent.) If Pg^c is allelic and codominant to Pg^a, there should be no B offspring of AC parents. This is in fact true for 31 offspring of 9 matings of AC parents. However, on the contrary assumption that Pg^c is not allelic to Pg^a and Pg^b, the expected number of B and BC offspring in the 8 AC backcross and intercross matings in only 1.4 each, compared to 9 BC (or C) and no B offspring observed. The results are consistent with allelism, but certainly not proof.

Another approach to the question of allelism is to examine the linkage relations of the various isozymes with other markers. Our preliminary data (Weitkamp, Townes, and May, unpublished) on the linkage relations of Pg with HL-A indicate a peak lod score of 2.9 at a recombination frequency of 18% with Pg 5. This is significant evidence for linkage with a

TABLE 9

SEGREGATION OF Pg A, Pg B, and Pg C[1]

Mating	Number	Offspring Phenotype A	B	AC	BC(C)
White					
A x A	20	57(53.6)	1(4.4)	0(0)	0(0)
AC x A	3	3	0(0)	1	3
AC x AC	1	2	0	2	1
AC x BC(C)	1	0	0(0)	0	1
A x BC(C)	7	3	3	6	2
Black					
A x A	30	71(82.2)	20(8.8)	0(0)	0(0)
A x B	12	19(26.9)	20(12.1)	0(0)	0(0)
B x B	1	0(0)	5	0(0)	0(0)
AC x A	4	10	0(0)	3	5
A x BC(C)	6	6	2	3	3

[1]Only families with unequivocal parental phenotypes were used. Five offspring with equivocal types (A or B) were omitted.

randomly selected locus at about the 5% level of confidence. On the other hand we can rule out linkage with Pg 2 or Pg 4 at less than 10% recombination (lod score<-2.0). Given our prior expectation that the genetic control of the pepsinogen isozymes may reside at a single locus, the evidence at the moment is still insufficient to conclude that there is more than one locus. The question can probably be resolved, however, with a reasonable amount of additional information.

ACKNOWLEDGEMENTS

We thank Ms. M. White for able technical assistance. This research was supported in part by NIH Research Career Development Award 5-K04-HD50248 (LRW), NIH Contract N01-HL-1-2404, and Grants AM-09247 and GM-19962.

REFERENCES

Bowen, P., W. Sissons, M. Beiner, H. Harris, and D. A. Hopkinson 1972. Pepsinogen polymorphism: Evidence for a third common allele. *Clin. Res.* 20: 929.

Samloff, I. M., W. M. Liebman, G. A. Glober, J. O. Moore, and D. Indra 1973. Population studies of pepsinogen polymorphism. *Amer. J. Hum. Genet.* 25: 178-180.

Samloff, I. M. and P. L. Townes 1970a. Electrophoretic heterogeneity and relationships of pepsinogens in human urine, serum and gastric mucosa. *Gastroenterology* 58: 462-469.
Samloff, I. M. and P. L. Townes 1970b. Pepsinogens: Genetic polymorphism in man. *Science* 168: 144-145.
Townes, P. L. and M. R. White 1974. Pepsinogen polymorphism frequencies in a Negro population. *Amer. J. Hum. Genet.* 26: 252-254.

EVOLUTIONARY ASPECTS OF ANIMAL GLUCOSE 6-PHOSPHATE DEHYDROGENASE ISOZYMES

SAMUEL H. HORI, SATOSHI YONEZAWA, YOSHIKATSU MOCHIZUKI, YOSHIKAZU SADO, and TSUTOMU KAMADA
Department of Zoology,
Faculty of Science,
Hokkaido University,
Sapporo, JAPAN

ABSTRACT. Rat liver G6PD exists in two molecular forms which differ strikingly in substrate specificity, sensitivity to inhibitors, inducibility, and molecular weight. In order to investigate whether or not these molecules arose from a common ancestral form, the properties of G6PD from a variety of animals were studied by means of polyacrylamide disc electrophoresis. The results indicate that although G6PDs have a wide range of substrate specificity, they can be classified into three major types with respect to their activity with G6P, Gal6P, and dG6P. The G6PDs of broader substrate specificity occur in higher animals, while G6PDs of narrow substrate specificity are widely distributed throughout the animal kingdom. This seems to suggest that G6PDs of broader substrate specificity might have evolved from those of narrower substrate specificity.

G6PD from *Asterias amurensis* was found to be peculiar among other invertebrate G6PDs in that its activity with Gal6P+NADP or +NAD was very high as compared with its activities with other substrates. In this respect, this enzyme appeared to resemble the vertebrate G6PD of broader substrate specificity commonly reported as hexose 6-phosphate dehydrogenase.

INTRODUCTION

Since the discovery by Warburg and Christian (1931) glucose 6-phosphate dehydrogenase (G6PD) has been extensively studied by numerous investigators. It is widely distributed throughout the animal and plant kingdoms and its main function is thought to be to provide reduced nicotinamide adenine dinucleotide phosphate (NADPH) for various reactions. G6PD as well as the other phosphogluconate pathway enzymes have been known to exist almost exclusively in cytosol (Glock and McLean, 1953; Newburgh and Cheldelin, 1956). However, recent studies demonstrated that these enzymes did exist also in a large particulate fraction, possibly in mitochondria and that the G6PD and phosphogluconate dehydrogenase of this fraction differed from those of the cytosol fraction in catalytic, immunological, and other properties (Bagdasarian and Hulanicka, 1965; Baquer and McLean, 1972; Baquer et al., 1972; Razumovskaya et al., 1970; Yamada and Shimazono, 1961; Zaheer et al., 1967). Furthermore, presence of G6PD in the microsomal fraction of mammalian livers has also been reported (Ohno et al., 1966; Shaw, 1966; Ruddle et al., 1968; Hori and Matsui, 1968). Ohno et al. (1966) designated the enzyme as hexose 6-phosphate dehydrogenase (H6PD) inasmuch as it was found to be as active on galactose 6-phosphate (Gal6P) as on G6P. Subsequent work on H6PD has revealed that this enzyme is resistant to inhibitors of G6PD, has a broad substrate and coenzyme specificity, is not precipitated by antibody against G6PD, and is present also in fishes (Beutler and Morrison, 1967; Srivastava and Beutler, 1969; Mandula et al., 1970; Shatton et al., 1971; Stegeman and Goldberg, 1971, 1972; Srivastava et al., 1972). Beutler and Morrison (1967) stated that H6PD might be identical with the enzyme formerly known as glucose dehydrogenase, since the H6PD activity could not be separated from the glucose dehydrogenase activity during purification of H6PD.

Whatever the designation of this enzyme having H6PD and glucose dehydrogenase activities may be, there is no doubt concerning the presence of two kinds of enzyme molecules having G6PD activity in vertebrates; one has a narrow and the other has a broad substrate specificity. Are they homologous or analogous molecules? This question would be answered by comparing the properties of as many G6PDs from both invertebrates and vertebrates as possible. Such comparative studies might provide some valuable information on the evolution of G6PD.

The present report is a summary of our previous studies on animal G6PDs which were initiated with the characterization of electrophoretically distinct molecular forms of rat G6PD, followed by comparison of G6PDs from a variety of animals.

METHODS AND RESULTS

PROPERTIES OF RAT G6PD

1. *Electrophoretic pattern.* Homogenates of rat liver exhibited seven bands of G6PD activity on polyacrylamide gels (Fig. 1). The pattern differed between male and female adults; relative activities of the D and F bands were greater in females than in males. Such a sex difference was not found in newborn rats. In earlier works, we described the D enzyme as more active in females than in males and reported that the relative activity of the D enzyme increased whenever the total G6PD activity was increased (Hori et al., 1966; Hori and Matsui,

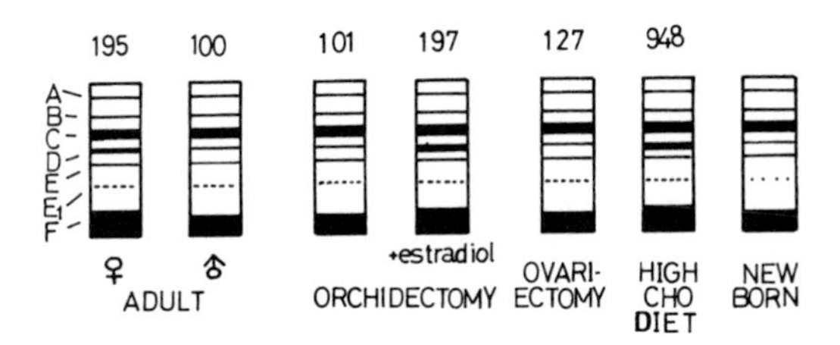

Fig. 1. Zymograms and relative activities (numbers) of G6PD from normal and treated rats (Hori et al., 1966; Hori and Matsui, 1967).

1967). It was, however, proved in recent studies that the increase of the D enzyme was rather a result of the increase of the F enzyme (Hori and Yonezawa, 1972).

2) Intracellular distribution, substrate specificity, and sensitivity to inhibitors. Bands A, C, E, and E_1 were located mainly in microsomes and were active toward G6P, Gal6P, and deoxyglucose 6-phosphate (dG6P), while bands D and F were found in mitochondria and cytosol and were specific for G6P. These two groups of enzymes also differed in sensitivity to inhibitors and heat; the former was resistant and the latter sensitive (Hori and Matsui, 1968; Kamada and Hori, 1970; Mochizuki and Hori, 1973).

3) Induction of the D and F enzymes. Glock and McLean (1953) have already reported that female rat livers have about twice as much G6PD activity as male rat livers. Zaheer, Tewari, and Krishnan (1967) also reported that the mitochondrial G6PD activity was greater in females than in males. The mechanism for such sex differences was, however, unknown.

If sex differences in total enzyme activity and in electrophoretic pattern are caused by sex hormones two possibilities might be considered; androgens inhibit or estrogens enhance the enzyme activity. We, therefore, tested effects of dehydroepiandrosterone and estradiol on castrated rats (Hori and Matsui, 1967). As expected, injection of estradiol benzoate into orchidectomized rats increased the total enzyme activity to the level of adult females and changed the zymogram from male to female type (Fig. 1). On the contrary, ovariectomy of young rats resulted in no increase of enzyme activity during maturation, and the zymogram remained the male type. Dehydroepiandrosterone had an inhibitory effect on hepatic G6PD. However, this hormone did not appear to be the main cause for the observed sex difference since castration and/or adrenalectomy did not increase the G6PD activity in male rats.

About a ten-fold increase of hepatic G6PD is possible by dietary treatment (Tepperman and Tepperman, 1958). Taking advantage of this phenomenon, we examined how the isozyme pattern changed after such a drastic increase of total enzyme activity. As a result, a marked increase of relative activity was observed only in the D and F enzymes (Fig. 1, Hori et al., 1966). In other words, it is the cytosol enzymes, but not the microsomal enzymes that respond to dietary and hormonal induction. This agrees with recent findings by other workers (Mandula et al., 1970; Kimura and Yamashita, 1972).

4) *Interconversion of the D and F enzymes.* As indicated in the above experiments, the D and F enzymes have quite similar properties in various respects. Are they products of a single gene or separate genes? The answer was given by the finding that the D enzyme could be formed in vitro from the F enzyme after treatment with cystamine HCl and had twice the molecular weight of the F enzyme. Inversely, conversion from the D to the F enzyme was induced with SH reagents (Hori and Yonezawa, 1972).

5) *Degradation of the C enzyme by trypsin.* In an attempt to characterize the A and C enzymes we happened to find that the C enzyme was degraded by the action of trypsin into two enzymatically active molecular forms which showed the same electrophoretic mobility as the E and E_1 enzymes and that lysosomal fractions could mimic the trypsin effect. The possibility was thus suggested that the E and E_1 enzymes might be degradation products of the C enzyme produced by lysosomal proteinases.

6) *Purification and properties of the purified C and F enzymes.* In order to obtain more precise information on the properties of these enzymes an attempt was made to purify them. As a result, about 3,300-fold purification of the C enzyme was achieved by means of cell fractionation, ammonium sulfate precipitation, hydroxylapatite chromatography, CM-Sepahdex and DEAE-Sepahdex column chromatography (final specific activity, 10 units/mg protein). On the other hand, the F enzyme was purified 380-fold from the cytosol fraction of rat livers which had been subjected to dietary induction, using DEAE-Sephadex and CM-cellulose (final specific activity, 122 units/mg protein) (Hori and Sado, 1974).

Properties of the purified C and F enzymes are summarized in Table 1 and Fig. 2. As is clear in these data, the C (microsomal) enzyme has properties closely similar to those of the H6PD of mouse liver reported by Beutler and Morrison (1967),

TABLE 1

PROPERTIES OF MICROSOMAL AND CYTOSOL G6PDs OF RAT LIVER

	pH optimum[1]	Km (M) (0.6 mM NADP as coenzyme)			Km (M) (6 mM G6P as substrate)		Molecular weight[2]	SH-inhibitors DEA and heat[3]	Mg^{++}
			pH 7.5	pH 10.0	pH 7.5	pH 10.0			
Microsomal (band C) enzyme	10.8	G6P	1.3×10^{-5}	1.3×10^{-4}	NADP 4×10^{-6}	3×10^{-6}	19×10^{4}	resistant	inhibited
		Gal6P	2.9×10^{-5}	5.8×10^{-4}					
		dG6P	1.1×10^{-3}	2.0×10^{-2}	NAD 1×10^{-6}	3×10^{-6}			
		glucose	5.0	too high to measure					
Cytosol (band F) enzyme	10.0	G6P	3.2×10^{-5}	4.0×10^{-4}	NADP 6×10^{-6}	12×10^{-6}	11×10^{4}	inhibited	activated
		Gal6P	5.9×10^{-3}	1.0×10^{-4}					
		dG6P	9.1×10^{-4}	2.1×10^{-2}					

1. Buffer used: Tris-HCl, I=0.07 for pH 7.5-8.9; glycine-NaOH, I=0.14 for pH 9.0-11.2. Values for G6P+NADP.

2. Determined by the methods of Zwaan (1967) and Shapiro et al., (1967).

3. DEA, dehydroepiandrosterone.

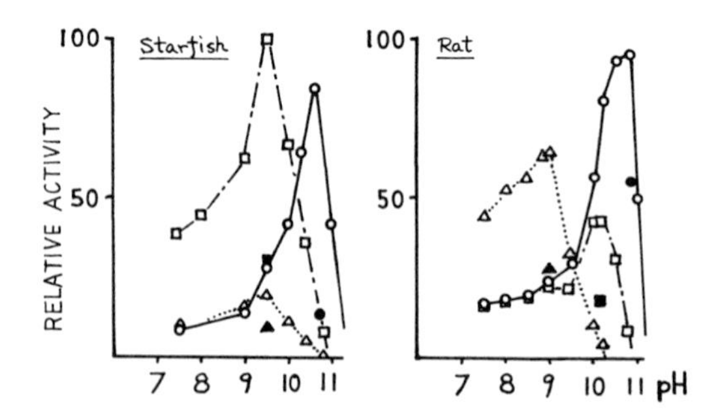

Fig. 2. Effect of pH on enzyme activity. Left, starfish, *Asterias amurensis*. Right, rat microsomes. o——o, G6P+NADP; □—·—□ , Gal6P+NADP; △-----△ , dG6P+NADP; ● , G6P+NAD; ■ , Gal6P+NAD; ▲ , dG6P+NAD. Substrate and coenzyme concentrations were 6 mM and 0.6 mM, respectively.

while the F (cytosol) enzyme resembles human erythrocyte G6PD reported by many workers (Marks et al., 1961; Kirkman, 1962; Yoshida, 1966, 1967).

Despite such striking difference in properties, there are some reasons to believe that both enzymes are functioning as G6PD in vivo at least in mammals (Srivastava and Beutler, 1969; Mandula et al., 1970); one is that Gal6P and dG6P have not been reported to occur in animals, so that Gal6PD and dG6PD activities of the microsomal enzyme might not be of physiological importance, and the second is that galactonate 6-phosphate formed from Gal6P by H6PD would not be a substrate for phosphogluconate dehydrogenase.

PHYLOGENIC STUDIES

A question thus arises whether the microsomal and cytosol enzymes are evolutionarily divergent or convergent molecules. To answer this question we have carried out comparative studies of G6PD from a wide variety of animals by means of polyacrylamide disc electrophoresis (Kamada and Hori, 1970; Mochizuki and Hori, 1973).

The results obtained by visual inspection and densitometric scanning of stained gels indicated that the enzymes from different species had a wide variety of substrate specificities. The enzymes were therefore temporarily classified into three types: type I enzymes which occur widely in the animal kingdom (present in cytosol) and are active on G6P; type II enzymes which occur in invertebrates and lower vertebrates, and which are probably present in membrane-bound organelles and are active on G6P and Gal6P; and type III enzymes, which occur only in vertebrates and which are present in microsomes and are active on G6P, Gal6P, and dG6P. A hypothesis was then proposed that type II and III enzymes might have evolved from type I enzymes (Kamada and Hori, 1970). A similar hypothesis was also put forward by Stegeman and Goldberg (1971) and Yamauchi and Goldberg (1973).

TYPE II G6PD

Type III enzymes are active with NAD+glucose, but not type I enzymes. In this respect, type II enzymes can be subtyped into: 1) invertebrate enzymes which are inactive with NAD and glucose as represented by hydra and squid enzymes; 2) invertebrate enzymes which are active with NAD, but not with glucose, as represented by scallop and sea cucumber enzymes; and 3) vertebrate enzymes which are active with NAD and also with glucose, though only slightly, as represented by *Xenopus* and sculpin enzymes (Table 2). These findings strongly favor the hypothesis that type II and type III enzymes have evolved from type I enzymes.

STARFISH G6PD

Further evidence suggesting a gradual broadening of the substrate specificity during G6PD evolution has been provided by the studies on starfish G6PD; the enzyme from *Asterias amurensis* had a very high activity with Gal6P, but was much less active with both G6P and dG6P and utilized NAD as well as NADP as electron acceptor when examined by gel electrophoresis. In this respect, starfish enzymes did not resemble any of the G6PDs so far studied. For further characterization, the enzyme was partially purified by ethanol precipitation and DEAE-cellulose column chromatography. As is clear in the pH-activity curves shown in Fig. 2, the enzyme was very active with Gal6P+NADP and its G6PD activity was low at pH 7.5-9.0, but increased markedly at pH 10.6, the phenomenon being comparable with that observed with rat microsomal G6PD (type III). The enzyme was inhibited 24% by 10 mM $MgCl_2$ and 80% by heat (5 min at 50° C), but not with 2 mM N-ethylmaleimide and 3 mM iodoacetamide. Details of these studies on starfish G6PD will be published elsewhere. These findings suggesting the resemblance of starfish G6PD with vertebrate type II or even type III enzymes indicate that starfish G6PD might be homologous to vertebrate enzymes of broader substrate specificity. The results of molecular weight estimation also favor this hypothesis, as will be described presently.

TABLE 2

SUBSTRATE SPECIFICITY OF ANIMAL GLUCOSE 6-PHOSPHATE DEHYDROGENASES.

Organism[1] Coenzyme	NADP				NAD				
Substrate	G6P	Gal6P	dG6P	G	G6P	Gal6P	dG6P	G	Type
Planaria	100	6	1	0	0	0	0	0	I
Eel (cytosol)	100	15	3	0	0	0	0	0	
Hydra	100	36	2	0	0	0	0	0	II
Squid	100	42	0	0	0	0	0	0	
Scallop	100	32	0	0	50	21	0	0	
Sea cucumber	100	68	10	0	21	38	4	0	
Starfish	100	256	52	7	48	206	51	8	?
Sculpin	100	97	12	2	28	36	19	12	II
Xenopus (Mc)	100	128	15	23	26	57	4	16	
Turtle (Mc)	100	178	330	45	46	105	363	88	III
Eel (Mc)	100	115	142	46	41	64	142	133	

Tissue homogenates were electrophoresed on thin polyacrylamide gels, stained, and scanned in a densitometer. The enzyme activity was expressed as the percentage of the activity with G6P+NADP (Mochizuke and Hori, 1973). (cytosol), cytosol enzyme; (Mc), microsomal enzyme.

[1] *Dugesia japonica; Anguilla japonica; Pelmatohydra robusta; Ommastrephes sloani pacificus; Patinopecten yessoensis; Sticopus japonicus; Asterias amurensis; Cottus sp.; Xenopus laevis; Pseudemys scripta elegans.*

MOLECULAR WEIGHT

The studies mentioned above are all concerned with the catalytic properties of G6PD molecules, but not with their general physical properties as proteins. We have therefore investigated the molecular weights of G6PDs from various animals in order to supplement the above data. The results are given in Table 3.

We initially thought that the type I G6PD gene was fairly stable so that it persisted throughout animal evolution without drastic changes, while type II and III G6PD genes might have evolved from type I G6PD genes by mechanisms such as gene duplication. However, as is clear in Table 3, the molecular weights of type I G6PDs from fish and amphibians are at least twice the molecular weights of type I G6PDs from reptiles, birds, and mammals. In addition, although not listed in the table, some invertebrate G6PDs so far studied have molecular weights of more than 20 x 10^4. On the other hand, there seems to be no doubt that the low molecular weight G6PD is the progenitor of the high molecular weight G6PD, since *Paramecium* (unpublished data), plants, and fungi have the low molecular weight G6PDs (Chilla et al., 1973; Engel et al., 1969; Muto and Uritani, 1970; Schnarrenberger et al., 1973; Scott, 1971; Yue et al., 1967). A question thus arises as to where the low molecular weight G6PDs of reptiles, birds, and mammals came from or why the lower animals possess the high molecular weight G6PDs. Unfortunately, data are still insufficient and more extensive studies with many other species are required to answer this question.

Contrary to such a range of molecular weights of type I enzymes, the molecular weights of type II and III enzymes fell within a relatively narrow range of 17-23 x 10^4 as far as the enzymes so far studied were concerned. Interestingly, the molecular weights of trypsinized enzymes were quite similar as shown in Table 3, thus strongly suggesting the structural resemblance among type II and III enzymes. Furthermore, the finding that the type I enzyme of the pond snail was converted by trypsin into a smaller, active molecule, just like type II and III enzymes was particularly interesting. This type I enzyme differed from the other type I enzymes in that its electrophoretic mobility was not affected by NADP added to gels, so that it was designated as type I-B in a previous study (Kamada and Hori, 1970). In contrast, the ordinary type I enzymes did not form smaller, active molecules upon trypsin digestion, and its mobility was reduced by excess NADP added to gels. In other words, some structural similarity might be present between this pond snail enzyme and type II and III

TABLE 3

MOLECULAR WEIGHTS OF G6PDs.[1]

Species[2]		rat	mouse	chicken	turtle	snake	*Rana*	*Xenopus*	salamander	funa	starfish	pond snail
Type 1 G6PD		11	12	11	10	11	26	25	21	28		14
Type II or III G6PD	intact	19	19	20	22	21	19	20	18	23	17	20[3]
	trypsin-treated[4]	12	13	12	12	11	12	13	12	10	10	10

[1]Determined by the method of Zwaan (1967). Molecular weight x 10^{-4}.

[2]*Rattus norvegicus; Mus musculus; Gallus domesticus; Clemmys japonica; Elaphe climacophora; Rana japonica; Xenopus laevis; Hynobius retardatus; Carassius carassius; Asterias amurensis; Radix auricularius japonicus.*

[3]This is type I G6PD, but differs from other type I G6PDs in that its electrophoretic mobility is not affected by NADP added to gels (Kamada and Hori, 1970).

[4]Molecular weights of trypsin-treated, enzymatically active molecules.

enzymes.

CONCLUSION

In our studies, animal G6PDs are classified into three types for convenience, but the definition of these types is by no means quantitative and hence should be revised in the future after G6PDs from a number of species have been isolated for quantitative analyses. However, such classification seems to be useful at present for speculating on the events which would have probably occurred during G6PD evolution.

The data presented in this report suggest the possibility that the evolution of G6PD might have occurred gradually from an enzyme of narrower substrate specificity to one of broader substrate specificity. However, the relationship between the evolutionary changes in substrate specificity and in molecular weight remains to be explained.

REFERENCES

Bagdasarian, G. and D. Hulanicka 1965. Changes of mitochondrial glucose-6-phosphate dehydrogenase and 6-phosphogluconate dehydrogenase during brain development. *Biochim. Biophys. Acta* 99: 369-371.

Baquer, N. Z. and P. McLean 1972. Evidence for the existence and functional activity of the pentose phosphate pathway in the large particle fraction isolated from rat tissues. *Biochem. Biophys. Res. Commun.* 46: 167-174.

Baquer, N.Z. , M. Sochor and P. McLean 1972. Hormonal control of the 'compartmentation' of the enzymes of the pentose phosphate pathway associated with the large particle fraction of rat liver. *Biochem. Biophys. Res. Commun.* 47: 218-226.

Beutler, E. and M. Morrison 1967. Localization and characteristics of hexose 6-phosphate dehydrogenase (glucose dehydrogenase). *J. Biol. Chem.* 242: 5289-5293.

Chilla, R., K. M. Doering, G. F. Domagk and M. Rippa 1973. A simplified procedure for the isolation of a highly active crystalline glucose-6-phosphate dehydrogenase from *Candida utilis*. *Arch. Biochem. Biophys.* 159: 235-239.

Engel, H. J., W. Domschke, M. Alberti and G. F. Domagk 1969. Protein structure and enzymatic activity. II. Purification and properties of a glucose-6-phosphate dehydrogenase from *Candida utilis*. *Biochim. Biophys. Acta* 191: 509-516.

Glock, G. E. and P. McLean 1953. Further studies on the properties and assay of glucose-6-phosphate dehydrogenase and 6-phosphogluconate dehydrogenase of rat liver. *Biochem.*

J. 55: 400-408.

Hori, S. H., T. Kamada and S. Matsui 1966. On the glucose 6-phosphate dehydrogenase in rat livers, with special regard to sex difference. *J. Fac. Sci. Hokkaido Univ.* Ser.6, *Zool.* 16: 60-66.

Hori, S. H. and S. Matsui 1967. Effects of hormones on hepatic glucose-6-phosphate dehydrogenase of rat. *J. Histochem. Cytochem.* 15: 530-534.

Hori, S. H. and S. Matsui 1968. Intracellular distribution of electrophoretically distinct forms of hepatic glucose 6-phosphate dehydrogenase. *J. Histochem. Cytochem.* 16: 62-63.

Hori, S. H. and Y. Sado 1974. Purification and properties of microsomal glucose 6-phosphate dehydrogenase (hexose 6-phosphate dehydrogenase) of rat liver. *J. Fac. Sci. Hokkaido Univ.* Ser. 6, *Zool.* (in press).

Hori, S. H. and S. Yonezawa 1972. The in vitro interconversion by mercaptans of type I glucose 6-phosphate dehydrogenase isozymes of rat. *J. Histochem. Cytochem.* 20: 804-810.

Kamada, T. and S. H. Hori 1970. A phylogenic study of animal glucose 6-phosphate dehydrogenases. *Japan. J. Genet.* 45: 319-339.

Kimura, H. and M. Yamashita 1972. Studies on microsomal glucose 6-phosphate dehydrogenase of rat liver. *J. Biochem.* 71: 1009-1014.

Kirkman, H. N. 1962. Glucose-6-phosphate dehydrogenase from human erythrocytes. I. Further purification and characterization. *J. Biol. Chem.* 237: 2364-2370.

Mandula, B., S. K. Srivastava and E. Beutler 1970. Hexose 6-phosphate dehydrogenase: Distribution in rat tissues and effect of diet, age and steroids. *Arch. Biochem. Biophys.* 141: 155-161.

Marks, P. A., A. Szeinberg and J. Banks 1961. Erythrocyte glucose-6-phosphate dehydrogenase of normal and mutant human subjects: properties of purified enzymes. *J. Biol. Chem.* 236: 10-17.

Mochizuki, Y. and S. H. Hori 1973. Further studies on animal glucose 6-phosphate dehydrogenases. *J. Fac. Sci. Hokkaido Univ. Ser.* 6, *Zool.* 19: 58-72.

Muto, S. and I. Uritani 1970. Glucose 6-phosphate dehydrogenase from sweet potato. *Plant and Cell Physiol.* 11: 767-776.

Newburgh, R. W. and V. H. Cheldelin 1956. The intracellular distribution of pentose cycle activity in rabbit kidney and liver. *J. Biol. Chem.* 218: 89-96.

Ohno, S., H. W. Payne, M. Morrison and E. Beutler 1966. Hexose-6-phosphate dehydrogenase found in human liver. *Science* 153: 1015-1016.

Razumovskaya, N. I., V. M. Pleskov, and T. L. Perova 1970.

Soluble and mitochondrial forms of pentose phosphate cycle dehydrogenases in skeletal muscle of rabbit. *Biochem.* 35: 196-201.

Ruddle, F. H., T. B. Shows and T. H. Roderick 1968. Autosomal control of an electrophoretic variant of glucose-6-phosphate dehydrogenase in the mouse (*Mus musculus*). *Genetics* 58: 599-606.

Schnarrenberger, C., A. Oeser and N. E. Tolbert 1973. Two isoenzymes each of glucose-6-phosphate dehydrogenase and 6-phosphogluconate dehydrogenase in spinach leaves. *Arch. Biochem. Biophys.* 154: 438-448.

Scott, W. A. 1971. Physical properties of glucose 6-phosphate dehydrogenase from *Neurospora crassa*. *J. Biol. Chem.* 246: 6353-6359.

Shatton, J. B., J. E. Halver and S. Weinhouse 1971. Glucose (hexose 6-phosphate) dehydrogenase in liver of rainbow trout. *J. Biol. Chem.* 246: 4878-4885.

Shaw, C. R. 1966. Glucose-6-phosphate dehydrogenase: Homologous molecules in deer mouse and man. *Science* 153: 1013-1015.

Srivastava, S. K. and E. Beutler 1969. Auxiliary pathways of galactose metabolism. *J. Biol. Chem.* 244: 6377-6382.

Srivastava, S. K., K. G. Blume, E. Beutler and A. Yoshida 1972. Immunological difference between glucose-6-phosphate dehydrogenase and hexose-6-phosphate dehydrogenase from human liver. *Nature* 238: 240-241.

Stegeman, J. J. and E. Goldberg 1971. Distribution and characterization of hexose-6-phosphate dehydrogenase in trout. *Biochem. Genet.* 5: 579-589.

Stegeman, J. J. and E. Goldberg 1972. Properties of hepatic hexose-6-phosphate dehydrogenase purified from brook trout and lake trout. *Comp. Biochem. Physiol.* 43B: 241-256.

Tepperman, H. M. and J. Tepperman 1958. The hexose monophosphate shunt and adaptive hyperlipogenesis. *Diabetes* 7: 478-485.

Warburg, O. and W. Christian 1931. Über Aktivierung der Robisonschen Hexose-Mono-Phosphorsäure in roten Blutzellen und die Gewinnung aktivierender Fermentlösungen. *Biochem. Z.* 242: 206-227.

Yamada, D. and N. Shimazono 1961. Recognition and solubilization of glucose 6-phosphate and 6-phosphogluconate dehydrogenases in the particle fraction of brain. *Biochim. Biophys. Acta* 54: 205-206.

Yamauchi, T. and E. Goldberg 1973. Glucose 6-phosphate dehydrogenase from brook, lake and splake trout: An isozymic and immunological study. *Biochem. Genet.* 10: 121-134.

Yoshida, A. 1966. Glucose 6-phosphate dehydrogenase of human erythrocytes. I. Purification and characterization of normal (B^+) enzyme. *J. Biol. Chem.* 241: 4966-4976.

Yoshida, A. 1967. Human glucose 6-phosphate dehydrogenase: Purification and characterization of Negro type variant (A^+) and comparison with normal enzyme (B^+). *Biochem. Genet.* 1: 81-99.

Yue, R. H., E. A. Noltmann, and S. A. Kuby 1967. Glucose 6-phosphate dehydrogenase (Zwischenferment) II. Homogeneity measurements and physical properties of the crystalline apoenzyme from yeast. *Biochemistry* 6: 1174-1183.

Zaheer, N., K. K. Tewari and P. S. Krishnan 1967. Mitochondrial forms of glucose 6-phosphate dehydrogenase and 6-phosphogluconic acid dehydrogenase in rat liver. *Arch. Biochem. Biophys.* 120: 22-34.

EVOLUTION OF GLUCOSE 6-PHOSPHATE DEHYDROGENASE AND HEXOSE 6-PHOSPHATE DEHYDROGENASE

AKIRA YOSHIDA
Department of Biochemical Genetics, City of Hope National Medical Center, Duarte, California 91010

ABSTRACT. Glucose 6-phosphate dehydrogenase (G6PD), which oxidizes only glucose 6-P, and hexose 6-phosphate dehydrogenase (H6PD), which oxidizes glucose 6-P, 2-deoxy glucose 6-P and galactose 6-P, have similar features, suggesting their common origin. Among more than 100 types of known human G6PD variants, G6PD Markham, G6PD Union and a few other variants can oxidize not only glucose 6-P but also 2-deoxy glucose 6-P and galactose 6-P as well. Therefore, single step mutations can induce an alteration of the substrate specificity; i.e., the differentiation of isozymes from their ancestral enzyme does not necessarily require an accumulation of a series of mutations.

Oxidation of the normal human G6PD by hydrogen peroxide in vitro changes the enzyme properties. The substrate specificity and kinetic properties of the oxidized enzyme are similar to those of G6PD Union. Thus, chemical modification of the enzyme in vitro mimics the mutational change of the enzyme.

Despite the fact that a mutation and chemical modification can change the substrate specificity, producing a modified G6PD with substrate specificity which is similar to that of H6PD, the present human H6PD is not structurally closely related to the G6PD. The divergence of the two enzymes presumably occurred early in evolution and they have separately accumulated evolutional modification for several hundred million years.

There is plenty of evidence which suggests that functionally similar enzymes were developed from a common ancestral protein in the process of evolution. For example, enzymes having the same co-factor also seem to have other features that indicate their common origins; such enzymes are transaminases requiring pyridoxal phosphate, dehydrogenases requiring NAD, NADP or flavine, esterases and proteases with different substrate specificity. Isozymes with homologous substrate specificity are clearly differentiated from a common protein. The process of such evolutional differentiation requires: a) duplication of a gene for an ancestral enzyme; and b) mutational change of kinetic properties and substrate specificity of an ancestral enzyme. The mechanism of mutational alteration of substrate specificity of an enzyme is not clear. Did a single step mutation induce significant alteration of substrate specificity of an enzyme, or, was an accumulation of a series of mutations required to induce significant change of the substrate specificity? Although many mutant enzymes were artificially produced in bacteria, information about the alteration of substrate specificity accompanying such mutations is obscure. Conversion of alanine dehydrogenase to glutamate-dehydrogenase by induced or spontaneous mutations in B. subtilis has been reported. However, it is not clear whether this is due to a mutation in the structural gene or in the regulatory gene, or due to other effects (Shen, et al., 1963; Hong, et al., 1963).

Human glucose 6-phosphate dehydrogenase (abbreviation: G6PD) has many advantages for the study of mutational alteration of substrate specificity. Firstly, the wild type of normal enzyme has a well-defined substrate specificity, i.e., NADP, but not NAD, is a co-enzyme and only D-glucose 6-phosphate is effectively oxidized by the enzyme. Secondly, the G6PD is located in the X-chromosome in man and other mammals, and therefore, the male always has only one type of the enzyme. Thus, one can avoid the confusing situation due to heterozygosity and hybrid enzyme molecules which usually occurs in the study of the variant enzymes located in autosomal chromosomes. Thirdly, more than 100 types of human G6PD variants have been reported at this time. And finally, an isozyme of G6PD, which can oxidize galactose 6-Phosphate and can use NAD as a co-enzyme, exists in man and in other vertebrates. This enzyme, designated as hexose 6-phosphate dehydrogenase (abbreviation: H6PD), is controlled by a gene located in an autosomal chromosome.

This paper discusses the nature of change of substrate specificity of human G6PD by mutation and by chemical modification, and relationships between G6PD and its isozyme, H6PD. A summary of the background knowledge about the structure of G6PD and H6PD is presented below, before the major topics are discussed.

Human G6PD is a predominantly dimeric form consisting of two identical subunits of molecular weight of about 55,000 in a reaction mixture in the presence of both glucose 6-phosphate and NADP (Yoshida and Hoagland, 1970). In the absence of the substrate, the human enzyme resumes tetrameric or dimeric form depending upon the pH and ionic strength of the solution (Cohen and Rosemeyer, 1969; Yoshida and Hoagland, 1970). G6PD from some organisms is mostly tetrameric (fish, bird) (Yamuchi and Goldberg, 1973; Yoshida, unpublished observation), and that from others is mostly dimeric (rat, deer mouse) (Matsuda and Yugari, 1967; Shaw and Koen, 1968). Basic structure of G6PD from various organisms seems to be similar.

Human G6PD associates with only one NADP, or NADPH, per active dimer at a time (Yoshida and Hoagland, 1970; Yoshida, 1973). The enzyme without bound NADP or the enzyme bound with NADPH cannot associate with glucose 6-phosphate; only the enzyme bound with NADP can associate with and oxidize glucose 6-phosphate (Yoshida, 1973). This would suggest that the two subunits cooperate to produce an NADP binding site and a glucose 6-phosphate binding site at a time. The normal human G6PD is highly specific for glucose 6-phosphate and NADP: and galactose 6-phosphate and 2-deoxy glucose 6-phosphate are not effectively oxidized by the enzyme. NAD is not effective as a co-enzyme.

In contrast to G6PD, H6PD can oxidize galactose 6-phosphate, 2-deoxy glucose 6-phosphate and glucose quite effectively (Table I). NAD is a fairly good co-enzyme particularly in the presence of galactose 6-phosphate, 2-deoxy glucose 6-phosphate or glucose as substrate.

TABLE I

KINETIC PROPERTIES OF HUMAN GLUCOSE 6-PHOSPHATE DEHYDROGENASE AND HEXOSE 6-PHOSPHATE DEHYDROGENASE

Enzyme	pH	Michaelis Constant (Km, mM)				
		G6P	2-deoxy	Gal 6P	NADP	NAD
Glucose-6-P** Dehydrogenase	8.0	0.039	6	8	0.0044	4
Hexose-6-P*** Dehydrogenase	7.1	0.005	0.066	0.007	-	-
	9.6	0.12	5.89	0.504	0.004	-

	pH	Substrate Specificity*		
		2-deoxy G6P/G6P	Gal 6P/G6P	NAD/NADP
Glucose-6-P Dehydrogenase	8.0	< 4	< 4	< 1
Hexose-6-P Dehydrogenase	7.1	110	83	22
	9.6	19	100	22

* Percentage of activity with glucose 6-phosphate plus NADP at the pH specified
** Yoshida (1966) ; *** Beutler & Morrison (1967)

H6PD is mostly located in the microsomal fraction of various tissues, while G6PD is in the cytoplasm (Beutler and Harrison, 1967). The electrophoretic mobilities of the two enzymes are different from each other.

H6PD is not yet purified in homogeneous form from humans and other organisms, and the structure of this enzyme is still obscure. However, from the data of gel filtration and zymogram patterns of the enzyme, it is likely that the enzyme from various organisms is dimeric, eg. deer mouse (Shaw and Koen, 1968), or tetrameric eg. fish (Stegman and Goldberg, 1971) and bird (Yoshida, unpublished), and its subunit molecular size is similar to that of G6PD. Therefore, structural homology exists between G6PD and H6PD.

1.) *Change of substrate specificity of human G6PD by mutation.* More than a hundred variants of G6PD have been found in man. These variants are distinguishable from each other by electrophoretic mobility on starch gel or by enzymatic characteristics, and therefore, all are individually different structural mutants. Most of them are rare, but some of them exist in high frequencies in certain populations. Several human G6PD variants listed in Table 2 have unusual substrate specificity, significantly differing from that of the normal wild-type enzyme. To compare their substrate specificity, the rates of utilization of substrate analogues are expressed, taking that of normal substrates, i.e., NADP and glucose 6-phosphate, as 100.

As shown in the Table, G6PD Markham, a common variant in a certain New Guinea population, can use NAD as a co-enzyme better than NADP, and it can efficiently oxidize 2-deoxy glucose 6-phosphate and galactose 6-phosphate. (Kirkman, et al., 1968). G6PD Union, a common variant in a Filipino population, (Yoshida, et al., 1970) and G6PD Benevento (McCurdy, pers. commun.) can also oxidize galactose 6-phosphate and 2-deoxyglucose 6-phosphate, while they cannot use NAD as a co-enzyme. Another interesting variant found in New Guinea is G6PD Munum (Yoshida, et al., 1974). This variant can use NAD as a co-enzyme, but cannot effectively oxidize 2-deoxy glucose 6-phosphate. For several variants, deamino NADP is a more efficient co-enzyme than NADP itself. Exact structural abnormality, i.e., specific amino acid substitution, of these variant enzymes has not yet been elucidated. However, considering the fact that most of the genetic variants are caused by a single step base substitution in a given gene, which has been demonstrated in many human hemoglobin variants and also in two human G6PD variants (Yoshida, 1967; Yoshida, 1970), most, if not all, of the G6PD variants listed in Table 2, are presumably associated with single amino acid substitutions. One can conclude that a slight change in the primary structure of the enzyme can induce

TABLE II

SUBSTRATE SPECIFICITY OF VARIANT G6PD

Variant	2 deoxy G6P	Gal-6-P	deamino NADP	NAD
Normal	< 4	< 4	55-60	< 1
Bat-Yam	40-45	-	-	-
Benevento	245	-	-	-
Hualien	72.2	-	-	-
Lifta	60	-	-	-
Manum	< 4	-	300-400	60-130
Markham	180	66	300	130
Mediterranean	25	20	350	-
Mexico	26-33	-	130-160	-
Orchomenos	105	-	350	-
Union	180	80	370	< 3
Worcester	< 4	-	21	-

Relative rate of substrate utilization in the reaction mixture containing 0.8 ~ 1 mM glucose 6-phosphate or 2-deoxy glucose 6-phosphate or galactose 6-phosphate, and 0.08 ~ 0.1 mM NADP or deamino NADP taking the utilization of the primary substrates as 100.

a significant alteration of its substrate specificity. As far as the substrate specificity is concerned, G6PD Markham and G6PD Union are more similar to H6PD than the wild type normal G6PD. The following facts can be deduced from these findings.

a) High 2-deoxy glucose 6-phosphate utilization is always accompanied with high galactose 6-phosphate utilization, presumably due to the structural similarity between the substrate analogues.

b) Certain variants (G6PD Union, G6PD Markham), with a high rate of utilization of 2-deoxy glucose 6-phosphate (or galactose 6-phosphate) have a high rate of utilization of deamino NADP, suggesting an interaction of the two binding sites, i.e., NADP binding site and glucose 6-phosphate binding site, of the enzyme.

The change of substrate specificity by mutation is not a very rare event, since more than ten G6PD variants which are associated with somewhat altered substrate specificity have been found to about one hundred human G6PD variants thus far reported.

It is of great interest to study whether or not somatic mutation in tissue culture could produce G6PD variants associated with altered substrate specificity. This can be examined by staining the cells with phenazine-methosulfate-tetrazolium dye using 2-deoxy glucose 6-phosphate as a substrate instead of glucose 6-phosphate, on the basis that normal cells cannot oxidize the substrate analogue and therefore cannot be stained, but the mutant cells associated with altered substrate specificity might oxidize the substrate analogue. Several investigators have carried out such experiments and found that most cells did not react with 2-deoxy glucose 6-phosphate, but only occasional cells did react with the substrate analogue (Sutton and Karp, 1971). However, it is not yet clear whether the altered substrate specificity was induced by somatic cell mutation, or it was simply due to secondary modification of the normal enzyme, since the inheritance of the abnormality has not been established. In this connection, it is important to examine whether or not chemical modification in vitro can change the substrate specificity of the enzyme.

2) *The change in substrate specificity of human glucose 6-phosphate by chemical modification.*

Kirkman and co-workers (1965) first observed a change in the substrate specificity of normal human G6PD by oxidation in vitro. Details of their study have not been published. We examined this phenomenon in more detail using a crystalline human G6PD (Yoshida, 1973).

The results of amino acid analysis and quantitation of sulfhydryl groups by chloromercuribenzoate indicate that the active dimeric enzyme (MW=110,000) contained about 18 cysteine residues and had no S-S bridges in the molecule (Table III and IV). Among these 18 sulfhydryl residues, about 8 residues are exposed to react with chloromercuribenzoate without denaturation and others can react with the SH-reagent only in the presence of urea (Table IV). A sulfhydryl group (or groups) which is exposed and reactive with chloromercuribenzoate is essential for the enzyme activity since the enzyme is inactivated by the SH-reagent. The enzyme inactivated by the SH-reagent can be fully reactivated by treatment with excess reducing reagents such as dithiothreitol.

The enzyme can be oxidized by H_2O_2 at pH 7.0, or by air at pH 8.0. After the oxidation of the enzyme by 1.5% H_2O_2 at 25° C for 2 hours, about 60% of the sulfhydryl groups were oxidized and the enzyme was strongly inactivated (Table V). Whereas, the oxidation of the enzyme by air induced only a slight decrease in the enzyme activity, although about 70% of the total sulfhydryl groups were oxidized (Table IV). This could be explained by the following mechanism: The enzyme has one tightly bound NADP per active dimer (Yoshida and Hoagland, 1969). When the enzyme was incubated with chloromercuribenzoate, NADP was released from the enzyme (Yoshida, unpublished observation) suggesting that a sulfhydryl group (or groups) is involved in

TABLE III

AMINO ACID COMPOSITION OF THE NATIVE AND OXIDIZED ENZYME

Amino Acid	Relative Molar Ratio (a) Native enzyme	Enzyme oxidized by H_2O_2	No. of residues per active dimer (b)
Cysteic acid	< 1	7.6	-
Aspartic acid	104	104	104
Threonine	45.0	45.1	43
Serine	46.8	46.5	52
Glutamic Acid	121.2	120.3	119
Proline	65.0	68.1	66
Glycine	72.3	76.6	75
Alanine	62.3	66.5	60
Cysteine	-	-	18
Valine	52.0	57.3	58
Methionine	22.1	19.6	24
Isoleucine	37.9	38.1	43
Leucine	86.1	85.6	88
Tyrosine	40.4	34.2	37
Phenylalanine	49.2	40.2	48
Lysine	50.3	55.8	54
Histidine	23.0	18.1	25
Arginine	53.4	51.6	50
Tryptophan	-	-	14

(a) Relative molar ratio found in acid hydrolysates (for 20 hrs) of the protein, taking aspartic acid as 104. Degradation of threonine and serine, and incomplete hydrolysis of valine, leucine, and isoleucine are not corrected.
(b) The nearest integral number of amino acid residues per active dimer, calculated from the previous multiple analysis (Yoshida, 1966).

NADP binding. Therefore, it is conceivable that such a functionally essential sulfhydryl group is protected from the air oxidation by bound NADP. In contrast, the oxidation of the enzyme by H_2O_2 could induce oxidation not only of sulfhydryl groups but also some other residues such as phenolic and indoyl groups, and some cysteine residues were oxidized to cysteic acid. In fact, amino acid analysis of the oxidized G6PD indicated that about 40% of the cysteine was oxidized to cysteic acid, and a significant amount of histidine and tyrosine were

TABLE IV

SULFHYDRYL CONTENT OF NATIVE AND OXIDIZED HUMAN GLUCOSE 6-PHOSPHATE DEHYDROGENASE

Enzyme	Enzyme Activity (% of Native Enzyme)	Sulfhydryl Content (μ mole/g)	
		No urea	8M urea
Native	100	71 (68-73)	168 (155-195)
Oxidized by Air	94 (93-95)	-	46 (40-49)
Oxidized by H_2O_2	23 (20-26)	-	62 (56-68)

were degraded (Table III).

A remarkable change in the substrate specificity was observed in the enzyme oxidized by H_2O_2 (Table V). The oxidized enzyme had a higher rate of utilization of deamino NADP, NAD and 2-deoxy glucose 6-phosphate. The altered substrate specificity is associated with a higher relative turnover rate of these substrate analogues and a higher affinity (lower K_m value) of the substrate analogues with the oxidized enzyme (Table VI).

Among the naturally existing G6PD variants, the substrate specificity, which is the higher utilization rate of 2-deoxy glucose 6-phosphate and deamino NADP, and the kinetic characteristics, which are the low K_m for 2-deoxy glucose 6-phosphate and the higher turnover rate of deamino NADP, of the Union variant, are similar to that of the oxidized normal enzyme. The substrate specificity of the oxidized enzyme is somewhat similar to that of H6PD (Table V). Thus, chemical modification of the enzyme in vitro mimics the mutational change of the enzyme.

3) *Divergence of glucose 6-phosphate dehydrogenase and hexose 6-phosphate dehydrogenase.*

The existence of the two forms of glucose 6-phosphate dehydrogenase, A and B, in deer mouse was first reported by Shaw and Barto (1965). Later, the existence of the two forms of enzyme was demonstrated also in other vertebrates (man, horse, bird, and fish).

In contrast to the A enzyme which is located in the X-chromosome, the B enzyme is located in an autosomal chromosome (Shaw and Barto, 1965) and is equally active toward glucose 6-phosphate and galactose 6-phosphate (Table I). The A enzyme (G6PD) was present in all tissues including erythrocytes, and the B enzyme (H6PD) was located in the microsomal fraction of various tissues but not in erythrocytes. H6PD does not seem to be present among the invertebrate species. Glucose 6-phos-

TABLE V

CHANGE OF SUBSTRATE SPECIFICITY OF HUMAN GLUCOSE 6-PHOSPHATE BY OXIDATION

Enzyme		Enzyme Activity (% of Native Enzyme)	Substrate Specificity Deamino NADP/NADP		NAD/NADP		2-Deoxy G6P/G6P		Gal 6P/G6P
			a	b	a	b	a	b	a
Normal Enzyme	Native	100	55 ± 4	57	0.3	4.7	3	12	4
	Oxidized by air	94 ± 2	59 ± 6	-	0.3	-	3	-	-
	Oxidized by H_2O_2	22 ± 9	222 ± 11	240	3.6	18	68 ± 13	110	-
G6PD Union		-	370	-	3	-	180	-	80
G6PD Markham		-	300	-	130	-	180	-	66
H6PD		-	-	-	22	-	19	-	100

a. Relative rate of substrate utilization determined in the reaction mixture containing 0.8 mM glucose 6-phosphate or 2-deoxy glucose 6-phosphate, and 0.08 mM NADP or deamino NADP or 0.4 mM NAD taking the utilization of the primary substrates as 100.

b. Relative V max.

TABLE VI

MICHAELIS CONSTANTS (μM) FOR PRIMARY SUBSTRATES AND ANALOGUES

Enzyme	Glucose 6-Phosphate	2-Deoxy Glucose 6-Phosphate	NADP	Deamino NADP	NAD
Native	39	6×10^3	3.5	6.1±0.6	4×10^3
Oxidized by H_2O_2	18 ± 2	530 ± 30	1.8±0.2	4.0±0.5	2×10^3
Union Variant	10 ± 2	640 ± 150	4.4±0.8	9.3±2	-

phate dehydrogenases isolated from bacteria (Bennerjee and Fraenkel, 1972), fungi (Scott and Tatum, 1971), yeast (Anderson, et al., 1968) and other invertebrates (Kamada and Hori, 1970) are similar to mammalian G6PD in respect to the substrate specificity and the effect of carbonate in the oxidation of glucose. An enzyme corresponding to the vertebrate's H6PD has not been found in these organisms. However, in these proceedings, Dr. Hori (1974) reported a finding which suggests an enzyme with broader substrate specificity in several asteroidae species. The nature of this enzyme and its evolutionary implications remain to be investigated.

The existence of two glucose 6-phosphate dehydrogenases, i.e., NADP-linked enzyme and NAD-linked enzyme in *Acetobactor xylinum* have been reported. The NAD-linked dehydrogenase can oxidize glucose as well as glucose 6-phosphate. This enzyme is presumably unique in this particular organism and it is not evolutionarily related to the vertebrate's H6PD, because of the different function role of the enzyme *Acetobactor xylinum*, as has been suggested by Benziman and Mazover (1973).

As it is shown in Table II, some of the G6PD variants can oxidize galactose 6-phosphate and their substrate specificity is similar to that of H6PD. It will be interesting to examine the structural similarity between the two enzymes. Since H6PD has not been obtained in homogeneous form, it was impossible to directly compare their structure at this time. The immunological properties of the two enzymes were compared using the partially purified human H6PD, and the partially purified and purified normal and variant human G6PD (Srivastava, et al., 1972).

G6PD activity was neutralized by anti-G6PD serum which was prepared by immunizing rabbit with crystalline human G6PD. However, the anti-serum, even at higher concentrations, had no effect on H6PD activity. It should be mentioned that several G6PD variants, including G6PD Union, a variant associated with substrate specificity which is similar to that of H6PD, were neutralized by the anti-serum (Yoshida, et al., 1970).

In double diffusion precipitation on agar gel, a precipitation line was observed with normal and variant human G6PD, but not with H6PD, even in the presence of higher concentrations of the anti-serum.

These results suggest that, despite the fact that a simple mutation can induce a significant change in the substrate specificity by producing a variant G6PD with substrate specificity similar to that of H6PD, the present human H6PD is not structurally closely related to the G6PD.

CONCLUSION

By examining many G6PD variants which are found in various human populations, one can conclude that mutations, most of them presumably due to single step base substitutions, can induce significant alteration of substrate specificity of the enzyme. Although it is not yet proven, the change of substrate specificity may also be produced by spontaneous or induced somatic cell mutations. Therefore, differentiation of isozymes and homologous enzymes from their ancestral enzyme does not necessarily require an accumulation of a series of mutations.

G6PD and H6PD presumably have been produced from a common ancestor by the mutational differentiation mentioned above. G6PD exists in all organisms from bacteria to man, while H6PD has been found only in vertebrates. This would suggest that H6PD has been differentiated from G6PD. Immunologic dissimilarity of the two enzymes in man suggests that the two enzymes diverged at an early phylogenic stage.

Recently Yamauchi and Goldberg (1973) observed some immunologic cross-reactivity between fish G6PD and fish H6PD. Although contamination in partially purified fish G6PD used for immunization was not ruled out, this may suggest structural similarity between the two enzymes. It is likely that divergence of G6PD and H6PD occurred early in evolution, several hundred million years ago, and the two enzymes have separately accumulated evolutionary modifications since then.

The chemical modification of the enzyme, in vitro, also changed the substrate specificity. At this time, the primary structure and the three dimensional structure of human G6PD is not yet known, and the exact structural abnormalities of the variant enzymes associated with altered substrate specificity, and the exact structural change associated with the chemical modification in vitro, are not elucidated. Such a study, which is now in progress, should lead to clear insight, not only into the structural requirement of G6PD and H6PD activity and specificity, but also of enzyme action, in general.

REFERENCES

Anderson, W.B., R.N. Horne, and R.C. Nordlie 1968. Glucose dehydrogenase activity of yeast glucose 6-phosphate dehydrogenase. II. Kinetic studies of the mode of activation by bicarbonate, phosphate, and sulfate. *Biochem.* 7:3997-4004.

Bennerjee, S. and D.G. Fraenkel 1972. Glucose 6-phosphate dehydrogenase from *Escherichia coli* and from a "high-level" mutant. *J. Bact.* 110:155-160.

Benziman, M. and A. Mazover 1973. Nicotinamide adenine dinucleotide and nicotinamide adenine dinucleotide phosphate-specific glucose 6-phosphate dehydrogenase of *Acetobactor xylinum* and their role in the regulation of the pentose cycle. *J. Biol. Chem.* 248:1603-1608.

Beutler, E. and M. Morrison 1967. Localization and characteristics of hexose 6-phosphate dehydrogenase (glucose dehydrogenase). *J. Biol. Chem.* 242:5289-5293.

Cohen, P. and M.A. Rosemeyer 1969. Subunit interactions of glucose 6-phosphate dehydrogenase from human erythrocytes. *Europ. J. Biochem.* 8:8-15.

Hong, M-M., W.C. Chen, and S.C. Shen 1963. Conversion of alanine dehydrogenase to glutamate dehydrogenase by nitrous acid-induced mutation in *Bacillus subtilis. II. Scientia Sinica* 12:557-564.

Kamada, T. and S.H. Hori 1970. A phylogenic study of animal glucose 6-phosphate dehydrogenase. *Japan J. Genet.* 45: 319-339.

Kirkman, H.N., S.A. Doxiadis, T. Valaes, N. Tassopoulos, and A.G. Bronson 1965. Diverse characteristics of glucose 6-phosphate dehydrogenase from Greek children. *J. Lab. & Clin. Med.* 65:213-221.

Kirkman, N.N., C. Kidson, and M. Kennedy 1968. Variant of human glucose 6-phosphate dehydrogenase. *Studies of samples from New Guinea in Hereditary Disorders of Erythrocyte Metabolism.* E. Beutler, Ed., Grune & Stratton, New York, pp. 126-145.

Matsuda, T. and Y. Yugari 1967. Glucose 6-phosphate dehydrogenase from rat liver. *J. Biochem.* 61:535-540.

Scott, W.A. and F.L. Tatum 1971. Purification and partial characterization of glucose 6-phosphate dehydrogenase from Neurospora crassa. *J. Biol. Chem.* 246:6347-6352.

Shaw, C.R. and E. Barto 1965. Autosomally determined polymorphism of glucose 6-phosphate dehydrogenase in Peromyscus. *Science* 148:1099-1100.

Shaw, C.R. and A.L. Koen 1968. Glucose 6-phosphate dehydrogenase and hexose 6-phosphate dehydrogenase of mammalian tissue. *Ann. N.Y. Acad. Sci.* 155:149-156.

Shen, S-C., M-M. Hong, and W-C. Chen 1963. Conversion of alanine dehydrogenase to glutamate dehydrogenase by nitrons and induced mutation in *Bacillus subtilis. I. Scientia Sinica* 12:545-556.

Srivastava, S.K., K.G. Blume, E. Beutler, and A. Yoshida 1972. Immunological difference between glucose 6-phosphate dehydrogenase and hexose 6-phosphate dehydrogenase from human liver. *Nature* 238:240-241.

Stegman, J.J. and E. Goldberg 1971. Distribution and characterization of hexose 6-phosphate dehydrogenase in trout. *Biochem. Genet.* 5:579-589.

Sutton, H.E. and G.W. Karp 1971. White blood cells variant for G6PD expression. *Excerpta Medica* 233:173.

Yamauchi, T. and E. Goldberg 1973. Glucose 6-phosphate dehydrogenase from Brook, Lake, and Splake trout: An isozymic and immunologic study. *Biochem. Genet.* 10:121-134.

Yoshida, A. 1966. Glucose 6-phosphate dehydrogenase of human erythrocytes. I. Purification and characterization of normal (B+) enzyme. *J. Biol. Chem.* 241:4966-4976.

Yoshida, A. 1967. A single amino acid substitution (asparagine to aspartic acid) between normal (B+) and the common Negro variant (A+) of human glucose 6-phosphate dehydrogenase. *Proc. Nat. Acad. Sci. U.S.A.* 57:835-840.

Yoshida, A. 1970. An amino acid substitution (histidine to tyrosine) in a glucose 6-phosphate dehydrogenase variant (G6PD Hektoen) associated with overproduction. *J. Mol. Biol.* 52:438-493.

Yoshida, A. and V.D. Hoagland 1970. Active molecular unit and NADP content of human glucose 6-phosphate dehydrogenase. *Biochem. Biophys. Res. Commun.* 40:1167-1172.

Yoshida, A., E. Baur, and A.G. Motulsky 1970. A Philippino glucose 6-phosphate dehydrogenase variant-G6PD Union-with enzyme deficiency and altered substrate specificity. *Blood* 35:506-513.

Yoshida, A. 1973. Hemolytic anemia and G6PD deficiency. *Science* 179:532-537.

Yoshida, A. 1973. Change of activity and substrate specificity of human glucose 6-phosphate dehydrogenase by oxidation. *Arch. Biochem. Biophys.* 159:82-88.

Yoshida, A., E. Giblette, L.A. Malcolm 1974. Heterogeneous distribution of glucose 6-phosphate dehydrogenase variants with enzyme deficiency in the Markham Valley area of New Guinea. *Ann. Hum. Genet.* 37:145-150.

GENETIC POLYMORPHISM AND DIFFERENTIATION IN *PARAMECIUM*

JULIAN ADAMS AND SALLY ALLEN
Division of Biological Sciences
The University of Michigan
Ann Arbor, Michigan 48104

ABSTRACT. *Paramecium aurelia* is a complex of fourteen syngens which are morphologically similar but between which no gene flow occurs. We have analyżed the isozymic variations in all fourteen syngens of *P. aurelia* and also in one syngen of *P. multimicronucleatum*. Isozyme variants at nine loci were utilized in this analysis: three esterase and, two isocitrate dehydrogenase loci, and one fumarase, β-hydroxybutyrate dehydrogenase, glutamate dehydrogenase and succinate dehydrogenase locus. Intrasyngenic variation in populations of the syngens is extremely low, with the exception of syngen 2 which possesses high levels of genetic variability. When variant alleles do appear within a syngen they are nearly always "private" variants occurring nowhere else. The virtual absence of geographic differentiation within the syngens strongly suggests that the alleles at the enzyme loci studied are selected for. Differences between the syngens are large, and there is no clear-cut distinction between the syngens of *P. aurelia* and *P. multimicronucleatum* on the basis of these data. Dendrograms constructed from the matrix of genetic distances portray a more complex set of relationships between the syngens than originally proposed by Sonneborn. From comparison of the genetic distances for *Paramecium* with similar values obtained from *Tetrahymena* and *Drosophila* we conclude that the syngens of *P. aurelia* are more realistically conceived of as species.

INTRODUCTION

The biological relationship of organisms which show few morphological differences may be grossly underestimated if only morphological criteria are used in assessing relationships. This problem is particularly acute in microorganisms where there is, in addition, lack of a fossil record and barriers to genetic recombination. The problem occurs in bacteria. It is also encountered in the ciliated protozoa. For example, all strains of *Paramecium aurelia* are morphologically quite similar; yet, between certain groups of strains gene flow does not occur. These breeding groups have been designated "biological species" or "syngens" by Sonneborn (1957a). They would be analogous to the sibling species described for other organisms such as in *Drosophila*.

Syngens occur in a number of familiar species-complexes of ciliated protozoa. In the "genus" *Paramecium*, for example, *P. aurelia* has 14 syngens, *P.multimicronucleatum*-5, *P.bursaria*-5, and *P.caudatum*-16 syngens (Sonneborn, 1957a, 1968, pers. com.; Rafalko and Sonneborn, 1959). For another familiar ciliate, *Tetrahymena pyriformis*, at least 12 syngens exist (see Elliott, 1973). Syngens have also been observed for several hypotrichs, *Euplotes patella*, *Oxytricha bifaria*, and *Stylonychia putrina* (see Sonneborn, 1957a). The existence of syngens is, therefore, a general phenomenon in the ciliated protozoa, and undoubtedly many more remain to be described. What is suggested by these few examples is that far greater diversification and differentiation has occurred during the evolution of these protozoa than is suspected from morphological criteria.

Biochemical comparisons, indeed, support this view. Both comparisons of nucleotide sequence divergence or of isozymes suggest that "vast molecular differences" separate the syngens (Allen and Li, 1974; Borden et al., 1974). The paper in Volume II reviews the isozyme data for 9 enzyme loci studied in the 14 syngens of *P. aurelia*. (Allen and Gibson, 1974). With one exception, each syngen appears to be a distinct entity in biochemical terms.

The few ecological studies that have been carried out suggest that the syngens of *P. aurelia* may coexist in natural populations, but that they may show unique distributions in their habitats (Hairston, 1958, 1967). In competition studies both in the laboratory and in natural populations syngen 2 was found to be more successful than other syngens with which it lives sympatrically (Gill, 1972; Gill and Hairston, 1972; Hairston and Kellerman, 1965).

Originally the first three syngens of *P. multimicronucleatum* were classified as syngens of *P. aurelia* (Sonneborn, 1957a, b). The "complex" *P. aurelia - multimicronucleatum* referred to "paramecia with vesicular micronuclei and cigar-shaped bodies" (Sonneborn and Dippell, 1957). Later these three syngens of *P. multimicronucleatum* were separated off since they regularly differed from the syngens of *P. aurelia* in the number of macronuclear anlagen which are formed following meiosis, two in *P. aurelia* and four in *P. multimicronucleatum* (Sonneborn and Dippell, 1957; Sonneborn, 1958). Comparison of three esterase loci in a stock of *P. multimicronucleatum*, probably belonging to syngen 3 with the syngens in *P. aurelia* revealed similarities in two of the loci to those found in syngens 7 and 13 of *P. aurelia;* the third locus was unique (Allen et al., 1973). Six other enzyme loci have now been examined in this stock of *P. multimicronucleatum,* and the results are reported here.

In this paper we reexamine the relationship between *P. aurelia* and *P. multimicronucleatum.* On the basis of the isozyme data we also present evidence that the syngens are unique in that they contain little genetic polymorphism, a low amount of geographical differentiation, and that the genetic distances separating syngens are large and are on the level of species.

MATERIALS AND METHODS

8MO is a stock of *P. multimicronucleatum* collected by John Vandermeer (University of Michigan) in Costa Rica. It was thought to be *P. aurelia* by Vandermeer but identified by Sonneborn and Dippell as *P. multimicronucleatum.* It is small, similar in size to syngens 4 and 8 of *P. aurelia,* and selfs extensively. Its small size, geographic origin and tendency to self are properties that suggest it belongs to syngen 3 (Sonneborn, personal communication). It was easily transferred to axenic culture where it reaches spectacular densities.

All enzyme comparisons utilized axenically grown stocks. The three esterases of 8MO were compared to those of stocks from each of the 14 syngens in *P. aurelia* using the electrophoretic procedures described in Allen et al. (1973). For the enzymes, succinate dehydrogenase (SDH), NAD-dependent glutamate dehydrogenase (GDH), β-hydroxybutyrate dehydrogenase (BDH), fumarase, and NADP-dependent isocitrate dehydrogenase ($IDH_{(s)}$ and $IDH_{(m)}$), the procedures referred to in Allen and Gibson (1974) were followed. The stocks of *P. aurelia* used in this analysis are listed along with their geographic origins in Table 1 in Allen and Gibson (1971) and Allen et al. (1973).

RESULTS

The electrophoretic mobilities of 9 enzymes from 8MO (*P. multimicronucleatum*) were compared to those of stocks in syngens of *P. aurelia* by starch gel electrophoresis. The same invariant mobility for SDH was observed in 8MO as found by Tait (1970) for the syngens of *P. aurelia.* The other 8 enzymes in 8MO showed variation: three had mobilities similar to stocks in one or more syngens of *P. aurelia* (esterase A, esterase C, and $IDH_{(s)}$) while the remaining five had unique

TABLE I

DISTRIBUTION OF ENZYME SUBTYPES AMONG 14 SYNGENS OF *PARAMECIUM AURELIA* AND *PARAMECIUM MULTIMICRONUCLEATUM*

Enzyme	Subtypes
Esterase A	(9) (10) (1,3,5,7,11,13,8MO) (6,12) (2) (4,8)
Esterase B	(2) (1,3,5) (6) (9) (7,13) (10,14) (12) (11) (8) (4) (8MO)
Esterase C	(8) (14) (1,2,3,4,5,6,9,10,12) (7,11,13,8MO)
Fumarase	(4,8,10,12,13) (2,3,7) (1,5,6,9,11,14) (8MO)
GDH	(10,11) (1,3,4,5,6,7,8,9,12,13,14) (2) (8MO)
BDH	(6,8) (7,14) (1,5,10,12) (2,3) (4,9,11) (13) (8MO)
$IDH_{(s)}$	(12) (13) (10) (11) (2,3) (1,5,6,7,8,9,14,8MO) (4)
$IDH_{(m)}$	(6) (4,10,12) (13) (1,2,3,5,7,8,9,11,14) (8MO)
SDH	(1,2,3,4,5,6,7,8,9,10,11,12,13,14,8MO)

Abbreviations: GDH=glutamate dehydrogenase; BDH= β-hydroxybutyrate dehydrogenase; $IDH_{(s)}$=isocitrate dehydrogenase (supernatant); $IDH_{(m)}$= isocitrate dehydrogenase (mitochondrial); SDH=succinate dehydrogenase.

mobilities different from stocks in any of the syngens of *P. aurelia* (esterase B, $IDH_{(m)}$, fumarase, BDH and GDH). Zymograms of the esterases and IDH isozymes in 8MO are compared to *P. aurelia* in Fig. 1 and 2.

Stocks within a syngen show little variability, with the exception of one syngen, although there is considerable differentiation between syngens. Table 1 shows the differences between the syngens of *P. aurelia* and *P. multimicronucleatum* for the nine enzymes utilized in this study. The low level of intrasyngenic variability is a striking aspect of these data and is all the more surprising when one considers that the stocks (strains) of (e.g.) syngen 1 of *P. aurelia* analyzed were obtained from locales as far apart as Japan and Scotland. The exception to this pattern of results

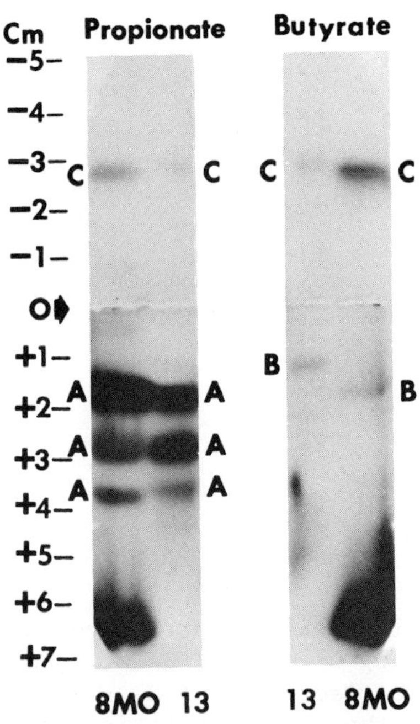

Fig. 1. Comparison of the esterases of 8MO (*P. multimicronucleatum*) with the esterases of syngen 13 of *P. aurelia*. Note the similarity of the A and C esterases and the difference in mobility of the B esterases. Distances in migration are marked off in centimeters from the origin (O) to the left of the photographs.

is syngen 2 which possesses substantial amounts of genetic variability. These results are reinforced by those of Tait (1970), Tait et al. (1971), Gibson and Adams (1974) and Gill and Hairston (personal communication) who also found that for a more restricted range of geographic locales and syngens, syngen 2 was the only syngen which contained appreciable amounts of genetic variability. The form in which we present the data, exhibiting no intrasyngenic differences, makes it particularly easy to calculate genetic distances between the syngens. Most of the arguments concerning the choice of an appropriate distance measure do not apply to these data, since many of the metrics give the same numerical values. Thus the euclidean distances based on the simple matching coefficient or mean character difference (Sneath and Sokal, 1973) are the same as the euclidean distances calculated using Nei's (1972) measure of genetic identity. Table 2

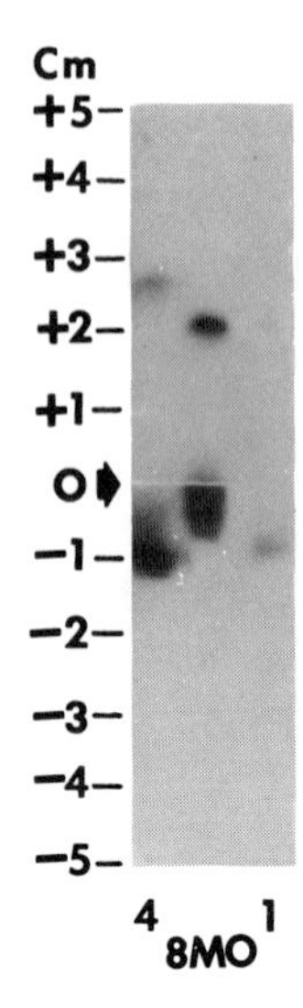

Fig. 2. Comparison of the isocitrate dehydrogenases of 8MO (*P. multimicronucleatum*) with the IDHs of syngens 4 and 1 of *P. aurelia*. The anodal isozyme is $IDH_{(s)}$ and the cathodal isozyme $IDH_{(m)}$. Note the electrophoretic similarity of the $IDH_{(s)}$ isozymes of 8MO and syngen 1 of *P. aurelia* and the unique $IDH_{(m)}$ present in 8MO.

---

shows the pairwise genetic distances between the 14 syngens of *P. aurelia* and *P. multimicronucleatum*. In general, the table shows that the distances between the syngens are large, although two of the syngens of *P. aurelia* , 1 and 5, are indistinguishable on the basis of the nine enzymes assayed in this study. The genetic distances in Table 2 show that the strain of *P. multimicronucleatum* cannot be distinguished with any confidence from the syngens of *P. aurelia*. The mean euclidean distance between all syngens is 0.81 whereas the mean distance between *P. multimicronucleatum* and the 14 syngens of *P. aurelia* is 0.87, both figures normalized to a 0-1 scale.

The pattern of distances between the syngens and species of *Paramecium* implies particular relationships between them. Accordingly dendrograms were constructed from the matrix of genetic distances. Fig. 3 shows the dendrograms for the syngens of *P. aurelia* with and without *P. multimicronucleatum*. The dendrograms represent the best estimate of the minimum length tree describing the relationships between the syngens

TABLE II

PAIRWISE DISTANCE MATRIX FOR THE 14 SYNGENS OF *P. AURELIA* AND THE SYNGEN OF *P. MULTIMICRONUCLEATUM* (8MO)

	2	3	4	1,5	6	7	8	9	10	11	12	13	14
3	0.577												
4	0.882	0.817											
1,5	0.817	0.577	0.817										
6	0.882	0.817	0.817	0.667									
7	0.817	0.667	0.882	0.667	0.817								
8	0.882	0.817	0.745	0.745	0.745	0.745							
9	0.817	0.745	0.745	0.577	0.667	0.745	0.745						
10	0.882	0.882	0.745	0.817	0.882	0.943	0.882	0.882					
11	0.882	0.817	0.882	0.745	0.882	0.745	0.882	0.745	0.882				
12	0.882	0.817	0.667	0.745	0.745	0.882	0.817	0.817	0.667	0.943			
13	0.943	0.817	0.817	0.817	0.882	0.667	0.817	0.882	0.882	0.817	0.817		
14	0.882	0.745	0.882	0.577	0.745	0.667	0.817	0.667	0.882	0.817	0.882	0.882	
8MO	0.943	0.882	0.943	0.817	0.882	0.745	0.882	0.882	0.882	0.817	0.943	0.817	0.817

The tabulated distance measures (d) are calculated using the relationship $d = \sqrt{1 - I}$. I is the measure of genetic identity of Nei (1972) calculated using the data for the nine isozyme loci described in the text.

and species. This tree was derived using the algorithm of Cavalli-Sforza and Edwards (1967) starting from an initial cluster obtained using the divisive algorithm of the same two workers (Edwards and Cavalli-Sforza, 1965). Although

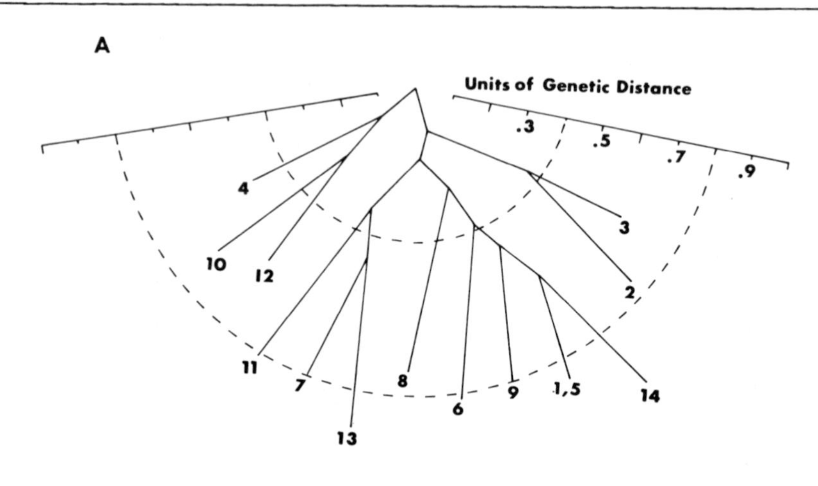

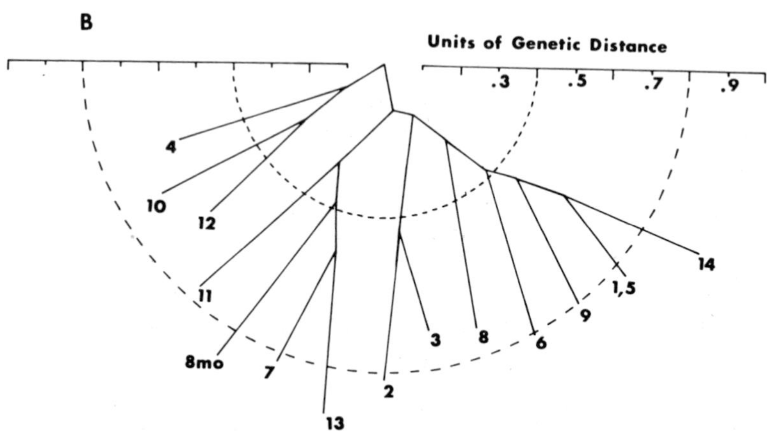

Fig. 3. A: Dendrogram of the 14 syngens of *P. aurelia*. B: Dendrogram of the 14 syngens of *P. aurelia* including *P. multimicronucleatum* (8MO). The dendrograms represent the "best" topology based on the criterion of minimum evolution. The networks are drawn to scale and plotted on polar coordinates so that distances between populations are read off along the radii. The trees are unrooted; however, the dendrograms are drawn such that the segment joining the results of the primary split passes through the origin.

there is no assurance that this topology generates the absolute minimum length, a variety of other topologies generated, using alternative clustering methods, all yielded trees with a longer net length. In so far as these trees represent the minimum length, under the assumption of parsimony of evolution, they therefore can be interpreted as representations of the evolution of this group of syngens. Thus they can be termed cladograms in the sense of Sneath and Sokal (1973). Although the interpretation of phenetic relationships as being phylogenetic implies many assumptions concerning the course of evolution (Fitch and Margoliash, 1967), the application of the minimum path method to other sets of data has been shown to give biologically reasonable results (Kidd and Sgaramella-Zonta, 1971; Ward and Neel, 1970). The dendrograms show that there are approximately four clusters of syngens. Syngens 4, 10 and 12 appear to be grouped together; 7, 11 and 13 group together; syngens 2 and 3 appear to be closely related, and syngens 1, 5, 6, 8, 9, and 14 form a cluster together. Sonneborn (1957a) has previously suggested on the basis of morphological and antigenic differences that syngens 2, 4 and 8 were closely related and that syngens 1, 3 and 9 were clustered together in a group. The analysis of the isozyme data presented in this paper suggest a much more complex pattern of relationships.

Fig. 3B shows the dendrogram of the 14 syngens of *P. aurelia* together with *P. multimicronucleatum*. The position of *P. multimicronucleatum* in the dendrogram confirms the earlier observation that *P. multimicronucleatum* cannot clearly be distinguished from the syngens of *P. aurelia* (Sonneborn and Dippell, 1957). The dendrogram shows that *P. multimicronucleatum* is closely related to syngens 7, 11, and 13 of *P. aurelia*.

DISCUSSION

The distribution of allozyme (allelic isozyme) frequencies between the syngens of *P. aurelia* and *P. multimicronucleatum* allows a distinction and identification of the syngens, which previously was only possible by a series of mating reactions, which were time-consuming, and tedious to perform. The use of allozyme variation in identifying populations, breeding groups or species is not new. Several previous studies (e.g. Ayala and Powell, 1972; Ayala et al., 1974; Dessauer and Nevo, 1969) have utilized these types of data successfully. There are however three aspects of our data which are unique and merit special consideration. First, the amount of intrasyngenic variation is extremely low; secondly, there is virtually no geographical differentiation within the syngens; and thirdly, the genetic distances between the syngens are very high.

Intrasyngenic variability: With the development of electrophoresis as a tool for identifying genetic variation, it has become apparent that the amount of polymorphism in natural populations is extraordinarily high. Surveys of genetic variation have now been carried out in practically every major group of plants and animals (see Lewontin, 1973, for a summary of the literature) and almost without exception the amount of genetic polymorphism has been found to be high. It is therefore of especial interest when a group of organisms is found to have a low amount of genetic polymorphism. Even higher amounts of genetic polymorphism are observed if thermal differences in enzymes with similar mobility are compared (Bernstein et al., 1973). Such also may be the case for *Paramecium* (Allen and Gibson, 1974). Natural populations of syngens 1, 2, 3, 4, 5 and 9, of *P. aurelia* have been studied in detail with regard to the criterion of mobility differences (Tait, 1970; Tait et al., 1971; Gibson and Adams, 1974; Gill and Hairston, personal communication). All these syngens have extensive geographic distributions and are often found living sympatrically. In general, natural populations of syngens 1, 3, 4, 5 and 9 show little polymorphism at the enzyme loci studied, although polymorphism has been observed for antigen loci (Pringle, 1956; Pringle and Beale, 1960). What few variants are found for the enzyme loci, are often "private" variants, found nowhere else. In contrast, natural populations of syngen 2 show a quite different pattern of variation compared to the other syngens and possess respectable amounts of heterozygosity, both for the enzyme as well as antigen loci (Gibson and Adams, 1974). It is unlikely that this pattern of results can be explained by the different features of the mating system of the syngens. Studies on natural populations of syngen 2 reveal that it is highly inbred (Gibson and Adams, 1974). However, there is no evidence that inbreeding is any more or less important in the other syngens. Sonneborn (1957a) has classified the syngens of *P. aurelia* as inbreeders or outbreeders on the basis of the length of the immature period following meiosis. According to his observations syngen 2 is neither an extreme inbreeder nor an extreme outbreeder. It is equally difficult to explain the presence of variability in syngen 2 and its absence in the other syngens by the action of selection. The extremely small amount of geographic differentiation that has occurred within the syngens argues for the action of strong selective forces on the loci studied. If this is true it is not immediately obvious why selection should maintain variability in one syngen and remove it in another syngen. It is of particular interest that syngen 2 is often found sympatrically with the other syngens, living in an apparently identical environment. Moreover, Gill and Hairston (1972) found that it was particularly successful in competition with syngen 5 and hypothesized that syngen 2 may be particularly well adapted for colonizing rare islands of favorable habitat in coarse-grained environments. It is hoped that more detailed studies in such populations can give us insight into the adaptive basis of variability.

Geographical differentiation within syngens: The second feature of our results with *P. aurelia* and *P. multimicronucleatum* is the virtual absence of geographical differentiation between the strains of any one syngen with regard to enzyme polymorphisms. The strains chosen for this study were purposely chosen to represent widely differing geographical locales and thus we can, with confidence, rule out the action of migration in retarding or preventing differentiation between the populations. The populations from which the strains used in this study were isolated are so far apart that they must be completely isolated genetically. Furthermore this lack of geographic differentiation cannot be explained by the absence of intrasyngenic variation, because variants <u>are</u> found in the syngens but in extremely low frequency. The results strongly suggest therefore that strong selective forces are acting to maintain the "wild-type" alleles in high frequencies and the variant alleles in low frequencies within the syngens. In the absence of selective forces far more differentiation would be expected, and at least some of the variant alleles would be found in intermediate and high frequencies.

Genetic distances between syngens: The allozyme (allelic isozyme) variation between the syngens of *P. aurelia* allows us to distinguish all syngens except 1 and 5 unequivocally. In fact such a distinction is possible with much less than the 9 enzymes reported in this communication. The distances between the syngens are therefore quite large. The magnitude of the distances is reflected in the structure of the dendrograms of Fig. 3. The lengths of all the terminal branches are long in relation to the size of the dendrogram itself. However, it is difficult to interpret the magnitude of these distances without reference to other groups of organisms. Table 3 shows three measures which can be used to compare the distances between the syngens of *Paramecium* with the distances between the syngens of another protozoan,

TABLE III

MEASURES OF DISTANCE

1) Mean Pairwise Euclidean Distance between Syngens

Paramecium	0.7887
Tetrahymena	0.9384

2) Sum of Squares of Distances

	S. Sqs.	No. of Syngens	"Mean Sq."
Paramecium	4.29	15	0.286
Tetrahymena	4.87	12	0.406

3) Length of Shortest Spanning Tree

	Span	N	Span/syngen
Paramecium	6.951	15	0.46
Tetrahymena	7.120	12	0.59

Tetrahymena pyriformis. Distance matrices and dendrograms were constructed using the same methods as for the *Paramecium* data. The *Tetrahymena* data were obtained from Borden et al. (1974). The mean pairwise distance between the syngens of *Tetrahymena* is even larger than that of *Paramecium*. The other two measures of dispersion, the total sum of squares of the euclidean distances, and the mean length of the tree also reflect the fact that the dispersion between the syngens of *Tetrahymena* is even greater than the dispersion between the syngens of *Paramecium*.

It is, for obvious reasons, impossible to directly estimate the amount of time elapsed since the biochemical divergence of the *Paramecium* syngens. Even if morphological divergence had occurred concurrently with biochemical divergence, the complete absence of a fossil record for Protozoa would still make the direct estimation of divergence time impossible. However Nei (1971) has developed a technique which allows the indirect estimation of divergence time using distance measures. The procedure assumes parsimony of evolution as well as the equality of the rate of gene substitution/year across different genes and species. This second assumption allows us to use the estimate of the rate of gene substitution/year estimated from vertebrate data by Dayhoff (1969). In practice this figure will vary with species and with loci, but as Nei's formulation shows this will tend to underestimate the divergence time. Table 4 shows the estimates of divergence

TABLE IV

ESTIMATED AVERAGE DIVERGENCE TIME

	Years
Paramecium	6.15×10^5
Tetrahymena (Borden et al., 1974)	1.23×10^6
Drosophila, sibling species (Nei, 1971)	5.7×10^5
Drosophila, non-sibling sp. (Nei, 1971)	1.4×10^6

time for *Paramecium*, *Tetrahymena* and *Drosophila*. The figures for *Paramecium* and *Tetrahymena* represent the mean divergence time for a pair of syngens. Although the calculation of these estimates involves a number of assumptions (see Nei, 1971) which would undoubtedly affect the estimates, a comparison of the *Drosophila* and Protozoan data is still valid. They suggest that on currently accepted criteria of species definition the syngens of *Paramecium* and *Tetrahymena* can be more realistically called species (see also Borden et al., 1974). Although the reliability of these estimates is suspect, we can see that the mean divergence time for the *Paramecium* syngens is similar to the figure for *Drosophila* sibling species and for the *Tetrahymena* syngens it is similar to the figure for *Drosophila* non-sibling species. This observation takes on more significance since a similar pattern of results is found in comparisons of nucleotide sequence divergence between *Drosophila* sibling and non-sibling species and *Paramecium* and *Tetrahymena* syngens (Allen and Li, 1974). The difference between *Paramecium* and *Tetrahymena* might, however, disappear if there are differences in generation time between these protozoan groups in natural populations, and if rates of evolution are generation dependent, as found by Kohne (1970) and Laird et al. (1969).

Regardless of the details, the distinction of the syngens using biochemical techniques is easy. Therefore, the last impediment to the elevation of protozoan syngens to formal species status is removed (Sonneborn, 1957a).

ACKNOWLEDGEMENTS

We thank I. Gibson, C.F. Sing and R.H. Ward for helpful discussions, and C.-I. Li for assistance in the conduct of experiments.

REFERENCES

Allen, S.L. and I. Gibson 1971. Intersyngenic variations in the esterases of axenic stocks of *Paramecium aurelia*. *Biochem. Genetics* 5:161-181.

Allen, S. L. and I. Gibson 1975. Syngenic variations for enzymes of *Paramecium aurelia*. *Isozymes IV. Genetics and Evolution*, C. L. Markert, editor, Academic Press, New York. pp. 883-899.

Allen, S.L. and C.I. Li 1974. Nucleotide sequence divergence among DNA fractions of different syngens of *Tetrahymena pyriformis*. *Biochem. Genetics*, in press.

Allen, S.L., S.W. Farrow,and P.A. Golembiewski 1973. Esterase variations between the 14 syngens of *Paramecium aurelia* under axenic growth. *Genetics* 73:561-573.

Ayala,F.J. and J.R. Powell 1972. Allozymes as diagnostic characters of sibling species of *Drosophila*. *Proc. Nat. Acad. Sci. U.S.A.* 69:1094-1096.

Ayala, F.J., M.L. Tracey, L.G. Barr, and J.G. Ehrenfeld 1974. Genetic and reproductive differentiation of the subspecies *Drosophila equinoxalis caribbensis*. *Evolution* 28:24-41.

Bernstein, S.C., L.H. Throckmorton, and J.L. Hubby 1973. Still more genetic variability in natural populations. *Proc. Nat. Acad. Sci. U.S.A.* 70:3928-3931.

Borden, D., E.T. Miller, G.S. Whitt, and D.L. Nanney 1974. Electrophoretic analysis of evolutionary relationships in *Tetrahymena*. *Evolution* (in press).

Cavalli-Sforza, L.L. and A.W.F. Edwards 1967. Phylogenetic analysis Models and estimation procedures. *Amer. J. Hum. Genet.* 19:233-257.

Dayhoff, M.O. 1969. *Atlas of Protein Sequence and Structure 1969*. National Biomedical Research Foundation, Silver Spring, Md. pp. 361.

Edwards, A.W.F. and L.L. Cavalli-Sforza 1965. A method for cluster analysis. *Biometrics* 21:362-375.

Elliott, A.M. 1973. Life cycle and distribution of *Tetrahymena*. In: *Biology of Tetrahymena*, A.M. Elliott (ed), Dowden, Hutchinson & Ross, Stroudsburg, Pa., pp. 259-286.

Fitch, W.M. and E. Margoliash 1967. Construction of phylogenetic trees. *Science* 155:279-284.

Gibson, I. and J. Adams 1974. Isozyme and antigen variation in a natural population of *Paramecium aurelia*. Manuscript in preparation.

Gill, D.E. 1972. Intrinsic rates of increase, saturation densities, and competitive ability. I. An experiment with *Paramecium*. *Am. Nat.* 106:461-471.

Gill, D.E. and N.G. Hairston 1972. The dynamics of a natural population of *Paramecium* and the role of interspecific competition in community structure. *J. Anim. Ecol.* 41:81-96.

Hairston, N.G. 1958. Observations on the ecology of *Paramecium*, with comments on the species problem. *Evolution* 12:440-450.

Hairston, N.G. 1967. Studies on the limitation of a natural population of *Paramecium aurelia*. *Ecology* 48:904-910.

Hairston, N.G. and S.L. Kellerman 1965. Competition between varieties 2 and 3 of *Paramecium aurelia:* the influence of temperature in a food limited system. *Ecology* 46: 134-139.

Kidd, K.K. and L.A. Sgaramella-Zonta 1971. Phylogenetic analysis: Concepts and Methods. *Amer. J. Hum. Genet.* 23:235-252.

Kohne, D.E. 1970. Evolution of higher-organism DNA. *Quart. Rev. Biophysics.* 3:327-375.

Laird, C.D., B.L. McConaughy,and B.J. McCarthy 1969. Rate of fixation of nucleotide substitutions in evolution. *Nature* 224:149-154.

Lewontin, R.C. 1973. Population Genetics. *Ann. Rev. Genet.* 7:1-17.

Nei, M. 1971. Interspecific gene differences and evolutionary time estimated from electrophoretic data on protein identity. *Am. Nat.* 105: 385-398.

Nei, M. 1972. Genetic distance between populations. *Am. Nat.* 106:283-292.

Pringle, C.R. 1956. Antigenic variation in *Paramecium aurelia*, variety 9. *Zeitschrift für Abstammungs-und Vererbungslehre* 87:421-430.

Pringle, C.R. and G.H. Beale 1960. Antigenic polymorphism in a wild population of *Paramecium aurelia*. *Genet. Res. Camb.* 1:62-68.

Rafalko, M. and T.M. Sonneborn 1959. A new syngen (13) of *Paramecium aurelia* consisting of stocks from Mexico, France and Madagascar. *J. Protozool.* 6 (Suppl.):30 (abst.)

Sneath, P.H.A. and R.R. Sokal. *Numerical Taxonomy*, Freeman, San Francisco. 573pp.

Sonneborn, T.M. 1957a. Breeding systems, reproductive methods and species problems in protozoa. In: *The Species Problem*, E. Mayr (ed.), Am. Assoc. Adv. Sci. Symp., Washington, D.C. pp. 155-324.

Sonneborn, T.M. 1957b. Varieties 13, 15 and 16 of the *P. aurelia-multimicronucleatum* complex. *J. Protozool.* 4 (Suppl.):21 (abst.).

Sonneborn, T.M. 1958. Classification of syngens of the *Paramecium aurelia - multimicronucleatum* complex. *J. Protozool.* 5(Suppl.): 17-18 (abst.).

Sonneborn, T.M. and R.V. Dippell 1967. The *Paramecium aurelia - multimicronucleatum* complex. *J. Protozool.* 4 (Suppl.): 21 (abst.).

Tait, A. 1970. Enzyme variation between syngens in *Paramecium aurelia. Biochem. Genetics* 4: 461-470.

Tait, A., G.H. Beale, and A.R. Oxbrow 1971. Enzyme polymorphism in a population of *Paramecium aurelia* in S.E. England. *J. Protozool.* 18(Suppl.): 26 (abst.).

Ward, R.H. and J.V. Neel 1970. Gene frequencies and microdifferentiation among the Makiritare Indians. IV. A comparison of a genetic network with ethnohistory and migration matrices; a new index of genetic isolation. *Amer. J. Hum. Genet.* 22: 538-561.

SYNGENIC VARIATIONS FOR ENZYMES OF *PARAMECIUM AURELIA*

SALLY ALLEN and IAN GIBSON
Division of Biological Sciences
The University of Michigan
Ann Arbor, Michigan 48104
and
School of Biological Sciences
University of East Anglia
Norwich, Norfolk, England

ABSTRACT. The distribution of isozymes observed for 9 different enzymes in the stocks of *Paramecium aurelia* reinforces the division of this species-complex into sibling species, or syngens. Each of the 14 syngens with the exception of syngens 1 and 5 can be separated clearly from each other in terms of these 9 enzymes. Each syngen thus appears to be a distinct entity in biochemical terms. Most syngens differ from each other in 6 out of the 9 enzymes with regard to electrophoretic mobility. This is probably a conservative estimate since temperature inactivation studies reveal additional differences between isozymes of similar mobility. Bacteria can contribute enzymes, and various effects of bacteria on paramecium esterases are described; it is, therefore, important to utilize axenically grown paramecia in making stock and syngen comparisons. With the exception of syngen 2, the frequency of intrasyngenic variants is low. These variants are specified by nuclear genes, even for mitochondrial enzymes. Natural populations may include several syngens living sympatrically. In one population 4 syngens were found, with syngen 2 dominant and exhibiting a high degree of polymorphism for both antigen and esterase loci. Gene frequencies appeared to fluctuate randomly, with one exception, which responded to changes in pond temperature. Although autogamy (self-fertilization) is the more common sexual process in natural populations, heterozygotes are detected, indicating that conjugation occurs in nature in appreciable frequency.

INTRODUCTION

Genetic research with the ciliated protozoan, *Paramecium aurelia*, has centered in the main on cytoplasmic inheritance and problems of nuclear differentiation. These studies have led to important considerations about individual characteristics which may or may not find their basis in nucleic acids and also about mechanisms of regulation of gene action. On the other hand there has been little contribution to the field of population genetics and indeed to the theory of evolutionary genetics. The most serious contribution to this field was in a paper by Sonneborn in 1957.

Of particular interest is the fact that *Paramecium aurelia* is a species complex and contains populations of animals between which gene flow does not occur. From the genetic point of view these breeding groups are sibling species. They have been referred to as "syngens" by Sonneborn (1957). At present 14 different syngens have been identified from various freshwater collections (Sonneborn, 1957, 1958; Sonneborn et al, 1959; Rafalko and Sonneborn, 1959). Some syngens are probably world-wide in distribution (syngens 1, 2, 4, 5, 6, 8, and 13) while others are more restricted in their distribution, such as syngen 3 (North America), syngen 9 (Europe), syngens 7, 10, 11, and 12 (Southern U.S.A.), and syngen 14 (Australia). A laboratory stock is established from a single autogamous individual and the stock is assigned to a particular syngen on the basis of its ability to exchange genes with other individuals in that syngen. Because of the procedure of selecting single autogamous cells, the standard laboratory stocks are homozygous for all their genes.

Although morphologically similar, the syngens vary in cell size, micronuclear chromosome number, and in the details of their breeding systems (Sonneborn, 1957, 1966). Phenotypic differences between syngens have also been recorded in their antigens, symbionts, mating types, temperature tolerance levels, salt tolerance levels, etc., but the mating isolating mechanism has been the one used in practice to distinguish between syngens. From general considerations involving these characteristics and others, Sonneborn proposed the grouping of the organisms into two major groups and suggested a bifurcate pathway of evolution from a common ancestor. The observation of mating between some syngens with the production of inviable F_2 cells even after exhaustive selective techniques (Haggard, 1974a, b) suggests an ancient evolutionary relationship.

Several considerations prompted us to examine the syngens of *P. aurelia* for enzymic differences. Already in another ciliated protozoan, *Tetrahymena pyriformis*, differences had been disclosed between syngens and strains within a syngen in the electrophoretic mobilities of esterases and phosphatases (Allen and Weremiuk, 1971) and more recently this has been shown for other enzymes (Borden et al, 1974). We felt that the characters under examination in *P. aurelia* at the time, i.e. cell surface antigens, cortical properties, symbionts, and mating types, might involve a highly selected group of characters in that they involved surface structure and not necessarily direct gene products. We thus started off seven years ago to compare (a) the enzymes and (b) the nucleotide sequences in the DNA of different syngens in *P. aurelia*. Another consideration which concerned us about the work done up until that point was the observation that in some cases the variation between stocks within a syngen appeared to be as great as that between syngens. Since enzyme studies with Tetrahymena had shown the ease with which enzyme variations could be found, we hoped they might also be readily found in Paramecium and the extent of intrasyngenic and intersyngenic variation assessed.

The experiments to be reported here identify both intra and intersyngenic variations in the enzymes of *P. aurelia* and the basis of the former. From these we have extended our studies to a pond in England where we have looked at certain gene frequencies with regard to enzyme and antigen loci and the environmental factors which may be involved in the process of natural selection of certain genotypes. In the accompanying paper the evolutionary relationships between the syngens based on these studies are assessed (Adams and Allen, 1974).

MATERIALS AND METHODS

Stocks of *Paramecium aurelia* have been obtained from several sources, and we are grateful for the support of the following individuals in providing them: T. M. Sonneborn, M. V. Schneller, G. H. Beale, W. van Wagtendonk, and N. G. Hairston. All cultures were maintained in bacterized medium, but some of them were transferred to axenic medium for reasons to be discussed. Some 60 stocks from the 14 syngens of *P. aurelia* have been examined for esterases in axenic medium, a fewer number for other enzymes. Lists of the axenic stocks used and the details of culture methods can be found in Allen and Gibson (1971) and Allen et al, (1973).

The normal method of examination of the stocks for enzymes has been to grow up 1-2 liters of cells to make extracts and to examine them by starch gel electrophoresis. For the esterases, extracts were made from frozen-thawed whole cells. For other enzymes, the cells were homogenized, centrifuged at high speeds, and the supernatants, or the supernatants from extracts of mitochondria, were used. In our laboratory lyphogel (Gelman) was used to concentrate these supernatants. For details of the methods of extraction and electrophoretic procedures see Allen et al, (1971), Borden et al (1974) and Tait (1970a).

RESULTS

Comparison of enzymes in syngens growing in bacterized medium

P. aurelia from all 14 syngens were examined for variation in six enzymes by Tait (1970a) following growth in bacterized medium. These six enzymes included: succinate dehydrogenase, glutamate dehydrogenase (NAD-dependent), β-hydroxybutyrate dehydrogenase, fumarase, and mitochondrial and supernatant isocitrate dehydrogenase (NADP-dependent). Succinate dehydrogenase showed no variation in any of the stocks and syngens. All other enzymes showed intersyngenic variation.

Intrasyngenic variation was observed for β-hydroxybutyrate dehydrogenase in syngens 2, 4, and 9 for mitochondrial isocitrate dehydrogenase in syngens 2 and 9, and for supernatant isocitrate dehydrogenase in syngen 2. In all but one case the pattern of intrasyngenic variation was different from the intersyngenic variations.

The esterases of the Paramecium stocks all react with the substrate, α-naphthyl propionate, but only some react with both α-naphthyl propionate and α-naphthyl butyrate. Some of these react to a greater extent with one than the other. Four classes of esterases could be distinguished: A esterases react with propionate, B react to a greater extent with butyrate, C to a greater extent with propionate, and D react equally well with both. At least three of these classes (A, B and cathodal C) may vary independently in the different syngens (Allen et al, 1971).

The initial studies with bacterized stocks indicated that only syngen 2 seemed to have any degree of intrasyngenic variation for the esterases. In general, these studies, like those of Tait (1970a), indicated that the frequency of intrasyngenic differences was low. Yet, differences could be observed between many of the syngens in the mobilities of either the A, B or cathodal C esterases.

Similar mobilities for the three classes of esterases were observed within some of the syngens; nevertheless, it was possible to separate many of the syngens from each other and to narrow down an unknown stock to a few syngens. Certain syngens proved difficult to separate, but others were different in all three classes of esterases. With 5 other enzymes, Tait (1970a) was able to define a stock to one of the 14 syngens apart from being able to separate syngen 1 from 5. This technique of enzyme analysis helped also to clarify differences between stocks which were considered to be the same in different laboratories. The most important conclusion from these studies, however, was that there were possibilities in defining the species-complex, *Paramecium aurelia* in biochemical terms and even relating different syngens using enzymic classification.

Effects of bacteria on enzyme patterns

A disturbing feature of the results just described was the difficulty in reproducing enzyme patterns between cultures of the same stock. Sometimes they looked to have as many differences as between some of the syngens. In Fig. 1 we show the propionyl esterases of stock 540, syngen 1, grown in the presence of bacteria and in their absence (Rowe et al, 1971). There are two bands (1 and 2) present in the stock when grown in bacterized medium which are absent from axenically grown cells. There is also a band (3) present in axenic cells which is absent from the bacterized culture. By growing the bacteria on their own, bands 1 and 2 can be attributed to bacterial sources. If live bacteria are added to an axenic culture, band 3 disappears after the cells have undergone approximately 13 fissions (7-10 days growth). Dead bacteria do not have the same effect. If bacteria are grown in a flask connected to, but separated by an ultrafine filter from a second flask containing an axenic culture of paramecia, band 3 disappears on the 6th day (Gibson and Cavill, 1973). If the bacterial culture is removed, then the esterase band reappears. Band 3 is a D esterase, but the bacterial bands 1 and 2, which are D in type, are different from the paramecium D esterases in their reaction to several inhibitors. Bacteria can positively contribute enzymes to the total array in an extract from a culture of paramecia even when the latter are starved. Furthermore, bacteria may inhibit the formation of particular paramecium esterases. Bacteria have also been shown to stimulate the activity of other paramecium esterases present in both bacterial and axenic media.

The results showed that comparison of stocks and syngens could be complicated by the contribution of enzymes from bacteria. These enzymes would interfere with the paramecium enzymes by overlapping with them on gels and therefore make comparisons between syngens difficult. For this reason early on in the studies we turned to the growth of Paramecium on the axenic medium developed by van Wagtendonk and his laboratory (Soldo et al, 1966). Any variations then would be due to genic differences attibutable to paramecia.

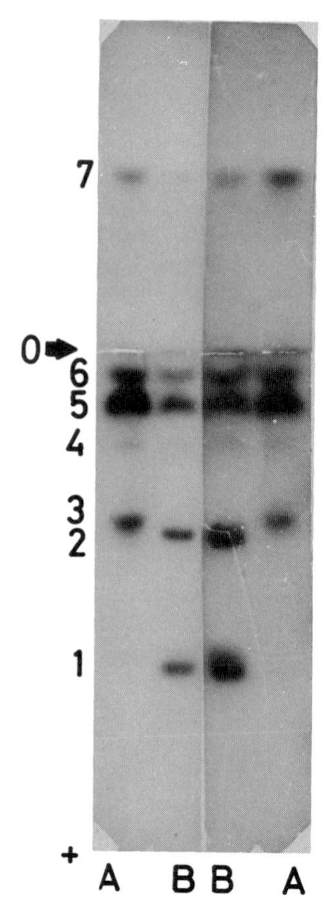

Fig. 1. Stock 540 (syngen 1) grown in axenic medium or with bacteria present. The substrate is α-naphthyl propionate. A (left) in axenic; B (right) in bacterized. B (left) in bacterized; A (right) in axenic. Electrophoresis was carried out for 5 hr.

Comparison of enzymes in axenic media

Some of the work on mitochondrial enzymes has been repeated on cells grown in axenic medium, and there is no evidence so far that mitochondrial preparations contain bacteria. Very similar electrophoretic patterns are observed for these enzymes when comparisons are made between bacterized and axenic paramecia. In the accompanying paper all comparisons were done on axenically grown cells (Adams and Allen, 1974).

The esterases, on the other hand, do show differences and the results are much clearer for paramecia grown in axenic medium. All four classes of esterases are found in each of the 14 syngens (Allen and Gibson, 1971; Allen et al, 1973). The A esterases are typically isozymic. Usually three isozymes are found, with the more cathodal isozyme being the most active. More than three isozymes may occur in stocks of syngen 2 with as many as five isozymes being observed in a single stock (see Fig. 3). A single B esterase is found in each stock and in each syngen. An exception may be syngen 4 where two isozymes are resolved in Electrostarch. The cathodal C esterase varies in activity in different stocks and in certain stocks appears as a doublet. There are, in addition, some rapidly migrating anodal C esterases which occur in certain syngens (2, 8, 10, 11 and 12). The D esterases are erratic in their behavior.

The mobilities of the A, B and cathodal C esterases are usually identical for stocks within the same syngen. Variant stocks have been observed only in syngens 1 (1/20), 2 (5/17) and 8 (2/13). Table I summarizes the enzymes and syngens which have been examined to date. Variant stocks for one of the 9 enzymes have been found in syngens 4, 8 and 9, for two of the 9 enzymes in syngen 1, and for 5 out of the 9 enzymes in syngen 2. For the most part too few stocks exist in the other syngens to test. Those that were tested, however, were derived from geographically widely-separated sources. It would appear then that the frequency of intrasyngenic variation is low, with the exception of syngen 2.

The mobilities of each of the types of esterases varies in different syngens (Fig. 2). As defined by differences in mobility, 6 subtypes exist for the A esterases, 10 subtypes for the B esterases, and 4 subtypes for the C cathodal esterases. The distribution of subtypes of these three esterases as well as that of the 6 other enzymes among the 14 syngens is shown in Table I. The syngens are definable in terms of the subtypes of these 9 enzymes except in the case of stocks of syngens 1 and 5. Some syngens differ for 8 of the 9 enzymes (2 and 13; 7 and 10; 11 and 12). Some differ for only three enzymes (1-5 and 3, 9 and 14; 2 and 3). Most syngens differ in 6 out of the 9 enzymes.

Further characterization of the axenic esterases

Comparisons of the esterases between syngens has also been carried out using inhibitors and temperature. The A esterases are all sensitive to eserine sulfate at a concentration of 10^{-3} to 10^{-4}M, the B esterases to 10^{-2}M, but the C and D esterases are resistant to 10^{-1}M (Allen and Gibson, 1971). These results are true for all the syngens suggesting some homology between the subtypes of each class (Allen et al, 1973).

TABLE I
DISTRIBUTION OF ENZYME SUBTYPES AMONG SYNGENS

Enzyme	Subtypes (Cathodal- - - - - - - - - -> Anodal)†
EsA	(9) (10) (1,3,5,7,11,13,14) (6,12) (2*) (4,8*)
EsB	(2*) (1*,3,5) (6) (9) (7,13) (10,14) (12) (11) (8) (4)
EsC	(8) (14) (1o,2,3,4,5,6,9,10,12) (7,11,13)
Fumarase+	(4,8,10,12,13) (2,3,7) (1,5,6,9,11,14)
GDH+	(10,11) (1,3,4,5,6,7,8,9,12,13,14) (2)
BDH+	(6,8)# (7,14) (1,5,10,12) (2⊛,3) (4⊛,9*,11) (13)
IDH_S+	(12) (13) (10) (11) (2⊛,3) (1,5,6,7,8,9,14) (4)
IDH_M+	(6) (4,10,12) (13) (1,2⊛,3,5,7,8,9⊛,11,14)
SDH+	(1,2,3,4,5,6,7,8,9,10,11,12,13,14)

† Using Electrostarch.

+ Data from Tait (1970a).

No migration into gel.

* Variant stocks show change in mobility.

⊛ Variant stocks show multiple zones of activity unlike "normal."

o Variant stock differs in activity.

Abbreviations: EsA = Esterase A; EsB = Esterase B;
EsC = Esterase C; GDH = glutamate dehydrogenase;
BDH = β-hydroxybutyrate dehydrogenase;
IDH_S = isocitrate dehydrogenase (supernatant);
IDH_M = isocitrate dehydrogenase (mitochondrial);
SDH = succinate dehydrogenase.

Studies on thermal stability, however, indicate differences in heat sensitivity between subtypes and, even more interesting, within a particular subtype. Table II summarizes the

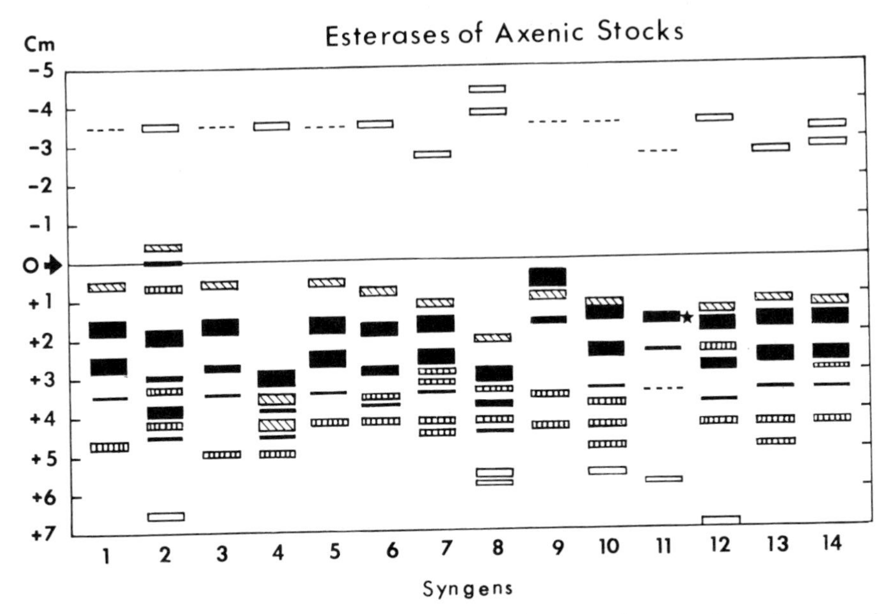

Fig. 2. Diagrams of the esterases in axenic stocks of the 14 syngens of *P. aurelia* in Electrostarch. Types of esterases are represented by the following symbols:

- A type, - B type, - C type, and

- D type. There is considerable variation between stocks of syngen 2 in the mobilities of the A esterases. Those shown are for stock 93. The * indicates the position of the B esterase in syngen 11 (hidden from view by an A esterase isozyme). The dashed lines indicate esterases of very low activities. Distances in migration are marked off in centimeters from the origin (0) on the margin in the diagram.

---

results so far on a few stocks from 7 syngens. Syngens 1, 3, 5 and 7 have the same subtype of A esterase with respect to mobility. Yet, differences in heat stability occur between as well as within syngens. This is also true for syngens 4 and 8 which have a common A esterase subtype and for syngens 1, 3 and 5 which have a common B esterase subtype. Four different variants (a-d) in terms of mobility are represented for the A esterases in the 5 syngen 2 stocks, but the two with a common mobility phenotype (2a) differ in thermal

TABLE II
THERMAL STABILITY OF "HOMOLOGOUS" ENZYMES

	A esterases		B esterases	
Syngen/Stock	Mobility Subtype	Stability to heat#	Mobility Subtype	Stability to heat#
1/60	1	-	1	+
1/90	1	-	1	+
1/513	1	+	1	-
1/551	1	-	1	-
2/562	2a	+		
2/He-1	2b	-		
2/He-2	2a	-		
2/He-3	2c	+		
2/He-4	2d	-		
3/92	1	+	1	-
4/51	4	-		
5/214	1	+	1	+
7/227	1	+		
8/138	4	+		

#Thermal inactivation was studied by placing the cell extract at 50°C for different lengths of time (25 minutes with these results) then running the extract by starch gel electrophoresis. The untreated control was run beside it on the same gel. - = unstable; + = stable.

stability. The important point to note is that differences in heat sensitivity occur for esterases of the same class with identical mobilities. Such a result suggests that homologous esterases even with the same mobility are not necessarily identical. Obviously then in making comparisons between syngens further chemical and physical studies on the esterases and other enzymes might prove greater differences than are suspected from electrophoretic studies.

Genes and enzyme variation in *P. aurelia*

Given the occurrence of variant stocks within a syngen, an analysis of the genetic basis of the phenotypic differences can be carried out. If a genetic basis were to be found, it would infer that genic differences might be responsible for the intersyngenic variations.

The genetic studies showed that each of the three classes

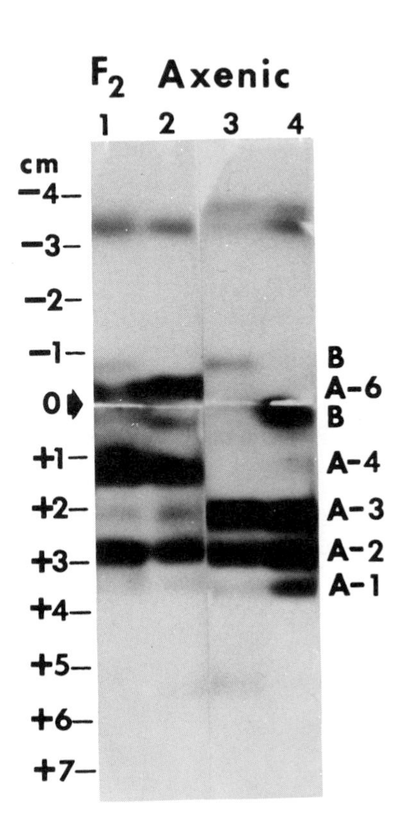

Fig. 3. Esterases of axenically grown F_2 exautogamous clones from the cross 93 x 305. 1 = 93A, 93B; 2 = 93A, 305B; 3 = 305A, 93B; 4 = 305A, 305B.

---

of esterases is coded by a separate gene, and the genes are unlinked (Allen and Golembiewski, 1972; Cavill and Gibson, 1972). Fig. 3 illustrates the appearance of both parental and recombinant classes in the F_2 exautogamous clones produced from crosses between two stocks in syngen 2 which differ in the two genes <u>esA</u> and <u>esB</u>. Tait (1968, 1970b) established a genetic basis for the intrasyngenic variation observed for β-hydroxybutyrate dehydrogenase and isocitrate dehydrogenase, both mitochondrial and supernatant. In all these cases the

TABLE III
GENES AND ENZYMES IN *P. aurelia*

Enzyme*	Syngen/Stock	Gene	Reference
EsA	2/50	esA^{50}	Allen and Golembiewski (1972)
"	2/93	esA^{93}	
"	2/305	esA^{305}	
EsB	2/50	esB^{50}	
"	2/93	esB^{50}	
"	2/305	esB^{305}	
EsA	8/138	esA^{138}	Cavill and Gibson (1972)
"	8/299	esA^{299}	
EsB	1/90	esB^{90}	
"	1/540	esB^{540}	
EsC	1/90	esC^{90}	
"	1/540	esC^{540}	
BDH	9/hu43	Hbd-1	Tait (1968, 1970a)
"	9/317	Hbd-2	
IDH_S	2/50	IDH_S-1	Tait (1970a, b)
"	2/du41	IDH_S-2	
IDH_M	2/50	IDH_M-1	
"	2/583	IDH_M-2	

* See legend to Table I for abbreviations.

enzymes appear to be coded by nuclear genes, even mitochondrial enzymes. A list of the genes and enzymes involved is shown in Table III.

The genetic basis for intrasyngenic differences presupposes

a genetic basis for intersyngenic differences. These results are important because they form the foundation for a genetics and ecology of populations of *P. aurelia*. Gene frequency changes can be measured within a closed environment containing different syngens and intrasyngenic variants, and the effects of environmental variation on the frequencies can be explored.

Studies on a population of *P. aurelia*

Within a pond at Hethersett, Norfolk, four syngens of *P. aurelia* have been found. By far the greatest number of isolates belong to syngen 2. Five esterase A alleles, two B esterase alleles, and 6 C and E antigen alleles were found in the syngen 2 isolates over a six-year period (Gibson and Adams, 1974). The number of esterase alleles observed in this one pond is very similar to the number so far sampled from isolates across the world. Within the pond, however, the frequency of the particular A and B esterase allele varies from month to month. One antigen allele increases in frequency dramatically in concert with the change in temperature of the pond. This temperature change occurs each year. With this exception, most of the frequencies seem to vary from month to month in an unpredictable way and also from year to year.

In other studies on natural populations variant stocks also appear to be more common in syngen 2 than in other syngens (Gill, 1972; Gill and Hairston, personal communication; Tait, 1970a; Tait et al, 1971).

Importance of conjugation and autogamy in nature

Genetic reorganization in *P. aurelia* involves conjugation or autogamy. In both cases new combinations of genes are set up. In the case of autogamy only homozygotes are produced but heterozygotes can appear following conjugation. Esterase and antigen loci were examined from the point of view of being homozygous or heterozygous. A esterase genes appear in heterozygous forms as do the c and e antigen genes. However, as Table IV shows they are always swamped by the numbers of homozygotes indicating the importance of autogamy. Although autogamy is common, nevertheless the frequency of heterozygotes is appreciable. This indicates that conjugation occurs in nature with appreciable frequency (Gibson and Adams, 1974).

DISCUSSION

The distribution of isozymes observed for 9 different enzymes in the stocks of *P. aurelia* reinforces the division of

TABLE IV

e antigen

Expected and observed frequency of heterozygotes and homozygotes for three loci among syngen 2 isolates from the Hethersett pond population.

	Heterozygotes	Homozygotes	Total
Observed	110	534	644
Expected *	415	229	644

c antigen

	Heterozygotes	Homozygotes	Total
Observed	28	475	503
Expected *	304	199	503

Esterase A

	Heterozygotes	Homozygotes	Total
Observed	16	297	313
Expected **	238	75	313

* Calculated from individual samples

** Calculated from yearly samples

this species-complex into sibling species, or syngens. Each of the 14 syngens with the exception of syngens 1 and 5 can be separated clearly from each other in terms of these 9 enzymes. Each syngen thus appears to be a distinct entity in biochemical terms.

Intersyngenic differences persist even when account is taken of intrasyngenic differences. The latter are infrequent and in the main are restricted to syngen 2 both when we look at isolates from world-wide locales and also in a restricted locale. The strict conservatism of the enzymes contrasts with the high degree of polymorphism observed for the immobilization antigens, which in all syngens examined vary genetically under a standard set of growth conditions.

A possible explanation for this dilemma may be that the

conservatism of the enzymes as judged by mobility differences may be more apparent than real. Temperature inactivation studies have revealed that esterases of the same class with similar mobilities in different syngens, or even within the same syngen, may in fact not be identical. This would imply a greater genetic heterogeneity both within and between syngens than is sensed by the electrophoretic procedures. Further physical and chemical characterization of these enzymes may indeed reveal greater differences between syngens and stocks within a syngen than heretofore suspected. Other studies have revealed differences in micronuclear chromosome numbers between syngens and stocks within a syngen, and studies in progress suggest DNA nucleotide sequence differences do occur between syngens and to a limited extent between stocks within a syngen.

Syngen 2 appears to be exceptional in the ease with which variant stocks can be identified. The reasons for this are not clear. This syngen shows no major differences from many of the other syngens in the way antigens and mating types are expressed nor in the amount of antigenic variation which can be detected. Stocks of syngen 2 seem to possess a larger array of different kinds of bacterial symbionts than other syngens (Beale et al, 1969) but how this fact fits with the observed enzyme variation is not apparent. In our population studies temperature changes are correlated with rapid changes in gene frequency in this syngen. Perhaps the response to environmental change results in selection for genetic variability to permit syngen 2 to occupy a wide range of habitats.

The relationships between the various syngens do not appear to be as straightforward as those deduced by Sonneborn (1957) from considerations of systems of mating type and antigen inheritance and other characters. These relationships will be considered in an accompanying paper (Adams and Allen, 1974) in which the isozyme data are used to measure genetic distances between the syngens.

ACKNOWLEDGMENTS

This research was supported by research grants from the National Institute of General Medical Sciences (GM-15879), U.S. Public Health Service, from the British Medical Research Council, and from the British Science Research Council.

REFERENCES

Adams, J. P. and S. L. Allen 1975. Genetic polymorphism and differentiation in *Paramecium*. *Isozymes IV. Genetics and Evolution*, C. L. Markert, editor, Academic Press, New York. pp. 867-882.

Allen, S. L. and I. Gibson 1971. Intersyngenic variations in the esterases of axenic stocks of *Paramecium aurelia*. *Biochem. Genet.* 5: 161-181.

Allen, S. L. and P. A. Golembiewski 1972. Inheritance of esterases A and B in syngen 2 of *Paramecium aurelia*. *Genetics* 71: 469-475.

Allen, S. L. and S. L. Weremiuk 1971. Intersyngenic variations in the esterases acid phosphatases of *Tetrahymena pyriformis*. *Biochem. Genet.* 5: 119-133.

Allen, S. L., B. C. Byrne, and D. L. Cronkite 1971. Intersyngenic variations in the esterases of bacterized *Paramecium aurelia*. *Biochem. Genet.* 5: 135-150.

Allen, S. L., S. W. Farrow, and P. A. Golembiewski 1973. Esterase variations between the 14 syngens of *Paramecium aurelia* under axenic growth. *Genetics* 73: 561-573.

Beale, G. H., A. Jurand, and J. R. Preer 1969. The classes of endosymbiont of *Paramecium aurelia*. *J. Cell. Sci.* 5: 65-91.

Borden, D., E. T. Miller, G. S. Whitt, and D. L. Nanney 1974. Electrophoretic analysis of evolutionary relationships in *Tetrahymena*. *Evol.* (in press).

Cavill, A. and I. Gibson 1972. Genetic determination of esterases of syngens 1 and 8 in *Paramecium aurelia*. *Heredity* 28: 31-37.

Gibson, I. and J. P. Adams 1974. Isozyme and antigen variation in a natural population of *Paramecium aurelia*. Manuscript in preparation.

Gibson, I. and A. Cavill 1973. Effects of bacterial products on a *Paramecium* esterase. *Biochem. Genet.* 8: 357-364.

Gill, D. E. 1972. Intrinsic rates of increase, saturation densities, and competitive ability. 1. An experiment with *Paramecium*. *Am. Nat.* 106: 461-471.

Haggard, B. W. 1974a. Interspecies crosses in *Paramecium aurelia* (syngen 4 by syngen 8). *J. Protozool.* 21: 152-159.

Haggard, B. W. 1974b. Interspecies transfer of marker genes in *Paramecium aurelia* (syngen 4 to syngen 8). *J. Protozool.* (in press).

Rafalko, M. and T. M. Sonneborn 1959. A new syngen (13) of *Paramecium aurelia* consisting of stocks from Mexico, France and Madagascar. *J. Protozool.* 6 (Suppl.): 30.

Rowe, E., I. Gibson, and A. Cavill 1971. The effects of growth conditions on the esterases of *Paramecium aurelia*. *Biochem. Genet.* 5: 151-159.

Soldo, A. T., G. A. Godoy, and W. J. van Wagtendonk 1966. Growth of particle-bearing and particle-free *Paramecium aurelia* in axenic culture. *J. Protozool.* 13: 494-497.

Sonneborn, T. M. 1957. Breeding systems, reproductive methods and species problems in protozoa. In *The Species Problem*, (E. Mayr, ed.). Am. Assoc. Adv. Sci. Symp., Washington, D.C., pp. 155-324.

Sonneborn, T. M. 1958. Classification of syngens of the *Paramecium aurelia - multimicronucleatum* complex. *J. Protozool.* 5 (Suppl.): 17.

Sonneborn, T. M. 1966. A non-conformist genetic system in *Paramecium aurelia*. *Am. Zoologist* 6: 589.

Sonneborn, T. M., M. V. Schneller, J. A. Mueller, and H. E. Holzman 1959. Extensions of the ranges of certain syngens of *Paramecium aurelia*. *J. Protozool.* 6 (Suppl.): 31.

Tait, A. 1968. Genetic control of β-hydroxybutyrate dehydrogenase in *Paramecium aurelia*. *Nature* 219: 941.

Tait, A. 1970a. Enzyme variations between syngens in *Paramecium aurelia*. *Biochem. Genet.* 4: 461-470.

Tait, A. 1970b. Genetics of NADP-dependent isocitrate dehydrogenase in *Paramecium aurelia*. *Nature* 225: 181-182.

Tait, A., G. H. Beale, and A. R. Oxbrow 1971. Enzyme polymorphism in a population of *Paramecium aurelia* in S.E. England. *J. Protozool.* 18 (Suppl.): 26.

MOLECULAR HETEROGENEITY AND THE STRUCTURE OF FEATHERS

ALAN H. BRUSH
Biological Sciences Group
University of Connecticut
Storrs, Connecticut

ABSTRACT. Hypotheses for the origin of molecular heterogeneity are reviewed with special regard to keratins. The nature of feather keratins and their organization into morphological entities can be explained by a number of mechanisms. The biochemical evidence for the structure of SCM-protein subunits of feathers, their potential control and interaction correlates protein structure with function. The heterogeneity of the keratins is compared to that of immunoglobulins and histocompatibility antigens. The selective pressures on systems which consist of relatively large numbers of protein subunits are discussed. Feather keratin heterogeneity is the product of selection for maximum morphological flexibility based on a minimum of genetic information. Evolution within the system appears to be conservative. The strategy has been one favoring the production of a relatively large number of small (MW = 11,000) proteins which interact under the influence of regulatory genes to produce the great range of morphological phenotypes.

INTRODUCTION

Molecular heterogeneity of genetic origin provides a basis for the evolution of structure and function at several levels of biological organization. Of special interest are proteins of similar sequence, presumably the products of related genes, whose function can be related directly to specific morphological structures. Protein heterogeneity can arise from translation errors such as relaxed vigor in sequence determination (Colvin, et al., 1954) or stepwise addition of peptides during synthesis (Lindley, et al., 1970); post-translational modifications (Midelfort and Mehler, 1972); preparative artifacts such as proteolytic enzyme activity, disruption of disulfide bonds (Hesbeeb, 1968) or subunit dissociation (Esterby and Rosemeyer, 1972); or conformational changes (Epstein and Schechter, 1968). Metastable proteins (Nickerson, 1973) may be singularly important in certain biochemical and physiological processes. Gene duplication followed by allelic mutations leads to production of different gene products and thus provides a significant source of apparent heterogeneity (Ohno, 1970). In addition to these mechanisms two other hypotheses have appeared to account for the heterogeneity associated with the keratin proteins of the vertebrate epidermis. One holds that keratinocytes accumulate mutations as they are dying cells and therefore insensitive to selection pressures (Mercer, 1961). The implication is that large numbers of mutated genes have accumulated in the keratinocyte line with no effect on viability. An alternative hypothesis is that the products of these cells are a family of closely related products with a positive selective advantage based on the functional requirements of the structures (e.g., feathers) involved. Two well documented examples of selection for functional heterogeneity are the immunoglobins and histocompatibility antigens.

The proximal sources capable of producing molecular heterogeneity, such as chemical or physical modifications, must be distinguished from ultimate sources such as selection. The former may act independently and capriciously. The latter provide explanations related to the requirements of the particular system, are capable of undergoing natural selection, and are of long term evolutionary significance.

PART 1

Feather keratins are a complex group of molecules produced by active cells in the vertebrate epidermis (Spearman, 1966; Flaxman, 1972). In birds the keratins are a family of closely related gene products which interact to ultimately produce the observed feather structure (See Part II for Discussion and Citations). The extraordinarily complex X-ray diffraction patterns of feathers, now known for over 40 years, indicate the complex nature of the subunit structure and their association. However, the precise nature of the interaciton among molecules or subclasses of molecules is largely unknown, despite the existence of detailed molecular models (Fraser, et al., 1972). The polypeptides are capable of self-assembly into larger units. The smallest of these, the keratin microfibrils are similar to the smallest units observable in electron microscopic studies of feathers. The molecular interactions which allow this organization are largely unknown (Fraser and McRae, 1973). The more complex assemblages of microfibrils into the pleated sheet structure associated with feathers are less well understood, and alternate models to explain the process exist (Lundgren and Ward, 1963; Fraser, et al., 1972). Finally, the entire morphological structure of the feather, the different types of feathers, and the difference between normal and mutant feathers must ultimately be accounted for by mechanisms of the chemical and genetic association of the keratin molecules.

Keratins are typical of the vertebrate integument. Keratins

have not been demonstrated among invertebrates where the functionally analogous molecule is chitin which has a strikingly different chemical composition. The epidermal cells which form the feather follicle produce a number of products during feather formation (Stettenheim, 1972). Histochemical evidence indicates that synthesis and keratinization occur simultaneously (Cane and Spearman, 1967, Matulionis, 1970). In feathers the specific gene products are distributed in both time and space to form the large variety of elements necessary to produce feathers (Kemp and Rogers, 1972; Brush, 1974b). Furthermore, they are under specific genetic control which results in the production of the morphological structures such as the variety of feathers, down, and scales. Some precedents exist for cases of a single cell or cell line producing a heterogeneous mixture of gene products. Multiple protein products are recognized in the hemoglobins, in the avian oviduct, which simultaneously synthesizes several eggwhite proteins, and in the widely distributed ability of tissue cells to produce isozymes.

PART II

The major protein constituent of feathers is a β-keratin, which contributes over 90% of the total weight. The difference among feather proteins, β-keratins and α-keratin, the main protein of hair, were first established through X-ray diffraction techniques in the 1930's. Feather keratins are not limited to feathers, but occur in the beak, claws, and scales as well. Phylogenetically, feather-like keratins are probably older than the birds themselves as structurally similar keratins are known from reptilian scales (Fraser, et al., 1971; Baden, et al., 1974). Although exact structural homologies have not been established, some analysis has been attempted (Alexander, 1970; Maderson, 1972a and b; Brush, 1974b) and it appears possible to study the evolutionary relationships of these structures on the molecular level (Brush, in prep.). Keratins and keratinization have been reviewed recently by Mercer (1961), Fraser, et al., (1972), Parakkal and Alexander (1972) and detailed aspects of the histological and histochemical processes of feather development are reviewed by Lucus and Stettenheim (1972).

The keratins from feathers are most easily and uniformly solubilized by preparing the S-carboxymethyl (SCM) derivatives of the polypeptide subunits produced by the cleavage of intermolecular disulphide bonds. This is accomplished by treatment with thioglycollic acid or mercaptoethanol and urea in alkaline solution (Harrap and Wood, 1964). The product is stable and gives complex patterns on polyacrylamide gel electrophoresis. The SCM-subunits have a molecular weight of approximately 11,000

(Jeffery, 1970; Brush, 1974a) and appear to be charge isomers of each other. Their pI was outside the range conveniently studied by iso-electric focusing (Brush, unpublished). Each of the SCM-polypeptide subunits consists of only 4-5 tryptic peptides (O'Donnell, 1973a) but detailed characterization of these products is needed urgently. Evidence now available suggests that each polypeptide is the product of a separate gene.

Amino acid compositional data is available for the different parts (e.g., shaft, vane) of adult contour feathers from several domestic species (summarized in Crewether, et al., 1965). In general, the acidic amino acids exceed tha basic residues and the overall pattern was qualitatively different from other animal fibers. The amino acid compositions are distinctly different in the component parts. Preliminary compositional analysis using various indices of amino acid similarity showed that differences in homologous feather parts were smaller between species than among different parts of the same feather (Brush, 1974a,c). The overall species differences are limited to only a few residues (mostly alanine, cystine, glycine, isoleucine, proline and tyrosine), which leads to the conclusion that selection for a particular function has occurred (Seifter and Gallop, 1966). No regular distribution exists for proline (Busch, 1970) despite the role it may play in structural determinations (Shor and Krimm, 1961). N-terminal sequences are available for several species. The terminal residue is acetylated in all cases. The sequence of the eight terminal amino acid residues in the turkey and goose calamus were identical.

Peptide maps of SCM-keratins and subunits of contour feathers have been prepared by column chromatography and high voltage electrophoresis of tryptic digests (O'Donnell, 1974a). The study was limited to a single species so no data are available on the differences of identifiable bands between species nor are there data on the degree of variation in individuals of differences of SCM-proteins from different feather parts. Thus we have relatively little information on the magnitude of primary structural differences responsible for the considerable morphological differences of the feather parts, the changes associated with feather mutations (Brush, 1972) or for comparative systematic use.

Only a single SCM-protein from feathers, band three, of the Emu (*Dromaius novae-hollandiae*) has been sequenced (O'Donnell, 1973b). The 102 amino acid residues were divided into five tryptic peptides. The largest one, T_3, with 65 residues contained large amounts of hydrophobic residues. It provides the crystalline portion of the molecule, but did not contain the cystine residues. About 30 residues are involved in the antiparallel pleated sheet (Fraser, et al., 1971). This may provide the site of intermolecular recognition. The cystine residues are restricted to the larger amorphous portions which may be the location of the stronger, intermolecular interactions. It is likely, but not yet proven, that the crystalline portion so important to the self assembly of the units is also the immunogenic site.

All SCM-polypeptides reacted in immunoelectrophoresis or double diffusion experiments with antisera to whole feather SCM-proteins. Similarly, the solubilized keratins from contour feathers of birds of other orders, down, scales, and skin will cross react with antisera to SCM-feather proteins of both chicken (*Gallus gallus*) and duck (*Anas platyrhynchos*). Often only a single precipitation band was present and no spurring was observed (Brush, 1974a). The broad immunological cross reactivity presumably reflects the chemical similarity and evolutionary conservativeness of the feather keratins.

The results of chemical and immunological studies can be related to protein heterogeneity and feather structure by a relatively simple hypothesis. Consider each SCM-polypeptide as a building block. Each consists of crystalline and amorphous portions. The former presumably resides in the T_3 tryptic peptide (O'Donnell, 1973b), is critical to the interaciton of these component units in self assembly (Harrap and Woods, 1964b), and therefore remains relatively intolerant of sequence change. The most active immunogenic sequences also lie in the area of the molecule, a proposition which agrees with the observed immunological conservatism. The remaining portions are the end peptides. They are amorphic in structure and contain the majority of the cystine residues. The strong intermolecular keratinizing bonds are formed here, but the important steric configurations are a function of the more conservative internal portions of the molecule. The N-terminal residues are acetylated. Whether this bears on the functional relationships of the molecule or simply represents the residue required for the initiation of synthesis is unclear. The existence of a relatively large number of gene products is explained by the stringent structural constraints on individual SCM-polypeptides to retain their functional identity and the structural requirements of the formation of the three-dimensional morphology of the feather. The spontaneous self-assembly in vivo must be under considerable regulatory control (Brush, 1972). By analogy the SCM-units are like gears in a watch, each differs only slightly from the other (e.g., in number and shape of teeth, overall diameter, spacing of teeth, thickness), but each is rigidly fixed into position for functional purposes.

By definition the genes for feather proteins are as old as birds themselves (ca. 135 million years). Their apparent conservativeness is a function of the requirements for their interactions on a molecular level. Their relatively small size (only 95-105 residues each) sets limits on the degree of change compatible with inter- and intra-molecular interactions. The resultant strategy has been one of an increased number of different types of building blocks with specific interacitons, as opposed to fewer, larger molecules each capable of a wider range of molecular interactions.

PART III

HETEROGENEITY AND FUNCTION

A major assumption in understanding, and thus explaining, the heterogeneity of enzymatic proteins is that each form is different functionally and therefore adapted to different micro-environmental conditions of cellular function. An alternative hypothesis holds that differentiation of alleles is not determined by selection but through the random fixation of neutral mutations. The selective hypothesis was first inferred in the explanation of isozymes where molecular types were differentiated on distributional, physical, and physiological grounds. Biochemical and physical differences were associated with such diverse physiological functions as temperature adaptation, various stratagies of anaerobic metabolism, cardiac malfunction, and torpor. The possible physiological roles of isozymes have been discussed by Hochachka and Somero (1973). The optional explanation, that the different forms of enzymatic molecules function similarly and are essentially neutral, is not disproven (Le Cam et al., 1972). It is possible that the selective pressures on alternative alleles varies in time and intensity. Thus allelic products of enzymes could be maintained in populations as the result of both selection and drift. Allelic proteins at a locus may or may not become fixed, but explain, at least in part, the unexpectedly high degree of genetic polymorphism in individuals and populations. The resolution of this problem is still uncertain (Lewontin, 1974).

The heterogeneity of feather SCM-proteins can be compared with that of other molecular systems. Unquestionably, the most variable group of proteins produced by animals is the immunoglobulins. Immunoglobulins have the dual functions of antibody recognition and initiation of cellular immune responses. The molecules are produced by a combination of both separate, non-allelic genes and other genes which may have several allelic forms. Whether the variation ultimately resides in a separate germ line for each gene or develops by hypermutation from only a few genes during somatic differentiation is still not settled (Galley and Edman, 1971). Nevertheless, the available evidence indicates that separate genes encode the variable (V) regions which have evolved independently, but at rates similar to those encoding constant (C) regions. The large number of genes involved in immunoglobulin production is undoubtedly the result of a series of duplication events followed by mutational changes. Since both conservative and variable regions exist, sequence variation was not random but the result of selective forces. The functional advantage of this arrangement is that a single antibody molecule can be specific for a great number of different antigens, with different affinities for each. Combined with the permutations possible among the V and C regions, a relatively small number of antibody genes could generate an enormous number of different antibody molecules (Smith, et al., 1971). At the same time the total number of genes involved would account for only 1% or less of the total genetic information in haploid cells (Hood and Talmage, 1970). Thus, immunoglobulin molecules are organized into separate regions, each with a different evolutionary history and each with a different function. The variable light chains are subject to the requirements of increasing diversity to increase antibody activity and yet must maintain constant regions for the critical interaction with the heavy chain. Similarly, the heavy chains must interact with the different light chains and yet be specific enough to influence cellular responses. bind complement, and perhaps eventually alter the antigen itself. Although much remains to be discovered about antibody function, their genetic control and interactions at the molecular level, the outline of their structure and evolutionary mechanisms are established (Porter, 1973; Edman, 1973). Similar arguments can be made regarding the heterogeneity of histocompatibility antigens, especially in regard to the generation of broad functional latitude (Burnet, 1973; Amos, 1972).

In many ways both the immunoglobulins and histocompatibility antigens are analogous to the model presented for feather proteins. A number of closely related genes evolved to produce a series of gene products whose structure and function are related intimately. Each polypeptide interacts with other molecules to form larger functional units. Several types of information are necessary in each system and both conservative and variable portions of each molecule are necessary. Isozymes are a specialized case since the interactions are between a large protein and a normally much smaller substrate molecule (Table 1).

The conditions producing heterogeneity in feathers may be different than in either isozymes or immunoglobulins, but combines features of both. The individual SCM-subunits do not have specialized functional roles as do enzymes in that they do not interact with a specific, smaller substrate molecule nor are they bound to membranes. Rather, intermolecular interactions

TABLE 1

COMPARISON OF ISOZYMES AND FEATHER KERATINS

Isozymes	Feather Keratins
Catalyze chemical reactions (regulatory function).	Building subunits of structural proteins.
Either hetero- or homo-polymers.	Heteropolymers
Involved in control of cellular functions and metabolism	Product of specialized cells which form larger structures.
Intracellular	Extracellular
Usually multigenic (isozymes) but may involve a heterozygous locus (allozymes).	Related family of proteins, multigenic.
Inducible, subject to feedback (and other) regulation.	Regulated by follicular activity.
Produce regular patterns of metabolic and biochemical adjustments to various parameters.	Produce structural elements under additional genetic control which form large third order structures of variable shape, color and function.
Responsive and adaptative to short term environmental and physiological condition.	Selection for long term stability. Functional differences related to structural changes.

appear to be limited to molecules of approximately equal size and shape. In this respect, the structural aspects determine function rather than an active site as is the case in enzymes. This condition is analogous to the interaction of the light and heavy chains of immunoglobulin. Since the interactions produce a non-living structure it is appropriate to consider the static effects of structure rather than the dynamic or kinetic effects characteristic of enzyme-substrate interactions, or the interactions among subunits in other types of multimeric proteins (such as facilitation by heme-heme interactions of hemoglobins). Structural proteins such as feather keratins represent a different strategy from that characteristic of the enzymes. Individual subunits are molded and produced under genetic control and may have different patterns in time. Intermolecular interactions are more permanent in the sense that disulphide bonds of keratinization require more energy to form and break than do the weaker forces typical of enzyme-substrate interactions. As is the case in immunoglobulins, sequence differences and similarities are important in both intra- and intermolecular interactions. Although feather keratins, unlike immunoglobulins, are not required to interact with a diverse range of substrates such as antigens, cell surfaces and complement, their interactions produce a large variety of morphologically diverse elements.

Feathers are produced by a set of proteins which presumably arose from a single ancestral gene. Morphological differences among the structures and the structural differences among the proteins are a reflection of their evolutionary history and a product of a variety of selective forces (Brush 1974c). The epidermal keratin structures differ in the number and types SCM subunits present, as well as their spatial organization. In feathers, the different morphological units share many of the same polypeptides but also possess identifiably unique ones. No reliable estimate exists for the total number of different molecules present nor do we have detailed models of the relationships of the primary sequences and the resultant functional differences. The mechanisms concerning the co-ordination of the behavior among individual polypeptides and other polymeric units are also essentially unknown.

A 12 S mRNA has been isolated from embryonic feather tissue (Partington, et al., 1973). In a cell-free system it is translated faithfully into feather keratin. The exact nature of this message is still unclear. It is not apparent, for example, why such a large mRNA is necessary to code for a relatively small molecule nor how the presumably single message produces the observed heterogeneity.

Discussion of the structure and function of feather keratins bears on the problem of the relative rates of evolution in proteins. While the rate of incorporation of sequence changes seems to be relatively constant for each molecule, different molecules appear to have evolved at quite different rates (Dickerson, 1971). It is possible that the consistency of molecular evolution has not been maintained in some lines, and a slow down in a variety of unrelated proteins in birds relative to other vertebrates has been observed (Prager, et al., in press). Furthermore, it appears that faster evolutionary rates are typical of extracellular proteins than those of intracellular molecules (Nolan, et al., in press). This is presumably a response to steric and other restrictions imposed by functions in the intracellular environment. It now appears that feather keratins have also evolved slowly, despite the fact that they are often classified as extracellular proteins. The explanation of this paradox is straightforward. Although the proteins function extracellularly, their successful organization into morphological elements is dependent on structural considerations and little variation is tolerated. Because of its relatively small size each subunit can undergo only limited sequence change and still retain all its functional aspects. Therefore, the evolutionary response was to produce an increased number of similar subunits each designed for a somewhat different, but related, purpose. Change in sequence occurred only in those regions of the molecule where function would be retained. Thus one would predict differences in rate between the folded and amorphous portion of the polypeptide. Molecular interaction must also be conserved, so conservativeness of the cystine-containing regions, as is the case in the V portion of light chain immunoglobulins, would also be predicted. Finally, the extremely conservative antigenic aspects of the feather SCM-subunits indicate that the immunogenetic portions must correspond with structurally conservative portions. Thus one would predict that the immunogenic sites and the sequences responsible for the highly organized portion of the molecule would coincide. All of these predictions remain to be tested.

Superimposed on the system of similar, but not quite identical, building blocks is a mechanism that determines the nature of the morphological superstructure of the feathers and the sequence of production. The nature of these regulatory elements is still not clear, as regulatory genes have not yet been demonstrated in eukaryotes (Goldberger, 1974). Nevertheless, such control elements must play an important role in translation of molecular information into the morphological phenotype.

Investigations on feather keratins reveal the nature of potential interactions among molecules. In addition to the specific design problem, consideration must be given to the evolutionary history, synthesis, and structure of the molecules. Keratinocytes of feather follicles are extremely complex cells which demonstrate a high degree of specialization with their ability to produce impressive amounts of specialized proteins. These proteins, in turn, are organized into highly structured units which provide an incredibly diverse array of biological and mechanical functions.

ACKNOWLEDGEMENTS

The preparation of this paper was supported by grants from the National Science Foundation (GB-35946X) and the University of Connecticut Research Foundation. I thank G. A. Clark and A. F. Scott for comments on the manuscript.

REFERENCES

Alexander, N. J. 1970. Comparison of α- and β-keratin in reptiles. *Z. Zellforsch.* 110: 153-165.

Amos, D. B. 1972. Biological significance of histocompatibility antigens. *Fed. Proc.* 31(3): 1087-1104.

Baden, H. S. Sviokla and I. Roth 1974. The structural protein of reptilian scales. *J. Exptl. Zool.* 187: 287-294.

Brush, A. H. 1972. Correlation of protein electrophoretic pattern with morphology of normal and mutant feathers. *Biochem. Genet.* 7: 87-93.

Brush, A. H. 1974a. Feather keratin-analysis of subunit heterogeneity. *Comp. Biochem. Physiol.* 48(4B): 661-670.

Brush, A. H. 1974b. Feather keratins. Ch. 4, In: *Chemical Zoology*, Vol. X. M. Florkin and B. Scheer, Eds., Academic Press, N. Y. (in press).

Brush, A. H. 1974c. *Some Taxonomic and Evolutionary Aspects of Feather Proteins.* Proc. 16th International Ornith. Congress, Canberra. (in press).

Burnet, F. M. 1973. Multiple polymorphism in relation to histocompatibility antigens. *Nature* 245: 359-361.

Busch, N. E. 1970. Investigations into the primary structure of feather keratin. Ph.D. Thesis, Iowa State U.; Univ. Microfilm #71-14, 211.

Cane, A. K. and R. I. C. Spearman 1967. A histochemical study of keratinization in the domestic fowl (*Gallus-gallus*). *J. Zool.* 153: 337-352.

Colvin, J. R., D. B. Smith, and W. B. Cook 1954. *Chem. Rev.* 54: 684-711.

Crewether, W. G., R. D. B. Fraser, F. G. Lennox and H. Lindley 1965. The chemistry of keratins. In: *Advances in Protein Chemistry*, C. B. Anfinsen, et al., (eds.) Academic Press, N. Y.

Dickerson, R. E. 1971. The structure of cytochrome *c* and the rates of molecular evolution. *J. Mol. Evol.* 1: 26-45.

Edelman, G. M. 1973. Antibody structure and molecular immunology. *Science* 180: 830-840.

Epstein, C. J. and A. N. Schechter 1968. An approach to the problem of conformational isozymes. In: *Multiple Forms of Enzymes*, E. S. Vesell (ed.). Annals N. Y. Acad. Sci. 151(1): 85-101.

Esterby, J. S. and M. A. Rosemeyer 1972. Purification and subunit interactions of yeast hexokinase. *Eur. J. Biochem.* 28: 241-245.

Flaxman, B. A. 1972. Cell differentiation and its control in the vertebrate epidermis. *Amer. Zool.* 12: 13-26.

Fraser, R. D. B. 1969. Keratins. *Sci. Amer.* 221:86-96.

Fraser, R. D. B., T. P. MacRae, D. A. D. Perry, and E. Suzuki 1971. The structure of feather keratin. *Polymer.* 12: 35-56.

Fraser, R. D. B., T. P. MacRae and G. E. Rogers 1972. *Keratins.* C. C. Thomas, Springfield, Ill.

Fraser, R. D. B. and G. E. MacRae,1973. *Conformation in Fiberous Proteins.* Academic Press, N. Y.

Goldberger, R. F. 1974. Autogenous regulation of gene expression. *Science* 183: 810-816.

Habeeb, A. F. S. A. 1968. Microheterogeneity of human serum albumin: Evidence for differences in reducibility in disulfide linkages. *Canad. J. Biochem.* 46: 789-795.

Harrap, B. S. and E. F. Woods 1964a. Soluble derivatives of feather keratin. I. Isolation, fractionation and amino acid composition. *Biochem. J.* 92: 8-18.

Harrap, B. S. and E. F. Woods 1964b. Soluble derivatives of feather keratins. II. Molecular weight and conformation. *Biochem. J.* 92: 19-26.

Hochachka, P. W. and G. N. Somero 1973. *Strategies of Biochemical Adaptation.* W. B. Saunders. Philadelphia, Pa.

Hood, L. and D. W. Talmage 1970. Mechanism of antibody diversity: Germ line basis for variability. *Science* 168: 325-334.

Jeffrey, P. D. 1970. The molecular weights of two reduced and carboxymethylated keratins by disk gel electrophoresis and a comparison of two methods of analyzing the results. *Aust. J. Biol. Sci.* 23: 809-819.

Kemp, D. J. and G. E. Rogers 1972. Differentiation of avian keratinocytes. Characterization of keratin proteins of adult and embryonic feathers and scales. *Biochem.* 11: 969-975.

LeCam, L. M., J. Neyman and E. L. Scott (Eds.) 1972. Darwinian, Neo-Darwinian, and Non-Darwinian Evolution. *Proc. 6th Berkeley Sym. on Math. Stat. and Prob.* Vol. V. U. Calif. Press, Berkeley, Ca.

Lewontin, R. C. 1974. *The Genetic Basis of Evolutionary Change.* Columbia University Press, New York.

Lindley, H., J. M. Gillespie and R. J. Rowlands 1970. A critique of some hypotheses relating to the heterogeneity of the high-sulphur proteins of wool. *J. Text. Inst.* 61: 157-165.

Lucus, A. M. and P. R. Stettenhiem 1972. *Avian Anatomy-Integument.* Agriculture Handbook 362. U. S. Gov. Printing Office, Washington, D. C.

Lundgren, H. P. and W. H. Ward 1963. The keratins. In: R. Borasky (ed.), *Ultrastructure of Protein Fibers.* Acad. Press, New York.

Maderson, P. F. A. 1972a. When? Why? and How?: Some speculation on the evolution of the vertebrate integument. *Am. Zool.* 12: 159-171.

Maderson, P. F. A. 1972b. On how an archosaurian scale might have given rise to an avian feather. *Amer. Nat.* 106: 424-428.

Matulionis, D. H. 1970. Morphology of the developing down feathers of chick embryos. *Z. Anat. Entwickl.-Gesch.* 132: 107-157.

Mercer, E. H. 1961. *Keratin and Keratinization: An Essay in Molecular Biology.* Pergamon, New York.

Midelfort, C. F. and A. H. Mehler 1972. Deamination in vivo of an aspargine residue of rabbit muscle aldolase. *Proc. Nat. Acad. Sci.* 69: 1816-1819.

Nickerson, K. W. 1973. Biological functions of multistable proteins. *J. Theor. Biol.* 40: 507-515.

Nolan, R. A., A. H. Brush, N. Arnhiem and A. C. Wilson 1974. An inconsistancy between protein resemblance and taxonomic resemblance: Immunological comparisons of diverse proteins from the chicken, pheasant and turkey. *Condor* (in press).

O'Donnell, I. J. 1973a. A search for a simple keratin - Fractionation and peptide mapping of proteins from feather keratins. *Aust. J. Biol. Sci.* 26: 401-413.

O'Donnell, I. J. 1973b. The complete amino acid sequence of a feather keratin from Emu (*Dromaius novae-hollandiae*). *Aust. J. Biol. Sci.* 26: 415-437.

Ohno, S. 1970. *Evolution by Gene Duplication.* Springer, New York.

Parakkal, P. F. and N. J. Alexander 1972. *Keratinization.* Acad. Press, New York.

Partington, G. A., D. J. Kemp and G. E. Rogers 1973. Isolation of feather keratin mRNA and its translation in a rabbit reticulocyte cell-free system. *Nature* 246: 33-36.

Porter, R. R. 1973. Structural studies of immunoglobulins. *Science* 180: 713-716.

Prager, E. M., A. H. Brush, R. A. Nolan, M. Nakanishi and A. C. Wilson 1974. Slow evolution of transferrin and albumin in birds according to micro-complement fixation analysis. *J. Mol. Evol.* (in press).

Seifter, S. and P. M. Gallop 1966. In: *The Proteins*, Vol. 4. H. Neurath, ed. Academic Press, New York.

Shor, R. and S. Krimm. 1961. Studies on the structure of feather keratin. II. A β-helix model for the structure of feather protein. *Biophys. J.* 1: 489-515.

Smith, G. P., L. Hurd and W. M. Fitch 1971. Antibody diversity. *Ann. Rev. Biochem.* 40: 969-1007.

Spearman, R. I. C. 1966. The keratinization of epidermal scales, feathers and hair. *Biol. Rev. Cambridge Phil. Soc.* 41: 59-96.

Spearman, R. I. C. 1973. *The Integument. A Textbook of Skin Biology*. Cambridge Univ. Press.

Stettenhiem, P. 1972. The integument of birds. *Avian Biology* 2: 1-63. D. S. Farner and J. R. King, eds., Academic Press, New York.

Suzuki, E. 1973. Localization of β-conformation in feather keratin. *Aust. J. Biol. Sci.* 26: 435-437.

ISOZYMES OF *AURELIA AURITA* SCYPHISTOMAE OBTAINED FROM DIFFERENT GEOGRAPHICAL LOCATIONS

PAUL L. ZUBKOFF AND ALAN L. LIN
Department of Environmental Physiology
Virginia Institute of Marine Science
Gloucester Point, Virginia 23062

ABSTRACT. The malate dehydrogenase (MDH, EC 1.1.1.37) and tetrazolium oxidase (TO) isozyme patterns of *Aurelia aurita* polyps show variations which are related to the geographical location from which these organisms were obtained. In contrast, glucose-6-phosphate dehydrogenase (G6PD, EC 1.1.1.49) from these *A. aurita* polyps shows a single band with identical electrophoretic mobilities.

On the basis of MDH and TO isozyme patterns, the polyps of *A. aurita* may be grouped into two, and possibly three categories: (a) Northern northwest Atlantic, including polyps from Massachusetts (United States), New Brunswick (Canada), and Prince Edward Island (Canada); (b) Middle northwest Atlantic, including polyps from Chesapeake Bay, Delaware Bay, and probably the Gulf of Mexico; (c) and Middle northeast Pacific, including polyps from Washington (United States). Polyps from Great Britain are more similar to those of category (a) than to any other.

Esterase isozyme patterns are too complex and variable for establishing the relationships of *A. aurita* polyps of different origin.

Isozyme patterns have been used to confirm the identity of polyps from New Brunswick and Prince Edward Island as *Aurelia aurita* and to show that these populations are identical.

INTRODUCTION

In the revised five kingdom systems of the living world based on three levels of organization, the *Cnidaria* are considered to be the simplest *Animalia* (Whittaker, 1969 and Margulis, 1971). Because the *Cnidaria* possess an elementary level of tissue organization (i.e., two-layered germ tissue, primitive nervous system), a high capacity for regeneration, and undergo metamorphosis, they are of particular interest as models for organization in the higher *Animalia*. In addition, nematocyst structure and function is a unique characteristic of this phylum. In order to develop a better understanding of the origin of these unusual features, investigation of the metabolic pathways, enzyme systems, and regulatory mechanisms of the Scyphozoan jellyfishes are considered worthy of continued analysis.

Reports from this laboratory have indicated differences between schyphistomae (polyps) of *Aurelia aurita* (Linnaeus, 1758) obtained from different locales. Polyps of *A. aurita* from Chesapeake Bay (*A.a C)* possess three types of nematocysts and have a significant quantity of β-alanine in their free amino acid pool. The polyspira nematocyst type and β-alanine are lacking in the polyps of *A. aurita* obtained from Massachusetts (*A.a M)* (Webb et al., 1972). In addition, the fully developed polyps of *A.a. C* have a circular mouth whereas those of *A.a. M* have a cruciform configuration (Morales-Alamo and Haven, 1974).

We report herein differences in the malate dehydrogenase (MDH, EC 1.1.1.37) and tetrazolium oxidase (TO) isozyme patterns of *A. aurita* scyphistomae obtained from different geographic locales. These differences in isozyme patterns further corroborate earlier observations that polyps from different locales have distinctive physiological and morphological features.

MATERIAL AND METHOD

The polyps of *A. aurita* used in this study were obtained from the VIMS culture collection,maintained in filtered York River water (salinity, 20.5 o/oo) and fed nauplii of newly hatched *Artemia salina* once a week (Table 1). The polyps of *A.a. GB* (Great Britian), *A.a. M*, *A.a. NB* (Bay of Fundy, New Brunswick, Canada), *A.a. PEI* (Malpeque Bay, Prince Edward Island, Canada), and *A.a. W* (Washington) were maintained at 12° or 15° C; polyps of *A.a. C*, *A.a. D* (Delaware Bay) and *A.a. GM* (Gulf of Mexico) were maintained at room temperature (19°-22° C). The identity of *A.a. C* polyps have been authenticated as *A. aurita* by life history analysis (Calder, 1971); *A.a. D.*, *A.a. GB*, *A.a. GM*, and *A.a. M* polyps have been confirmed in a similar manner; *A.a. W* polyps have been identified as *Aurelia sp* only, while those of *A.a. NB* and *A.a. PEI* are assumed to be *A. aurita* because of the seasonal abundance of the medusae in those areas (D.R. Calder, personal communication).

For MDH and TO, polyacrylamide gel electrophoresis using 7.5% gels, Tris-HCl-TEAMED buffer, pH 8.9, and staining procedures were performed as described previously (Lin and Zubkoff, 1973). In the cases of *A.a. M* and *A.a. GM*, extracts containing 5-15 μg of protein were used because of a scarcity of polyps.

TABLE I

HISTORY OF *AURELIA AURITA* SCYPHISTOMAE CULTURES

Abbreviation	Location	Life Stage Isolated	Culture Established	Source
A.a. C	Chesapeake Bay	medusae*	1968	Morales-Alamo
A.a. D	Delaware Bay	medusae*	1970	Calder
A.a. GB	Great Britain	culture*	1923	Lambert
A.a. GM	Gulf of Mexico	medusae*	1962	Spangenberg
A.a. M	Massachusetts	polyps*	1969	Anonymous
A.a. NB	New Brunswick	podocysts	1971	Calder
A.a. PEI	Prince Edward Island	podocysts	1971	Calder
A.a. W	Washington	polyps	1972	Calder

*Authenticated as *A. aurita* by life history studies (D.R. Calder, Personal Communication).

In order to determine whether the observed differences were in mitochondrial or cytoplasmic MDH, mitochondria were prepared from *A.a. C* medusae which have an MDH isozyme pattern very similar to *A.a. C* polyps. The tentacles and oral arms of the fresh medusae were mixed with 0.6 M sucrose-Tris-glycine buffer, pH 8.3 (1:1, w/v). After the homogenate was centrifuged at 2000 x g for five minutes, the supernatant was removed and the precipitate discarded. The 2000 x g supernatant was centrifuged in a Sorvall centrifuge at 7000 x g for 30 minutes, the supernatant was discarded, and the precipitate (mitochondria fraction) resuspended in fresh sucrose buffer and centrifuged again. After this washing process had been repeated four times, the mitochondria fraction was homogenated with 3% sucrose-Tris-glycine buffer, and the homogenate was applied directly to the top of a 7.5% gel without the aid of a spacer gel. Electrophoresis was carried out as usual.

The detection of G6PD was essentially the same as that for MDH and the staining solution was prepared according to Brewer (1970).

For the detection of esterases, 40-60 μg of protein were applied to 7.5% gels, the staining solution contained 50 ml of NaH_2PO_4 (0.2 M), 10 ml Na_2HPO_4 (0.2 M), 40 ml H_2O, 1.5 ml of 1% α-naphthyl acetate and 40 mg of Fast Blue RR salt (Sigma), pH 6.0 (Brewer, 1970). The gels were incubated in the dark for 1-2 hours.

RESULTS

Malate dehydrogenase. The MDH isozyme patterns when 20-35 μg of proteins were applied to each gel are shown in Figure 1. The MDH isozyme pattern of polyps of *A.a. C* are consistent with the previous report in which four bands (0.29, 0.33, 0.39, 0.43) were observed (Lin and Zubkoff, 1973); the major band was identified as a fast moving band (0.43).

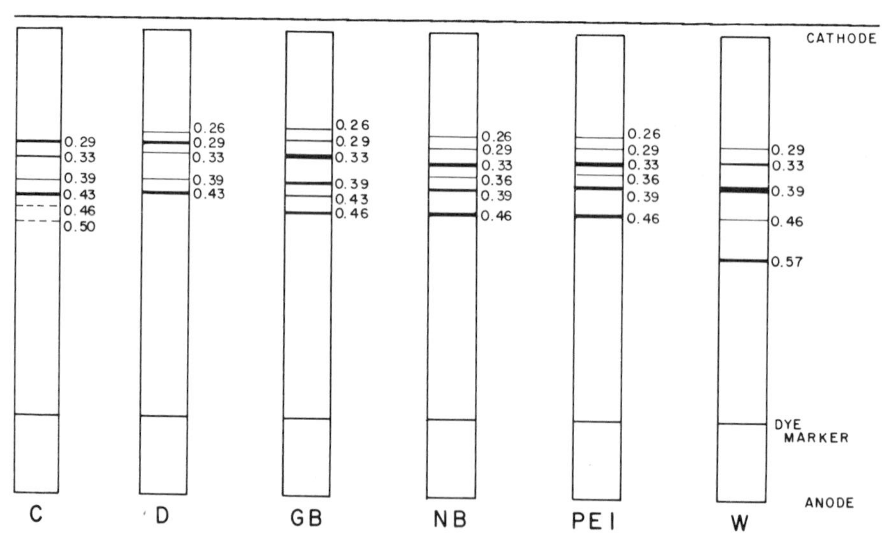

Fig. 1. Malate dehydrogenase isozymes of *Aurelia aurita* polyps obtained from different locales.

The MDH isozyme patterns of *A.a. C* medusae and scyphistomae are quite similar (Figures 1 and 2). When mitochondria are harvested and washed free of the soluble fraction of the cell, the faster migrating MDH bands (0.39, 0.43) are detected, but the slower moving bands are not (Figure 2).

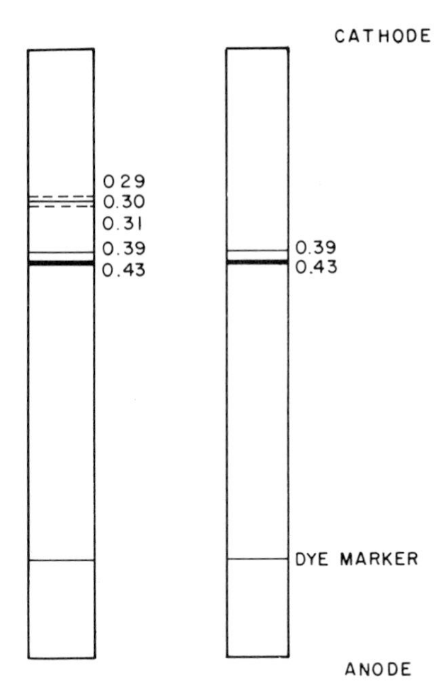

Fig. 2. MDH isozymes of *Aurelia aurita* medusae from the Chesapeake Bay. Left: whole tissue homogenate; Right: mitochondrial fraction.

These results are similar to those of Grimm and Doherty (1961) who demonstrated that the MDH of beef heart exists in two forms: cytoplasmic and mitochondrial.

Of the seven other populations studied (Table 2), the MDH isozymes of *A.a. D* and *A.a. GM* polyps show a pattern similar to that of *A.a. C:* all have the same fast moving major band (0.43). *A.a. GB*, *A.a. M*, *A.a. NB*, and *A.a. PEI* have a slower moving major band (0.33) and a distinct band at 0.39. Three of these (*A.a. GB*, *A.a. NB*, and *A.a. PEI)* have a relatively fast moving band (0.46), and two trace bands (0.26, 0.29). The MDH isozyme patterns of *A.a. NB* and *A.a. PEI* are identical. The *A.a. W* MDH isozyme pattern is unique: three major bands with diminishing degrees of intensity (0.39 > 0.33 = 0.57) and two trace bands (0.29, 0.46).

Tetrazolium oxidase. As previously reported (Lin and Zubkoff, 1973), the TO of *A.a. C* polyps have a major band (0.51) and two other bands (0.39, 0.62); it is the same as that of the *A.a. D* polyps (Figure 3, Table 3). The TO isozyme patterns of *A.a. GB*, *A.a. NB*, and *A.a. PEI* are very similar (10 bands). The TO isozyme bands *A.a.NB* and *A.a.PEI* are identical. In the

TABLE II

MDH ISOZYMES OF POLYPS OF *AURELIA AURITA* OBTAINED FROM DIFFERENT GEOGRAPHIC LOCATIONS

Relative Mobility[1]	Location and Relative Intensity							
	Chesapeake	Delaware	Great Britain	Gulf of Mexico[2]	Massa-chusetts[2]	New Brunswick	Prince Edward Island	Washington
	C	D	GB	GM	M	NB	PEI	W
0.26		+	+			+	+	
0.29	++	+++	+			+	+	+
0.33	++	+	++++	+[3]	+++	++++	++++	+++
0.36						+	+	
0.39	+	++	+++		+	++	++	++++
0.43	++++	++++	++	+	+			
0.46	+[3]		++			+++	+++	+
0.50	+[3]							
0.57								+++

[1] Relative Mobility = $\frac{\text{Distance from origin to band}}{\text{Distance from origin to dye marker}}$

[2] Observations at low concentrations of extracts only

[3] Trace band observed occasionally

TABLE III

TO ISOZYMES OF POLYPS OF *AURELIA AURITA* OBTAINED FROM DIFFERENT GEOGRAPHICAL LOCATIONS

Location and Relative Intensity

Relative Mobility	Chesapeake C	Delaware D	Great Britain GB	Massa-chusetts M	New Brunswick NB	Prince Edward Island PEI	Washington W
0.12			+	+	+	+	
0.16			+	+	+	+	
0.23			++		++	++	
0.37			++		+++	+++	+++++
0.39	+++	++		++			
0.42			++	++	+	+	+
0.45				++			
0.51	++++	++++					
0.58			+		+	+	
0.62	++	+					
0.64			+		+	+	+++
0.68			+		+	+	+
0.74							++
0.82			++	++	++	++	
0.88			+		+	+	

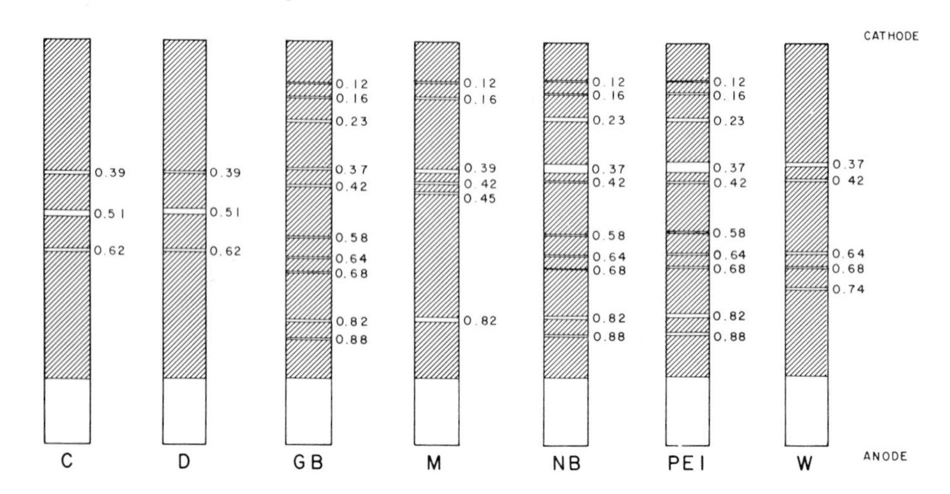

Fig. 3. Tetrazolium oxidase isozymes of *Aurelia aurita* polyps from different locales.

A.a.W polyps,one extremely strong TO band(0.37),two faint bands (0.64, 0.74) and two trace bands (0.42, 0.68) are observed.

Glucose-6-phosphate dehydrogenase. The G6PD of all the *A. aurita* polyps differs from MDH, TO, and esterases which have multiple bands; G6PD has a single band with a relative mobility of 0.20.

Esterases. Unlike the MDH and TO isozyme patterns which show trends, esterase isozyme patterns of *A. aurita* polyps are too different for direct interpretation (Table 4). In order to detect the extremely small amount of esterases present in the polyps, extracts with high protein concentrations (40-60 μg) from 15 polyps of *A.a. C* and *A.a. D* and 1-2 polyps of *A.a. GB, A.a. W*, and *A.a. NB* were applied to each gel. Although polyps of *A.a. C* and *A.a. D* are very closely related with respect to MDH and TO isozyme patterns, they have completely different esterase patterns.

DISCUSSION

On the basis of the MDH isozyme patterns of *A. aurita* polyps studied from eight geographical regions, there are at least two, and possibly a third, distinct groups coincident with three geographic groupings:

1. Northern northwest Atlantic (major bands=0.33, 0.39, 0.46) *A.a. GB, A.a. NB, A.a. PEI*

TABLE IV

ESTERASE ISOZYMES OF POLYPS OF *AURELIA AURITA* OBTAINED FROM DIFFERENT GEOGRAPHIC LOCATIONS

	Location and Relative Intensity				
Relative Mobility	Chesapeake C	Delaware D	Great Britain GB	New Brunswick NB	Washington W
0.44		+		+++	
0.48			+++		
0.50		++			
0.54-0.56				++++	+
0.60-0.61			++++	++	+++
0.67					++++
0.70		+++	++		
0.74	++				
0.78					++
0.80					++
0.82	+				
0.83	+				

2. Middle northwest Atlantic (major band=0.43) *A.a. C, A.a. D*
3. Middle northeast Pacific (major bands=0.33, 0.39, 0.57) *A.a. W*

A.a. GM is more similar to the second group (*A.a. C* and *A.a. D*) while *A.a. M* is more similar to the first (*A.a. GB, A.a. NB, A.a. PEI*).

These observations on MDH isozyme patterns of the six collections of *A. aurita* from the North American Atlantic coast are consistent with the suggestion that northern and southern varieties may exist, with one group acclimated to colder waters and the other to warmer waters. Spangenberg (1964) suggested that subspeciation of North American *A. aurita* may be appropriate because *A.a. GM* scyphistomae differ significantly from those of *A.a. GB* with respect to nematocyst complement and mechanism of locomotion. Other observations show significant differences among polyps from various locations in (1) nematocyst complement, (2) mouth shape, (3) podocyst formation, and (4) the presence of β-alanine in the free amino acid pool (Table 5). Our observations on MDH isozyme patterns corroborate these studies.

TABLE V

CHARACTERISTICS OF *AURELIA AURITA* SCYPHISTOMAE

	Chesapeake Bay C	Delaware Bay D	Germany	Great Britain GB	Gulf of Mexico GM	Massachu-setts M	New Brunswick NB	Sweden	Washington W	Reference
Nematocyst Complement										a,b,c,d
Atrichous isorhiza						+				
"a" Atrichs	+					-				
polyspira	+									
Microbasic heterotrichous										
euryteles	+					+				
Podocyst Formation	-			+		+		+		d,e,f
Mouth Shape of Mature Polyps										g,h,i,j,k
circular	+	+			+					
cruciform			+	+		+	+			
β-alanine	+					-				c
Isozymes										
(No. of Bands)										
MDH	4	5		6			6		5	
G6PD	1	1		1	1	1			1	
TO	3	3		10		6	10		5	
Esterase	3	3		3			3		5	
VIMS Cultures										
size of polyp (mm)	1-2	1-2		4-8	0.5-1	4-8	3-6		3-6	
temperature	19-22°	19-22°		12-15°	19-22°	12-15°	12-15°		12-15°	

a, Spangenberg, 1964; b, Calder, 1971; c, Webb et al., 1972; d, Calder, Personal Communication; e, Chapman, 1968;

f, Chapman, 1970; g, Thiel, 1962; h, Smith, 1964; i, Chapman, 1965; j, Russell, 1970; k, Morales-Alamo and Haven, 1974.

In addition, the analysis of the TO isozyme patterns is consistent with the above three groupings: *A.a. GB*, *A.a. NB*, and *A.a. PEI* have ten TO isozymes; *A.a. C* and *A.a. D* have three. *A.a. M* TO isozymes (Table 3) are very similar to the first group. Unfortunately, TO of *A.a. GM* could not be determined because of a lack of material. Although it is possible that the laboratory temperatures at which these polyps were maintained could influence the isozyme patterns, it seems unlikely that the holding temperature would consistently bias the isozyme patterns and morphological characteristics to the same extent (Table 5).

As mentioned previously, the *A.a. W* has been identified only as belonging to the genus *Aurelia*. The nematocyst complement has not been confirmed nor a detailed life history followed. The MDH isozyme patterns, with the exception of the major band (0.57), indicate that these polyps are similar to the first group (Northern northwest Atlantic) and the TO isozyme pattern indicates the same, although there are differences in several minor bands. The 0.33 and 0.39 or 0.43 MDH bands, common to all polyps of *A. aurita*, serve to distinguish *Aurelia* from polyps of *Chrysaora quinquecirrha* (Desor, 1848) and *Cyanea capillata* (Linnaeus, 1758): *Chrysaora* have doublets (0.21, 0.22 and 0.53, 0.54) whereas *Cyanea* has a single band (0.50) and a triplet (0.25) (Lin and Zubkoff, 1973). It is remotely possible that this population from the Pacific belongs to another species, i.e. *A. limbata* (D.R. Calder, personal communication). The *A.a. NB* and *A.a. PEI* are identical with respect to three enzymes (MDH, TO, and esterases) studied and are closely related to *A.a. M* and *A.a. GB*. Although a detailed analysis of the *A.a. NB* and *A.a. PEI* by life history studies or nematocyst complement has not been undertaken, these polyps are undoubtedly *A. aurita*.

In contrast to both the MDH and TO isozyme patterns, the esterase isozymes of the polyps of the *A. aurita* obtained from 5 locales do not indicate common electrophoretic behavior (Table 4). Although all esterase isozyme patterns indicate at least three components, the differences of these patterns are far too great and the number of observations are too few to interpret in even a qualitative manner. α-napthyl acetate, the substrate for the non-specific esterases, is hydrolyzed by any of the four classes of esterases: carboxyesterase (EC 3.1.1.1), arylesterase (EC 3.1.1.2), acetylesterase (EC 3.1.1.6) and cholinesterase (EC 3.1.1.8) (Holmes and Masters, 1967). It is possible that the use of other substrates and inhibitors which permit the detection of esterases on a more selective basis may be useful.

Although numerous studies have been reported on the MDH

isozymes of a wide variety of organisms, few have been undertaken on a single species of the lowest *Animalia* from varying geographic locations. Conde-del-Pino et al. (1966) reported two MDH isozymes in the flatworm, *Schistosoma mansoni*, from Puerto Rico while Coles (1971) observed two or four MDH isozymes and variations in the non-specific esterases of the same species isolated from different regions of East Africa. In their survey of 46 animals, Thornber et al. (1968) using starch gel electrophoresis (pH 7.0) found that the sea anemone, *Oulactia tenebrosa*, had 11 MDH isozyme bands, the most complex pattern of all organisms studied. However, conformational changes or generation of artifacts through the electrophoresis procedures were not entirely ruled out.

Johnson (1971, 1974) hypothesized that enzyme polymorphisms are often associated with regulatory reactions in metabolism; that is, regulatory enzymes usually at branch points in metabolic pathways, have many isozymes, but non-regulatory enzymes do not. He developed this hypothesis by classifying enzymes as either regulatory or non-regulatory based on several criteria including equilibrium product : substrate ratio. If the equilibrium product : substrate ratio deviates by more than an order of magnitude from its predicted thermodynamic equilibrium, it is taken as *prima facie* evidence that the reaction rate is either regulated or is of regulatory importance in the overall pathway. On this basis, the data of three groups of highly developed organisms, insects, vertebrates, and humans, indicate that MDH is a non-regulatory enzyme and G6PD is a regulatory enzyme. However, evidence that MDH is subject to allosteric control has been reported by Cassman and Vetterlein (1974). Cytoplasmic MDH from beef heart exists in two forms, with one form (dehydrogenase b) showing a protein concentration dependence in binding of NADH, cooperative kinetics with respect to NADH, and having fructose-1,6-biphosphate as an allosteric inhibitor.

The MDH and G6PD of the *Cnidaria* do not conform to Johnson's classification: G6PD of *A. aurita* from different locations and *C. quinquecirrha* has a single band (Zubkoff and Lin, 1973), whereas MDH has as many as seven bands on polyacrylamide gels (Lin and Zubkoff, 1973). In *Hydra eutonina*, G6PD exists as a single band and MDH has three bands (C. Sassaman, personal communication); in both *H. magnipapillata* and *Pelmatohydra robusta*, G6PD exists as two bands but the MDH pattern has not been determined (Kamada and Hori, 1970). Using starch gels, Thornber et al. (1968) reported as many as 11 MDH and 2 G6PD bands for the anemone, *Oulactis tenebrosa*. Kamada and Hori (1970) observed only one G6PD band in two anemones, *Anthopleura midori* and *A. japonica*, and we observed one G6PD and 2 MDH bands in another sea anemone, *Aiptasiomorpha luciae* (unpublished). Beattie and O'Day (1971) observed three major and four minor MDH bands in the estuarine anemone, *Diadumene leucolena*. A shift in the amount of the major bands occurred when the organisms were maintained under conditions of temporary anaerobiosis.

The metabolic patterns of the *Cnidaria* are undoubtedly of significance: the aquatic polyps are sessile and quite primitive when compared to the highly developed and dynamic fruit flies and vertebrates. The metabolism of glucose to the level of phosphoenolpyruvate (PEP) occurs via the glycolytic pathway; in the presence of oxygen, PEP is converted to pyruvate which is completely oxidized to CO_2 and H_2O via the tricarboxylic acid cycle. However, in many invertebrates, under anoxic conditions, PEP is carboxylated to oxaloacetate by PEP carboxykinase, and the oxaloacetate reduced to malate by MDH and subsequently to fumarate and succinate. Malic enzyme (EC 1.1.1.40) is also important for the conversion of pyruvate to malate. Thus, malic enzyme functions in the same capacity as lactate dehydrogenase in glycolysis of higher organisms (Hochachka and Mustafa, 1973; Hochachka et al., 1973). Although MDH may catalyze the same reaction in higher and lower *Animalia*, the overall pathways of carbohydrate metabolism may differ significantly. In view of the lack of specific information about the metabolism of the lower *Animalia*, it is conceivable that MDH may respond to mechanisms of regulation which are as yet unidentified and do not conform to Johnson's criteria.

In summary, MDH and TO isozyme patterns provide additional corroborative evidence that polyps with different morphological characteristics obtained from widely separated geographic locales also differ in physiological characteristics. In addition, two groups of unauthenticated polyps have been clearly identified as *A. aurita*.

ACKNOWLEDGEMENTS

We thank Dr. D.R. Calder for the polyps, Dr. R.E. Black for suggesting the mitochondria separation experiment and for reviewing the manuscript, and Mrs.P. Crewe for maintaining the cultures. This study was conducted as part of the Virginia Jellyfish Research Program under Public Law 89-720, The Jellyfish Act, Contract No. N-043-226-72(G) from the National Marine Fisheries Service of the National Oceanic and Atmospheric Administration, U.S. Department of Commerce and the Commonwealth of Virginia. Contribution No. 656 from the Virginia Institute of Marine Science.

REFERENCES

Beattie, C.W. and D.H. O'Day 1971. Alternations in coelenterate malic dehydrogenase isoenzymes during temporary anaerobiosis. *Comp. Biochem. Physiol.* 40B:917-922.

Brewer, G.J. 1970. *An Introduction to Isozyme Techniques.* Academic Press, New York. 186 pp.

Calder, D.R. 1971. Nematocysts of polyps of *Aurelia, Chrysaora* and *Cyanea,* and their utility in identification. *Trans. Amer. Micros. Soc.* 90:269-274.

Cassman, M. and D. Vetterlein 1974. Allosteric and nonallosteric interactions with reduced nicotinamide adenine dinucleotide in two forms of cytoplasmic malic dehydrogenase. *Biochemistry* 13:684-689.

Chapman, D.M. 1965. Co-ordination in a scyphistoma. *Amer. Zool.* 5:455-464.

Chapman, D.M. 1968. Structure, histochemistry and formation of the podocyst and cuticle of *Aurelia aurita.* *J. Mar. Biol. Ass. U.K.* 48:187-208.

Chapman, D.M. 1970. Further observations on podocyst formation. *J. Mar. Biol. Ass. U.K.* 50:107-111.

Coles, G.C. 1971. Variations in malate dehydrogenase isoenzymes of *Schistosoma mansoni.* *Comp Biochem. Physiol.* 38B: 35-42.

Conde-del-Pino, E., M. Perez-Villar, A.A. Cintron-Rivera, and R. Seneriz 1966. Studies on *Schistosoma mansoni.* I. Malic and lactic dehydrogenase of adult worms and cercariae. *Exptl. Parasit.* 18:320-326.

Grimm, F.C. and D.G. Doherty 1961. Properties of the two forms of malic dehydrogenase from beef heart. *J. Biol. Chem.* 236:1980-1985.

Hochachka, P.W., J. Field, and T. Mustafa 1973. Animal life without oxygen: Basic biochemical mechanisms. *Amer. Zool.* 13:543-555.

Hochachka, P.W. and T. Mustafa 1973. Invertebrate facultative anaerobiosis. *Science* 178:1056-1060.

Holmes, R.S. and C.J. Masters 1967. The developmental multiplicity and isoenzyme status of cavian esterases. *Biochim. Biophys. Acta.* 132:379-399.

Johnson, G.B. 1971. Metabolic implications of polymorphism as an adaptive strategy. *Nature (Lond.)* 232:347-349.

Johnson, G.B. 1974. Enzyme polymorphism and metabolism. *Science* 184:28-37.

Kamada, T. and S.H. Hori 1970. A phylogenic study of animal glucose-6-phosphate dehydrogenases. *Japan J. Genetics* 45:319-339.

Lin, A.L. and P.L. Zubkoff 1973. Malate dehydrogenase and tetrazolium oxidase of scyphistomae of *Aurelia aurita, Chrysaora quinquecirrha,* and *Cyanea capillata* (Scyphozoa: Semaeostomeae). *Helogoländer wiss. Meeresunters.* 25: 206-213.

Margulis, L. 1971. Whittaker's five kingdoms of organisms: Minor revisions suggested by considerations of the origin of mitosis. *Evolution* 25:242-245.

Morales-Alamo, R. and D.S. Haven 1974. Atypical mouth shape of polyps of the jellyfish, *Aurelia aurita,* from Chesapeake Bay, Delaware Bay, and Gulf of Mexico. *Ches. Sci.* 15:22-29.

Russell, F.S. 1970. *The Medusae of the British Isles. II. Pelagic Scyphozoa.* Cambridge Univ. Press, London, 284 pp + XV pl.

Smith, R.I. (ed.) 1964. *Keys to Marine Invertebrates of the Woods Hole Region. Contrib. 11, Syst.-Ecol. Progr.,* Mar. Biol. Lab., Woods Hole, Mass. 208 pp.

Spangenberg, D.B. 1964. New observations of *Aurelia. Trans. Amer. Micros. Soc.* 83:448-455.

Thiel, H. 1962. Untersuchungen über die strobilisation von *Aurelia aurita* Lam. an einer population der Kieler Forde. *Kieler Meeresforchungen* 18:198-230.

Thornber, E.J., I.T. Oliver, and P.B. Scutt 1968. Comparative electrophoretic patterns of dehydrogenases in different species. *Comp. Biochem. Physiol.* 25:973-987.

Webb, K.L., A.L. Schimpf, and J. Olmon 1972. Free amino acid composition of scyphozoan polyps of *Aurelia aurita, Chrysaora quinquecirrha* and *Cyanea capillata* of various salinities. *Comp. Biochem. Physiol.* 43B:653-663.

Whittaker, R.H. 1969. New concept of kingdoms of organisms. *Science* 163:150-160.

Zubkoff, P.L. and A.L. Lin 1973. Dehydrogenases of the Cnidarian, *Chrysaora quinquecirrha* (Scyphozoa: Semaeostomeae). 9th Inter. Conf. Biochem., *Abstr. 2d20,* Stockholm, Sweden, 1-7 July 1973.

PHOSPHOGLUCOSE ISOMERASE IN MARINE MOLLUSCS

NOËL P. WILKINS
Department of Zoology
University College
Galway, Ireland

ABSTRACT. A high degree of genic variability is observed at the PGI locus in marine species of molluscs. Individual species have between one and nine alleles at this locus and the effective number of alleles is positively correlated with the actual number of alleles observed. These findings contrast with those of vertebrate species in which polymorphism is usually low at this "group I" enzyme locus. Actual and effective numbers of alleles are higher in intertidal molluscs than in sublittoral, and higher in epifaunal species than in infaunal. The results suggest that polymorphism is greater at this locus in those species which experience high environmental variability.

INTRODUCTION

The extent of genetic variation at some structural gene loci is considered, from a selectionist viewpoint, to be related to the degree of environmental heterogeneity experienced by the members of a species. Species inhabiting highly variable or ecologically unpredictable environments are postulated to exhibit greater genetic diversity than those inhabiting more homogeneous environments.

Evidence is accumulating that many marine species of molluscs exhibit polymorphism at a number of enzyme loci (Milkman and Beaty, 1970; Koehn and Mitton, 1972; Berger, 1973; Levinton, 1973; Snyder and Gooch, 1973). Among these, the locus encoding phosphoglucose isomerase is notable in that up to nine alleles are segregating in appreciable proportions in some species (Levinton, 1973; Wilkins and Mathers, 1974). This paper considers the variability at this locus in marine molluscs, and its relationship to environmental heterogeneity.

MATERIALS AND METHODS

Animals were collected from wild populations in natural habitats around the coast of Ireland. Tissues were excised from living individuals, homogenized in 50 mM tris-5mM magnesium sulphate, subjected to electrophoresis and stained as previously described (Wilkins and Mathers, 1974).

RESULTS

In all species each individual exhibits either one or three major bands of enzyme activity. The patterns observed, together with their frequency distributions, are consistent with an hypothesis of genetic control of a dimeric molecule at a single autosomal locus in each species. Table 1 lists the actual and effective (Crow and Kimura, 1970) numbers of alleles in each species, together with similar data available for other marine molluscs. Detailed treatment of individual allele frequencies and phenotype distributions within populations of the different species is given elsewhere (see references) or is in preparation. In general, variations within species involve, at most, different allele frequencies in spatially isolated wild populations. Table 2, for example, summarizes the data for the scallop *Pecten maximus* in which nine alleles are segregating at this locus. All the alleles are represented in the populations from the three geographically separated areas, situated on the west, southwest, and east coasts of Ireland. Gene flow between these populations through larval dispersal or adult migration appears unlikely on hydrographical and behavioral considerations. Clearly, the occurrence of nine alleles in this species, and also of multiple alleles in other species (Snyder and Gooch, 1973; Levinton, 1973) is not attributable to genetic phenomena occurring exclusively in a single isolated population. Neither are the alleles only of rare occurrence. The effective number of alleles ranges from 1.1 to 5.6, and there is a significant positive correlation between actual and effective numbers of alleles over the eighteen species ($r = 0.813$, $P < 0.001$). The effective number of alleles rather than the actual number determines the amount of heterozygosity at any locus; species with a high effective number of alleles will contain a larger proportion of heterozygous individuals than one with a lower effective number, irrespective of their actual number of alleles. The positive correlation between actual and effective numbers of alleles in molluscs indicates that heterozygosity is greater in species with higher actual numbers of alleles.

The high degree of genetic variability at this locus in marine molluscs generally can be assessed by comparison with Table 3, which presents the corresponding data for some vertebrate species including man. In most vertebrates, mutant alleles are rare at this locus. In man, for instance, the nine mutant alleles are extremely rare (effective number

TABLE 1

ACTUAL AND EFFECTIVE NUMBERS OF ALLELES AT THE PGI LOCUS IN MARINE MOLLUSCS

Species	No of Individuals Analyzed	Actual No of Alleles	Effective No of Alleles ($1/\Sigma p^2$)	Reference
Littorina saxatilis	466	4	1.9	Snyder & Gooch 1973
Nucula annulata	71	2	1.2	Levinton 1973
Nuculana pontonia	13	3	1.2[A]	Gooch & Schopf 1973
Malletia sp.	8	2	1.1[A]	Gooch & Schopf 1973
Mytilus edulis	70	7	3.9	Levinton 1973
Modiolus demissus	62	6	2.6	Levinton 1973
Pecten maximus	356	9	5.6	This study
Chlamys varia	194	5	2.8	This study
Chlamys opercularis	115	4	2.1	This study
Ostrea edulis	316	2	1.1	Wilkins & Mathers 1973
Crassostrea gigas	48	5	1.4	Mathers et al 1974
Crassostrea angulata	50	4	1.2	Mathers et al 1974
Tapes decussata	89	5	3.2	Wilkins & Mathers 1974
Venus verrucosa	19	5	4.6	This study
Mercenaria mercenaria	55	6	2.5	Levinton 1973
Macoma balthica	85	3	2.1	Levinton 1973
Ensis ensis	46	3	2.6	Wilkins & Mathers 1974
Mya arenaria	58	3	1.7	Levinton 1973

A = Minimum value

TABLE 2

FREQUENCIES OF THE NINE ALLELES (A - J) IN THREE POPULATIONS OF THE SCALLOP *PECTEN MAXIMUS*

Sample	No	A	B	C	D	E	F	G	(H&J)	Effective No of Alleles
[1]GALWAY	69	.022	.130	.239	.254	.232	.036	.058	.028	5.06
KERRY	196	.051	.140	.186	.286	.161	.105	.054	.018	5.59
DUBLIN	91	.039	.132	.269	.203	.187	.093	.055	.022	5.57
TOTAL	356	.042	.136	.218	.258	.181	.089	.055	.021	5.60

1 From Wilkins & Mathers 1974.

TABLE 3
ACTUAL AND EFFECTIVE NUMBERS OF ALLELES AT THE PGI LOCUS
IN SOME VERTEBRATE SPECIES

Species	No of Individuals Analyzed	Actual No of Alleles	Effective No of Alleles	Reference
FISH				
Anguilla rostrata	731	3[C]	1.5	Williams et al 1973
Zoarces vivipara	888	3[D], 3.	1,1	Yndgaard 1972
Salmo salar	50	1[D], 1.	1,1	Burnell & Wilkins unpublished
MAMMALS				
Mus musculus	30[A]	2	1.1	Selander et al 1969
Hystricomorph rodents (7 species)	11-92	1-3	1-1.2	Carter et al 1972
Rabbit		2		Welch, Fitch, and Parr 1970
Sheep		2		Welch, Fitch, and Parr 1970
Pig	200	2	1.6	Tariverdian 1970
Man[B]	3397	10	1.01	Detter et al 1968

A = Sub-species *M.m. domesticus*

B = erythrocytes

C = Two other monomorphic, non-interacting loci present.

D = PGI is encoded at 2 interacting loci in these species.

of alleles = 1.01). Sufficient data are not available to indicate the effective number of alleles in rabbit and sheep, but it cannot exceed two in either case. In none of the vertebrate species does the effective number of alleles exceed 1.65, whereas 66% of the molluscs exceed this value.

The data for the three species of fish are interesting: in two of these (*Zoarces vivipara* and *Salmo salar*) the PGI locus is duplicated, the functional isozymes being formed by the assortment of the products of the duplicate loci. In the eel *Anguilla rostrata* on the other hand, three separate non-interacting loci appear to be operative.

The molluscan species listed in Table 1 occupy different depth zones in the marine environment, varying from the high intertidal to the profundal region. Figure 1 indicates broadly their zonation. Eight of the species occur either exclusively or predominantly in the intertidal zone, above the low water spring tide level. The remaining ten species are typically found only in the sub-littoral zone (below the low water spring tide level), and two of them (*Nuculana*; *Malletia*) are extremely deep water species. In general, actual and effective numbers of alleles are higher in organisms living high on the shore than in sub-littoral species; the mean values tend to decrease from the high intertidal species through the low intertidal to the sub-littoral (Fig. 1). The unweighted mean number of alleles observed in the total intertidal group is 4.88; the mean effective number of alleles is 2.81. In the sub-littoral group the corresponding values are 3.90 and 2.03. While the values for the sub-littoral group are lower than those for the intertidal, the difference between them is not statistically significant. If, however, the data for *Pecten maximus* and *Chlamys opercularis* are omitted from the former group (see below), the values are 3.25 and 1.58. These values do differ significantly from those of the intertidal group ($t_{16^o} = 2.368$, $P < 0.05$ for actual alleles; $t_{16^o} = 2.831$, $P < 0.02$ for effective alleles).

In their natural habitat, whether intertidal or sub-littoral, the post-settlement (juvenile and adult) individuals take up *in situ* postures which are characteristic of each species. In some, the individuals live on, or attached to, the substrate (epifaunal species), whereas other species burrow into the substrate (infaunal species). In both cases individuals spend the whole of their post-settlement life in a relatively restricted locality which is "selected" irreversibly at settlement. Members of the Pectinidae, (*Pecten*, *Chlamys*), however, have developed considerable power of swimming off the bottom (Morton, 1964) and are known to swim freely from area to area. Locality selection is therefore

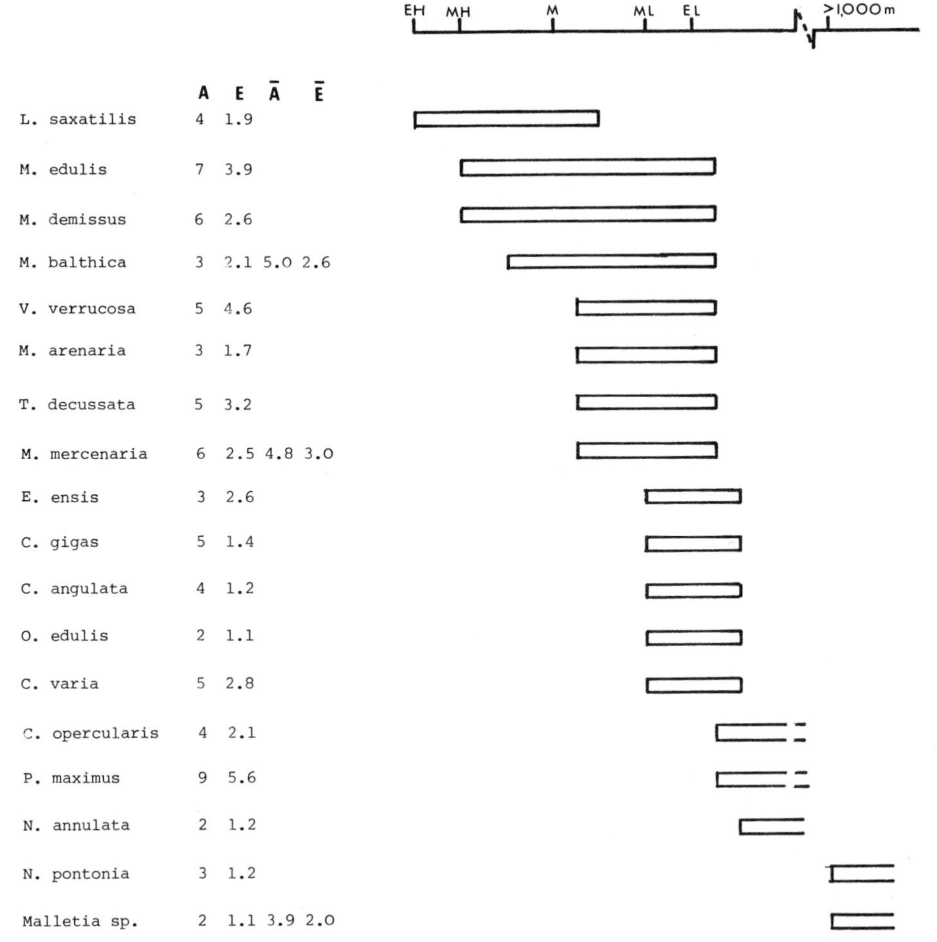

Fig. 1. Zonation of molluscs on the shore and sub-littorally, together with the actual and effective numbers of PGI alleles in each species. A = Actual number of alleles; E = Effective number of alleles; EH = Extreme high water spring tide level; MH = Mean high water spring tide level; M = Mid-shore; ML = Mean low water spring tide level; EL = Extreme low water spring tide level.

---

a reversible ongoing process in these species. It is reasonable to suppose that these swimming-epifaunal species experience a wider degree of environmental heterogeneity than the relatively non-mobile infaunal and normal epifaunal species. In Table 4 the species are grouped according to their level on the shore (intertidal or sub-littoral) and their characteristic *in situ* posture. Among the non-swimming species, there is a steady reduction in the mean actual and effective numbers of alleles in the series intertidal epifaunal, intertidal

TABLE 4
ACTUAL AND EFFECTIVE NUMBERS OF ALLELES
AT THE PGI LOCUS IN MARINE MOLLUSCS[a]

	IN SITU POSTURE OF ADULTS					
	EPIFAUNAL		INFAUNAL		SWIM/EPIFAUNAL	
	A[b]	E[c]	A[b]	E[c]	A[b]	E[c]
INTERTIDAL						
L. saxatilis	4	1.9				
M. edulis	7	3.9				
M. demissus	6	2.6				
MEAN	5.67	2.80				
T. decussata			5	3.2		
V. verrucosa			5	4.6		
M. mercenaria			6	2.5		
M. balthica			3	2.1		
M. arenaria			3	1.7		
MEAN			4.40	2.82		
SUBLITTORAL						
O. edulis	2	1.1				
C. angulata	4	1.2				
C. gigas	5	1.4				
MEAN	3.67	1.23				
E. ensis			3	2.6		
N. annulata			2	1.2		
N. pontonia			3	1.2		
Malletia sp.			2	1.1		
MEAN			2.50	1.52		
C. varia					5	2.8
C. opercularis					4	2.1
P. maximus					9	5.6
MEAN					6.00	3.50

[a]Species are grouped according to their zone on the shore (intertidal or sub-littoral) and their characteristic *in situ* posture.

[b]Actual number of alleles.

[c]Effective number of alleles.

infaunal, sub-littoral epifaunal, sub-littoral infaunal. The difference between the extremes of this series (intertidal epifaunal and sub-littoral infaunal) is statistically significant ($t_{50} = 3.895$, $P < 0.02$). The differences between the other groups, when any two of them are compared, are not statistically significant in "student's t" tests, although the difference in effective allele number between the epifaunal groups of the intertidal and sub-littoral zones approaches statistical significance ($t_{40} = 2.738$, $0.06 > P > 0.05$). The difference between the total intertidal and the non-swimming sub-littoral species is, as previously stated, statistically significant, whereas the combined infaunal species do not differ significantly from the combined epifaunal. The swimming-epifaunal group differs significantly from the other sub-littoral groups, but not from the intertidal.

DISCUSSION

As pointed out by Selander and Kaufman (1973) a major difference exists in levels of genetic variability between vertebrates and invertebrates. The proportion of polymorphic loci ranges from 10 to 20% in local populations of vertebrates, and at more than half of the polymorphic loci only two common alleles are segregating. In contrast, three common alleles, on average, segregate per polymorphic locus in invertebrate species, and 20 to 50% of their loci are polymorphic. In so far as these findings cannot be related to differences in dispersal ability or total species size, they lend support to Levins' theory (1968) of adaptive strategies and they suggest that isozyme variability is maintained in wild populations by natural selection (Selander and Kaufman, 1973).

Considering the PGI locus in molluscs on its own, it is clear from a comparison of Tables 1 and 3 that this locus typifies the situation reported by Selander and Kaufman (1973). The mean number of alleles observed at this locus in molluscs is 4.3 and the effective number is 2.4. In the vertebrate species listed, the corresponding values are 2.3 and 1.2. That such a high level of variability is maintained at this particular locus in molluscs is unexpected in the light of patterns of enzyme variability emerging in other species. Studies on suites of enzymes in other species indicate that genes coding for enzymes involved in critical metabolic reactions (Group I), in particular those concerned with glucose metabolism, are generally less variable than genes coding for enzymes (Group II) catalyzing more peripheral reactions (Gillespie and Kojima, 1968; Selander and Yang, 1969; Kojima et al, 1970; Ayala and Powell, 1972; Cohen et al, 1973; Harris et al, 1974). No firm generalizations can be drawn for molluscs from the variability observed in the single glycolytic enzyme discussed here. Nevertheless, eight of the eighteen molluscan species (~45%) have five or more alleles encoding this Group I enzyme whereas not one of the twelve Group I enzymes in three species of *Drosophila* has five alleles (Ayala and Powell, 1972) and only two of sixteen (~13%) of Group II enzymes have a maximum of five alleles. Moreover, the variability observed at the PGI locus in these molluscan species is as great or greater than that observed at their own Group II loci viz. LAP, TO and, Est. (Koehn and Mitton, 1972; Berger, 1973; Koehn et al, 1973; Levinton, 1973; Mathers et al, 1974) and at least one other Group I enzyme locus (PGM^1) is highly polymorphic in some of these species (unpublished results).

If this high level of variability at the PGI locus in molluscs is maintained by natural selection, as might be inferred from Selander and Kaufman (1973), then it is worthwhile to inquire whether differences in levels of genetic variability at this locus are evident between individual molluscan species occupying ecologically differing habitats.

The marine environment (from extreme high water, spring tides to profundal depths) is extremely heterogeneous. Environmental conditions are highly variable tidally, diurnally and seasonally in the intertidal zone, and highly invariable in the deep sub-littoral zone. A sharp discontinuity occurs at the level of low water spring tides. Organisms living higher on the shore experience large-scale variations in physical environmental conditions such as temperature, salinity and humidity which to a great extent reflect terrestrial climatic and seasonal conditions complicated by the twice-daily tidal submersion. Organisms living below the low water spring tide level are only occasionally, if ever, exposed to ambient climatic conditions (and then only for brief periods), and the sea-water medium severely dampens seasonal and other fluctuations in physical conditions. In both zones, organisms which burrow into the substrate may also experience a lesser degree of environmental variability than epifaunal organisms. Burrowing can be interpreted as an ethological homeostatic mechanism.

The actual and effective numbers of PGI alleles are less in sub-littoral molluscs than in intertidal. This difference is statistically significant when *Pecten maximus* and *Chlamys opercularis* are excluded from the normal sub-littoral group. The exclusion of these two species is justified on the basis of their well developed swimming ability and migratory behavior. *Chlamys varia* has not been excluded from the normal sub-littoral group, although it, too, can and does swim. In the samples analyzed, however, all the *C. varia* individuals were byssally attached to the substrate, whereas those of *C. opercularis* and *P. maximus* were not. For this reason *C. varia* was treated as a normal sub-littoral species when these were compared with the intertidal. Their exclusion from the normal sub-littoral group of species would, in any case, tend to increase rather than decrease the difference observed between the groups ($t_{130} = 3.282$, $P < 0.01$ for effective alleles).

The actual and effective numbers of alleles are less in infaunal than in epifaunal species. This agrees with the earlier conclusion of Levinton (1973). While this reduction in variability is not, on its own, statistically significant, taken in conjunction with the increased variability observed in swimming species it indicates the importance of *in situ* posture in comparing species groups: swimming species are more variable than epifaunal species, and these in turn are more variable than infaunal.

The results suggest, therefore, that the extent of genetic variation at the PGI locus in these species is influenced both by habitat zone (intertidal or sub-littoral) and *in situ* posture, and the effects of these are additive : variability is greatest in intertidal epifaunal species and least in sub-littoral infaunal species. Since these species groups experience, respectively, the maximum and minimum degree of relative environmental variability, the results suggest that increased genetic variability at this locus has adaptive value in species experiencing greater environmental heterogeneity.

ACKNOWLEDGMENT

This work was supported by a grant from the National Science Council of Ireland.

REFERENCES

Ayala, F. J. and J. R. Powell 1972. Enzyme variability in the *Drosophila willistoni* group. VI. Levels of polymorphism and the physiological function of enzymes. *Biochem. Genet.* 7: 331-345.

Berger, E. M. 1973. Gene-enzyme variation in three sympatric species of *Littorina*. *Biol. Bull.* 145: 83-90.

Carter, N. D., M. R. Hill, and B. J. Weir 1972. Genetic variation of phosphoglucose isomerase in some hystricomorph rodents. *Biochem. Genet.* 6: 147-156.

Cohen, T. P. W., G. S. Omenn, A. G. Motulsky, S.-H. Chen, and E. R. Giblett 1973. Restricted variation in the glycolytic enzymes of human brain and erythrocytes. *Nature New Biol.* 241: 229-233.

Crow, J. F. and M. Kimura 1970. *An introduction to population genetics theory.* Harper and Row, New York, 591.

Detter, J. C., P. O. Ways, E. R. Giblett, M. A. Baughan, D. A. Hopkinson, S. Povey, and H. Harris 1968. Inherited variations in human phosphohexose isomerase. *Ann. Hum. Genet. Lond.* 31: 329-338.

Gillespie, J. H. and K. Kojima 1968. The degree of polymorphism in enzymes involved in energy production compared to that in non-specific enzymes in two *Drosophila ananassae* populations. *Proc. Natl. Acad. Sci.* U.S.A., 61: 582-585.

Gooch, J. L. and T. J. M. Schopf 1973. Genetic variability in the deep sea: relation to environmental variability. *Evolution* 26: 545-552.

Harris, H., D. A. Hopkinson, and E. R. Robson 1974. The incidence of rare alleles determining electrophoretic variants: data on 43 enzyme loci in man. *Ann. Hum. Genet. Lond.* 37: 237-253.

Koehn, R. K. and J. B. Mitton 1972. Population genetics of marine pelecypods. I. Ecological heterogeneity and evolutionary strategy at an enzyme locus. *Amer. Nat.* 106: 47-56.

Koehn, R. K., F. J. Turano, and J. B. Mitton 1973. Population genetics of marine pelecypods. II. Genetic differences in microhabitats of *Modiolus demissus. Evolution* 27: 100-105.

Kojima, K., J. H. Gillespie, and Y. N. Tobari 1970. A profile of *Drosophila* species' enzymes assayed by electrophoresis. I. Number of alleles, heterozygosities and linkage disequilibrium in glucose-metabolising systems and some other enzymes. *Biochem. Genet.* 4: 627-638.

Levins, R. 1968. *Evolution in changing environments.* Princeton University Press, Princeton, N. J., 120.

Levinton, J. 1973. Genetic variation in a gradient of environmental variability. *Science* 180: 75-76.

Mathers, N. F., N. P. Wilkins, and P. R. Walne 1974. Phosphoglucose isomerase and esterase phenotypes in *Crassostrea angulata* and *Crassostrea gigas (Bivalvia: Ostreidae). Biochemical Systematics and Ecology* 2: 93-96.

Milkman, R. and L. D. Beaty 1970. Large scale electrophoretic studies of allelic variation in *Mytilus edulis. Biol. Bull.* 139: 430.

Morton, J. E. 1964. Locomotion. in *Physiology of Mollusca,* edited by K. M. Wilbur and C. M. Yonge. Academic Press, London, 473.

Selander, R. K., W. G. Hunt, and S. Y. Yang 1969. Protein

polymorphism and genic heterozygosity in two European sub-species of the house mouse. *Evolution* 23: 379-390.

Selander, R. K. and S. Y. Yang 1969. Protein polymorphism and genic heterozygosity in a wild population of the house mouse (*Mus musculus*). *Genet.* 63: 653-667.

Selander, R. K. and D. W. Kaufman 1973. Genic variability and strategies of adaptation in animals. *Proc. Nat. Acad. Sci.* U. S. A. 70: 1875-1877.

Snyder, T. P. and J. L. Gooch 1973. Genetic differentiation in *Littorina saxatilis* (*Gastropoda*). *Mar. Biol.* 22: 177-182.

Tariverdian, G. 1970. Zur populationsgenetik der Phosphohexoseisomerasen (E.C.5.3.1.9) beim Schwein. *Humangenetik* 9: 110-112.

Welch, S. G., L. I. Fitch, and C. W. Parr 1970. A variant of rabbit phosphoglucose isomerase. *Biochem. J.* 117: 525-531.

Wilkins, N. P. and N. F. Mathers 1973. Enzyme polymorphisms in the European oyster, *Ostrea edulis* L. *Anim. Blood Grps. Biochem. Genet.* 4: 41-47.

Wilkins, N. P. and N. F. Mathers 1974. Phenotypes of phosphoglucose isomerase in some bivalve molluscs. *Comp. Biochem. Physiol.*, 48: 599-612.

Williams, G. C., R. K. Koehn, and J. B. Mitton 1973. Genetic differentiation without isolation in the American eel *Anguilla rostrata*. *Evolution* 27: 192-204.

Yndgaard, C. F. 1972. Genetically determined electrophoretic variants of phosphoglucose isomerase and 6-phosphogluconate dehydrogenase in *Zoarces viviparus L.* *Heredites* 71: 151-154.

MIGRATION AND POPULATION STRUCTURE IN THE PELAGICALLY DISPERSING MARINE INVERTEBRATE, *MYTILUS EDULIS*

RICHARD K. KOEHN
Department of Ecology and Evolution
State University of New York
Stony Brook, New York 11790

ABSTRACT. *Mytilus edulis* is a widely distributed sedentary intertidal marine invertebrate with a pelagically dispersing larval stage of from 21 to 55 days. Allozyme (allelic isozyme) variation at three poly-allelic loci was analyzed in order to discover the relative contribution of interpopulation migration and/or natural selection in the maintenance of patterns of spatial variation. Patterns of geographic variation at the three loci were discordant and no evidence for isolation by distance could be demonstrated. Allele frequencies at one locus were uniform over large portions of the species range while abrupt and gradual transitions of frequencies were observed at other loci over great, moderate, and very small distances. A net heterozygote deficiency was observed in many samples with largest deficiencies being associated with populational samples contiguous between large areas significantly different in the genetic composition of their populations. Correlations were tested between the magnitudes of spatial variation in allele frequency and heterozygote deficiency and found to be nonsignificant. Additionally, loci exhibiting significant geographic variation in allele frequency did not exhibit greater heterozygote deficiency than a locus at which no significant geographic variation occurred. Individuals in some areas possessing an allele at one locus common to distant geographic regions, did not exhibit frequencies at other loci typical of those regions. The absence of data suggesting a significant role of interpopulation migration, in addition to the discordant among-loci patterns of macrogeographic variation, suggests a minor role for migration in all local populations except those adjacent to populations of large differences in genetic composition. These data demonstrate that natural selection is the principal force affecting patterns of spatial variation and that selection is qualitatively and quantitatively different for each of the studied loci.

INTRODUCTION

This study is an attempt to characterize the relative

contributions of migration and selection to spatial genetic variation in an intertidal marine invertebrate with a typical life cycle of extended larval dispersal and sedentary adulthood. Natural selection and interpopulational migration are important in all natural populations for the maintenance of intrapopulation variation, but their relative roles are difficult to estimate and therefore poorly understood. Migration has received much theoretical consideration of its effects on gene frequency variance in structured populations (Wright, 1951; Kimura and Weiss, 1964; Holgate, 1964; Deakin, 1966; Gadgil, 1971; Rohlf and Schnell, 1971; Endler, 1973) but empirical investigations have failed to demonstrate that the absence of migration is important to differentiation among populations (Kettlewell, 1958; Ehrlich and Raven, 1969; Bradshaw, 1971; Endler, 1973; Williams et al., 1973). The relative effects of migration and selection upon spatial patterns of variation are difficult to decipher, since they may operate in either complementary or antagonistic ways. Geographically uniform frequencies may be due to substantial area-independent stabilizing selection or to high effective migration rates in combination with negligible area-specific directional selection. In contrast, geographic heterogeneity of allele frequencies may be due to either low interpopulational migration rates in combination with high or low area-specific selection or to high effective interpopulational migration rates in combination with high area-specific selection. Hence any observed pattern of interpopulation variation has a variety of explanations encompassing various magnitudes of migration and selection. In populations of marine invertebrates, some authors have attempted to unravel the joint effects of migration and selection by comparing spatial heterogeneity at two or three loci between two or three species differing in dispersal characteristics (see, for example, Berger, 1973; Snyder and Gooch, 1973). Although these authors observed highest indices of genetic similarity (Hedrick, 1971) among population samples of pelagically dispersed species, geographical homogeneity need not be due to large effective migration rates. Whereas the influence of migration might be expected to diminish with increased distance, both magnitudes and modes of selection are also expected to change with distance. Thus, rank order correlations among species, between the degree of dispersal and genetic spatial heterogeneity, are not conclusive evidence that migration significantly retards differentiation, particularly when species are compared at different loci and from different selective regimes.

The results presented here will demonstrate that significant spatial differentiation can occur in a species of extended larval dispersal and that different loci may exhibit discordant patterns of differentiation. In *Mytilus edulis*, interpopulational migration is detectable only in areas that are intermediate to populations of large genetic differences, and no correlations between genetic difference and distance can be demonstrated.

The extent to which migration and natural selection are the the principal determinants of spatial patterns of genetic variation is closely related to the life cycle of the organism being studied. As such, a brief outline of the life cycle of *Mytilus edulis* is provided below.

LIFE CYCLE OF MYTILUS EDULIS

Mytilus edulis is normally a dioecious species. Postlarval individuals are sedentary, though a well-developed foot is present and movement of small adults occurs (Field, 1922; Milkman, unpublished). Adults live to about six years in eastern North America. Spawning occurs from late spring to early fall (Field, 1922; Engle and Lossanoff, 1944; Chipperfield, 1953; Lubinsky, 1958). Reproductive females are extraordinarily fecund, liberating about 12 million eggs in a single spawn and perhaps 25 million in a spawning season (Field, 1922). Gametes are liberated into the surrounding water mass where fertilization takes place (Chipperfield, 1953). A variety of larval stages occur, each differing in response to light and gravity (Bayne, 1964a). The most prolonged stage is the veliconcha, the feeding larva, lasting for a minimum period of twenty-two days. In the absence of suitable settling sites, veliconchas may delay metamorphosis for about 55 days from fertilization (Bayne, 1965). Larvae settle on a number of substrates (Engle and Loosanoff, 1944; Chipperfield, 1953), but Bayne (1964b) described primary settlement upon filamentous algae followed by a secondary migratory phase which terminated by settlement in adult mussel beds. The extended larval dispersal state suggests an important role for interpopulational migration while sedentary adulthood should make the impact of differential selection unavoidable.

METHODS AND MATERIALS

Detailed methods for collection, preservation, and preparation of samples have been described (Koehn et al., in prep.). These authors have additionally described the electrophoretic conditions for optimal resolution of several loci, including the three discussed here. Leucine aminopeptidase (L-Leucyl-Peptide-Hydrolase; E.C.3.4.1.1; LAP), aminopeptidase (Amino-

Acyl-Dipeptide-Hydrolase; E.C.3.4.1.3.; AP), and glucose-phosphate isomerase (D-Glucose-6-Phosphate Keto-isomerase; E.C. 5.3.1.9; GPI) were resolved by horizontal starch gel electrophoresis with the discontinuous pH 8.4 lithium hydroxide buffer of Selander et al. (1969).

RESULTS

Approximately 130 samples of *Mytilus edulis* were taken throughout the geographic range on the east coast of North America ranging from Virginia, their approximate southern limit of distribution, to Newfoundland. Detailed descriptions of geographic locations of these samples, as well as their genetic composition are given by Koehn et al (in prep.), though complete data will be furnished upon request. All alleles at all loci are designated according to their relative electrophoretic mobility. These designations differ from those used previously to describe variation at the LAP locus (Milkman and Beaty, 1970; Koehn and Mitton, 1972; Mitton et al., 1973). Modifications in nomenclature are given by Koehn et al. (in prep)).

Leucine aminopeptidase. (LAP).-- As has been previously noted (Milkman and Beaty, 1970; Milkman, 1971), three common electrophoretic variants were observed (Fig. 1) in all samples and consist of LAP^{98}, LAP^{96}, and LAP^{94}. Two additional rare alleles were observed, LAP^{100} and LAP^{92}, at frequencies less than .050. For statistical reasons, rare alleles were pooled within samples with common alleles of most similar electrophoretic mobility. Because of the different pattern of geographic variation, samples from Long Island Sound are discussed separately. The LAP^{94} allele ranged from 0.50 to 0.60, with an average value of 0.56, in all samples from Virginia to Cape Cod (Fig. 2). In samples north of Cape Cod, the frequency of this allele abruptly decreased to approximately 0.10 and remained low throughout the Gulf of Maine. Intermediate frequencies were observed in the Cape Cod Canal. In samples from Nova Scotia and Newfoundland the frequency increased to approximately 0.35. Geographic variation in LAP^{98} and LAP^{96} allele frequencies was approximately the complement of the LAP^{94} allele from Virginia to the northern Gulf of Maine. However, in Nova Scotia, the LAP^{94} allele was more common and the LAP^{98} less common than samples from the Gulf of Maine. The frequency of LAP^{96} was also greater in Nova Scotia when compared to the Gulf of Maine (Fig. 2).

There were significant differences in the frequencies of LAP alleles between Atlantic Coast samples and those within Long Island Sound. Since LAP^{98} and LAP^{96} positively co-vary, this difference is most conspicuous for the LAP^{94} allele. The

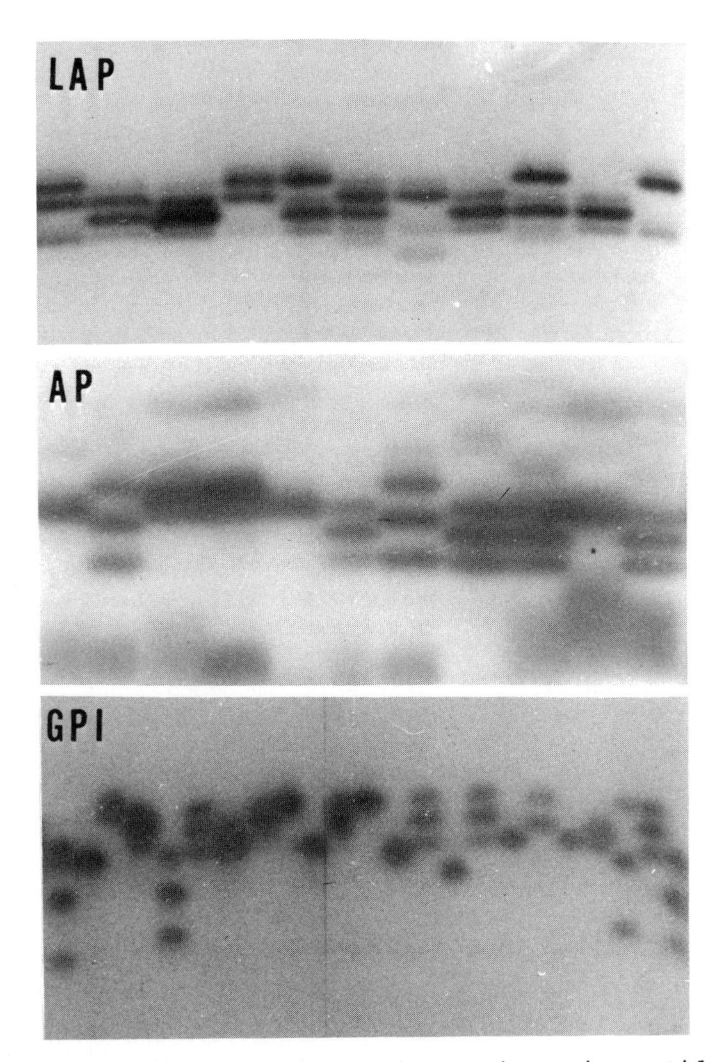

Fig. 1. Polymorphic variation at the Leucine Aminopeptidase (LAP), Aminopeptidase (AP), and Glucose Phosphate Isomerase (GPI) loci. Genotypes, from left to right: (LAP) 98/96, 96/94, 94/94, 98/96, 98/94, 96/94, 96/96, 94/94, 98/94, 94/94, 98/98; (AP) 92/92, 97/82, 97/92, 97/92, 92/92, 92/82, 97/82, 92/82, 92/82, 92/92, 92/82; (GPI) 95/72, 93/93, 105/100, 100/95, 93/82, 100/93, 95/93, 100/95, 100/100, 93/93, 100/95, 100/100, 93/89, 100/93, 89/89, 100/93, 93/93, 100/-, 93/93, 95/93, 100/82, 100/93, 93/82.

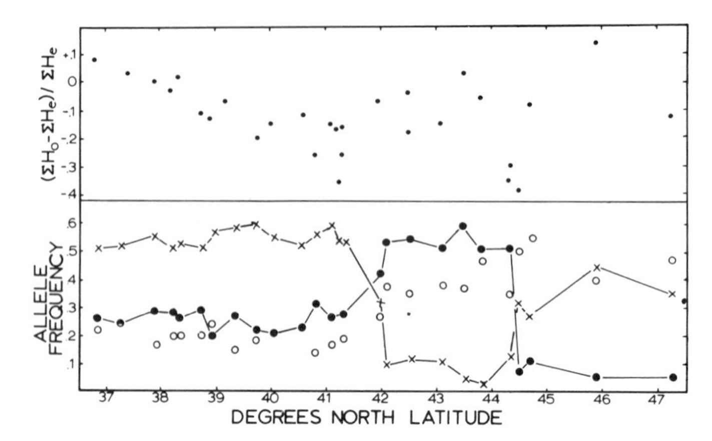

Fig. 2. Geographic variation of the three most common LAP alleles: LAP^{98} (●), LAP^{96} (o), and LAP^{94} (x).

frequency of the LAP^{94} at New London, Connecticut was 0.48, statistically like samples from outside Long Island Sound. This allele steadily decreased in frequency with progressive sampling away from the entrance into Long Island Sound. A minimum frequency of 0.11 occurred at Glen Island, New York. No change in allele frequency was observed on the Long Island shore, but samples were statistically homogeneous and ranged between 0.14 at the western end of Long Island Sound to 0.20 at Inlet Point, the most eastern locality on Long Island.

Many samples deviated significantly from Hardy-Weinberg expectations and in each case resulted from a deficiency of heterozygotes. The life cycle as presently understood would suggest much interlocality mixing of zygotes each generation and individuals at a particular point could have originated from a source population of considerable geographic distance. In order to investigate the possibility of interpopulation mising, geographic patterns and magnitudes of heterozygote deficiency were estimated. $\underline{D_t}$ was computed as a $\Sigma H_o - \Sigma H_e / \Sigma H_e$, where ΣH_o is the total number of heterozygotes observed and ΣH_e the total expected by the expanded Hardy-Weinberg binomial. For simplicity in illustration, values of $\underline{D_t}$ for all samples except those of Long Island Sound and the Cape Cod area are given in Figure 2. The most southern samples exhibited a small but statistically non-significant excess of heterozygotes. The relative frequency of observed heterozygotes declined in more northern samples. A maximum deficiency (significantly different from zero) was observed in two regions, near the entrance to Long Island Sound and in northern Gulf of Maine and southern

Nova Scotia. These two areas of dramatic (35%) deficiency of heterozygotes are spatially associated with regions of significant variation of allele frequency (Fig. 2).

Aminopeptidase (AP).-- A total of seven different alleles were observed at the AP locus (Fig. 3). However, four of these

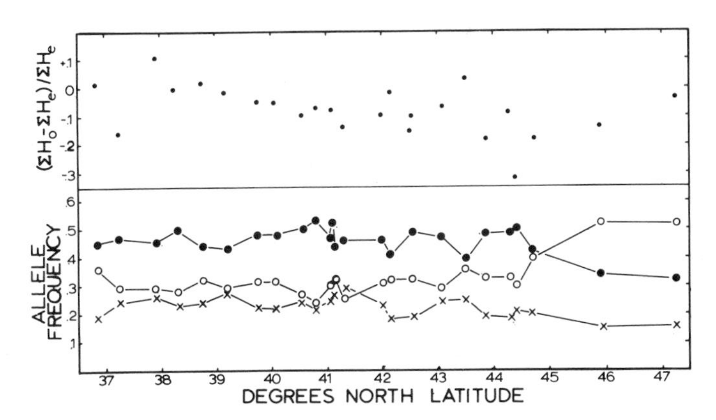

Fig. 3. Geographic variation of the three most common AP alleles: GPI^{97} (●), GPI^{92} (o), and GPI^{82} (x).

(AP^{105}, AP^{100}, AP^{72} and AP^{87}) were extremely rare and therefore pooled with alleles of more similar electrophoretic mobility.

In contrast to variation of the LAP locus, frequencies of AP alleles were relatively constant from Virginia to southern Nova Scotia (Fig. 3). All samples from Long Island Sound exhibited frequencies within the range of variation observed in coastal samples. At 44.5° latitude (Fig. 3), the AP^{92} allele increased significantly from 0.32 to 0.52 while the AP^{97} allele concomitantly decreased. A statistically significant deficiency of heteroyzgotes (D_t) was associated with the change in AP frequency in this region.

Glucose phosphate isomerase (GPI).-- A total of nine different electrophoretic alleles was observed at the GPI locus (Fig. 1) and up to seven within a single population sample. Three alleles were common (GPI^{100}, GPI^{93}, and GPI^{89}), while other alleles were observed only rarely in various areas of the species range (GPI^{107}, GPI^{105}, GPI^{95}, GPI^{84}, GPI^{82}, GPI^{72}). All samples corresponded with Hardy-Weinberg expectations and therefore D_t values are not given.

There was significant geographic variation in the frequency of the three most common GPI alleles (Fig. 4). Estimated

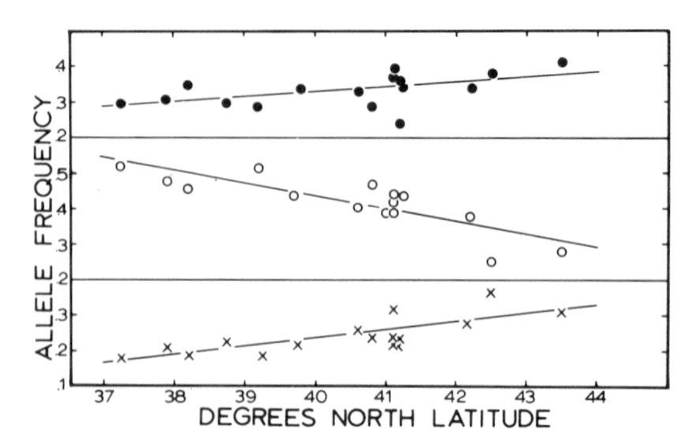

Fig. 4. Geographic variation of the three most common GPI alleles: GPI^{100} (●), GPI^{93} (o), and GPI^{89} (x).

frequencies were linearly dependent upon the latitude from which samples were taken. The GPI^{100} allele varied from 0.30 to 0.40 between latitudes 37° and 45°, respectively (β = .0125 ± .0058; $F_{1,14}$ = 4.690; P<.05). Likewise, the frequencies of the GPI^{89} and GPI^{93} alleles varied clinally with latitude (β = .027 ± .0052; $F_{1,14}$ = 19.017; P<.001 and β = -.0353 ± .0063; $F_{1,14}$ = 30.888; P<.001; respectively). Samples from Long Island Sound were statistically homogeneous at frequencies expected by the regressions on latitude of coastal samples.

For migration to be detectable, effective migration rates must be of great enough magnitude to counterbalance the differentiating effects of natural selection. An extended dispersal stage, significant geographic variation in the frequency of LAP alleles and the frequent deficiency of heterozygotes at this locus, all suggest that the last may be generated in local samples by the Wahlund Effect -- in this case, the settling at a locality of individuals from parental populations of significantly different genetic compositions. The geographic origins of individuals in local samples is not known. Such information would provide a thorough understanding of the role of migration. The influence of migration upon patterns of allele frequence variation was examined by testing two predictions: (1) significant interpopulation migration should generate a correlation between magnitudes of spatial variance of allele frequency and heterozygote deficiency, and (2) loci at which alleles exhibit significant geographic variation should, on the average, exhibit greater heterozygote deficiencies in samples than loci at which allele frequencies are geographically homogeneous.

Correlation will be observed between spatial variance and heterozygote deficiency only when samples included in a variance computation proportionally represent the parental populations (i.e., appropriate variance) that have contributed immigrants to an individual sample. Since this cannot be done, individual localities were grouped so as to represent a range from the minimum to the maximum possible variance, thus avoiding the arbitrary nature of a single correlation computation. In other words, depending on the true effective migration patterns, individuals at a locality could have minimally originated at that point or maximally from any place within the range of the species distribution in North America. The minimum spatial variance would thus be represented by replicated samples taken from the immediate vicinity while the maximum possible variance contribution of immigrants to a local sample would be the maximum observed variance among all samples along the North American continent. A series of variance computations were made where samples were grouped so as to represent progressively larger geographic areas (Table 1). Within each group, the weighted among-sample variance was computed for arc-sin transformed allele frequencies. Additionally, the average deficiency of each heterozygote class as well as the net heterozygote deficiency ($\underline{D_t}$) were computed (Table 1). Among-group heterogeneity of within group variance was tested by Bartlett's test of homoscadasticity and groups were significantly different for all alleles ($\underline{LAP^{98}}$, $\chi^2{}_{(8)} = 37.05$, $P<<.001$; $\underline{LAP^{96}}$. $\chi^2{}_{(8)} = 26.32$, $P<.003$; $\underline{LAP^{94}}$, $\chi^2{}_{(8)} = 79.94$, $P<<.001$). Among-group average heterozygote deficiency was also significantly heterogeneous ($F_{9,120} = 2.321$, $P<.03$). Product-moment correlation coefficients (Sokal and Rohlf, 1969) were determined between the among-group variance estimates for any two alleles and the average deficiency exhibited by the heterozygote class involving those two alleles. In addition, all possible among-group pairwise comparisons were made of allele frequency variance with the average heterozygote deficiency. For example, the among-group variation of the arc-sin transformed variances of $\underline{LAP^{98}}$ and $\underline{LAP^{96}}$ were compared for the among-group variation of $\overline{D}_{98,96}$, but no significant correlation was found ($\underline{r} = -.191$, df = 18, $P>.80$). All other correlation tests gave non-significant coefficients with the exception of among-group $\underline{D_t}$ and variance of $\underline{LAP^{98}}$ and $\underline{LAP^{96}}$ ($\underline{r} = -.536$, df = 16, $P<.03$).

Migration would generate larger deficiencies of heterozygotes at spatially heterogeneous loci than spatially homogeneous loci. This is more easily tested than the foregoing analysis, since an arbitrary group of samples is equally representative of migration for each locus. Four areas were used in the among-loci variation of among sample variance and

TABLE 1. A comparison of the within geographic area arc-sin transformed allele frequency variance and average deficiency of heterozygotes at the LAP locus. See text for explanation.

Area	Allele Variance			Heterozygote Deficiency				
	LAP^{98}	LAP^{96}	LAP^{94}	$\bar{D}_{98,96}$	$\bar{D}_{98,94}$	$\bar{D}_{96,94}$	$\bar{D}_{t}$	k*
Stony Brook Beach	1.32	1.23	4.03	-.129	-.122	-.135	-.132	9
Nissequogue (Subtidal)	4.59	8.28	5.90	-.125	-.263	-.293	-.177	10
Nissequogue (Beach)	5.35	7.25	5.42	-.095	-.057	-.087	-.077	21
Nissequogue (Estuary)	12.54	13.33	18.91	-.062	-.167	-.252	-.159	10
Northport	4.33	5.60	2.51	-.102	-.252	-.073	-.125	11
Connecticut Shore	23.41	27.81	99.02	-.199	-.273	-.143	-.215	6
Long Island Shore	1.30	1.50	2.59	-.101	-.177	-.190	-.133	6
Atlantic Coast								
North Cape Cod	27.37	17.14	94.76	-.092	-.338	-.153	-.152	16
South Cape Cod	8.87	10.45	21.86	-.152	-.148	-.036	-.097	35
Nfld. and Nova Scotia	9.10	11.37	11.76	-.292	-.442	-.197	-.192	4

*Number of Samples

heterozygote deficiency (Table 2): replicated samples at two localities, samples of the Connecticut shore of Long Island Sound, and coastal samples south of Cape Cod. No significant differences among loci were observed in either allele variance or $\overline{D}_t$ for Nissequogue estuary and subtidal samples. Among-loci variances were different among Atlantic Coast samples (Allele 1, $\chi^2_{(2)} = 8.16$, $P<.025$; Allele 2, $\chi^2_{(2)} = 6.65$, $P<.04$; Allele 3, $\chi^2_{(2)} = 22.91$; $P<.001$), but $\overline{D}_t$ values were homogeneous ($F_{2,63} = 2.46$, $P>.07$). Among loci variances and average heterozygote deficiencies were both heterogeneous for the Connecticut samples, but this heterogeneity was due to the LAP locus alone (Table 2).

DISCUSSION

Various analyses failed to demonstrate a significant role for migration in the observed patterns of spatial variation. The magnitudes of heterozygote deficiency are not correlated with regional among-sample allele frequency variance, and loci which exhibit significant geographic variation in allele frequency (large within-area variance) do not exhibit a greater average heterozygote deficiency than more geographically homogeneous loci. Additionally, the discordant patterns of geographic variation among the LAP, AP, and GPI loci (Figs. 2, 3, and 4) militate against a significant role for migration and suggest both different sources and magnitudes of natural selection affecting each of the three loci. It is clear from these patterns that intra-regional migration cannot explain large areas of homogeneity in genetic composition without postulating strong selection at other loci. For example, if the geographic homogeneity at the AP locus is due to large effective migration, variation at the LAP and GPI loci must be due to selective coefficients of a sufficient magnitude to totally obscure the effects of interregional migration. Although there is a geographical association between largest magnitudes of heterozygote deficiency and spatial variance in allele frequency at the LAP and AP loci (Figs. 2 and 3), this localization suggests only slight importance of migration relative to selection and only in populations that are intermediate between areas differing greatly in the genetic composition of their populations. The change in LAP allele frequencies along the Connecticut Shore in Long Island Sound is a dramatic demonstration of the differentiation that can occur among areas for which no barriers to migration can be postulated. Whatever the selective factor may be, this pattern of spatial variation is reminiscent of data typically seen for organisms of low dispersal ability for which no

TABLE 2. Comparisons among loci within four geographic areas of arc-sin transformed allele frequency variances and average heterozygote deficiencies at the LAP, AP and GPI loci. See text for explanation.

Area	Allele Variance 1	2	3	$\bar{D}_t$	k*
Nissequogue (Subtidal)					
LAP[1]	4.59	8.28	5.90	-.177±.0663	8
GPI[2]	.69	2.71	3.22	-.101±.0577	6
AP[3]	2.19	1.50	1.33	-.092±.0939	8
Nissequogue (Estuary)					
LAP	12.54	13.33	18.91	-.168±.1507	10
GPI	18.60	4.18	13.93	-.049±.0982	8
AP	6.43	7.92	6.94	-.049±.0813	10
Connecticut Shore					
LAP	23.41	27.81	99.02	-.207±.0527	7
GPI	.59	3.43	2.93	-.107±.0894	6
AP	2.32	5.46	2.65	-.080±.0527	6
Atlantic Coast-S. Cape Cod					
LAP	8.86	10.45	21.83	-.098±.1014	35
GPI	5.27	5.04	5.71	-.044±.0973	13
AP	2.25	3.37	2.16	-.050±.0652	10

[1] Alleles LAP^{98}, LAP^{96}, LAP^{94}; [2] Alleles GPI^{100}, GPI^{93}; GPI^{89};
[3] Alleles AP^{97}, AP^{92}, AP^{82}; * Number of samples.

significant interpopulational movement can be demonstrated (Koehn and Rasmussen, 1967; Selander et al., 1969; Merritt, 1972).

Although significant inter-regional movement of larvae must be occurring, there is little evidence that this movement is genetically important. It is quite obvious that any of the above loci taken singly, would give a totally different impression of the probable role of migration. Hence, rank order correlations among species, between the degree of dispersal and genetic spatial heterogeneity, are not conclusive evidence that migration is important. However, there is not yet available sufficient comparative information to know how the problem of understanding the effects of migration on genetic composition of natural marine invertebrate populations might be fruitfully approached. If species with great dispersal ability are typically observed to be more geographically homogeneous in their compositions than species with more restricted dispersal, it will be difficult to dismiss the role migration may play in affecting patterns of spatial variation. However, since interpopulational migration rates for individual loci are constant but selective coefficients can be significantly heterogeneous, it is not likely that even if a relatively clear among-species pattern is observed that comparative studies will provide quantitative information on migration.

ACKNOWLEDGEMENTS

Interpretations of these data were greatly influenced by discussions with Dr. Roger Milkman, who collected some of the data discussed here and with whom these results will be jointly described elsewhere in more detail. This study was supported by National Science Foundation Grant GB-25343 and USPHS Career Award GM-28963. This is contribution number 54 from the Program in Ecology and Evolution, State University of New York Stony Brook, New York 11790.

REFERENCES

Bayne, B. L. 1964a. The responses of the larvae of *Mytilus edulis* (L.) to light and to gravity. *Oikos*. 15: 162-174.

Bayne, B. L. 1964b. Primary and secondary settlement in *Mytilus edulis* L. (Mollusca). *J. Anim. Ecol.* 33: 513-523.

Bayne, B. L. 1965. Growth and the delay of metamorphosis of the larvae of *Mytilus edulis* (L.). *Ophelia*. 2: 1-47.

Berger, E. M. 1973. Gene-enzyme variation in three sympatric species of *Littorina*. *Biol. Bull.* 145: 83-90.

Bradshaw, A. D. 1971. Plant evolution in extreme environments. pp. 20-30, In: *Ecological Genetics and Evolution*. R. Creed (ed.), Blackwell Sci. Publ., Oxford. 391 pp.

Chipperfield, P. N. J. 1953. Observations on the breeding and settlement of *Mytilus edulis* (L.) in British waters. *J. Mar. Biol. Ass. U. K.* 32: 449-476.

Deakin, M. A. B. 1966. Sufficient conditions for genetic polymorphism. *Amer. Natur.* 100: 690-692.

Endler, J. A. 1973. Gene flow and population differentiation. *Science*. 179: 243-250.

Engle, J. B. and V. L. Loosanoff 1944. On season of attachment of larvae of *Mytilus edulis* Lin. *Ecology*. 25: 433-440.

Ehrlich, P. R. and P. H. Raven 1963. Differentiation of populations. *Science*. 165: 1228-1232.

Field, I. A. 1922. Biology and economic value of the sea mussel *Mytilus edulis*. *Bull. U. S. Bureau Fish.* 38: 127-260.

Gadgil, M. 1971. Dispersal: Population consequences and evolution. *Ecology*. 52: 253-261.

Hedrick, P. W. 1971. A new approach to measuring genetic similarity. *Evolution*. 25: 276-280.

Holgate, P. 1964. Genotype frequencies in a section of a cline. *Heredity*. 19: 507-509.

Kettlewell, H. B. D. 1958. A survey of the frequencies of *Biston betularia* (l.)(Lep.) and its melanic forms in Great Britain. *Heredity*. 12: 51-72.

Kimura, M. and G. H. Weiss 1964. The stepping stone model of population structure and the decrease of genetic correlation with distance. *Genetics* . 49: 561-576.

Koehn, R. K. and D. I. Rasmussen 1967. Polymorphic and monomorphic serum esterase heterogeneity in catostomid fish populations. *Biochem. Genet.* 1: 131-144.

Koehn, R. K. and J. B. Mitton 1972. Population genetics of marine pelecypods. I. Ecological heterogeneity and adaptive strategy at an enzyme locus. *Amer. Natur.* 106: 47-56.

Koehn, R. K., R. Milkman, J. B. Mitton, and John F. Boyer. (in prep.) Population genetics of marine pelecypods. IV. Selection, migration and genetic differentiation in the Blue Mussel, *Mytilus edulis*.

Lubinsky, I. 1958. Studies on *Mytilus edulis* L. of the "Calanus" expeditions to Hudson Bay and Ungava Bay. *Canad. J. Zool.* 36: 869-881.

Merritt, R. B. 1972. Geographic distribution and enzymatic properties of lactate dehydrogenase allozymes in the Fathead Minnow, *Pimephales promelas*. *Amer. Natur.* 196: 173-184.

Milkman, R. 1971. Genic polymorphism and population dynamics in *Mytilus edulis*. *Biol. Bull.* 141: 397.

Milkman, R. and L. D. Beaty 1970. Large-scale electrophoretic studies of allelic variation in *Mytilus edulis*. *Biol. Bull.* 139: 430.

Mitton, J. B., R. K. Koehn, and T. Prout 1973. Population genetics of marine pelecypods. III. Epistasis between functionally related isoenzymes of *Mytilus edulis*. *Genetics*. 73: 487-496.

Rohlf, F. J. and G. D. Schnell 1971. An investigation of the isolation-by-distance model. *Amer. Natur.* 105: 295-324.

Scheltema, R. S. 1971. Larval dispersal as a means of genetic exchange between geographically separated populations of shallow-water benthic marine gastropods. *Biol. Bull.* 140: 284-322.

Selander, R. K., W. G. Hunt, and S. Y. Yang 1969. Protein polymorphism and genic heterozygosity in two European subspecies of the House Mouse. *Evolution* 23: 379-390.

Snyder, T. P. and J. L. Gooch 1973. Genetic differentiation in *Littorina saxatilis* (Gastropoda). *Marine Biol.* 22: 177-182.

Sokal, R. R. and F. J. Rohlf 1969. *Biometry*. W. H. Freeman and Co., San Francisco. 776 pp.

Williams, G. C., R. K. Koehn, and J. B. Mitton 1973. Genetic differentiation without isolation in the American Eel, *Anguilla rostrata*. *Evolution* 27: 192-204.

Wright, S. 1951. The genetical structure of populations. *Ann. Eugen.* 15: 323-354.

Index

I

J

K

L

M

N

O

P

R

S

T

U

A 5
B 6
C 7
D 8
E 9
F 0
G 1
H 2
I 3
J 4